AF839721

J. Wolfgang Wägele, Thomas Bartolomaeus (Eds.)

Deep Metazoan Phylogeny: The Backbone of the Tree of Life

Also of Interest

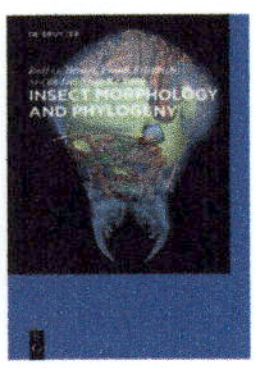

Insect Morphology and Phylogeny. A textbook for students of entomology
Rolf G. Beutel, Frank Friedrich, Xing-Ke Yang, Si-Qin Ge, 2013
ISBN 978-3-11-026263-6, e-ISBN 978-3-11-026404-3

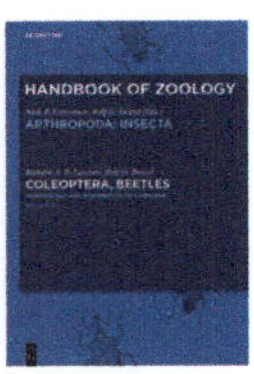

Handbook of Zoology/Handbuch der Zoologie
Arthropoda: Insecta: Coleoptera
Volume 3: Morphology and Systematics (Phytophaga)
Richard A.B. Leschen, Rolf G. Beutel (Eds.), 2013
ISBN: 978-3-11-027446-2, e-ISBN 978-3-11-027446-2

Deep Metazoan Phylogeny: The Backbone of the Tree of Life

New Insights from Analyses of Molecules, Morphology, and Theory of Data Analysis

Edited by
J. Wolfgang Wägele
Thomas Bartolomaeus

DE GRUYTER

Editors
Professor Dr. J. Wolfgang Wägele
Stiftung Zoologisches Forschungsmuseum
Alexander Koenig (ZFMK)
Leibnitz-Institut für Biodiversität der Tiere
Adenauerallee 160
53113 Bonn
w.waegele.zfmk@uni-bonn.de

Professor Dr. Thomas Bartolomaeus
Universität Bonn
Institut für Evolutionsbiologie und Zooökologie
An der Immenburg 1
53121 Bonn
tbartolomaeus@evolution.uni-bonn.de

ISBN 978-3-11-027746-3
e-ISBN 978-3-11-027752-4

Library of Congress Cataloging-in-Publication Data
A CIP catalog record for this book has been applied for at the Library of Congress.

Bibliographic information published by the Deutsche Nationalbibliothek
The Deutsche Nationalbibliothek lists this publication in the Deutsche Nationalbibliografie; detailed bibliographic data are available in the Internet at http://dnb.dnb.de.

© 2014 Walter de Gruyter GmbH, Berlin/Boston
Cover image: L. Willatt, East Anglian Regional Genetics Service/Science Photo Library/Agentur Focus
Typesetting: PTP-Berlin Protago-TEX-Production GmbH, Berlin
Printing and binding: Hubert & Co. GmbH & Co. KG, Göttingen
♾ Printed on acid-free paper

Printed in Germany

www.degruyter.com

List of Contributing Authors

Thomas Bartolomaeus
Institute of Evolutionary Biology and Animal Ecology
University of Bonn
Bonn, Germany
e-mail: tbartolomaeus@evolution.uni-bonn.de

Tjard Bergmann
ITZ, Division of Ecology and Evolution
Stiftung Tierärztliche Hochschule Hannover
Hannover, Germany
e-mail: tjard.bergmann@ecolevol.de

Matthias Bernt
Faculty of Mathematics and Computer Science
University of Leipzig
Leipzig, Germany
e-mail: bernt@informatik.uni-leipzig.de

Christoph Bleidorn
Institute of Biology II
University of Leipzig
Leipzig, Germany
e-mail: bleidorn@uni-leipzig.de

Janus Borner
Institute of Zoology and Zoological Museum
University of Hamburg
Hamburg, Germany
e-mail: janus.borner@uni-hamburg.de

Iris Bruchhaus
Bernhard Nocht Institute for Tropical Medicine
Hamburg, Germany
e-mail: bruchhaus@bni-hamburg.de

Thorsten Burmester
Institute of Zoology and Zoological Museum
University of Hamburg
Hamburg, Germany
e-mail: thorsten.burmester@uni-hamburg.de

Karolin von der Chevallerie
ITZ, Division of Ecology and Evolution
Stiftung Tierärztliche Hochschule Hannover
Hannover, Germany
e-mail: karolin.chevallerie@ecolevol.de

Alexander Donath
Stiftung Zoologisches Forschungsmuseum
Alexander Koenig –
Leibniz-Institut für Biodiversität der Tiere (ZFMK)
Bonn, Germany
and
Department of Computer Science
University of Leipzig
Leipzig, Germany
e-mail: a.donath@zfmk.de

Janina Dordel
FB05 Biology/Chemistry
University of Osnabrück
Osnabrück, Germany
e-mail: janina_dordel@web.de

Jason Dunlop
Museum für Naturkunde
Leibniz Institute for Research on Evolution and Biodiversity
Berlin, Germany
e-mail: jason.dunlop@mfn-berlin.de

Ingo Ebersberger
Center for Integrative Bioinformatics Vienna
Medical University of Vienna
Vienna, Austria
and
Institute for Cell Biology and Neurosciences
Goethe University
Frankfurt, Germany
e-mail: ebersberger@bio.uni-frankfurt.de

Igor Eeckhaut
Biology of Marine Organisms and Biomimetism
University of Mons-Hainaut
Mons, Belgium
e-mail: Igor.EECKHAUT@umons.ac.be

Carina Eisenhardt
Department of Computer Science, and Interdisciplinary Center for Bioinformatics
and
Institute of Biology II
University of Leipzig
Leipzig, Germany
e-mail: carina@eisenhardt.tk

Michael Eitel
ITZ, Division of Ecology and Evolution
Stiftung Tierärztliche Hochschule Hannover
Hannover, Germany
and
The Swire Institute of Marine Science
The University of Hong Kong
Hong Kong, China
e-mai: meitel@hku.hk

Frauke Diersing
FB05 Biology/Chemistry
University of Osnabrück
Osnabrück, Germany
e-mail: fffisse@gmx.de

Martin Fritsch
Institute of Biosciences
University of Rostock
Rostock, Germany
e-mail: martin.fritsch@uni-rostock.de

Peter Grobe
Stiftung Zoologisches Forschungsmuseum
Alexander Koenig –
Leibniz-Institut für Biodiversität der Tiere (ZFMK)
Bonn, Germany
e-mail: p.grobe@zfmk.de

Heike Hadrys
ITZ, Division of Ecology and Evolution
Stiftung Tierärztliche Hochschule Hannover
Hannover, Germany
and
Division of Invertebrate Zoology
American Museum of Natural History
New York, NY, USA
and
Department of Molecular, Cellular and Developmental Biology
Yale Univeristy
New Haven, USA
e-mail: heike.hadrys@ecolevol.de

Thomas Hankeln
Institute of Molecular Genetics
Johannes Gutenberg-University Mainz
Mainz, Germany
e-mail: hankeln@uni-mainz.de

Stefanie Hartmann
Institute of Biochemistry and Biology
University of Potsdam
Potsdam, Germany
e-mail: stefanie.hartmann@uni-potsdam.de

Steffen Harzsch
Zoological Institute and Museum
Ernst-Moritz-Arndt University Greifswald
Greifswald, Germany
e-mail: steffen.harzsch@uni-greifswald.de

Bernhard Hausdorf
Zoological Museum
University of Hamburg
Hamburg, Germany
e-mail: hausdorf@zoologie.uni-hamburg.de

Conrad Helm
Institute of Biology
University of Leipzig
Leipzig, Germany
e-mail: helm@rz.uni-leipzig.de

Martin Helmkampf
Zoological Museum
University of Hamburg
Hamburg, Germany
e-mail: martin.helmkampf@asu.edu

Holger Herlyn
Institute of Anthropology
Johannes Gutenberg-University Mainz
Mainz, Germany
e-mail: herlyn@uni-mainz.de

Jana Hertel
Department of Computer Science
University of Leipzig
Leipzig, Germany
e-mail: jana.hertel@bioinf.uni-leipzig.de

Natascha Hill
Institute of Biochemistry and Biology
University of Potsdam
Potsdam, Germany
e-mail: natascha.hill@uni-potsdam.de

Christoph Hösel
FB05 Biology/Chemistry
University of Osnabrück
Osnabrück, Germany
e-mail: c.hoesel@web.de

Wolfgang Jakob
ITZ, Division of Ecology and Evolution
Stiftung Tierärztliche Hochschule Hannover
Hannover, Germany
e-mail: wolfgang.jakob@ecolevol.de

Markus Koch
Institute of Evolutionary Biology and Animal Ecology
University of Bonn
Bonn, Germany
Email: mkoch@evolution.uni-bonn.de

Veiko Krauss
Department of Computer Science and Interdisciplinary Center for Bioinformatics
University of Leipzig
Leipzig, Germany
e-mail: veiko@bioinf.uni-leipzig.de

Patrick Kück
Stiftung Zoologisches Forschungsmuseum Alexander Koenig –
Leibniz-Institut für Biodiversität der Tiere (ZFMK)
Bonn, Germany
e-mail: patrick_kueck@web.de

Deborah Lanterbecq
Biology of Marine Organisms and Biomimetism
University of Mons-Hainaut
Mons, Belgium
e-mail: deborah.lanterbecq@umons.ac.be

Jörg Lehmann
Department of Computer Science and Interdisciplinary Center for Bioinformatics
University of Leipzig
Leipzig, Germany
e-mail: joe@bioinf.uni-leipzig.de

Peter Lesný
Institute of Evolutionary Biology and Animal Ecology
University of Bonn
Bonn, Germany
e-mail: plesny@evolution.uni-bonn.de

Harald Letsch
Department für Tropenökologie und Biodiversität der Tiere
Vienna, Austria
e-mail: harald.letsch@univie.ac.at

Rudi Loesel
Institute for Biology II (Zoology)
RWTH Aachen University
Aachen, Germany
e-mail: loesel@bio2.rwth-aachen.de

Daniel Merkle
Department of Mathematics and Computer Science
University of Southern Denmark
Odense, Denmark
Denmark
e-mail: daniel@imada.sdu.dk

Karen Meusemann
Stiftung Zoologisches Forschungsmuseum
Alexander Koenig –
Leibniz-Institut für Biodiversität der Tiere (ZFMK)
Bonn, Germany
e-mail: mail@karen-meusemann.de

Martin Middendorf
Faculty of Mathematics and Computer Science
University of Leipzig
Leipzig, Germany
e-mail: middendorf@informatik.uni-leipzig.de

Bernhard Misof
Stiftung Zoologisches Forschungsmuseum
Alexander Koenig –
Leibniz-Institut für Biodiversität der Tiere (ZFMK)
Bonn, Germany
e-mail: b.misof.zfmk@uni-bonn.de

Burkhard Morgenstern
Institute of Microbiology and Genetics
University of Göttingen
Göttingen, Germany
e-mail: bmorgen@gwdg.de

Carsten H.G. Müller
Zoological Institute and Museum
Ernst-Moritz-Arndt University Greifswald
Greifswald, Germany
e-mail: carstmue@uni-greifswald.de

Adina Mwinyi
LGC Genomics
Berlin, Germany
e-mail: Adina.Mwinyi@lgcgenomics.com

Maximilian P. Nesnidal
Zoological Museum
University of Hamburg
Hamburg, Germany
e-mail: nesnidal@gmail.com

Tetyana Nosenko
Department of Earth and Environmental
Sciences, Palaeontology & Geobiology
Ludwig-Maximilians-Universität München
Munich, Germany
e-mail: nosenko@bio.lmu.de

Hans-Jürgen Osigus
ITZ, Division of Ecology and Evolution
Stiftung Tierärztliche Hochschule Hannover
Hannover, Germany
e-mail: h.j.osigus@ecolevol.de

Omid Paknia
ITZ, Division of Ecology and Evolution
Stiftung Tierärztliche Hochschule Hannover
Hannover, Germany
e-mail: omid.paknia@ecolevol.de

Christiane Paul
Institute of Biochemistry and Biology
University of Potsdam
Potsdam, Germany
e-mail: cpaul@uni-potsdam.de

Yvan Perez
Institut Méditerranéen de Biodiversité et
d'Ecologie „Evolution Genome Environment"
Aix-Marseille Université
Marseille, France
e-mail: perezyvan@hotmail.com

Lars Podsiadlowski
Institut für Evolutionsbiologie und Ökologie
University of Bonn
Bonn, Germany
e-mail: lars@cgae.de

Günter Purschke
FB05 Biology/Chemistry
University of Osnabrück
Osnabrück, Germany
e-mail: purschke@biologie.uni-osnabrueck.de

Björn Quast
Institut für Evolutionsbiologie und Ökologie
University of Bonn
Bonn, Germany
Email: bquast@evolution.uni-bonn.de

Björn M. von Reumont
Department of Life Sciences
Natural History Museum London
London, UK
and
Stiftung Zoologisches Forschungsmuseum
Alexander Koenig –
Leibniz-Institut für Biodiversität der Tiere (ZFMK)
Bonn, Germany
e-mail: bmvr@arcor.de

Stefan Richter
Institute of Biosciences
University of Rostock
Rostock, Germany
e-mail: stefan.richter@uni-rostock.de

Birgen Holger Rothe
Zoological Museum
University Hamburg
Hamburg, Germany

Bernd Schierwater
ITZ, Division of Ecology and Evolution
Stiftung Tierärztliche Hochschule Hannover
Hannover, Germany
and
American Museum of Natural History
New York, NY, USA
and
Department of Ecology and Evolutionary Biology
Yale University
New Haven, CT, USA
e-mail: bernd.schierwater@ecolevol.de

Martin Schlegel
Institute of Biology
University of Leipzig
Leipzig, Germany
e-mail: schlegel@rz.uni-leipzig.de

Andreas Schmidt-Rhaesa
Zoological Museum
University of Hamburg
Hamburg, Germany
e-mail: andreas.schmidt-rhaesa@uni-hamburg.de

Gerhard Scholtz
Institute of Biology
Humboldt University Berlin
Berlin, Germany
e-mail: gerhard.scholtz@rz.hu-berlin.de

Fabian Schreiber
Wellcome Trust Sanger Institute
Wellcome Trust Genome Campus
Hinxton, Cambridgeshire, UK
e-mail: fs@ebi.ac.uk

Michael Schrödl
Zoologische Staatssammlung München
Munich, Germany
e-mail: schroedl@zi.biologie.uni-muenchen.de

Joachim Selbig
Institute of Biochemistry and Biology
University of Potsdam
Potsdam, Germany
e-mail: jselbig@uni-potsdam.de

Sabrina Simon
ITZ, Division of Ecology and Evolution
Stiftung Tierärztliche Hochschule Hannover
Hannover, Germany
and
Sackler Institute for Comparative Genomics
American Museum of Natural History
New York, NY, USA
e-mail: ssimon@amnh.org

Thomas Stach
Institute of Biology
Humboldt University Berlin
Berlin, Germany
e-mail: tstach@zoosyst-berlin.de

Peter F. Stadler
Institute of Computer Science
and
Interdisciplinary Center for Bioinformatics
University of Leipzig
and
Max Planck Institute for Mathematics in the Sciences
and
Fraunhofer Institute for Cell Therapy and Immunology IZI
Leipzig, Germany
and
Department of Theoretical Chemistry
University of Vienna, Austria
and
Center for non-coding RNA in Technology and Health, University of Copenhagen, Denmark
and
Santa Fe Institute, NM, USA
e-mail: Peter.Stadler@bioinf.uni-leipzig.de

Martin E.J. Stegner
Institute of Biosciences
University of Rostock
Rostock, Germany
e-mail: martin.stegner@uni-rostock.de

Roman R. Stocsits
Research Institute of Molecular Pathology
Vienna, Austria
and
Stiftung Zoologisches Forschungsmuseum
Alexander Koenig –
Leibniz-Institut für Biodiversität der Tiere (ZFMK)
Bonn, Germany
e-mail: roman.stocsits@imp.ac.at

Torsten H. Struck
FB05 Biology/Chemistry
University of Osnabrück
Osnabrück, Germany
and
Stiftung Zoologisches Forschungsmuseum
Alexander Koenig –
Leibniz-Institut für Biodiversität der Tiere (ZFMK)
Bonn, Germany
e-mail: torsten.struck.zfmk@uni-bonn.de

Hakim Tafer
Department of Computer Science
University of Leipzig
Leipzig, Germany
e-mail: htafer@gmail.com

Ralph Tiedemann
Institute of Biochemistry and Biology
University of Potsdam
Potsdam, Germany
e-mail: tiedeman@uni-potsdam.de

Lars Vogt
Institute of Evolutionary Biology and Animal Ecology
University of Bonn
Bonn, Germany
e-mail: lars.m.vogt@gmail.com

J. Wolfgang Wägele
Stiftung Zoologisches Forschungsmuseum
Alexander Koenig –
Leibniz-Institut für Biodiversität der Tiere (ZFMK)
and
Lehrstuhl für Spezielle Zoologie
Rheinisch Friedrich-Wilhelms-Universität Bonn
Bonn, Germany
e-mail: w.waegele.zfmk@uni-bonn.de

Mathias Weber
Institute of Molecular Genetics
Johannes Gutenberg-University Mainz
Mainz, Germany
e-mail: weber011@uni-mainz.de

Michael Weidhase
Institute of Biology
University of Leipzig
Leipzig, Germany
e-mail: michael.weidhase@uni-leipzig.de

Anne Weigert
Institute of Biology
University of Leipzig
Leipzig, Germany
e-mail: anne.weigert@uni-leipzig.de

Alexandra R. Wey-Fabrizius
Institute of Molecular Genetics
Johannes Gutenberg-University Mainz
Mainz, Germany
e-mail: wey@uni-mainz.de

Alexander Witek
Institute of Molecular Genetics
Johannes Gutenberg-University Mainz
Mainz, Germany
e-mail: witek.alexander@gmail.com

Gert Wörheide
GeoBio-Center
and
Department of Earth and Environmental Sciences, Palaeontology & Geobiology
Ludwig-Maximilians-Universität München
and
Bayerische Staatssammlung für Paläontologie und Geologie
Munich, Germany
e-mail: woerheide@lmu.de

Content

Part II: New Tools and Methods

J. Wolfgang Wägele and Thomas Bartolomaeus

1 Introduction

Animal phylogeny: riddles and answers

This book summarizes major results of a **priority program** funded by the German Research Foundation during a period of six years (2005–2011) to study the backbone of the metazoan tree of life. More than 30 teams from different institutes and universities contributed new data and innovative phylogenetic analyses. New bioinformatics tools were developed to overcome some methodological problems. During this time a large number of young scientists were trained in phylogenetics, working in an unprecedented multidisciplinary network of specialists.

This major effort was motivated by mutually **incompatible phylogenetic hypotheses** published recently, which left at a loss all those who write textbooks, teach systematics, or need a phylogeny as a basis to understand evolution. The contradictions between **morphology-based hypotheses** and **molecular phylogenies**, and the incompatibility of various "well supported" molecular analyses called for a major endeavor. Also, a closer look at poorly studied **morphological characters** was commendable to get independent evidence. Therefore, the morphological part of the priority program had a focus on the anatomy of brains, nervous systems, and mesoderm derivatives.

The reader will find here several **up-to-date reviews** which give an overview over various hypotheses. We did not aim at a harmonization of these hypotheses to produce one fully resolved tree: available data and current methods of data analysis do not allow the selection of a "best" hypothesis covering all nodes in the backbone of the animal tree of life. This book rather cautions against accepting the latest ideas, even when they are published in high ranking journals and adopted by a large audience, and it offers starting points for additional research.

Starting the priority program, one major assumption was that by increasing the data sets (number of sequenced genes, number of taxa) the phylogenetic signal in molecular data sets would become more distinct. While this is in general true for the signal-to-noise ratio, we had to realize that in molecular phylogenetics **hidden systematic errors** can occur independently of the quantity of sampled data (see the chapters on theory and methods). These errors are not caused by simple noise or by absence of signal in the data. They are probably the reason why morphological evidence strongly contradicts molecular phylogenies (see e.g. the arthropod chapter by Wägele and Kück). To better understand these sources of error, several theoretical studies were conducted. Their results are also presented in this book.

It became clear that new tools needed to avoid systematic errors still have to be developed. Therefore current users of popular software (see also examples in this book) must be aware of some dramatic pitfalls which remained unnoticed until recently.

The book is divided into two major parts: the first part contains all chapters with phylogenetic analyses of different taxa of Metazoa, the second part is dedicated to methods and tools.

The first two chapters discuss the origin of and relationships between **diploblastic metazoans**. Both studies are based on phylogenomic data, but they differ in the composition of data sets and in the way data were analyzed. While Wörheide et al. conclude that sponges are monophyletic and represent among the extant fauna the earliest lineage of Metazoa, a concept also supported by functional and morphological characters (e.g. retention of the feeding mechanism also seen in Choanoflagellata, absence of digestive exoenzymes etc.), Eitel et al. favor a clade Diploblasta, composed of Cnidaria, Ctenophora, Porifera and Placozoa as sister-taxon to Bilateria, because this clade got high support in their analyses of sequence data. This conflict is very typical for the current situation of phylogenetics. The solution is to search for **sources of error**, to better understand the footprints left by evolution in genomes at the level of site patterns. We have to strike out on new paths of data analysis. We also have to check the **plausibility of hypotheses** derived from phylogenetic graphs, which are supposed to represent the historical course of macroevolution. Plausibility means that a scenario based on a tree hypothesis must also be able to explain the evolution of modes of life, transitions between body constructions and ecological specializations, also including all available morphological and fossil evidence. Such considerations are lacking in the literature of the past 20 years.

In some cases it is more honest to state that we do not have sufficient evidence. An example is the analysis of the origin of **Chaetognatha**, discussed by Perez, Müller, and Harzsch. They consider all available data (sequences, HOX genes, embryonic development, anatomy, new neuroanatomical data), point out to possible contradictory interpretations and tentatively propose that chaetognaths may be a separate lineage which evolved parallel to other protostomes.

Further neuroanatomical studies of protostomes provided surprising evidence for the homology of **brain characters** present in **annelids and arthropods,** including Onychophora. Loesel demonstrates that these brains have – with their mushroom bodies, central body and glomeruli – a complexity not seen elsewhere. The homology of this type of brain is highly probable. Schmidt-Rhaesa and Rothe present new data on the brains of **Gastrotricha** and **Cycloneuralia** (Scalidophora and Nematoida). These brains are quite simple. Assuming that molecular phylogenies are without error, the Ecdysozoa hypothesis would imply a drastic reduction of brain structure (and of other anatomical features) in Gastrotricha and Cycloneuralia in comparison with arthropods, which is unusual in non-sessile animals. Alternatively, one has to assume a parallel evolution of highly similar brains in annelids and arthropods, which also is very unlikely. The search for systematic errors in molecular phylogenies (chapter by Kück, Misof, and Wägele) shows a way out of this dilemma.

Hankeln et al. focused on the **Platyzoa** hypothesis (Platyhelminthes + Gastrotricha + Gnathifera). They analyzed molecular data sets and review the literature. Their Figure 7.3 represents a typical result obtained with currently available methods, where Platyzoa are monophyletic (though not strongly supported) and related to a large group containing the Lophotrochozoa. The authors caution that in inferred tree graphs many lineages exhibit long branches and that these graphs may still contain artifacts. On the other hand, some details like the close relationship between parasitic **Acanthocephala** and **Seisonidea** seem to be plausible in view of the implied evolution of anatomy and modes of life.

The **Lophophorata** (= Tentaculata) hypothesis (Phoronida, Brachiopoda, Ectoprocta) was studied by Nesnidal et al. using phylogenomic data. The Lophophorata is a typical example for the careening of molecular phylogenies. While many molecular phylogeneticists agree that these taxa belong to a clade that also contains annelids and mollusks (the Lophotrochozoa), many previous studies rejected the monophyly of Lophophorata. Published ideas include paraphyly of Brachiopoda or a closer relationship between Ectoprocta (Bryozoa) and Entoprocta (Kamptozoa). Nesnidal et al. confirm the monophyly of Lophophorata and of the studied Brachiopoda. Kamptozoa are not included in Lophophorata.

Struck et al. provide a deeper inside into the phylogeny of **Annelida**, also based on molecular data. Zoologists must get used to the idea that other taxa with the traditional rank of phyla are now considered to be annelids, namely the **Sipuncula**, **Echiura**, and **Myzostomida**. Also, it is clear now that the traditional separation of Polychaeta and Oligochaeta is an artifical, non-phylogenetic classification. Arguments for the placement of Myzostomida are reviewed in detail by Bleidorn et al. Studies into the fate of the embryonic coeloms in Annelida and **Panarthropoda** revealed differences in almost all details of coelomogenesis except for the fact that, if present, these structures are topologically identical in both groups. Since no obvious function could be assigned to these organs in arthropods, Koch, Quast, and Bartolomaeus conclude that most likely historical constraints underlie their existence. In this chapter they modify the traditional concept on coelomic sacs as mesoblastic organs and demonstrate the significance of high-resolution electron microscopy in studying coelomic cavities and nephridia.

Arthropods are the most species rich taxon of Metazoa and their phylogeny is especially difficult to understand due to the old age and strong divergence of many lineages. Wägele and Kück focus on the **Mandibulata** and summarize morphological evidence (including fossils) that strongly supports the idea that insects are crustaceans. Crustacea are paraphyletic, as suggested by many previous molecular studies. However, in contrast to molecular phylogenies, the available morphological information shows that **Remipedia** (relic cave crustaceans) conserve a bauplan with many derived characters seen also in **Myriapoda** and **Hexapoda**. A new evolutionary scenario suggests that **Tracheata** (= Atelocerata) are monophyletic and that they evolved from a remiped-like marine ancestor. An explanation for the failure of molecular phylogenies that do not recover these relationships is suggested. These observations would suggest that **Pancrustacea** is a junior synonym of Mandibulata.

This case is similar to the contradiction between those neuroanatomical data that support the Articulata hypothesis, and the molecular phylogenies that suggest a clade Ecdysozoa. Life would be easier if we ignored the morphological evidence, but this attitude will not uncover the history of life.

Simon and Hadrys present a closer look at the phylogeny within **Pterygota** using molecular data. They do not propose a final solution but discuss analyses of their own new data in the light of the recent literature. Some of the currently supported ideas are the monophyly of Chiastomyaria (Ephemeroptera and Neoptera), with Odonata as a separate lineage at the base of the pterygote tree. The monophyly of Polyneoptera (orthopteroid taxa, Dictyoptera, Dermaptera and others) is confirmed, and within Holometabola the sister-group relationship between Hymenoptera and the remaining taxa.

The phylogeny within **Crustacea** is still a riddle. Stegner, Fritsch, and Richter provide new neuroanatomical data. They studied the central complex in the brain of representatives of several major taxa and conclude that this is a homologous part of the brain inherited from the last common ancestor of all crustaceans and insects. This brain complex has been reduced subsequently in several crustacean lineages. Further evidence for relationships cannot be derived from these data. The molecular analyses presented by von Reumont and Wägele confirm several results of previous publications and suggest that stability can be achieved when data and substitution models are selected carefully. For example, in recently published tree graphs Remipedia are the crustacean taxon closest to insects, an idea that is also partly compatible with the scenario presented by Wägele and Kück. However, stability does not imply absence of systematic errors. A suspicious detail seen in molecular phylogenies is that the morphologically highly evolved Malacostraca, which share e.g. several special brain characters with insects, are placed near Cirripedia and Copepoda.

The remaining last arthropod clade, the **Chelicerata**, is discussed by Dunlop, Borner and Burmester. Their updated phylogeny is basically consistent with previous morphology-based hypotheses, however, the available molecular data do not allow a finer resolution of relationships within Arachnida.

Due to the comparatively good congruence in published literature, the phylogeny of **Deuterostomia** was not a major theme in our program. However, Schlegel, Weidhase and Stadler conducted an analysis of new molecular data to understand the major nodes in the deuterostome tree. The authors provide a review of recent and some new hypotheses, like the subdivision into the clades Xenacoelomorpha, Ambulacraria and Chordata, and within Chordata into the Cephalochordata and Olfactores. The corresponding morphological perspective is summarized by Stach in a critical way. In comparison with the molecular trees, the conclusions are not always the same.

The first part of the book concludes with a review on the information contained in **mitochondrial gene arrangements** (Podsiadlowski et al.). Gene order is quite variable and not devoid of misleading parallel mutations. It is however possible to infer some node patterns for higher taxa.

The second part of this book begins with the presentation of a new tool that is especially valuable for the management of morphological evidence (Grobe and Vogt). **MorphDBase** is a database for morphological data and metadata. It implements for the first time standards for content, formats, nomenclature, terminology, and concepts. This web-based system can be used by any registered user, its data pool is growing rapidly. The necessity to standardize the terminology and to conceptualize characters used in phylogenetic analyses is elaborated for a case study on neuroanatomical characters. Together with many colleagues, Loesel and Richter worked out definitions and concepts that are the basis for **neurophylogeny**, the study of the evolution of nervous systems along the metazoan tree of life.

Misof et al. explain new tools useful for the assessment of molecular data quality prior to tree inference. These include **measures of tree-likeness**, the search for **noise in alignments**, or the selection of **informative data subsets** in large phylogenomic data matrices. Bernt et al. give an overview on computational methods for the **analyses of gene order rearrangements**, which are mostly applied for mitochondrial genomes.

The **RNAsalsa** software package is presented by Stocsits et al. This software is relevant to consider secondary structure information of RNA molecules in phylogeny inference. RNAsalsa infers the secondary structure for each sequence and builds alignments based on this information. The results can be used for phylogeny inference with RNA-specific substitution models.

Searching for additional types of molecular data that might be useful for phylogenetic analyses, Lehmann et al. tested **near intron pairs** (NIPs). The authors explain the workflow, summarize hitherto published results, and present new analyses. Donath and Stadler provide a wider overview on higher order molecular characters (**molecular morphology**) such as repetitive elements, NUMTs, or microRNAs. Though this type of characters is not free of homoplasies, it is obvious that NIPs and other "rare genomic events" bear relevant phylogenetic information.

Kück, Misof and Wägele caution against a too credulous acceptance of results of maximum likelihood analyses (ML). Though in theory the ML method is infallible when certain assumptions are fulfilled, the reality of incompatible results and the implausibility of some hypotheses prove that there exist several **undetected sources of error** besides the well-known model misspecifications and weak data sampling. Simulations help to identify the conditions that produce systematic errors as a basis for the development for new tree building algorithms.

Since the period of single gene sequencing is rapidly being succeeded by genome sequencing, growing masses of data have to be handled. The thorough review by Ebersberger and von Haeseler is dedicated to **challenges and pitfalls of phylogenomics**, pointing to the near future of our science.

The main progress that had been achieved during the "Deep Metazoan Phylogeny" project are detailed morphological data based on a broader taxon sampling including previously neglected groups, and a critical evaluation of sequence data. If it were possible to infer phylogeny from those molecular and morphological traces that during the course of evolution remained in all extant taxa, one would expect that identical phylogeny hypotheses should result from analyzing these traces. As long as this is not the case, it is much more the structure of the data that poses conflicts on the analyses than priority of one data source over the other. This book actually presents some examples where additional data and modified analytical tools reveal congruence between molecular and morphological studies, occasionally corroborating traditional views (e.g., monophyly of the Lophophorata) or weakening classical ideas (e.g., Myzostomida as isolated phylum). This book shows the present progress on our long way to a uniform picture of metazoan phylogeny.

Part I: **New Data and Phylogenies**

Gert Wörheide, Tetyana Nosenko, Fabian Schreiber, and Burkhard Morgenstern

2 Progress and perspectives of the deep non-bilaterian phylogeny, with focus on sponges (Phylum Porifera)

Abstract: Sponges (Phylum Porifera) are benthic aquatic animals of great ecological and pharmaceutical importance. The phylum is arguably the earliest branching metazoan taxon and of considerable significance for the reconstruction of early metazoan evolution. However, the deep phylogeny of sponges is still debated, as the exact branching pattern of their main clades and their relationships to the other non-bilaterian animals is not unequivocally resolved. Here, the current state of the debate about the deep phylogeny of sponges and other non-bilaterians is reviewed and we highlight some issues that need to be considered when using phylogenomic data to reconstruct these relationships. Several studies, mainly based on single or a few genes, have suggested that sponges are a paraphyletic assemblage that share a grade of construction rather than common ancestry. However, recent phylogenomic analyses have suggested that sponges are monophyletic, consistent with earlier cladistic analyses based on morphological characters. This finding has many implications for the evolutionary interpretation of early animal traits. However, we conclude that further evidence from sequence-independent characters needs to be sought to test existing phylogenetic hypotheses. Only by cross-corroboration with independent data might we come closer to fully resolving the deepest branches of the animal tree of life.

2.1 Introduction

Phylum Porifera (sponges) contains organisms that live exclusively in aquatic habitats, mostly marine, but some also colonized freshwater environments. All adult sponges are sessile, and are, mainly due to their filter-feeding lifestyle, of great ecological importance (Bell, 2008). Furthermore, sponges are among the most toxic animals in the ocean and are consequently of great interest for the development of drugs (Faulkner, 2002). However, the systematic and phylogenetic relationships within and among poriferans have proven quite difficult to reconstruct (reviewed in Erpenbeck and Wörheide, 2007).

Taxonomy and systematics of recent (and fossil) sponges have, until recently, mainly relied on morphological characters, such as the shapes of spicules and their arrangement in the sponge body (see, e.g., Boury-Esnault and Rützler, 1997). Characteristics of the soft tissue are only of prime importance in those taxa that do not have spicules at all (e.g., keratose sponges, see Hooper and Van Soest, 2002) or as additional characters in class Calcarea (see, e.g., Wörheide and Hooper, 1999; 2003). Sponges often lack complex morphological characters. In addition, sponge morphological characters are often prone to homoplasies (Erpenbeck et al., 2006) and can vary depending on local environmental conditions (Maldonado et al., 1999). For these reasons, phylogenies based on morphological traits often lack sufficient resolution (Manuel et al., 2003; Dohrmann et al., 2008). The advent of molecular systematics alleviated this problem by providing large morphology-independent data sets and by doing so revived interest in sponge phylogeny (see review by Wörheide et al., 2012).

Sponges are arguably the earliest branching metazoan lineage (Philippe et al., 2009; Pick et al., 2010), and are therefore pivotal to our understanding of the early evolution of animals and the reconstruction of our last common ancestor (Taylor, Thacker, and Hentschel, 2007; Larroux, 2011). The relationships of sponges to the other non-bilaterian taxa (Placozoa, Cnidaria, and Ctenophora) as well as the relationships of those to the Bilateria are of similar fundamental importance to our understanding of the evolution of key bilaterian traits, such as the nervous system (Nickel, 2010; Conaco et al., 2012) and muscles (Steinmetz et al., 2012).

Here we summarize the progress and perspectives on the deep phylogeny of sponges that have been made in the last decade. We concentrate on the relationships of the main sponge lineages ("classes") to each other and to the other metazoans. For an overview of the internal relationships of the main sponge lineages we refer to Wörheide et al. (2012). We focus on the results of phylogenomic[1] analyses to resolve the deepest branches in the animal tree of life and point to limitations of this approach.

2.2 The challenge of reconstructing non-bilaterian relationships

Higher-level, non-bilaterian relationships remain controversial and highly topical, as judged by the many conflicting hypotheses published in recent years in leading science journals (see reviews by Edgecombe et al., 2011; Philippe et al., 2011a). These hypotheses are based on phylogenomic data sets that contain increasing amounts of sequence data generated in the last few years. Sequences are commonly drawn from transcriptome sequencing projects (e.g., expressed sequence tag (EST) data; due to lower costs) and also increasingly from fully sequenced genomes (Philippe and Telford, 2006). Phylogenomics should be distinguished from multi-gene analyses (e.g., Sperling, Peterson, and Pisani, 2009), which typically include fewer than 30 genes that are selected before rather than after sequencing.

Rockas, Kruger, and Carroll (2005) were among the first to use multiple protein-coding genes (50) in a phylogenomic study but were unable to resolve relationships

1 Phylogenomics was originally described by Eisen and Fraser (2003) as the "intersection of evolution and genomics" and is now mainly applied to the analysis of large multi (> 30) gene alignments. (Eisen and Fraser, 2003)

between the major non-bilaterian lineages, which most likely diverged in a relatively short time span before the “Cambrian Explosion” (Peterson et al., 2008). The authors concluded that these relationships are difficult, if not impossible, to resolve using sequence data from crown-group taxa (Rokas and Carroll, 2006). Another attempt to resolve the higher rank metazoan relationships using a large sequence data set (150 genes) has been made by Dunn et al. (2008). This study focused on bilaterian relationships. While the addition of large amounts of novel EST data from bilaterians improved the resolution of their relationships, the non-bilaterian part of their tree was not well represented. The most astonishing result of this study was perhaps the alleged sister-group relationship of the Ctenophora to the remaining Metazoa, i.e. a “basal” position within the Metazoa as the earliest branching taxon. A subsequent study by the same group of scientists included more non-bilaterians and reconstructed basically the same topology with Ctenophora-basal (Hejnol et al., 2009).

However, Pick et al. (2010) further improved the non-bilaterian taxon sampling of the Dunn et al. (2008) data set and convincingly showed that the “basal” position of the Ctenophora in Dunn et al. (2008) was an artifact of long branch attraction[2] (LBA, see, e.g., Bergsten, 2005; Wägele and Mayer, 2007) by distantly related outgroups. Philippe et al. (2011a) supported this conclusion and, in addition, discovered several other shortcomings of the Dunn et al. (2008) study, including contaminations and alignment errors caused by frameshifts.

Another scenario for early metazoan evolution, a “modernized Urmetazoa hypothesis”, was proposed by Schierwater et al. (2009a) based on the analysis of a “total evidence” data set that included nuclear protein-coding and mitochondrial genes as well as morphological characters. This study reconstructed monophyletic “Diploblasta” (i.e., non-bilaterian) animals with “basal” Placozoa as the sister-group to the Triploblasta (Bilateria). However, it was recently shown that the supermatrix constructed by Schierwater et al. contained many errors such as questionable orthology of some genes, frameshifts, and other biological and in silico contaminations (Philippe et al., 2011a). Analyses of the error-corrected data set recovered a different tree topology where the Diploblasta clade no longer received support.

The data set constructed by Philippe et al. (2009) included the most comprehensive sampling of non-bilaterian taxa to date. They analyzed 128 genes to recover highly supported monophyletic Porifera as the sister-group to the remaining Metazoa, and revived the Coelenterata hypothesis (Hyman, 1940) (sister-group relationship of Ctenophora and Cnidaria). The Placozoa was the sister to the Eumetazoa clade. This tree topology was the only one that withstood a recent re-analysis of the aforementioned phylogenomic studies (Philippe et al., 2011b). However, while a more recent study by Nosenko et al. (2013) again supported the “Coelenterates” using their full gene-set alignment, the position of the ctenophore and placozoan branches were less

2 LBA, an artifact where taxa with long branches are artificially attracted to each other in phylogenetic analyses without being truly related (Felsenstein, 1978).

stable when two non-overlapping gene-partitions of this large matrix were analyzed. Therefore, we take a conservative approach here and consider the relationships of Ctenophora and Placozoa to other metazoans as currently unresolved, but with an early branching position of Porifera (as in Dohrmann and Wörheide, 2013) (see Figure 2.1).

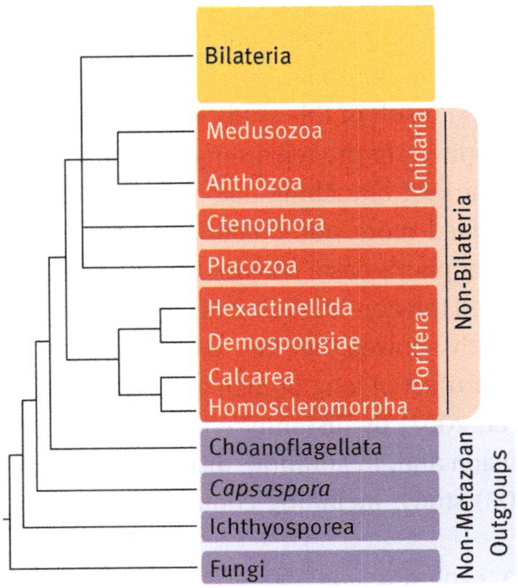

Figure 2.1: Consensus of higher-level phylogenetic relationships of Metazoa, including relationships to non-metazoan relatives (modified from Dohrmann and Wörheide, 2013).

2.2.1 Some issues to consider when reconstructing deep metazoan phylogeny

To resolve higher-level, non-bilaterian relationships, potential sources of error need to be eliminated whenever possible from phylogenomic data sets. There are two main categories of sources of errors that can lead to artifactual phylogenies inferred from molecular data: (1) *stochastic error* caused by a limited amount of sequence data, (2) *systematic error* caused by non-phylogenetic signal and/or methodological shortcomings such as model-misspecification and paralogy (Philippe et al., 2011a).

Stochastic error is probably the easiest to deal with as it often disappears when the length of the data set (alignment) used for phylogenetic inference is increased. In one of the earliest phylogenomic studies based on genome sequences from a few species of yeast, Rokas et al. (2003a) showed that stochastic error often results in conflicting phylogenies when individual genes are analyzed, but its effect largely disappears when > 20 genes or > 8000 randomly selected nucleotides are used for the analysis. However, the reduction of stochastic error does not automatically result in well-supported phylogenies. Bootstrap support values for some phylogenetic relationships can be low even in the absence of stochastic error if the phylogenetic signal is weak and non-phylogenetic signal resulting from systematic error is present (Philippe et al., 2011a).

In contrast to stochastic error, which is a function of the sample size, systematic errors can result from misspecifications in the model of sequence evolution (model violations), which often do not take into account aspects such as compositional heterogeneity, heterotachy (shifts in site-specific evolutionary rates over time), site non-independence, and site-heterogeneous nucleotide or amino acid replacement (Rodriguez-Ezpeleta et al., 2007). This type of error can be minimized by using evolutionary models that account for these evolutionary patterns and/or by excluding fast-evolving sites from analyses (Naylor and Brown, 1997; Rodriguez-Ezpeleta et al., 2007). Recently, it was demonstrated that a switch from the WAG (Whelan and Goldman, 2001) to the CAT (Lartillot and Philippe, 2004) model, which eliminates the likely erroneous assumption of identical distribution of substitutions among sites and accounts for site-heterogeneous amino acid replacement, can reconcile some phylogenomic inferences of non-bilaterian animal relationships (Philippe et al., 2011a). Further, if the orthology assumption is violated, the history of the gene is different from the history of the organisms from which it was obtained, resulting in a gene tree–species tree conflict.

To minimize systematic error, several main issues should be considered before the analysis of phylogenetic/phylogenomic data: (a) orthology assignment and alignment construction, (b) gene marker selection, (c) choice of substitution models, and (d) taxon sampling, including outgroup selection (Philippe et al., 2001a).

Although it might sound trivial to the experienced, care must be taken when assembling supermatrices from EST and genome data to select only orthologous sequences, since the inclusion of paralogs can result in erroneous tree topologies. Although it is sometimes expected that this effect will be buffered in genome-scale analyses, this is not always the case, especially when the true phylogenetic signal is weak (Philippe et al., 2011a).

Today's approaches for inferring orthology relationships can be roughly divided into graph-based and tree-based methods. Given a set of proteins from the species of interest, graph-based methods start with a similarity search using tools like BLAST (Altschul et al., 1997) and use bit score or E-value as a proxy for the evolutionary distance between protein pairs. From the distances between all protein pairs between complete proteomes, graph-based methods build ortholog groups using a variety of clustering criteria: best reciprocal hits, e.g. OMA (Altenhoff et al., 2010), OrthoInspector (Linard et al., 2011), InParanoid (Ostlund et al., 2010); best triangular hit, e.g. COG, eggNOG (Powell et al., 2011); or Markov clustering, e.g., OrthoMCL (Chen et al., 2006). In contrast, tree-based approaches use reconstructed protein family trees to infer orthologs. The inference is done by a tree reconciliation or mapping of the protein (gene) tree to the corresponding species tree (Mirkin, Muchnik, and Smith, 1995; Page and Charleston, 1997), which labels the internal nodes of the protein trees as either speciation or duplication events (Zmasek and Eddy, 2001). Based on this labeling, orthology and paralogy relationships are inferred. Most existing methods rely on full proteomes as input and cannot use nucleotide, e.g., non-translated EST, sequences.

The programs OrthoSelect (Schreiber et al., 2009) and HaMsTR (Ebersberger, Strauss, and von Haeseler, 2009) have been developed to fill this gap. OrthoSelect uses an initial BLAST search to assign sequences to KOG groups and employs different tools (ESTScan, standard 6-frame translation) to find the most correct translation for an EST. HaMsTR is based on a set of core orthologs and uses HMMer for group assignment of a set of translated EST and protein sequences. For the full list of orthology methods, see http://questfororthologs.org/orthology_databases.

Both approaches outlined above have certain advantages and disadvantages. Tree-based approaches are perhaps a more intuitive way of assigning orthology. They make use of a multiple sequence alignment, which is normally more reliable than pairwise alignments. Unfortunately, the high computational complexity and reliance on correct multiple sequence alignments and protein trees make tree-based methods unsuitable for large protein families or large numbers of proteins (Liu et al., 2009; Thompson et al., 2011). Graph-based methods are computationally less demanding and easier to automate. Implementations of these methods handling large sets of sequences already exist. However, they are not hierarchical and group proteins from different species into a single group, which is not a natural representation. Furthermore, although graph-based methods are computationally less expensive than most tree-building methods, they still pose a problem when large numbers of taxa (hundreds of species) are analyzed.

Accurate alignment of the chosen orthologous sequences is also critical, as it is well known that the results of phylogenetic analyses can be dramatically affected by the underlying sequence alignments. For example, Wong et al. (2008) have shown how different methods for multiple alignment can lead to contradicting phylogenetic trees derived from these alignments by standard phylogeny methods. The problem is that multiple alignments often contain regions of low sequence similarity where alignment quality is questionable. Consequently, several computational approaches have been developed to identify non-informative or poorly aligned regions within multiple alignments. The program Gblocks (Castresana, 2000) automatically selects conserved, gap-free blocks from multiple alignments; more recently, the program TrimAl has been proposed that is based on similar ideas (Capella-Gutiérrez, Silla-Martínez, and Gabaldón, 2009). By contrast, Noisy (Dress et al., 2008) allows the identification of phylogenetically uninformative homoplastic sites in multiple alignments using circular split systems. Aliscore (Misof and Misof, 2009) detects random similarity in multiple alignments using Monte Carlo resampling within a sliding window. All these methods are used to remove unreliable regions from sequence alignments prior to phylogenetic analysis. Even though removing positions from an alignment reduces the amount of information available for phylogeny inference, it has been shown that the quality of the resulting trees can be significantly improved by restricting phylogenetic analyses to reliable and informative alignment regions (Talavera and Castresana, 2007).

Although several automated pipelines have been developed to perform some or all of these steps (e.g., Roure, Rodríguez-Ezpeleta, and Philippe, 2007; Ebersberger, Strauss, and von Haeseler, 2009; Kumar et al., 2009; Schreiber et al., 2009), supermatrix assembly and alignment must include a quality control step that often – if matrix size is not prohibitive – involves manual inspection and curation. However, with ever increasing sizes of phylogenomic matrices this will not always be feasible in future and automated approaches to matrix assembly, such as AGALMA (Howison, Sinnott-Armstrong, and Dunn, 2012), might be the way forward, also to allow for objective repetition of experimental procedures. As for the evolutionary model, it has been shown that the CAT model (Lartillot and Philippe, 2004), which accounts for the variation of evolutionary patterns across sites, is currently the best-fitting model for phylogenomic data sets that contain taxa from all major bilaterian and non-bilaterian metazoan lineages (Philippe et al., 2009; Philippe et al., 2011a). However, even the CAT model does not account for all peculiarities of amino acid changes during evolution. Therefore the development of more sophisticated evolutionary models is of prime importance in future.

One aspect that has not received much attention yet is the effect of gene sampling on the resulting deep-level metazoan phylogeny. A common practice to date is to include as many relatively conserved genes into a phylogenetic alignment as the data allow (e.g., Gatesy and Baker, 2005; Dunn et al., 2008; Hejnol et al., 2009; Meusemann et al., 2010; Srivastava et al., 2010). This "random-gene" approach minimizes cognitive biases associated with a priori choice of target genes and reduces the impact of gene-specific signals on phylogenetic inferences (reviewed in Philippe et al., 2001a). However, genes selected using this method demonstrate a diverse spectrum of evolutionary rates and, therefore, have different phylogenetic utilities (Tekle et al., 2010; Ebersberger et al., 2011; Fong and Fujita, 2011). Phylogenetic utility (or informativeness) of each gene marker is not a constant value; it depends on the divergence times of interest (Graybeal, 1994; Giribet, 2002). It has been demonstrated that sites informative for resolving relationships between terminal branches can be homoplasious (noisy) at deeper levels of the same tree (Wenzel and Siddall, 1999). Due to the uneven distribution of the true phylogenetic signal-to-noise ratio across time, additional gene selection criteria are required when a phylogenetic problem is associated with deep-in-time rapid radiation events, such as gene-specific evolutionary rates linked to gene functions, history (e.g., recent gene duplications or horizontal gene transfer events), and its position in metabolic networks (Hudson and Conant, 2011). Although several methods of quantifying the phylogenetic informativeness have been proposed and tested (Townsend and Lopez-Giraldez, 2010), there are as yet no clear thresholds for assessing the goodness of a given gene or gene set for resolving a particular phylogenetic problem (Philippe et al., 2011a). Therefore, a consistency of a tree topology inferred from different data sets remains the main criterion for verifying the validity of a phylogenetic hypothesis (Comas, Moya, and Gonzalez-Candelas, 2007a; Ebersberger et al., 2011).

To assess the effect of gene sampling on resulting deep-level metazoan phylogeny, Nosenko et al. (2013) performed analyses of two large supermatrices (> 9000 amino acid characters each) that had identical taxon sampling, but non-overlapping gene content. These analyses reconstructed conflicting phylogenies. Relationships between major sponge lineages and the position of ctenophores relative to other metazoans were in the center of this conflict. Remarkably, the two partitions showed different saturation levels. The less saturated data set composed of genes involved in translation supported monophyletic Porifera and Coelenterata as the sister-group to the Bilateria (similar to the traditional morphology-based metazoan phylogeny supported also by the molecular phylogenomic study by Philippe et al., 2009). On the contrary, the more saturated data set that included genes from various functional categories ("random" genes) favored paraphyletic Porifera and ctenophores at the base of the Metazoa (similar to Dunn et al., 2008). This result violates the consistency criterion (Comas, Moya, and Gonzalez-Candelas, 2007a; Ebersberger et al., 2011) and reflects the overall conflicting results of the higher-level metazoan phylogenomic studies described above. Since the two supermatrices were constructed and analyzed using the same methods, they were expected to have similar levels of non-phylogenetic signal associated with ortholog selection, aligning, and the choice of an evolutionary model. Therefore, different levels of saturation between the two data sets and the observed inconsistency of resulting phylogenies should be attributed to their different gene compositions.

Considering the potential sources of error outlined above, conflicting results among studies addressing higher-level metazoan phylogeny are not surprising, since these studies attempt to reconstruct cladogenetic events that occurred hundreds of millions of years ago (Ma), possibly as early as the Cryogenian (~ 650 Ma, Peterson et al., 2008). Resolving such ancient splits with molecular sequence data is difficult due to a relatively low phylogenetic signal at short internal branches (see Rokas and Carroll, 2006) and phylogenetic signal erosion caused by multiple substitutions along long terminal branches (saturation) ("non-phylogenetic signal", see Philippe et al., 2011a). Felsenstein (1985) showed that a minimum of three non-conflicting apomorphies (substitutions) are necessary on each branch to achieve high-level node support (95 % bootstrap support).

Finally, taxon sampling and outgroup selection can also have a significant effect on tree topology (Pick et al., 2010; Philippe et al., 2011a) and branch support (Philippe et al., 2009), respectively, as limited or uneven across-lineage sampling of ingroup taxa and too distant outgroups may enhance LBA artifacts (Bergsten, 2005). One should also try to avoid missing data, i.e., genes should be selected that are available for as many taxa as possible, although it has been suggested that missing data per se are not a problem as long as the present data are sufficiently informative and the appropriate method of analysis is chosen (Hartmann and Vision, 2008; Wiens and Moen, 2008; Wiens and Morrill, 2011). However, Roure et al. (2012) have recently shown that large amounts of missing data frequently present in large phylogenomic data sets (e.g., Hejnol et al., 2009) may lead to phylogenetic artifacts because systematic errors might be exacerbated by "reducing the number of species effectively available for the detection of multiple substitutions". Roure et al. (2012) consequently argued to use smaller but less gappy data sets with a relatively small number of genes (about 50), preferably the highly conserved and least saturated ones. Genes involved in translation might be perfect candidates here, as they are highly expressed and ubiquitously present in transcriptome data sets.

In summary, resolving higher-level metazoan relationships is still a difficult and challenging endeavor. However, the increased availability of massive genome-scale data coupled with continuous improvement of analytical approaches have the potential to eventually achieve a well-supported backbone phylogeny of the animal tree of life.

2.2.2 Are sponges paraphyletic (or monophyletic after all), and why is this important?

Porifera as a monophyletic group is well supported by morphological character analyses (e.g., Böger, 1983; Ax, 1995; Reitner and Mehl, 1996). Within the phylum, the four extant classes Demospongiae, Hexactinellida, Calcarea and Homoscleromorpha are well defined morphologically (Hooper and Van Soest, 2002; Gazave et al., 2012) as well as by molecular data (see reviews by Erpenbeck and Wörheide, 2007; Wörheide et al., 2012). However, the phylogenetic relationships between those four taxa and their placement relative to the other non-bilaterians are still debated. Different hypotheses about the relationships within monophyletic Porifera have been postulated based on cladistic analyses of morphological characters (see Erpenbeck and Wörheide, 2007 for a summary). With the introduction of molecular systematics to the study of deep sponge relationships in the early 1990s, the hypothesis that sponges might be a paraphyletic assemblage sharing a grade of construction rather than common ancestry was proposed. The history of this debate has recently been reviewed in detail by Wörheide et al. (2012). Briefly, according to the hypothesis of paraphyletic Porifera, either Calcarea (earlier studies), or Homoscleromorpha (later studies), or both are more closely related to the rest of the Metazoa than the other sponge clades Demospongiae and Hexactinellida. However, the majority of molecular studies that found sponges to be paraphyletic either did not provide significant support for the relevant nodes, and/or were based on limited amounts of sequence data, did not include all relevant taxa and/or used simple (distance) methods to reconstruct the phylogeny (see summary by Wörheide et al., 2012). At present, analyses of the largest phylogenomic data sets (Philippe et al., 2009; Pick et al., 2010) using the best available methods (Philippe et al., 2011a) agreed with the results of cladistic analyses of morphological data (e.g., Böger, 1983; Ax, 1995; Reitner and Mehl, 1996) on sponge monophyly. Congruence of phylogenetic hypotheses resulting from independent data is among the strongest supporting evidence for a particular hypothesis (Comas, Moya, and Gonzalez-Candelas, 2007a; Pisani, Benton, and Wilkinson, 2007). Consequently, sponge monophyly should be regarded at present as the best possible working hypothesis, until rejected by an independent data set. So, why is the phylogenetic status of Porifera as either monophyletic, or paraphyletic important for our understanding of early animal evolution?

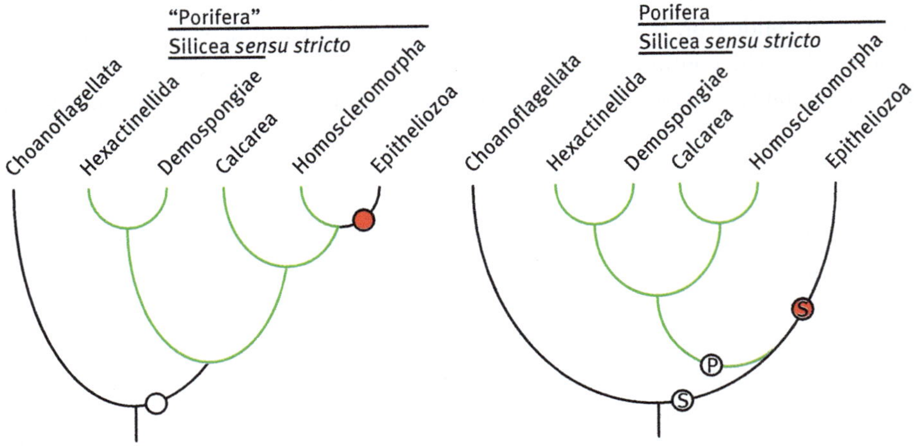

Figure 2.2: Alternative scenarios for the higher-level relationships of extant sponges. **Left:** sponge paraphyly (e.g., Sperling, Peterson, and Pisani, 2009). According to this scenario, a sponge-like body plan (white circle) was acquired in the last common ancestor of Porifera and Eumetazoa (= Cnidaria, Ctenophora, Placozoa, Bilateria) and subsequently lost (red circle) from the last common ancestor of Eumetazoa. Alternative paraphyly scenarios exist mainly in earlier studies, where homoscleromorphs were often not included (see text for details). **Right:** sponge monophyly (e.g., Philippe et al., 2009). According to this scenario, the sponge-like body plan (white circles) was acquired either in the stem lineage of Porifera (P), or, if choanocytes are considered homologous to choanoflagellate cells as judged by outgroup comparison to the well-established sister-group of the Metazoa, the Choanoflagellata (see text for details), in stem-group metazoans (S). The latter scenario would require one gain and one loss (indicated by white/red circles marked with S), as in the paraphyly hypothesis (modified from Wörheide et al., 2012).

If sponges are indeed monophyletic and represent the earliest branch in the metazoan tree, some inferences can be made about the last common ancestor of all animals. However, these partly hinge on the homology of sponge choanocytes and choanoflagellate cells. Choanocytes, flagellate filtering cells that line the sponge's aquiferous system, are one of the most remarkable features of this group of animals. Water is pumped from the external medium through a system of internal canals to chambers lined by those flagellate cells. Since Choanoflagellata is well established as the sister-group to the Metazoa (King et al., 2008), homology of sponge choanocytes and choanoflagellate cells appears obvious. Nonetheless, it has been proposed that the overall morphological similarities of both cell types could result from convergent evolution (see discussion in Woollacott and Pinto, 1995; Karpov and Leadbeater, 1998; Philippe et al., 2009). If choanocytes and choanoflagellate cells are not homologous, then the poriferan aquiferous system may have evolved independently in the sponge stem group (Figure 2.2) and, therefore, must not necessarily have been present in the last common ancestor of all animals. If the opposite is true and sponge choanocytes are homologous to choanoflagellate cells, then the last common ancestor of all animals probably was a filter-feeding sponge-like organism. Another important issue is the evolution of a basement membrane and, consequently, "true" epithelia as classically recognized. In sponges, a type-IV collagen-containing basement membrane (Boute et al., 1996) is only present in class Homoscleromorpha. If sponges are monophyletic, then there are two possible scenarios of the metazoan basement membrane evolution: (a) it was present in the last common ancestor of the Metazoa and has been subsequently lost from all sponge lineages except the Homoscleromorpha (Figure 2.2; see also discussion in Lavrov, 2007), or (b) it convergently evolved in homoscleromorph sponges and Eumetazoa (see Fig. 3 in Philippe et al., 2009). Scenario (a) implies widespread loss of a basement membrane in several main sponge lineages. Loss of this trait is not uncommon as it apparently also occurred in some Turbellaria (Platyhelminthes, Brusca and Brusca, 2003). Furthermore, Leys and Riesgo (2011) recently indicated that a major constituent of a basement membrane, type-IV collagen, is also present in sponge lineages other than the Homoscleromorpha. It appears therefore plausible that stem group metazoans already possessed type-IV collagen (as suggested by Aouacheria et al., 2006). However, it is unanswered at present whether the basement membrane is a convergently evolved (and more recent) autapomorphy of homoscleromorphs and eumetazoans, which involved an independent co-option of type-IV collagen into this structure, or autapomorphic for the Metazoa.

A few studies suggested a sister-group relationship between the Homoscleromorpha (formerly included in Demospongiae) and the Calcarea within monophyletic Porifera (e.g., Philippe et al., 2009; Pick et al., 2010). Unequivocal morphological synapomorphies are difficult to identify, but recent ultrastructural studies suggested that similarities in the kinetid of the choanocytes of calcisponges and homoscleromorphs can be considered as such (Pozdnyakov and Karpov, 2013). This sister-group relationship has novel implications for the evolution of some of the key traits of sponges, e.g., spiculogenesis. Spicules are one of the taxonomic hallmarks of sponges, present in most species and used for systematics (Hooper and Van Soest, 2002). A sister-group relationship of Calcarea (calcareous spicules) and Homoscleromorpha (siliceous spicules) would imply that silica spiculogenesis either evolved several times independently in sponges (see also Maldonado and Riesgo, 2007) or is autapomorphic for Porifera secondarily lost from Calcarea. Further, in demosponges and hexactinellids spicules are produced around an axial filament (Uriz et al., 2003), which in demosponges contains silicatein (Shimizu et al., 1998). Apparently silicatein was characterized in a single hexactinellid species, *Crateromorpha meyeri* (Müller et al., 2008).

However, a subsequent study suggested that classical silicateins are not necessarily involved in the formation of spicules in other hexactinellids (Veremeichik et al., 2011). The situation in Homoscleromorpha is less clear as they seem to build their silica spicules differently than Demospongiae (Maldonado and Riesgo, 2007). However, recent paleontological work (Botting et al., 2012) suggested that both calcareous and siliceous spicules originated from a common bimineralic ancestor, i.e., are homologous. Obviously, more effort is required to answer whether spiculogenesis in sponges is homologous or a convergently evolved trait.

If sponges are paraphyletic, then choanocytes and the aquiferous system are most parsimoniously interpreted as being ancestral for the Metazoa. This implies that our most recent common ancestor was a sponge-like organism (Figure 2.2). If true, then non-sponge metazoans are derived from sponge-like organisms. The loss of poriferan attributes could have involved either neotenic evolution from "an early developmental stage of ancient sponges" (Maldonado, 2004), or a homoscleromorph-like larva becoming sexually mature (Nielsen, 2008).

Turning back to the evolution of the basement membrane, the alleged closer relationship of Homoscleromorpha with the Eumetazoa in the sponge paraphyly scenario is important, because the presence of a basement membrane in both, traditionally considered to define "true" epithelia, has been understood as a synapomorphy of Homoscleromorpha and Eumetazoa (Sperling, Pisani, and Peterson, 2007). Thus, Sperling et al. (2009) included Homoscleromorpha in the Epitheliozoa (Ax, 1995), a clade combining Eumetazoa and Placozoa.

2.3 Conclusions and outlook

Deep sponge phylogenetics has come a long way in recent years. Large-scale phylogenomic analyses have so far rejected the hypothesis that sponges are paraphyletic; results of several studies are consistent with the notion of monophyletic Porifera. However, while sponge monophyly should be regarded as the best possible working hypothesis at present, it is imperative to further test this hypothesis with additional data. Because of the caveats of sequence-based analyses, as outlined above, further evidence from sequence-independent markers should be sought. However, the presence/absence of microRNAs, a set of sequence-independent markers proven useful to corroborate phylogenomic analyses in the Bilateria (Philippe et al., 2011a; Rota-Stabelli et al., 2011), have failed as yet to provide conclusive evidence to reject either sponge monophyly or paraphyly (Sperling et al., 2010). Other sequence-independent markers such as the not yet fully explored "rare genomic changes" (Rokas and Holland, 2000), including transposable elements (Piskurek and Jackson, 2011), might become promising targets for such endeavors once more sponge and non-bilaterian nuclear genomes have fully been sequenced. With those additional novel data sets at hand we might come closer to fully resolve the deepest branches of the animal tree of life.

Acknowledgments

We acknowledge funding from the German Research Foundation (DFG) through the Priority Program 1174 “Deep Metazoan Phylogeny” (Project Wo896/6).

Michael Eitel, Wolfgang Jakob, Hans-Jürgen Osigus, Omid Paknia, Karolin von der Chevallerie, Tjard Bergmann, and Bernd Schierwater

3 Phylogenetics and phylogenomics at the root of the Metazoa

Abstract: Knowledge of the phylogeny and evolution of the non-bilaterian phyla is critical for reconstructing the origin and diversification of animal life. At the base of metazoan evolution four extant phyla are found, Placozoa, Cnidaria, Porifera, and Ctenophora. The phylogenetic relationships between these phyla, class-level relationships within phyla, the unicellular ancestor and the split off of bilaterians have been subject of controversial discussion for more than a century. We have generated and analyzed a variety of molecular markers to resolve deep phylogenetic relationships in early metazoans and learned that we might not be able to resolve the base of the metazoan Tree of Life (ToL) by molecular sequence markers alone. Thus we also included characters from morphology, rRNA molecule morphology, mitochondrial genomes, and the genomic organization of *Antp*-class genes. Our analyses of single character-type data sets as well as total evidence analyses suggest a position for the Placozoa close to the last common metazoan ancestor (LCMA).

3.1 Introduction

At the base of metazoan evolution phylogenetic relationships between the Placozoa, Porifera, Cnidaria, and Ctenophora have been highly controversial (Collins et al., 2005; DeSalle and Schierwater, 2008; Philippe et al., 2011a). As a consequence there has been no agreement on the origin of Metazoa (the "urmetazoan" bauplan) and the root for bilaterian animals (Miller and Ball, 2008; Blackstone, 2009). Different studies have created different hypotheses on early animal relationships. Critical reevaluation and reanalyses of existing data sets using different algorithms have shown the error-prone nature of large data sets and analytical methods (Philippe et al., 2011a) and have increased the variety of evolutionary scenarios. Identifying stable hypotheses on early animal evolution is a hard task using current (limited) data sets, phylogenetic inference methods, and unsolved problems in character weighting (particularly when combining morphological and molecular data). Almost all possible evolutionary scenarios, no matter how meaningful, have been suggested at least once (see DeSalle and Schierwater, 2008; Schierwater et al., 2009b for overview). Among the meaningful scenarios there are only three alternatives that we want to discuss here. Either Placozoa or Porifera are sister-groups to all other metazoans or there was an early split of non-bilaterian animals and Bilateria. In the latter case the Urmetazoon might have been derived from a placula-like bauplan, which was closely related to the bauplan of the recent Placozoa (Syed and Schierwater, 2002a; Schierwater et al., 2009a).

The phylum Placozoa is represented by the most simply organized animals, which are missing most of the metazoan-typical characters. They lack nerve or muscle cells, organs, an extracellular matrix, a basal lamina, any kind of symmetry and anterior-posterior axis (Syed and Schierwater, 2002a, 2002b; Schierwater et al., 2009a for details). Traditionally an early branching position for the Placozoa has been proposed on the grounds of morphological simplicity (Grell and Benwitz, 1971; Ivanov, 1973; for controversial views see: Ax, 1995; Peterson and Eernisse, 2001; Miller and Ball, 2008). Based on molecular data (mostly ribosomal RNA sequences), a number of phylogenetic analyses have shaken the phylogenetic position of the placozoan *Trichoplax adhaerens* and all possible evolutionary scenarios (apart from Placozoa being within Bilateria) have been suggested; the most often cited views place Placozoa in a clade close to the Cnidaria (e.g. Wainright et al., 1993; Aleshin et al., 1995; Odorico and Miller, 1997). The mitochondrial genome, however, clearly rejects a close relationship to Cnidaria and suggests an early branching of Placozoa in a non-bilaterian clade (Ender and Schierwater, 2003; Dellaporta et al., 2006; Signorovitch et al., 2007). It is unclear whether this branching is the result of long branch attraction in the mitochondrial data (as discussed in Dellaporta et al., 2006). However, this split was also recovered using nuclear genomic data at least in some most recent analyses (Nosenko et al., 2013).

The Porifera are more complexly organized than the Placozoa. For example, the number of cell types is two to five-fold higher than in Placozoa and a defined body axis (foot-osculum) is often present (e.g. Kilian, 1993). Most molecular studies that include Placozoa suggest Porifera as the first branching extant metazoan phylum (e.g. Philippe et al., 2009; Pick et al., 2010).

The phylum Cnidaria shows a more complex bauplan compared to the Placozoa and Porifera and it is particularly notable also for its complexity in life history. Primitive organs appear as gonads and eyes, and a well-defined oral-aboral (head-foot) axis defines the bauplan (Werner, 1993). Metagenesis, i.e. the regular alternation between a bisexual (usually medusa) and vegetative (usually polyp) generation, in the later branching cnidarian classes Staurozoa, Hydrozoa, Scyphozoa, and Cubozoa (so called "Medusozoa"), reaches an unrivaled diversity of appearances like in no other phylum. Monophyly of the phylum Cnidaria and a position of Anthozoa as sister-clade to other cnidarians have virtually been uncontested (Bridge et al., 1992; Collins, 2002). Relationships within and among classes, however, remain problematic. Organismal data have revealed conflicting scenarios about sister- and molecular group relationships (Collins, 2002; Marques and Collins, 2004). The presently best-supported trees confirm an early branching position for monophyletic Anthozoa, provide some resolution for most of the Medusozoa groups and suggest an evolutionary later appearance of Cubozoa (cf. Collins et al., 2006a). Recent mitogenomic data, however, question a monophyletic Anthozoa, claiming octocorallian Anthozoa to form a sister-clade to Medusozoa (Kayal et al., 2013).

The Ctenophora show superficial similarity to cnidarian medusae but completely lack any kind of cnidocytes or metagenesis. Eight "comb rows" of fused cilia arranged along the sides of the animal represent a unique ctenophoran character. The phylogenetic position of ctenophores and all other non-bilaterian phyla, however, has been and still is under dispute (Podar et al., 2001; Aleshin and Petrov, 2002; Dunn et al., 2008). Historically, Ctenophora has generally been found as a sister-group to the Bilateria in cladistic analyses of morphological characters (e.g. Zrzavý et al., 1998; Peterson and Eernisse, 2001; Martindale et al., 2002; Collins et al., 2005). Phylogenomic studies that have given weak support for a sister-group relationship to all other metazoans (Dunn et al., 2008; Hejnol et al., 2009) have been shown to be the result of long-branch attraction artifacts (Philippe et al., 2009; Pick et al., 2010; Siddall, 2010; Philippe et al., 2011; Nosenko et al., 2013).

Radiation within the non-bilaterian phyla varies greatly. For Placozoa only one extant species has yet been described (see however Voigt et al., 2004; Eitel and Schierwater, 2010; Eitel et al., 2013 for undescribed species lines). Some 8,500 valid species of Porifera, 190 species of Ctenophora, and approximately 10,200 accepted extant species of Cnidaria are known (Appeltans et al., 2012a). In Placozoa no higher taxonomic levels could be erected yet, because of the lack of species descriptions. Porifera were divided into four different classes (Gazave et al., 2012; Van Soest et al., 2012) and Ctenophora have traditionally been divided into two well-defined classes (see e.g. Liley, 1958), while morphological (Harbison, 1985) and molecular (Podar et al., 2001) data reject this scenario. The most species rich phylum Cnidaria includes five classes, Antho-, Stauro-, Scypho-, Hydro-, and Cubozoa. Based on mtDNA data and ribosomal gene sequences, class-level relationships have begun to clear up (Bridge et al., 1992; Bridge et al., 1995) but are not completely resolved yet. This is highly unfortunate, since assembling the cnidarian tree of life will allow us to better understand basic principles of life cycle evolution and morphological diversity as well as the origin of body axes in triploblasts.

In sum, phylogenetic relationships near the base of the metazoan ToL have become a mess with the advent of molecular systematics and remain a big phylogenetic conundrum (cf. Blackstone, 2009). Open questions include: (i) a possible sister-clade relationship between non-bilaterians and bilaterians (triploblastic animals), (ii) the relationships among non-bilaterian phyla, (iii) the relationships of Staurozoa, Cubozoa, Hydrozoa and Scyphozoa within Cnidaria, (iv) the systematics of Placozoa, (v) monophyly vs. paraphyly in the "Eumetazoa" and Porifera. The often found term "Eumetazoa" for animals with true epithelia that perform gastrulation leading to the establishment of distinct germ layers, and which includes all metazoans except Porifera and Placozoa is most likely outdated. Porifera have been shown to possess true epithelia at least in some species and in some life stages (Humbert-David and Garrone, 1993; Boute et al., 1996; Boury-Esnault et al., 2003; Ereskovsky and Tokina, 2007; Ereskovsky et al., 2010), and – although widely neglected – gastrulation was identified at least in hexactinellid and calcareous sponges (Boury-Esnault et al., 1999; Leys, 2004;

Leys and Eerkes-Medrano, 2005; Ereskovsky et al., 2013). The Placozoa have true epithelia although gap-junctions and a basement membrane are missing. So far, the placozoan embryonic development has not been clarified (see Eitel et al., 2011) and gastrulation remains unknown. The monophyly of the sponges has been questioned in several studies (Medina et al., 2001; Peterson and Eernisse, 2001; Rockas, Kruger, and Carroll, 2005; Dunn et al., 2008; Sperling et al., 2009; Erwin et al., 2011; Nosenko et al., 2013) but has been supported in several other studies (e.g. Philippe et al., 2009; Pick et al., 2010; Nosenko et al., 2013). The term "Eumetazoa" is no longer helpful.

Obviously, morphological characters are very limited in number, DNA sequence markers seem to be insufficient yet, and other markers not yet satisfactorily explored. In addition, the relative weighting of different character sets in a total evidence (concatenated) data matrix has not been addressed yet for the base of the metazoan ToL. It will be crucial to explore new data sets, such as gene arrangement data (e.g. in mitochondrial genome and Antp gene clusters) and RNA molecular morphology data, and to find a valuable way to score them. It seems that in the given context and near future, data analysis and interpretation will become at least as crucial as data collection and taxon sampling.

3.2 Project data

In the course of our research project we generated different data sets that contribute to class-level relationships in Cnidaria (Collins et al., 2006b; Collins et al., 2006a; Nosenko et al., 2013), phylum-level phylogenetic signals in ANTP-class genes (Cartwright et al., 2006; Kamm and Schierwater, 2006; Kamm et al., 2006; Monteiro et al., 2006; Schierwater et al., 2008; Schierwater and Kamm, 2010), the phylogenetic position of the Placozoa (Dellaporta et al., 2006; Srivastava et al., 2008; Schierwater, de Jong, and DeSalle, 2009; Schierwater et al., 2009a), the diploblast-triploblast split (Schierwater et al., 2009b) and total evidence analysis approaches (Schierwater et al., 2009a). In the following we will briefly summarize the importance of the different data sets for a discussion on the phylogenetic relationships at the base of the Metazoa.

3.2.1 ANTP superclass genes

The structure and genomic organization of ANTP superclass genes support an early split of Hox gene evolution in Bilateria and Hox-like genes in non-bilaterian animals. This suggested split is seen at different levels, but the ANTP superclass, particularly the Hox/ParaHox-like gene repertoire in the hypothetical "Urmetazoon" remains speculative. The evolutionary scenarios suggested below are consistent with the idea of Placozoa as the earliest branching phylum in a non-bilaterian clade.

Although EvoDevo data on gene expression have not entered data matrices for phylogenetic analyses yet, these data appear nonetheless phylogenetically informative. Parsimonious interpretations of comparative expression patterns again support an early split at least of Cnidaria and Bilateria (Cartwright et al., 2006; de Jong et al., 2006; Kamm et al., 2006; Jakob and Schierwater, 2007).

Several gene families have diversified during animal evolution, concurrently with the radiation of the Metazoa. The homeobox genes are a key example (e.g. Holland, 2001; Jakob et al., 2004; Holland and Takahashi, 2005; Monteiro et al., 2006). They play important roles during the development and cell proliferation in animals, plants, and fungi (Ferrier and Holland, 2001). The homeobox genes are characterized by the presence of a conserved 180 bp sequence encoding a DNA binding motif, the homeobox. In the Metazoa these genes could be divided into the superclasses ANTP, PRD, POU, LIM, SIX and TALE (Galliot et al., 1999; Gauchat et al., 2000; Banerjee-Basu and Baxevanis, 2001). Most of them appear to be restricted to the animal kingdom. The highest gene diversity by far has been observed in the ANTP superclass, which seems to represent a vital factor for the evolution of metazoan bauplans, by driving the development of body axes and segment identities. The ANTP superclass includes the Hox/ParaHox, extended Hox, and NKL genes. All Bilateria possess Hox/ParaHox, NKL and extended Hox genes, serving conserved functions. Clustered Hox genes, which provide positional information along the anterior-posterior body axis across Bilateria, are the best-known examples. With fundamental roles and high diversity among metazoans, the ANTP genes are suitable candidates for elucidating the relationship between genetic and phenotypic change in early evolution. Among extant lower metazoan phyla, ANTP genes have been studied fairly well, within Placozoa (*Trichoplax*), Porifera (*Amphimedon*), and Cnidaria (*Nematostella*, *Eleutheria*, and *Clytia*) (Chourrout et al., 2006; Larroux et al., 2007; Schierwater et al., 2008). Comparing the ANTP gene composition of these three taxa shows that the Placozoa possess less ANTP diversity than Cnidaria but a higher diversity than found in Porifera (Table 3.1).

Nematostella has the largest ANTP superclass gene repertoire with 78 genes and *Amphimedon* has the smallest repertoire with only eight genes. While an earlier study on *Trichoplax* revealed a low diversity of ANTP genes (Jakob et al., 2004; Hadrys et al., 2005; Monteiro et al., 2006), a more detailed investigation using genome data of *Trichoplax* showed that the *Trichoplax* genome harbors at least 14 ANTP class genes (Schierwater et al., 2008) (Table 3.1). Phylogenetic analyses conducted to infer relationships of these 14 genes to known ANTP families suggest that most of the *Trichoplax* ANTP genes (11 out of 14 genes) are clustered in three groups representing Hox/ParaHox-like, extended Hox and NKL genes (Figure 3.1), suggesting the presence of a complex ANTP repertoire in *Trichoplax adhaerens*.

The most striking difference among the non-bilaterian phyla is the lack of any Hox/ParaHox-like and extended Hox genes in Porifera. All *Amphimedon* ANTP genes fall within the NK-related group. For some researchers this finding suggests that the

Table 3.1: Comparison of homeobox genes between three taxa of different early branching metazoan phyla (*Trichoplax adhaerens*, Placozoa; *Nematostella vectensis*, Cnidaria; *Amphimedon queenslandica*, Porifera). The homeobox gene complement of *Trichoplax adhaerens* shows a slightly higher diversity of the homeobox classes ANTP, PRD and LIM. Orthology assignment of the *Trichoplax* homeobox genes is based upon BLAST hits in the case of non-ANTP genes. ANTP genes have been classified according to the phylogenetic analysis shown in Figure 3.1. Modified after Schierwater et al. (2008).

	Trichoplax	Nematostella	Amphimedon
ANTP	14	78	8
NKL	NK2, NK5, NK6, Hex, Dlx, Dbx/Hlx + 2 NK-related genes	all (bilaterian) families except Tlx	NK2(2/3), NK6, Msx, BarH-related, Hex-related, Tlx-related
Ext. Hox	Not, Mnx	all families except Eng and possibily Vax	–
Hox/ParaHox	Gsx + 1 gene with some affinity to ext. Hox and Hox/ParaHox	Gsx, anterior Hox, posterior Hox/Cdx-like+several genes with unclear relation to bilaterian Hox/ParaHox genes	–
PRD	9(Arx, Ebx/Arx-like, Pax3-like, PaxB, Prd/Pax-like, Pitx, Gsc, Otp)	33 (PaxA, B, C, Pax6-like, Arx, Rx, Pitx, Otx, Otp, Gsc & several unassigned)	9 (Arx, PaxB, Rx, OG12)
POU	2 (Pou 3, 4)	5 (POU 1, 3, 4, 6)	4 (POU 1, 6, 2–5)
LIM-HD	5 (LIM 1, 1/5, 3/4, 2/9, Isl)	6 (Lim 1, 2, 3/4 Awh, Isl)	3 (LIM 3, Lin-11, Isl)
SIX	2 (Six 1/2, Six 3/6)	5 (Six 2, 3, 4, 4/5)	1 (Six 1/2)
TALE	4 (PBX/PBC, Pknox, Irx, Meis)	7 (PBC, TGIF, Meis, Irx)	6 (Meis, PBC, Irx)
HNF	1	1	–
Total	37	134	31

Figure 3.1: Bayesian analysis shows supported grouping for 11 out of 14 placozoan ANTP genes to known ANTP genes from Cnidaria, Porifera, and Bilateria. Genes involved in the analysis represent Placozoa (*Trichoplax adhaerens* Ta), Cnidaria (*Nematostella vectensis* Nv, *Eleutheria dichotoma* Ed, *Acropora formosa* Af), Porifera (*Amphimedon queenslandica* Aq), protostomes (*Drosophila melanogaster* Dm, *Anopheles gambiae* Ag, *Discocelis tigrina* Dt, *Schistocerca gregaria* Sg), and deuterostomes (*Branchiostoma floridae* Bf, *Danio rerio* Dr, *Homo sapiens* Hs, *Oikopleura dioica* Od, *Strongylocentrotus purpuratus* Sp, *Xenopus laevis* Xl). Genes of the NKL subclass are framed in yellow, extended Hox genes in blue and Hox/ParaHox-like genes in red. *Amphimedon*, *Nematostella* and *Trichoplax* genes are color-coded. Tree rooted on midpoint. Modified after Schierwater et al. (2008). ►

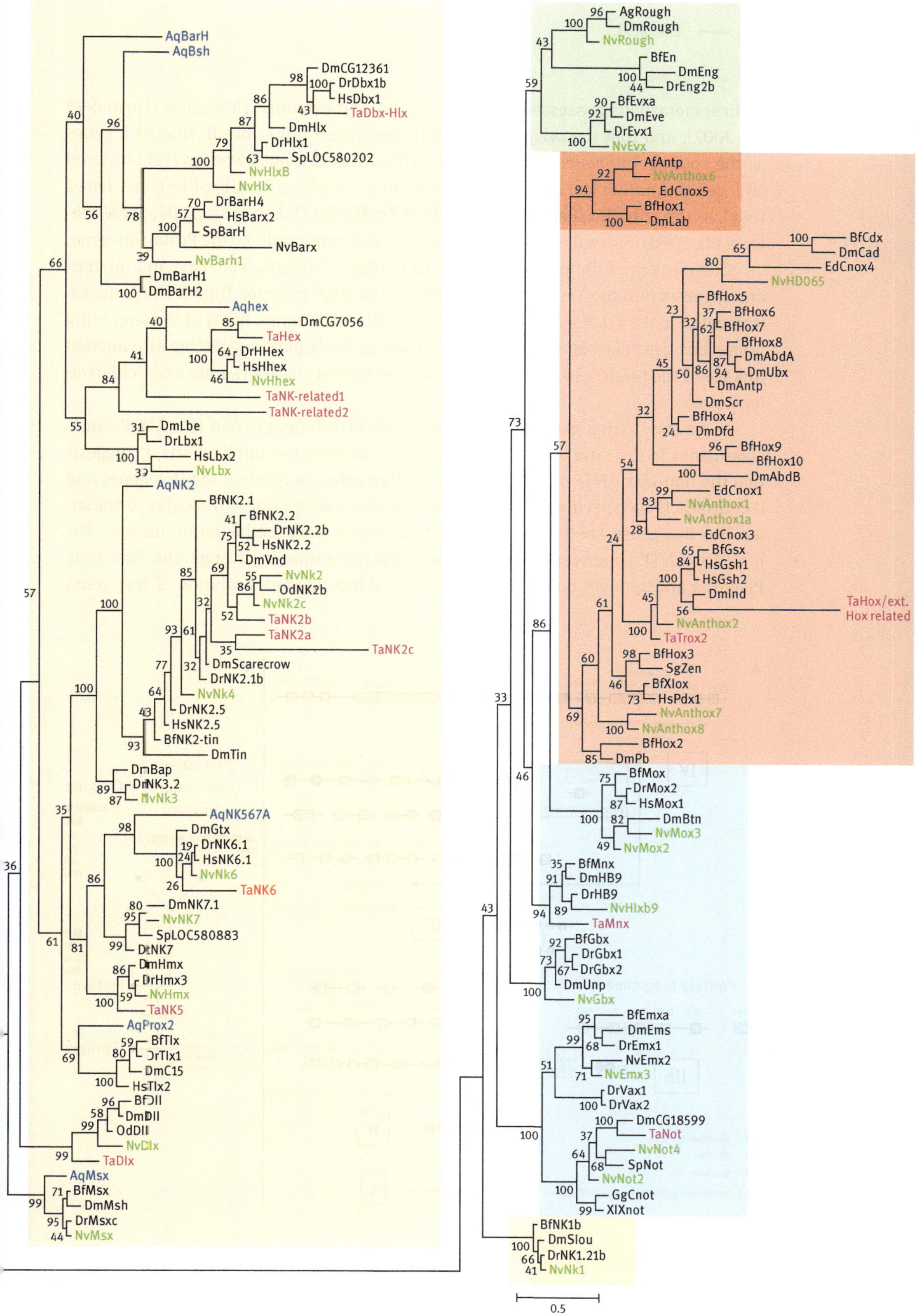

AqBarH
AqBsh
DmCG12361
DrDbx1b
HsDbx1
TaDbx-Hlx
DmHlx
DrHlx1
SpLOC580202
NvHlxB
NvHlx
DrBarH4
HsBarx2
SpBarH
NvBarx
NvBarh1
DmBarH1
DmBarH2
Aqhex
DmCG7056
TaHex
DrHHex
HsHhex
NvHhex
TaNK-related1
TaNK-related2
DmLbe
DrLbx1
HsLbx2
NvLbx
AqNK2
BfNK2.1
BfNK2.2
DrNK2.2b
HsNK2.2
DmVnd
NvNk2
OdNK2b
NvNk2c
TaNK2b
TaNK2a
TaNK2c
DmScarecrow
DrNK2.1b
NvNk4
DrNK2.5
HsNK2.5
BfNK2-tin
DmTin
DmBap
DrNK3.2
NvNk3
AqNK567A
DmGtx
DrNK6.1
HsNK6.1
NvNk6
TaNK6
DmNK7.1
NvNK7
SpLOC580883
DrNK7
DmHmx
DrHmx3
NvHmx
TaNK5
AqProx2
BfTlx
DrTlx1
DmC15
HsTlx2
BfDll
DmDll
OdDll
NvDlx
TaDlx
AqMsx
BfMsx
DmMsh
DrMsxc
NvMsx
AgRough
DmRough
NvRough
BfEn
DmEng
DrEng2b
BfEvxa
DmEve
DrEvx1
NvEvx
AfAntp
NvAnthox6
EdCnox5
BfHox1
DmLab
BfCdx
DmCad
EdCnox4
NvHD065
BfHox5
BfHox6
BfHox7
BfHox8
DmAbdA
DmUbx
DmAntp
DmScr
BfHox4
DmDfd
BfHox9
BfHox10
DmAbdB
EdCnox1
NvAnthox1
NvAnthox1a
EdCnox3
BfGsx
HsGsh1
HsGsh2
DmInd
TaHox/ext.
Hox related
NvAnthox2
TaTrox2
BfHox3
SgZen
BfXlox
HsPdx1
NvAnthox7
NvAnthox8
BfHox2
DmPb
BfMox
DrMox2
HsMox1
DmBtn
NvMox3
NvMox2
BfMnx
DmHB9
DrHB9
NvHlxb9
TaMnx
BfGbx
DrGbx1
DrGbx2
DmUnp
NvGbx
BfEmxa
DmEms
DrEmx1
NvEmx2
NvEmx3
DrVax1
DrVax2
DmCG18599
TaNot
NvNot4
SpNot
NvNot2
GgCnot
XlXnot
BfNK1b
DmSlou
DrNK1.21b
NvNk1
0.5

earliest metazoans possessed no Hox, ParaHox and extended Hox genes (Larroux et al., 2007), and that the origin of the ANTP superclass rather lies in true NKL genes as the sponge *Amphimedon* lacks any Hox/ParaHox-like but owns several clustered NKL genes. Another key issue is the identification of a Hox/ParaHox-like gene, *Trox-2* (*Gsx*), as the only Hox/ParaHox type gene in *Trichoplax* (Jakob et al., 2004; Schierwater et al., 2008). Several evidences suggest that this gene could be the Proto-Hox gene.

Comparison of the physical linkage in Porifera, Cnidaria, Placozoa and bilaterians suggests that most ANTP gene families could have emerged from a single ancestral region in the *Trichoplax* genome (Figure 3.2). Furthermore, many of the remaining homeobox gene classes are reciprocally linked in *Trichoplax*, suggesting also ancient linkage of the predecessors of all metazoan homeobox classes (Kamm and Schierwater, 2006).

In contrast to Hox-like genes, the assignment of orthology to non-Hox ANTP superclass genes in Cnidaria is straightforward. This fact has key implications, it suggests that the non-Hox ANTP superclass gene families originated before the Hox genes and it contradicts the speculation that the difficulties to assign true orthologies to cnidarian Hox-like genes are the result of early divergence from the bilaterian lineage. The non-Hox ANTP genes are likely older and had more time to diverge. The data from Placozoa and sponges better suit the concept of independent evolution of Hox genes

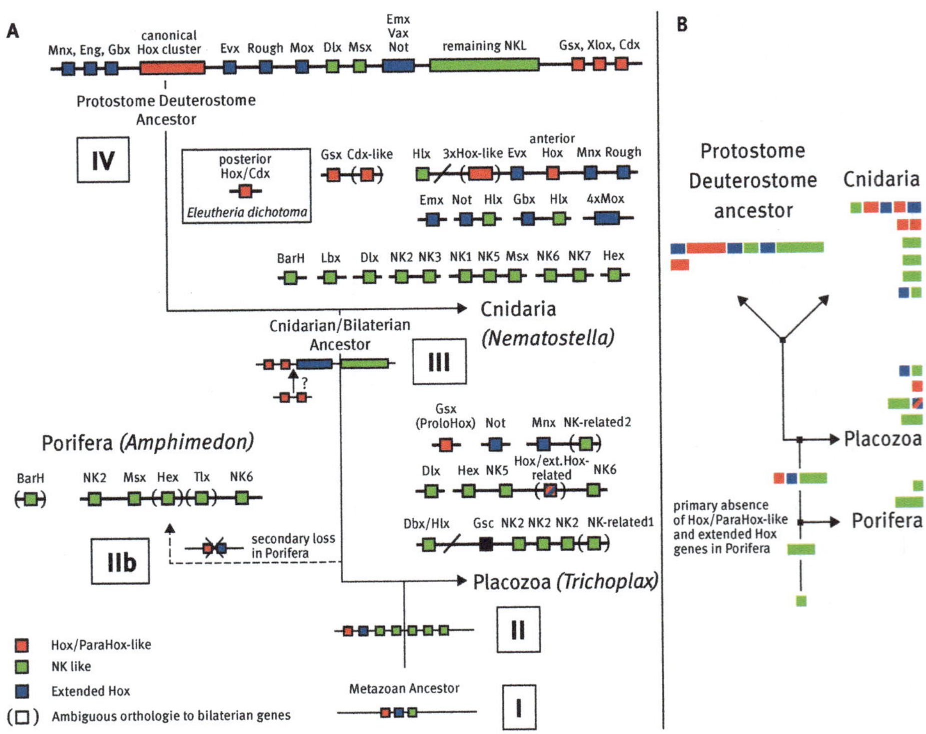

in Bilateria and Hox-like genes in non-bilaterians. In Placozoa we see a low diversity of ANTP class genes, *Gsx* (*Trox-2*; Jakob et al., 2004) as the only (Para/) Hox-like gene and two ANTP class homeobox genes of the extended Hox family, *Not* and *Mnx* (Monteiro et al., 2006). In Porifera we only find non-Hox ANTP genes. In Cnidaria, we find an almost complete non-Hox ANTP gene repertoire plus some orthologs to bilaterian Hox/ParaHox genes. *Trichoplax* possesses another eight NKL genes (*Dlx*, 3 copies of *NK2*, *NK5*, *NK6*, *Hex* and *Dbx/Hlx*; Kamm and Schierwater, 2006). In other words, the structure and genomic organization of non-Hox ANTP superclass genes support an early split of Hox gene evolution in Bilateria and Hox-like genes in non-bilaterian animals.

Cnidaria possess most of the bilaterian non-Hox ANTP genes. Combined evidence from phylogenetic analysis and linkage data shown in Kamm and Schierwater (2006) suggest a common origin for the ANTP superclass genes. Most likely the last common ancestor of cnidarians and bilaterians had a gene region harboring the almost full non-Hox ANTP superclass gene complement that must have undergone several rounds of rearrangements, duplications and gene losses in different lineages.

To clarify the evolution of the Hox system in animals, the Hox-like genes in two representative cnidarians were characterized (Kamm and Schierwater, 2006): *Eleutheria dichotoma*, a typical cnidarian with polyp and medusae life cycle stages (Kuhn

◂ **Figure 3.2:** Two hypothetical scenarios for the evolution of the ANTP superclass in Metazoa. **A.** The Placozoa first scenario. (I) The metazoan ancestor may have already possessed clustered single representatives of NKL, extended Hox and Hox/ParaHox-like genes. (II) Expansion of the NKL subclass. This situation is exemplified by the ANTP gene complement in *Trichoplax adhaerens*, which harbors several clustered NKL genes, two genes of the extended Hox and a single Hox/ParaHox-like gene which is a true *Gsx* ortholog and may be the founder of the Hox/ParaHox-like subclass of ANTP genes. (IIb) Based on the assumption that Placozoa are basal within extant metazoan phyla, Porifera must have lost extended Hox, Hox/ParaHox-like and some NKL genes. An alternative scenario that Porifera are most basal would in turn require loss of some NKL genes in Placozoa (e.g. *Msx* and *Bar/Bsh* related genes). (III) Further expansion of ANTP genes in a cnidarian-bilaterian ancestor leading to almost complete repertoires of NKL and extended Hox genes. Occurrence of anterior Hox and posterior Hox/*Cdx*-like genes. The assumption that a *Gsx* type gene was the founder of Hox/ParaHox-like genes requires that *Gsx* and a posterior Hox/*Cdx*-like gene were part of an ancestral ANTP mega-array for which we still lack any evidence in extant species (question mark in III). In urochordates, however, we see one example of a ParaHox gene (*Cdx*), which is tightly linked to a Hox gene (Seo et al., 2004). (IV) Occurrence of a true canonical Hox system and complete ParaHox cluster in Bilateria. Linkages other than for *Trichoplax adhaerens* are based upon Chourrout et al., 2006; Kamm et al., 2006; Larroux et al., 2007; Ryan et al., 2007. In the case of *Nematostella*, some linkages have been deduced from the current JGI genome assembly. The proposed ANTP mega-array in a cnidarian-bilaterian ancestor has been deduced from Pollard and Holland, 2000; Castro and Holland, 2003; Luke et al., 2003; Kamm and Schierwater, 2006; Ryan et al., 2007. Remark: For completeness the black square in the *Trichoplax* genome represents the PRD class gene *Gsc*, which is tightly linked to NKL genes. **B.** A Porifera first scenario (simplified) for the evolution of ANTP genes leading to crown group non-bilaterians and a putative protostome-deuterostome ancestor: starts with a single (proto-)NK gene and an initial expansion of the NKL subclass. Hox/ParaHox-like and extended Hox genes are an invention of the lineage leading to Placozoa, Cnidaria and Bilateria. Modified after Schierwater et al. (2008).

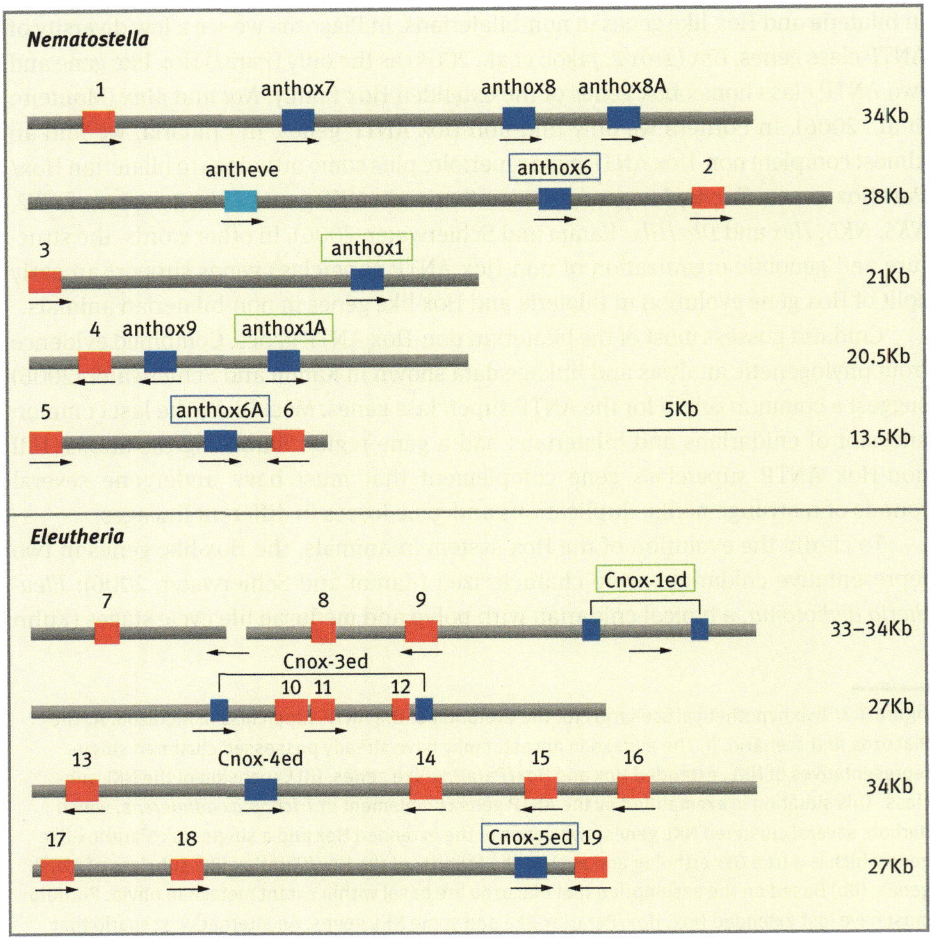

Figure 3.3: The Hox-like genes of *Nematostella* and *Eleutheria* are organized differently and do not reflect the cluster patterns of true Hox genes. The figure summarizes the genomic organization of Hox-like genes in *Nematostella vectensis* and *Eleutheria dichotoma*. Rectangles show the position of the genes in the genomic context. Arrows show transcriptional orientation of the genes. Hox-like genes are in dark blue, Hox-related genes are in light blue and non-Hox genes are in red and numbered: 1: retrotransposon; 2: *Rho*/*Rsa*-related gene; 3: putative metalloproteinase inhibitor; 4: putative ANF-receptor; 5: Phosphatidylinositol phosphate kinase; 6: putative lam_G domain; 7, 8, 10–12: fragmented reverse transcriptases; 9, homolog to Danio rerio putative protein; 13: reverse transcriptase; 14: resembles metabotropic glutamate receptor; 15: two or more ORFs resembling Homo sapiens put. protein; 16: last exon of RRN3; 17: Dfp domain gene; 18: POU gene; 19: 1.2 Kb ORF resembling *Plasmodium* MAEBL (interrupted by vector). Modified after Kamm et al. (2006).

et al., 1996), and the cnidarian *Nematostella*, a basal cnidarian (anthozoan) with a polyp stage only (Darling et al., 2005). Some cnidarian Hox-like genes clearly are relatives to anterior Hox or posterior Hox/*Cdx* type genes of Bilateria but "intermediate" genes are missing. Cnidarian Hox-like genes are not organized in clusters, as this is the case for Bilateria. Comparison of gene arrangement in *Eleutheria* and *Nematostella* revealed a different organization, i.e. organization of Hox-like genes are different within the Cnidaria. Additionally, the Hox-like genes are not linked and are flanked by non-Hox genes (Figure 3.3). Expression patterns of Hox-like genes in Cnidaria are heterogenous and do not conform to colinearity rules (Figure 3.4), a finding that is also supported by another study on the hydrozoan model system *Clytia hemisphaerica* (Chiori et al., 2009). Thus, the hypothesis of a fragmented Hox cluster in Cnidaria can be rejected, as this would imply a conservation of spatial colinearity.

Analyses of cnidarian Hox-like genes have highlighted the importance to study non-bilaterian models to understand the evolution of bilaterian gene families (Technau and Steele, 2011). The phylum Cnidaria harbors genes related to anterior Hox and posterior Hox/*Cdx* genes, but most of the Hox-like genes present are likely to postdate divergence with the bilaterian line, accounting for their unclear relationships to true Hox classes. The Hox cluster presumably arose from the outside in – from a two gene state via a series of unequal crossing-over events – and the cnidarian "anterior Hox-like" and "posterior Hox/*Cdx*-like" types may be derived from these ancestral two outer genes. Contrary to expectations, a definitive Hox system is obviously not essential for axial patterning in non-bilaterians. Moreover, the spectacular range of morphological variation across the Cnidaria demonstrates that the canonical Hox system is not mandatory for the elaboration of a wide range of variations of a basic body plan. Current knowledge implies that true clustered Hox genes evolved in Urbilateria after the cnidarian branch (Kamm et al., 2006; Chiori et al., 2009). Data from sponges further predate the evolution of Hox-like genes and support the organization in clusters to be a bilaterian invention (Larroux et al., 2007).

Inhibition of Hox-like genes in the hydrozoan *Eleutheria* create substantial bauplan alterations, including the formation of multiple oral poles ("heads") by *Cnox*-2 and *Cnox*-3 inhibition, deformation of the main body axis by *Cnox*-5 inhibition and duplication of tentacles by *Cnox*-1 inhibition (Jakob and Schierwater, 2007). Some of the observed gene inhibition phenotypes look like direct links to bauplan patterns that have evolved in other hydrozoan taxa, while others seem to have derived functions. *Cnox* genes in Cnidaria may provide a very efficient means for macro-evolutionary bauplan alterations, for example by multiplication of body parts (Figure 3.5). Unfortunately, comparative RNA-Interference studies from other marine invertebrates are missing, but increasing knowledge in applying this technique will soon bring new insights into ANTP-class gene function in other early branching metazoans (Rivera et al., 2011).

It was first pointed out by Pollard and Holland (2000) for vertebrates, and subsequently later for other lineages (Coulier et al., 2000; Castro and Holland, 2003; Garcia-Fernandez, 2005) that the linkage pattern of ANTP genes in metazoan genomes strongly points to a common origin of ANTP superclass genes from the same genomic region through several series of cis-duplications.

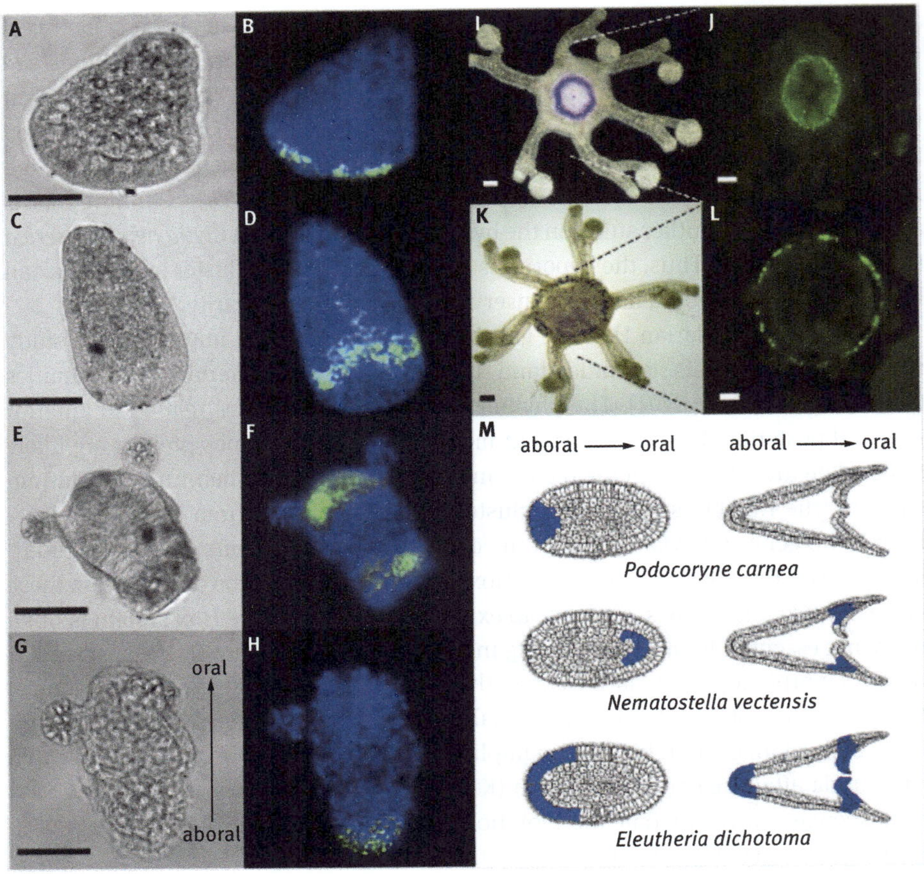

Figure 3.4: Expression patterns of related Hox-like genes are heterogenous among cnidarians and do not conform to colinearity rules. In planulae, the "anterior Hox-like" genes (= *Cnox*-5ed in *Eleutheria*, anthox6 in *Nematostella*) are expressed ectodermally and aborally (*Eleutheria*; [A–D, M]), entodermally and orally (*Nematostella*; [M]), or aborally in both ecto- and endoderm (*Podocoryne* (Yanze et al., 2001); [M]). In polyps, the corresponding genes are not expressed in *Podocoryne*, but are expressed both orally and aborally in *Eleutheria* (E, F, M) and at the oral end only in *Nematostella* (M). (A–F) *Cnox*-5ed in 3-day-old planula larvae (A, B), 4- to 5-day-old planula larvae (signal moves toward the center as an ectodermal ring; [C, D]), and in young primary polyps (oral and aboral; [E, F]). (G and H) Aboral expression of *Cnox*-4ed in a 10-day-old primary polyp. (I and J) Ectodermal oral expression of *Cnox*-3ed around the manubrium of a medusa. (K and L) Ectodermal oral *Cnox*-1ed expression in the "Cnidoblast channel" of the medusa stage. Scale bar equals 50 mm. Modified after Kamm et al. (2006).

Also the combined evidence from phylogenetic analyses and linkage data clearly suggest a common origin for the ANTP superclass genes. Most likely the last common ancestor of cnidarians and bilaterians had a gene region harboring the almost full ANTP superclass gene complement. Recent evidence also suggests that cnidarians

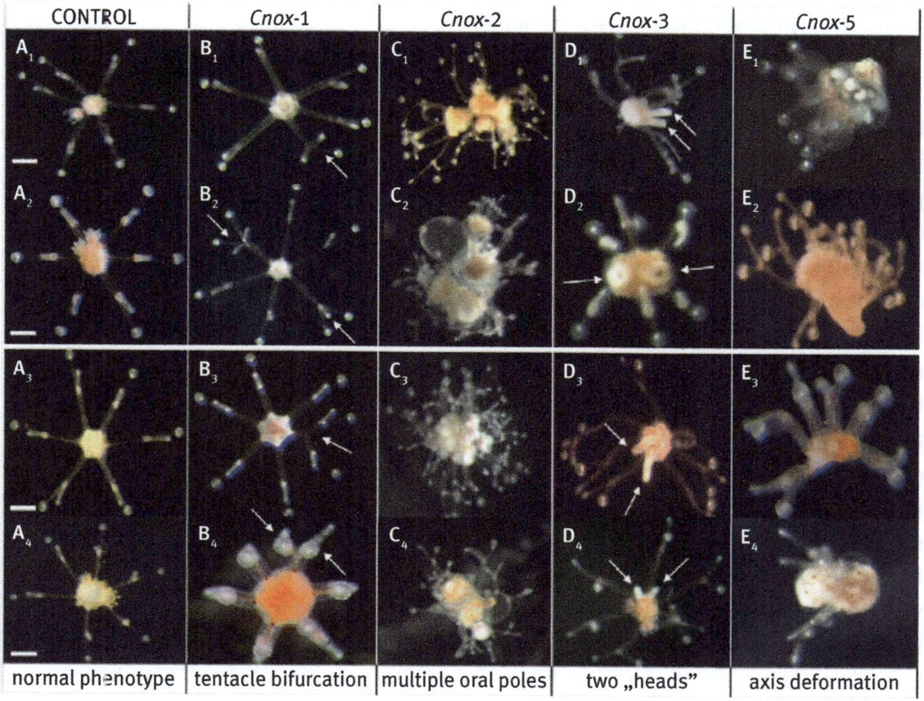

Figure 3.5: Phenotypic changes in knock down experiments of Hox-like genes in the hydrozoan *Eleutheria dichotoma*. Four of the five Hox-like genes produced phenotypically abnormal medusae, main features of bauplan change relate to additional tentacle bifurcation (*Cnox*-1; see arrows), multiple oral poles or heads (*Cnox*-2), head duplication (*Cnox*-3; see arrows), and oral-aboral body axis deformation (*Cnox*-5). The two upper panels show life pictures of medusae transfected with double stranded RNA ($B_{1,2}$, $C_{1,2}$, $D_{1,2}$, $E_{1,2}$), and the lower panel medusae transfected with morpholino oligos ($B_{3,4}$, $C_{3,4}$, $D_{3,4}$, $E_{3,4}$). $A_{1,2}$ and $A_{3,4}$ are controls, scale bar is 100 mm. Modified after Jakob and Schierwater (2007).

represent a pre-Hox state, with only anterior and posterior Hox-like genes (Kamm et al., 2006). We therefore conclude that the expansion of the non-Hox ANTP gene families took place before a true Hox system evolved. Moreover, the Hox system is clearly not mandatory for axial patterning in non-bilaterians, neither Hox genes nor Hox clusters are strictly required for the elaboration of morphological diversity. Comparative expression patterns and functional studies as well as the structure and genomic organization of ANTP superclass genes in non-bilaterians again provide strong support for an early split at least of Cnidaria and Bilateria.

3.2.2 Intra-phylum relationships in Cnidaria

The phylum Cnidaria currently comprises more than 10,000 accepted marine species (Appeltans et al., 2012a; Appeltans et al., 2012b), which display a remarkable morphological diversity (Daly et al., 2007 for overview). The classification of taxa within this phylum up to class-level relationships is controversially discussed (see e.g. Bridge et al., 1995; Collins, 2002; Marques and Collins, 2004). Whereas there is strong support for Anthozoa as the earliest branching class, the relationships between the remaining cnidarian classes Staurozoa, Cubozoa, Scyphozoa and Hydrozoa remain unclear. We have addressed this phylogenetic problem and analyzed large molecular sequence data from a representative choice of cnidarian taxa to further clarify relationships between and within classes. We have incorporated new sequence data of the large and small nuclear ribosomal RNA subunit (28S rRNA and 18S rRNA, respectively) in different data subsets and analyzed them using different phylogenetic methods (i.e. maximum likelihood, maximum parsimony, minimum evolution and Bayesian analysis). The results of molecular analyses as well as morphological and developmental/ life cycle characters have been combined by the authors (Collins et al., 2006a) in a working hypothesis on cnidarian relationships which is shown in Figure 3.6. According to this working hypothesis the class Anthozoa (Hexacorallia and Octocorallia) is a sister-group to Medusozoa (comprising Staurozoa, Cubozoa, Scyphozoa and Hydrozoa). Within the medusozoan clade the distinct class Staurozoa is the earliest branching lineage. Another remarkable point is the sister-group relationship between Cubozoa and Scyphozoa, forming the "Acraspeda" (Gegenbaur, 1857) clade as a sister-group to Hydrozoa. Although alternative scenarios within Medusozoa might be arguable (especially with respect to the position of Staurozoa), a recent study on complete cnidarian mitochondrial genomes by Kayal and co-workers gave further support for the robustness of the proposed scenario by Collins et al. (2006a). The observed mitochondrial gene order within Medusozoa has been mapped on the previously mentioned cnidarian phylogeny resulting in an overall conclusive picture of medusozoan mitochondrial genome evolution (Kayal et al., 2012).

3.2.3 Systematic composition of the phylum Placozoa using mitochondrial genomes

Since the description of the only described placozoan species, *Trichoplax adhaerens* (Schulze, 1883), a number of molecular studies (mainly 16S, 18S or 28S rRNA sequence analyses) have provided compelling evidence for an unexpected molecular diversity within this so far monotypic phylum (e.g. Voigt et al., 2004; Eitel and Schierwater, 2010). The genetic distances between identified placozoan lineages are comparable to distances observed between genera or even families in the Porifera or Cnidaria. Early studies on complete animal mitochondrial (mt) genomes have shown

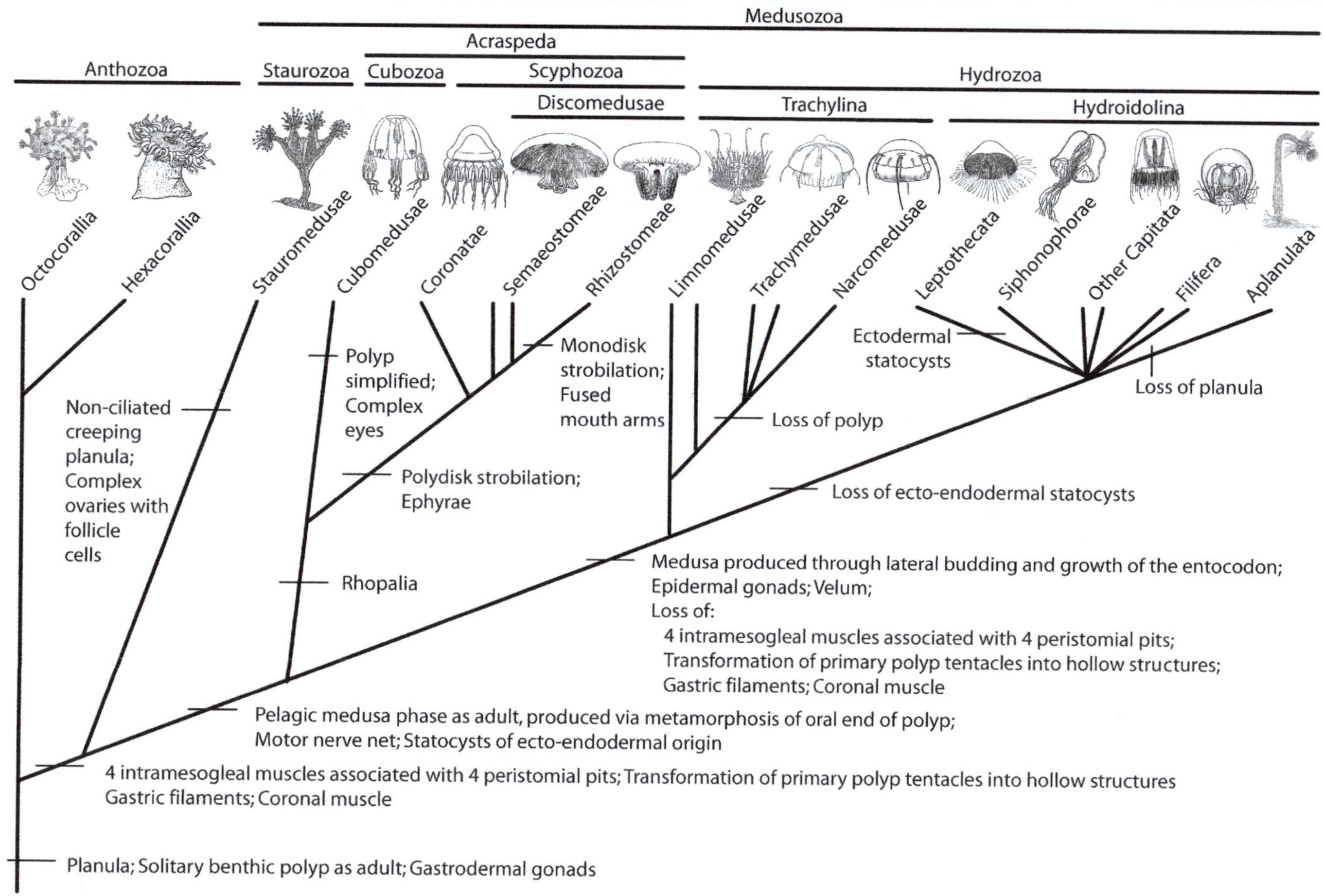

Figure 3.6: Phylogenetic relationships among cnidarian classes after Collins et al. (2006a). Ancestral characters of selected clades are mapped at nodes. From Collins et al. (2006a).

that whole mt genome data can be suitable to infer phylogenetic relationships within and in many cases between animal phyla (e.g. Medina et al., 2006; Wang and Lavrov, 2008). Besides sequence data for phylogenetic analyses mitochondrial genomes provide a variety of phylogenetic informative characteristics such as genome size/architecture, number/order of encoded genes or proportion of non-coding sequence (see e.g. Osigus et al., 2012). The analysis of complete mitochondrial genomes from various placozoan lineages therefore must help to further characterize the diversity within this phylum and also should give important insights into the evolution of early animal mitochondrial genomes.

In 2006, we published the first complete placozoan mitochondrial genome. The *Trichoplax adhaerens* mitochondrial genome is the largest metazoan circular mt genome with a size of 43,079 bp. It encodes 12 respiratory chain genes (nad1–6, nad4L, cox1–3, cob and atp6, but atp8 and atp9 are missing), two rRNA genes (rnS and rnL) and 24 tRNAs. Beside the large proportion of non-coding sequences, the mitochondrial genome contains eight additional open reading frames as well as several group I/II introns (Dellaporta et al., 2006; Burger et al., 2009). Introns can be found in three genes: nad5, 16S and cox1. One intron is found in nad5, two in the fragmented 16S rRNA gene. The fragmented cox1 gene harbors seven introns. The *Trichoplax* mitochondrial genome therefore displays a unique combination of unusual animal mitochondrial genome characteristics, underlining the exceptional status of this organism for comparative genomic studies. The insights gained from the *Trichoplax* mt genome substantially contributed to two subsequent comparative studies on mitochondrial genome diversity in Placozoa. The *Trichoplax adhaerens* and other placozoan mt genomes (see below) show some general characters that are shared between placozoan and non-metazoan mitogenomes and should thus be considered as plesiomorphic for the Metazoa: a generally large circular mt genome, presence of large non-coding regions, and genes that are interrupted by introns.

Mitochondrial genomes of four additional placozoan lineages (Placozoa spp. H3, H4, H8 and H15) have been analyzed (Signorovitch et al., 2007; Miyazawa et al., 2012). The characterized lineages display a remarkable variation with respect to mt genome size (from 32 kb to more than 37 kb) and number of encoded open reading frames (from two to eight). With respect to relationships within the Placozoa, gene arrangement as well as protein coding sequence data strongly support a subdivision of lineages in two groups, 'A' and 'B'. Both data sources also support a close relationship between *Trichoplax adhaerens* and Placozoa sp. H3 (group B) as well as Placozoa spp. H4, H8 and H15 (group A) on the other hand. Members from both groups are clearly distinguishable from each other by an inversion of an approximately 20 kb mt genome fragment. Even within groups A and B, each placozoan lineage displays a unique gene arrangement due to translocation and/or inversion events. The proposed subdivision of placozoan lineages into two groups is also strongly supported by several phylogenetic analyses based on protein coding sequence data (see Signorovitch et al., 2007). While most recent studies only comprise members of four out of seven known placozoan clades, the phylogeny in Figure 3.7 comprises unpublished data from members of two additional clades (Placozoa sp. H5, clade IV and Placozoa sp. H12, clade VII). The observed topology is in agreement with previous analyses supporting the well-known subdivision into two distinct groups (Signorovitch et al., 2007; Burger et al., 2009). A total of 9 complete or almost complete mitochondrial genomes have been sequenced so far. From these we gained some basic insights into the major splits within the Placozoa. Given that the number of genetic lineages is expected to increase probably more than tenfold (Eitel and Schierwater, 2010; Eitel et al., 2013), a large number of placozoan taxa are still awaiting mitogenomic characterization.

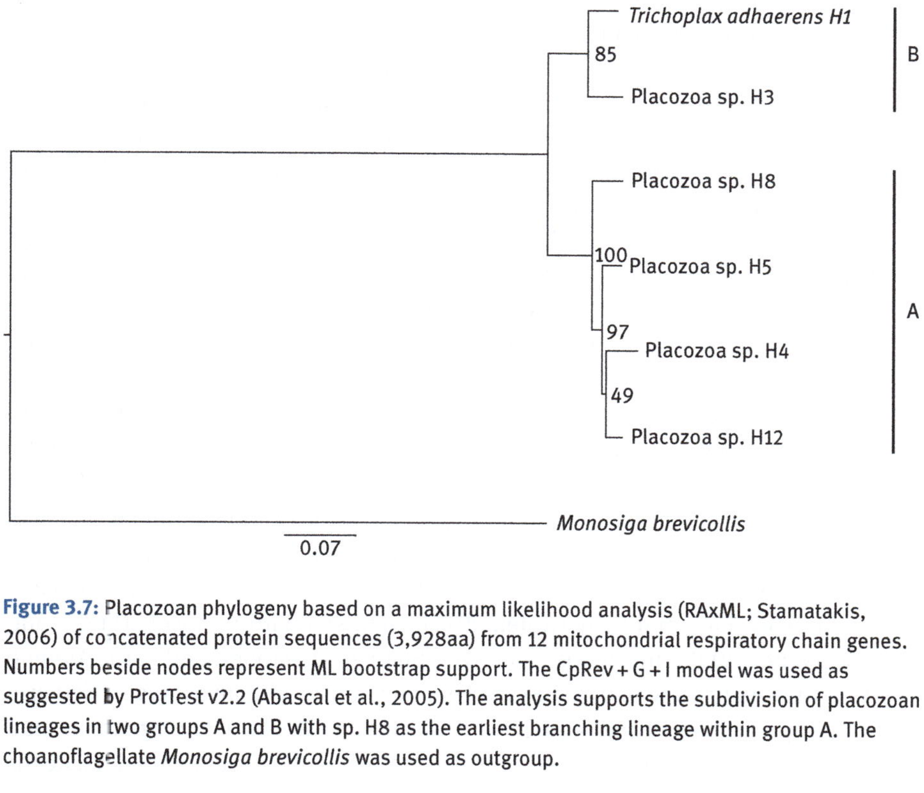

Figure 3.7: Placozoan phylogeny based on a maximum likelihood analysis (RAxML; Stamatakis, 2006) of concatenated protein sequences (3,928aa) from 12 mitochondrial respiratory chain genes. Numbers beside nodes represent ML bootstrap support. The CpRev + G + I model was used as suggested by ProtTest v2.2 (Abascal et al., 2005). The analysis supports the subdivision of placozoan lineages in two groups A and B with sp. H8 as the earliest branching lineage within group A. The choanoflagellate *Monosiga brevicollis* was used as outgroup.

3.2.4 Enlarged nuclear data sets to infer inter- and intra-phylum relationships

The main problems of previous analyses on early metazoan relationships relate to insufficient taxon sampling and a limited number of molecular characters. Both issues need to be addressed in future studies. To increase the taxon sampling we have sequenced ESTs from one placozoan representative (Placozoa sp. H4) and genomic

DNA from two more placozoans (Placozoa spp. H2 and H16). In addition, 454 sequencing was performed on a *Vallicula multiformis* (Ctenophora) cDNA pool from various developmental stages as well as on cDNA from an adult *Beroe* sp. (in collaboration with Gert Wörheide, LMU Munich, Germany).

The inter-phylum analyses include more than 100 taxa with large transcriptome and whole genome data sets and will be addressed elsewhere. The focus here is on preliminary analyses of the intra-phylum relationships in the Placozoa using EST and whole genome data from four placozoan species-lineages.

(1) *Trichoplax adhaerens:* Sequences derived from the *Trichoplax* nuclear genome project where 11,520 genes were predicted (Srivastava et al., 2008).

(2) Placozoa sp. H2: This is an ongoing sequencing project and the final results will be presented elsewhere (Schierwater et al., in prep.). In brief, genomic DNA was isolated from 4,000 healthy growing animals of the undescribed Placozoa sp. H2 (Eitel and Schierwater, 2010) using standard protocols (Ender and Schierwater, 2003). In collaboration with Rob DeSalle and Apurva Narechania (American Museum of Natural History, New York, USA), a genomic library with an insert size of 300 bp was created and sequenced at Ambrygen (California, USA). More than 80 million reads were created by Illumina sequencing and successively assembled using the Abyss assembler (Simpson et al., 2009). Afterwards, all contigs were re-assembled using CAP3 (Huang and Madan, 1999). In total, the final contigs spanned about 100 Mb with an average 57x coverage.

(3) Placozoa sp. H4: A cDNA library was constructed for the undescribed Placozoa sp. H4 using an RNA pool from several developmental stages and from cultures kept under different environmental conditions (for details, see Eitel et al., 2011). 4,015 clones were randomly sequenced, quality clipped and assembled to a total of 2,194 unique clusters.

(4) Placozoa sp. H16: This is an ongoing sequencing project and the final results will be presented elsewhere (Schierwater et al., in prep.). In brief, genomic DNA was isolated from more than 3,000 healthy growing animals of the undescribed Placozoa sp. H16 (Eitel and Schierwater, 2010) using standard protocols (Ender and Schierwater, 2003). In collaboration with Stephen Dellaporta and Maria Moreno (Yale University, New Haven, USA), a genomic library with an average insert size of 600 bp was created and sequenced. Two runs on a Roche 454FLX genomic analyzer yielded more than 293,000 reads and more than 96 Mb sequence data. Reads were assembled using Newbler (Roche) and the contigs re-assembled using CAP3.

For the annotation of the final sequences (either ESTs cluster or genomic contigs), the HaMStR (Ebersberger et al., 2009) hidden Markov model based search approach was applied (supported by I. Ebersberger). A set of core orthologs was therefore created (kindly provided by I. Ebersberger) based on the proteomes of five completely sequenced species: the fungus *Phycomyces blakesleeanus*, the choanoflagellate *Monosiga brevicollis*, the placozoon *Trichoplax adhaerens*, as well as the cnidarians *Nematostella vectensis* and *Hydra magnipapillata*. A total of 1,117 core orthologs were selected and used as input for the HaMStR analyses with the parameters set to '-eval_limit=1e-20' and '-representative'. *Trichoplax* was used as the reference species for reciprocal blast searches. The analyses resulted in 938, 287 and 291 orthologs from the Placozoa spp. H2, H4 and H16 data sets, respectively (Figure 3.8A). 74 complete or partial genes were found in all four placozoans. Using a perl script (kindly provided by Sabrina Simon), these were aligned at the amino acid level together with according orthologs from *Monosiga brevicollis*, which served as the outgroup. Alignments were concatenated and all ambiguously aligned positions as well as gap positions were removed manually. This process resulted in a data set of 3,120 amino acid positions (1,978 invariable sites). Based on this limited proportion of the four placozoan proteomes, mean uncorrected pairwise distances were calculated using PAUP v.4b10 (Swofford, 2003). The protein sequence distance between the closely related placozoan species *Trichoplax adhaerens* and Placozoa sp. H2 (see Eitel et al., 2013 for a current placozoan 16S-based phylogeny) is 0.1 %, while that to the other placozoans is up to 6.7 % (Figure 3.8B, lower panel). These differences support the previous assumption that the Placozoa harbor taxa on various taxonomic levels (Voigt et al., 2004; Eitel and Schierwater, 2010; Eitel et al., 2013). Phylogenetic studies within the Placozoa were previously based on mitochondrial genome data only. To compare the findings from the nuclear genes with those from mitochondrial genomes, protein sequences of twelve mitochondrial protein coding genes were aligned (nad1–6, nad4L, cox1–3, cytb and atp6) from the same four placozoans (Dellaporta et al., 2006; Signorovitch et al., 2007 and own, unpubl. data). *Monosiga brevicollis* was again used as outgroup. Alignments were concatenated and all ambiguities and gap positions removed, as was done for the nuclear data set. This resulted in 3,795 amino acid positions (2,010 invariable sites). The sequence divergence for this mitochondrial data set is higher compared to the nuclear set (Figure 3.8B, upper panel), but the order of magnitude between the various placozoans is comparable.

While the genome analyses are still in progress, phylogenetic inferences on preliminary data sets yield highly congruent topologies, showing a deep split between the formerly described placozoan subgroups 'A' and 'B' (Figure 3.8B).

In sum, the preliminary results from nuclear and mitochondrial data support a high diversity within the Placozoa and different taxa at least at the genus and family rank.

3.2.5 Studying placozoan development to identify early metazoan traits

Unraveling animal life cycles and embryonic development is basic to understanding animal biology and regularly sheds light on phylogenetic relationships. Despite big efforts and over a hundred years of research, the life cycle of the Placozoa is yet unresolved. The *Trichoplax adhaerens* genome has been sequenced and several more

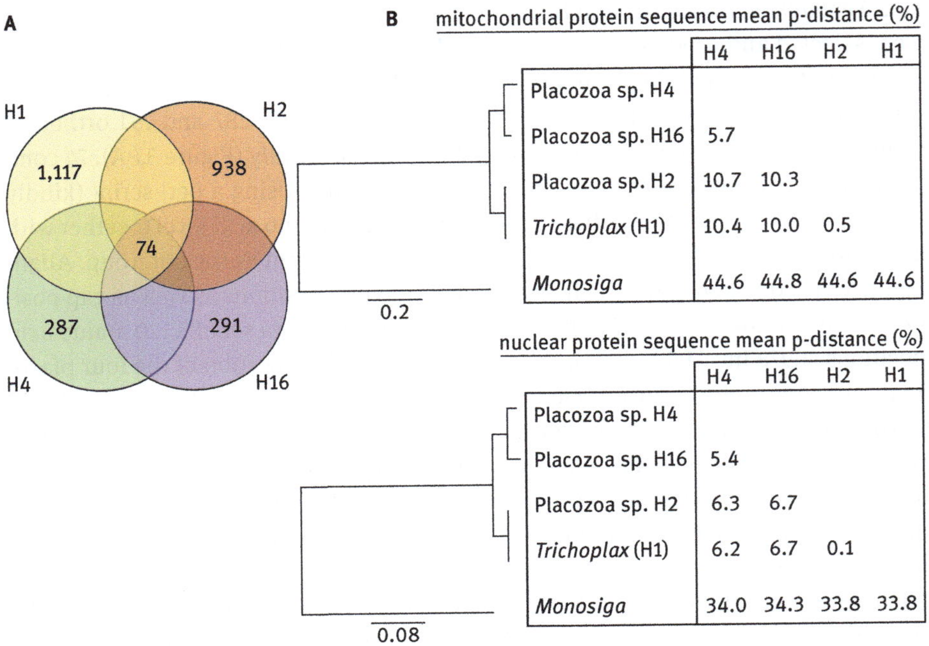

Figure 3.8: The *Trichoplax adhaerens* (16S haplotype H1) nuclear genome was the first placozoan genome to be sequenced (Srivastava et al., 2008). Ongoing transcriptome and whole genome sequencing produced large amounts of sequence data from three additional species lineages, Placozoa spp. H2, H4 and H16 (Schierwater et al., unpubl. data). Based on a set of 1,117 core ortholog proteins from *Trichoplax*, the raw sequence data was searched for orthologs. A variable number of orthologs was found, shown in (A), and 74 orthologs were identified in all four placozoans in the preliminary data set. These were used for preliminary ML tree reconstructions (B, lower panel). In addition, mitochondrial protein sequences were used to show genetic distances (B, upper panel) in comparison to the nuclear proteins. For methods see main text.

placozoan genomes are on their way before we know about the basic biology of this group. A few previous studies have given some first insights into the development of oocytes (Grell, 1971, 1972; Grell and Benwitz, 1974; Grell, 1984) and one genetic study gives support for sexual reproduction in natural populations (Signorovitch et al., 2005).

By addressing the sexual reproduction and early development of different placozoan lineages, we have generated comparative data allowing identification of general morphological traits in the development of this phylum. By optimizing culture conditions we could reproducibly induce sexual reproduction – as seen by oocyte maturation and early embryonic development – in three placozoan species lineages: *Trichoplax adhaerens*, Placozoa sp. H2 and Placozoa sp. H16 (Eitel et al., 2011). The water temperature was shown to be a critical element for the induction of sexual reproduction similar to other early metazoan animals like sponges (e.g. Simpson, 1984; Witte et al., 1994; Mercurio et al., 2007; Whalan et al., 2007; Abdo et al., 2008) and cnidarians (e.g. Jennison, 1979; Chen et al., 1995; Lucas and Lawes, 1998; Chiaverano et al., 2004; Bavestrello et al., 2006; de Putron and Ryland, 2009).

At least five conserved sperm markers are expressed in healthy growing placozoans. This and the regular observation of oocytes support that placozoans reproduce sexually in the field in addition to the standard mode of vegetative reproduction. Furthermore, the maturation of oocytes as well as the expression of sperm-associated genes provided evidence that the life cycle includes bisexual reproduction (Eitel et al., 2011). The sum of observations indicates a simple life cycle with an alternation between bisexual and vegetative reproduction. The identified putative sperm protein Dnajb13 is associated with the sperm flagellum in other animals indicating flagellated sperms in the Placozoa. This is a presumed ancestral feature of metazoans (Adiyodi and Adiyodi, 1994). As shown for *Trichoplax*, cortical granules were also found in Placozoa sp. H2 oocytes. Such granules have been known from oocytes across different metazoan phyla (Nosek, 1984; Cran, 1989; Velilla et al., 2004; Eckelbarger and Blades-Eckelbarger, 2005; Swiatek, 2006; Bembenek et al., 2007; Hamel et al., 2007; Mei et al., 2008; Wong and Wessel, 2008; Zhang et al., 2010a). Shortly after fertilization these granules fuse to create the fertilization membrane, which prevents polyspermy (e.g. Anderson, 1968; Dandekar and Talbot, 1992) and protects the embryo.

The fertilization membrane is two layered in *Trichoplax adhaerens* at least in the observed stages, while it is three layered in Placozoa sp. H2 from the 4-cell stage onward. Future studies will have to show if this feature is a fixed morphological difference or related to developmental stage. Unfortunately all embryos observed so far died sooner or later not developing beyond a 128-cell stage. Reasons are unknown but obviously some fundamental environmental needs are not met as yet. Further culture improvements are needed to see the progress of embryonic development.

Despite the above shortcoming, we can summarize observations on placozoan sexual reproduction and the very early developmental traits as follows:

(1) Placozoan reproduce bisexually, with sperms possibly being created and stored before the oocyte maturation starts. Nothing is known so far on the onset of oocyte generation.
(2) With 50–100 µm, oocytes are relatively large, given the small size of the animal.
(3) The expression of the flagellum-associated protein Dnajb13 indicates that placozoans build flagellated sperm cells, an ancestral metazoan characteristic.
(4) Fertilization in the Placozoa is internal. Shortly after the (not yet observed) fertilization, a fertilization membrane is built. This protective eggshell resembles fertilization membranes of other metazoans.
(5) Cleavage is total and equal, at least until the 128-cell stage.

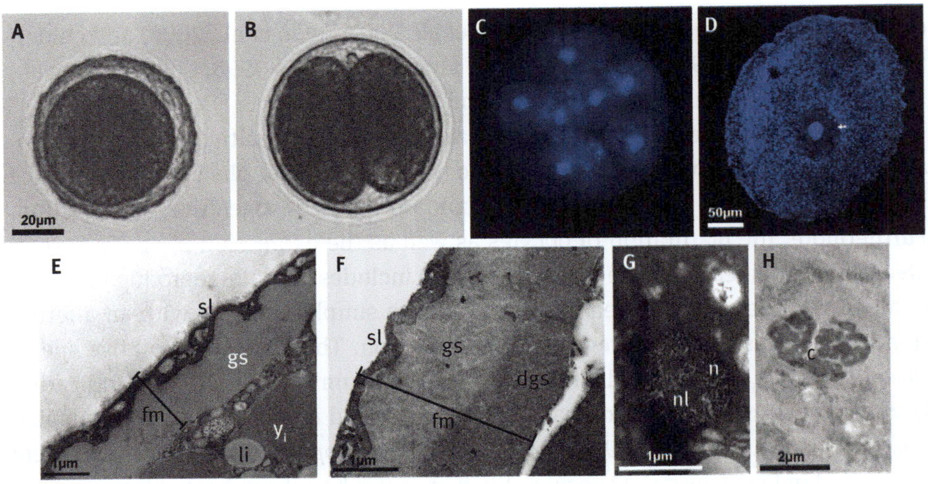

Figure 3.9: Embryonic development in Placozoa sp. H2. Shown are embryos at the zygote, 2-cell, and 8-cell stage inside the fertilization membrane (A–C). Nucleic acid staining with DAPI shows a direct correlation of blastomer number and fluorescent signals under standard fluorescent microscopy (C). Maturing oocytes and zygotes show a large nucleus compared to somatic cells of the mother animal (arrow in D). Ultrastructural analyses of developing Placozoa sp. H2 oocytes and embryos using TEM imaging shows that the early 'fertilization membrane' is made up of two layers as reported for *Trichoplax* (E), whereas three layers are distinguishable in later stages (F) in Placozoa sp. H2 only. In some sections, nuclei (G) and chromosomes (H) were found in blastomers, indicating a normal cell cycle in developing embryos that has never been observed before. yi = yolk inside oocyte, li = lipid droplet, fm = fertilization membrane, sl = striped layer, gs = ground substance, dgs = dense ground substance, n = nucleus, nl = nucleolus, c = metaphase chromosomes. The scale bar of (A) also applies to (B) and (C). Modified from Eitel et al. (2011).

3.2.6 Total evidence analysis

While generating sequence data was a hurdle twenty years ago, today next generation sequence techniques allow for fast access to large amounts of sequence data. The amount of sequence data currently available has spawned a revolution in the understanding of biological diversity (Graur and Li, 2000). As a result, the use of single genes for resolving taxonomic relationships, which can be biased in several ways, has been more and more replaced by total evidence analyses of concatenated data sets (Delsuc et al., 2003; Hoofer et al., 2003; Teeling et al., 2003; Hedges et al., 2004; Wolf et al., 2004). Total evidence analyses have proven to overcome the shortcomings of single gene phylogenies by diminishing weakly supported data while summing up shared features (Pupko et al., 2002; Gadagkar et al., 2005). Total evidence data sets can be based upon (1) several genes, (2) gene positions, (3) amino acids and (4) morphological characters.

Single-gene data sets have frequently led to contradicting phylogenies because not all genes are equally suited to resolve phylogenies of closely related species (Mason-Gamer, 1996; Kopp and True, 2002), relationships between major classes (Giribet et al., 2001; Hwang et al., 2001), or deep metazoan splits (Baldauf and Palmer, 1993; Rokas et al., 2003b). For phylogenetic reconstruction it is crucial to consider two factors before choosing a single or multiple genes (Rokas et al., 2003a; 2003b). The first is the analytical factor involving choice of optimality criterion (Huelsenbeck, 1995), data availability (Philippe, 1994; Cummings et al., 1995), taxon sampling (Graybeal, 1993) and specific assumptions in the modeling of sequence evolution (Yang et al., 1994). The second is the biological factor such as action of natural selection or genetic drift (Maddison, 1997; Satta et al., 2000; Martin and Burg, 2002).

To overcome the shortcomings of previous approaches on phylogenetic relationships between recent non-bilaterian groups we employed different analytical and biological factors. We combined morphological with molecular data (sequence as well as ribosomal RNA structure) for 24 representative metazoan taxa (Schierwater et al., 2009; Schierwater et al., 2009b). The total number of characters included was 17,664 from 51 partitions, giving 7,822 phylogenetically informative characters. In addition, a matrix was constructed with a larger number of taxa based on the Dunn et al. (2008) study with 73 taxa for the same gene partitions. This matrix had 17,637 total characters and 9,421 phylogenetically informative characters. Furthermore, Hox gene expression was compared for a placozoan and a cnidarian bauplan to test predictions from the placula hypothesis (Syed and Schierwater, 2002a).

The combination of parsimony, maximum likelihood (with morphological characters removed), and mixed Bayesian analysis of the smaller concatenated matrix support the hypothesis of a sister-clade relationship between bilaterians and non-bilaterians (Porifera, Ctenophora, Placozoa and Cnidaria) with the choanoflagellate *Monosiga brevicollis* as a sister-taxon to these two groups (Figure 3.10). Our analyses place the Placozoa at the earliest branching position in a pure non-bilaterian clade. It strongly supports Porifera, Bilateria, and Fungi as monophyletic groups. The four cnidarian classes together with the Ctenophora form the "Coelenterata". The grouping of Porifera, Cnidaria and Ctenophore is only weakly supported in the analyses, but the Placozoa are clearly excluded from a Porifera-Coelenterata group. As a conclusion of these findings, a new placula hypothesis was formulated (Schierwater et al., 2009a).

3.2.7 Conclusions

In an attempt to combine different data sets, from morphology to sequence data, we performed the first comprehensive total evidence analyses to address relationships among early metazoans. The surprising, but not at all new result is the hypothesis of an early split of Bilateria and non-bilaterian animals. However, while we addressed a number of analysis options and data and taxon sampling, the crucial issue of charac-

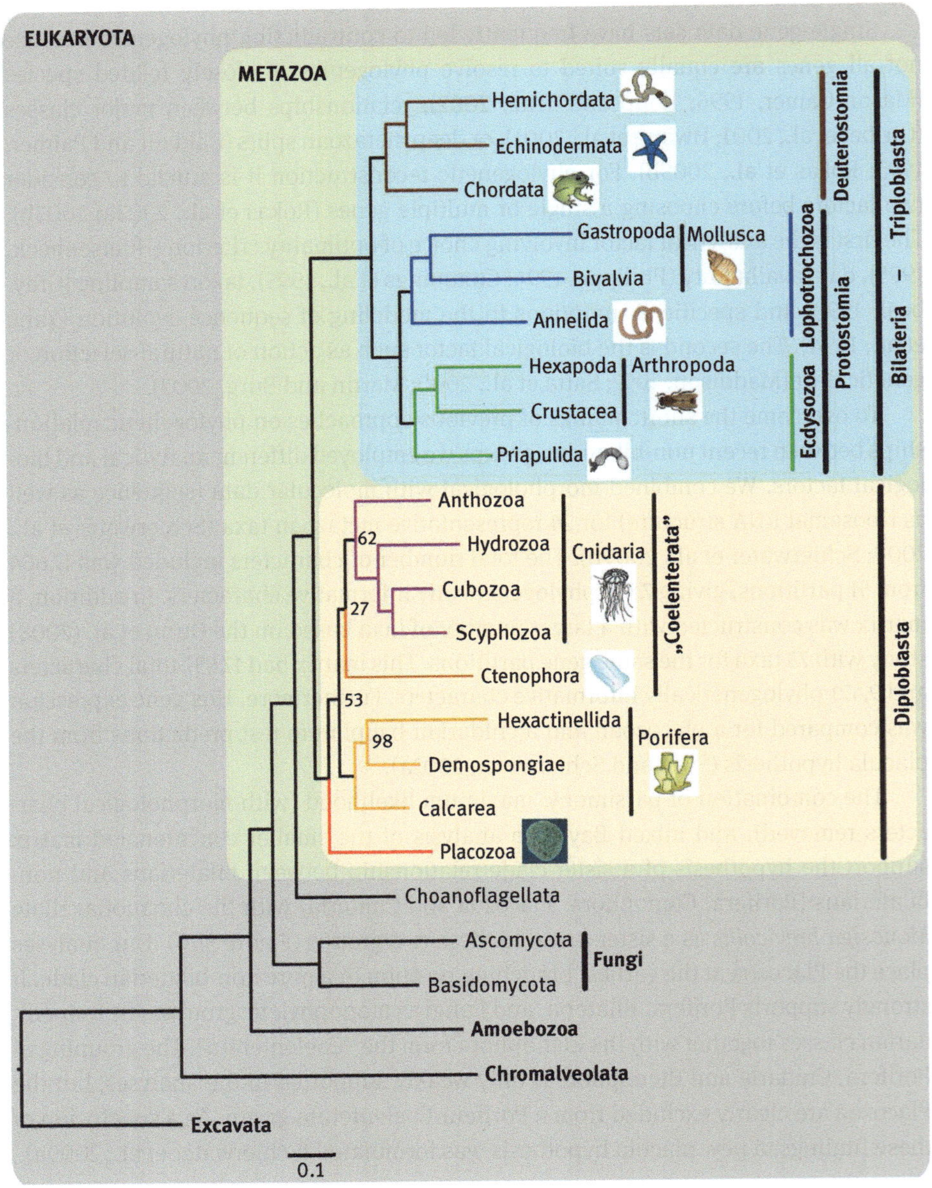

ter weighting has not been addressed yet. At present it seems that a massive increase in sequence data alone is not enough to resolve the base of the metazoan Tree of Life. It almost looks like the opposite, i.e. as more sequence data become available more topological scenarios appear. Thus, at the base of metazoan phyla we need to strike a blow for the quality – in addition to the quantity – of data (cf. Osigus et al., 2013). It is hard to see, however, how this issue can be addressed in practice. For example, the mitochondrial genome size in combination with gene loss versus gene gain probabili-

◄ **Figure 3.10:** Maximum likelihood phylogenetic tree of metazoan relationships using the concatenated data matrix. Node support is based on the best ML tree filtered through 1,000 rapid bootstrap replicates. Only support values below 100 % are shown. Bayesian inference supported strongly (posterior probability 1.0) all nodes with the exception of monophyly of Cnidaria. The maximum a posteriori and the Bayesian 50 % majority-rule consensus trees disagreed with the best ML tree in supporting a Ctenophora–Anthozoa clade with posterior probability of 0.98. Please note that "Coelenterata" is not a taxonomic unit, but rather it is a traditional grouping for reasons of convenience. The alpha shape parameters of the gamma distribution were 0.507454 and 0.651659 for the nucleotide and amino acid partitions, respectively. Log-likelihood = −261429.821426. From Schierwater, de Jong, and DeSalle, 2009.

ties makes it *cum grano salis* impossible not to regard the features of the placozoan mitochondrial genome as plesiomorphic for the Metazoa. But this is only a handful of high-quality data, which in practice get easily overrun by thousands or even hundreds of thousands of sequence characters of extremely poor quality per unit. This is without mentioning the large degrees of freedom in the analytical tools.

Acknowledgments

The study was supported by German Research Foundation grants SCHI 277/20-1, SCHI 277/20-2, SCHI 277/20-3, SCHI 277/21-1, SCHI 277/24-2, SCHI 277/24-3, SCHI 277/26-1 to B. Schierwater. M. Eitel was kindly supported by a doctoral fellowship of the Evangelisches Studienwerk e.V. Villigst and by a fellowship within the Postdoc-Program of the German Academic Exchange service. H.J. Osigus acknowledges a doctoral fellowship of the Studienstiftung des deutschen Volkes, K. v.d. Chevallerie a fellowship of the Evangelisches Studienwerk e.V. Villigst and T. Bergmann funding through a H. Wilhelm Schaumann Stiftung fellowship. We gratefully acknowledge bioinformatic support by Ingo Ebersberger and Sabrina Simon.

Yvan Perez, Carsten H.G. Müller, and Steffen Harzsch

4 The Chaetognatha: An anarchistic taxon between Protostomia and Deuterostomia

Abstract: The phylogenetic position of the chaetognaths within the Bilateria is heavily debated but this taxon so far has resisted all serious attempts to pin down its phylogenetic affinities. Despite the ever increasing number of molecular phylogenetic studies and an emerging consensus for protostome affinities, we must conclude that still the position of the Chaetognatha is among the most enigmatic issues of metazoan phylogeny. Here, we review the current molecular phylogenetic studies and discuss morphological features that play important roles in the discussion of chaetognath affinities including the integument, musculature and neuromuscular innervation, sensory systems, the nervous system and aspects of embryology. The chaetognath genome is likely the product of a unique evolutionary history and witnesses a long isolation of this group. Furthermore, morphological characters provide evidence for the long evolutionary distance that separates the Chaetognatha from its closest (unknown) metazoan relative and suggest that this taxon in many aspects seems to have explored its own evolutionary pathways in generating tissue and organ diversity. Both the genome and morphological characters include many autapomorphies of this group in addition to a character mix of protostome and deuterostome features. However, despite intense analyses of their genome and tissue ultrastructure, so far we do not know of any convincing synapomorphies that clearly would suggest a sister-group relationship to another metazoan taxon. These facts are most parsimoniously interpreted such that Chaetognatha are likely an early offshoot of the protostome lineage but so far it is not possible to precisely say whether they are basal lophotrochozoans, basal ecdysozoans or sister-group to all others protostomes. All together, we currently tend to suggest a new basal rooting of the Bilateria, which considers the Chaetognatha as the sister-group to the Lophotrochozoa and Ecdysozoa (soft polytomy). This assemblage may be called Hyponeuria to indicate the ventralization of the nervous system as one of the most important steps in metazoan evolution.

4.1 Who are the Chaetognatha?

The taxon Chaetognatha (arrow worms) represents a group of small marine predators that are major components of the zooplankton throughout the world's oceans and comprise more than 150 described species from all geographical and vertical ranges of the ocean (Figures 4.1, 4.2). With a body length between just a few millimeters up to 120 mm, these glassily transparent carnivores are among the most abundant pelagic organisms, but several epibenthic species are also known (Bieri, 1991; Shinn, 1997; Nielsen, 2001; Mutlu, 2006; Kapp, 2007). According to palaeontological evidence, the

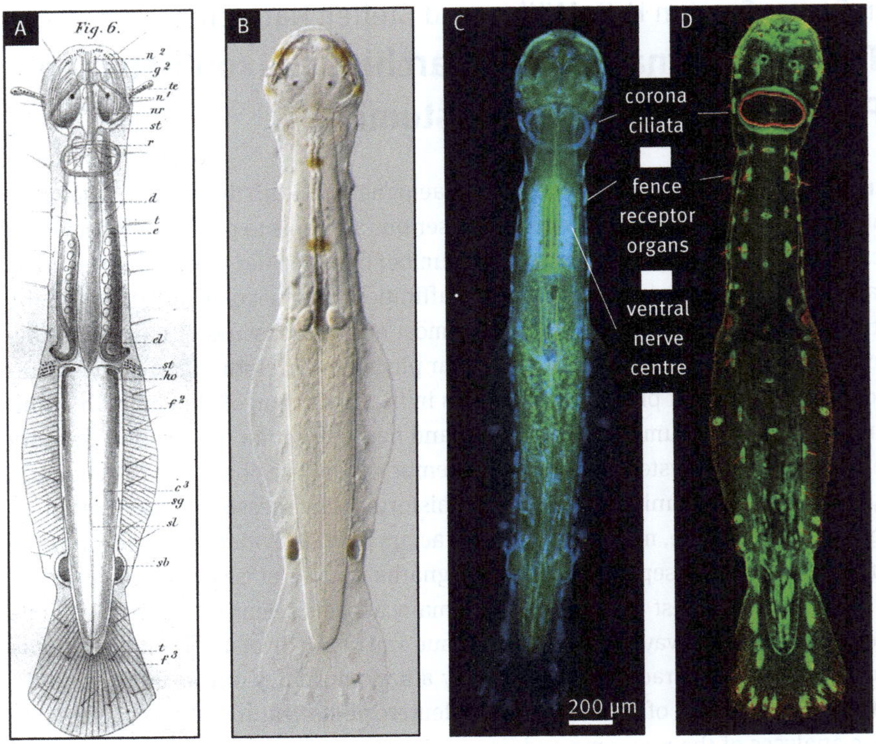

Figure 4.1: A. *Spadella cephaloptera*; drawing from Hertwig (1880) with the author's original nomenclature (from top to bottom; note that the current identification of the various organs may be different; consider e.g. the "Riechorgan"): Abbreviations: *n2* Nerv zwischen Bauchganglion und Schlundganglion (frontal connective between brain and vestibular ganglion); *g2* Oberes Schlundganglion (brain); *te* Tentakelartiger Fortsatz der Kopfkappe (tentacular head process); *n1* Commissur zwischen Bauchganglion und Schlundganglion (main connective between brain and ventral nerve center); *nr* Riechnerv ("olfactory nerve", nerve that links the corona ciliata to the brain); *st* Querseptum (transverse septum); *r* Riechorgan ("olfactory organ", the corona ciliata); *d* Darm (gut); *t* Tastorgan (mechanosensory organ); *e* Eierstock (ovary); *el* Eileiter (oviduct); *st* Querseptum (transverse septum); *ho* Hoden (testes); *c3* Schwanzhöhle (Coelom); *sg* Samengang (seminal duct); *sl* Längsseptum (longitudinal septum); *sb* Samenblase (seminal vesicle); *t* Tastorgan (mechanosensory organ); *f3* Schwanzflosse (tail). **B.** light microscopic image of *S. cephaloptera*. **C.** *S. cephaloptera*; immunohistochemical localization of RFamide-related petides (*green*) and histochemistry to label cell nuclei (*blue*; reprinted with permission from Harzsch et al., 2008). **D.** *S. cephaloptera*; immunohistochemical localization of acetylated alpha-tubulin (*red*) and histochemistry to label nuclei (*green*; reprinted with permission from Harzsch and Wanninger, 2010).

Chaetognatha were present in the Early Cambrian (approx. 540–520 Myr years ago) Chengjang biota (Vannier et al., 2007), as well as in the Middle Cambrian Burgess Shale (Szaniawski, 2005). Vannier et al. (2007) suggested placing them among the earliest active predator metazoans and argued that the ancestral chaetognaths were planktonic with possible ecological preferences for hyperbenthic niches close to the

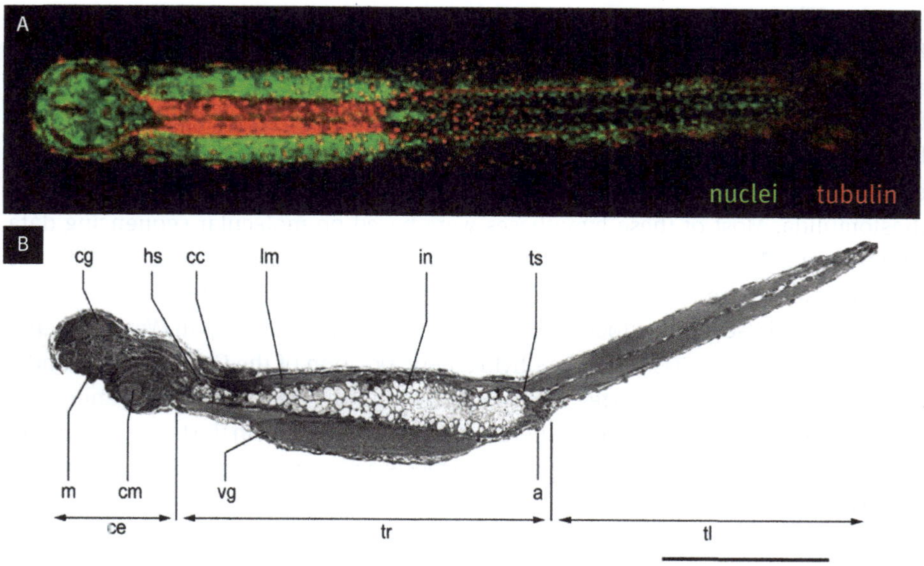

Figure 4.2: A. Two-day old hatchling of *S. cephaloptera*; immunohistochemical labeling of tubulin (*red*) and histochemical labeling of nuclei (*green*; reprinted with permission from Rieger et al., 2011). **B.** Light micrograph of a longitudinal section through a juvenile of *Spadella cephaloptera* (Perez, unpublished data). *a* anus; *cc* corona ciliata; *ce* cephalic segment; *cg* cephalic ganglion/brain; *cm* cephalic muscles; *hs* head-trunk septum; *in* intestine; *lm* longitudinal muscles; *m* mouth; *tl* tail segment; *tr* trunk segment; *vg* ventral nerve center. Scale bar 0.25 mm.

sea bottom. Recent Chaetognatha are characterized by the presence of horizontally projecting fins and, at the anterior end, two groups of moveable, cuticularized grasping spines used in capturing prey such as copepods. The so-called protoconodonts, spine-like small shelly microfossil elements that were abundant in the Cambrian, today are convincingly interpreted as parts of the chaetognath grasping apparatus or, at least, as belonging to protoconodont animals most closely related to Chaetognatha (e.g. Szaniawski, 1982; 2002; 2005; Vannier et al., 2007; but see Conway-Morris, 2009; Szaniawski, 2009 for a controversial discussion). The presence of protoconodonts in the lowermost Cambrian and the complexity of their feeding apparatus points to a Precambrian origin of these animals (Vannier et al., 2007).

4.2 Phylogenetic relationships and insights from molecular approaches

The phylogenetic position of the chaetognaths within the Bilateria is heavily debated but this taxon so far has resisted all serious attempts to pin down its phylogenetic affinities on the basis of traditional characters. In the last 200 years, various phylogenetic affiliations have been proposed (summarized e.g. by Ghiradelli, 1968; 1995; Bone, Kapp, and Pierrot-Bults, 1991; Telford, 2004; Harzsch and Müller, 2007; Harzsch and Wanninger, 2010; Edgecombe et al., 2011; Nielsen, 2012). For instance, close relationships to Ecdysozoa as a whole, or, among them, to Arthropoda and Nematoda have been assumed. In contrast, chaetognaths were thought to be the sister-group of Lophotrochozoa, or, further up the lophotrochozoan clade, closely related to Gnathostomulida. Most of these hypotheses were based on molecular sequencing data (Harzsch and Müller, 2007; for further details see below). A placement within the deuterostomes based on their embryonic development (Kuhl, 1938; Hyman, 1959) was accepted in several textbooks (Brusca and Brusca, 1990; Kapp in Westheide and Rieger, 1996) but Ax (2001) left the phylogenetic position of the Chaetognatha undecided. Nielsen (2001), suggested a placement within the Protostomia and united the Chaetognatha together with the Rotifera and Gnathostomulida in the taxon Gnathifera but this author is more cautious now and suggests a placement as a basal protostome group (Nielsen, 2012).

4.2.1 Single gene analysis and total evidence approach

4.2.1.1 Nuclear ribosomal genes

Over the last 20 years, phylogenetic studies based on single-gene analyses, sometimes combined with morphological data, also failed to come up with a stable hypothesis on the phylogenetic positions of Chaetognatha. Wada and Satoh (1994) clearly excluded the Chaetognatha from the Deuterostomia in an analysis based on a 18S rRNA sequence, a phylogenetic marker widely used to infer the new metazoan phylogeny. In a study also based on this gene, Telford and Holland (1993) proposed to consider the chaetognaths descendants from an early metazoan branch possibly originating prior to the radiation of the major coelomate groups. Halanych (1996a; 18S rRNA) suggested a chaetognath-nematode relationship but did not include Nematomorpha and Gastrotricha into his analysis. He recognized long-branch attraction as a possible source of error but tried to minimize this error by a four-taxon analysis. Interestingly, the ribosomal cluster of Chaetognatha is duplicated, leading to two classes of paralogous 18S rRNA (Papillon et al., 2006) and 28S rRNA (Telford and Holland, 1997) genes both of which diverge extremely from other Metazoa. Giribet et al. (2000) combined 18S rRNA data of 145 terminal taxa with 276 morphological characters in a total evidence approach. This analysis yielded a sister-group relationship of Chaetognatha with Nemertodermatida, which the authors qualified as unstable and difficult to justify on the basis of morphological/anatomical characters. Peterson and Eernisse (2002) analyzed with maximum parsimony 138 morphological characters from 40 metazoan groups and 304 18S rRNA sequences. Their analyses placed the Chaetognatha within the Ecdysozoa and here within the Nematoida similar to the result of Halanych (1996a). However, this position was unstable so that the authors suspected that this placement is potentially prone to long-branch attraction, and are unaware of any morphological synapomorphies shared exclusively by chaetognaths and either nematodes or nematomorphs. Mallatt and Winchell (2002) combined the analysis of 28S and 18S rRNA sequences. Chaetognath sequences grouped with those of an onychophoran, but this was unstable and qualified as artifactual by these authors due to long-branch attraction. In a recent attempt to improve rRNA data sets used in previous studies, Mallat, Craiga, and Yoderb (2012) constructed a taxon-rich alignment of rRNA genes from most metazoans (371 taxa from 33 of the 36 metazoan phyla). However, because of base-compositional heterogeneity and several highly divergent sequences used in this study, phylogenetic reconstructions did not support the bilaterian clades of lophotrochozoans and ecdysozoans. Among these highly divergent taxa, chaetognath sequences grouped with a mongrel and implausible assemblage including the cephalopods, onychophorans, and pauropod myriapods.

4.2.1.2 Intermediate filaments

Erber et al. (1998) conducted a comparative molecular genetic analysis of intermediate filament proteins which are major components of the cytoskeleton. Within a field of ca. 20 representatives of the protostomes and deuterostomes, the novel intermediate filament sequence of *Sagitta elegans* was very unusual which the authors attributed to a particularly high evolutionary drift of *Sagitta*. In principle, however, Erber et al. (1998) found that many of the intermediate filament proteins obtained from the various different protostomian taxa investigated (including the Chaetognatha) display a "lamin homology segment in their tail domains" (p. 754). Among Protostomia, moreover, chaetognath intermediate filament sequences show closest similarities with members of the Lophotrochozoa. Lophotrochozoans have in common L-type intermediate filaments which "share with the nuclear lamins the long version of the coil 1b subdomain and, in most cases, a homology segment of 120 residues in the carboxy-terminal tail domain" (p. 751).

4.2.1.3 Tropomyosin

Matus et al. (2006) analyzed the phylogenetic position of the chaetognaths using a new data set from the tropomyosin gene of 56 metazoans including the sequence of *Flaccisagitta enflata*. Bayesian and maximum likelihood analyses recovered the clades of Ecdysozoa and Lophotrochozoa as well as the deuterostomes and showed a sister-group relationship of Chaetognatha and Lophotrochzoa with high support values.

4.2.2 Multiple gene analysis and phylogenomics

Owing to the increasing sequencing effort, a considerable amount of molecular data is now available to evolutionary biologists. Recently, the metazoan phylogeny has been strengthened by additional molecular evidences, in particular from the 'phylogenomic approach', which entails the use of growing databases to build large data sets of concatenated genes in a large number of species. According to several authors (Delsuc, Brinkmann, and Philippe, 2005; Philippe and Telford, 2006), collecting a large number of orthologous genes from numerous species is a good strategy to avoid stochastic and systematic errors when inferring a phylogenetic hypothesis. The generation of Expressed Sequence Tags (EST) is then a fruitful method for gathering many sequences from many taxa. So far, EST libraries are available in public databases for two species of chaetognaths, *Spadella cephaloptera* (Marlétaz et al., 2006) and *Flaccisagitta inflata* (Matus et al., 2006).

In order to resolve the affinities of chaetognaths, several studies used genomic data. All confirmed that the chaetognaths are not deuterostomes, although there is no consensus as to whether they are lophotrochozoans or diverged before the lophotrochozoan/ecdysozoan split. Indeed, molecular analyses using Bayesian inference and maximum likelihood methods for analyzing broad data sets supported either a sister-group relationship with Lophotrochozoa (Matus et al., 2006; Dunn et al., 2008; Philippe et al., 2011b) or a position as a sister-group to all Protostomia (Marlétaz et al., 2006; Philippe et al., 2007; Marletaz et al., 2008; Marlétaz and Le Parco, 2008). By using the same methods and a broader taxon sampling, Dunn et al. (2008) suggested a sister-group relationship of Chaetognatha to the Lophotrochozoa, whereas Marlétaz et al. (2008) again proposed a most likely position of the Chaetognatha as a sister-group to all Protostomia.

Finally, in an analysis of nuclear-coded housekeeping genes, Chaetognatha emerged within the Ecdysozoa (Helmkampf, Bruchhaus, and Hausdorf, 2008a) but the authors did not qualify this position as robust and excluded alternative hypotheses concerning their relationship. Nevertheless, placement within the Ecdysozoa was also suggested in the study of Paps, Baguñà, and Riutort (2009a). Possible roles of long-branch attraction effects in chaetognath phylogenomic studies were discussed by Marlétaz and Le Parco (2008). Due to the lack of sequences from additional protostome data, long-branch attraction has not been conclusively ruled out and all analyses emphasized that new genomic data and adequate methods of reconstruction are required to challenge the assumption that chaetognaths have a basal position within the protostomes. In conclusion, even robust approaches provided by broad taxon and gene sampling in recent phylogenomic studies failed to deliver a stable hypothesis on the phylogenetic position of Chaetognatha. Despite the ever increasing number of such studies and a consensus for protostome affinities, we must conclude that today the position of the Chaetognatha is still among the most enigmatic issues of metazoan phylogeny.

4.3 Peculiarities of *Hox* genes, the mitochondrial genome, and a transcriptome

As there appear to be specific diagnostic amino acid motifs in the *Hox* genes of the three main lineages of bilaterians, this family of homeotic genes can be an important and informative field of investigations. Papillon et al. (2003) isolated the homeodomains of six *Hox* genes from *Spadella cephaloptera*, one belonging to the paralogy group 3 and four to the median class. Yet, these authors were not able to identify any sequence belonging to the posterior *Hox* genes. One important result was the discovery of a new homeodomain with a unique set of signature amino acid motifs shared both with median and posterior *Hox* proteins of protostomes and deuterostomes. This unique mosaic organization suggests that at least some of the median genes in extant metazoans may have derived from tandem duplication of an ancestral median/posterior one and that the Chaetognatha might be an early off-shoot of the triploblastic lineage that predates the deuterostome/protostome split. However, the authors also noted carefully that this mosaic gene could be highly derived and only present in Chaetognatha. In 2007, eight *Hox* genes and one *ParaHox* gene were isolated from *F. enflata* (Matus, Halanych, and Martindale, 2007). The presence of the mosaic median/posterior gene was confirmed in this pelagic species and, additionally, two posterior *Hox* genes were isolated. The finding of posterior *Hox* genes in Chaetognatha supports the hypothesis that the mosaic median/posterior gene is actually an apomorphy of all extent chaetognaths inherited from their last common ancestor. Yet, the careful analysis of homeodomains showed that the posterior *Hox* genes of chaetognaths possess both ecdysozoan and lophotrochozoan signature amino acid motifs, while the central class *Hox* genes lack the diagnostic amino acid motifs used to assign lophotrochozoan or protostomes affinities (e.g. Lox5 spiralian parapeptide and Ubd-A peptide found in both ecdysozoans and lophotrochozoans). Thus, this last study did not allow the authors to decide between the two hypotheses tested of a sister-group relationship of Chaetognatha to all other Protostomia versus to all Lophotrochozoa (Matus, Halanych, and Martindale, 2007).

So far, five complete mitochondrial genomes (mtDNA) are available, two in benthic species (Helfenbein et al., 2004; Papillon et al., 2004) and three in pelagic ones (Miyamoto, Machida, and Nishida, 2010). Chaetognatha display a very unusual mtDNA both in size and composition. It is one of the smallest known in metazoans and contains only 14 of the 37 usual genes. Most surprising was the inability to isolate 21 transfer RNA from the 22 usually present in the mtDNA of others metazoans. Of the 13 common genes encoding proteins, the ATP synthetase subunit 6 and 8 are missing. Helfenbein et al. (2004) analyzed the mtDNA of the benthic species *Paraspadella gotoi*. Comparisons of amino acid sequences from mitochondrial encoded proteins using a parsimony-based approach yielded a single most parsimonious tree that suggests a position of the Chaetognatha as a sister-group to all other Protostomia. Yet, Papillon et al. (2004) characterized the mtDNA of another benthic species, *S. cepha-*

loptera, and provided a more complex analysis using several approaches including a phylogenetic analysis of primary sequences, a gene order comparison and the diagnosis of signature amino acid residues in gene encoding for the NAD5 protein. Their analyses placed the chaetognaths within the protostomes in a clade that includes molluscs and annelids. However, the authors cautiously limited their conclusions and suggested only that chaetognaths are linked in some way to the protostomes, arguing that the features shared between annelids, molluscs and chaetognaths are actually bilaterian plesiomorphies.

Few nuclear genomic data are available in Chaetognatha and little is known about the structure and composition of their genome. Important questions remain about the organization of their *Hox* cluster and duplication events, insertions, deletions and copy number. What is the level of polymorphism within populations? Do chaetognaths regulate their genome in a unique and original manner? Two recent studies, however, offered first opportunities to explore some of these questions in the species *S. cephaloptera*. An overview of chaetognath genomics given by a detailed survey of transcriptomic data showed that mRNA of *S. cephaloptera* undergoes trans-splicing maturation (Marlétaz et al., 2008; Marlétaz and Le Parco, 2008). A partial genome duplication which occurred early in the lineage was also reported for 30 % of genes, followed by a high conservation of duplicated genes (Marlétaz et al., 2008). Phylogenetic analysis of cytochrome b, cytochrome oxydase I and two ribosomal protein (L36a and L40a) genes showed that *S. cephaloptera* displays a fast molecular rate and a high level of genetic polymorphism at the population level (Marlétaz et al., 2008; Marlétaz, Gyapay, and Le Parco, 2010). This strong genetic variation was confirmed by a structural analysis of four bacterial artificial chromosomes clone inserts corresponding to four different alleles of the *Hox1* region in which mobile element activity mainly mediated by multiple families of retrotransposons and DNA transposons should have played a prominent role in structural variations (Marlétaz, Gyapay, and Le Parco, 2010). Nevertheless, while none of the polymorphic insertions induced modification of the gene order in the genomic region considered, the authors wondered how selective processes cope with intense sequence turnover to preserve the integrity of functional genomic regions and suggested that the Chaetognatha genome is likely the product of a very original evolutionary history (Marlétaz et al., 2008).

4.4 Unusual features: The role of morphology in our understanding of chaetognath phylogeny

There is a long-standing interest into the body organization of Chaetognatha. General information on their morphology and anatomy has been summarized in the contributions by Hertwig (1880), John (1933), Hyman (1959), Goto and Yoshida (1987), Bone and Goto (1991), Kapp (1991), Nielsen (2001; 2012), and Ax (2001); the most detailed review of their microscopic anatomy is that of Shinn (1997). Much effort has been directed towards understanding the structure, development and physiology of the muscles in the body wall and trunk (Duvert and Salat, 1979; 1980; Duvert, Gros, and Salat, 1980; Duvert and Savineau, 1986; Savineau and Duvert, 1986, Dress and Duvert, 1994; Duvert and Dress, 1994; Royeula et al., 2003), the visceral muscles (Duvert and Salat, 1995), and muscles in the appendages (Casanova, Duvert, and Goto, 2003) of chaetognaths. Furthermore, the ultrastructural and cellular organization of the connective tissue (Duvert and Salat, 1990), the mesoderm (Shinn and Roberts, 1994), the coelomic epithelia (Welsch and Storch, 1982; Ghirardelli, 1995), the digestive system (Arnaud et al., 1996; Perez et al., 1999; Perez, Casanova, and Mazza, 2000; 2001), and of the reproductive organs (Shinn, 1992; Carré, Djediat, and Sardet, 2002) have been explored (review Shinn, 1997). In the debate on chaetognath phylogenetic relationships, several important morphological characters have emerged that have been discussed repeatedly. In the following, we will discuss some of these characters with respect to their phylogenetic significance.

4.5 Unique features of chaetognath development

Aspects of chaetognath embryology and neurogenesis were recently reviewed by Rieger et al. (2011) and the following section was reprinted with permission from this contribution. The embryonic development of Chaetognatha is known from the fundamental studies by Hertwig (1880), Doncaster (1902), and John (1933) on *Sagitta* and *Spadella*. Chaetognaths are hermaphroditic and develop directly, so that the larvae at hatching display a body organization that is in most respects similar to the adult. More recently, the embryology and various aspects of growth and reproduction of Chaetognatha have been explored in laboratory cultures (Kuhl and Kuhl, 1965; Reeve, 1970; Kuhlmann, 1977; Goto and Yoshida, 1997; Shimotori and Goto, 1999, 2001; Carré, Djediat, and Sardet, 2002; review Pearre, 1991; Kapp, 2000) including the mating sequence (Goto and Yoshida, 1985), the fertilization process (Goto, 1999; Carré, Djediat, and Sardet, 2002), neurogenesis (Rieger et al., 2011, Perez et al., 2012) and aspects of regeneration (Duvert, Perez, and Casanova, 2000). Furthermore, the embryonic expression patterns of actin genes (Yasuda et al., 1997), the brachyury gene (Takada, Goto, and Satoh, 2002) and a median *Hox* gene (Papillon et al., 2005) have been analyzed by whole mount in situ hybridization. The studies on embryology have used time-lapse imaging (Kuhl and Kuhl, 1965) and injections of fluorescent tracer molecules (Shimotori and Goto, 1999; 2001) to examine developmental fates of the blastomeres and the establishment of axial properties of the embryos. Carré, Djediat, and Sardet (2002) observed the fate of a *Vasa*-positive germ granule in eggs and traced these through the development from embryo to adult. Thus, as in *Drosophila melanogaster*, *Vasa*-proteins here seem to be involved in germ cell specification.

Embryonic cleavage in chaetognaths is total and equal, and sufficiently close to the radial cleavage pattern of deuterostomes to suggest a deuterostome affinity (discussed in Shimotori and Goto, 2001). This placement was also based on the fact that differentiation of the coelom seemingly resembles enterocoely (reviewed by Salwini-Plawen, 1986; Ghirardelli 1995; Kapp, 2000; 2007). However, a recent analysis of the developmental fate of the first four blastomeres led Shimotori and Goto (2001) to propose that chaetognaths are more similar to protostomes in their developmental program. In addition, their embryology includes some features that were suggested to be unusual and difficult to relate to other Bilateria so that Kapp (2000) coined a new term, heterocoely, to describe their mode of coelom formation during gastrulation. This author, reviewing the classical studies by Hertwig (1880), Doncaster (1902), and John (1933) has concluded that many other aspects of chaetognath embryogenesis, especially the process of mesoderm formation, seem to be unique to this taxon and not suggestive of a close affinity to any other bilaterian phylum. However, we feel that beyond taking these classical studies as a foundation, chaetognath embryogenesis and especially germ layer and coelom formation urgently need to be reanalyzed by contemporary methods including electron microscopy (see e.g. Shinn and Roberts, 1994) in order to allow for a phylogenetic comparison of chaetognath development.

Gastrulation describes the process of germ layer formation and is intimately linked to the formation of the digestive system that includes mouth opening, gut and anus. Gastrulation patterns play essential roles in discussions on bilaterian evolution (e.g. Arendt and Nübler-Jung, 1997; Nielsen, 2001; 2005a; 2005b; Hejnol and Martindale, 2009; Martindale and Hejnol, 2009; Lacalli, 2010; Nielsen, 2010). The blastopore is the site of tissue internalization, and two major phylogenetic concepts are based on the fate of the blastopore. In protostomes, the blastopore becomes the mouth, and the anus (if present) is formed secondarily at a different site (reviewed in Hejnol and Martindale, 2009; Martindale and Hejnol, 2009). In deuterostomes, the blastopore becomes the anus while the mouth is formed secondarily at a different site in the animal hemisphere of the embryos. In Chaetognatha, an invagination of the blastoporus initiates the formation of the entoderm and subsequently, a second opening forms the stomodaeum opposite to the blastoporus. For a short period the embryo possesses both a stomodaeum and a blastopore. The mouth then closes, followed shortly afterwards by the blastopore, so that young hatchlings for a time have neither a mouth opening nor an anus. Both openings are re-established some days after hatching. Hence, Chaetognatha conform to the developmental pattern of deuterostomes as laid out above. However, Hejnol and Martindale (2009) have emphasized that, on closer examination, a considerable variation in gastrulation patterns is evident between bilaterian subtaxa, especially in protostomes, which leads them to stress the extreme variability in blastopore fate. In ctenophores and cnidarians, the mouth and the blastopore have a common origin and these animals gastrulate at the animal pole. Nevertheless, bilaterians gastrulate at the vegetal pole (Hejnol and Martindale, 2009; Martindale and Hejnol, 2009). The authors propose that in Bilateria a separation of the signaling centers that determine the sites of mouth formation versus site of germ layer specification has taken place and that this ancient separation explains the variation of the spatial relation of blastopore and mouth in Bilateria. Furthermore, they critically review the concept of amphistomy, which refers to the lateral closure of a slit-like, elongate blastopore. The latter, by staying open at both ends, can account in principle for the formation of both the bilaterian mouth and the anus (e.g. Arendt and Nübler-Jung, 1997; Nielsen, 2001). Amphistomy appears, however, to occur in this idealized way in only a few protostomian taxa.

In summary, chaetognath gastrulation displays a mosaic of both protostome and deuterostome features, which undoubtedly accounts for the abundance of hypotheses on their phylogenetic relationships. The mode of gastrulation has been one of the key reasons for allying them to deuterostomes (summarized in Takada, Goto, and Satoh, 2002), but this link should not necessarily be taken at face value. In a recent review, Lacalli (2010) raises the point that deuterostomy could have emerged more than once in the taxa conventionally assigned to the Deuterostomia. The condition is linked to a delay in mouth formation in the developing dipleurula larvae of one subgroup, the Ambulacraria (echinoderms plus hemichordates), and this could be associated with their distinctive mode of feeding, which relies on a ciliated food-collecting domain that surrounds the mouth. In contrast, there is little evidence as yet that the other subgroup (chordates) ever had feeding larvae of this type (Lacalli, 2005) but the delay in mouth formation (and hence deuterostomy) in chordates may instead be more a consequence of the reorganization of the head region required by dorsoventral inversion, a feature of chordates but not the Ambulacraria. In chordates, in fact, the embryonic anus closes as well, so both the mouth and the definitive anus are formed quite late, and neither occupies the precise site of the blastopore. Since the hatching stages of chordates rely on a substantial store of yolk, there is a greater freedom to rearrange the openings to the gut without compromising function.

It is possible, in this way, to view deuterostomy as a secondary change that may have occurred several times and for different adaptive reasons. If, in fact, the mouth has been the subject of evolutionary experiments of this type in taxa conventionally assigned to the deuterostomes, it is a short step to suggesting that chaetognaths may simply represent a third such experiment in which the fate of the blastopore is clearly distinct from protostomes, but without necessarily meaning that chaetognaths are taxonomically close to deuterostomes. This conclusion, that chaetognaths display a third variant of deuterostomy, is substantiated by an analysis of the expression pattern of the brachyury gene in a closely related chaetognath species, *Paraspadella gotoi* (Takada, Goto, and Satoh, 2002). These authors conclude that the expression pattern of this gene in the embryonic chaetognath blastopore and mouth resembles that of hemichordates and echinoderms, whereas the pattern in the region of the new mouth opening of the hatchling appears to be novel.

4.6 Integument: multilayered epidermis and intra- and basiepidermal plexus

The integument of the Chaetognatha is complex and contains a multilayered epidermis (synonymous term: (pluri-)stratified epidermis/epithelium: e.g., Furnestin, 1967; Duvert and Salat, 1979; Welsch and Storch, 1983; Duvert, Boulignand, and Salat, 1984; Ahnelt, 1984, Shinn, 1997) that has attracted numerous zoosystematists and evolutionary biologists as it somehow resembles the vertebrate epidermis, but contrasts with other invertebrates that are usually equipped with a single-layered epidermis (Welsch and Storch, 1983). Even though there are exceptions to this rule, as for instance the spherically aggregated iridophore (epidermal) cells in the interambulacral zone of some diadematid sea urchins (review by Holland, 1984), a multilayered epidermis is only present in some specialized areas of the body. In contrast, a multilayered, non-cuticularized epidermis of taxon-specific thickness is predominant in Chaetognatha. Only at the inner side of the hood (praeputium), the ventral cephalic hemisphere around the mouth opening (covered by a cuticularized prismatic epithelium), in some parts of the lateral and tail fins, and on the eyes the chaetognath epidermis remains monolayered (e.g., Ahnelt, 1984; Kapp, 1991; Shinn, 1997). Based on the examination of histological and ultrathin sections, previous authors noticed the epidermal cells to become flattened from apical to proximal layers, but assigned them to a single type of epithelial cells (e.g. Hertwig, 1880: "mehrschichtiges Pflasterepithel", Kapp, 1991). Ahnelt (1984), Duvert, Boulignang, and Salat (1984), Shinn (1997) and Harzsch et al. (2009) however distinguished a single distal layer of polygonal, relatively voluminous cells with numerous large vacuoles and secretory granules ("surface cells", "superficial cells", "outer epidermal cells", "distal epidermal cells") from a subjacent multilayer of elliptical to extremely flattened cells ("compact interdigitating cells", "filament-rich cells", "inner epidermal cells", "proximal epidermal cells") which are strongly intertwined by desmosome-like columnar junctions and are equipped with a massive apparatus of tightly accumulated tonofilaments (see scheme in Figure 4.3 and Figure 4.4B,D). The massive filamentous apparatus in proximal epidermal cells is presumed to antagonize coelomic pressure (Shinn, 1997). Furthermore, if further elaborated and compressed, it provides functional specializations of this cell type, as for example observed in the sheath cells wrapping the various elements of the central and peripheral nervous system (Shinn, 1997; Harzsch et al., 2009; see also our Figure 4.4B–C,E). As a third cell type, "vacuolated cells" (sensu Ahnelt, 1984 or Shinn, 1997) may occur locally in the chaetognath epidermis (e.g., collarette region) which are considered derivatives of the proximal epidermal cells. The chaetognath integument also includes an extensive basi- and intraepidermal (neuronal) plexus (Figures 4.3, 4.4A–B,D) which displays two horizontal layers of concentration. The first is found close by the thick and fibrillar extracellular matrix (ECM) and comprises transverse, oblique and longitudinal neurite bundles whose connection pattern somewhat resembles an orthogon (compare Fig. 4 of Ahnelt, 1984; cf. Figures 4.3, 4.4B in this paper). The second plexus aggregation is found crammed into the tran-

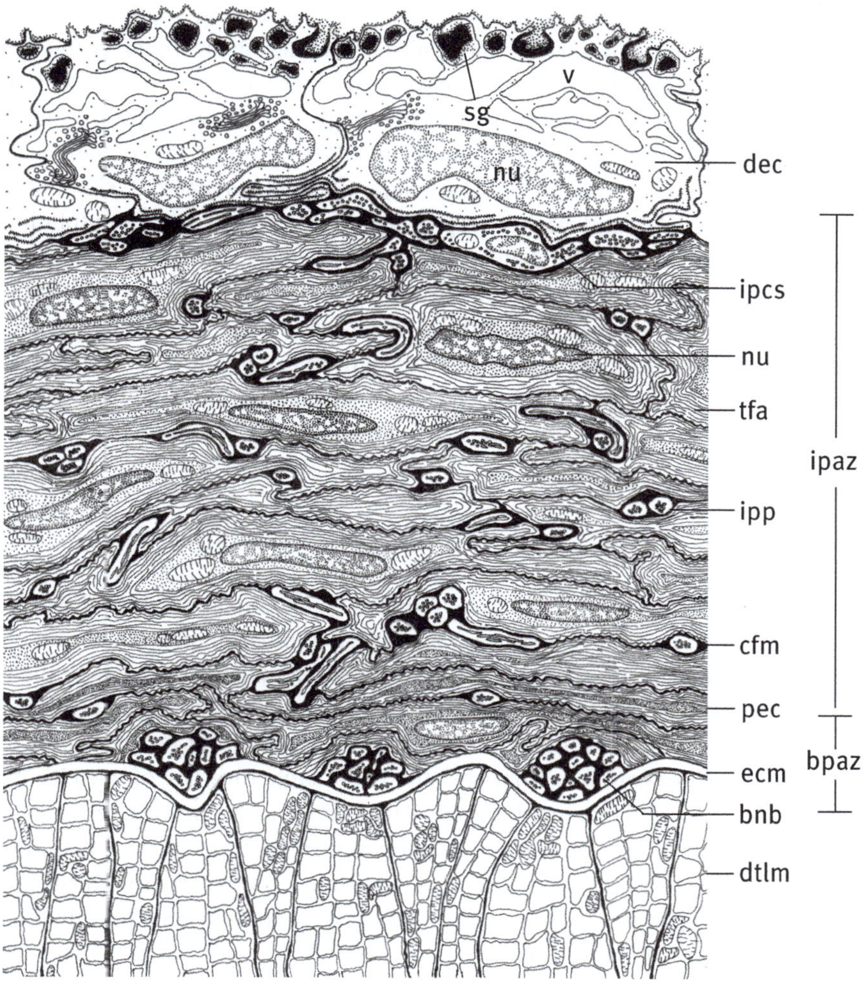

Figure 4.3: Semischematic reconstruction of the chaetognath integument using the example of the dorsolateral trunk region of *Spadella cephaloptera* (cross section; C. H. G. Müller, unpublished figure). The integument exhibits an apical layer of cuboidal distal epidermal cells, a subjacent multi-layer of horizontally flattened proximal epidermal cells, a massive extracellular matrix (ECM), and an extensive intra- and basiepidermal plexus. Note the two aggregation levels of the neuronal plexus, one below the distal epidermal layer (containing also interneuronal cell somata) and the other directly above the ECM, here in the shape of distinct neurite bundles. The plexus profiles are often surrounded by a filamentous, extracellular matrix that has been blackened (*dark matter* between neurites) for providing a better contrast. Original drawing. *bpaz* basiepidermal plexus aggregation zone; *bnb* basiepidermal neurite bundle (proximal aggregation zone of the neuronal plexus); *cfm* condensed filamentous mass (in-between pec's, surrounding ipp); *dltm* dorsal longitudinal trunk musculature; *dec* distal epidermal cells; *ecm* extracellular matrix; *ipcs* interneuronal cell soma (in the intraepidermal plexus); *ipaz* distal intraepidermal plexus aggregation zone; *ipp* intraepidermal plexus profiles (neurites); *nu* nucleus; *pec* proximal epidermal cell; *sg* secretory granule; *tfa* tonofilament apparatus (in pec); *v* vacuole.

sition zone of distal and proximal epidermal cells without showing any kind of geometric arrangement (Figures 4.3, 4.4D). There, also neuronal somata are noticeable by their characteristic shape and cytoplasmic appearance (cf. Shinn, 1997). The distal and proximal aggregation zone is connected by a highly complex network of plexus profiles (neurites or neurite bundles) which are crammed into the tiny interspaces of the proximal epidermal cells and surrounded by extracellular filamentous material (simplified as dark matter in Figure 4.3) and can be comprehensively visualized using anti-tubulin immunhistochemistry (Harzsch and Müller, 2007 and our Figure 4.4A). Whereas extensive intraepidermal plexus are common in various metazoan phyla (review of Richter et al., 2010), the combination of a mainly orthogonal basiepidermal plexus and a heavily diversified intraepidermal plexus may be inevitably linked to a multilayered epidermis which is as thick as found in recent Chaetognatha. Consequently, this combination may be interpreted as an additional apomorphy of this animal group.

4.7 Muscle ultrastructure and neuromuscular innervation

Chaetognatha include many taxa which are ferocious predators and the mechanical performance of their locomotory muscles was suggested to be comparable to that of the fastest invertebrate muscles (Duvert, 1989). Muscles have been intensively studied in Chaetognatha and it is well documented that their locomotory muscles of the cross-striated type display a high level of variation at the ultrastructural level (Casanova and Duvert, 2002). These authors compared the remarkable diversity of muscles in Chaetognatha to that found e.g. in arthropods suggesting a convergence constrained by the need to construct a contractile apparatus in response to similar functional needs. However, chaetognaths lack oblique-striated muscles which are typical for many invertebrates capable of super-contractions such as e.g. Nematoda or Annelida and Mollusca (reviewed in Casanova and Duvert, 2002).

In representatives of the genus *Sagitta*, the body wall comprises two types of muscles called primary and secondary muscles. The primary musculature accounts for ca. 80 % of the trunk wall volume and is of the typical cross-striated type (Bone and Duvert, 1991). However, it has an unusual pattern of neuromuscular innervations that will be discussed below. The secondary muscles account for less than 1 % of the body wall musculature but their structure is unique within the animal kingdom and hence considered an autapomorphy of this taxon (Duvert, 1991) and a subtype of this musculature was proposed to have super-contracting properties (Casanova and Duvert, 2002). It forms two pairs of longitudinal bands, two medio-dorsal and ventral and two lateral ones suggesting an antagonistic functional role (Duvert, 1991). In the genus *Sagitta*, this cross-striated muscle is composed of two different types of sarcomeres, S1 and S2, which alternate along the muscle's longitudinal axis separated by Z-discs and hence this muscles type was termed heterosarcomeric muscle (Casanova and Duverts, 2002). S1 sarcomeres show structural features typical of invertebrate slow fibers including long sarcomeres in which a thick filament is surrounded by 8–10 thin filaments (Duvert, 1991; Casanova and Duvert, 2002; Royela et al., 2002). In contrast, S2 sarcomeres do not show a cross-striation but possess parallel arrays of thicker (10–15 nm) and thinner (ca. 6 nm) filaments that seem to differ from the characteristic thick and thin myofilaments and do not display A and I bands. The molecular composition of these filaments is largely unknown but they were suggested to be cytoskeletal filaments with contractile properties (Duvert, 1991). However, in an immunohistochemical study with antisera against various muscle proteins, Royela et al. (2003) found that the S2 sarcomeres do possess actin (probably corresponding to the thinner filaments) but that they lack myosin and also paramyosin suggesting that the thicker filaments may represent a type of intermediate filament. Hence, these authors suggested that the S2 sarcomeres are not true muscular structures but rather "pseudosarcomeres" of unknown physiological role. While these heterosarcomeric secondary muscles are typical for the genus *Sagitta*, other taxa show variations on this theme. In the genera *Spadella* and *Paraspadella*, the secondary muscles are homosarcomeric and composed only of S1 sarcomeres with a classical cross-striation (S2 sarcomeres are absent) but with discontinuous Z-discs (Casanova and Duvert, 2002). In *Xenokrohnia*, the secondary muscles display a hybrid architecture of the types described for the genera *Sagitta* versus *Spadella*.

In conclusion, whereas oblique-striated musculature, as is typical for many other invertebrates, is absent in Chaetognatha, this taxon has experimented with various types of cross-striated muscles. We agree with Duvert (1991) that ultrastructural features of muscle architecture at the moment do not help us to specifically relate Chaetognatha to other metazoan phyla but we nevertheless can conclude that this taxon seems to have explored its own evolutionary pathways in generating muscle diversity.

The same holds true for types of neuromuscular innervation. Chaetognatha show a great variety in the ultrastructure of neuromuscular junctions that was suggested to match the diversity seen e.g. in arthropods (Duvert and Barets, 1983). At least three types of neuromuscular synapses have been described so far, two related to muscles in the head and one related to the trunk muscles. In the head, nerves emerging from the brain are embedded in supporting structures, being either ramifications of the extracellular matrix (ECM) or sheets of extremely flattened, glial-like proximal epidermal cells (Figure 4.4E, for details read Section 4.6 on epidermis). These nerves, which target the esophageal and somatic head muscles, have conventional nerve endings and neuromuscular junctions that display ultrastructural features similar to classic motor end plates (Duvert and Barets, 1983; Shinn, 1997). Hence, the first type of synapses is related to the esophageal muscles which are innervated by more or less classical neuromuscular contacts from extensions of nerves penetrating the muscle tissue (Duvert and Barets, 1983). The second type of synapses is related to muscles of the head, where extensions of muscle cells protrude through the ECM and extend towards axons to form synaptic contacts with these. Contrarily, the primary muscle of the trunk (see above) is completely separated from the epidermis and the nervous system by a thick ECM which is neither crossed by extensions of muscle fibers nor by neurites of neurons (Figure 4.4). This means that in this third type of synaptic contacts, neurites do not have direct synapses with underlying locomotory muscle fibers, but that the presynapses are sepa-

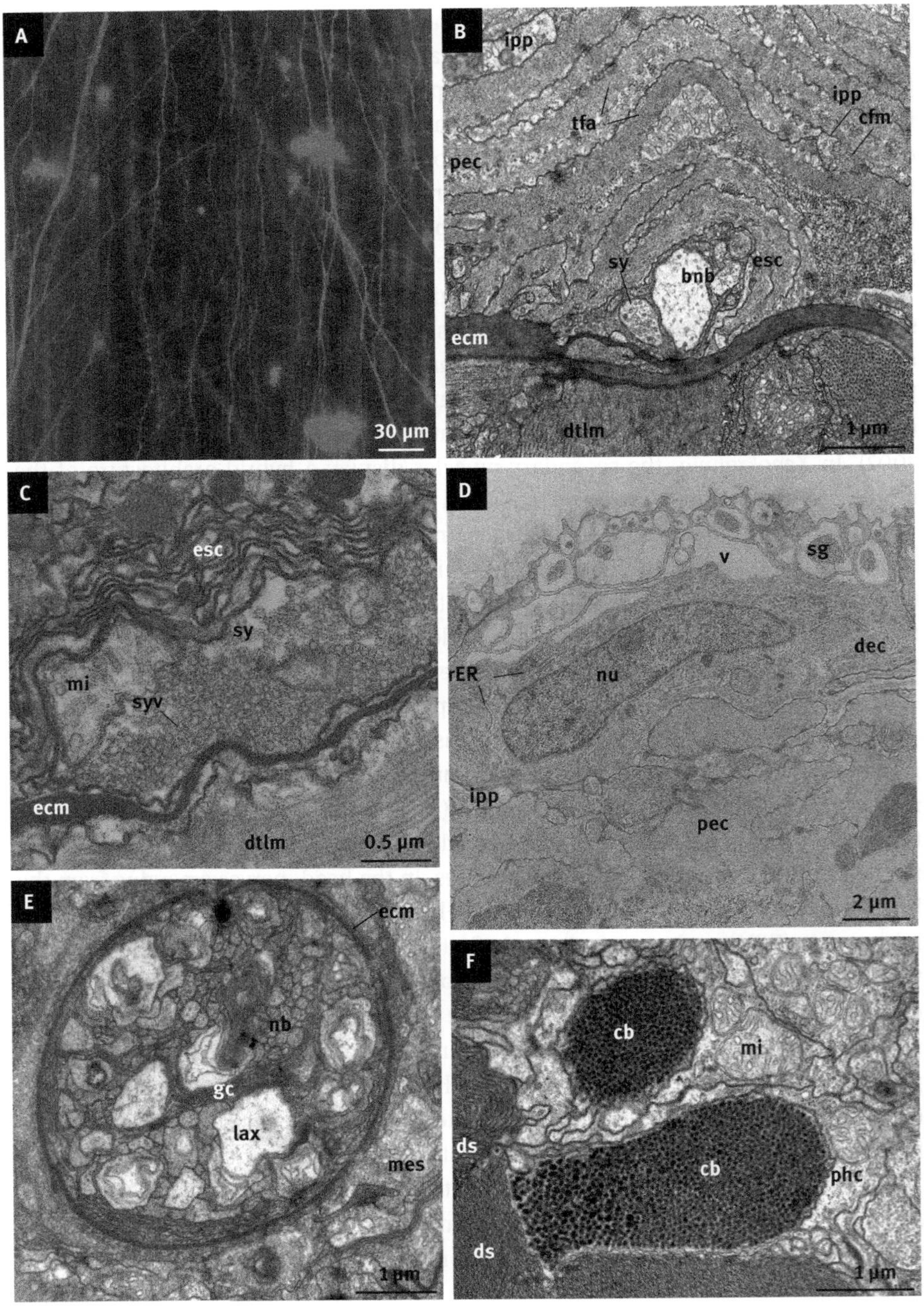

rated from underlying muscles by this thick, fibrillar ECM (Figure 4.4B–C) which was also termed "connective stratum" by Duvert and Barets (1983). This very specialized type of neuromuscular innervations is mirrored in the arrangement of acetylcholinergic fibers which deliver their transmitter by presynaptic endings that terminate on the epidermal side of the ECM and therefore diffusely bathe the muscle cells with acetylcholine (Bone et al., 1987; Duvert et al., 1997). In Chaetogntha, acetylcholine is known to be the major neuromuscular transmitter and it is obvious now that it reaches muscle cells in the trunk primary musculature by diffusion through the ECM to elicit muscle contraction (Bone and Goto, 1991; Shinn, 1997). The cholinergic neurites are part of a profuse intra- and basiepithelial neuronal network (Duverts and Barets, 1983; Bone and Pulsford, 1984; Duvert and Salat, 1990), the peripheral, exclusively epidermal plexus (Bone et al., 1987; Bone and Goto, 1991; Duvert et al., 1997; Shinn, 1997; Harzsch and Müller, 2007; Figure 4.4A in this paper). This extensive intra- and basiepidermal plexus both in the head region and in the trunk can also be visualized by anti-aspartate immunohistochemistry (Duvert et al., 1997). It is linked to the ventral nerve center but mechanosensory input from the numerous ciliary fence receptors in the epidermis also feeds into the plexus (Bone and Pulsford, 1984; Bone and Goto, 1991; Shinn, 1997), which seems to act as a functionally semiautonomous system that mediates sensory-motor integration.

Duvert and Barets (1983) discuss whether this unusual type of neuromuscular innervation of the chaetognath primary trunk musculature may have structural counterparts in other animals and evaluate evidence from Pogonophora and Annelida.

◂ **Figure 4.4:** Immunohistochemistry and ultrastructure of the multilayered epidermis and associated intra- and basiepidermal plexus in *Parasagitta setosa* (A,C) and *Spadella cephaloptera* (B,D–F). Whole mount of an adult specimen labeled for acetylated α-tubulin. **A.** High power magnification of a dorsal trunk section showing the extensive spatial fiber network of the intra- and basiepidermal plexus (modified from Harzsch and Müller, 2007). **B–F.** Transmission electron micrographs (C. H. G. Müller, unpublished data). **B.** Cross section through the basal region of the dorsal trunk epidermis showing the basiepidermal plexus aggregation zone exemplified by one basiepidermal neurite bundle sitting in a depression of the ECM. **C.** Close-up of a basiepidermal neurite bundle of the ventral trunk region containing one large neuromuscular synapse filled with numerous synaptic vesicles. **D.** Distal part of the multilayered epidermis of the laterodorsal region of the head. Note distal layers of filamentous proximal epidermal cells and overlaying vacuolated distal epidermal cells that include many secretory granules. Plexus profiles are recognized between the proximal epidermal cells. **E.** Cross section through fully encapsulated dorsal nerve in the head embedded into the musculus expansus superior. **F.** Oblique section of the eye illustrating the apical regions of several neighbored photoreceptor cells. The micrograph shows the transition to highly modified ciliary process comprising a proximal core body and a lamellar distal segment. *bnb* basiepidermal neurite bundle (proximal aggregation zone of the neuronal plexus); *cb* conical body; *cfm* condensed filamentous mass (in-between *pec*'s, surrounding ipp); *dltm* dorsal longitudinal trunk musculature; *dec* distal epidermal cells; *ds* distal segment; *ecm* extracellular matrix; *esc* epidermal sheath cell; *gc* glial cell (extremely flattened epidermal sheath cell); *ipp* intraepidermal plexus profiles (neurites); *lax* large axon; *mes* dorsal head muscle (musculus expansus superior); *mi* mitochondrion; *nb* neurite bundle; *nu* nucleus; *pec* proximal epidermal cell; *phc* photoreceptor cell; *rER* rough endoplasmatic reticulum; *sg* secretory granule; *sy* synapse; *syv* synaptic vesicle; *tfa* tonofilament apparatus (in *pec*); *v* vacuole.

However, more ultrastructural data are necessary to draw a meaningful conclusion in this issue but such comparisons for the future may hold some potential for phylogenetic insights. At the moment we conclude that, together with the secondary musculature and the other unusual morphological, embryological and genomic features, the uncommon third type of chaetognath neuromuscular contact may be taken as evidence for the long evolutionary distance that separates the Chaetognatha from its closest (unknown) metazoan relative.

4.8 The visual system

The structural diversity of chaetognath eyes has not been fully explored yet but the available studies indicate significant taxon-specific variations of eye design (Ducret, 1978; Goto and Yoshida, 1981, 1983, 1984; Goto, Terazaki, and Yoshida, 1989; reviews Bone and Goto, 1991; Shinn, 1997). Chaetognath eyes belong to the type of ciliated photoreceptors. They are located within the dorsal epidermis of the head, where they are encapsulated by the ECM and several sheath cells. Based on the orientation of the photoreceptor cells, chaetognath eyes are classified as "indirect" eyes in the sagittids and spadellids, and as "direct" eyes in the genus *Eukrohnia*. In the direct eyes of *Eukrohnia*, the apices of the photoreceptor cells are directed distally and the eyes appear to have dozens of closely packed facets which consist of a clear lens-like structure in the apex of the ciliary processes of the photoreceptor cells (Ducret, 1978; Bone and Goto, 1991). In this eye type, the plasmalemma of the ciliary processes is elaborated into a complex array of microvilli called "rhabdome" and probably the light-sensitive part of the cell. However, within the genus *Eukrohnia* there is considerable variation in the architecture of the photoreceptive membranes and in some species of this genus, the eyes are not facetted (Bone and Goto, 1991). It appears that even within this genus so much structural variation is present that exploring eye diversity in other chaetognath genera seems to be a warranting enterprise. The Chaetognatha seem to have thoroughly experimented with eye design so that we expect that many surprises await discovery.

The well-studied "indirect" or "inverted" eyes of the sagittids exhibit one central pigment cell that is surrounded by 70 to 600 photoreceptors cells per eye depending on the species (Figure 4.5). The receptor cell somata are located in the unpigmented peripheral region of the eye and extend axons into the optic nerve towards the posterior compartment of the brain (Rieger et al., 2010). Towards the pigment cell, the photoreceptor cells extend a receptoral process, the "distal segment" which most likely houses the rhodopsin-like photopigments (Goto and Yoshida, 1988). The distal segment is anchored to an axial centriole by a connecting piece with a $9 \times 2 + 0$ microtubular configuration so that the entire receptoral process is considered a highly modified cilium (Shinn, 1997). The "conical body" of the receptoral process (Figure 4.4F), which adjoins the distal segment proximally, is considered unique among metazoan photoreceptor cells (Eakin and Westfall, 1964; Goto and Yoshida, 1984). It has a clear

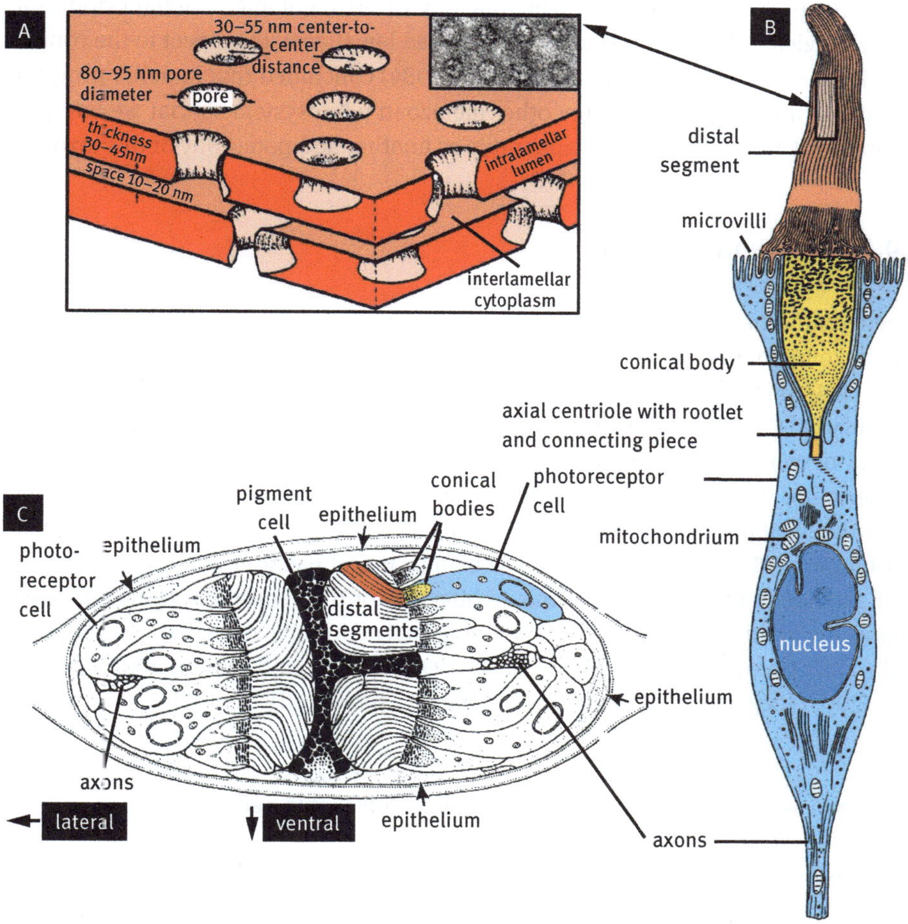

Figure 4.5: **A.** *Spadella schizoptera*, photoreceptor cell of the eyes, schematic representation of the lamellar structure of the distal segment (modified from Goto, Takasu, and Yoshida, 1984). **B.** Photoreceptor cell from the eye of *Sagitta scripsae* (modified from Eakin and Westfall, 1964). **C.** Diagram of a cross section through an indirect chaetognath eye (modified from Shinn, 1997).

and refractive appearance in live animals so that a dioptric function was discussed. Ultrastructural studies showed that its proximal part is composed of dense granules, whereas its distal part contains irregular cords of undefined material (Eakin and Westfall, 1964). A possible role of this structure as participating in the organization of the photoreceptor membranes in the distal segment was discussed by Goto, Takasu, and Yoshida (1984). These are arranged as stacks of lamellae which enclose a 30–45 nm wide intralamellar lumen (Figures 4.4F, 4.5). Apposing membranes of a lamella are continuous across this space at numerous pores. These pores have an 80–95 nm center-to-center distance and are regularly aligned in a grid-like pattern of almost crystalline consistency. The lamellae are separated by 10–20 nm of interlamellar cytoplasm. The cytological origin of the stacks of perforated lamellae with respect to the conical body is unclear. Goto, Takasu, and Yoshida (1984) concluded that lamellar arrays of this nature are not present in any other metazoan photosystem so that we interpret the entire ciliary photoreceptive apparatus as another autapomorphy of this taxon.

4.9 The nervous system

Chaetognaths have a complex nervous system that comprises intra- and basiepidermal components (Ahnelt, 1980; 1984; Salvini-Plawen, 1988; reviews of Bone and Goto, 1991; Shinn, 1997; see also Section 4.6 on the epidermis). The studies by Bone and Pulsford (1984) and Goto and Yoshida (1984; 1987) are the most thorough ones on general nervous system organization and internal brain anatomy of Chaetognatha. Furthermore, the fine structure of the brain has received much attention (Scharrer, 1965; Ahnelt, 1980; 1984; Rehkämper and Welsch, 1985; Goto and Yoshida, 1987; Salvini-Plawen, 1988). Recent immunohistochemical und electron microscopic studies have extended our knowledge of the architecture of the ventral nerve center, sense organs, the brain and the intra- and basiepidermal neuronal plexus in the trunk (Bone et al., 1987; Goto et al., 1992; Duvert et al., 1997; Harzsch and Müller, 2007; Harzsch et al., 2009; Rieger et al., 2010; 2011; Müller et al., submitted).

4.9.1 The ventral nerve center and individually identifiable neurons

The central nervous system consists of six ganglia in the head region, one ventral nerve center in the body, tracts interconnecting these ganglia, and peripheral nerves connecting these ganglia with the intra- and basiepidermal neuronal plexus (Figure 4.6; Goto and Yoshida, 1987; Bone and Goto, 1991; Shinn, 1997; Harzsch and Müller, 2007; Harzsch et al., 2009; Rieger et al., 2010). In addition to the central part of the nervous system, Chaetognatha have a peripheral, exclusively epidermal, plexus-like part (see above and Section 4.6 on epidermis). The basiepidermal ventral nerve center (syn.: "ventral ganglion") is an elongate structure lying in the ventral hemisphere of the trunk, sitting on the ECM and surrounded by the epidermis (Bone and Pulsford, 1984; Goto and Yoshida, 1987; Harzsch and Müller, 2007; Harzsch et al., 2009). Together with the intra- and basiepidermal neuronal plexus, the ventral nerve center controls swimming by initiating contractions of the body wall musculature and coordinating mechanosensory input from the numerous ciliary fence and ciliary tuft receptors in the epidermis (Bone and Pulsford, 1984; Bone and Goto, 1991; Shinn, 1997). Similar to the brain, the ventral nerve center consists of a central fibrillar neuropil core, flanked by lateral clusters of cell somata (Figure 4.7; Goto and Yoshida, 1987; Bone and Goto, 1991; Shinn, 1997). The neuropil core displays an unexpected degree of internal organization being composed of dozens of serially repeated synaptic microcompartments (Harzsch and Müller, 2007). Furthermore, immunolocalization of RFamide-related molecules provides evidence for the presence of serially arranged, individually identifiable neurons that are homologous between four chaetognath species examined (Fig. 8 in Harzsch et al., 2009), suggesting that the potential to generate serially arranged neurons with individual identities is part of the chaetognath ground pattern. This view is supported by the finding of numerous septum-like protrusions of glial-like proximal epidermal cells that compartmentalize the lateral clusters of neuronal somata into distinct serial packages along the anterior-posterior axis (Harzsch et al., 2009).

In addition to the Chaetognatha, individually identifiable neurons are present in the nervous systems of all other major taxa within the Protostomia but also in some basal deuterostomes (discussed in Harzsch and Müller, 2007; Harzsch et al., 2009). Therefore, these authors suggested that developmental programs generating neurons with an individual identity may date back to the ground pattern of Protostomia, perhaps even that of Bilateria.

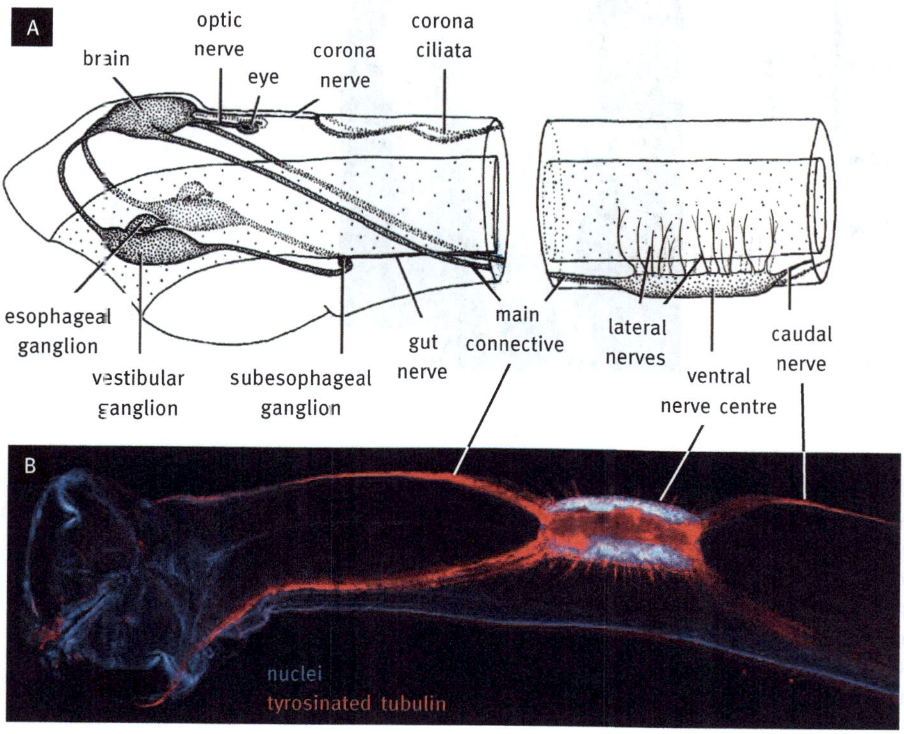

Figure 4.6: **A.** Schematic diagram of the chaetognath nervous system (modified from Kapp, 1996). **B.** *Sagitta bipunctata,* obliquely ventral of the ventral nerve center, immunohistochemical labeling of tubulin (*red*) and histochemical labeling of nuclei (*green*; Rieger, Perez, Müller, Harzsch; unpublished data).

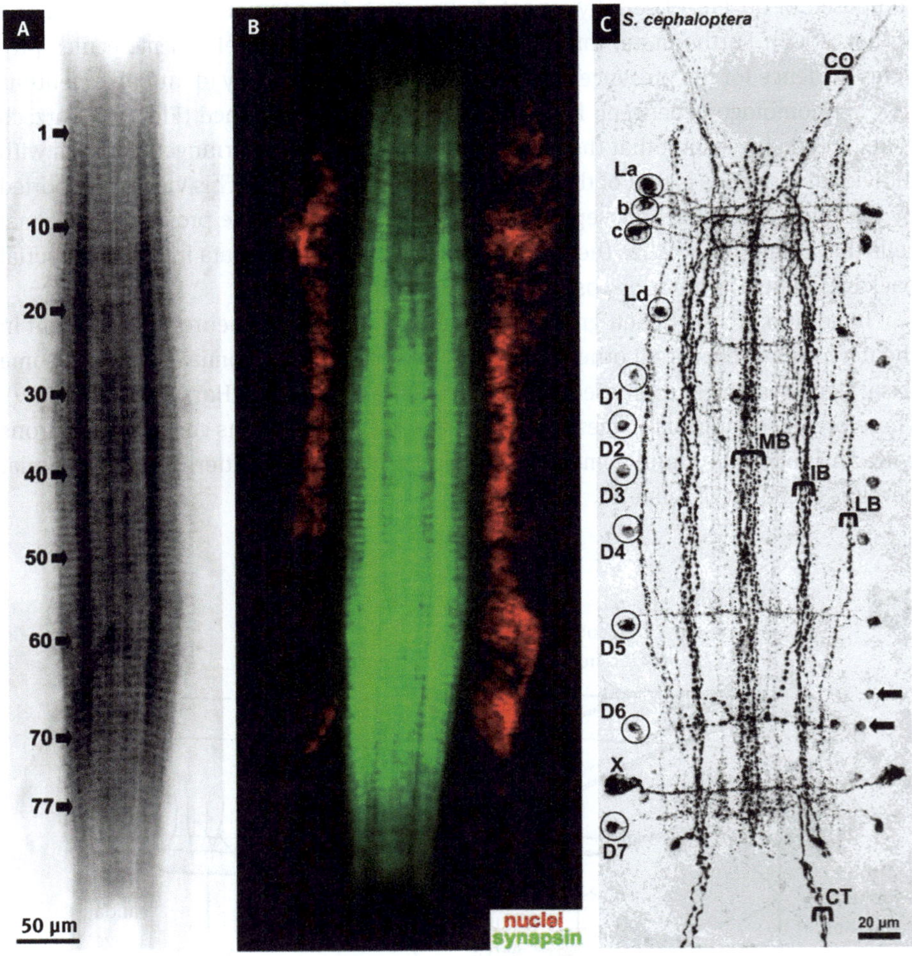

Figure 4.7: **A, B.** Immunolocalization of synapsin (SYNORF 1; *green*) in a whole mount of the ventral nerve centee of adult *Parasagitta setosa* combined with a nuclear marker (*red*; ventral views; A is black-white inverted; reprinted from Harzsch and Müller (2007), copyrights kept by the authors). Synaptic contacts are confined to the central neuropil core which shows a highly ordered subdivision into ca. 80 serially arranged microcompartments. Scale bar: 50 µm. **C.** Localization of RFamide-like immunoreactivity in the ventral nerve center of *Spadella cephaloptera* (ventral view of whole mount specimens, black-white inverted confocal laser-scan images). Identifiable neurons are encircled and labeled with letters (reprinted with permission from Harzsch et al., 2009). *Arrows* point to additional, unidentified neurons. *CO* main connective; *CT* caudal tract; *IB* intermediate bundle; *LB* lateral bundle; *MB* medial bundle.

4.9.2 The cephalic nervous system

In Chaetognatha, the cephalic nervous system consists of the unpaired cerebral complex, including the brain (syn.: "cerebral ganglion") and the retrocerebral organ according to Salvini-Plawen (1988), as well as a pair of vestibular ganglia, a pair of esophageal ganglia, and a subesophageal ganglion from which the stomatogastric nerve emerges (Figures 4.6, 4.8; Bone and Pulsford, 1984; Goto and Yoshida, 1984; 1987; reviews Bone and Goto, 1991; Shinn, 1997; Rieger et al., 2010). Except for the elaborate intra- and basiepidermal neuronal plexus, all components of the cephalic nervous system are deeply sunk into the musculature that moves the grasping spines and the esophagus. Despite their clear subepidermal location, looking embedded into the head musculature, the cephalic nervous system is completely basiepidermal from a typological point of view, as all associated ganglia and connectives (see below) are with few exceptions totally ensheathed by the likewise invaginated ECM. The antero-posteriorly extending bundles which connect the brain and the ventral nerve center are called the main connectives. Connectives also link the brain and the vestibular ganglia (frontal connectives), the vestibular ganglia and the esophageal ganglia, as well as the vestibular ganglia and the subesophageal ganglion (Goto and Yoshida, 1987; Shinn, 1997; partim Rieger et al., 2010). The vestibular ganglia presumably control the operation of the grasping spines and receive sensory input from the vestibular ridge papillae surrounding the mouth via the frontal nerves (e.g. John, 1933). The esophageal and subesophageal ganglia exclusively consist of nerve cell bodies (Goto and Yoshida, 1987). The esophageal ganglion gives rise to the medially directed ventral esophageal nerve that seems to target the esophageal muscles (Shinn, 1997). The slightly sunk brain is situated immediately at the base of the multilayered epidermis of the head and consists of a neuropil core with numerous synapses surrounded by clusters of neuronal cell somata (Rehkämper and Welsch, 1985; Goto and Yoshida, 1987; Bone and Goto, 1991; Rieger et al., 2010). Its basiepidermal location is given by encapsulation by the ECM and specialized sheath cells produced by the proximal epidermis cells (Salvini-Plawen, 1988; Bone and Goto, 1991; Shinn, 1997). Immunohistochemical studies have examined the distribution of serotonin and RFamide-like immunoreactive neurons (Bone et al., 1987, Goto and Yoshida, 1987; Goto et al., 1992; Rieger et al., 2010) and aspartate immunoreactivity (Duvert et al., 1997) in the brain. Rieger et al. (2011) suggested a structural and functional subdivision of the rather complex chaetognath brain into two domains, one posterior domain that may be primarily involved in the integration of sensory input, and the anterior domain that may be involved in the control of the mouthparts and the anterior part of the digestive system. Nerves emerging from the brain target esophageal and somatic head muscles as described above. Peripheral organs associated with the brain are a pair of eyes (Goto and Yoshida, 1984; 1988; Goto, Takasu, and Yoshida, 1984; Goto, Terazaki, and Yoshida, 1989), the corona ciliata (syn.: "ciliated loop") which is localized in the dorsal part of the head (Bone and Goto, 1991; Shinn, 1997; Müller et al., 2013), and the retrocerebral organ, a structure with an unknown putative sensory function, possibly a baroreceptor (Shinn, 1997).

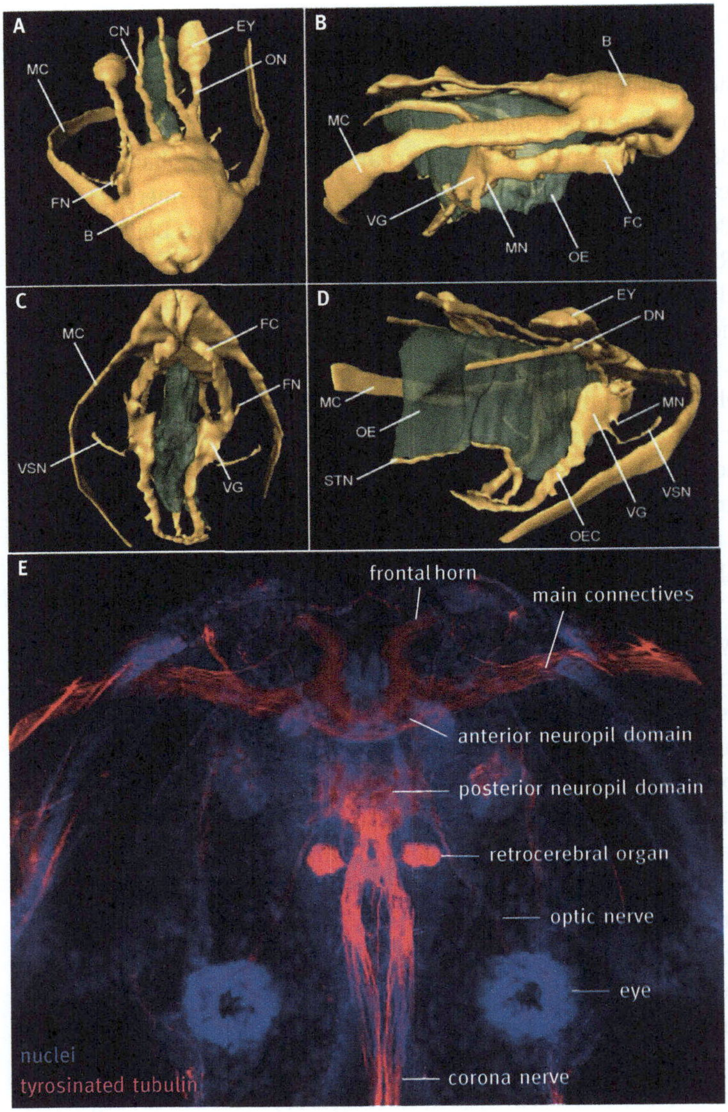

Figure 4.8: 3D reconstruction of the cephalic nervous system of *Ferrosagitta hispida* in relation to the gut based on a series of approximately 400 transverse semithin sections through the head; structures of the nervous system are colored yellow, the gut is stained green (reprinted with permission from Rieger et al., 2010). **A.** Fronto-dorsal view; **B.** Lateral view; **C.** Fronto-ventral view; **D.** Posterio-lateral view. *B* brain; *CN* coronal nerve; *DN* dorsal nerve; *EY* eye; *FC* frontal connective; *FN* frontal nerve; *MC* main connective; *MN* mandibulare nerve; *OE* esophagus; *OEC* esophageal commissure; *ON* optic nerve; *STN* stomatogastric nerve; *VG* vestibular ganglion. **E.** Components of the cephalic nervous system of *Spadella cephaloptera*, immunolocalization of tyrosinated tubulin (*red*) and histochemical labeling of nuclei (*blue*; confocal laser-scan microscopy; reprinted with permission from Rieger et al., 2010).

4.9.3 Brain structure and development in chaetognaths, stomatogastric innervation and phylogenetic considerations

For the Protostomia, Nielsen (2001; 2005a; 2011) suggested that a perioral/circumesophageal brain in addition to the intraepidermal neuronal plexus characterizes their ground pattern. Such a circumoral brain can be recognized in most protostomian groups (e.g. Nielsen, 2001, his Fig. 12.2; and Harzsch, 2006) and is by Nielsen (2001; 2005a) considered to be one of the key apomorphies of the Protostomia that may have evolved from a circumoral concentration of nerve cells around the mouth as present in Cnidaria. The brain components in Chaetognatha are also arranged in such a typical circumoral pattern so that we have suggested that the arrangement of the chaetognath cephalic nervous system is suggestive of protostome affinities (Harzsch and Müller, 2007; Rieger et al., 2010).

Developmental data provided some further insight into this issue (Rieger et al., 2011). These authors showed that the brain of chaetognath hatchlings is delayed in development when compared to the ventral nerve center. At hatching, the brain is represented simply by a fiber loop that extends anteriorly from the ventral nerve center and surrounds the area where a few days later the new mouth opening and its connection to the gut will form (Figure 4.9). This primary brain, which topologically can be viewed as a circumoral nerve ring, will develop into the posterior brain domain of the adult (see above). Acting as a kind of sensory brain, it receives input from the sensory organs and may be involved in the modulation of motor behaviors in response to changing sensory input (Rieger et al., 2010). Slightly later, a second brain component develops in hatchlings that topologically can also be viewed as being arranged in a circumoral pattern, and this develops into the anterior brain domain. This anterior domain gives rise to the stomatogastric nervous system which, in the adult, is involved in controlling the activity of the grasping spines, the innervation of the teeth, and the musculature responsible for opening and closing the mouth (Rieger et al., 2010). The circumesophageal arrangement of the adult cephalic nervous system including, in addition to the brain, vestibular and subesophageal ganglion has already been recognized by Nielsen (2001). However, the situation in the hatchlings clearly shows that we do not only face one but even two brain components that have a basically circumoral arrangement.

As mentioned above, in other bilaterian invertebrates, the ventral cords, with suitable cross-connections, typically exhibit a circumoral arrangement as well (or circum-blastoporal, Nielsen's equivalent, in some cases). Rieger et al. (2012) pointed out that together with the oral and/or stomatogastric innervation, one then has much the same condition as in chaetognaths, a CNS comprised of two essentially separate circumoral subsystems. It appears that among ecdysozoans and spiralian protostomes, there is a well-defined stomodeal derivative (e.g. a foregut or similar) supplied with stomatogastric nerves and ganglia quite separate from the ventral cords supplying the body wall and muscles of the trunk. Taking into account the secondary modifications seen in many larvae, this layout may nevertheless be important for phylogenetic

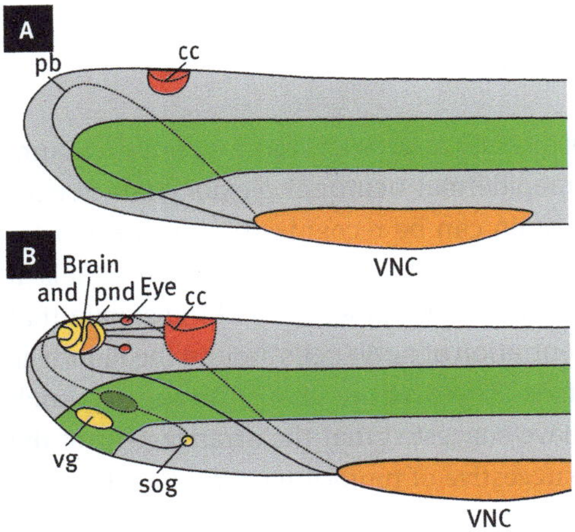

Figure 4.9: Schematic drawings of the nervous system in a hatchling (**A**) and an adult (**B**) chaetognath to show the development of the double circumoral brain (reprinted from Rieger et al., 2011). *and* anterior neuropil domain of the brain; *cc* corona ciliata; *pb* primordial brain; *pnd* posterior neuropil domain of the brain; *sog* sub-esophageal ganglion; *vg* vestibular ganglion; *VNC* ventral nerve center.

considerations. Hence, the widespread occurrence among invertebrates of a CNS consisting of two essentially separate circumoral components, one for oral innervation and one for the rest of the body, was taken as a starting point by Rieger et al. (2011) to discuss how this situation might have arisen. They suggested that the axial nerves devoted to locomotory control, whether arranged in a circumoral fashion or not, may have evolved before an oral innervation of any complexity was established. In such scenarios, a well-defined stomatogastric innervation might well be a quite late innovation but the question of whether stomatogastric innervation in bilaterians might be older than, contemporary with, or more recent in origin than their trunk innervation needs further discussion.

4.10 Conclusion: Chaetognatha on the playground of metazoan evolution

Despite the ever increasing number of molecular phylogenetic studies and an emerging consensus for protostome affinities, we must conclude that today the position of the Chaetognatha is still among the most enigmatic issues of metazoan phylogeny. As outlined above, the chaetognath genome is likely the product of a unique evolutionary history and shows the long isolation of this group. Furthermore, morphological characters provide evidence for the long evolutionary distance that separates the Chaetognatha from its closest (unknown) metazoan relative and suggest that this taxon in many aspects seems to have explored its own evolutionary pathways in generating tissue and organ diversity. Both the genome and morphological characters include many autapomorphies of this group in addition to a character mix of protostome and deuterostome features. However, despite intense analyses of their genome and tissue ultrastructure, so far we do not know of any convincing synapomorphies that clearly would suggest a sister-group relationship to another metazoan taxon.

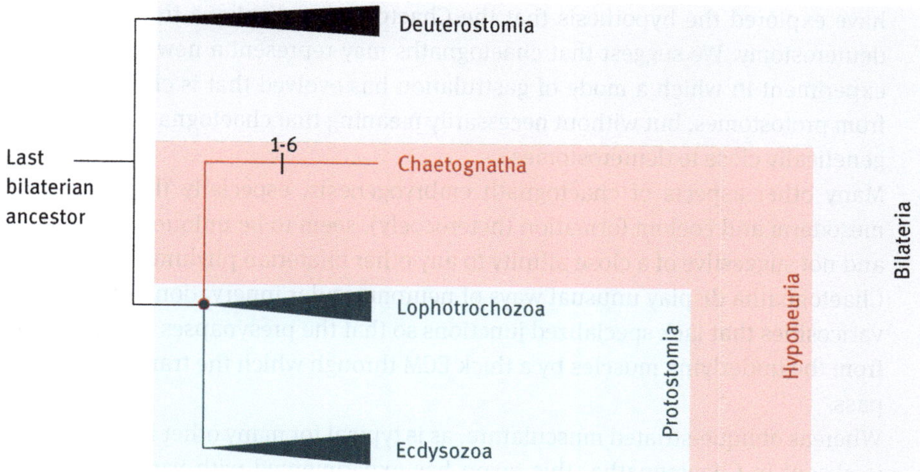

Autapomorphies of Chaetognatha

1 heterocoely
2 specific Median/posterior Hox gene
3 specific neuromuscular innervation through basement membrane
4 secondary, heterosarcomeric muscles
5 multistratified epidermis
6 parallel arrangement of perforated membranes in outer segment of photoreceptor cells

Figure 4.10: Phylogenetic hypothesis proposed in this contribution and by Nielsen (2012). Chaetognatha are likely an early branching protostome lineage but so far it is not possible to precisely say whether they are basal lophotrochozoans, basal ecdysozoans or sister-group to all others protostomes. Hence, we suggest Chaetognatha as the sister-group to the Lophotrochozoa plus Ecdysozoa (soft polytomy), a new assemblage which we suggest to call "Hyponeuria" to indicate the ventralization of the nervous system as one of the most important steps in metazoans' evolution.

- The ribosomal cluster of Chaetognatha is duplicated so that two classes of paralogous 18S rRNA and 28S rRNA genes are present both of which diverge extremely from other Metazoa.
- The sequence of the intermediate filament protein (nuclear lamins) gene of *Sagitta elegans* is very unusual compared to a field of ca. 20 protostome and deuterostome sequences suggesting a particularly high evolutionary drift of the chaetognath sequence.
- The unique mosaic organization of a posterior/median *Hox* gene shows a long, isolated evolution of the *Hox* complex.
- Chaetognatha display a very unusual mitochondrial genome which is one of the smallest known in metazoans and contains only 14 of the 37 usual genes.
- Chaetognath gastrulation displays a mosaic of protostome features plus distinct aspects conventionally assigned to the deuterostomes. In this contribution, we

have explored the hypothesis that the Chaetognatha display a third variant of deuterostomy. We suggest that chaetognaths may represent a new evolutionary experiment in which a mode of gastrulation has evolved that is clearly distinct from protostomes, but without necessarily meaning that chaetognaths are phylogenetically close to deuterostomes.

- Many other aspects of chaetognath embryogenesis, especially the process of mesoderm and coelom formation (heterocoely), seem to be unique to this taxon and not suggestive of a close affinity to any other bilaterian phylum.
- Chaetognatha display unusual ways of neuromuscular innervations with axonal varicosities that lack specialized junctions so that the presynapses are separated from the underlying muscles by a thick ECM through which the transmitter must pass.
- Whereas oblique-striated musculature, as is typical for many other invertebrates, is absent in Chaetognatha, this taxon has experimented with various types of cross-striated muscles and evolved a unique type of secondary muscles with two different kind of sarcomeres that alternate within the muscles.
- The chaetognath integument mainly consists of a multilayered, non-cuticularized epidermis that is made up by two different cell types (single layer of vacuolated and secretory distal epidermal cells and multilayer of flattened, filamentous proximal epidermal cells). This multilayered, dual-type epidermis is a unique, hence apomorphic character of Chaetognatha.
- The intra- and basiepidermal neuronal plexus is a plesiomorphic character retained from metazoan or bilaterian ground pattern. Due to thickness of the epidermis, however, this intraepidermal aspect of the chaetognath plexus is particularly complex. Therefore, the combination of an orthogonal basiepidermal neuronal plexus and a heavily diversified intraepidermal plexus may represent a further derived, apomorphic state characterizing the Chaetognatha, however inevitably linked to the possession of a multilayered epidermis.
- The structure of the ciliated photoreceptors in the eyes shows a unique architecture with a basal body, a conical body, an outer segment with parallel stacks of perforated membrane lamellae.
- The circumesophageal arrangement of the adult cephalic nervous system including, in addition to the brain, vestibular and subesophageal ganglia is suggestive of protostome affinities. However, the situation in the hatchlings clearly shows that we do not only face one but even two brain components that have a basically circumoral arrangement.

The facts summarized in this and previous reviews are most parsimoniously interpreted such that Chaetognatha are likely an early offshoot of the protostome lineage (compare Nielsen, 2012) but so far it is not possible to precise whether they are basal lophotrochozoans, basal ecdysozoans or sister-group to all others protostomes. All together, we currently tend to suggest a new basal rooting of the Bilateria (Figure 4.10), which considers the Chaetognatha as the sister-group to the Lophotrochozoa and Ecdysozoa (soft polytomy), and this assemblage may be called **Hyponeuria** to indicate the ventralization of the nervous system as one of the most important steps in metazoans' evolution.

Acknowledgments

We would like to thank Verena Rieger for her contribution to the experimental work that is the basis of this report's section on the nervous system. We gratefully acknowledge the exchange with Thurston Lacalli on metazoan gastrulation patterns, Carolin and Joachim Haug on fossil chaetognaths, and with Günter Purschke on photoreceptor structure. Research on arrow worms by SH was supported by grants HA 2540/7-1, 2, 3 in the DFG focus program "Metazoan Deep Phylogeny".

Rudi Loesel

5 Brain complexity in protostomes

Abstract: During the last few years, the use of molecular methods for deriving phylogenetic relationships has dramatically rearranged the protostome tree. Traditional taxa such as the segmented "Articulata" have been refuted in favor of the "Ecdysozoa/ Lophotrochozoa" hypothesis, resulting in new challenges in interpreting the evolution of protostome neuronal complexity. This chapter first gives an overview of our current knowledge of the ancestral neuroarchitecture of those protostome clades that exhibit the most complex brain morphologies, namely arthropods, onychophorans, annelids, nemerteans, and mollusks. We will then discuss competing ideas of protostome evolution in the light of new anatomical findings and highlight some of the novel concepts that have been suggested to reconcile the conflict between morphology and molecules.

5.1 Introduction

Recent hypotheses on metazoan phylogeny have recognized three major clades of nephrozoan[1] animals: Deuterostomia, Ecdysozoa, and Lophotrochozoa (Jondelius et al., 2002). Usually, Ecdysozoa and Lophotrochozoa are grouped together, forming the taxon Protostomia (Hickman et al., 2008). While there are only some 70,000 described species of deuterostomes, the number of protostome species is estimated to be in the range of several million. This huge amount of species is reflected by a vast variety of protostome body plans, ranging from the relatively simple flatworms to highly complex forms such as insects and cephalopods. Is the apparent diversity of protostome body organizations, reproduction strategies, developmental patterns, and occupied ecological niches accompanied by an equally diverse organization of the one organ that best echoes the lifestyle of an animal, its brain? We will start our comparison with the arthropods, which are – according to the number of described species – the most successful animal group on this planet.

5.2 Arthropoda

Within the arthropods, insects are the most thoroughly investigated group. This is in part due to the fact that many insect species serve as model organisms for various aspects of fundamental research in neurobiology. Figure 5.1A illustrates the mor-

1 The taxon Nephrozoa comprises all bilaterian animals excluding the basal Nemertodermatida and Acoela.

phology of a generalized insect brain. It is composed of three neuromeres, the protocerebrum, the deutocerebrum and the tritocerebrum. Like all arthropod brains, it is divided into an outer cell rind or cortex, where the neuronal cell bodies reside. Embedded in this cortex we find a number of so-called neuropils, areas in the central nervous system that exclusively contain dendritic and axonal ramifications of neurons but never cell bodies. Individual neuropils are characterized by their specific neuroarchitecture, i.e. their particular internal "wiring" and arrangement of neuronal arborizations. This neuropil-specific scheme does not make it necessarily an easy, but generally a feasible task to homologize neuropils over a wide taxonomic range.

In insects, the most prominent neuropils of the protocerebrum are the unpaired central body and the bilaterally paired mushroom bodies. Several lines of evidence suggest that the central body serves as a higher command center for limb coordination, locomotor control, and navigation (Strausfeld, 1999; Strauss, 2003; Heinze and Homberg, 2007). Its internal architecture is characterized by a highly complicated but stereotypic pattern of perpendicularly interweaving columnar and tangential fibers thus presumably facilitating interhemispherical information exchange (Loesel, Nässel, and Strausfeld, 2002).

The mushroom bodies are the main sensory integration modules of the insect brain. They are important in memory formation and are believed to be the neuronal basis for associative behaviors (Heisenberg, 2003; Farris and Roberts, 2005). Mushroom bodies consist of up to several thousand parallel axons of intrinsic neurons, called globuli cells. The somata of these globuli cells can easily be distinguished from other somata in the cell cortex since they are relatively small, contain a minimum amount of cytoplasm and form a dense cluster around the anterior margins of the mushroom bodies. In many insects this anterior region is cup-shaped and represents the main input area to the mushroom bodies. The most prominent inputs to the mushroom bodies originate in the deutocerebrum. Here, a cluster of small glomerular neuropils receives terminals from olfactory receptor neurons of the antennae. This information is then relayed to the mushroom bodies and lateral aspects of the protocerebrum via the antennocerebral tract (Loesel and Heuer, 2010).

Another sensory modality that feeds into the protocerebrum is visual information. The optic lobes, which contain at least three optic neuropils, the lamina, the medulla, and the lobula, that gradually process information provided by photoreceptors of the compound eye, are attached to the protocerebrum. In several insect orders, the medulla as well as the lobula are split into two separate neuropils. Apart from this, deviations from the basic scheme of the optic lobe are extremely rare (Loesel, 2006). Taking all the evidence together it becomes apparent that the anatomy of the insect brain is highly conserved. Differences in the behavioral repertoire between species are only reflected in differences in the relative size and connectivity of individual neuropils but not in the general architecture of the brain. For a more detailed overview of insect brain morphology see Loesel (2011) or Strausfeld (2012).

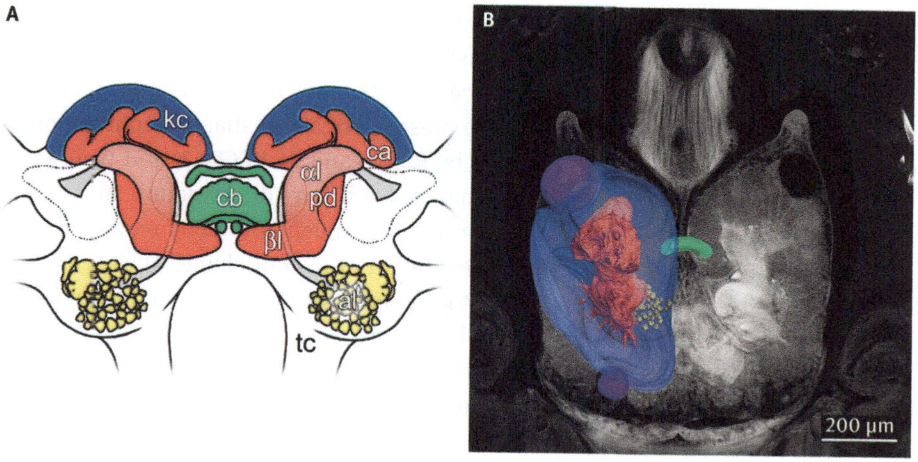

Figure 5.1: **A.** Internal anatomy of a generalized insect brain. The protocerebrum contains the central body (*cb*) and the paired mushroom bodies (*in red*). A cluster of densely packed globuli cells which for historical reasons in insects are called Kenyon cell (*kc*) engulf the primary input region of the mushroom body, the calyx (*ca*). Bundles of Kenyon cell axons form the peduncle (*pd*) that splits up into the alpha-, and beta-lobes (*αl, βl*). The mushroom body receives input from olfactory glomeruli (*in yellow*) of the deutocerebrum or antennal lobe (*al*) via the antennocerebral tract (*in grey*). The tritocerebrum (*tc*) represents the third neuromere of the insect brain (modified from Strausfeld, 1998). **B.** Three-dimensional reconstruction of major brain compartments superimposed onto a section through the head of *Harmothoe areolata*. Brain areas that correspond to neuropils in the insect brain are color-coded in the same way. Globuli cell somata (*blue*), mushroom body neuropil (*red*), central body (*green*), olfactory glomeruli (*yellow*). Eyes are shown in *magenta* (from Heuer and Loesel, 2009).

The closest relatives of hexapods are the crustaceans. In fact, Hexapoda is most likely a taxon within Crustacea, rendering the latter an invalid paraphyletic group. Therefore the term Tetraconata or alternatively Pancrustacea has been suggested to unify hexapods and crustaceans (Dohle, 2001; Regier, Schultz, and Kambic, 2005). As a matter of fact, the crustacean brain does exhibit many similarities when compared to the insect's.[2] The major cerebral neuropils of insects that were described above are also present in those crustacean species that have been studied thoroughly enough. Especially the optic lobes show a high degree of resemblance (Sinakevitch et al., 2003). Differences are mainly found in the outward shape of neuropils but not with respect to their neuronal architecture. Compared to insects, the crustacean central body is more elongated and spindle-shaped (Loesel, Nässel, and Strausfeld, 2002) and primary olfactory neuropils are often not glomerular but wedge-shaped (Krieger et al., 2012). The presence of mushroom bodies in crustaceans has long eluded neu-

2 The question of which crustacean group is closest to insects is under debate and is discussed in Chapter 21 by Loesel and Richter.

roanatomists since these associative centers display a radically different external morphology when compared to insect mushroom bodies: the so-called hemiellipsoid bodies found in the optic stalk of crustaceans lack many of the anatomical characteristics of mushroom bodies such as the presence of a cup-shaped input region or elongated lobes. However, recent studies by Brown and Wolff (2012) and Wolff et al. (2012) have demonstrated that the neuronal organization, ultrastructure, and immunoreactivity of hemiellipsoid bodies correspond to mushroom body architecture in insects, despite their obvious divergent evolution of overall shape. These common morphological traits (central body, olfactory glomeruli, mushroom body/hemiellipsoid body) are also found in small and basal crustacean representatives (Andrew, Brown, and Strausfeld, 2012) suggesting that the structure of the tetraconate brain is highly conserved and might be a plesiomorphic feature.

This idea is supported by studies on the brains of the two remaining arthropod taxa, the Myriapoda and Chelicerata. In myriapods, the supraesophageal ganglia of all investigated species contain olfactory glomeruli and mushroom bodies (Strausfeld, 2012). The presence of a central body has been demonstrated in Chilopoda (Sombke et al., 2012) but not in Diplopoda. The apparent lack of a central body in millipedes has been interpreted as a secondary reduction, since the need for a higher cerebral control center to orchestrate complicated limb movements might have been lost in these animals that use their legs in a highly stereotypical fashion (Loesel, Nässel, and Strausfeld, 2002).

In chelicerates, descriptions of the brain have revealed an unpaired midline neuropil that unlike the central body of Chilopoda and Tetraconata is situated at the posterior rim rather than in the center of the brain (Loesel, Nässel, and Strausfeld, 2002; Loesel et al., 2011). For historical reasons this body is called arcuate and it is relatively large and spans the entire width of the cerebral ganglion. However, analyses of the neuroarchitecture of the arcuate body have demonstrated a number of shared characters with the central body. Today, the arcuate body is viewed as possibly homologous to the central body by most specialists. Moreover, the arcuate body seems to play a similar role in leg coordination, since those chelicerate species with the highest locomotor abilities have the largest and most elaborate arcuate bodies (Loesel, 2004; Strausfeld et al., 2006a; Loesel et al., 2011; Strausfeld, 2012).

The mushroom bodies of chelicerates as well as those in Tetraconata and Myriapoda are characterized by a system of parallel fibers that originate from clustered globuli cells. Mushroom bodies are present in all chelicerate species except the Acari (mites and ticks) with their significantly reduced nervous systems. Unlike in Mandibulata, where mushroom bodies are mainly innervated by olfactory interneurons, the mushroom bodies of chelicerates are supplied by additional massive inputs from the visual system. Mushroom bodies in chelicerates show a remarkable diversity in relative size, ranging from slender structures in scorpions to the enormous multilobed neuropils that occupy the greater part of the brain in *Limulus polyphemus* or in representatives of the amblypygids. These differences in mushroom body size and elaboration seem to be correlated with the number of olfactory glomeruli present in a given species. Olfactory glomeruli of chelicerates can be positioned in whatever part of the central nervous system that lies close to the main olfactory receptors, whereas olfactory glomeruli in mandibulates are always located in the deutocerebrum (Strausfeld, 2012).

Even though it is unknown which number of olfactory glomeruli, which elaboration of the mushroom bodies, and which relative size of the central body represent the ancestral condition, the basic neuroarchitecture of chelicerates most likely represents the ancestral arthropod brain. This is supported by an outgroup comparison with the sister-taxon of arthropods, the onychophorans, whose brains are remarkably chelicerate-like. In representatives of the Onychophora we find the same neuropils that are characteristic for all arthropod central brains, namely olfactory glomeruli, mushroom bodies, and a central body/arcuate body (Strausfeld et al., 2006a; 2006b). Assuming that these prominent cerebral structures were inherited from a common ancestor they might even be plesiomorphic, i.e. older than the Onychophora/Arthropoda clade. If true, similar neuroarchitectural features should be found in the brains of other protostome taxa as well. As a matter of fact this is the case.

5.3 Annelida

Early accounts on annelid brain anatomy have identified neuronal arrangements that are reminiscent of arthropod mushroom bodies and were termed *Globulusfasersystem* (system of globuli cell fibers, Holmgreen, 1916) or *corpora pedunculata* (Hanström, 1926). At the time these findings were published, annelids were viewed as close relatives to the arthropods (Articulata hypothesis) and therefore the presence of mushroom bodies in both taxa was to be expected. The view on protostome phylogeny changed dramatically with the advent of molecular methods. Today, most molecular trees place annelids and arthropods far apart in two different protostome clades, the Lophotrochozoa and Ecdysozoa, respectively (Aguinaldo et al., 1997; Eernisse, 1997). Since then the early neuroanatomical findings of Holmgreen and Hanström have largely been ignored and did not make a difference in discussions on the early evolution of the nervous system. This might in part be due to the fact that since annelids and arthropods are no longer viewed as sister-groups, the presence of a shared neuronal complexity had to be interpreted as homoplasy. Another problem concerning these early anatomical accounts was that they were sometimes done in a rather superficial way and in many cases only hand-drawn reconstructions and not the original data were published. This made it hard to assess the usefulness of the data for modern comparative studies.

The use of novel anatomical techniques such as immunocytochemistry and confocal laser-scanning microscopy has opened the opportunity to reevaluate early morphological descriptions. Recent studies on the brain of annelids (Figure 5.1B) actually show an almost identical neuroarchitecture of polychaete mushroom bodies when compared to insects (where these brain centers are described in most detail).

In a publication on the cerebral ganglion of *Nereis diversicolor*, Heuer and Loesel (2008) found that the mushroom body of this species is composed of thousands of small diameter globuli cells that engulf the principal input region. From here, an elongated peduncle extends posteriorly and splits into a number of output lobes, such mirroring the anatomy of the insect mushroom body in the greatest detail. However, one difference with regard to the sensory input was found. While in insects olfactory information is relayed to the mushroom body only indirectly via the olfactory glomeruli and the antennocerebral tract, a dual pathway was found in *N. diversicolor*: the first one resembles the one found in insects; the second provides input via fibers of presumably olfactory receptor neurons that directly innervate the mushroom body. While this might be viewed as a major deviation from the arthropod olfactory system, findings in a study by Rösler et al. (1999) offer another interpretation. In this study, the development of olfactory glomeruli in the moth *Manduca sexta* was pharmacologically prevented during ontogenesis. Ingrowing axons of olfactory receptors therefore did not find their original target neuropils but kept on growing into the protocerebrum and directly innervated the mushroom bodies. This demonstrates that in insects a targeting machinery, which guides olfactory receptors to the mushroom bodies, is existent, but suppressed in the presence of olfactory glomeruli. This targeting mechanism might be interpreted as an evolutionary relict dating back to deep-time ancestors of arthropods that did not have glomeruli as part of their olfactory pathway.

Until now, mushroom bodies have been described and three-dimensionally reconstructed in additional polychaete species such as *Harmothoe areolata* (Heuer and Loesel, 2009) and *Lepidonotus clava* (Heuer et al., 2010). These studies have revealed the presence of an additional cerebral neuropil that had not been described in earlier accounts. It is a small, arcuated, and unpaired midline neuropil superficially resembling the central body/arcuate body of arthropods. Due to its minute size it has up to now eluded all attempts to analyze its neuroarchitecture in detail. However, columnar fibers that enter the annelid central body from anterior and horizontal fibers that connect this neuropil to lateral aspects of the brain have been identified thus exhibiting a connectivity similar to what is found in insects. The annelid central body is only present in those species that have well-developed parapodia and can use these limb-like structures for rapid maneuvering during hunting. In this respect the annelid central body corresponds to the locomotor ability of a given species as does the central body/arcuate body of arthropods.

Such an elaborate architecture matching the complexity of arthropod brains, however, has not been found in all annelids investigated. In a comprehensive survey of the cerebral anatomy of 22 species representing different annelid taxa and a wide range of lifestyles, elaborate mushroom bodies, olfactory glomeruli, and central bodies have only been described in a few vagile, predominantly predatory polychaetes. In some polychaete species only clusters of densely packed globuli cells which are usually indicative of mushroom bodies have been found but no discernible mushroom body neuropil per se. In sedentary polychaetes as well as in clitellate representatives, brain complexity is generally impoverished, exhibiting no anatomical subdivision of the uniform brain neuropil (Heuer et al., 2010).

What then is the ancestral architecture of the annelid brain? To resolve this question a robust phylogeny of annelids is much needed. Several lines of evidence suggest that errant polychaetes most likely represent the ancestral condition (summarized by Bartolomaeus, Purschke, and Hausen, 2005). The latest phylogenetic tree published by Struck et al. (2011) rescues the traditional Errantia and Sedentaria (which includes Clitellata) as sister-groups. According to the evolutionary scenario supported by this phylogeny, the more complex brains of vagile polychaetes could be an apomorphy for the Errantia and thus do not necessarily represent the ancestral condition in annelids. As yet, the competing phylogenetic hypotheses do not allow a conclusion whether the first annelids were equipped with elaborate brains. This, however, would be a prerequisite to homologize mushroom bodies and central bodies of annelids and arthropods.

In a different approach, Tomer et al. (2010) analyzed local gene expression patterns, the so-called molecular fingerprints, of certain brain areas. They found that the mushroom bodies of the polychaete *Platynereis dumerilii* and the fly *Drosophila melanogaster* share identical fingerprints found in no other compartment of the brain, suggesting that a certain level of brain complexity was already present in the last common ancestor of annelids and arthropods. Considering the phylogenetic distance of these two phyla one must then assume that complex cerebral architectures should be likewise present in the ground pattern of additional protostome groups.

5.4 Nemertea

In comparison to their vermiform unsegmented body organization, ribbon worms possess an astonishingly elaborate brain. Despite their simple outward morphology, nemerteans are active hunters that pursue their prey animals by following them in their tracks (Thiel and Kruse, 2001). They actually feed on the predatory polychaetes mentioned in the previous section (Nordhausen, 1988). For this purpose nemerteans use a number of different sensory organs which are mainly situated in the frontal region of the animal's body, such as a number of eyes and the chemosensory cerebral organs. The information from these receptors is then relayed to the brain.

With respect to neuroanatomy, the green ribbon worm *Lineus viridis* is the most thoroughly investigated species of nemerteans. Its brain measures approximately half of the head's width. In the living animal it can be identified as a reddish structure in the shape of an inverted U that shines through the semitransparent tissue of the body wall. In all nemerteans the brain consists of four elongated lobes, two ventral and two dorsal ones that are anteriorly interconnected. In this way the frontal part of the brain forms a ring which surrounds the rhynchocoel and the proboscis. From this ring the four lobes extend posteriorly and the two ventral ones give rise to the medullary cords while the two dorsal ones terminate in the cerebral organ, the main chemosensory receptor of nemerteans. All parts of the brain are composed of neuropil surrounded by a massive layer of densely packed neuronal somata of various sizes. It is still unknown to what extent the brain neuropil is regionalized. While glial sheaths, which subdivide the cerebral neuropil, have been identified in Azan stained sections, a partition of the neuropil is not reflected in the branching patterns of individual neurons. The only clearly discernible subunits are the cerebral organs, chemosensory structures that form the posterior part of the dorsal brain lobes. The center of the cerebral organ is characterized by a cluster of densely packed small diameter somata which resemble the globuli cells that form the annelid and arthropod mushroom bodies. However, a neuropil that is similar to the typical mushroom body peduncle is not present. Based on the current neuroanatomical data (all information summarized in this paragraph according to Beckers, Faller, and Loesel, 2011) no phylogenetic affinities of Nemertea with other lophotrochozoan groups can be postulated (see also Beckers, Loesel, and Bartolomaeus, 2013).

5.5 Mollusca

With over 100,000 species the mollusks are the second largest animal phylum, exceeded only by arthropods. The anatomical variety of mollusks extends from relatively simple wormlike body shapes to some of the most complex forms of invertebrates. It is generally accepted that Mollusca comprises eight higher taxa whose phylogenetic relationships are under dispute. Practically no information on the internal architecture of the nervous system is available for the deep sea dwelling Monoplacophora. Due to their lifestyle the largely sessile Bivalvia exhibit a derived central nervous system architecture and will not be considered in the following. Out of the remaining six taxa either the aplacophoran Solenogastres and Caudovofeata or the Polyplacophora constitute the most basal extant mollusk lineages and presumably represent the ancestral condition (Salvini-Plawen, 2003; Haszprunar, Schander, and Halanych, 2008; Todt et al., 2008).

All of these basal mollusks are equipped with a relatively simple brain that does not match the complexity found in arthropods and errant polychaetes. The cerebral ganglion of Solenogastres is composed of an outer cell body cortex and a central neuropil that contains no discernible subcompartments (Faller et al., 2012). The brain of Caudovofeata is a little more elaborate since the central neuropil is divided into an anterior and a posterior portion. An additional feature of Caudovofeata, which has not been found in Solenogastres, is the presence of so-called precerebral ganglia, spherical clusters of neuronal somata that are attached to the anterior surface of the brain. The number of precerebral ganglia varies from three to six pairs between species but intraspecific variability has also been observed. Nerve fibers connect the precerebral ganglia to anterior body parts, possibly conveying sensory information. In this respect, the precerebral ganglia are somewhat similar to mushroom bodies but do not show the same elaboration and neuronal complexity as mushroom bodies in annelids and arthropods (Salvini-Plawen, 1972; Shigeno, Sasaki, and Haszprunar, 2007; Faller et al., 2012).

In polyplacophorans, the brain has the shape of a loop that encircles the mouth anteriorly and is therefore referred to as cerebro-buccal ring. Lateral to the mouth opening, the cerebro-buccal ring splits and gives rise to the paired ventral and lateral nerve cords. The polyplacophoran brain is basically a mere elongation and anterior connection of the four main nerve cords. This is reflected by its internal neuroanatomy that does not significantly differ from the neuronal complexity found in these nerve cords. The cerebro-buccal ring is only slightly thicker in diameter. Its central neuropil, which is sandwiched in-between an anterior and posterior layer of neuronal cell bodies, shows no subdivisions or specialized areas (Faller et al., 2012).

The central nervous system of the benthic Scaphopoda represents the basic tetraneurous system of Conchifera with two pairs of nerve cords and a distinct brain that is composed of two fused cerebral ganglia (Reynolds and Steiner, 2008). The brain consists of a central neuropil that is surrounded by a cortex of neuronal somata. Cell bodies in this cortex have been found to be uniformly distributed in *Entalina quinquangularis*. In *Antalis entalis*, two clusters of cell bodies that are situated in the posterior cortex are discernible from neighboring somata because they are more tightly packed and contain less cytoplasm. In arthropods and certain annelids such clusters of condensed cell bodies are indicative of the presence of a mushroom body. In *A. entalis*, however, no corresponding specialization of the underlying neuropil was observed (Faller et al., 2012).

Gastropods possess a more elaborate central nervous system than the mollusk groups discussed so far. Their four nerve cords connect a number of well-developed ganglia that usually occur pairwise. In many advanced gastropods the nervous system is condensed anteriorly accompanied by a shortening of nerve cords and fusion of certain ganglia. The cerebral ganglia are commonly the largest and are often fused to form a continuous brain (Brusca and Brusca, 2003).

The gastropod brain is highly lateralized and composed of two hemiganglia that are connected via the prominent cerebral commissure. Each hemiganglion is subdivided into three sections, the procerebrum, the mesocerebrum, and the postcerebrum. The best studied system of the brain is the olfactory pathway which in the snail *Achatina fulica* consists of approx. 100,000 sensory neurons and 25,000 interneurons per hemisphere, surpassing the rest of the cerebral ganglion's number of neurons. Fibers of receptor neurons either terminate in olfactory glomeruli that are situated within the tentacle or they directly innervate the procerebrum where they contact a system of parallel fibers that emanate from densely packed, small diameter cell bodies. From there, the information is relayed to other parts of the central nervous system (Chase and Tolloczko, 1989; 1993). In this respect the neuronal organization is similar to what is found in arthropods and annelids, where the olfactory pathway also consists of receptors, olfactory glomeruli and the parallel fiber system of the mushroom bodies. Within Mollusca, the complexity of the gastropod nervous system is only exceled by the brain of cephalopods.

The nervous system of the Cephalopoda exhibits a degree of centralization and complexity that is unique in protostomes and equals the complexity found in vertebrates. In cephalopods, the ganglia that are distributed throughout the body in other mollusk groups are concentrated by a shortening of commissures and nerve cords to form a massive brain that surrounds the esophagus (Figure 5.2).

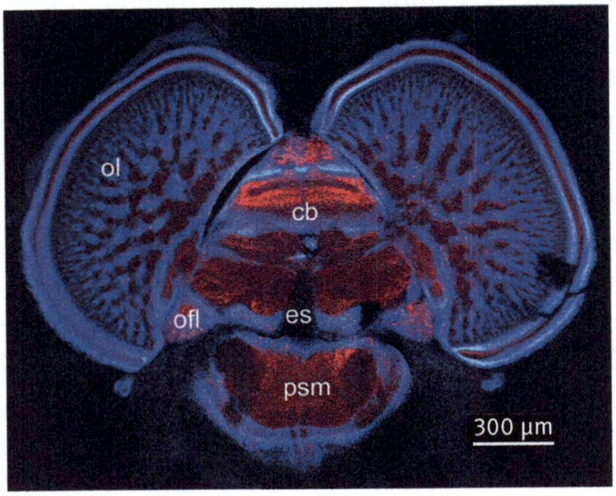

Figure 5.2: Section through the brain of the pygmy squid *Idiosepius notoides* labeled with the nuclear marker DAPI (*blue*) and an antiserum directed against FMRFamid (*red*) showing the central brain (*cb*), the esophagus (*es*), the olfactory lobes (*ofl*), the optic lobes (*ol*), and the posterior subesophageal mass (*psm*).

Detailed descriptions of the external anatomy of the cephalopod brain have been published (Young, 1971) and a number of behavioral studies have demonstrated various intelligent behaviors in these animals (Hanlon and Messenger, 1996). The internal architecture of certain brain areas has been analyzed in detail and wiring schemes have been reconstructed (e.g. Young, 1991 for the learning system). Despite the vast amount of information on individual species and selected brain areas astonishingly little has been published on the ancestral pattern of the cephalopod brain and the phylogentic origin of its circuitry. Along with the relatively small number of neuroanatomists who have worked with these animals, this might also be due to the fact that the sheer size of cephalopod brains makes it hard to reconstruct its basic connectivity. The approach that has been taken by Wollesen et al. (2008; 2009), who used the pygmy squid *Idiosepius notoides*, a species that is small enough to actually place sections of the entire brain under a confocal microscope, might help to work out its basic architecture in the future. Until we have enough data to possibly homologize brain centers of cephalopods with neuronal assemblies in other clades we have to assume that the neuroarchitecture of the cephalopod brain has evolved completely independently from simple forms that may be represented by non-conchiferan mollusks.

5.6 Evolutionary origin of complex brains in protostomes

We have now taken a look at those protostome clades that exhibit a high degree of brain complexity (brain comprises significantly more than just a few hundreds of neurons). Let us recapitulate the facts: arthropods and onychophorans have a highly conserved cerebral anatomy. In both groups, the olfaction system contains olfactory glomeruli and mushroom bodies. The central body of hexapods, crustaceans and myriapods is most likely homologous to the arcuate body of chelicerates and onychophorans and plays a role in limb coordination. A similar degree of complexity with the presence of glomeruli, mushroom bodies, and a central body is found in annelids but only in certain errant taxa. Nemerteans and basal mollusks do not show any of these complex brain characters. Only clusters of glomeruli cells that in arthropods and annelids are indicative of mushroom bodies but no mushroom body neuropil per se has been described in these groups. Gastropods possess an olfactory system that is somewhat similar to arthropods and annelids. Finally, cephalopods exhibit a complexity that is not matched by any other protostome clade. The brains of all other protostomes are comparatively simple and do not contain the sheer number of neurons that is necessary to form complex traits. For a more detailed description of the remaining protostomes see the chapter 6 by Schmidt-Rhaesa and Rothe in this book.

How can these findings be interpreted? There are three conflicting evolutionary scenarios that explain the distribution of complex cerebral traits; each has its strengths and its drawbacks. (1) If one were to ignore the molecular trees of Metazoa that have been published lately, then the neuroanatomical data itself would clearly speak for a close phylogenetic relationship between arthropods and annelids, i.e. for the traditional Articulata hypothesis. This, however, would imply that recent molecular phylogenies, which are in support of the Ecdysozoa/Lophotrochozoa, are mistaken. (2) The alternative to the first scenario then would be that complex brain morphologies have evolved several times independently. With respect to the close similarity of arthropod and annelid brains this seems unlikely. Even though it could be argued that once a brain reaches a particular degree of complexity, certain neuroarchitectural features have to evolve in a similar way due to computational needs. This view is not supported by the anatomical data that clearly demonstrate that the elaborate brains of cephalopods or nemerteans do not possess mushroom bodies and a central body. There seem to be different solutions to the problem of how to 'construct' a complex brain. Moreover, it does not explain the accumulating evidence from molecular developmental (evo-devo) studies that suggest a deep-time origin of complex neuroanatomical traits. This leads to a third scenario that incorporates the available morphological as well as the molecular data and does not conflict with the modern view on animal phylogeny.

(3) In 2005 Lichtneckert and Reichert published a review that summarized conserved genetic patterning mechanisms in insect and vertebrate brain development. The authors concluded that gene expression patterns in the developing brain supply sufficient evidence that (i) Geoffroy St Hilaire's inversion hypothesis is correct, that (ii) the brains of protostomes and deuterostomes are homologous, and (iii) that the last common ancestor of protostomes and deuterostomes (PDA) was already equipped with a tripartite brain. All this supports the idea that the "urbilaterians" might have been fairly complex animals with an accordingly complex brain. Since then a number of evo-devo studies have pointed in the same direction. Tomer et al. (2010), for example, claimed a homology of mushroom bodies in insects and annelids based on molecular fingerprint studies. A comparison of ParaHox gene expression patterns lead Hui et al. (2009) to the conclusion that the last PDA was complex, at least in terms of those ParaHox components that are important in the development of a variety of nervous system components (eyes, neurosecretory cells in the hindbrain, and the anterior-posterior axis of the nerve cord). Combining the data in these and in several other publications (Tautz, 2004; Dennes et al., 2007 to name just a few) the PDA would have possessed serially iterated body segments, a mesodermally lined coelomic space, a throughgut with separate mouth and anus, lateral appendages, heart, eyes and a tripartite brain with a centralized nervous system. According to these hypotheses, the PDA was already equipped with a complex brain, which must have undergone a severe reduction in complexity several times throughout bilaterian radiation to explain the simplicity of the nervous system of many recent phyla. From the morphological point of view, this might not be the parsimonious assumption but it is a plausible one, nevertheless.

There are a number of examples where a secondary reduction of the central nervous system has demonstrably occurred, e.g. in barnacles (Callaway and Stuart, 1999) or sessile annelids (Heuer et al., 2010). Neuroanatomical characters might be especially susceptible to secondary reductions in complexity since nervous tissue is particularly costly in development and maintenance due to its high metabolic requirements (Laughlin, 2001). Regressive changes in neural complexity are commonly attributed to a narrowing of the ecological niche in a monotonous and extremely simplified environment and a cessation of mobile life (Kaufman, 2008), and can also be observed in sessile, semi-sessile, and parasitic species (Bullock, 2002). The nervous system in none-parasitic cirripedes, for example, is much simpler than that of free-swimming crustaceans, lacking a distinctive tripartition and comprising only an estimated 200 neurons in *Balanus nubilus* (Gwilliam, 1987). Similarly, the nervous system of the highly derived, parasitic rhizocephalic cirripeds is largely reduced. Comparable trends can also be observed in bivalve mollusks, in brachiopods, and in parasitic cestodes, illustrating that comparatively simple neuroarchitectures do not necessarily reflect ancestral conditions.

So far we cannot verify this third scenario. Some authors caution against the idea that complexity in gene expression patterns necessarily reflects morphology (Hejnol and Martindale, 2008a). Moreover, there is no hard evidence that the majority of bilaterian phyla has undergone a sessile or infaunal 'bottleneck' that would explain the multiple reductions which are postulated by this hypothesis. It all comes down to the question of whether recent molecular phylogenies that support the Ecdysozoa/ Lophotrochozoa are valid. These trees contradict classical morphological phylogenies that explain the emergence of complex brain morphologies with fewer problems. Some authors claim that a systematic error in analyzing molecular data might lead to a repeatedly high support for certain topologies in the tree that are nevertheless wrong (see chapter 27 by Kück and Wägele in this book). As long as this discussion is unresolved none of the tree hypotheses on the emergence of complex brain architectures that were outlined above should be disregarded.

Our knowledge on nervous system anatomy has significantly increased during the last few years, but our understanding of brain evolution still represents a rough stone that gradually gains shape as it is worked on. Future studies that map the molecular identity and ontogenesis of defined neuronal assemblies might help to identify additional features of the protostome brain that are homologous but escaped morphological comparisons because their outward appearance has changed. Of special interest in this respect are the clusters of globuli cells that do not give rise to a 'classical' mushroom body. These clusters have been described in a number of protostome clades. Should it be demonstrated that these cells are nevertheless homologous to mushroom body neurons, the neuroarchitecture of the brain in many protostome clades would have to be viewed as much more conserved and complex than previously assumed.

Andreas Schmidt-Rhaesa and Birgen Holger Rothe

6 Brains in Gastrotricha and Cycloneuralia – a comparison

Abstract: Four taxa, Nematoda, Priapulida, Kinorhyncha and Loricifera share unique characters in their nervous system. Their brain is composed of a ring neuropil with an anterior and posterior position of neuronal somata. Together with Nematomorpha, which are related based on other characters but show a derived nervous system, all taxa are regarded as monophyletic in a taxon Cycloneuralia and the characteristic form of the brain is called cycloneural. We review here the diversity present in the nervous systems of nematodes, nematomorphs, priapulids, kinorhynchs and loriciferans with regard to function and phylogeny. Additionally, the taxon Gastrotricha is often regarded as closely related to Cycloneuralia due to some morphological correspondence, but molecular analyses do not support this relationship. The brain structure of gastrotrichs is in the form of a dorsal commissure with lateral somata and therefore unique among bilaterians.

6.1 Introduction

The taxon Cycloneuralia includes the five taxa Nematoda, Nematomorpha, Priapulida, Kinorhyncha and Loricifera. Their name refers to a characteristic, cycloneuralian, construction of the brain. The cycloneuralian brain is described as a ring neuropil around the anterior part of the intestinal tract with a ventral origin of the paired or unpaired nerve cord. Somata of the ring neuropil are present anterior and posterior of the neuropil (see e.g. Fig. 47.1 in Nielsen, 2012). This characteristic distribution of somata is regarded as characteristic for a cycloneuralian brain, as well as its ring-shaped (circumoral or circumintestinal) nature (see Richter et al., 2010). Cycloneuralia are potentially related to the taxa Gastrotricha (as Nemathelminthes) and/or to Arthropoda (as Ecdysozoa). Because molecular analyses do often not show monophyletic Cycloneuralia and because the same analyses always show Gastrotricha as not related to Ecdysozoa, it is our aim to reinvestigate and compare the brain structure of cycloneuralian taxa and try to use the brain structure for phylogenetic questions.

6.2 Phylogenetic background

Monophyly and phylogenetic relationships of Cycloneuralia were not always clear. Together with taxa such as Gastrotricha, Rotifera and Acanthocephala they were often lumped together in a grouping named Aschelminthes or Pseudocoelomates (see e.g. Schmidt-Rhaesa, 2012a for summary). It was quite clear that such a group was not monophyletic, but well-argued hypotheses on phylogenetic relationships were developed comparably late (Lorenzen, 1985; Ahlrichs, 1995; Nielsen, 1995; Wallace, Ricci, and Melone, 1995; Ehlers et al., 1996). According to these authors, Gastrotricha, Nematoda, Nematomorpha, Priapulida, Kinorhyncha and Loricifera are closely related and, with the exception of Lorenzen (1985), Gastrotricha are the first taxon to branch off, being the sister-group to a monophylum covering the remaining taxa.

Unfortunately, different names were introduced. Ahlrichs (1995) and Ehlers et al. (1996) chose the "old" name Nemathelminthes for the monophylum including the six taxa, while Nielsen (1995) named it Cycloneuralia. The name Nemathelminthes has a long tradition with a changing content of taxa (see Schmidt-Rhaesa, 2012a). The name Cycloneuralia appears, in retrospective, as unfortunate, because the cycloneuralian condition of the brain is not present in gastrotrichs (see below). Additionally, Ahlrichs (1995) had used the same name, Cycloneuralia, for the subtaxon (Nematoda + Nematomorpha + Priapulida + Kinorhyncha + Loricifera), in reference to the particular organization of the brain. Nielsen (1995) named this group Introverta, a name that infers homology of mobile anterior regions of the body. Such regions are present in priapulids, kinorhynchs and loriciferans as an introvert, in nematomorph larvae such a region is often named proboscis and the homology to the introvert is questionable (see Schmidt-Rhaesa, 1998). In nematodes, a mobile anterior region is only present in the species *Kinonchulus sattleri* (Riemann, 1972), but homology to an introvert is, in our opinion, not supported. These names have been used in parallel for some time, but it appears that Cycloneuralia for (Nematoda + Nematomorpha + Priapulida + Kinorhyncha + Loricifera) is becoming established, especially since Nielsen (2012) adopted the name in this sense in the third edition of his book *Animal Evolution*. Cycloneuralia is used in this sense throughout this chapter. The taxon (Gastrotricha + Cycloneuralia) may be polyphyletic (see below).

Internal relationships within Cycloneuralia appear to be ((Nematoda + Nematomorpha)(Priapulida + Kinorhyncha + Loricifera)) (Ahlrichs, 1995; Nielsen, 1995; Wallace, Ricci, and Melone, 1996; Ehlers et al., 1996). Some (superficial?) resemblance of the nematomorph larva with priapulids, kinorhynchs and loriciferans led to the alternative hypothesis that these taxa might be closer related to each other in a taxon Cephalorhynchia (Malakhov, 1980; Adrianov and Malakhov, 1995; 1999). Nevertheless, there are a number of potential synapomorphies supporting a (Nematoda + Nematomorpha) relationship (see Ahlrichs, 1995; Schmidt-Rhaesa, 1998) and this sister-group relationship is widely accepted under the name Nematoida. The three taxa Priapulida, Kinorhyncha and Loricifera are very likely monophyletic, due to a corresponding body organization composed of an introvert with scalids and a trunk. Again, two different names were used for this taxon. Lemburg (1995) introduced the name Scalidophora, while Nielsen (1995) used Cephalorhyncha. As this name originally refers to the inclusion of Nematomorpha, its restriction to a smaller taxon appears problematic and Nielsen (2012) therefore adopted Scalidophora. Internal relationships between the three scalidophoran taxa remain not clearly resolved, with a set of characters being available for each of the three possible sister-group relationships (see Lemburg, 1999).

Cycloneuralia became important when analyses of DNA sequences stared to be used in systematics. Aguinaldo et al. (1997) were the first to make sense of the clustering of sequences from arthropod and cycloneuralian representatives that repeatedly had occurred in the analyses and proposed the monophyly of a taxon Ecdysozoa that included arthropods and cycloneuralians. Such a hypothesis contradicts a closer relationship between arthropods and annelids, which was regarded to be well supported by the segmental body organization of both taxa. Nevertheless, Ecdysozoa is supported meanwhile by numerous analyses of different genes, of genome and other data (see e.g. Giribet, 2003; Telford et al., 2008; Edgecombe et al., 2011) and with a corresponding cuticle ultrastructure and growth by molting, there is also support from morphology (see e.g. Schmidt-Rhaesa et al., 1998).

Although supporting monophyly of Ecdysozoa, molecular analyses do not support monophyly of Cycloneuralia. There is no uniform signal in the many available molecular analyses. Generally, nematodes and nematomorphs cluster together (supporting monophyletic Nematoida) and priapulids and kinorhynchs cluster together as well in most analyses (see Aguinaldo et al., 1997; Dunn et al., 2008; Hejnol et al, 2009; Paps, Baguñà, and Riutort, 2009a as a few selected examples). However, in the few analyses including Loricifera, Scalidophora do not appear as a monophyletic taxon (Park et al., 2006; Sørensen et al., 2008). The analyses by Dunn et al. (2008) and Hejnol et al. (2009) imply monophyletic Cycloneuralia under inclusion of tardigrades. Gastrotricha never occur close to any ecdysozoan taxon in molecular analyses, instead, gastrotrich taxa always come up within Spiralia (Lophotrochozoa), usually close to flatworm taxa.

In summary, the structure of the nervous system is the most important morphological character supporting monophyletic Cycloneuralia, but molecular analyses do not support this monophyly. Therefore, a closer look at the nervous system, in particular the structure of the brain, is essential to resolve the question of cycloneuralian monophyly and the relationship of gastrotrichs.

6.3 Brain organization in Cycloneuralia and Gastrotricha

In general, comparatively few species have been investigated concerning their brain structure. This applies in particular to immunostainings. Particular focus is put on the shape of the neuropil, the presence and distribution of somata associated with the neuropil and the origin of the ventral nerve cord(s).

6.3.1 Nematoda

In nematodes, the brain surrounds the elongate pharynx somewhere in its median region. A proper comparison of the brain structure has not been made and knowledge is mainly based on the detailed investigations on model species, *Ascaris lumbricoides* and *A. megalocephala* (Goldschmidt, 1908; 1909; 1910) and *Caenorhabditis elegans* (White et al., 1986). Further data are found in several descriptions of nematode species (see, e.g., Chitwood and Chitwood, 1950 for a summary). Ultrastructural investigations supplement these data (e.g. Roggen, Raski, and Jones, 1967; Strote, Bonow, and Attah, 1996; summary in Wright, 1991).

Several immunohistochemical stainings were made in nematodes, including antibodies which are broadly used to document the nervous system. FMRFamides are present in several components of the nervous system (e.g. Leach, Trudgill, and Gahan, 1987 for *Goodeyus ulmi*; Atkinson et al., 1988 for *Panagrellus redivivus*, *Caenorhabditis elegans*, and *Heterodera glycines*; Geary et al., 1992a for *Panagrellus redivivus*; Schinkmann and Li, 1992 for *Caenorhabditis elegans*; Crowden et al., 1993 for *Ascaris suum*; Jagdale and Gordon, 1994 for *Romanomermis culicivorax*; Kimber et al., 2001 for *Globodera* species). Serotonin, in contrast, is present only in few neurons (Horvitz et al., 1982 for *Caenorhabditis elegans*; Johnson et al., 1996 for *Ascaris suum*). Tubulin has been shown from nematodes as a molecule (Dawson, Guttteridge, and Gull, 1983) and as β-tubulin genes (Geary et al., 1992b; Kwa et al., 1995), but has not been used to reveal the structure of the nervous system.

The emerging picture of the structure of the nervous system in nematodes is diverse. All nematodes have a brain in the form of a circumpharyngeal ring neuropil, from which the ventral longitudinal nerve cord, further longitudinal neurite bundles and neurites innervating sensory structures in the head region arise. Not so clear is the distribution of neuronal somata contributing to this neuropil. Goldschmidt (1908; 1909) documented from *Ascaris* species somata anterior and posterior of the ring neuropil and these somata are concentrated in certain clusters he called ganglia. Such "ganglia" (better: clusters of somata) have also been reported from other species, e.g., from *Anthraconema* species (Zur Strassen; 1904), *Leptosomatum acephalatum* (Timm, 1953) or from *Xiphinema index* (Roggen, Raski, and Jones, 1967). Other authors report a more homogeneous distribution of somata anterior and posterior of the ring neuropil, e.g., Türk (1903) for *Thoracostoma strasseni* or Strote, Bonow, and Attah (1996) for *Onchocerca volvlus*. Malakhov (1978; 1982, summarized in 1994) observed in *Pontonema vulgare* a homogeneous distribution of somata around the ring neuropil (with not even a distinction between an anterior and a posterior cluster) and he regarded such a distribution as ancestral for nematodes. Additional peculiar patterns may exist, such as the presence of somata only posterior of the ring neuropil in *Romanomermis culicivorax* (Jagsdale and Gordon, 1994).

Concluding from these data that there is some variation in the brain structure in nematodes, at present it is not possible to conclude whether a "typical" cycloneuralian brain is even ancestral for nematodes.

6.3.2 Nematomorpha

In freshwater nematomorphs, there are some descriptions of the brain as ring-like (Rauther ,1904; 1905; 1930; Švábenik 1909; 1925) and other descriptions as being mainly subesophageal (a muscular pharynx is not present in nematomorphs) with a fine supraesophageal commissure (Vejdovsky, 1886; 1894; Montgomery, 1903). This latter structure is also present in juveniles of the marine *Nectonema munidae* (Schmidt-Rhaesa, 1996), the dorsal commissure is reduced in adults of this species. In the freshwater species *Paragordius varius*, serial histological sections show a ring-like organization in the anterior-most region, close to the mouth opening, which extends dorsal of the esophagus (Schmidt-Rhaesa, 2012b). The ventral nerve cord is in this species very voluminous in its anterior-most part, it is connected to the dorsal brain by lateral connections. The distribution of somata has not been investigated. As a typical ring-like neuropil is not present, nematomorphs do not show a cycloneuralian brain. The detection of the brain in larvae has not been successful, neither in ultrastructural studies (Zapotosky, 1974; Jochmann and Schmidt-Rhaesa, 2007) nor with immunohistochemical stainings (Schmidt-Rhaesa and Rothe, unpublished results).

6.3.3 Priapulida

The brain surrounds the intestinal tract in the anterior part of the introvert, anterior of or in the anterior region of the pharynx. It is composed of a ring neuropil giving rise to the unpaired ventral nerve cord. In the small *Tubiluchus* species and in *Meiopriapulus fijiensis*, the brain is cycloneural in possessing somata anterior and posterior of the neuropil (Figure 6.1B–E; Rehkämper et al., 1989; Storch et al. 1989; Rothe and Schmidt-Rhaesa, 2010a). In the macroscopic species (data for *Halicryptus spinulosus*, *Priapulus caudatus* and *Priapulopsis bicaudatus*), however, somata are present only anterior of the neuropil (Horst, 1881; Scharff, 1885; Joffe and Kotikova, 1988; Adrianov and Malakhov, 1996). *Maccabeus tentaculatus* appears to deviate from this pattern, somata are found here medial of the neuropil (Por and Bromley, 1974). Adrianov and Malakhov (1996) describe local thickenings in the mass of anterior somata. There are eight such thickenings, called ganglia by Adrianov and Malakhov (1996). This number corresponds to the number of scalids in the first ring. In *Tubiluchus troglodytes* nuclei are also not homogeneously distributed, but are concentrated in 20–25 clusters in the anterior ring (Rothe and Schmidt-Rhaesa, 2010a). This number corresponds to the arrangement of scalids posterior of the first ring. In the posterior somata, clustering of nuclei is also present, but much less distinct compared to the anterior somata (Rothe and Schmidt-Rhaesa, 2010a).

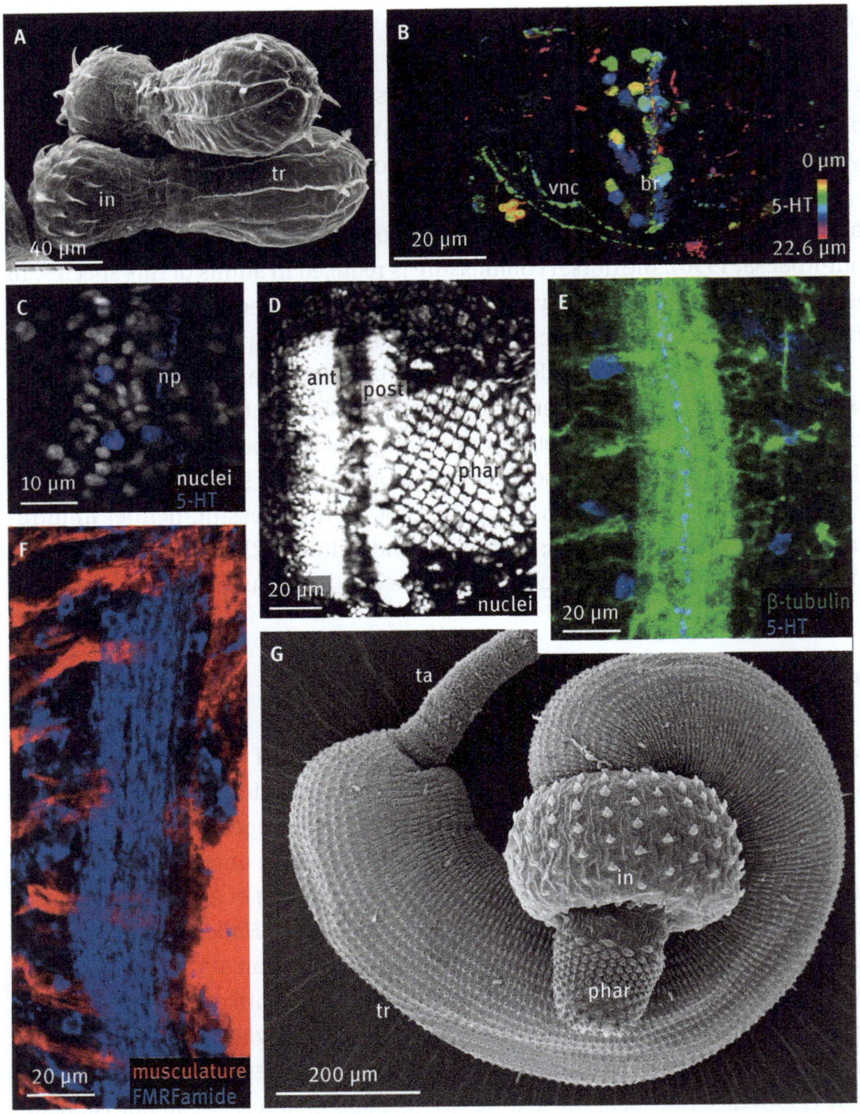

6.3.4 Kinorhyncha

The circumintestinal and cycloneural brain in kinorhynchs is about two thirds anterior of the pharynx, one third overlaps with the anterior part of the pharynx. A ring neuropil and anterior as well as posterior somata are present. This has been investigated in different species (*Echinoderes aquilonius*, *E. capitatus*, *Pycnophyes dentatus*, *P. greenlandicus*, *P. kielensis* and *Zelinkaderes floridensis*, see Kristensen and Higgins, 1991; Neuhaus, 1994; Nebelsick, 1993). According to these descriptions, the anterior

◂ **Figure 6.1:** Priapulida. Anterior is always to the left (except in G). **A.** Two early larvae of *Priapulus caudatus* showing introvert (*in*) and trunk (*tr*). **B.** Whole mount of *P. caudatus* larva, staining against serotonin (5-HT). The introvert is withdrawn, causing a bend in the course of the ventral nerve cord (*vnc*) and a deep position of the brain (*br*). **C.** Some of the nuclei of the brain (nuclei in *white*) are serotonin-immunoreactive (5-HT in *blue*), *np* = neuropil. **D.** Anterior end of *Tubiluchus troglodytes* shows accumulation of nuclei in two portions anterior (*ant*) and posterior (*post*) of the brain, *phar* = pharynx. **E.** Double staining shows serotonin-immunoreactive fiber and nuclei (*blue*) among a broader neuropil shown by β-tubulin-immunoreactivity (*green*). **F.** FMRFamide-immunoreactivity is stronger in the neuropil compared to 5-HT. More FMRFamide-immunoreactive nuclei are present anterior of the neuropil than posterior. **G.** *Tubiluchus* sp. from the Canary Islands, showing everted pharynx (*phar*), introvert (*in*), trunk (*tr*) and tail (*ta*). A,G with SEM. B–F with confocal laser scanning microscopy, A depth color coded, B,D maximum projections, C,E,F single planes.

somata form ten clusters (lobes), while the distribution of posterior somata is more or less homogeneous. A pair of ventral nerve cords originates from the brain. The cords are very close together (Kristensen and Higgins, 1991). The previously mentioned data were obtained with TEM investigations; they were confirmed by Herranz, Pardos, and Boyle (2013) and own unpublished immunohistochemical investigations (Figure 6.2B–G). In addition, the ventral part of the brain, between the origin of the ventral nerve cords, appears thinner than the remaining part, as most of the fibers seem to proceed into the nerve cords.

6.3.5 Loricifera

Very few TEM investigations from *Nanaloricus mysticus* and *Pliciloricus enigmaticus* describe the structure of the brain (Kristensen, 1991a), no immunohistochemical data are available. The large brain of Loricifera is in the introvert, around the buccal tube in *N. mysticus* or around the esophagus in *P. enigmaticus*. It is cycloneural in structure; the somata are clustered into lobes. Anterior of the ring neuropil, eight lobes of somata are present, posterior of the neuropil there are ten lobes (Kristensen, 1991a).

6.3.6 Gastrotricha

The structure of the brain in gastrotrichs has been reconstructed in different forms. "Old" descriptions, e.g. by Zelinka (1889) and Remane (1936) reported the brain as a dorsal commissure with lateral "ganglia" composed of cell somata. A reconstruction based on ultrathin sections stated the brain in *Turbanella cornuta* to have a circumoral neuropil and anterior and posterior accumulations of somata (Teuchert, 1977). This reconstruction has some resemblance to a cycloneural brain, although the neuropil is not completely ring-shaped, i.e. distinctly uneven in proportion in the dorsal and ventral part. Another reconstruction of the brain in *Cephalodasys maximus* by

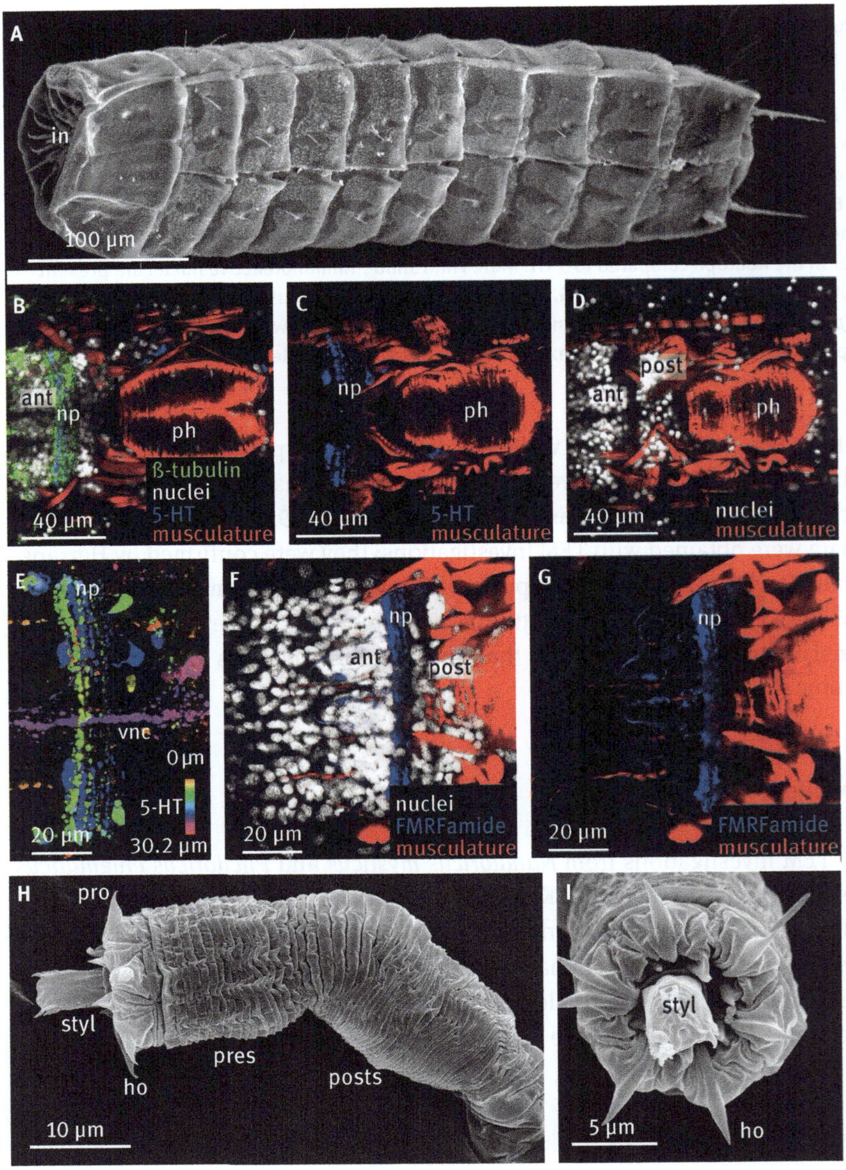

Figure 6.2: Kinorhyncha (A–G) and Nematomorpha (H,I). Anterior is always to the left. **A.** Ventral view on a female of *Pycnophyes kielensis* showing segmented trunk and withdrawn introvert (*in*). **B–D.** Different stainings of comparable sections through the brain/pharynx region of *P. kielensis* showing the arrangement of anterior somata (*ant*), neuropil (*np*) and posterior somata (*post*) (nuclei in *white*, musculature [phalloidin] in *red*, neurons in *green* [β-tubulin] and *blue* [5-HT]), *ph* = pharynx. **E–G.** Brain of *P. kielensis* shown with different markers (see labels in images), abbreviations as in B–D, *vnc* = ventral nerve cord. **H,I.** Larvae of *Gordius aquaticus* showing proboscis (*pro*) with hooks (*ho*) and stylets (*styl*), preseptum (*pres*) and postseptum (*posts*). A,H,I with SEM, B–G with confocal laser scanning microscopy, B–D,F,G single planes, E maximum projection.

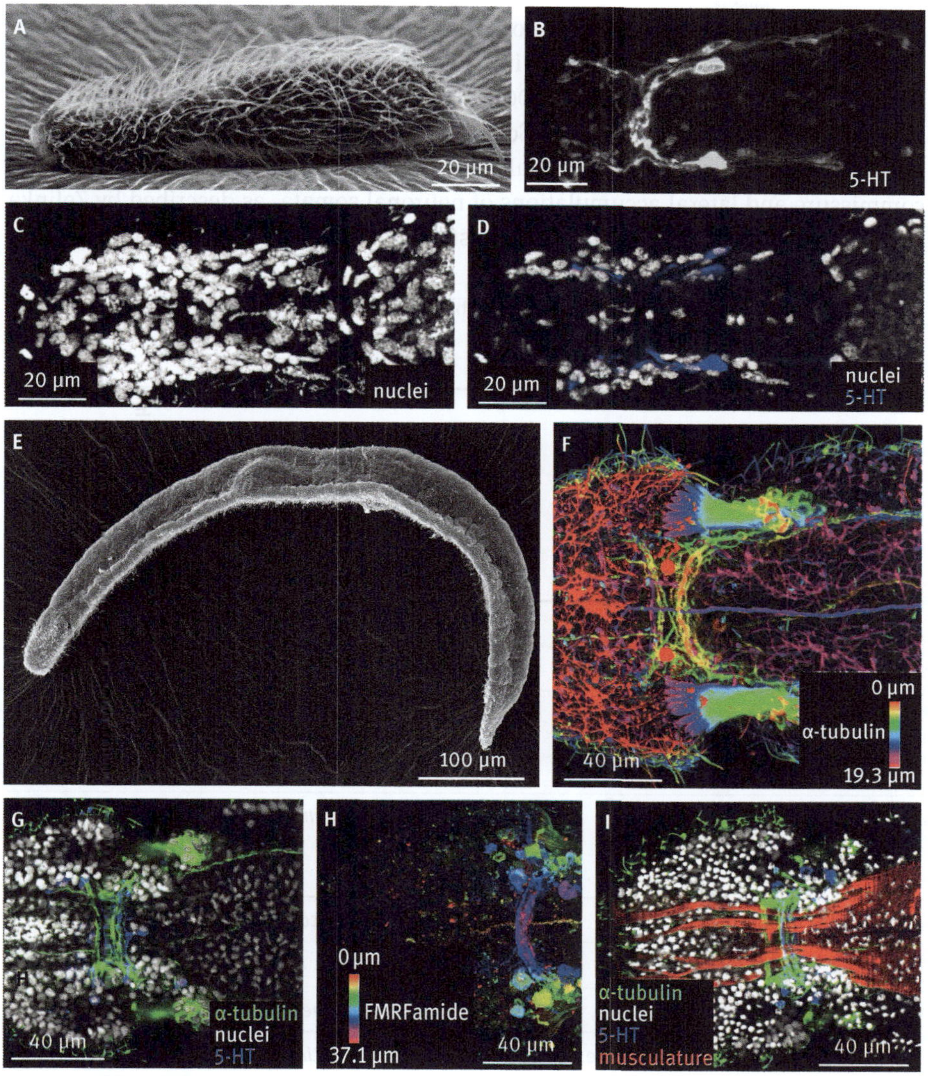

Figure 6.3: Gastrotricha. Anterior is always to the left. **A.** Dorsal view on *Chaetonotus maximus* (Chaetonotida, Paucitubulatina). **B.** Serotonin (5-HT)-immunoreactive components of the nervous system in the anterior body region of *Chaetonotus maximus*: dorsal commissure (*dc*), anterior projections (*ap*), lateroventral nerve cords (*lvnc*) and lateral neuronal somata (*lns*). **C.** General distribution of somata in the head region. **D.** Additional staining against serotonin (*blue*) identifies neuronal somata. **E.** *Cephalodasys maximus* (Macrodasyida). **F–I.** Different stainings showing the dorsal commissure (*dc*) of the brain and the presence of lateral neuronal somata (*lns*)in *Cephalodasys maximus*, fcot in F are anterior feet-like structures carrying adhesive tubes. A,E with SEM, B–D,F–I with confocal laser scanning microscopy, D,G–I single planes, B,C,F maximum projection.

Wiedermann (1995) reported a more or less complete covering of the entire neuropil with somata. Immunohistochemical investigations (Figure 6.3; Table 6.1; Hochberg, 2007; Rothe and Schmidt-Rhaesa, 2008; 2009; 2010b; Rothe, Schmidt-Rhaesa, and Kieneke, 2011; Rothe, Kieneke, and Schmidt-Rhaesa, 2011) confirmed that the brain in gastrotrichs is composed of a dorsal commissural neuropil with a very fine ventral commissure and lateral accumulations of neuronal somata, as described by Zelinka (1889) and Remane (1936). The brain of gastrotrichs is therefore not cycloneural.

Table 6.1: Overview of the stainings used in published investigations on the nervous system in Gastrotricha.

		serotonin	RF-like amides	acetylated α-tubulin	β-tubulin	histamin	SCP_B
Joffe and Wikgren 1995	*Turbanella cornuta*	X					
Liesenjohann et al. 2006	*Dactylopodola baltica*			X			
Hochberg 2007	*Neodasys cirritus* *Turbanella cf. hyalina* *Xenodasys riedli*		X				
Rothe and Schmidt-Rhaesa 2008	*Turbanella ambronensis* *T. cornuta* *T. hyalina*	X		X			
Rothe and Schmidt-Rhaesa 2009	*Dactylopodola baltica*	X	X	X	X	X	
Rothe and Schmidt-Rhaesa 2010a	*Oregodasys cirratus*	X	X	X			
Hochberg and Atherton 2011	*Lepidodasys worsaae*		X				X
Rothe, Schmidt-Rhaesa, and Kieneke, 2011	*Neodasys chaetonotideus*	X	X	X	X		
Rothe, Kieneke, and Schmidt-Rhaesa, 2011	*Xenotrichula intermedia* *X. velox*	X	X	X			

6.4 Functional aspects of the cycloneuralian brain

The anterior end of cycloneuralian taxa expresses radial symmetrical patterns, especially concerning sensory structures. These patterns may be responsible for the shape of the brain as a circumoral ring. The mouth opening is terminal and opens at the anterior tip. In nematodes, there are three rings of head sensory structures with hexaradial symmetry. All realized arrangements go back to a basic scheme with six head sensilla in the first ring, six sensilla in the second ring and four sensilla plus two amphids in the last ring (see, e.g., Malakhov, 1994). In scalidophorans, a pentaradial symmetry of scalids on the introvert is common (see e.g., Adrianov and Malakhov, 1996; Schmidt-Rhaesa, 2012a for Priapulida; Kristensen and Higgins, 1991; Sørensen and Pardos, 2008; Neuhaus, 2012 for Kinorhyncha; Kristensen, 1991a; Heiner, Schmidt-Rhaesa, and Kristensen, 2012 for Loricifera). Nematomorphs are an exception, because in adults no sensory structures are externally recognizable and the mouth opening is closed in some species (Schmidt-Rhaesa, 2012b). Larvae have a hexaradial symmetry of hooks and three stylets, but their nervous system is unknown.

The ring-shape of the cycloneuralian brain may be explained by the radial pattern of sensory structures around the terminal mouth opening. In other taxa, a ventral mouth opening and a non-radial distribution of sensory structures is present and this may explain the location of the brain in the dorsal region.

6.5 Conclusions, comparison within protostomes and evolutionary scenarios

6.5.1 Are Cycloneuralia monophyletic?

Despite the fact that the cycloneuralian brain is a unique and complex structure, it is not realized in all members of the taxon. No nematomorph species has a cycloneuralian brain and within Nematoda and Priapulida there are some species that have a distribution of nuclei around the ring neuropil different from the cycloneuralian condition. This can be explained in two ways: either the cycloneuralian brain is an autapomorphy of Cycloneuralia and all deviating states are derived, or a cycloneuralian condition evolved several times in parallel. As cycloneuralian brains are distributed broadly among Cycloneuralia and as they are complex structures composed of several substructures (ring neuropil, somata anterior and somata posterior of the neuropil), a unique origin in the cycloneuralian ancestor appears likely.

6.5.2 How are cycloneuralian taxa related to Arthropoda?

Numerous analyses indicate a close relationship of cycloneuralian taxa with arthropods in a common taxon Ecdysozoa, but the brain structure of cycloneuralians and arthropods is completely different. Arthropods have, very much simplified, a brain structure as in spiralians (lophotrochozoans) with a central neuropil and a sheath of somata. Scenaria deriving the cycloneuralian and the arthropod brains from a common ancestral structure or deriving one type from the other are hard to imagine. Therefore, the brain structure does not help very much to discuss the relationship between arthropods and cycloneuralians among Ecdysozoa. As many analyses reveal,

Cycloneuralia as not monophyletic, it would be elegant if the brain structure could help to evaluate alternative relationships, but due to the reasons named above, the brain structure does not help much.

6.5.3 Are Gastrotricha related to cycloneuralian taxa or to Ecdysozoa?

The gastrotrich brain strongly differs from the cycloneuralian brain in structure. It is possible to imagine the evolution of a ring-shaped brain with a paired or unpaired ventral nerve cord from a dorsal commissure with laterally emerging nerve cords by a ventral shift of the ends of the dorsal commissure until they meet ventrally, but in this scenario the transition from lateral somata (as in gastrotrichs) to anterior and posterior somata (as in the cycloneuralian brain) is difficult to explain. Nevertheless, the gastrotrich brain structure is also not indicative of a closer relationship to another taxon, especially a spiralian (lophotrochozoan) relationship as suggested by molecular data. In spiralian (lophotrochozoan) taxa, a central compact neuropil surrounded by a sheath of somata prevails (see, e.g., Morris et al., 2007; see also chapter by R. Loesel). This also applies to a comparison of the gastrotrich brain with the arthropod brain.

In some respects the gastrotrich brain may be compared with brains in acoels, which, at least in some species, form a dorsal commissure (Raikova et al., 1998; Reuter et al., 2001a; Reuter, Raikova, and Gustafsson, 2001b). The distribution of neuronal somata is unknown in acoels, but there is no indication that they are clustered lateral of the commissure. Therefore, the gastrotrich brain may be regarded as a mosaic of plesiomorphic (the dorsal commissure) and apomorphic features (the peculiar distribution of neuronal somata).

6.5.4 Conclusions

Both Cycloneuralia and Gastrotricha each have a unique brain structure. In Cycloneuralia there are some deviations in some species and more data are needed especially from nematodes. The brain structure supports monophyly of each Cycloneuralia and Gastrotricha, but unfortunately does not help much to resolve the relationship of these taxa to other taxa and test phylogenetic hypotheses based on molecular analyses.

Acknowledgments

We thank our colleagues from the project Deep Metazoan Phylogeny, especially the colleagues working on nervous systems, for numerous constructive discussions and an excellent working atmosphere.

Thomas Hankeln, Alexandra R. Wey-Fabrizius, Holger Herlyn, Alexander Witek, Mathias Weber, Maximilian P. Nesnidal, and Torsten H. Struck

7 Phylogeny of platyzoan taxa based on molecular data

Abstract: In this chapter we discuss the concept of Platyzoa (Platyhelminthes + Gastrotricha + Gnathifera) with a focus on recent molecular phylogenetic evidence. Besides an introduction to the taxa, we present an up-to-date summary of molecular phylogenetic analyses and highlight the results of studies generated within the "Deep Metazoan Phylogeny" priority project 1174 of the German Research Foundation. While several aspects of internal phylogenies of Platyzoan subgroups (Syndermata, Acanthocephala) could already successfully be addressed, the question for monophyly of Platyzoa is awaiting further analysis. In this context, we provide novel data on the position of Gastrotricha and discuss potential long-branch artifacts in molecular phylogenetic analyses of platyzoan taxa.

7.1 Introduction

Molecular data have profoundly changed our view of the metazoan tree of life. For example, within Bilateria three major taxa, Deuterostomia, Ecdysozoa and Lophotrochozoa have been recognized, the latter two essentially unrecognized until the mid-1990s (for review see Halanych, 2004). Lophotrochozoa is defined by the last common ancestor of Annelida, Mollusca and all three lophophorate taxa (Brachiopoda, Phoronida and Ectoprocta), plus all the descendants of that ancestor (Halanych et al., 1995). Lophotrochozoan taxa thus show a wide variety and plasticity in body plans and in developmental, embryonic and morphological characters. The concept of Lophotrochozoa was first proposed based on 18S rRNA data (Halanych et al., 1995) and confirmed by evidence from other single genes as well as by larger-scale expressed sequence tag (EST) data (Dunn et al., 2008; Halanych, 2004; Hausdorf et al., 2007; Passamaneck and Halanych, 2006; Struck and Fisse, 2008). Additionally, Cavalier-Smith (1998) proposed Platyzoa as a taxon comprising Platyhelminthes and Acanthognatha, which consisted of Gastrotricha, "Rotifera", Acanthocephala and Gnathostomulida (Figures 7.1 and 7.2). Lophotrochozoa, in the way defined by Halanych (2004), either contains Platyzoa or, alternatively, Platyzoa is the sister-group to Lophotrochozoa.

Platyzoan taxa are generally direct developers, lack a vascular system, and their worm-shaped body is usually ciliated, flat and unsegmented. However, morphological synapomorphies supporting the monophyly of Platyzoa are lacking (Giribet 2008).

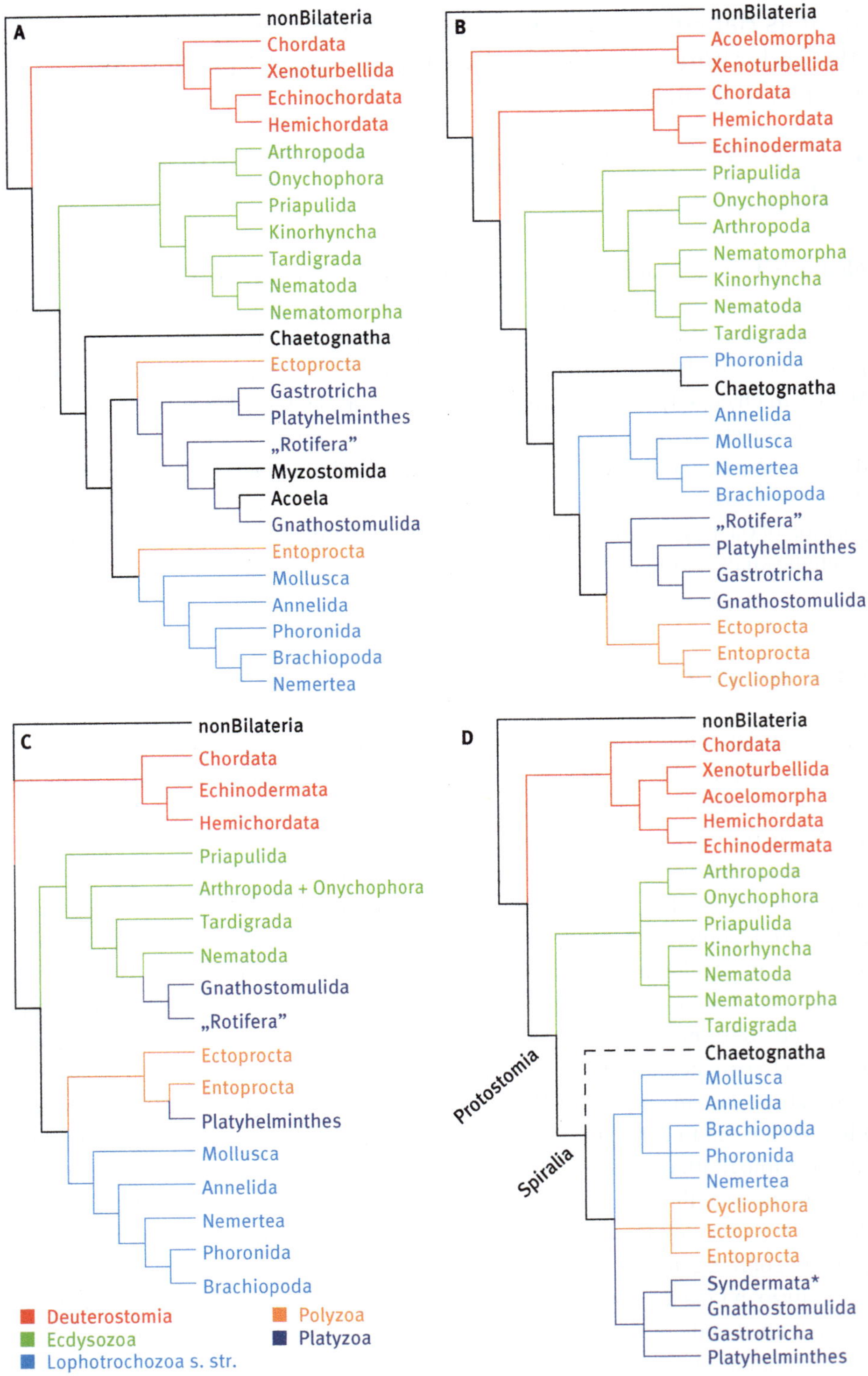
A
nonBilateria
Chordata
Xenoturbellida
Echinochordata
Hemichordata
Arthropoda
Onychophora
Priapulida
Kinorhyncha
Tardigrada
Nematoda
Nematomorpha
Chaetognatha
Ectoprocta
Gastrotricha
Platyhelminthes
„Rotifera"
Myzostomida
Acoela
Gnathostomulida
Entoprocta
Mollusca
Annelida
Phoronida
Brachiopoda
Nemertea
B
nonBilateria
Acoelomorpha
Xenoturbellida
Chordata
Hemichordata
Echinodermata
Priapulida
Onychophora
Arthropoda
Nematomorpha
Kinorhyncha
Nematoda
Tardigrada
Phoronida
Chaetognatha
Annelida
Mollusca
Nemertea
Brachiopoda
„Rotifera"
Platyhelminthes
Gastrotricha
Gnathostomulida
Ectoprocta
Entoprocta
Cycliophora
C
nonBilateria
Chordata
Echinodermata
Hemichordata
Priapulida
Arthropoda + Onychophora
Tardigrada
Nematoda
Gnathostomulida
„Rotifera"
Ectoprocta
Entoprocta
Platyhelminthes
Mollusca
Annelida
Nemertea
Phoronida
Brachiopoda
Deuterostomia
Ecdysozoa
Lophotrochozoa s. str.
Polyzoa
Platyzoa
D
nonBilateria
Chordata
Xenoturbellida
Acoelomorpha
Hemichordata
Echinodermata
Arthropoda
Onychophora
Priapulida
Kinorhyncha
Nematoda
Nematomorpha
Tardigrada
Chaetognatha
Mollusca
Annelida
Brachiopoda
Phoronida
Nemertea
Cycliophora
Ectoprocta
Entoprocta
Syndermata*
Gnathostomulida
Gastrotricha
Platyhelminthes
Protostomia
Spiralia

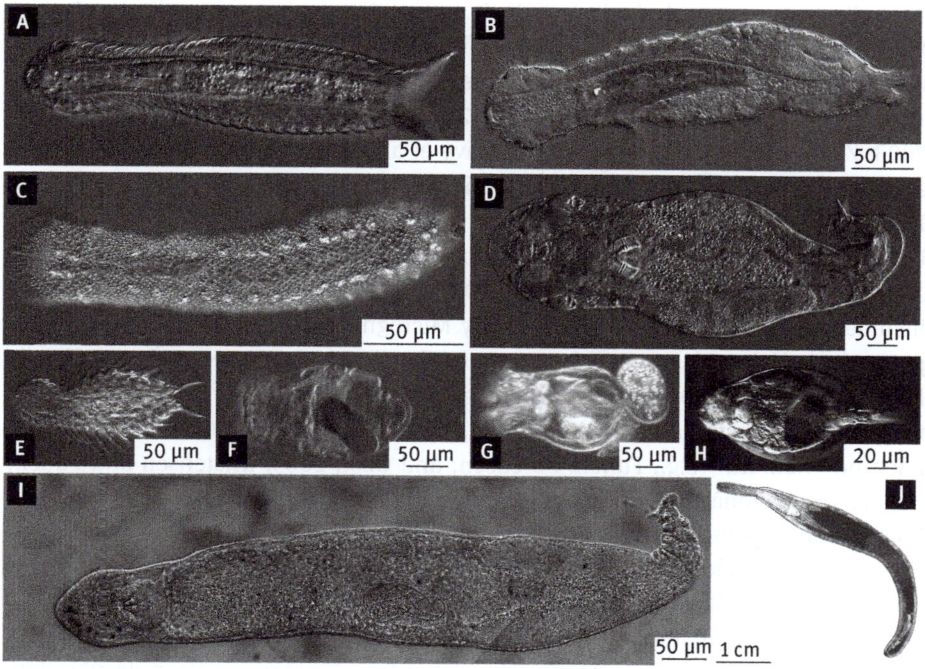

Figure 7.2: Photos of representatives of Gastrotricha, Syndermata and Gnathostomulida. **A.** *Lepidodermella squamata* (Gastrotricha, Chaetonotida). **B.** *Dactylopodola baltica* (Gastrotricha, Macrodasyida). **C.** *Tetranchyroderma megastoma* (Gastrotricha, Macrodasyida). **D.** *Philodina roseola* (Syndermata, Bdelloida). **E.** *Halichaetonotus schromi* (Gastrotricha, Chaetonotida). **F.** *Brachionus calyciflorus* (Syndermata, Monogononta). **G.** *Brachionus plicatilis* (Syndermata, Monogononta). **H.** *Monostyla sp.* (Syndermata, Monogononta). **I.** *Gnathostomula paradoxa* (Gnathostomulida, Bursovaginoidea). **J.** Male *Paratenuisentis ambiguus* (Syndermata, Acanthocephala). For photos A–D and F–I we used a Zeiss Axioskop microscope equipped with an AxioCamMRc5 digital camera. Picture E was obtained with a standard optical microscope and a standard digital camera. Picture J was obtained using a Zeiss Axiophot microscope and an analog camera.

◂ **Figure 7.1:** Current views of the metazoan tree of life based on molecular data. Tree topologies are redrawn from the original works of [A] Dunn et al., 2008 (Fig. 1 therein), [B] Hejnol et al., 2009 (Fig. 2 therein) and [C] Hausdorf et al., 2010 (Fig. 1B therein). Part [D] illustrates a consensus view on the metazoan tree. This consensus topology is mainly based on the works mentioned above and built according to the authors' subjective views. The position of Xenoturbellida and Acoelomorpha is derived from the work of Philippe et al. (2011b), which is in contrast to most other studies regarding the position of Acoelomorpha. Multifurcations illustrate open questions in metazoan phylogeny. Colors highlight putative monophyla taking the consensus [D] as reference. Asterisk (*) in [D] indicates where the taxon "Rotifera" is replaced with Syndermata ("Rotifera" including Acanthocephala).

Accordingly, most morphological-cladistic and total evidence analyses find platyzoan taxa as paraphyletic or polyphyletic assemblages within Lophotrochozoa or Protostomia, although a few total evidence analyses support monophyletic Platyzoa (Eernisse, Albert, and Anderson, 1992; Giribet et al., 2000; Peterson and Eernisse, 2001; Zrzavý, 2003; Zrzavý, Hypsa and Tietz, 2001; Zrzavý et al., 1998). Molecular analyses using single genes were also inconsistent in this respect. Some supported monophyly of Platyzoa, whereas others found these taxa to be paraphyletic within Lophotrochozoa, but always with low support (see Halanych, 2004; Paps, Baguñà, and Riutort, 2009a; 2009b; Passamaneck and Halanych, 2006). Moreover, additional taxa have been included within Platyzoa as well, namely Micrognathozoa, Cycliophora and Entoprocta (e.g., Halanych, 2004; Kristensen and Funch, 2000; Winnepenninckx, Van de Peer and Backeljau, 1998). Whereas inclusion of Micrognathozoa is substantiated by their anatomical similarities to the other jaw-bearing platyzoan taxa Gnathostomulida and "Rotifera" (e.g., Funch, Sørensen, and Obst, 2005; Kristensen and Funch, 2000; Paps, Baguñà, and Riutort, 2009a; Worsaae and Rouse, 2008), Cycliophora and Entoprocta are possibly more closely related to other lophotrochozoans such as Ectoprocta (e.g., Hausdorf et al., 2007; Hejnol et al., 2009; Helmkampf, Bruchhaus, and Hausdorf, 2008b; Struck and Fisse, 2008).

Recently several attempts have been made to resolve the phylogeny of Bilateria or Protostomia using phylogenomic approaches based on large scale molecular data sets derived from ESTs (Figure 7.1). Many of these studies included several of the platyzoan taxa so that the monophyly of Platyzoa could be tested at least in part. Some analyses based on up to 79 ribosomal proteins (RP) recovered monophyletic Platyzoa with strong support (Hausdorf et al., 2007; Hausdorf et al., 2010; Helmkampf, Bruchhaus, and Hausdorf, 2008b; Struck and Fisse, 2008), but others found paraphyletic assemblages of platyzoan taxa (Hausdorf et al., 2007; Witek et al., 2008; 2009) or polyphyletic positions within Protostomia, with Platyhelminthes nested within Lophotrochozoa and "Rotifera", Acanthocephala and/or Gnathostomulida close to ecdysozoan taxa (Hausdorf et al., 2010; Helmkampf, Bruchhaus, and Hausdorf, 2008b; Nesnidal et al., 2010; Struck and Fisse, 2008). Thus, with respect to the monophyly of Platyzoa, phylogenies based on RP data are as inconsistent as the analyses of single or only a few genes and likewise depend on the taxon sampling and reconstruction methods used. Other phylogenomic studies with larger, mixed data sets of hundreds of different genes and at least one representative of each platyzoan taxon except for Acanthocephala and Micrognathozoa found monophyletic Platyzoa placed within Lophotrochozoa (Dunn et al., 2008; Hejnol et al., 2009; see Figure 7.1). Unfortunately, platyzoan taxa were among the most unstable taxa in these analyses. Support for the monophyly of Platyzoa as well as their position within Lophotrochozoa was only strong if certain platyzoan taxa such as Gnathostomulida or "Rotifera" were excluded. Moreover, except for Platyhelminthes, gene coverage of platyzoan taxa within data matrices was low throughout these analyses (Dunn et al., 2008; Hejnol et al., 2009). Another reason for the inconsistent results in the molecular analyses and the low support values for the phylogenetic placements of platyzoan taxa is probably the problem of long-branch attraction as platyzoan taxa are characterized by long branches (Edgecombe et al., 2011). We will discuss this problem in more detail at the end of this chapter.

Likewise, studies using complete mitochondrial (mt) genome data have not yet indicated if Platyzoa are monophyletic or not, because these data are available for only a few of the platyzoan subtaxa (Platyhelminthes, "Rotifera", and Acanthocephala, respectively). At least the mtDNA results consistently show Platyhelminthes close to Acanthocephala (or "Rotifera" and Acanthocephala, if "Rotifera" was included in the analyses) and therefore suggest monophyletic Platyzoa (Gazi et al., 2012; Min and Park, 2009; Steinauer et al., 2005; Weber et al., 2013). However, this Platyzoa clade is sometimes placed close to Ecdysozoa (probably due to long-branch attraction (LBA) artifacts, see below). Thus, analytic methods to reduce LBA should join the sequencing of more platyzoan mitochondrial genomes in future assessments of the Platyzoa concept. Moreover, the phylogenetic signal of the mitochondrial gene order should also to be taken into account more thoroughly and systematically.

In summary, monophyly of Platyzoa comprising Platyhelminthes, Gastrotricha, Gnathostomulida, Micrognathozoa, "Rotifera" and Acanthocephala has so far gained some support from molecular studies, but this support is still weak and eventually due to long-branch attraction. The same is true for placing Platyzoa either within Lophotrochozoa or as their sister-group. In the following sections, we will discuss the progress in molecular phylogenetic reconstructions of specific platyzoan taxa in more detail.

7.2 The phylogenetic position of Platyhelminthes

Platyhelminthes are acoelomate worms, which are dorsoventrally flattened and lack an anus and segmentation. Presently about 13,000 living species are described. The vast majority are parasites well known as trematodes or flukes and cestodes or tapeworms, several species of which infest humans or our livestock and pets. Their free-living relatives, paraphyletic "Turbellaria", occur in all marine and terrestrial habitats. Small interstitial platyhelminths can be as small as 1 mm, whereas some fresh water species can be up to 0.5 m long and parasitic forms can be considerably longer (e.g., the 25 m fish tapeworm, *Diphyllobothrium latum*) (Westheide and Rieger, 1996). Traditionally Platyhelminthes were said to consist of Acoelomorpha, Catenulida and Rhabditophora (Westheide and Rieger, 1996). However, monophyly is not well supported by morphological data, and is sharply refuted by molecular data (e.g., Dunn et al., 2008; Haszprunar, 1996a; 1996b; Hejnol et al., 2009; Ruiz-Trillo et al., 2002; Telford et al., 2003; Zrzavý, 2003). Specifically, the molecular analyses indicate that Acoelomorpha are not closely related to the other Platyhelminthes, but are either sister to all other bilaterians (e.g., Hejnol et al., 2009; Paps, Baguñà, and Riutort, 2009b; Ruiz-

Trillo et al., 1999; Wallberg et al., 2007) or go within deuterostomes (Philippe et al., 2011b). Hence, the exclusion of Acoelomorpha from Platyhelminthes is now generally well accepted. For a more detailed discussion of relationships within Platyhelminthes see the review by Riutort et al. (2012).

Due to their simple acoelomate body organization, Platyhelminthes once occupied a central role in considerations about the early evolution of Bilateria. Together with Nemertea, which are also generally dorsoventrally flat and show a habitus similar to "turbellarian" flatworms, Platyhelminthes were formerly seen as the sister-group to all other bilaterian taxa (see Brusca and Brusca, 1990; Halanych, 2004; Hyman, 1951). However, both Nemertea and Platyhelminthes show spiral cleavage during their development and, hence, closer affinities to spiralian taxa (Lophotrochozoa) have been proposed for both (Westheide and Rieger, 1996). Within a clade of spiralian taxa, different phylogenetic positions for Platyhelminthes have been put forward based on morphological data. Some authors still maintained a close relationship of Nemertea to Platyhelminthes, based on such similarities as a reduced hyposphere in the larva, the lack of an anus and the shape of the larval ciliary bands (Nielsen, 2001; Nielsen, Scharff and Eibye-Jacobsen, 1996; Sørensen et al., 2000). Several other morphology-based studies placed Platyhelminthes within Platyzoa or at least close to platyzoan taxa such as Gnathostomulida and Gastrotricha (e.g., Ax, 1995; Cavalier-Smith, 1998; Garey et al., 1998; Giribet et al., 2000; Meglitsch and Schram, 1991; Peterson and Eernisse, 2001; Zrzavý et al., 1998), whereas others just placed them within a clade of spiralian taxa like Annelida or Nemertea (e.g., Brusca and Brusca, 2003; Haszprunar, 1996a; 1996b; Rouse and Fauchald, 1995; Zrzavý, 2003; Zrzavý, Hypsa and Tietz, 2001). For a detailed review of these morphological analyses, see Jenner (2004a). Molecular data from single nuclear genes or mt genomes so far support only a placement of Platyhelminthes within Lophotrochozoa (see de Rosa et al., 1999; Halanych, 2004; Paps, Baguñà, and Riutort, 2009a; 2009b; Passamaneck and Halanych, 2006; Podsiadlowski et al., 2009). Some of these analyses suggested a position within monophyletic Platyzoa, whereas others placed Platyhelminthes as sister to several or all other lophotrochozoan taxa. However, none of these relationships was substantially supported, and none of the analyses found a close relationship of Nemertea and Platyhelminthes.

A phylogenomic study of 60 RP genes as part of the "Deep Metazoan Phylogeny" (DMP) priority project (Struck and Fisse, 2008) found Platyhelminthes either as sister to the other platyzoan taxa, "Rotifera" and Acanthocephala, which were the only other representatives of Platyzoa in that study, or placed them within Lophotrochozoa as part of a basal polytomy. Again, a close relationship to Nemertea was not found and topology testing significantly rejected it. In subsequent phylogenomic analyses, most with a more comprehensive sampling of genes and platyzoan taxa, Platyhelminthes appeared as part of a monophyletic Platyzoa clade, either as sister to Gastrotricha, to Gnathifera or to Gastrotricha/Gnathostomulida (Dunn et al., 2008; Hausdorf et al., 2010; Hejnol et al., 2009) or part of a polytomy (Witek et al., 2009). However, nodal support values were low in all cases. Hence, at present, Platyhelminthes have yet to be placed precisely. Within Lophotrochozoa, they are most likely part of Platyzoa, but this awaits further study.

7.3 Gastrotricha: Phylogenetic case study using four genes

Gastrotricha are members of the marine interstitium and their acoel body is generally slender and dorsoventrally flattened. Gastrotrichs are among the smallest multicellular animals, ranging from only 60 µm to 1.5 mm in length (Brusca and Brusca, 2003). This means that some unicellular eukaryotes are substantially larger than some gastrotrich species. Gastrotricha is split into two major groups, Macrodasyida and Chaetonotida. The Chaetonotida, with a small body size of usually less than 0.5 mm, have amazingly intricate cuticular scales and spines. Though Macrodasyida and Chaetonotida differ in the orientation of their Y-shaped pharyngeal lumen or in the possession of elaborate cuticular structures, several characters support monophyly of Gastrotricha. For example, their cilia, both locomotory and sensory, are entirely covered by the cuticle, which is unique in Bilateria (Westheide and Rieger, 1996). The cuticle also covers the outlet of the duo-gland adhesive system (Todaro et al., 2006). Because gastrotrichs do not possess an anchor cell, the tension during adhesion at sand grains is transmitted via this cuticle (Ruppert, 1991a; Tyler and Rieger, 1980). Moreover, all gastrotrichs investigated so far show a 'muscular double helix'; that is, a muscle arrangement in two spirals similar to the DNA backbone organization (Hochberg and Litvaitis, 2001a; Hochberg and Litvaitis, 2001b; Hochberg and Litvaitis, 2003; Todaro et al., 2006). Studies based on 18S rRNA with a comprehensive taxon sampling consistently recovered monophyly of Gastrotricha (Petrov et al., 2007; Todaro et al., 2006) in line with morphological-cladistic analyses (Kieneke, Riemann, and Ahlrichs, 2008).

Traditional morphological studies placed Gastrotricha within a group comprising all taxa with a pseudocoelomate organization; i.e., with Nematoda, Priapulida, Kinorhyncha and "Rotifera" (Hyman, 1951). Or they were joined with Nematoda, Priapulida and others in a group of thread-like worms, the hypothesized Nemathelminthes, which share a two-layered cuticle, cuticle present in the pharynx and esophagus, and a terminal mouth opening (Ax, 2001; Westheide and Rieger, 1996). In more recent morphological studies, Gastrotricha were placed as sister to the ecdysozoan taxon Introverta, comprising Nematoda, Nematomorpha, Priapulida, Kinorhyncha and Loricifera, a topology known as the Cycloneuralia hypothesis (Nielsen, 1995; Nielsen, 2001; Sørensen et al., 2000). This placement is in part congruent with the Pseudocoelomates or Nemathelminthes hypotheses, because all these taxa were also members of Nemathelminthes. Others proposed a sister-group relationship of Gastrotricha to all Ecdysozoa (Peterson and Eernisse, 2001; Schmidt-Rhaesa, 2002; Schmidt-Rhaesa et al., 1998; Zrzavý, 2003). This gastrotrich-ecdysozoan link was based on such shared traits as a collar-shaped circumpharyngeal brain, lack of an apical organ in developmental stages, Y-shaped cylindrical pharynx and a two-layered cuticle (Zrzavý, 2003).

Alternatively, Gastrotricha has been related to taxa with spiral cleavage like Platyhelminthes, Annelida or Mollusca (e.g., Giribet et al., 2000; Halanych, 2004; Petrov et al., 2007). Based on morphological and/or molecular data, Gastrotricha has been proposed to be sister to all or most other lophotrochozoan taxa or to be within Platyzoa (e.g., Cavalier-Smith, 1998; Giribet et al., 2000; Halanych, 2004; Paps, Baguñà, and Riutort, 2009a; 2009b; Petrov et al., 2007). More specifically, based on similarities of protonephridial ultrastructure, the presence of monociliated epidermal cells and 18S rRNA data, a sister-group relationship to Gnathostomulida, the Monokonta or Neotrichozoa hypothesis has been suggested (Cavalier-Smith, 1998; Rieger, 1976; Sterrer, Mainitz, and Rieger, 1985; Zrzavý, Hypsa, and Tietz, 2001; Zrzavý et al., 1998). A total evidence analysis found a sister-group relationship to Platyhelminthes (Giribet et al., 2000) and other studies a sister-group relationship of Gastrotricha to "Rotifera" (Todaro et al., 2006). Finally, Petrov et al. (2007) significantly rejected a closer relationship of Gastrotricha to Ecdysozoa or a basal position of Gastrotricha within Bilateria using topology tests.

To circumvent possible pitfalls associated with 18S rRNA phylogeny in the Gastrotricha case, we conducted a multiple-gene approach based on three additional genes, namely an approximately 3,600 bp fragment of 28S rRNA mimicking Struck et al. (2007; 2008), a 970 or 820 bp fragment of the nuclear protein-coding gene methionine adenyltransferase (MAT) inspired by Peterson et al. (2004) and a 760 or 676 bp fragment of myosin heavy chain type II (MHC) such as used by Ruiz-Trillo et al. (2002) of four gastrotrich species (two Chaetonotida and two Macrodasyida), three rotifers, and one gnathostomulid (our GenBank numbers are listed on the tree of Figure 7.3). Gene amplification and sequencing generally followed Dordel et al. (2010). After aligning each gene with MUSCLE (Edgar, 2004) and masking ambiguously aligned positions with AliScore (Kück et al., 2010; Misof and Misof, 2009), we used different partitioned maximum likelihood (ML) analyses employing RAxML 7.2.8 (Stamatakis, 2006) to analyze the effect of the following analytical strategies on the phylogenetic results obtained from the same data set: (1) all taxa, all data as nucleotides (6,250 total nt); (2) all taxa, protein-coding genes as amino acids instead of nucleotides (5,063 nt, 457 amino acids); (3) reducing the number of outgroup taxa, and (4) excluding all taxa showing significant heterogeneity in their nucleotide composition based on posterior predictive tests (Lartillot and Philippe, 2004) except for the essential "Rotifera" and/or Gnathostomulida. Finally, we conducted Bayesian inferences (BI) using PhyloBayes 3.2f (Lartillot and Philippe, 2004) with both the time-homogeneous CAT model and the time-heterogeneous Bayesian non-parametric version of the mixture of branch lengths (MBL) model (Kolaczkowski and Thornton, 2008; Zhou et al., 2007).

In the partitioned ML analysis of the complete nucleotide data set a clade comprising Platyhelminthes, Gastrotricha and "Rotifera" was recovered with a bootstrap support (BS) of 72, which is equivalent to the taxon Platyzoa without Gnathostomulida (Figure 7.3). Gnathostomulida was found as sister to Nematoda probably due to long-branch attraction (BS = 72). Within the supported clade comprising Platyhelminthes, Gastrotricha and "Rotifera", Gastrotricha was more closely related to "Rotifera" than to Platyhelminthes (Figure 7.3), but bootstrap support was low. Gastrotricha did not appear as a monophyletic taxon, but formed a paraphyletic assemblage with respect to "Rotifera" (Figure 7.3). The chaetonotidan taxa were more closely related to "Rotifera" than the macrodasyidan taxa. However, bootstrap support for these relationships was again generally low. Monophyly of the gastrotrichan subtaxa Macrodasyida (i.e. *Tetranchyroderma* and *Dactylopodola*) and of Chaetonotida (i.e. *Halichaetonotus* and *Lepidodermella*) were always recovered with strong nodal support.

As mentioned above we tried different strategies that eventually reduce the potential impact of LBA on the phylogenetic reconstructions. Excluding long-branched outgroup taxa such as the nematodes found a sister-group relationship of Gnathostomulida to the other lophotrochozoan taxa and the relationships of the other platyzoan taxa remained unaltered. Excluding taxa with a heterogeneous base composition such as the nematodes placed Gnathostomulida within the clade of the other platyzoan taxa in Figure 7.3 as sister to Syndermata congruent with the Gnathifera hypothesis. However, in both analyses support for these relationships remained low. Employing time-heterogeneous substitution models in a Bayesian context we were also not able to significantly increase support for the phylogenetic positions of platyzoan taxa.

The topology tests significantly rejected all hypotheses constraining closer relationships of Gastrotricha to ecdysozoan taxa (i.e., the hypotheses Ecdysozoa, Pseudocoelomates, Nemathelminthes and Cycloneuralia) as well as a basal position of Gastrotricha within Bilateria (Table 7.1). An inclusion of Entoprocta within Platyzoa ("Platyzoa expanded") was rejected. The relationship of Gastrotricha to Gnathostomulida (Monokonta or Neotrichozoa hypothesis) was not quite rejected with the data set comprising all data as nucleotides, but with the data set comprising both nucleotide and amino acid data (Table 7.1). On the other hand, closer relationships of Gastrotricha either to Platyhelminthes or to Gnathifera (Acanthognatha hypothesis) were not rejected. Finally, monophyly of Gastrotricha, inclusion of Gnathostomulida in Lophotrochozoa and monophyly of Gnathifera were not rejected either.

These results obtained with sequences from our four genes strengthen those of past phylogenomic studies, in which Gastrotricha were only represented by the macrodasyidan species *Turbanella ambronensis* and Chaetonotida were unrepresented (Dunn et al., 2008; Hejnol et al., 2009). In those studies, Gastrotricha was sister to Platyhelminthes in Platyzoa, but nodal support for this placement was low, and branch attachment studies by Dunn et al. (2008) showed that high branch attachment frequencies also existed to other platyzoan taxa and, to a low degree, to ecdysozoan taxa. In addition, in these phylogenomic analyses *Turbanella ambronensis* was among the taxa that exhibited only low matrix coverage, because it had only 74 out of 1487 genes present in the supermatrix of Hejnol et al. (2009). Now, considering all evidence, the morphology-based hypotheses of a relationship of Gastrotricha to ecdysozoan taxa can be refuted, just as the gastrotrich-platyhelminth link is strengthened.

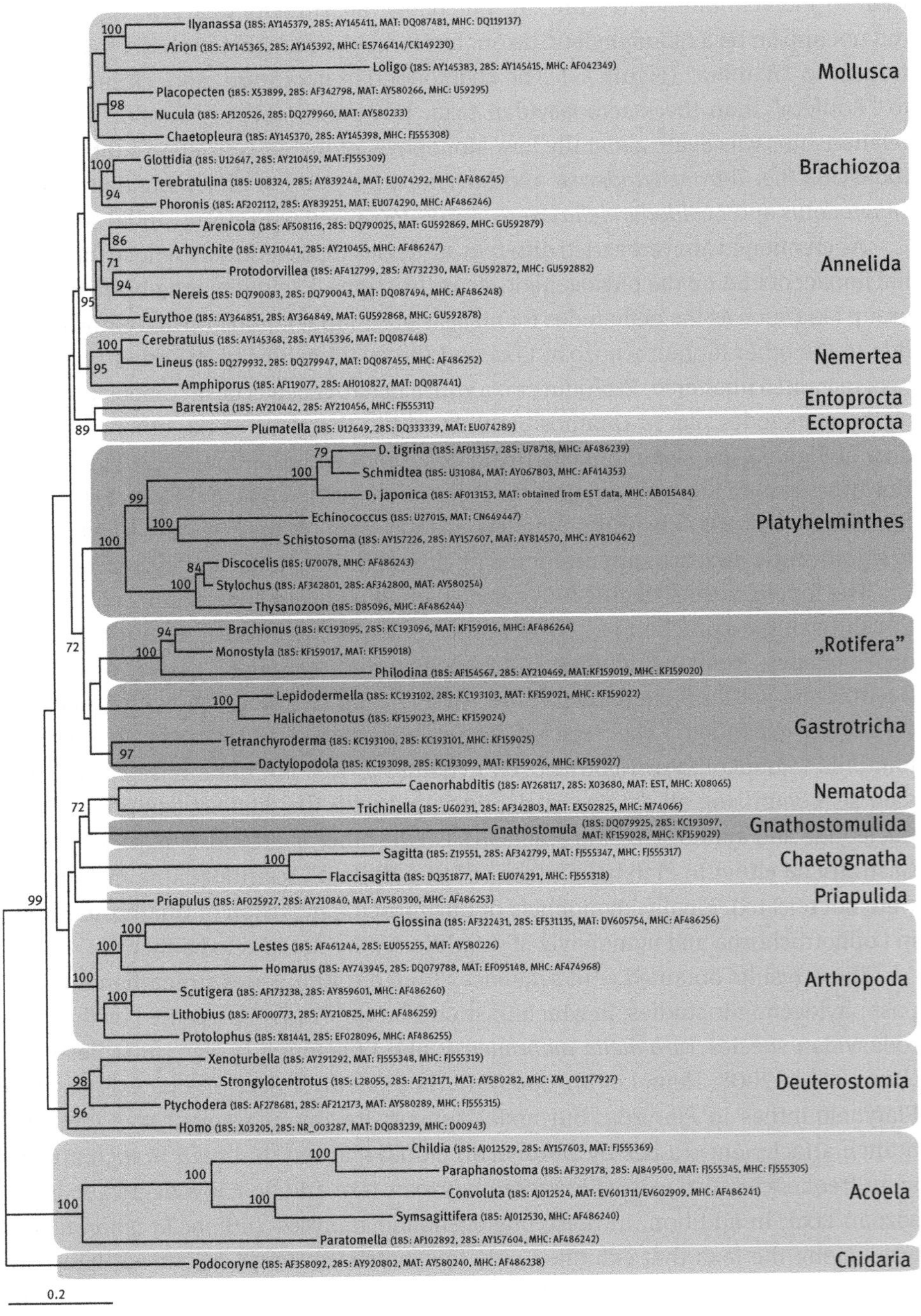

With respect to the evolution of morphological traits, the unrelatedness of gastrotrichs to ecdysozoans has profound impact. Especially, the Y-shaped cylindrical pharynx and collar-shaped circumpharyngeal brain must have evolved independently in Gastrotricha and ecdysozoans. The Y-shaped pharynx is a sucking pharynx and hence similar selection pressures for effective sucking might have resulted in its convergent evolution, which is further supported by the fact that Hirudinea and Pantopoda also

Table 7.1: Testing hypotheses about platyzoan relations that differ from those in the ML tree of Figure 7.3. Results of the approximately unbiased (AU) test (Shimodaira, 2002) for the data sets based on all taxa and all positions as either all nucleotides (AllNuc) or nucleotides for the rRNA and amino acids for protein-coding genes (NucAA). On the right, significant p values are in bold.

Hypothesis	Applied constraint	AllNuc	NucAA
"Rotifera"	("Rotifera", Gastrotricha)	0.755	0.907
Monophyly of Gastrotricha	(Gastrotricha)	0.488	0.356
Lophotrochozoa including Gnathostomulida	(Nemertea, Annelida, Mollusca, Brachiozoa, Ectoprocta, Entoprocta, Platyhelminthes, "Rotifera", Gnathostomulida, Gastrotricha)	0.463	0.284
Platyhelminthes	(Platyhelminthes, Gastrotricha)	0.067	0.228
Gnathifera	("Rotifera", Gnathostomulida)	0.118	0.078
Acanthognatha	("Rotifera", Gnathostomulida, Gastrotricha)	0.167	0.103
Monokonta	(Gnathostomulida, Gastrotricha)	0.087	**0.012**
Platyzoa expanded	(Entoprocta, Platyhelminthes, "Rotifera", Gnathostomulida, Gastrotricha)	**0.022**	**0.018**
Basal bilaterians	(Cnidaria, Acoela, Gnathostomulida, Gastrotricha)	**0.009**	**0.004**
Ecdysozoa	(Arthropoda, Nematoda, Priapulida, Gastrotricha)	**2e-06**	**2e-07**
Pseudocoleomates	(Nematoda, Priapulida, "Rotifera", Gnathostomulida, Gastrotricha)	**6e-06**	**1e-45**
Nemathelminthes	(Nematoda, Priapulida, "Rotifera", Gastrotricha)	**5e-06**	**0.004**
Cycloneuralia	(Nematoda, Priapulida, Gastrotricha)	**0.001**	**9e-61**

◂ **Figure 7.3:** Partitioned ML reconstructions of four genes with all taxa included and all genes coded as nucleotides (lnL = −111,893.999, 6,250 positions in total: 1,271 for 18S rRNA, 3,091 for 28S rRNA, 799 for methionine adenyltransferase (MAT), 472 for myosin heavy chain type II (MHC)). Accession numbers are provided after the genus name. EST = data obtained from expressed sequence tags libraries. Phylogenetic analyses conducted with RAxML 7.2.8 (Stamatakis, 2006) applying the GTR+Γ+I models for each gene and 10 replicates. Bootstrap values (Felsenstein, 1985) were computed from 500 replicates. Values above 70 % are given at branches. Higher taxonomic units are indicated.

show such a Y-shaped pharynx. Moreover, within Gastrotricha it is even not certain if the *inverted* Y-shaped pharynx of Macrodasyida is the plesiomorphic condition or if the upright Y-shaped pharynx of Chaetonotida is (Kieneke, Riemann, and Ahlrichs, 2008). Only the latter would support a relationship to ecdysozoan taxa such as Nematoda.

Recent studies showed that the ground pattern of the gastrotrich nervous system consists of a brain with a solid arch-like dorsal commissure with laterally positioned cell somata and a fine ventral commissure as well as a pair of longitudinal, lateroventral nerve cords, which join posteriorly (Rothe and Schmidt-Rhaesa, 2009). Though superficially resembling a ring, the gastrotrich brain is different from the cycloneuralian ring-shaped brain, which consists of a characteristic position of the neuropil between the anterior and posterior pericarya as well as the ventral origin of either paired ventral nerve cords or a single nerve cord (Schmidt-Rhaesa, 2007). The gastrotrich brain is most similar to the nervous system of Acoela (Rothe and Schmidt-Rhaesa, 2009), which could represent the basal-bilaterian line (see above). Thus, this feature of gastrotrichs could conceivably be the primitive state: "Compared to non-bilaterian animals with the net-like plexus and no cerebral ganglion, the acoel brain fits in the probable basal position of this taxon by expressing a certain grade of condensation of the nerve plexus in the anterior end to form a more or less condensed commissural brain" (Rothe and Schmidt-Rhaesa, 2009). However, neither morphological nor molecular data support such a basal-bilaterian position for Gastrotricha (herein, as well as Dunn et al., 2008; Hejnol et al., 2009; Peterson and Eernisse, 2001; Petrov et al., 2007; Schmidt-Rhaesa et al., 1998; Todaro et al., 2006; Zrzavý, 2003) and the few molecular studies based on 18S rRNA data which recovered such a position were clearly affected by LBA problems (e.g., Peterson and Eernisse, 2001). Hence, there are two possibilities regarding the evolution of the organization of the nervous system in Gastrotricha. Either Gastrotricha still exhibit the plesiomorphic condition as it is also found in Acoela (Rothe and Schmidt-Rhaesa, 2009) or the organization was secondarily derived or simplified.

In summary, Gastrotricha is not closely related to Ecdysozoa (Table 7.1) and we obtained some support for the relation to Rotifera and Platyhelminthes (Figure 7.3). However, their placement within Platyzoa or Lophotrochozoa is not entirely conclusive, in part because our findings misplaced gnathostomulids (see below) and are thus not fully trustworthy. Neither can certain affiliations be rejected by topology tests as shown by the present results for four genes as well as by other molecular studies (e.g., Dunn et al., 2008; Hejnol et al., 2009; Petrov et al., 2007; Todaro et al., 2006). Within the DMP priority project, we have generated deeply covered EST libraries of three macrodasyidan and one chaetonotidan gastrotrichs, increasing both the taxon and data coverage of Gastrotricha. These data are now being analyzed by sophisticated methods that ameliorate the misleading effect of biases such as LBA (see below).

7.4 The Gnathifera concept: Support from phylogenomic data

The taxon Gnathifera originally was assumed to include Gnathostomulida and Syndermata, the latter comprising "Rotifera" and Acanthocephala. The taxon name, which means "jaw-bearing", is derived from the pharyngeal jaws that all included taxa possess, except for the highly specialized endoparasitic Acanthocephala (Ahlrichs, 1995; Ahlrichs, 1997).

Gnathostomulids were originally regarded as a "turbellarian" and hence platyhelminth subtaxon (Ax, 1956), until being reclassified as a gnathiferan subtaxon (Riedl, 1969). The around 100 described species belong to the meiobenthos. Gnathostomulids are hermaphrodites with a direct development and a monociliary epithelium, which they share with Gastrotricha (Rieger, 1976).

Two other taxa have been assigned to Gnathifera, Micrognathozoa and Cycliophora (see Sørensen, Sterrer, and Giribet, 2006 and references therein). Micrognathozoa is an especially good candidate. The only micrognathozoan species described so far (Limnognathia maerski; Kristensen and Funch, 2000) has a complicated jaw apparatus showing ultrastructural similarities to the jaws of gnathostomulids, monogononts, bdelloids and seisonids (Kristensen, 2002; Sørensen, 2003). On the other hand, first analyses of molecular data (one mitochondrial and three nuclear loci) could neither confirm nor reject a phylogenetic position of Micrognathozoa within Gnathifera (Giribet et al., 2004). More comprehensive molecular data are thus needed to finally settle the phylogenetic position of Micrognathozoa. Turning to Cycliophora, more recent analyses of molecular data suggest they are closer to Entoprocta than to Gnathifera (Fuchs et al., 2010; Hejnol et al., 2009; Paps, Baguñà, and Riutort, 2009a).

The concept of a close relationship between Gnathostomulida and Syndermata in Gnathifera is based on the likely homology of structures such as supportive rods in gnathostomulid jaws and rotiferan trophi (e.g., Ahlrichs, 1995; Ahlrichs, 1997; Haszprunar, 1996b; Herlyn and Ehlers, 1997; Rieger and Tyler, 1995). However, first analyses of molecular data (single or few loci as well as broad scale EST data) did not support the validity of a monophylum Gnathifera (Dunn et al., 2008; Littlewood et al., 1998; Paps, Baguñà, and Riutort, 2009a; 2009b). Our own study within the DMP priority project yielded a more substantiated view: Witek et al. (2009) recovered Gnathifera (Syndermata + Gnathostomulida) based on > 11,000 amino acid positions of RP sequences derived from EST projects, with high support (100 ELW value using TreeFinder, > 89 bootstrap using TreeFinder and PhyML, 0.98 posterior probability using PhyloBayes). Gnathiferan monophyly was subsequently upheld by another study using RP sequences (Hausdorf et al., 2010). On the other hand, several studies questioned the suitability of RP data for phylogenetic reconstructions (e.g., Bleidorn et al., 2009a; Nesnidal et al., 2010), and an analysis of broad-scale EST data (a 1,487 gene matrix) recovered Gnathifera as polyphyletic, with Gnathostomulida and Gastrotricha closer to Platyhelminthes than to Syndermata (Hejnol et al., 2009). However, this observed polyphyletic origin of Gnathifera could be a species-specific effect, because the sister-groups Gnathostomulida and Gastrotricha were represented by just a single species each. To attain a convincing conclusion, a broader taxon and data sampling within Gnathifera, especially within Gnathostomulida is certainly desirable. To this end, a deeply covered EST library of another gnathostomulid species has been generated as part of this project and is now being analyzed together with new gastrotrich and syndermatan data obtained in this project.

7.4.1 Phylogenomics support monophyletic Syndermata and paraphyletic "Rotifera"

The taxon Syndermata (Ahlrichs, 1997) includes Monogononta, Bdelloidea and Seisonidea (traditionally summarized as "Rotifera" or "wheel animals") as well as Acanthocephala ("thorny-headed worms"). The name, which means "joined skin", refers to a characteristic syncytial epidermis (Ahlrichs, 1995) that probably evolved as a novelty in the stem lineage of Syndermata. Other authors conceive Acanthocephala as a subtaxon of "Rotifera" (e.g., Garey et al., 1996). However, due to the absence of a wheel organ in Acanthocephala we prefer the name Syndermata over the alternative "Rotifera".

The species grouped within Syndermata are very diverse in lifestyle. While most monogononts and bdelloids are free living, seisonids live epizoically on leptostracan crustaceans of the genus *Nebalia*. Finally, acanthocephalans are obligate endoparasites infesting a wide range of arthropod intermediate- and vertebrate definite-hosts (Ax, 2003). These different lifestyles go along with differing morphological features. The wheel organ, for instance, is well developed in free-living monogononts and bdelloids, where it enables swimming and feeding. A homologous structure might furthermore be present in the epizoically living seisonids (Nielsen, 2012; see also Leasi, Rouse, and Sørensen, 2011), but is absent from the endoparasitic Acanthocephala. Acanthocephalans are characterized by a retractable and hooked proboscis with specific musculature, which anchors the animal to the host's intestine wall (e.g. Herlyn and Ehlers, 2001; and references therein). Most strikingly, acanthocephalans lack a digestive tract and therefore also the pharyngeal hard parts representing the eponymous evolutionary novelty of Gnathifera (see, e.g., Schmidt-Rhaesa, 2007).

The phylogenetic relationships between acanthocephalans and the other syndermatan subgroups are still a matter of debate (Figure 7.4; see Witek et al., 2008 and references therein), partly due to the disputed homology of diverse morphological characters (see Clement, 1993; Markevich, 1993; Melone et al., 1998; Ricci, 1998; Sørensen, 2002). Within the DMP priority project, major progress has been made on this question (Weber et al., 2013; Wey-Fabrizius et al., submitted; Witek et al., 2008; 2009). Thus, Witek et al. (2008) could for the first time provide significant support for the paraphyly of Eurotatoria on the basis of ESTs encoding 79 RPs. That is, the results consistently supported Figure 7.4A; a closer relationship of Bdelloidea and Acanthocephala to each other than each of the groups had to Monogononta, with moderate support values (76 ELW support using TreeFinder, 78 % bootstrap support using PhyML, 0.83 posterior probability using PhyloBayes; Figure 7.5A). This topology conflicts with three of the debated hypotheses which favor a sister-group relationship of Monogononta and Bdelloidea (Figure 7.4B–D). Other authors meanwhile upheld the paraphyly of Eurotatoria on the basis of complete mt genome data from one monogonont, one bdelloid and up to two acanthocephalan species (Figure 7.5B1, B2; Gazi et al., 2012; Min and Park, 2009), but with regards to the scarce taxon sampling in all these studies, paraphyly of Eurotatoria was still not conclusively settled. Major improvements in terms of taxon sampling within Syndermata allowed for a more reliable analysis of mt genome data, again revealing paraphyletic Eurotatoria (Figure 7.5B3; Weber et al., 2013).

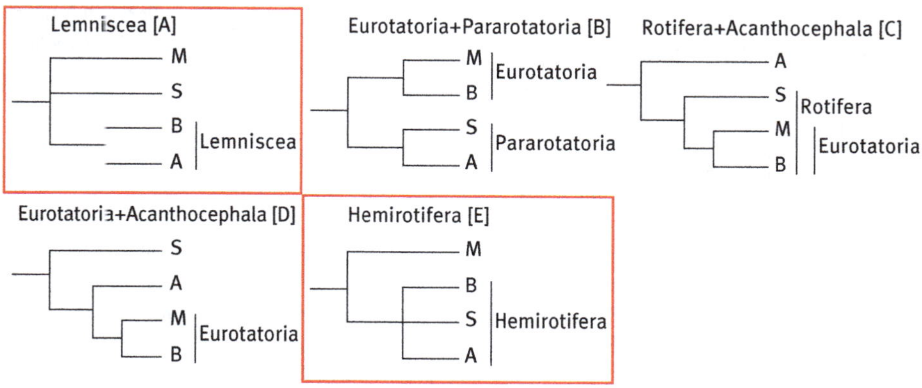

Figure 7.4: Competing hypotheses on the internal phylogeny of Syndermata, as discussed in the literature. Recent molecular analyses favor topologies showing paraphyletic Eurotatoria (see red boxes). A = Acanthocephala, B = Bdelloidea, M = Monogononta, S = Seisonidea.
[A] Lemniscea hypothesis. The hypothesis is primarily based on the hypothesized homology of paired epidermal intrusions in the neck region of bdelloids and acanthocephalans (called lemnisci in the latter; Lorenzen, 1985: 210-223), and supported by molecular studies using 16S rRNA, 18S rRNA, 28S rRNA, histone *H3* and *cox1* sequences (García-Varela and Nadler, 2006; Garey et al., 1996; 1998; Giribet et al., 2000; 2004; Mallatt, Craig, and Yoder, 2012).
[B] Eurotatoria + Pararotatoria hypothesis. The proposed synapomorphy of Eurotatoria is the highly developed wheel organ, while Pararotatoria share spermatozoa with "dense bodies" (Ahlrichs, 1995; Ahlrichs, 1997; Ahlrichs, 1998).
[C] "Rotifera" + Acanthocephala hypothesis. This traditional hypothesis is derived from morphological data including nervous system, toes, sensory and masticatory organs (Melone et al., 1998). Support comes from analyses of 18S rRNA data (García-Varela et al., 2000).
[D] Eurotatoria + Acanthocephala hypothesis. The hypothesis is derived from *hsp82* sequences (Mark Welch, 2000). The lack of acrosomal structures has been proposed as a potential synapomorphy of Eurotatoria and Acanthocephala (Sørensen et al., 2000).
[E] Hemirotifera hypothesis. This hypothesis was established by analyzing a combination of four molecular loci (18S rRNA, 28S rRNA, histone *H3* and *cox1* sequences) and 74 morphological characters (Sørensen and Giribet, 2006).

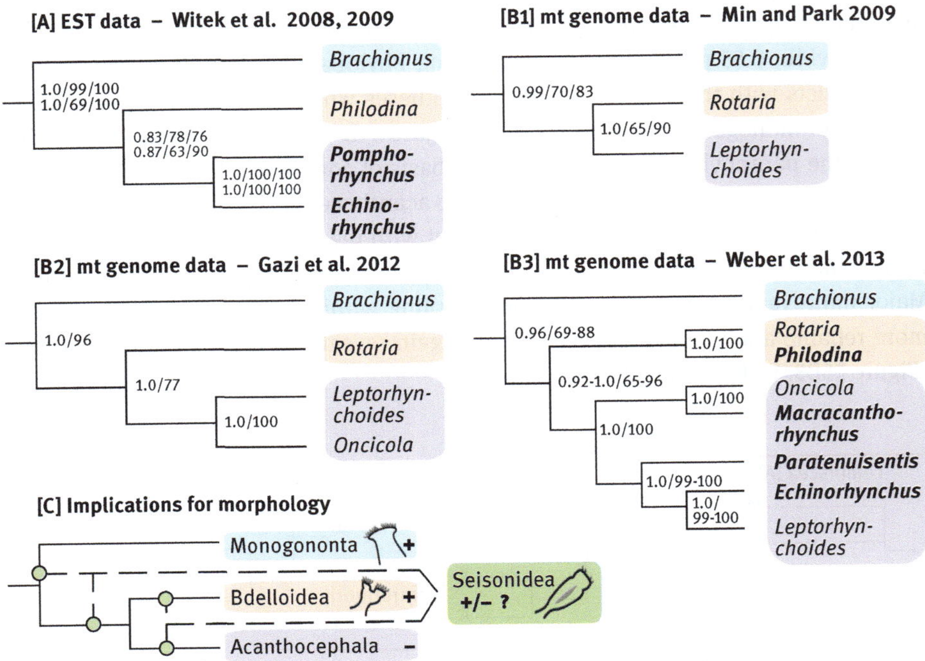

Figure 7.5: Paraphyletic Eurotatoria as derived from broad-scale molecular data sets. Box colors indicate taxa belonging to Monogononta (blue), Bdelloidea (orange) or Acanthocephala (purple). Paraphyletic Eurotatoria with Bdelloidea+Acanthocephala was obtained based on analyses of EST data [A] and mt genome data [B1–3]. Topologies and support values are redrawn from original studies. Support values represent Bayesian posterior probabilities (PP), bootstrap support (BS) and, if applicable, expected likelihood weights (ELW) in the given order (support values part [A]: upper row = Witek et al., 2008; lower row = Witek et al., 2009; BS support part [D]: only minimum and maximum BS values of the analyses are denoted as described in the original work). Data of the taxa in bold were introduced by DMP funded studies. Implications for the evolution of the wheel organ within Syndermata are illustrated in [C]. Pictograms illustrate the degree of wheel organ development: fully developed in Bdelloidea and Monogononta (+), reduced or absent in Seisonidea (+/– ?), absent in Acanthocephala (–). The position of Seisonidea within the tree has not yet been determined based on broad-scale molecular data. Green dots illustrate probable branching points of Seisonidea as derived from morphological and small-scale molecular analyses (see Witek et al., 2008 and references therein). Molecular evidence for monophyletic Pararotatoria (Seisonidea + Acanthocephala) is provided by analyses of 18S rRNA (Herlyn et al., 2003), 18S rRNA + *hsp82* data (Zrzavý, 2001a) as well as by EST data (Wey-Fabrizius et al., submitted).

Paraphyly of Eurotatoria raises the question of how the wheel organ evolved within Syndermata (Figure 7.5C). The most likely alternatives are (i) the wheel organ emerged in the syndermatan stem lineage followed by a partial or complete reduction of the trait within Seisonidea and Acanthocephala, or (ii) the convergent evolution of a wheel organ in the stem lineages of monogononts, bdelloids and possibly also seisonids. The situation was additionally complicated by the unsettled position of

Seisonidea within Syndermata (Figure 7.5C), though ultrastructural (Ahlrichs, 1997) as well as first molecular data (Herlyn et al., 2003) suggest a sister-group relation of Seisonidea and Acanthocephala. A recent phylogenomic study within the DMP priority project for the first time included broad data for Seisonidea and corroborates this proposed sister-group relation of Seisonidea and Acanthocephala (=Pararotatoria) with maximum support values (Wey-Fabrizius et al., submitted). Such a topology implies that the acanthocephalan endoparasitism evolved via an epizootic stage and that early acanthocephalans parasitized arthropods before a host change to arthropod intermediate- and vertebrate definite-hosts evolved (Herlyn et al., 2003; see also Near, 2002). Paraphyly of Eurotatoria is maintained in that study, with Bdelloids being more closely related to Pararotatoria. The wheel organ most likely evolved once on the syndermatan stem lineage and was reduced during the establishment of biological interactions with hosts in the pararotatorian ancestor resulting in a complete loss during the evolution of endoparasitism in Acanthocephala.

7.4.2 Phylogeny of Acanthocephala – from mitochondrial genes to morphology

As mentioned above, acanthocephalans are obligate endoparasites with highly derived morphological characters. The more than 1,100 described species infest a wide range of aquatic and terrestrial arthropod (intermediate hosts) and vertebrate species (definite and paratenic hosts), including human food sources such as commercial poultry, fish and pigs.

Four higher-ranked taxa are currently distinguished within Acanthocephala, namely Archiacanthocephala, Eoacanthocephala, Palaeacanthocephala and Polyacanthocephala (Amin, 1987; Garcia-Varela et al., 2002). Analyses of 18S rRNA, 28S rRNA and COI sequences suggest monophyletic Archiacanthocephala as the sister-group to the remaining Acanthocephala (García-Varela and Nadler, 2006; García-Varela et al., 2000; Near, 2002; Near, Garey, and Nadler, 1998). Polyacanthocephala appeared as sister to monophyletic Eoacanthocephala, together building the sister-group to monophyletic Palaeacanthocephala (Garcia-Varela et al., 2002; García-Varela and Nadler, 2006). In contrast to these findings, other 18S rRNA-based findings questioned the monophyly of Palaeacanthocephala, in particular of the palaeacanthocephalan subtaxon Echinorhynchida (Garcia-Varela et al., 2002; Herlyn et al., 2003; Near, 2002; Verweyen, Klimpel, and Palm, 2011), and an analysis of 138 morphological characters suggested archiacanthocephalans was a paraphyletic assembly at the base of the acanthocephalan clade (Monks, 2001). As far as the above-mentioned molecular studies are concerned, it is very likely that scarce taxon representation and scarce gene coverage hampered the resolution of the phylogenetic relationships within Acanthocephala (e.g., Fontaneto and Jondelius, 2011; Gazi et al., 2012; Herlyn et al., 2003; Verweyen, Klimpel, and Palm, 2011).

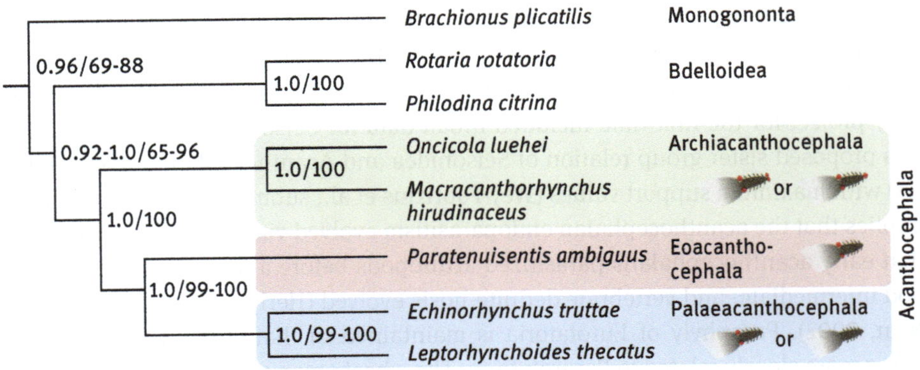

Figure 7.6: Phylogeny within Acanthocephala as inferred from complete mt genome data (Weber et al., 2013). Topology and support values are redrawn from the original work (support values represent Bayesian posterior probabilities (PP) and bootstrap support (BS); only minimum and maximum BS values of the analyses are denoted as described in the original work). Pictograms illustrate body and hooked proboscis of an acanthocephalan. The distribution of apical and lateral sensory organs is indicated by red dots on the proboscis (dots on apical proboscis = apical sensory organs, dots on basal proboscis: lateral sensory organs).

Our analyses within the DMP framework helped to answer some of the questions regarding relationships within Acanthocephala. Based on complete mt genomes of 26 species including those of four newly sequenced syndermatans (Weber et al., 2013), we recovered the expected paraphyly of Eurotatoria and moreover obtained a highly robust topology within Acanthocephala. According to these data, monophyletic Archiacanthocephala represents the sister taxon to a clade uniting at least monophyletic Eo- and Palaeacanthocephala (Figure 7.6). Polyacanthocephala, though not sampled in the present study, most probably are the sister-group to Eoacanthocephala (see Garcia-Varela et al., 2002).

Taking sensory organs as an example, we demonstrated that the character distribution within Acanthocephala does not necessarily follow a parsimonious pattern, but rather involved gain and loss events (Weber et al., 2013). This is shown on the right side of Figure 7.6. In particular, the obtained monophyly of Palaeacanthocephala implies a loss of lateral sensory organs (which are absent in some Palaeacanthocephala) shortly after their emergence in the acanthocephalan stem lineage. Moreover, the observed topology suggests that apical sensory organs evolved once within Acanthocephala, on the stem lineage of Archiacanthocephala.

Questions remain about relations within Syndermata. Besides the phylogenetic positions of Seisonidea and Polyacanthocephala, future studies should focus on the monophyly of Archi-, Eo- and Palaeacanthocephala. All these aspects require a denser taxon sampling.

7.5 Hypothetical Platyzoa and the long-branch problem

Increased substitution rates, displayed by long branches in phylogenetic reconstructions, may result in artificial taxon groupings through "long-branch attraction" (Felsenstein, 1978). The theoretical and methodical aspects of the so-called "long-branch attraction" are discussed in more detail by Kück and Wägele (this book). In molecular phylogenetic reconstructions, many platyzoan taxa exhibit long branches, as do a few other lophotrochozoans (e.g., Myzostomida) and some Ecdysozoa (e.g., Nematoda). This poses a severe problem in phylogenetic reconstructions including these taxa. An example is seen in the four-gene analyses presented here (Figure 7.3): Considering that so much morphological and molecular data support an inclusion of Gnathostomulida within Gnathifera (e.g., Ahlrichs, 1997; Witek et al., 2009), the deviating association of Gnathostomulida with nematodes in the tree most likely represents such a long-branch artifact. In line with this, topology testing (Table 7.1) does not reject the inclusion of Gnathostomulida in Gnathifera.

Besides Gnathostomulida, endoparasitic Acanthocephala and Platyhelminthes also exhibit comparatively long branches in most molecular phylogenetic analyses. First analyses of the newly sequenced gastrotrich EST data (see above) revealed long-branched species within Gastrotricha, as well. Hence, it might even be that monophyly of Platyzoa as well as their phylogenetic position within Lophotrochozoa (e.g., Dunn et al., 2008; Hejnol et al., 2009; Struck and Fisse, 2008) is just an artifact resulting from LBA (Edgecombe et al., 2011). On the other hand, it cannot be excluded that increased substitution rates are an apomorphic feature of a true clade Platyzoa.

To overcome the LBA problem and to explore whether Platyzoa is an artificial group or a monophyletic taxon, some important points have to be considered. First, the selection of the data set seriously influences the results of phylogenetic reconstructions. The mt genome data, for example, provide congruent and highly supported trees within lower-order clades, such as the internal phylogeny of Syndermata and Acanthocephala (Figures 7.5, 7.6, and Weber et al., 2013), Cestoda (Waeschenbach, Webster, and Littlewood, 2012), or birds (Pacheco et al., 2011), but exhibit severe LBA artifacts when reconstructing high-order metazoan-wide trees (Bernt et al., 2013; Gazi et al., 2012; Perseke et al., 2013). Moreover, certain categories of nuclear genes such as RP data can introduce bias in phylogenetic reconstructions (e.g., Bleidorn et al., 2009a). Therefore, to address the Platyzoa question, extracting and combining a broad range of different genes obtained by transcriptome sequencing would be the preferred approach.

Second, different methodological strategies to address the LBA problem have to be applied. Several methods for detecting and attenuating LBA artifacts have been proposed, and they also alleviate other misleading signals in the data such as compositional biases, heterogeneous rates of certain substitutions, or mutational saturation (Foster, 2004; e.g., Foster and Hickey, 1999; Kubatko and Degnan, 2007; Whelan, 2008). Such methods include separate partition analyses, long-branch extraction, outgroup removal, adding artificial LB sequences, parametric simulation, spectral analyses, different compositional vectors along the tree, and a temporal hidden Markov model (for reviews see Bergsten, 2005; Foster, 2004; Whelan, 2008). Altogether, these approaches should be able to reveal the phylogenetic signal within phylogenomic data sets, although some of the methods are still computationally intensive. Despite their promise, depending on the specific data set, these methods can sometimes fail. For example, in our four-gene analyses (Figure 7.3), we employed time-heterogeneous substitution models or excluded certain taxa in order to reduce the potential impact of LBA on the phylogenetic reconstructions. However, none of these approaches effectively increased the support for a specific topology.

Finally, improved taxon sampling and data coverage will be crucial in platyzoan phylogeny. The importance of thorough taxon coverage is already well documented. For example, the EST study of Struck and Fisse (2008) found that including the long-branched ecdysozoan taxa Tardigrada and Nematoda makes Platyzoa become sister to these taxa, while excluding them results in a more reasonable sister-group relationship of Platyzoa with Lophotrochozoa. In the EST-based study of Dunn et al. (2008) Platyzoa strangely contained Acoelomorpha and Myzostomida, but increasing the taxon sampling and data coverage of acoelomorphan taxa lead to Acoelomorpha branching off first in the bilaterian tree and positioned Myzostomida as part of the annelid radiation (Hejnol et al., 2009). The examples illustrate the findings of Philippe et al. (2004), that a more complete representation of the biodiversity of a taxon promises to generate a more reliable placement of that taxon by breaking up long branches (see also Bergsten, 2005 for review).

In Platyzoa, some taxa such as cestodes (Platyhelminthes) are currently well represented in larger phylogenomic studies with respect to the numbers of genes, but other platyzoans such as gastrotrichs and gnathostomulids are still heavily underrepresented (Hejnol et al., 2009). Furthermore, several candidate platyzoan groups are so far only represented by a single species such as *Turbanella ambronensis* for Gastrotricha or *Gnathostomula peregrina* for Gnathostomulida. Traditionally, the generation of EST libraries for Gastrotricha, "Rotifera" or Gnathostomulida has been cumbersome due to their small body size (see Figure 7.2) and their usually patchy habitat distribution. However, novel sequencing technologies, ("next generation sequencing", NGS; see, e.g., Metzker, 2010) promise to make substantial progress in sequence data acquisition, even when only limited amounts of DNA or RNA are available (Ramsköld et al., 2012). Within the DMP priority project, we were able to generate additional NGS libraries of two turbellarian platyhelminths, four gastrotrichs, one gnathostomulid, and eight syndermatans, with some of the data yet unpublished and currently being analyzed. This broader coverage of taxa and sequences already enabled us to address phylogenetic questions within platyzoan subgroups and, more importantly, will facilitate examination of the yet unanswered question of the monophyly of Platyzoa and the phylogenetic position of such a monophylum.

Acknowledgments

The authors gratefully thank the German Research Foundation (DFG) for generous funding of the priority project 1174 "Deep Metazoan Phylogeny" (Ha-2103/4-1, Ha-2103/4-3, STR-683/3-1, Pu-84/5-1, STR-683/5-2). T.H. acknowledges additional support by the Johannes Gutenberg University Center for Computational Sciences (SRFN).

Maximilian P. Nesnidal, Martin Helmkampf, Iris Bruchhaus,
Ingo Ebersberger, and Bernhard Hausdorf

8 Lophophorata monophyletic – after all

Abstract: The lophophorates are mainly marine, sessile invertebrates that are characterized by a lophophore, a horseshoe-shaped tentacular filter apparatus surrounding the mouth opening. Lophophorata comprises three divergent lineages, Phoronida, Brachiopoda and Ectoprocta (Bryozoa). Their phylogenetic relationships to other metazoan phyla as well as to each other have long remained unclear. Based on ontological and morphological evidence, Lophophorata was long considered to be the sister or paraphyletic stem-group of Deuterostomia. In contrast, phylogenomic analyses indicated that the lophophorates are more closely related to trochozoans than to deuterostomes in accordance with previous molecular analyses. The phylogenomic studies supported monophyletic Brachiozoa including Brachiopoda and Phoronida as well as Polyzoa including Ectoprocta, Entoprocta and Cycliophora, thus rendering Lophophorata polyphyletic. In contrast to previous molecular phylogenetic studies, the phylogenomic analyses presented here indicate that Lophophorata are monophyletic and that phoronids are the sister-group of ectoprocts. The conflicting results might be due to compositional bias resulting in systematic errors in former phylogenetic studies. The monophyly of Lophophorata would imply that the lophophore and the epistome are synapomorphies of the lophophorate lineages. Radial cleavage may be another apomorphy of Lophophorata. However, among phoronids spiral cleavage is known as well, indicating that the cleavage pattern has changed several times within lophophorates.

8.1 Introduction

Lophophorates are mainly marine, sessile invertebrates that are characterized by a three-part body consisting of a protosome forming the epistome, a circumoral mesosome with the lophophore and a large metasome, the trunk. The lophophore is a horseshoe-shaped tentacular filter apparatus surrounding the mouth opening, while the epistome is a muscular lobe that is used to push the infiltrated particles into the mouth opening. Lophophorata in the sense of Hyman (1959) (= Tentaculata Hatschek 1891) comprises three divergent lineages, Phoronida, Brachiopoda and Ectoprocta (Bryozoa). Phoronida or horseshoe worms are marine wormlike animals living in upright tubes made of their own secretion including chitin (Figure 8.1). Diverse trace fossils have been interpreted as tubes of phoronids, but reliable records are known from the Devonian on. There are only about 10 recent species. Brachiopoda or lamp shells are also exclusively marine animals that resemble bivalves in possessing two shells (Figure 8.2). However, in contrast to the lateral valves of mussels, there is a dorsal and a ventral valve in brachiopods. Usually, a stalk-like pedicle projects through an opening in the ventral valve and attaches the animal to the substrate. From the Cambrian on brachiopods dominated seafloors till the Permian extinction. Today there are only about 380 living species known. With more than 4,500 extant species, Ectoprocta (Bryozoa) or moss animals are much more diverse than the two other lophophorate lineages in the recent fauna, and the only lophophorate group with representatives in freshwater, although the highest diversity is found in tropical seas. Ectoprocts form colonies (Figure 8.3) consisting of a few to thousands of individual zooids that are generated by asexual budding from a founding individual. The zooids remain interconnected with each other and can become highly specialized for reproductive, defensive or other functions. They form exoskeletons that may be strengthened by chitinous and in some groups calcareous components. Such exoskeletons are known from the Cambrian on.

Figure 8.1–3: Representatives of the three lineages of Lophophorata. **1.** Phoronida: *Phoronis muelleri* Selys-Longchamps, 1903 from Helgoland, Germany (photo courtesy of Peter Grobe). **2.** Brachiopoda: *Terebratulina retusa* (Linnaeus, 1758) from Skarnsundet, Trondheim Fjord, Norway (photo courtesy of Kåre Telnes, www.seawater.no). **3.** Ectoprocta: *Flustra foliacea* (Linnaeus, 1758) from Middelfart, Denmark (photo courtesy of Dirk Schories).

Based on ontological and morphological data, the lophophorate lineages were initially considered the sister or paraphyletic stem-group of Deuterostomia (Ax, 1989; 2001; Brusca and Brusca, 2003; Lüter and Bartolomaeus, 1997; Sørensen et al., 2000). However, molecular phylogenetic analyses altered our views about the relationships of the lophophorate lineages dramatically. Based on analyses of 18S rDNA sequences, Halanych et al. (1995) were the first to recognize that the lophophorate lineages are more closely related to trochozoan phyla, namely Annelida and Mollusca, than to deuterostomes. As a consequence, they united Lophophorata and Trochozoa to form Lophotrochozoa. Since then, the monophyly of Lophotrochozoa has been confirmed by further analyses of rDNA sequences (Mackey et al., 1996; Mallatt and Winchell, 2002; Mallat, Craig, and Yoder, 2010; Paps, Baguñà, and Riutort, 2009a; Passamaneck and Halanych, 2006; Peterson and Eernisse, 2001), single nuclear protein-encoding genes (e.g., Anderson, Cordoba, and Thollesson, 2004; Ruiz-Trillo et al., 2002), mitochondrial protein sequences (Helfenbein and Boore, 2004; Jang and Hwang, 2009; Nesnidal et al., 2011; Stechmann and Schlegel, 1999; Sun et al., 2009; 2011; Waeschenbach et al., 2006), and the presence of specific *Hox* genes (de Rosa et al., 1999; Passamaneck and Halanych, 2004). Finally, the monophyly of Lophotrochozoa has been further corroborated by analyses of data sets including several nuclear protein-encoding sequences (Helmkampf, Bruchhaus, and Hausdorf, 2008a) as well as by phylogenomic approaches (Bleidorn et al., 2009a; Dunn et al., 2008; Hausdorf et al., 2007; 2010; Hejnol et al., 2009; Helmkampf, Bruchhaus, and Hausdorf, 2008b; Nesnidal et al., 2010; Struck and Fisse, 2008; Witek et al., 2008; 2009).

While the monophyly of the Lophotrochozoa has meanwhile been widely accepted, the phylogenetic relationships of the lophophorate lineages within Lophotrochozoa are still not robustly resolved. Halanych et al. (1995) suggested that lophophorates are polyphyletic, because ectoprocts formed the sister-group of all other lophotrochozoans in their tree. Moreover, they proposed that phoronids are the sister-clade of articulate brachiopods, making brachiopods paraphyletic. However, it turned out that the clustering of phoronids and articulate brachiopods was an artifact probably caused by a chimeric sequence (Cohen, Gawthrop, and Cavalier-Smith, 1998). The monophyly of Brachiozoa (=Phoronozoa) including brachiopods and phoronids was later independently corroborated by analyses based on rDNA (Cohen, 2000; Cohen, Gawthrop, and Cavalier-Smith, 1998; Cohen and Weydmann, 2005; Mackey et al., 1996; Mallat, Craig, and Yoder, 2010; Paps, Baguñà, and Riutort, 2009a; Zrzavý et al., 1998), sodium-potassium ATPase α-subunit (Anderson, Cordoba, and Thollesson, 2004), and multiple nuclear protein-encoding sequences (Paps, Baguñà, and Riutort, 2009b; Sperling, Pisani, and Peterson, 2011) as well as total evidence analyses including morphological characters (Giribet et al., 2000; Peterson and Eernisse, 2001; Zrzavý et al., 1998). Finally, the monophyly of Brachiozoa has also been confirmed by phylogenomic approaches (Helmkampf, Bruchhaus, and Hausdorf, 2008b; Nesnidal et al., 2010) in the framework of the Deep Metazoan Phylogeny network. The relationships within Brachiozoa are, however, in dispute.

Whereas some rDNA analyses indicate that brachiopods are paraphyletic and phoronids are the sister-group of inarticulate brachiopods (Cohen, 2000; 2013; Cohen, Gawthrop, and Cavalier-Smith, 1998; Cohen and Weydmann, 2005), brachiopods emerged as monophyletic in analyses of morphological data (Ax, 2001; Carlson, 1995; Giribet et al., 2000; Nielsen, 2001; Peterson and Eernisse, 2001; Sørensen et al., 2000; Williams et al., 1996; Zrzavý et al., 1998), multiple nuclear protein-encoding sequences (Sperling, Pisani, and Peterson, 2011), and phylogenomic data sets (Helmkampf, Bruchhaus, and Hausdorf, 2008b; Nesnidal et al., 2010).

Analyses of EST data gathered in the Deep Metazoan Phylogeny network had one further interesting implication, namely that ectoprocts are the sister-group of entoprocts (and cycliophorans). As a consequence, the old Polyzoa (=Bryozoa sensu lato) hypothesis was revived (Bleidorn et al., 2009a; Hausdorf et al., 2007; 2010; Hejnol et al., 2009; Helmkampf, Bruchhaus, and Hausdorf, 2008b; Struck and Fisse, 2008; Witek et al., 2008; 2009). This hypothesis has previously been supported only by few morphologists (Nielsen, 1971; 1985; 2001), but has recently also been corroborated by analyses of rDNA sequences (Mallat, Craig, and Yoder, 2010; Paps, Baguñà, and Riutort, 2009a), albeit with weak support.

While phylogenomic studies are likely to reduce the influence of random errors and gene-specific influences on phylogenetic inference, they cannot cope with the fact that model violations such as compositional biases in the data can confound accurate tree reconstruction (Delsuc, Brinkmann, and Philippe, 2005; Jermiin et al., 2004; Rodríguez-Ezpeleta et al., 2007). We recently demonstrated that a biased amino acid composition in a ribosomal protein data set indeed affects phylogenetic analyses of the metazoan phyla (Nesnidal et al., 2010). Several phylogenomic analyses of the relationships of lophotrochozoan lineages relied exclusively on ribosomal proteins (Bleidorn et al., 2009a; Hausdorf et al., 2007; 2010; Helmkampf, Bruchhaus, and Hausdorf, 2008b; Struck and Fisse, 2008; Witek et al., 2008; 2009). We therefore compiled a new data set using all protein classes to reduce the influence of bias in ribosomal proteins (Bleidorn et al., 2009a) on the analyses of the phylogenetic relationships of lophophorates.

8.2 Materials and Methods

8.2.1 Data sources and orthology assignment

In the framework of the Deep Metazoan Phylogeny network, our group has generated EST data from the following lophophorates: *Alcyonidium diaphanum* (Ectoprocta), *Flustra foliacea* (Ectoprocta), *Tubulipora* sp. (Ectoprocta), *Novocrania anomala* (Brachiopoda) and *Phoronis muelleri* (Phoronida) (Hausdorf et al., 2007; 2010; Helmkampf, Bruchhaus, and Hausdorf, 2008b). The EST data used in our analyses have been deposited in the NCBI EST database (http://www.ncbi.nlm.nih.gov/projects/dbEST)

under accession numbers (*Alcyonidium diaphanum*: LIBEST_025703; *Flustra foliacea*: LIBEST_028288; *Tubulipora* sp.: LIBEST_025704; *Novocrania anomala*: LIBEST_028289; *Phoronis muelleri*: LIBEST_028290). Additional data were extracted from publicly available EST and genome projects.

The data set for tree reconstruction was compiled in a two-step procedure. For the initial ortholog search, we first defined a set of seven species with completely sequenced genomes, the so-called primer taxa: *Caenorhabditis elegans* (Nematoda), *Daphnia pulex* (Crustacea), *Apis mellifera* (Insecta), *Schistosoma mansoni* (Platyhelminthes), *Capitella capitata* (Annelida), *Helobdella robusta* (Annelida) and *Lottia gigantea* (Mollusca). We then used InParanoid-TC (Ebersberger, Strauss, and von Haeseler, 2009) to identify genes for which an ortholog was present in each of the seven primer taxa. Finally, we extended the resulting 1,297 ortholog groups with sequences from 51 additional taxa using HaMStR (Ebersberger, Strauss, and von Haeseler, 2009). Ortholog groups with less than 30 taxa were discarded.

8.2.2 Alignment, alignment masking and protein selection

The amino acid sequences of the proteins were aligned with MAFFT using the most accurate option L-INS-i (Katoh and Toh, 2008a). To increase the signal-to-noise ratio, sections with random sequence similarity were identified with ALISCORE version 1.0 (Kück et al. 2010; Misof and Misof, 2009) and subsequently excluded with ALICUT (http://www.utilities.zfmk.de). We constructed individual trees for each protein using RAxML version 7.3.5 (Ott et al., 2007; Stamatakis, 2006) with the LG+G+F model to check for unusual topologies and long branches that might indicate hidden paralogy and contaminations. One gene tree shows a very long, highly supported branch separating a clade including Deuterostomia and Ecdysozoa, into which the three nemertean representatives were nested. This topology is inconsistent with the position of Nemertea within Spiralia inferred in other analyses and our own analyses, if this protein is excluded. This topology indicates probably a paralogy. Thus, we excluded this protein. We also inspected each protein alignment manually for contaminant sequences and poorly conserved motives. Problematic sequences were excluded from the individual unmasked alignments and all single protein data sets were re-aligned with MAFFT and masked using ALISCORE. All masked alignments that were at least 100 amino acids long and contained at least 30 taxa after the various preprocessing and filtering steps were subsequently concatenated. The resulting superalignment has been deposited at TreeBASE (http://www.treebase.org, accession number S14244).

8.2.3 Phylogenetic analyses

We performed Maximum Likelihood analyses using a parallel Pthreads-based version of RAxML version 7.3.5 (Ott et al., 2007; Stamatakis, 2006) with the LG+G+F model (Le and Gascuel, 2008). Based on the alignments, we computed 10 maximum likelihood trees using 10 distinct randomized maximum parsimony starting trees and chose the tree with the highest likelihood. Rapid bootstrapping (Stamatakis, Hoover, and Rougemont, 2008) was used to assess the statistical branch support in the reconstructed phylogeny. The number of necessary replicates was inferred using the extended majority-rule consensus tree criterion (Pattengale et al. 2009; 250 replications inferred using the option autoMRE in RAxML).

We performed Bayesian inference analyses with the CAT model that adjusts for site-specific amino acid frequencies (Lartillot et al., 2013) as implemented in PhyloBayes MPI version 1.3b (http://megasun.bch.umontreal.ca/People/lartillot/www/index.htm). Four independent chains were run for 20,000 points, of which 5,000 were discarded as burn-in. The largest discrepancy observed across all bipartitions (maxdiff) was 0.29. Taking every 10th sampled tree, a 50 %-majority rule consensus tree was computed using all chains.

To test predefined phylogenetic hypotheses, we calculated the maximum-likelihood tree for a specified hypothesis by resolving multifurcations in a constrained tree with RAxML. Then we computed per-site log likelihood scores for the global and the constrained maximum-likelihood trees with RAxML and performed an approximately unbiased test (Shimodaira, 2002) using CONSEL (Shimodaira and Hasegawa, 2001) to investigate whether the data can be equally well explained with the alternative hypothesis compared to the maximum-likelihood tree.

8.2.4 Influence of compositional heterogeneity among lineages on the phylogenetic analyses

Effects of compositional heterogeneity between lineages on the phylogenetic analyses can be mitigated by the exclusion of data partitions that show compositional heterogeneity between lineages (Collins, Fedrigo, and Naylor, 2005; Nesnidal et al., 2010). To facilitate the identification of such partitions, we scored the compositional heterogeneity of all single protein alignments using relative composition variability, the average variability in composition between taxa (Phillips and Penny, 2003), but using frequencies instead of absolute numbers (Zhong et al., 2011). We adapted the formula of relative composition frequency variability (*RCFV*) to amino acids instead of nucleotides:

$$RCFV = \sum_{i=1}^{20} \sum_{j=1}^{n} \left| A_{ij} - A_i \right| / n$$

where A_{ij} is the frequency of amino acid i in taxon j, and A_i is the average frequency of amino acid i across the n taxa. Constant sites were excluded for the *RCFV* calculations. To evaluate the influence of compositional heterogeneity among lineages on the phylogenetic analyses, we constructed data sets by excluding the 10 % and the 33 % most heterogeneous protein sequences.

8.3 Results and Discussion

8.3.1 Deuterostome versus protostome relationships of the lophophorate lineages

We compiled a data set including 29,896 amino acid positions from 155 proteins of which orthologs were found in at least 30 the 58 included taxa (percentage of missing data 37 %). Both a Maximum Likelihood analysis with the LG model (Figure 8.4) and a Bayesian inference analysis with the CAT model (Figure 8.5) based on this data set revealed a subdivision of Bilateria into deuterostomes (bootstrap support 66 %; posterior probability 0.78) and protostomes (100 %/1.00). Within protostomes, nemertodermatids are sister to Ecdysozoa+Spiralia (including the lophophorates). The monophyly of Spiralia including the lophophorates is strongly supported in both trees (92 %/1.00). Thus, our phylogenetic analyses corroborate the hypothesis that the lophophorate lineages are more closely related to mollusks, annelids and relatives than to deuterostomes as first postulated by Halanych et al. (1995).

Traditionally Lophophorata were considered the sister or paraphyletic stem-group of Deuterostomia (Ax, 1989; 2001; Brusca and Brusca, 2003; Lüter and Bartolomaeus, 1997; Sørensen et al., 2000). This placement was based on the following characters: a body organization with three distinct coelomic cavities, namely protocoel, mesocoel and metacoel (archimery); a mesosomal tentacular apparatus; entomesoderm derived from the archenteron by enterocoely, larvae with upstream-collecting ciliary bands; heterogeneously assembled metanephridia. However, the hypothesis that the lophophorate lineages are more closely allied to Deuterostomia than to annelids and mollusks could be rejected by an approximately unbiased test based on our data set (Table 8.1, Hypothesis 1). Nielsen (1985, 2001) argued that ectoprocts show no trace of archimery and that only Brachiopoda and Phoronida form a monophyletic group with Deuterostomia sensu stricto (his Neorenalia). Lüter (2000) suggested that the origin of the coelomic anlage from differentiated archenteral epithelium, which he defined as enterocoely, is a synapomorphy of Brachiopoda and Deuterostomia; he therefore considered these two taxa as sister-groups. Consequently, we tested the hypotheses that Brachiozoa or Brachiopoda alone are the sister-group of Deuterostomia. Both possibilities were rejected (Table 8.1, Hypotheses 2–3).

The conflicting results concerning the phylogenetic relationships of the lophophorates is a major incongruity between morphological and molecular phylogenetic approaches for resolving the relationships of metazoan phyla. However, in the last

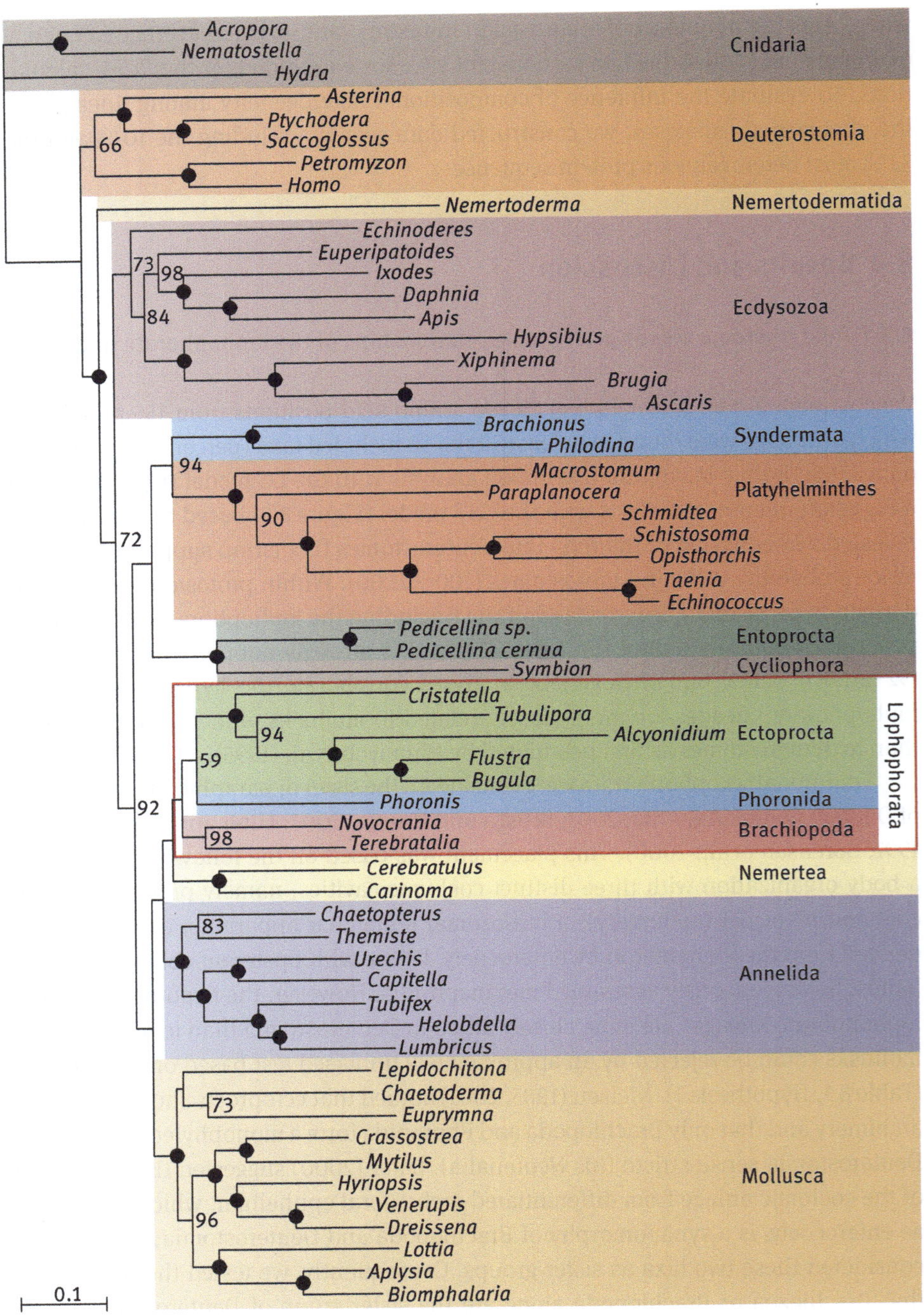

Figure 8.4: Maximum-likelihood tree calculated with the LG+G+F model based on 29,896 aligned amino acid positions derived from 155 proteins of 58 taxa. Bootstrap support values larger than 50 % are shown. Nodes with 100 % bootstrap support are indicated by black circles.

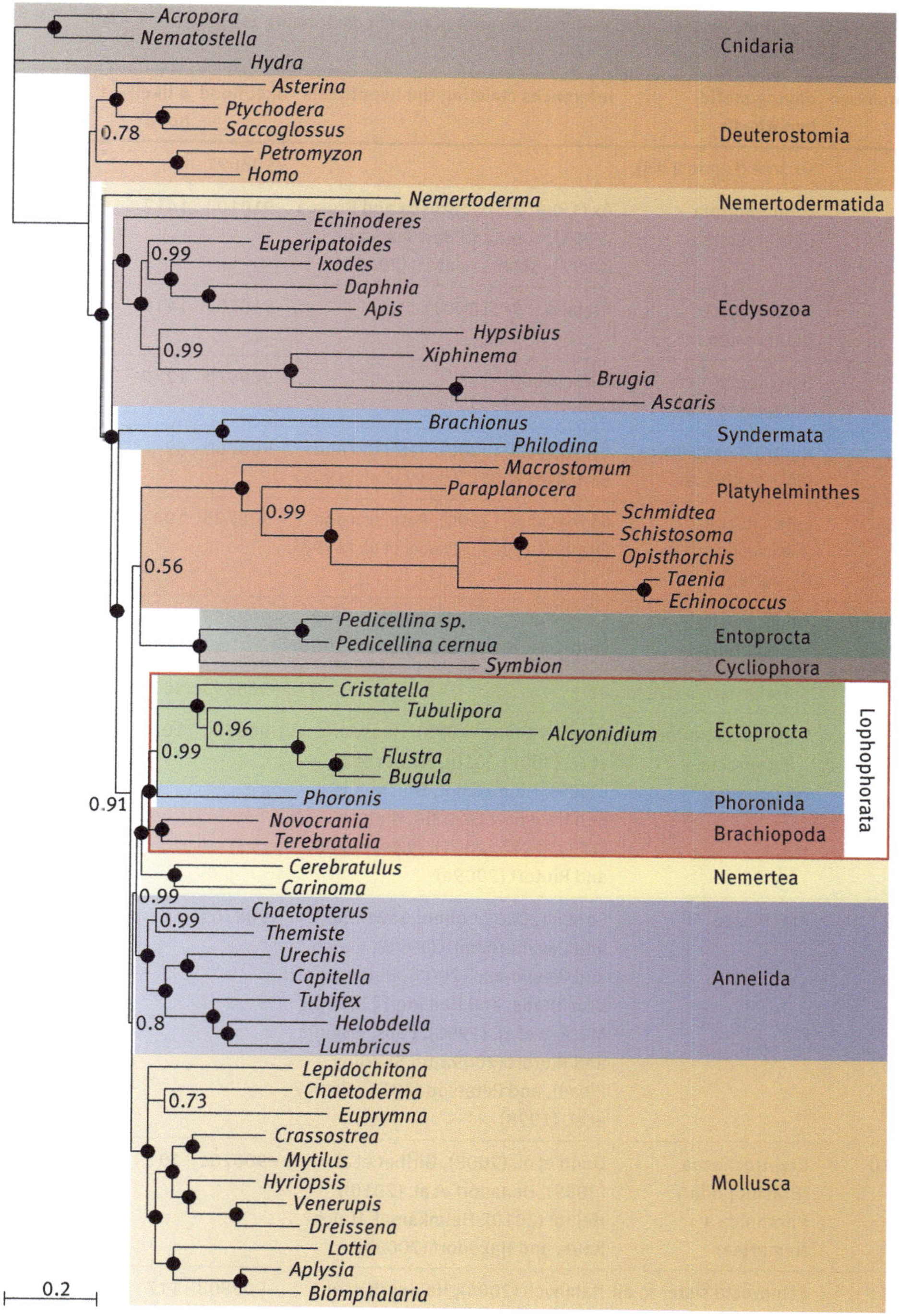

Figure 8.5: Bayesian inference reconstruction with the CAT model. Bayesian posterior probabilities are shown. Posterior probabilities equal to 1.0 are indicated by black circles.

Table 8.1: Topology test results based on the superalignment comprising 155 genes.

number[a]	phylogenetic hypothesis	references claiming the hypothesis	likelihood	Δ likelihood[b]	AU[c]
	ML tree (Figure 8.4A)		−908691	0	0.749
1	Lophophorata + Deuterostomia	Ax (1989, 2001), Brusca and Brusca (2003), Lüter and Bartolomaeus (1997), Sørensen et al. (2000)	−910103	1412	0.000*
2	Brachiozoa + Deuterostomia	Nielsen (1985; 2001)	−910102	1410	0.000*
3	Brachiopoda + Deuterostomia	Lüter (2000)	−909970	1279	0.000*
4	Neotrochozoa (Annelida + Mollusca)	Giribet et al. (2000), Peterson and Eernisse (2001), Zrzavý et al. (1998)	−908760	69	0.012*
5	Eutrochozoa (Neotrochozoa + Nemertea)	Giribet et al. (2000), Peterson and Eernisse (2001), Zrzavý et al. (1998)	−908795	104	0.005*
6	Trochozoa (Entoprocta + Eutrochozoa)	Giribet et al. (2000), Peterson and Eernisse (2001), Zrzavý et al. (1998)	−908855	164	0.003*
7	Spiralia sensu stricto		−908993	302	0.000*
8	Polyzoa (Ectoprocta + Entoprocta + Cycliophora)	Cavalier-Smith (1998), Hausdorf et al. (2007, 2010), Hejnol et al. (2009), Helmkampf, Bruchhaus, and Hausdorf (2008b), Nielsen (1971, 1985, 2001), Paps, Baguñà, and Riutort (2009a)	−908701	10	0.481
9	Brachiozoa	Cohen (2000), Cohen, Gawthrop, and Cavalier-Smith (1998), Cohen and Weydmann (2005), Helmkampf, Bruchhaus, and Hausdorf (2008b), Mackey et al. (1996), Paps, Baguñà, and Riutort (2009a,b), Sperling, Pisani, and Peterson (2011), Zrzavý et al. (1998)	−908701	10	0.480
10	Kryptrochozoa (Brachiopoda + Phoronida + Nemertea)	Dunn et al. (2008), Giribet et al. (2009), Hausdorf et al. (2010), Hejnol (2010), Helmkampf, Bruchhaus, and Hausdorf (2008b)	−908701	10	0.480
11	Ectoprocta sister to all other Lophotrochozoa inclusive Platyzoa	Halanych (2004), Halanych et al. (1995), Passamaneck and Halanych (2006)	−908808	117	0.003*

number[a]	phylogenetic hypothesis	references claiming the hypothesis	likelihood	Δ likelihood[b]	AU[c]
12	Ectoprocta sister to all other protostomes	Giribet et al. (2000)	−909054	362	0.000*
13	Conchozoa (Brachiopoda + Phoronida + Mollusca)	Cavalier-Smith (1998), Halanych (2004)	−908826	134	0.000*
14	Phoronida + Rhynchonelliformea	Halanych et al. (1995); Zrzavý et al. (1998)	−908747	56	0.034*
15	Phoronida + Craniformea + Linguliformea	Cohen, Gawthrop, and Cavalier-Smith (1998); Cohen (2000, 2013); Cohen and Weydmann (2005)	−908745	54	0.032*
16	Phylactolaemata + Phoronida	Mundy, Taylor, and Thorpe (1981)	−908973	282	0.000*
17	Phylactolaemata + Stenolaemata	Anstey (1990); Tsyganov-Bodounov et al. (2009)	−908714	23	0.134
18	Cheilostomata + Phylactolaemata + Stenolaemata	Anstey (1990)	−909349	658	0.000*
19	Stenolaemata + Cheilostomata	Cuffey and Blake (1991)	−909110	419	0.000*
20	Phylactolaemata + Eurystomata	Cuffey (1973)	−908722	31	0.043*
21	Phylactolaemata + Ctenostomata	Bassler (1953)	−909333	642	0.000*

[a] Numbers refer to the order of appearance in the text.
[b] Δ likelihood: differences between the likelihood of a constrained tree and the maximum-likelihood tree.
[c] AU: approximately unbiased test (*p*-values). Values for topologies significantly rejected at the 0.05 level are indicated by an asterisk.

decade the morphological evidence for a close relationship between the lophophorate groups and the deuterostomes has become weaker by careful re-examinations of the characters in question. It has been shown that neither brachiopods nor phoronids possess three coelomic cavities, because a protocoel is lacking in all lophophorate groups (Bartolomaeus, 2001; Lüter, 2000). Thus, the archicoelomate concept (Siewing, 1980) uniting Lophophorata and Deuterostomia, founded on the similarities of three distinct coelomic cavities, lost its basis. Additionally, the finding that Pterobranchia may nest within the enteropneusts (Cameron, Garey, and Swalla, 2000; Cannon et al., 2009; Peterson and Eernisse, 2001; Winchell et al., 2002) suggests that the ancestral deuterostome more closely resembled a mobile worm-like enteropneust than a sessile colonial pterobranch. This means that the similar tentacular feeding apparatuses of lophophorates and pterobranchs are not a synapomorphy of lophophorates and deuterostomes as supposed previously (Ax, 1989; 2001; Lüter and Bartolomaeus, 1997), but evolved independently as convergent adaptations to a sessile lifestyle (Halanych, 1996b). Moreover, Lüter (2000) argued that the mesoderm does not originate by enterocoely in Ectoprocta and Phoronida, but that this is the case in Brachiopoda and Deuterostomia only. However, the origin of the mesoderm of brachiopods is also in dispute. Jenner (2004a) tentatively concluded that reports of true enterocoely, i.e., mesoderm origin by epithelial folding, in brachiopods appear unsupported and that no fundamental difference in the source of mesoderm and mode of coelomogenesis exists between brachiopods and various protostomes. Thus, few morphological characters argue against protostome affiliations of brachiopods and phoronids.

The relationships of Lophophorata within Spiralia need further corroboration. According to our phylogenomic analyses (Figure 8.4–8.5), Lophophorata is the sister-group of Nemertea and both form the sister-group of Annelida. This result is in conflict with three nested hypotheses, namely the Neotrochozoa hypothesis that suggests that Annelida is the sister-group of Mollusca, the Eutrochozoa hypothesis according to which Nemertea is the sister-group of Neotrochozoa, and the Trochozoa hypothesis proposing that Entoprocta is the sister-group of Eutrochozoa (Giribet et al., 2000; Peterson and Eernisse, 2001; Zrzavý et al., 1998). All three hypotheses could be rejected by an approximately unbiased test based on our data set (Table 8.1, Hypotheses 4–6). If the trochophora larvae of annelids, mollusks and entoprocts are homologous, the position of the lophophorates in our trees (Figure 8.4–8.5) indicates that the lophophorates would be derived from ancestors with a trochophora larva.

8.3.2 Monophyly of Lophophorata and Ectoprocta+Phoronida

Both the Maximum Likelihood analysis with the LG model (Figure 8.4) and the Bayesian inference analysis with the CAT model (Figure 8.5) revealed monophyletic Lophophorata (42 %/1.00) as sister-group of Nemertea. Furthermore, they indicate a sister-group relationship between ectoprocts and phoronids (59 %/0.99) within Lophophorata. Interestingly, a sister-group relationship between ectoprocts and phoronids has been proposed based on morphological data (Jägersten, 1972; Lüter et al., 2007; Temereva and Malakhov, 2011). The eversion of a ventral invagination (the metasomal tube in phoronids and the ventral sac in some ectoprocts) at the beginning of metamorphosis (Jägersten, 1972; Lüter et al., 2007; Temereva and Malakhov, 2011; but see Grobe, 2008) and the loss of setae (Temereva and Malakhov, 2011) might be synapomorphies of ectoprocts and phoronids.

The monophyly of the lophophorates implies that the lophophore of ectoprocts, phoronids and brachiopods is homologous, despite some differences in the structure between these groups (Nielsen, 2001), and that this character and the epistome are synapomorphies of the lophophorate lineages (Lüter, Grobe, and Bartolomaeus, 2007). These innovations are the morphological prerequisites for the transition from a vagile to the sessile, filter-feeding lifestyle of the lophophorates.

Similar to the lophophore, radial cleavage was formerly considered a symplesiomorphy of lophophorates and deuterostomes (Ax, 1989; 2001; Brusca and Brusca, 2003). However, there are no doubts about the homology of the spiral cleavage of platyhelminths, entoprocts, nemerteans, annelids, and mollusks, the closest relatives of the lophophorates. Given our tree (Figure 8.4–8.5), parsimony would suggest that radial cleavage evolved secondarily in lophophorates. An alternative hypothesis would be that Lophophorata is the sister-group to the mentioned phyla with spiral cleavage (and the syndermates). However, this hypothesis could be rejected based on the phylogenomic data (Table 8.1, Hypothesis 7). The finding that early cleavage displays spiral features in at least some phoronids (Pennerstorfer and Scholtz, 2012; Rattenbury, 1954) shows how variable cleavage patterns are and corroborates the hypothesis that the radial cleavage of lophophorates is probably secondarily derived from spiral cleavage. It remains unclear whether radial cleavage can be considered a synapomorphy of the lophophorate lineages (with atavistic tendencies in phoronids) or whether the ancestral cleavage pattern has been modified independently in brachiopods, phoronids and ectoprocts.

The monophyly of Lophophorata and a sister-group relationship between ectoprocts and phoronids are in conflict with the Polyzoa (= Bryozoa) hypothesis suggesting that ectoprocts are the sister-group of entoprocts (and cycliophorans), and the Brachiozoa hypothesis, according to which Brachiopoda and Phoronida form a monophylum, as well as with the Kryptrochozoa hypothesis that suggests that brachiopods and phoronids form a clade with nemerteans. The Polyzoa hypothesis (Cavalier-Smith, 1998; Nielsen, 1971; 1985; 2001) has been supported by previous phylogenomic analyses (Bleidorn et al., 2009a; Hausdorf et al., 2007; 2010; Hejnol et al., 2009; Helmkampf, Bruchhaus, and Hausdorf, 2008b; Struck and Fisse, 2008; Witek et al., 2008; 2009) and also by some analyses of rDNA sequences (Mallat, Craig, and Yoder, 2010; Paps, Baguñà, and Riutort, 2009a). The Brachiozoa hypothesis has originally been based on analyses of rDNA sequences (Cohen, 2000; Cohen, Gawthrop, and Cavalier-Smith, 1998; Cohen and Weydmann, 2005; Mackey et al., 1996; Mallat, Craig, and Yoder, 2010; Paps, Baguñà, and Riutort, 2009a; Zrzavý et al., 1998), but has also been corroborated by analyses of sodium-potassium ATPase α-subunit sequences (Anderson, Cordoba, and Thollesson, 2004), multiple nuclear protein-encoding sequences (Paps, Baguñà, and Riutort, 2009b; Sperling, Pisani, and Peterson, 2011), and phylogenomic data sets (Helmkampf, Bruchhaus, and Hausdorf, 2008b; Nesnidal et al., 2010). The Kryptrochozoa hypothesis was based on results of phylogenomic studies (Dunn et al., 2008; Hausdorf et al., 2010; Hejnol, 2010; Helmkampf, Bruchhaus, and Hausdorf, 2008b; Giribet et al., 2009).

Resolving the multifurcations in the trees constraining monophyletic Polyzoa, Brachiozoa or Kryptrochozoa resulted in the same topology that could not be rejected with an approximately unbiased test based on the new phylogenomic data set (Table 8.1, Hypotheses 8–10).

The approximately unbiased test showed that the support for a tree in which ectoprocts and entoprocts are sister-groups in accordance with the Polyzoa (= Bryozoa) hypothesis and in which phoronids and brachiopods are sister-groups in accordance with the Brachiozoa hypothesis is almost as strong as the support for the maximum-likelihood tree with monophyletic lophophorates. Previous analyses of ribosomal protein data sets already indicated that there are signals for the conflicting hypotheses in the genomic data (Nesnidal et al., 2010). Nesnidal et al. (2010) attributed the conflicting signal to compositional bias in the ribosomal protein data set. To assess the influence of compositional bias on the phylogenetic analysis of the current data set, we ordered the used proteins according to relative compositional variability and excluded those with the highest heterogeneity. If we exclude the 10 % most heterogeneous proteins (remaining alignment length 27,602 amino acids), the bootstrap support for the ectoproct-phoronid clade will increase to 78 %, and if we exclude the 33 % most heterogeneous proteins (remaining alignment length 22,029 amino acids), the bootstrap support for this clade will increase even to 90 %. This indicates that both the Polyzoa (=Bryozoa) hypothesis and the Brachiozoa hypothesis are actually artifacts resulting from systematic errors in tree reconstruction due to compositional bias. The support for the monophyly of Lophophorata increased slightly to 60 % if the 10 % most heterogeneous proteins were excluded, but collapsed if the 33 % most heterogeneous proteins were excluded. In this case, the relationships between Ectoprocta + Phoronida, Brachiopoda, Nemertea and Entoprocta + Cycliophora remained unresolved in a bootstrap 50 % majority rule tree. We suppose that this is an artifact resulting from the compositional similarity of brachiopods and nemerteans. If nemerteans are excluded, the bootstrap support for the monophyly of Lophophorata increased to 70 %.

The Bayesian inference analysis with the CAT model (Figure 8.5) provided stronger support for the monophyly of Lophophorata (posterior probability 1.00) and Ectoprocta+Phoronida (posterior probability 0.99) than the Maximum Likelihood analysis with the LG model, even if no heterogeneous proteins are excluded. This might indicate that the CAT model is a better approximation to the evolutionary process in the investigated data set than the LG model.

Some other hypotheses concerning the phylogenetic position of the three lophophorate lineages within protostomes are also in conflict with the monophyly of Lophophorata. Ectoprocts were supposed to represent the sister-group to all other Lophotrochozoa including Platyzoa, i.e., Platyhelminthes, Syndermata and allies (Halanych, 2004; Halanych et al., 1995; Passamaneck and Halanych, 2006), or even to all other protostomes except chaetognaths (Giribet et al., 2000). Both hypotheses could be rejected based on the phylogenomic data presented here (Table 8.1, Hypotheses 11–12). The Conchozoa hypothesis suggesting that brachiopods and phoronids form a clade with mollusks (Cavalier-Smith, 1998; Halanych, 2004) could also be rejected based on the current data (Table 8.1, Hypotheses 13).

Furthermore, the results that phoronids are the sister-group of ectoprocts (59 %/0.99) and that brachiopods are monophyletic (98 %/1.00) are in conflict with hypotheses supposing that phoronids are sister to subgroups of paraphyletic brachiopods. The hypotheses that Brachiopoda is paraphyletic and that phoronids are the sister-group of articulate brachiopods represented by *Terebratalia* (Halanych et al., 1995; Zrzavý et al., 1998) or inarticulate brachiopods represented by *Novocrania* (Cohen, Gawthrop, and Cavalier-Smith, 1998; Cohen, 2000; 2013; Cohen and Weydmann, 2005) have been significantly rejected based on the current data (Table 8.1, Hypotheses 14–15). This indicates that phoronids did not evolve from brachiopods and have not secondarily reduced the shell.

8.3.3 Phylogenetic relationships within Ectoprocta

There is a variety of hypotheses concerning the phylogenetic relationships of the four recent major groups of Ectoprocta, i.e., Phylactolaemata (represented in our data set by *Cristatella*), Stenolaemata (*Tubulipora*), Ctenostomata (*Alcyonidium*) and Cheilostomata (*Bugula* and *Flustra*). As our previous analyses based on ribosomal proteins (Hausdorf et al., 2010), the phylogenetic analyses of the new data set (Figure 8.4–8.5) showed the monophyly of Ectoprocta and a sister-group relationship between Phylactolaemata, the freshwater bryozoans, and the predominantly marine Gymnolaemata (= Stenolaemata + Ctenostomata + Cheilostomata; as originally defined by Allman (1856) and used, e.g., by Ax, 2001; Bassler, 1953; Jebram, 1973; Todd, 2000), confirming morphological classifications of Jebram (1973), Cuffey and Blake (1991), Todd (2000), and Ax (2001). Within Gymnolaemata, Ctenostomata and Cheilostomata form a monophyletic group, Eurystomata (as proposed by Marcus, 1938; and equivalent to Gymnolaemata as used by Cuffey, 1973, Fuchs, Obst, and Sundberg, 2009; Tsyganov-Bodounov et al., 2009; Waeschenbach, Taylor, and Littlewood, 2012). These relationships were also corroborated by recent analyses of Fuchs, Obst, and Sundberg (2009) and Waeschenbach, Taylor, and Littlewood (2012) based on rDNA and mitochondrial genes.

Diphyletic Ectoprocta with phylactolaemates forming a clade with phoronids as hypothesized by Mundy, Taylor, and Thorpe (1981) could be significantly rejected by topology tests (Table 8.1, Hypothesis 16). Thus, the similarities of phylactolaemates and phoronids concerning lophophore development, intrinsic body wall muscles and adoral budding direction noted in this study are either symplesiomorphies or convergences. The hypotheses that phylactolaemates and stenolaemates constitute a monophyletic group (Anstey, 1990; Tsyganov-Bodounov et al., 2009) could not be ruled out (Table 8.1, Hypothesis 17), although there is hardly any morphological evidence for it. However, the hypothesis that this clade is the sister-group of Cheilostomata (Anstey,

1990) could be rejected by topology tests (Table 8.1, Hypothesis 18), as could the hypothesis that Stenolaemata constitutes a monophyletic group with Cheilostomata (Cuffey and Blake, 1991) (Table 8.1, Hypothesis 19). This implies that the similarities between Stenolaemata and Cheilostomata concerning the laminated wall, the presence of avicularia, peristomes, intrazooecial pores and pseudopores noted by Cuffey and Blake (1991) are either symplesiomorphies or convergences. In contrast, the lack of musculature on the body wall, parietal muscles that pass through the cystid controlling its extension, cellular rosettes occupying the pores in the cystid wall between zooids, and branched funiculus vessels without longitudinal muscles that contact the rosettes and effect transfer of fluids through the colony via these pores may be autapomorphies of Eurystomata (Ax, 2001). The classification of ectoprocts into Pyxibryozoa or "box-like bryozoans" including phylactolaemates and eurystomates and Tubulobryozoa (= Stenolaemata) or "tubular bryozoans" (Cuffey, 1973) could also be excluded by topology tests (Table 8.1, Hypothesis 20) indicating that the form of the zooecium has been convergently modified in the major ectoproct groups. Finally, the hypothesis that Phylactolaemata are derived from Ctenostomata (Bassler, 1953) has been significantly rejected (Table 8.1, Hypothesis 21).

Some results of the analyses of the new sequence data sets like the monophyly of Lophophorata and the sister-group relationship of Ectoprocta and Phoronida are in conflict with previous phylogenomic studies. Analyses reducing the amount of compositional heterogeneity in the sequence data indicate that these conflicts are due to compositional bias. There is variety of such systematic errors potentially affecting phylogenetic analyses based on sequence data, which are difficult to detect and to eliminate. Therefore, we suggest to confirm the monophyly of Lophophorata and their relationships by additional sequence-independent evidence like clade-specific microRNAs (Wheeler et al., 2009), near-intron positions (Krauss et al., 2008) or other rare genomic events (Rokas and Holland, 2000) in future studies.

Acknowledgments

We are grateful to Michael Kube and Richard Reinhardt (Max Planck Institute for Molecular Genetics, Berlin) for the construction and sequencing of cDNA libraries and the working groups of Thomas Hankeln, Bernhard Lieb and Torsten Struck for the collaboration within the "Deep Metazoan Phylogeny" network. We thank Peter Grobe, Dirk Schories and Kåre Telnes for providing photos and two anonymous referees for helpful comments. This study was funded by the priority program "Deep Metazoan Phylogeny" of the German Research Foundation (Deutsche Forschungsgemeinschaft) (HA 2763/5).

Torsten H. Struck, Günter Purschke, Janina Dordel, Christoph Hösel, Maximilian P. Nesnidal, Frauke Diersing, Christoph Bleidorn, Christiane Paul, Natascha Hill, Ralph Tiedemann, Joachim Selbig, and Stefanie Hartmann

9 Phylogeny and evolution of Annelida based on molecular data

Abstract: Annelids inhabit diverse habitats from the deep sea to terrestrial soils. Being one of the three animal taxa exhibiting true segmentation, deciphering the evolution of Annelida is crucial to our understanding of the evolution of segmentation. Traditionally, Annelida comprised polychaetes (e.g., lugworms, fanworms, fireworms) and clitellates (e.g., earthworms, leeches). In this chapter, we summarize results regarding the phylogeny and evolution of Annelida that have been obtained in the last years using different sources of molecular data. Results to date support hypotheses in which non-segmented Sipuncula (peanut worms) and Echiura (spoon worms) are part of the annelid radiation, and Clitellata are placed among taxa like the lugworms within polychaetes. The majority of annelid taxa is found within one large clade that split into Errantia and Sedentaria and thus was named Pleistoannelida. On the other hand, the most recent analyses included support for placements of Chaetopteridae (parchment worms), Oweniidae or small, interstitial polychaete families as part of the basal annelid radiation. Based on these results and the fossil record, biodiversity of Annelida most likely originated from an early annelid resembling a crawling form with grooved palps and parapodia with simple chaetae.

9.1 Introduction

Annelida, the segmented worms, is an ancient and ecologically important animal taxon with over 16,500 described species that occupy terrestrial and aquatic habitats worldwide. Annelids range in length from < 1 mm to > 3 m and from less than 10 segments to up to more than 1,000 segments (Figure 9.1). As bioturbators, scavengers, and predators they are important members of all ecosystems. They are the most abundant macrofaunal organisms in the largest habitat on the planet, the deep sea (Grassle and Maciolek, 1992; Hessler and Sanders, 1967). Several annelids are used as model organisms (Shain, 2009), but their positions within the annelid tree of life are uncertain. Annelida is one of three major animal taxa exhibiting true segmentation and, thus, is critical in any investigation regarding body plan evolution. Therefore, we need to understand the composition and branching history of groups within the annelid radiation if we aim to reconstruct the last common ancestor of bilaterians

and to understand evolution of segmentation and other morphological character complexes.

Annelida has traditionally been classified into Polychaeta and Clitellata (Fauchald and Rouse, 1997; Grube, 1850; Michaelsen, 1928–1932; Rouse and Fauchald, 1997; Rouse and Pleijel, 2001). However, two non-segmented lophotrochozoan taxa,

◂ **Figure 9.1:** Diversity of Annelida. **A–F** Members of basal annelid radiation. **G–T** Pleistoannelida. **G–H** Orbiniida. **I–M** Errantia. **N–T** Sedentaria. **A.** *Owenia fusiformis* (Oweniidae), anterior end, length about 100 mm, **B.** *Phascolion strombus* (Sipuncula), introvert protruding from gastropod shell, length about 50 mm, **C.** *Eurythoe complanata* (Amphinomidae), length about 140 mm, **D.** *Protodrilus* sp. (Protodrilidae), length about 5 mm, **E.** *Trilobodrilus axi* (Dinophilidae), length about 1 mm, **F.** *Nerilla antennata* (Nerillidae), length about 1.5 mm, **G.** *Stygocapitella subterranea* (Parergodrilidae), length about 2.5 mm, **H.** *Scoloplos armiger* (Orbiniidae), juvenile, length about 120 mm, **I.** *Nematonereis* sp. (Eunicidae), length about 200 mm, **J.** *Protodorvillea kefersteini* (Dorvilleidae), anterior end, length about 15 mm, **K.** *Tomopteris helgolandica* (Tomopteridae), length about 25 mm, **L.** *Myzostoma cirriferum* (Myzostomida) on *Antedon bifida*, length about 4 mm, **M.** *Ophiodromus* sp. (Hesionidae), length about 70 mm, **N.** *Pygospio elegans* (Spionidae), length about 25 mm, **O.** *Fabricia stellaris* (Sabellidae), length about 4 mm, **P.** *Aricidea suecica* (Paraonidae), length about 20 mm, **Q.** *Eupolymnia nebulosa* (Terebellidae), length about 200 mm, **R.** *Euclymene* sp. (Maldanidae), length about 100 mm, **S.** *Enchytraeus bigeminus* (Clitellata, Enchytraeidae) length about 15 mm, **T.** *Acanthobdella peledina* (Clitellata, Hirudinea) length about 40 mm. B, C, I, L, M, Q–T courtesy of Prof. Westheide, Osnabrück.

Echiura and Sipuncula, have been discussed to be within Annelida, indicating a much higher evolutionary variability of segmentation than expected. Echiura, a clade of large, non-segmented marine worms with chaetae, usually live in burrows of soft marine sediments (Purschke, 1996). First support from molecular data for an inclusion of Echiura into Annelida was derived from elongation factor 1α data (McHugh, 1997) and, specifically, a sister-group relationship to Capitellidae was proposed based on 18S rRNA data (Bleidorn, Vogt, and Bartolomaeus, 2003a; 2003b). Moreover, the development of the nervous system in Echiura also reveals traces of an ancestral segmentation pattern (Hessling, 2002; 2003; Hessling and Westheide, 2002). Sipuncula are relatively large, non-segmented worms without chaetae that burrow in marine sediments, live in crevices or inhabit abandoned tubes and shells (Schulze, Cutler, and Giribet, 2005). Based on morphology, sipunculan origins are controversial: some morphological data suggest a molluskan affiliation (Rouse, 1999; 2000; Rouse and Fauchald, 1995; Scheltema, 1993), whereas others indicate an annelid relationship (e.g., Bartolomaeus and Quast, 2005; Kristof, Wollesen, and Wanninger, 2008; Rice, 1985; Tzetlin and Purschke, 2006). Early molecular and total evidence studies placed Sipuncula within Lophotrochozoa in close relationship to Mollusca and/or Annelida, but without strong support (e.g., Giribet et al., 2000; Halanych, 2004; Schulze, Cutler, and Giribet, 2007; Winnepenninckx, Van de Peer, and Backeljau, 1998). While the position of Echiura as sister to Capitellidae is well supported by 18S rDNA data, the position of Sipuncula within Annelida is generally inconsistent and lacks strong support.

Analyses of morphological and molecular data support clitellate monophyly and have provided robust hypotheses of relationships within this taxon (Erséus, 2005; Erséus and Kallersjo, 2004; Purschke, 2002). On the other hand, monophyly of Polychaeta lacks support, and resolving polychaete relationships was difficult (McHugh, 2000; 2005). Polychaetous annelids have been classified into approximately 80 fami-

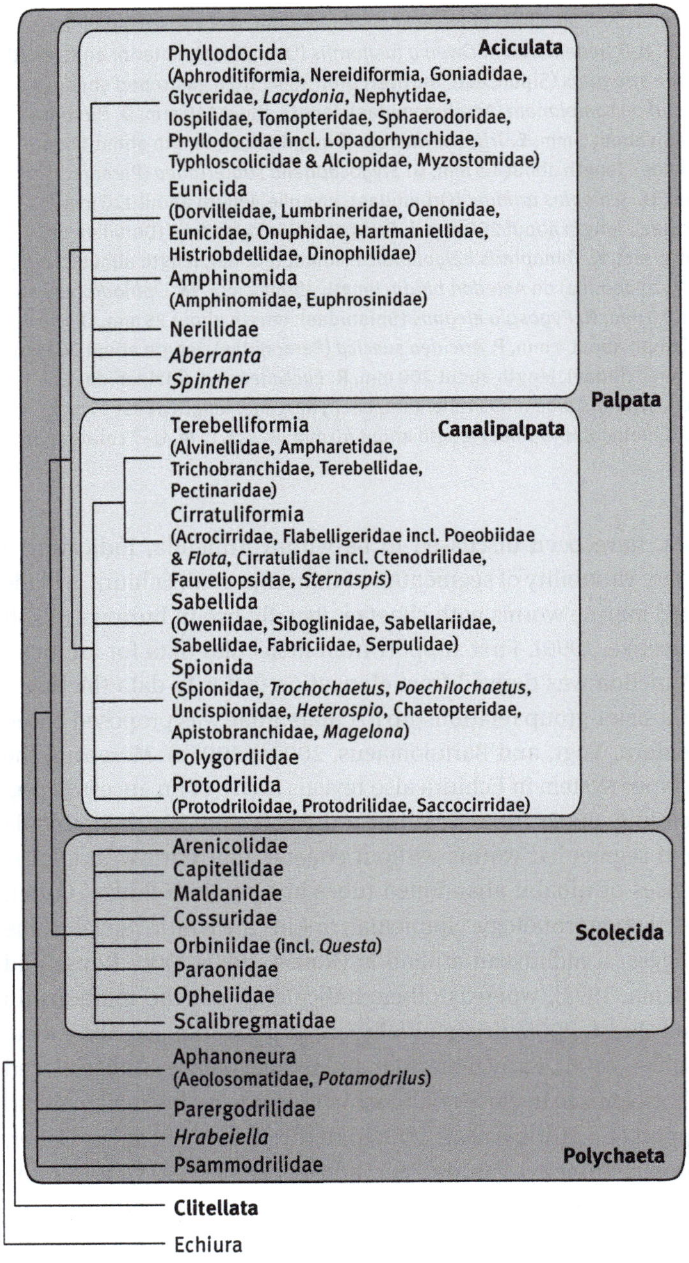

Figure 9.2: Schematic depiction of the phylogeny of Annelida as proposed by the morphological cladistic analyses of Rouse and Fauchald (1997). Grey boxes indicate higher taxonomic units.

lies, each generally supported as monophyletic, but more inclusive nodes were not well supported (Rouse and Pleijel, 2001). Until the middle of the last century, Polychaeta was divided into two major groups, Errantia (Aciculata in Figure 9.2) and Sedentaria (Scolecida and Canalipalpata in Figure 9.2) (e.g., Fauvel, 1923; 1927; 1953). This division is based on a classification proposed by de Quatrefages (1866). De Quatrefages (1866) suggested the names Annelidae erraticae, which was later named Errantia, and Annelidae sedentariae, which was later named Sedentaria. These names referred to names already being used by Audouin and Milne Edwards (1834) as Annélides errantes and Annélides tubicoles (ou sédentaires) or Lamarck (1818) as Sédentaires. These names reflect the mode of living as either errant, more vagile worms, or as sedentary, more sessile ones, but the system itself was based on the presence or absence of distinct body regions also known as tagmatization (de Quatrefages, 1866). However, Dales (1963; 1977) rejected this systematization and argued that the groupings reflect the mode of living rather than evolutionary history and are only valuable for practical purposes of identification. Fauchald (1977) and others followed Dales and rejected the division into two clades by grouping families only into equally ranked orders.

Rouse and Fauchald's (1997) morphological cladistic analyses have provided the most taxonomically inclusive hypothesis of polychaete relationships based on morphology to date. They proposed Polychaeta to be monophyletic and to consist of two major clades, Scolecida and Palpata, the latter divided into Canalipalpata and Aciculata (Figure 9.2). Subsequent morphological cladistic analyses by Rouse (1999; 2000) supported these results. Within a short time, the systematic scheme of Rouse and Fauchald (1997) was adopted in textbooks or reviews of polychaetes (e.g., Glasby et al., 2000; Rouse and Pleijel, 2001; 2003; 2006; 2007). However, some morphologists questioned the monophyly of Scolecida, Palpata, Aciculata, and Canalipalpata (Bartolomaeus, 1998; Bartolomaeus, Purschke, and Hausen, 2005; Hausen, 2005; Westheide et al., 1999).

Molecular analyses do not support the classification scheme developed by Rouse and Fauchald (1997) (see Hall, Hutchings, and Colgan, 2004; McHugh, 2000; 2005; Struck and Purschke, 2005). While the analyses of Rouse and Fauchald (1997) provide objective and testable hypotheses of polychaete relationships and thus significantly moved the field forward, many aspects of annelid phylogeny are still poorly understood. Therefore, reconstruction of the annelid ground pattern, as well as the placement of their root, are major problems in annelid systematics and have considerable implications for understanding annelid evolution (McHugh, 2000; 2005; Rouse and Pleijel, 2003).

For example, it is uncertain whether the ancestral annelid was a burrowing form (Clark, 1969; Rouse and Fauchald, 1997) or a surface-crawling errant form (Conway Morris and Peel, 1990; Westheide, 1997), drawing into question the use of model organisms as representatives of the ancestral annelid (e.g., Arendt et al., 2002; Denes et al., 2007). The first hypothesis about annelid evolution (e.g., Fauchald, 1974; 1977) proposes a progression from an earthworm-like marine ancestor of relatively simple organization (e.g., without head appendages, parapodial lobes, or pygidial cirri) to more complex forms, and it was based on the hydrostatic theory of the evolution of segmented coeloms by Clark (1964; 1969). In this theory, the primary function of segmented coelomic cavities is that of a hydrostatic skeleton, aiding peristaltic burrowing by means of strong, complete muscular septa. In addition to a well-developed body wall musculature comprised of circular and longitudinal muscles, this system allows an effective generation of peristaltic waves that push the body forward even in comparatively firm substrates such as terrestrial soils. Because earthworms most closely reflect this condition, oligochaete-like polychaetes like Capitellidae, Opheliidae, Paraonidae, or Arenicolidae were thought to represent ancestral conditions in Annelida, whereas polychaetes like Sabellidae or Nereididae were considered as more derived (Fauchald, 1974; 1977). Morphological cladistic analyses (Rouse, 1999; 2000; Rouse and Fauchald, 1997) are in agreement with this hypothesis (Figure 9.2).

However, others have argued for the opposite direction of annelid evolution, going from more complex forms like Amphinomidae or Nereididae to simple forms (e.g., Conway Morris and Peel, 1995; Storch, 1968; Westheide, 1997; Westheide and Watson Russell, 1992). The segmented body organization is thought to have arisen from chaetae-bearing parapodia as protective and locomotory devices. In this scenario, the segmental organization of the coelom occurred secondarily as an adaptation to the need to transport blood to and from external appendages such as the parapodia and the musculature (Westheide, 1997). This view is in accordance with the observation that in invertebrates the vessels of the blood vascular system are spaces between the basal extracellular matrices of adjacent epithelia and hence between adjacent coelothelia accomplishing blood transport from and to the external periphery of the body (e.g., Bartolomaeus, 1994; Rieger and Lombardi, 1987; Ruppert, 1991b; Westheide, 1997). Moreover, the first evolutionary scenario has been criticized, because most oligochaete-like, burrowing polychaetes lack muscular septa similar to the ones of earthworms (Mettam, 1985; Westheide, 1997). In addition, the absence of parapodia and other appendages in Clitellata could also be regarded as a primary adaptation to their highly specialized reproductive biology and only secondarily to locomotion (Westheide, 1997; Westheide and Müller, 1996). In line with this hypothesis, Clitellata are placed within polychaetes (Conway Morris and Peel, 1995; Westheide, 1997). However, strong support was lacking for both hypotheses.

Therefore, as part of the DFG priority program SPP 1174: "Deep Metazoan Phylogeny – Stammesgeschichte der Großgruppen der Tiere" we as well as others addressed the phylogeny and evolution of Annelida using different molecular approaches. First, a targeted-gene approach was employed using either complete mitochondrial genomes or nuclear genes such as 18S and 28S rRNA, protein-coding genes like aldolase, ATP synthase b chain, catalase, methionine adenosyltransferase, and myosin II heavy chain. Second, we determined expressed sequence tags (EST) libraries of various annelid taxa to carry out phylogenomic analyses. Third, we explored the utility of novel gene structure marker systems such as introns for their probable phylogenetic signal. These approaches will be described in detail next; they also allow conclusions about the evolution of Annelida.

9.2 Phylogenetic analyses of Annelida using targeted genes

Early molecular studies of Annelida predominantly used either nuclear 18S rRNA or elongation factor 1α data, but resolution of and support for deep-level annelid relationships was weak and contradictory in these analyses and did not provide substantial progress towards a robust annelid phylogeny (for reviews see McHugh, 2000; 2005). Therefore, other nuclear and mitochondrial genes have been explored in this context. The first such study was carried out by Brown et al. (1999), who used fragments of three other genes (U2 snRNA, histone H3 and the D1 and D9–10 expansion elements of the 28S rRNA). Other genes that were used to infer annelid phylogeny are the mitochondrial cytochrome c oxidase subunit I and 16S rRNA and nuclear 28S rRNA, aldolase, ATP synthase b chain, catalase, methionine adenosyltransferase, and myosin II heavy chain (Colgan, Hutchings, and Braune, 2006; Dordel et al., 2010; Rousset et al., 2007; Struck et al., 2008; 2007). However, although up to four genes were used in each of these studies, the actual number of nucleotide positions was not much more than twice the number of a typical 18S rRNA study (e.g., 1,770 in Rota, Martin, and Erséus, 2001). Only one study used 11 genes and substantially increased the number of positions by three to five times in comparison to a typical 18S rRNA study (Dordel et al., 2010). Finally, Zrzavý et al. (2009) were the first to conduct a total evidence analysis of morphological data and sequence information of six available genes (histone H3, elongation factor 1α, 18S rRNA, 28S rRNA, and the mitochondrial 16S rRNA and cytochrome oxidase I) totaling up to 3,903 cladistically informative characters.

A different approach was taken that exploited the information present in complete mitochondrial genomes. In addition to sequence information of rRNA and protein-coding genes, gene order rearrangement and duplication events can be used for phylogenetic inferences as well. The first studies focused on the position of Annelida within Protostomia (e.g., Boore and Brown, 2000; Helfenbein and Boore, 2004), but subsequent studies steadily increased the number of complete or nearly complete mitochondrial genomes of annelid taxa (e.g., Bleidorn et al., 2007; Bleidorn et al., 2009b; Bleidorn, Podsiadlowski, and Bartolomaeus, 2006; Mwinyi et al., 2009; Shen et al., 2009; Struck et al., 2007; Vallès and Boore, 2006; Vallès, Halanych, and Boore, 2008; Wu et al., 2009; Zhong et al., 2011; Zhong, Struck, and Halanych, 2008). A data set of mitochondrial rRNA and protein-coding genes typically contains about 6,200 nucleotide positions (e.g., Zhong et al., 2011). However, the number of analyzed mitochondrial genomes from Annelida is currently still too low to address the phylogeny of Annelida in general. Nonetheless, some progress could be achieved by all these analyses.

The most substantial progress was achieved with respect to the taxon composition of Annelida. As mentioned above, molecular data of both elongation factor 1α data (McHugh, 1997) and 18S rRNA (Bleidorn, Vogt, and Bartolomaeus, 2003a; 2003b) supported the hypothesis of an inclusion of Echiura within Annelida; a taxonomy corroborated by several other analyses (Bleidorn et al., 2007; Bleidorn, Podsiadlowski, and Bartolomaeus, 2006; Brown et al., 1999; Colgan, Hutchings, and Braune, 2006; Dordel et al., 2010; Mwinyi et al., 2009; Rousset et al., 2007; Shen et al., 2009; Struck et al., 2008; 2007; Wu et al., 2009; Zhong, Struck, and Halanych, 2008; Zrzavý et al., 2009). More specifically, topology tests significantly rejected an exclusion of Echiura from Annelida (Struck et al., 2007). Based on 18S rRNA data, a sister-group relationship of Echiura to Capitellidae was proposed and was strongly confirmed by the largest studies by number of nucleotide positions from nuclear genes (Struck et al., 2008; 2007; Zrzavý et al., 2009). Based on these data it can be concluded that in all probability Echiura is the sister-group of Capitellidae and hence a subtaxon of Annelida.

The taxonomic affinity of Sipuncula is less clear in these analyses. Most of these recent studies consistently recover Sipuncula either as sister to Annelida or within Annelida, depending on the taxon composition of the analysis (Bleidorn, Podsiadlowski, and Bartolomaeus, 2006; Colgan, Hutchings, and Braune, 2006; Dordel et al., 2010; Mwinyi et al., 2009; Rousset et al., 2007; Shen et al., 2009; Struck et al., 2008; 2007; Zrzavý et al., 2009). In the analyses of mitochondrial data, Sipuncula were usually recovered as sister to Annelida, but crucial taxa such as Chaetopteridae, which eventually branch off before Sipuncula, are still lacking in these analyses. Nodal support for a close affinity of Sipuncula and Annelida was more or less strong in these analyses, but not for the inclusion of Sipuncula into Annelida. Mitochondrial gene order data also strongly support an annelid affinity of Sipuncula, because their gene order is very similar to the annelid gene order that is unique within Lophotrochozoa (Vallès and Boore, 2006). Moreover, recent studies of microRNAs also supported a close relationship of Sipuncula and Annelida (Helm et al., 2012; Sperling et al., 2009), and topology tests based on a targeted-gene approach significantly rejected an exclusion of Sipuncula from Annelida (Dordel et al., 2010; Mwinyi et al., 2009; Struck et al., 2007). Therefore, based on these studies a close affinity of Sipuncula was supported, but their position within Annelida remained uncertain. Several analyses placed Sipuncula in the basal part of the annelid tree or as sister to Annelida (Colgan, Hutchings, and Braune, 2006; Dordel et al., 2010; Mwinyi et al., 2009; Sper-

Figure 9.3: Schematic depiction of Fig. 10 in Struck et al. (2008) based on Maximum Likelihood analyses of 18S and 28S rRNA data. Species of their analysis have been collapsed to the family level. Nodal support values of the different analyses are given at the branches as indicated in the inset (MP BS = Maximum Parsimony bootstrap support, NJ BS = Neighbour Joining bootstrap support, ML BS = Maximum Likelihood bootstrap support, PP = Posterior probability). For branch length refer to the original study. Vertical bars to the right indicate higher taxonomic units. ▸

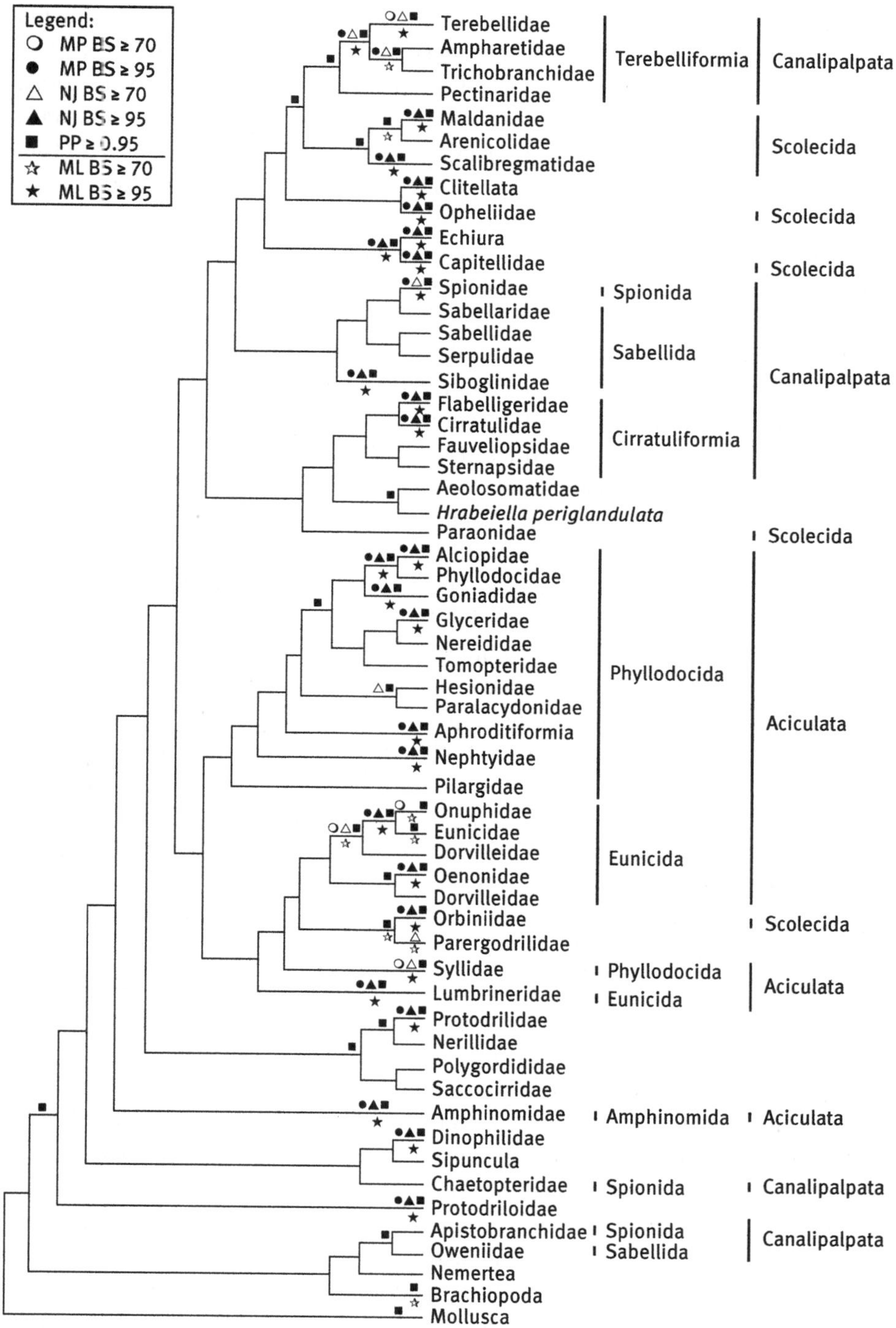

Legend:
MP BS ≥ 70
MP BS ≥ 95
NJ BS ≥ 70
NJ BS ≥ 95
PP ≥ 0.95
ML BS ≥ 70
ML BS ≥ 95
Terebellidae
Ampharetidae
Trichobranchidae
Pectinaridae
Terebelliformia
Canalipalpata
Maldanidae
Arenicolidae
Scalibregmatidae
Scolecida
Clitellata
Opheliidae
Scolecida
Echiura
Capitellidae
Scolecida
Spionidae
Spionida
Sabellaridae
Sabellidae
Serpulidae
Siboglinidae
Sabellida
Canalipalpata
Flabelligeridae
Cirratulidae
Fauveliopsidae
Sternapsidae
Cirratuliformia
Aeolosomatidae
Hrabeiella periglandulata
Paraonidae
Scolecida
Alciopidae
Phyllodocidae
Goniadidae
Glyceridae
Nereididae
Tomopteridae
Hesionidae
Paralacydonidae
Aphroditiformia
Nephtyidae
Pilargidae
Phyllodocida
Aciculata
Onuphidae
Eunicidae
Dorvilleidae
Oenonidae
Dorvilleidae
Eunicida
Orbiniidae
Parergodrilidae
Scolecida
Syllidae
Phyllodocida
Lumbrineridae
Eunicida
Aciculata
Protodrilidae
Nerillidae
Polygordiidae
Saccocirridae
Amphinomidae
Amphinomida
Aciculata
Dinophilidae
Sipuncula
Chaetopteridae
Spionida
Canalipalpata
Protodriloidae
Apistobranchidae
Spionida
Oweniidae
Sabellida
Canalipalpata
Nemertea
Brachiopoda
Mollusca

ling et al., 2009; Struck et al., 2008; Zrzavý et al., 2009), but others found Sipuncula with weak support as sister to Terebelliformia (Colgan, Hutchings, and Braune, 2006; Struck et al., 2007), to Serpulidae/Flabelligeridae/Orbiniidae (Rousset et al., 2007), to Orbiniidae (Bleidorn, Podsiadlowski, and Bartolomaeus, 2006; Struck et al., 2007; Zhong et al., 2011; Zhong, Struck, and Halanych, 2008), to Dinophilidae (Zrzavý et al., 2009), or to Nereididae (Shen et al., 2009; Zhong, Struck, and Halanych, 2008).

With respect to the general phylogeny of Annelida some progress could be achieved, especially considering the last three comprehensive studies based on targeted genes (Struck et al., 2008; 2007; Zrzavý et al., 2009) (Figure 9.3). Several analyses recovered the monophyly of Annelida including Echiura and Sipuncula (Colgan, Hutchings, and Braune, 2006; Struck, Halanych, and Purschke, 2005; Struck et al., 2007). All analyses placed Clitellata clearly within polychaetes and rendered "Polychaeta" paraphyletic. Topology testing significantly rejected a sister-group relationship of Clitellata to polychaetes (Struck et al., 2007). More specifically, Clitellata was part of a larger clade comprising Terebelliformia, Arenicolidae, Maldanidae, Opheliidae, Scalibregmatidae, Echiura, and Capitellidae (Struck et al., 2008; 2007). Within this clade a close relationship of Terebelliformia and Arenicolidae/Maldanidae was also present (Dordel et al., 2010; Rousset et al., 2007; Struck et al., 2008; Zrzavý et al., 2009) as well as monophyly of Terebelliformia (Rousset et al., 2007; Struck et al., 2008; 2007; Zrzavý et al., 2009). Spionidae was placed among the sabellidan taxa Sabellidae, Serpulidae, and Sabelleriidae and usually as sister to the latter (Colgan, Hutchings, and Braune, 2006; Dordel et al., 2010; Struck et al., 2008; 2007; Zrzavý et al., 2009). Moreover, Struck et al. (2008) inferred monophyly of Cirratuliformia as proposed by Rouse and Fauchald (1997). On the other hand, a clade of Phyllodocida, Eunicida, and Orbiniida was found in the analyses of Struck et al. (2008; 2007) and Zrzavý et al. (2006). Finally, Zrzavý et al. (2009) and Struck et al. (2008) found basal positions of Oweniidae, Apistobranchidae, Chaetopteridae, Amphinomidae, Sipuncula, Myzostomida, Protodrilidae, Protodriloidae, Saccocirridae, Polygordiidae, Nerillidae, and Dinophilidae.

Although congruence between these multiple-gene analyses increased relative to the previous single-gene analyses, support values for most basal nodes were still very low and topology testing could only reject certain phylogenetic hypotheses with statistical significance. This is in agreement with simulation studies by Struck et al. (2008), who showed that an increasing amount of sequence information will more likely recover the phylogeny of Annelida using Maximum Likelihood approaches and also result in strong nodal support. Moreover, Dordel et al. (2010) found that amino acid data are better suited than nucleotide data, probably due to the higher number of possible character states (20 amino acids compared to four nucleotides), which decreases the likelihood of convergence. In addition, amino acid alignments allow to robustly reconstruct deeper nodes because they are generally less variable than the corresponding nucleotide alignments and therefore less affected by saturation. Hence, a robust phylogeny of Annelida based on molecular data may be obtained by increasing the number of protein-coding genes in phylogenetic analyses of Annelida. However, using a targeted-gene approach is not the most effective method to achieve this goal. For a targeted-gene approach suitable primers have to be determined and tested for each gene, the amplification has to be optimized for each gene individually for all annelids or – even worse – for different annelid subgroups to obtain single gene fragments, and then each gene for each single species has to be sequenced individually. Therefore, to determine hundreds of protein-coding genes a targeted-gene approach is very time consuming and expensive.

9.3 Phylogenomic analyses of Annelida

Due to the limitations of the targeted-gene approach we tested new methods to robustly reconstruct a phylogeny of Annelida. Phylogenomic approaches using EST libraries have previously been successfully used to infer lophotrochozoan relationships (e.g., Dunn et al., 2008; Hausdorf et al., 2007; Hejnol et al., 2009; Struck and Fisse, 2008). In contrast to targeted-gene approaches, transcriptomic data are randomly sequenced, and no gene-specific primers are necessary. At the same time, many more genes are sequenced than in targeted-gene approaches and more information can therefore be gathered. However, due to the randomness any two EST libraries only partially overlap, but the number of common genes increases with the number of sequences present in each. This makes this approach especially suitable for next-generation sequencing strategies. The first studies using such an approach with respect to Annelida were still based on classical Sanger sequencing of about 1,400 clones (= sequences) on average.

The first transcriptomic analysis addressed the phylogenetic position of Myzostomida with respect to Annelida (Bleidorn et al., 2009a; see corresponding chapter on Myzostomida in this book). Dordel et al. (2010) explored the utility of this kind of data to investigate the affiliation of Sipuncula with Annelida. From EST libraries of 15 annelid, 2 sipunculan, and 31 lophotrochozoan taxa they retrieved the sequence information of 78 ribosomal protein genes to build a data matrix of 11,827 amino acid positions, which corresponds to 35,481 nucleotide positions. The data set was 3.5 times larger than the nucleotide data matrices, which Dordel et al. (2010) assembled from 11 targeted genes. The resulting tree was congruent with previous studies based on targeted genes (e.g., Struck et al., 2008) and bootstrap support values of basal nodes moderately increased. More importantly, only this ribosomal data set of Dordel et al. (2010) significantly rejected a sister-group relationship of Sipuncula and Annelida in favor of a placement of Sipuncula within Annelida, while the other smaller data sets based on 11 targeted genes resulted in non-significant results. Dordel et al. (2010) demonstrated the potential of a phylogenomic approach. Given a limited taxon sampling, their analyses could not yet resolve the phylogeny of Annelida in general.

A year later Struck et al. (2011) provided a phylogenomic study with a more comprehensive taxon sampling covering 20 of the ~ 80 traditional polychaete 'families', Myzostomida, Echiura, Sipuncula, and Clitellata. From the EST libraries they were able to compile a data set of 47,953 amino acid positions derived from 231 gene fragments, a data set comprising four times more positions and twice as many annelid taxa compared to Dordel et al. (2010). While generally congruent with the analyses of Dordel et al. (2010), Zrzavý et al. (2009) and Struck et al. (2008), bootstrap support was substantially higher at basal nodes and reached 100 at some nodes. Struck et al. (2011) were able to provide the best supported phylogeny based on molecular data to date (Figure 9.4). Chaetopterida, Myzostomida and Sipuncula were part of the first annelid radiation in their analyses. The remaining annelids split into two major well-supported clades with bootstrap support ranging from around 80 up to 100 for both clades. One clade comprised Eunicida, Phyllodocida, Amphinomidae, and Orbiniidae, and the other one Sabellidae, Spionidae, Siboglinidae, Cirratuliformia, Terebelliformia, Arenicolidae, Opheliidae, Capitellidae, Echiura, and Clitellata. These two clades were also recovered in the analyses of Struck et al. (2008) and Zrzavý et al. (2009), except for the position of Amphinomida, which was placed in the basal part of the annelid tree in those analyses. Moreover, the higher-level clade containing these two clades was also well supported (Struck et al., 2011).

The taxon composition of each clade in Figure 9.4 is very similar to previously proposed groups of Annelida or "Polychaeta", namely Errantia and Sedentaria. Errantia traditionally comprised Amphinomida, Phyllodocida and Eunicida (Fauvel, 1923) and, interestingly, Orbiniidae were regarded as an intermediate stage between Errantia and Sedentaria (Fauvel, 1927). On the other hand, Sedentaria comprised all remaining polychaete groups, of course with the exclusion of Clitellata, which were considered outside polychaetes at that time (Fauvel, 1927). Given the overall agreement in taxon composition, Struck et al. (2011) resurrected the names Errantia and Sedentaria for these two clades revealed by their analyses, albeit without formal definitions. The name Pleistoannelida for the well-supported higher-level clade containing Errantia and Sedentaria was suggested in a subsequent study (Struck, 2011).

Within Errantia, the most surprising result of the phylogenomic study was the non-monophyly of Phyllodocida (Struck et al., 2011). Aphroditiformia (scale worms) were sister to Orbiniidae, and Eunicida was sister to a clade comprising the other phyllodocidan taxon in their analyses (Figure 9.4). Within Sedentaria, a well-supported clade that consisted of Sabellidae and Spionidae was found. Also a clade of Terebelliformia, Arenicolidae, Opheliidae, Capitellidae, Echiura, and Clitellata received strong support after the exclusion of unstable taxa in Struck et al. (2011). Within this clade the sister-group relationship of Arenicolidae and Terebelliformia is well supported. Bootstrap support values for all other relationships within Sedentaria were poorly supported with nodal support below 70, but congruence to previous analyses is high.

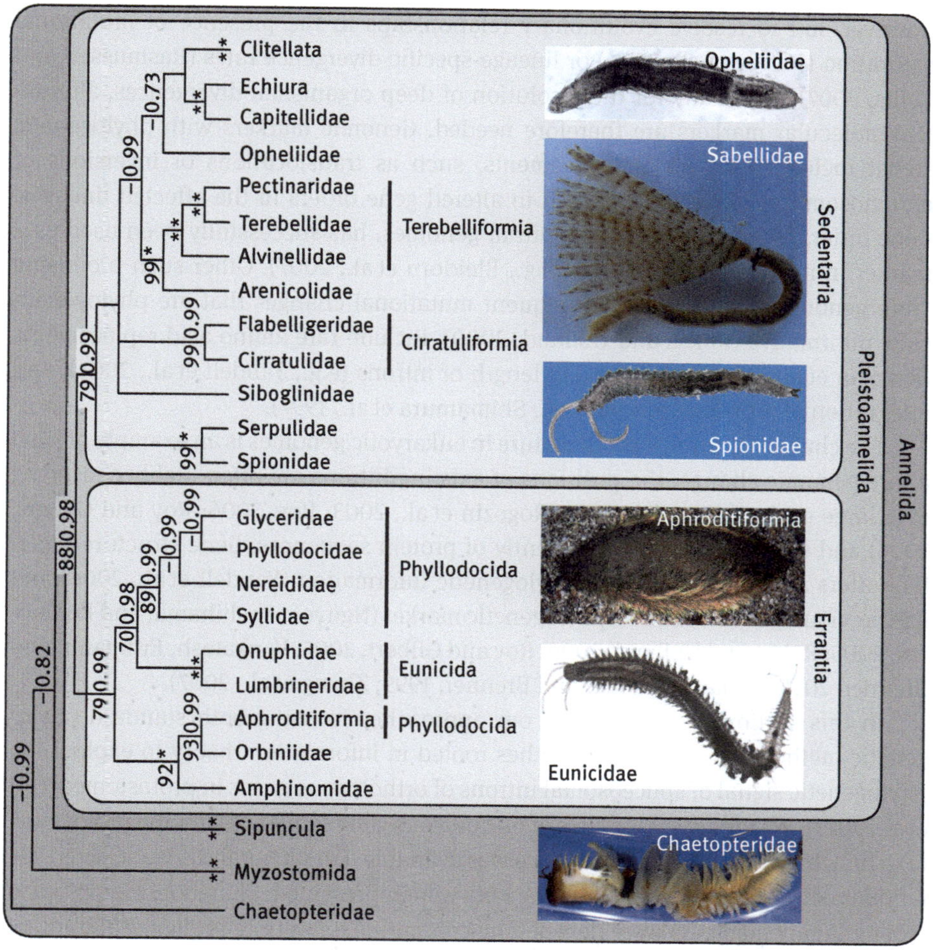

Figure 9.4: Schematic depiction of Fig. 1 in Struck et al. (2011) based on Bayesian inferences of phylogenomic data of 231 genes derived from expressed sequence tags (EST) libraries. Species of their analysis have been collapsed to the family level. Nodal support values are given at the branches and only bootstrap support values ≥ 70 and posterior probability ≥ 0.70 are shown (– = value below threshold, * = 100 or 1.00). For branch length refer to the original study. Grey boxes indicate higher taxonomic units. Pictures of Aphroditiformia and Chaetopteridae are a courtesy of Prof. Westheide, Osnabrück.

9.4 Gene structure data as phylogenetic markers

Phylogenomic approaches as described in the previous section have clear advantages over using targeted genes. For example, they can overcome problems resulting from biases that lead to non-phylogenetic signal or stochastic error related to gene length (Jeffroy et al., 2006). Large-scale phylogenetic analysis of sequence data may, however, fail to resolve evolutionary relationships in the presence of mutational saturation (Jeffroy et al., 2006) or lineage-specific divergence rates (Rasmussen and Kellis, 2007). Especially for the resolution of deep organismal divergences, alternative molecular markers are therefore needed. Genomic markers with phylogenetic signal include genomic rearrangements, such as translocations or inversions of chromosomal regions, which result in altered gene orders in the affected lineages. Gene order, especially on mitochondrial genomes, has successfully been used as a marker in molecular systematics (e.g., Bleidorn et al., 2007). Other such promising "rare genomic changes", i.e., infrequent mutational changes that are phylogenetically informative (Rokas and Holland, 2000), include rare amino acid replacements (Rogozin et al., 2007), position and length of introns (e.g., Yandell et al., 2006), and integration of retrotransposons (e.g., Shimamura et al., 1997).

The change in intron-exon structure in eukaryotic genomes is an example of such a rare genomic change: the positions of existing introns are often highly conserved over large evolutionary distances (Rogozin et al., 2003; Roy, 2006; Roy and Gilbert, 2006) and evolve largely independently of protein sequences. Gene structure therefore offers a valuable source of phylogenetic information (Yandell et al., 2006) and has previously been used as a phylogenetic marker (Nguyen, Yoshihama, and Kenmochi, 2005; Rokas and Holland, 2000; Roy and Gilbert, 2005; Venkatesh, Erdmann, and Brenner, 2011; Venkatesh, Ning, and Brenner, 1999; Zheng et al., 2007).

In this section, we summarize our approaches to using both standard phylogenetic methods as well as approaches rooted in information theory to explore the phylogenetic signal of spliceosomal introns of orthologous genes in protostomes (Hill et al., 2013). We identified families of orthologous gene sequences in nine protostome taxa for which a full genome sequence was available. For our analysis we selected 324 families of orthologs in which all nine taxa were represented. Shared intron presence among family members were detected by computing a multiple sequence alignment for each family and mapping intron positions onto the aligned sequences. This information was coded in a matrix for phylogenetic analyses.

Dollo Parsimony, which assumes an ancestral state (here: absence of an intron) and a derived state (here: presence of an intron), was applied to our intron absence/presence matrix. The resulting phylogeny strongly supported the monophyla Arthropoda, Insecta, and Trochozoa, but annelids were paraphyletic, with *Capitella* and *Lottia* forming a sister-clade to *Helobdella*. Ecdysozoa were not recovered in this analysis, due to a clade consisting of *Caenorhabditis* and *Schistosoma* as sister-group of Arthropoda. This unexpected result could be due to two reasons. First, Ecdysozoa in its present taxa composition could be an artificial grouping that resulted from biases in other molecular phylogenetic studies such as long branches (Kück et al., 2012; Wägele and Misof, 2001). Second, as Dollo parsimony does not allow convergent evolution of characters and essentially minimizes the degree of reductions, it could be misled by biases in the data set. Because *Caenorhabditis* and *Schistosoma* showed high proportions of single-taxon introns, they are assumed to have higher rates of intron gain compared with the other taxa. If intron evolution is faster in *Caenorhabditis* and *Schistosoma*, the close relationship between these two taxa may be an analytical artifact equivalent to long-branch attraction, caused by multiple intron loss and/or gain events at the same intron positions in parallel.

While the absence/presence of introns can be used to cluster taxa using standard phylogenetic algorithms, especially when the data have been carefully filtered, this type of data is also well suited for an analysis using information theoretic measures. Information theory is a branch of mathematics that was first developed in the context of communication systems (Shannon, 1948), and it has since found application in many fields, including molecular biology (Adami, 2004). Its central goal is the quantification of information, and one such measure is Mutual Information. Mutual Information quantifies the mutual dependence, or shared information, of two variables, such as intron absence/presence and evolutionary relationship, and the resulting distances can be used for clustering data. Mutual Information has been used to analyze primary sequence data (Penner, Grassberger, and Paczuski, 2011) and to detect correlations of different types of biological variables (Hummel et al., 2005; Weckwerth and Selbig, 2003), including species-specific DNA signatures (Bauer, Schuster, and Sayood, 2008). We computed Mutual Information-based distances for ingroup taxa of the 342 families of orthologs and clustered the data hierarchically according to these distances (Figure 9.5). The result strongly supports Lophotrochozoa, Ecdysozoa, Arthropoda, Insecta, and Trochozoa. Annelida is recovered as a monophyletic group.

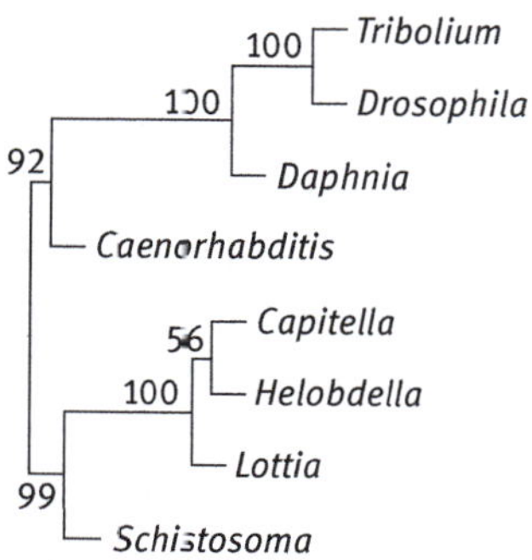

Figure 9.5: Mutual Information-based distance measures (modified from Fig. 4A of Hill et al., 2013). Mutual Information-based distances were clustered using average linkage. Support values were obtained by changing 5 % of the character states at random (1,000 replicates).

These results demonstrate that Mutual Information can be used to analyze the phylogenetic signal present in gene structure. Moreover, Mutual Information is independent of explicit model assumptions, meaning that no assumptions about the character polarity and about ancestral and derived character states are required. Outgroups can therefore be omitted, and Mutual Information analyses may be restricted to ingroup taxa. As the choice of inadequate outgroup taxa may cause tree reconstruction artifacts like long-branch attraction (Holton and Pisani, 2010), Mutual Information may also be useful when appropriate outgroups are not available.

We have shown that gene structure data can be used for phylogenetic questions related to Annelida, and this kind of information may be used in the future to address the phylogeny of Annelida with a much better representation of their biodiversity. The computational approach described here for genomic data can be augmented by molecular genetic work. For this purpose, orthologs of genes with putatively informative introns, which were identified in the course of our computational work, can be PCR-amplified from genomic DNA, cloned, and sequenced from annelids and other lophotrochozoan taxa, for which genomes are not yet available. This experimentally derived data can then be used to complement the data matrix based on our genome-wide scans in future studies.

9.5 Evolution of Annelida

In summary, our recent efforts as well as those of others achieved some substantial progress regarding the phylogeny of Annelida based on molecular data. These results clearly support a placement of Chaetopteridae, Sipuncula, and Myzostomida as well as eventually other taxa such as Oweniidae, Magelonidae, Apistobranchidae or the interstitial polychaete families like Polygordiidae or Dinophilidae as part of the basal radiation of Annelida (Struck, 2011; 2012). Most of these are species-poor taxa. The majority of Annelida is split into two taxa, Sedentaria and Errantia, in these analyses. Struck (2012) formally defined the composition of the clade Errantia by the last common ancestor of Eunicida, Phyllodocida and all organisms or species that share a more recent common ancestor with these two taxa than with Clitellata and Spionidae. Similarly, the composition of the clade Sedentaria is defined by consisting of Clitellata, Echiura, Spionidae and all organisms or species that share a more recent common ancestor with these three taxa than with Phyllodocida or Eunicida. The clade comprising Errantia and Sedentaria was named Pleistoannelida and defined by the last common ancestor of Sedentaria and Errantia, and all the descendants of that ancestor.

This annelid phylogeny based on molecular data also supports some hypotheses about character evolution in Annelida. Character loss is confirmed as a widespread evolutionary pattern in annelids, as previously postulated (Bleidorn, 2007; Purschke, Hessling, and Westheide, 2000; Struck, 2006). For example, segmentation has been independently lost in Sipuncula and Echiura, according to molecular phylogeny. Similarly, Clitellata as well as several earthworm-like polychaetes like Capitellidae, Opheliidae, or Arenicolidae occupy highly derived positions within the annelid tree of life and hence exhibit reductions of parapodia and palps as an adaptation to a burrowing mode of living. Interstitial polychaete taxa like Dinophilidae might be secondarily reduced due to paedomorphosis rather than reflecting an ancestral, primitive condition of Annelida (Struck, 2006; Westheide, 1987; Worsaae and Kristensen, 2005).

Albeit given a still limited taxon sampling Struck et al. (2011) also used the molecular phylogeny to reconstruct the ancestral conditions of different annelid character traits. In this reconstruction the last common ancestor of Annelida most likely possessed only a pair of grooved palps, bicellular eyes, nuchal organs, and biramous parapodia that bear only simple chaetae as well as internalized supporting chaetae (Figure 9.6).

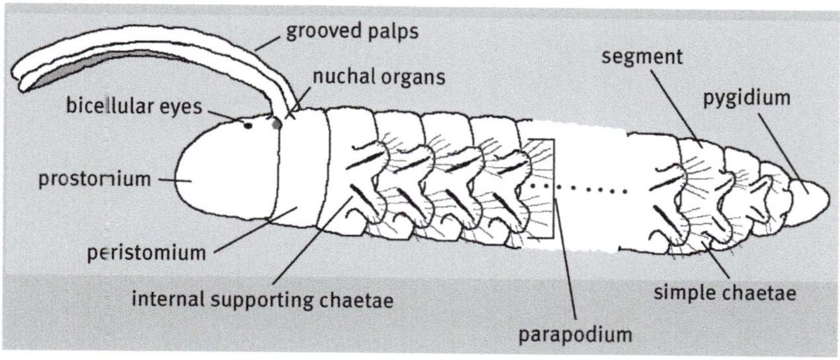

Figure 9.6: Drawing of the ancestral state reconstruction of Annelida modified from Fig. 2 in Struck et al. (2011).

This reconstruction resembles specimens found in the sparse fossil record of early annelids (Conway Morris and Peel, 2008; Eibye-Jacobsen, 2004; Struck, 2011; Vinther, Eibye-Jacobsen, and Harper, 2011). These fossils also show only one pair of anterior appendages, most likely palps, and parapodia with only simple chaetae. However, internalized supporting chaetae are not present in the fossils known to date (Eibye-Jacobsen and Vinther, 2012). Neither the molecular phylogeny, the ancestral reconstruction based on this phylogeny nor the fossil record are in agreement with the evolutionary scenario that the ancestral annelid was a burrowing form resembling an earthworm-like marine ancestor of relatively simple organization (e.g., without head appendages, nuchal organs, parapodial lobes, or pygidial cirri) (Clark, 1969; Rouse and Fauchald, 1997). In contrast, based on these results the last common ancestor had a more complex organization with palps and parapodia and is more in agreement with an ancestral surface-crawling form and the hypothesis that the segmentation of the coelom followed the evolution of parapodia, rather than the other way round (Conway Morris and Peel, 1990; Westheide, 1997). However, this latter scenario regarded the protective function of the chaetae as the driving force in the evolution of parapodia in early annelids, but the results of the ancestral reconstruction and the fossil record highlight that the locomotory aspect of parapodia might be more likely to be the driving force in the evolution of parapodia and their chaetae (Struck, 2011). Given these results and considerations, the early annelid probably resembled a crawling form with grooved palps and parapodia with simple chaetae. From this starting point, the amazing diversity of Annelida evolved, resulting in burrowing earthworms, blood-sucking leeches, filter-feeding fan- and Christmas tree-worms, predatory fire- and bobbit-worms, tiny interstitial worms, non-segmented peanut- and spoon-worms, and sediment-ingesting lug- and bamboo-worms (Rouse and Pleijel, 2001). There are only few other Metazoan taxa with such an extraordinary diversity, which clearly demonstrates the high evolutionary potential of the annelid bauplan.

Christoph Bleidorn, Conrad Helm, Anne Weigert, Igor Eeckhaut, Deborah Lanterbecq, Torsten H. Struck, Stefanie Hartmann, and Ralph Tiedemann

10 From morphology to phylogenomics: Placing the enigmatic Myzostomida in the tree of life

Abstract: Myzostomida are enigmatic protostome animals and their phylogenetic position has puzzled researchers for nearly 200 years. Here we summarize the history of the classification of these animals and introduce their biology and anatomy. Synthesizing recent analyses of ultrastructural, immunohistochemical, and genomic data we show strong support for placing myzostomids within Annelida. As such, especially rare genomic changes markers like microRNA content and mitochondrial gene order strongly confirm an annelid ancestry. Due to a long history as commensals/parasites of crinoid echinoderms, a mixture of character conditions plesiomorphic for annelids and apomorphic for myzostomids is hardly surprising and made phylogenetic studies difficult.

10.1 From Leuckart to Nansen – discovery and early classification of Myzostomida

Myzostomida were first described by Friedrich Sigismund Leuckart (Leuckart, 1827) in a treatise of helminth relationships. He mentioned an ectocommensal species repeatedly found on a Mediterranean feather star (*Antedon mediterranea*) and classified this new genus (initially named *Myzostomum*, now referred to as *Myzostoma*) as trematode flatworms. Ectocommensal myzostomids are disc-shaped, acoelomate worms bearing five pairs of parapodia equipped with chitinous chaetae (Figure 10.1). Not everybody was convinced with this kind of classification. John V. Thompson was the first to report about the existence of an unusual parasite of comatulid crinoids (Thompson, 1836), which was later formally described as *Myzostoma cirriferum* (Leuckart, 1836), by far the best investigated myzostomid species to date. In the light of the century long phylogenetic controversy it is interesting to read Thompson's first account on these animals: "Connected with the natural history of the comatula is that of a nondescript parasite, which appears to be a complete zoological puzzle, as it is not possible to determine from its figure and structure to what class it ought to be referred." And he ends with: "Query, Is it a perfect animal or a larva, and does it belong to the Crustacea, Annelides, or what?"

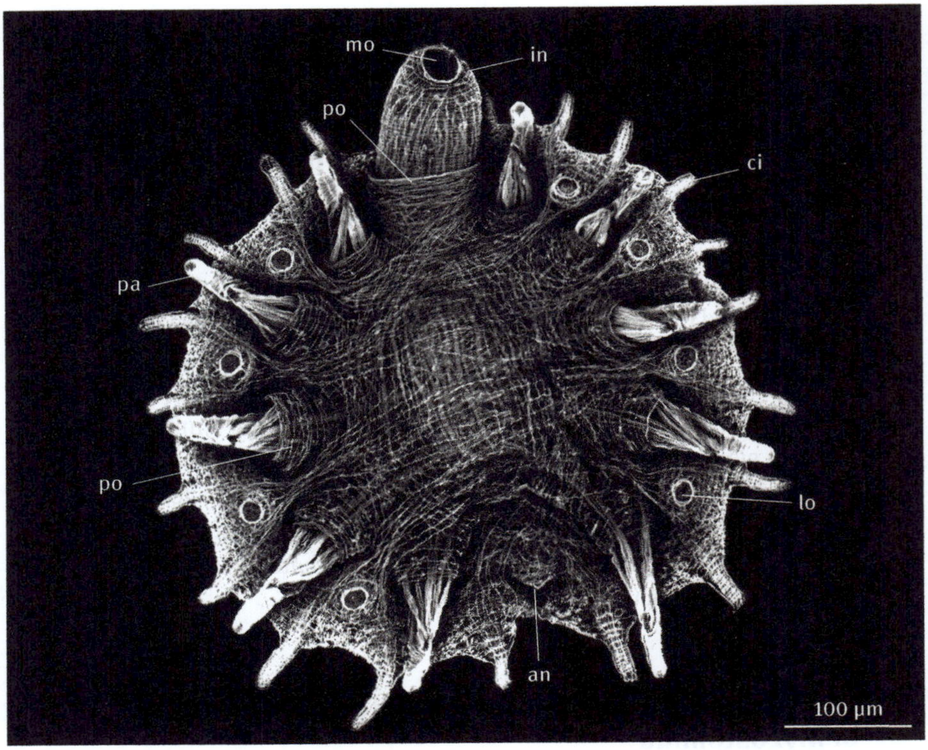

Figure 10.1: Ventral view of *Myzostoma cirriferum* with Phalloidin-rhodamine labeling; confocal micrographs. Note the sets of serial structures: five pairs of parapodia (*pa*), four pairs of lateral organs (*lo*), and ten pairs of cirri (*ci*). *an* anus, *ci* cirrus, *in* introvert, *po* pouch of introvert respectively parapodium, *lo* lateral organ, *mo* mouth opening, *pa* parapodium.

The main character that convinced Leuckart to place myzostomids within trematodes was the assumed presence of several pairs of suckers on the ventral side. Also, the branching gut with its diverticulae resembles that of some flatworm species. Even though, in contrast to these, myzostomids clearly have an anus. The so-called suckers were later shown to be sense organs which might be homologous to the lateral organs of polychaetes (Wheeler, 1896). A first detailed description of the muscular and gonadal systems was given by Carl Semper, who also discovered the first juvenile stage of a myzostomid (Semper, 1858). His initial goal was to find characters which would help to classify this enigmatic group. Yet, in his introductory words he admitted that he had to abandon his hopes. However, especially the first description of a juvenile stage inspired him and others to at least speculate about the phylogenetic relationships. Semper already wrote about a rather superficial resemblance of the two leg-paired juvenile stage to tardigrades. Later on, others also pointed to this affinity (Carus, 1863), an interpretation which goes back to Johannes Müller (Müller, 1841). Semper himself rather speculated that myzostomids might be relatives of parasitic crustaceans, as he saw a resemblance with their larvae. However, he also mentioned that it would be most consequent to erect a new class for these so-hard-to-classify animals. The erection of a new taxon was then proposed by Ludwig von Graff who grouped myzostomids together with tardigrades and pentastomids into Stelechopoda (Graff, 1877).

The first thorough investigations of myzostomid development were conducted by Mecznikow (1866), followed by Beard (1884). Myzostomid larvae are of the trochophora type and bear a metatroch and a telotroch. Moreover, they are equipped with two pair of long chaetae and the remarkable overall similarity led both authors to the conclusion that myzostomids must be related to polychaetes annelids. This view was also adopted by Benham (1896) who classified them as order Myzostomaria within Annelida. However, there was still no consensus in placing myzostomids with polychaetes. The famous Nobel laureate Fridtjof Nansen highlighted in his dissertation that the myzostomid larvae lack the prototroch (Nansen, 1885), which later became the defining feature of the trochophora larvae (sensu Rouse, 1999). Instead, he described them as intermediate forms between annelids and arthropods, which, translated into a cladistic framework, would mean to regard them as the sister group of Articulata.

10.2 Biology of Myzostomida

The life-cycle of myzostomids is usually connected to crinoid echinoderms. Some species are endoparasites of other echinoderms and even a few records of infection of other marine invertebrates (a hexactinelid sponge and a coral) have been verified (Grygier, 2000). Again, by far best known is the biology (Figure 10.2) of the European myzostomid *Myzostoma cirriferum*, a protandric hermaphroditic ectocommensal symbiont of crinoids of the genus *Antedon* (Eeckhaut and Jangoux, 1993). Female stages are fertilized by a spermatophore and subsequent intradermic penetration of sperm cells (Eeckhaut and Jangoux, 1991). Mature eggs are released in the water column and pass through a spiralian pattern of cleavage. Four different larval stages are distinguished and late stages are eventually caught by their host *Antedon*, which treat them as food particles. Due to the possession of long chaetae, they can attach to the host's tube feet and enter the ambulacral groove, where they undergo metamorphosis. Juvenile stages and early males remain in cysts of the wall of the ambulacral groove and grow in this place, often leading to local deformations of pinnulae. Late males leave the place of infection, start to move freely on the host's arms and calyx, and live on stealing food particles from the ambulacral groove. While growing, these animals become more flattened and disc-shaped, and individuals pass through a functional male stage, via a hermaphroditic stage to become functional females (Wheeler, 1896). It has been shown that endoparasitism evolved multiple times from such ectocommensal ancestors, including forms infesting the integument, gonads, coelom, or digestive system of echinoderms (Lanterbecq et al., 2006).

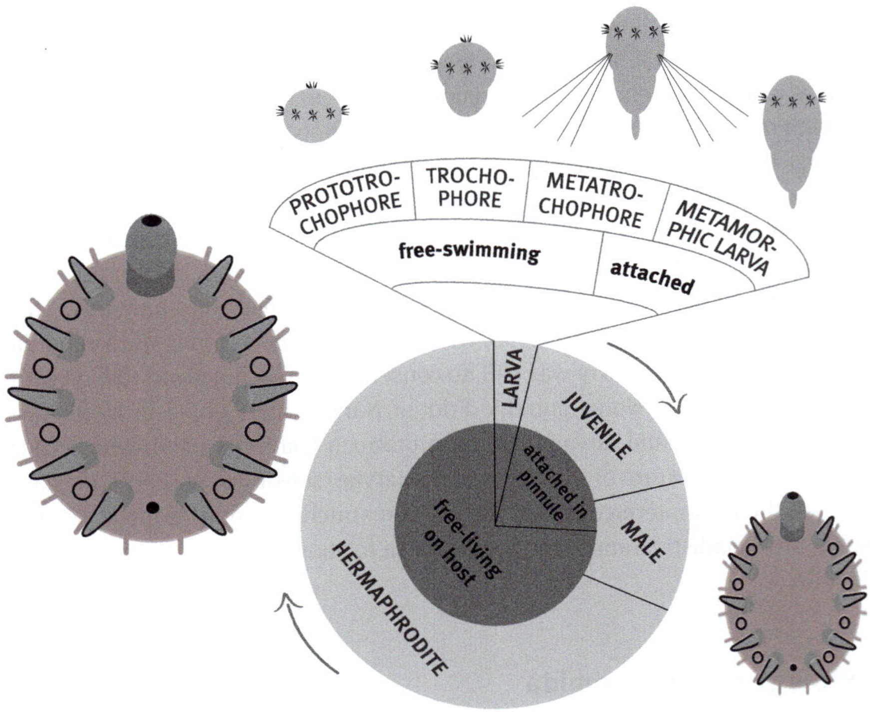

Figure 10.2: Life cycle of *Myzostoma cirriferum* according to Eeckhaut and Jangoux (1991).

10.3 Cladistic analyses of morphological and molecular data – setting up a controversy

The placement of Myzostomida as part of the Annelida remained stable for a hundred years. This changed after myzostomids were included in cladistic analyses of morphological characters to explore their affinities. Within a few years four large analyses were published, two placing myzostomids as sister group to annelids and allies (Haszprunar, 1996a; Zrzavý et al., 1998), one recovering them nested within annelids (Rouse and Fauchald, 1997) and one uniting Myzostomida with Syndermata and Cycliophora in the new taxon Promastigozoa (Zrzavý, Hypsa, and Tietz, 2001). It comes without surprise that there are huge differences in the character selection of these four disagreeing cladistics analyses (Jenner, 2003). For example, the outcome of the analyses by Zrzavý, Hypsa, and Tietz, (2001) was heavily influenced by coding the aberrant sperm morphology of myzostomids using several characters for all the substructures. The myzostomid spermatozoon is rather unusual for metazoans, as it is of the filiform type with the flagellum at the anterior end and lacking the accessory centriol (Mattei and Marchand, 1987). A similar condition has been reported for syndermatan spermatozoa (Mattei and Marchand, 1987; Melone and Ferraguti, 1994;

Ahlrichs, 1998; Foata et al., 2004). However, several characters which may support an annelid affinity (e.g. the presence of a nectochaeta larvae, aciculae, and cirri) have been neglected by Zrzavý, Hypsa, and Tietz (2001).

In the meantime the first molecular systematic analyses including myzostomids popped up and tantalizingly supported a close relationship to either flatworms or syndermatans (Eeckhaut et al., 2000; Zrzavý, Hypa, and Tietz, 2001). Even though the long branch leading to myzostomids in those analyses made the results at least suspect, good bootstrap support (Eeckhaut et al., 2000) and congruence with a morphological character matrix (Zrzavý, Hypa, and Tietz, 2001) strengthened the hypothesis. The presence of many characters which are typically found in annelids or other trochozoans (e.g. serial protonephridia, chaetae, trochophora larvae, etc.), but never in platyzoans (flatworms and relatives), were at least confusing. Consequently, myzostomids were treated as missing links between platyzoans and annelids, and numerous reductions of the above mentioned characters were assumed in the platyzoan lineage (Eeckhaut et al., 2000; Zrzavý, 2001a).

Both available molecular analyses addressing the phylogenetic position of myzostomids suffered from either restricted taxon sampling (Eeckhaut et al., 2000) or long branches (Zrzavý, Hypa, and Tietz, 2001), which may positively mislead the analysis. Later on, an analysis of lophotrochozoan relationships based on the 28S rRNA gene came up with a tree uniting myzostomids with bryozoans – a result that was immediately regarded as an artifact by the authors (Passamaneck and Halanych, 2006). Readdressing those three data sets, which were based on 18S rRNA, 28s rRNA and elongation factor-1α (EF-1α), showed that all analyses were likely biased by systematic errors or a limited taxon sampling (Bleidorn et al., 2007). As such, after analyzing an expanded EF-1α data set, no support for a myzostomid-flatworm clade could be shown. Moreover, for both ribosomal genes, parametric bootstrap analyses and spectral analyses gave evidence that the position of myzostomids is likely an artifact due to long-branch attraction.

10.4 Phylogenomics and rare genomic changes – phylogenetic analyses of long-branched taxa

It has been shown that molecular phylogenetic analyses based on one or few genes (like those mentioned above) were often biased due to sampling errors. It is not surprising that incongruent results are widespread in the literature, even when addressing quite similar data sets (Rokas et al., 2003a). The first larger data sets available for non-model organism phylogenetics were based on complete mitochondrial genomes or have been derived from expressed sequence tags (ESTs). Bilaterian mitochondrial genomes are typically circular extrachromosomal molecules of around 16 kb coding 37 densely packed genes, of which two are ribosomal genes, 13 protein-coding genes and 22 tRNAs (Boore, 1999). Analyzing the concatenated amino acid sequences of the first nearly complete myzostomid mitochondrial genome revealed a clade uniting myzostomids with annelids and allies (Bleidorn et al., 2007). This result was later supported by including data from two additional myzostomids (Bleidorn et al., 2009a), including taxa from the two major lineages of myzostomids as shown by Lanterbecq et al. (2006). Much larger data sets became available by sequencing EST-libraries of focal taxa. For this approach total mRNA is directionally cloned and a certain amount of clones are randomly sequenced (Philippe and Telford, 2006). As this is a top-down approach a comparable set of highly expressed housekeeping genes will be recovered across taxa. Even more data became available using RNAseq based on the new sequencing techniques including 454- and Illumina sequencing. Several EST-based phylogenomic studies which included myzostomids have been published and disagreed in the placement of this enigmatic taxon (Figures 10.3 and 10.4). Whereas in Dunn et al. (2008) and Bleidorn et al. (2009a) myzostomids were grouped with platyzoan taxa, Hejnol et al. (2009) and Hartmann et al. (2012) recovered them as annelid relatives. In a fifth study, the placement of myzostomids is still ambiguous and depends on the choice of amino acid substitution model (Struck et al., 2011). Interestingly, when using the CAT-model, which has been shown to be less prone to long-branch attraction, myzostomids group within annelids in this latter study.

An additional source of phylogenetic information comes from the so-called rare genomic changes marker, which became a valuable tool to place problematic taxa (Rokas and Holland, 2000). Especially the analyses of mitochondrial gene order and the absence/presence pattern of microRNAs have been shown to be useful to unravel deep metazoan phylogenetic relationships (Boore and Brown, 1998; Boore, Lavrov, and Brown, 1998; Sperling and Peterson, 2009). Mitochondrial gene order is highly conserved in annelids (Jennings and Halanych, 2005; Bleidorn, Podsiadlowski, and Bartolomaeus, 2006) and the myzostomids investigated so far show an order of protein coding genes which is identical to the one in most annelids (Bleidorn et al., 2007; Bleidorn et al., 2009a). MicroRNAs are involved in the regulation of gene-expression and were shown to be highly conserved across animal taxa. It has been demonstrated that microRNA families are continuously added through bilaterian evolution, but are thought to be only rarely lost secondarily (Hertel et al., 2006; Wheeler et al., 2009). These properties make them a potential powerful phylogenetic tool and they have been included in several phylogenetic analyses (Sperling et al., 2009; Heimberg et al., 2010). By analyzing small RNA libraries of *Myzostoma cirriferum* several microRNA families have been detected which are exclusively shared by annelids and myzostomids (Helm et al., 2012). Summarizing, both data sets based on rare genomic changes marker systems strongly support a close relationship of annelids and myzostomids.

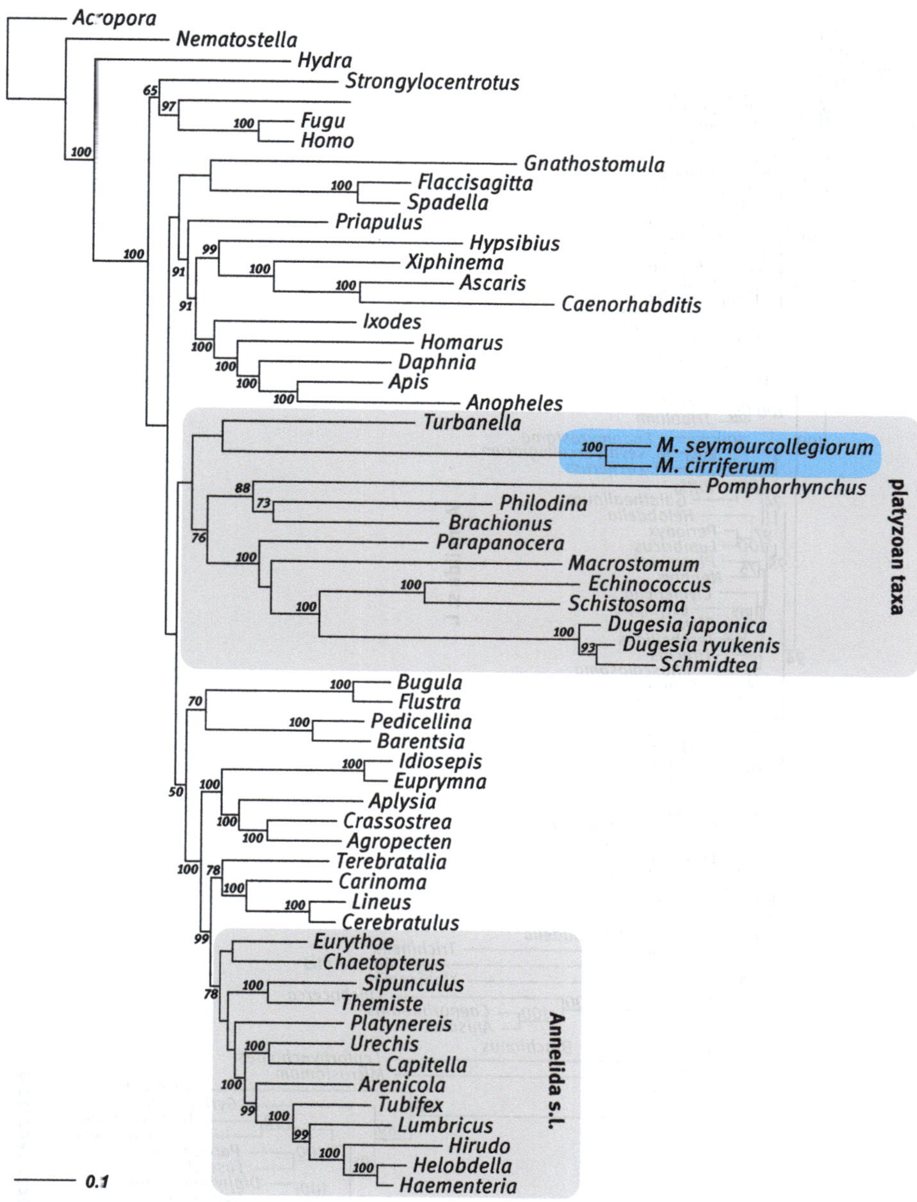

Figure 10.3: Maximum Likelihood analysis of a concatenated amino acid supermatrix based on 77 ribosomal proteins. Myzostomida group together with a couple of long-branched Platyzoa. Analysis details are available in Bleidorn et al. (2009a).

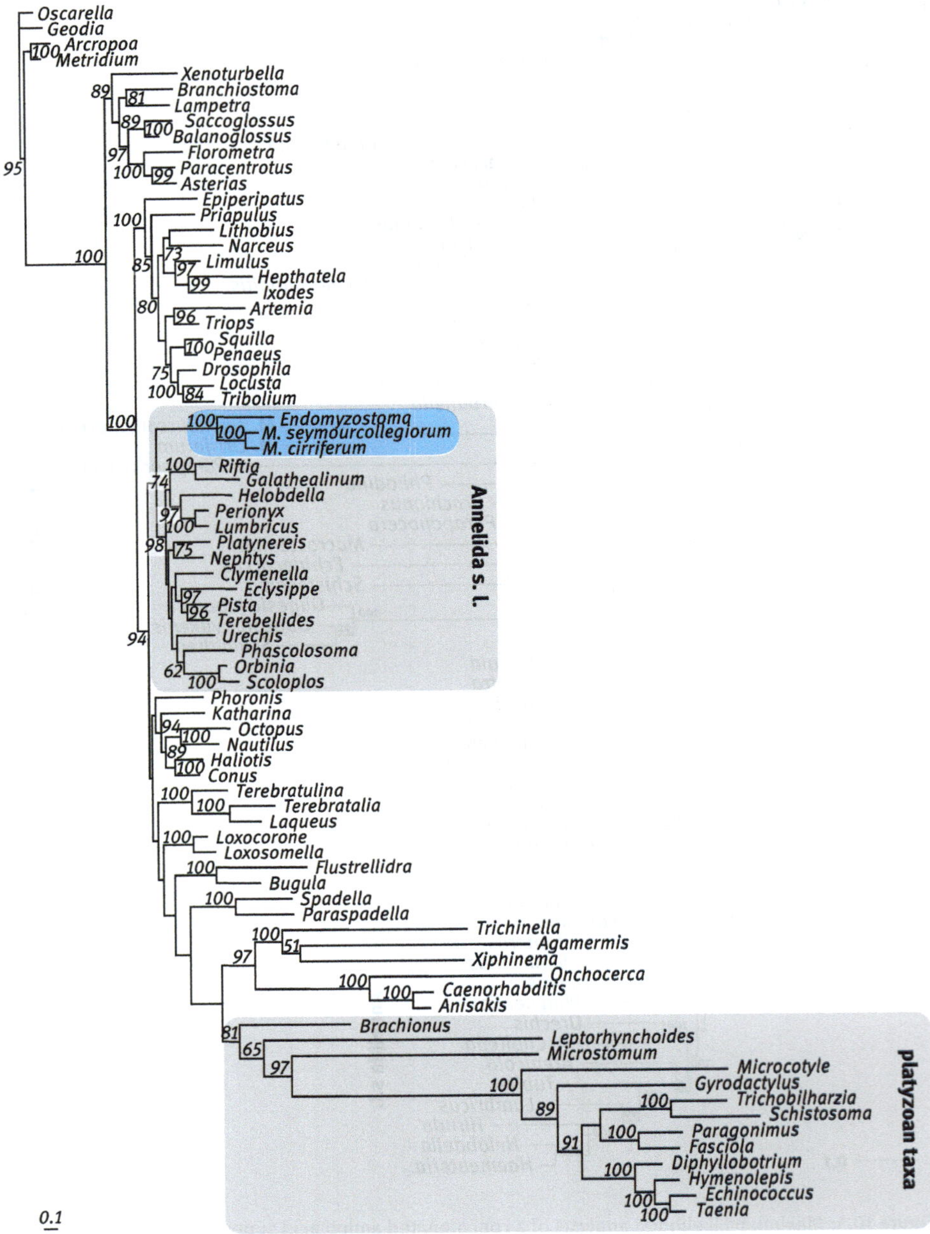

Figure 10.4: Maximum Likelihood analysis of a concatenated amino acid supermatrix based on 12 mitochondrial proteins. Myzostomida group as sister taxon of Annelida. Analysis details are available in Bleidorn et al. (2009a).

10.5 Morphological and evolutionary developmental studies of myzostomids – towards understanding the evolution of a highly adapted body plan

Myzostomids have a long and well-documented history as commensals or parasites of crinoids and few other echinoderm species. The oldest unequivocal traces of myzostomid infestation stem from the Carboniferous (Eeckhaut, 1998) and the obligate symbiosis with crinoids may even date back to the Ordovician (Warn, 1974). Obviously, myzostomids are highly adapted to their hosts and a previous study has shown that diversification in this group is achieved due to co-speciation (Lanterbecq, Rouse, and Eeckhaut, 2010). Using a well-supported molecular phylogeny, Lanterbecq, Rouse, and Eeckhaut (2009) conducted an ancestral character state reconstruction to elucidate the myzostomid ground pattern. Based on their analysis, the ancestor was an ectocommensal species bearing six trunk segments and five pairs of polychaete-like parapodia. The parapodia of myzostomids are in ventral position and the animals use them to walk on their host. These parapodia are often so highly adapted to a given host that they are useless when they are forced to walk on other substrates. Myzostomid parapodia are uniramous and equipped with chitinous chaetae which are ultrastructurally identical to those of other annelids (Lanterbecq et al., 2008). The five pairs of parapodia show a serially repeated set of muscles, including hook retractor muscles, spindle muscles, short protractor muscles, stronger flexor muscles, extensor muscles, and the acicular retractor (Figure 10.5D). Similar arrangements can also be found in other annelids. The musculature of the myzostomid body comprises besides an outer transverse muscle layer and an inner longitudinal one, numerous prominent dorsoventral muscles (Figure 10.5C), which is a rather unusual condition for annelids. Interestingly, due to the convex, respectively concave, arrangement of the body wall muscles a radial symmetry overlays the originally bilateral organization (Helm et al., 2013). Such a superficial radial symmetry can be found in other organs and muscular substructures as well. As many ectocommensal myzostomid species are disc-shaped a similar anatomy can be expected for all of those. However, many endoparastic myzostomids are rather worm shaped (longer than wide) and their detailed anatomy remains to be investigated.

Given their annelid ancestry, myzostomids seem to have lost coelomic cavities of their segments. Even their number of segments is unclear, as these animals possess six pairs of serotonergic perikarya and protonephridia, five pairs of parapodia, and four pairs of ventral organs (Pietsch and Westheide, 1987; Müller and Westheide, 2000). The presence of at least six segments has been assumed for their last common ancestor. Consequently, the nervous system of myzostomids shows some derivations from the rope-ladder-like nervous system which is found in many (but not all) other annelids. For myzostomids, its structure has only been investigated in *M. cirriferum* with help of confocal laser microscopy (Müller and Westheide, 2000). The nervous system consists of a compound brain, connected by a dorsal commissure, a ventral nerve

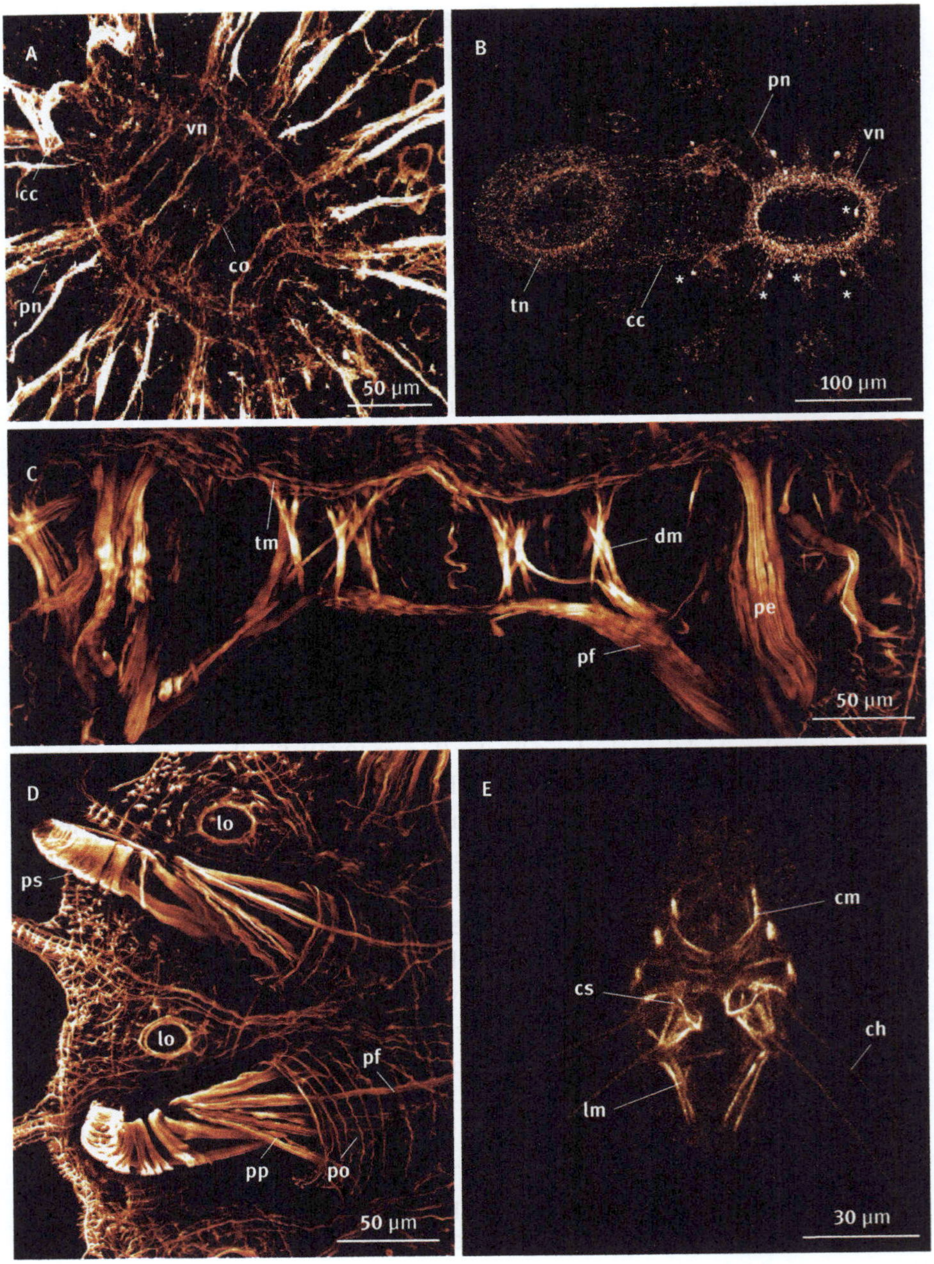

mass and a pair of anterior connectives joining the former to the latter (Figure 10.5B). Two ventral main nerve cords are joined anteriorly and posteriorly. They are connected by twelve commissures resembling a ladder-like nervous system. Moreover, twelve peripheral nerves arise from each cord (Figure 10.5A). Two of them innervate

◄ **Figure 10.5:** Neuronal and myoanatomical characters of *Myzostoma cirriferum*. **A.** anti-tyrosinated α-tubulin immunoreactivity; **B.** anti-FMRFamide immunoreactivity; **C–E.** Phalloidin-rhodamine labeling; confocal micrographs. **A.** Dorsal view, anterior is left. Adult myzostomids have twelve commissures within the ventral nerve mass. Note that the median nerve is not visible in the staining shown. **B.** Dorsal view, anterior is left. The ventral nerve mass, the trunk innervations and parapodial nerves are visible. Note the five pairs of somata (indicated by *asterisks*) at the base of each parapodial nerve. The posterior-most pair is fused. **C.** Cross section, dorsal is up. The entire body is stiffened by dorsoventral musculature. **D.** Detailed ventral view of the parapodia. Myzostomids parapodia are uniramous and embedded in a muscular pouch. **E.** Myzostomids metatrochophore, anterior is up. Note the weak circular musculature. The prominent chaetae exhibit autofluorescence. *cc* circumesophageal connective, *co* commissure, *cm* circular muscle fibers, *ch* chaetae, *cs* chaetal muscle sack, *dv* dorsoventral muscle fibers, *lm* longitudinal muscle fibers, *lo* lateral organ, *pe* parapodial extensor muscle, *pf* parapodial flexor muscle, *pn* parapodial nerve, *po* parapodial pouch, *pp* parapodial protractor muscle, *ps* parapodial spindle muscle, *tm* transversal muscle fibers, indicate FMRFamidergic somata.

the anterior body regions. The remaining nerves form five groups of two nerves, each innervating a pair of parapodia, the corresponding two cirri, and the lateral organs. Additionally, a single median nerve is visible midventrally, representing a common annelid characteristic (Müller and Westheide, 2000). Within and ahead of the ventral nerve mass, serotonergic somata are situated, forming six distinct groups of cells. Furthermore, five pairs of unipolar FMRFamidergic somata are visible at the base of each parapodial nerve. The arrangement of the peripheral nerves, the commissures within the ventral nerve mass and the order of serotonergic and FMRFamidergic somata therefore strongly support a segmented ancestry of Myzostomida.

Other annelids usually exhibit a posterior growth zone and new segments are generated from posterior to anterior (Shankland and Seaver, 2000). In myzostomids it seems to be another mechanism, at least regarding the appearance of parapodia, as it has been observed that they appear in the order 3-4-5-2-1 (Jägersten, 1940). As expected, at least in adult myzostomids no growth zone could be identified by analyzing the expression pattern of a stem-cell marker (Weigert et al., 2013). The genetic mechanism underlying this process remains to be elucidated. Genes which might be involved in the process of segmentation in annelids are *Hox* genes. Myzostomids seem to exhibit the set of *Hox* genes typical for Lophotrochozoa and even show signatures of conserved amino acids in the flanking region of the homeobox in some *Hox* genes, which are otherwise only found in annelids (Bleidorn et al., 2009c). Initial studies suggest that these genes might be organized in a cluster in myzostomids, but a possible colinearity of their expression has not been tested yet. Interestingly, these genes and other transcription factors typically involved in larval development are found to be highly expressed in mRNA-libraries of adult specimens (Hartmann et al., 2012). It remains to be investigated if these genes are involved in the sexual transformation from functional males to females in these organisms.

Whereas the similarity of adult morphology to annelids is for some character complexes hardly visible due to a long evolutionary history of adaptation to a symbiotic/parasitic lifestyle, the anatomy of myzostomid larvae clearly indicate the annelid roots of this taxon. The myzostomid larvae (Figure 10.5E) strongly resemble the trochophora type with the exception that the diagnostic prototroch (sensu Rouse, 1998) is missing (Eeckhaut, Fievez, and Müller, 2003). Later stages bear two pairs of long chaetae, similar to other annelid nectochaeta larvae, e.g., those of Sabellariidae.

10.6 Integration of molecules and morphology to place an enigmatic animal taxon

The case of placing Myzostomida in the tree of life shows the value of integrating data from multiple sources (e.g., morphology, genomics, neurobiology, etc.) to reach a synthesis of available evidence. Genomic approaches made it possible to analyze rare genomic changes markers as absence/presence patterns of microRNA families and mitochondrial gene order which clearly favor an annelid ancestry for Myzostomida. Single gene analyses have been found to be affected by systematic biases leading to putative long-branch attraction artifacts in phylogenetic analyses. Instead, concatenated supermatrix and supertree analyses of the largest available genetic data sets are congruent with the hypothesis of an annelid affinity (Hartmann et al., 2012). Cladistic analyses of morphological data sets were shown to suffer from miscoding and biases in character selection. However, ultrastructural and immunohistological investigations of prominent character complexes (e.g., muscular system, nervous system, chaetae, larval forms) are largely congruent with comparable investigations from annelids. Due to a long history as commensals/parasites of crinoid echinoderms a mixture of character conditions plesiomorphic for annelids and apomorphic for myzostomids is hardly surprising and made phylogenetic studies difficult. Given the organization of the parapodial musculature and the presence of acicula, the myoanotomy of the papilla bearing axial pharynx (Helm et al., 2013), and serial protonephridia in adult individuals, a closer relationship with errant polychaetes (e.g., Phyllodocida) seems probable. Notably, several parasitic and commensal species can be found in this annelid lineage. Placing the myzostomids in the Annelida is the best supported and most plausible hypothesis in the light of the current data.

Acknowledgments

The authors gratefully acknowledge support through the priority program "Deep Metazoan Phylogeny" of the German Research Foundation (DFG) (BL787/2-1 and BL787/2-2 to CB, STR 683/5-2 to THS, TI349/4-1 to RT). CB was supported by the European Union due to ASSEMBLE grant agreement no. 227799. Collection of *Myzostoma cirriferum* was kindly supported by Station Biologique Roscoff, France.

Markus Koch, Björn Quast, and Thomas Bartolomaeus

11 Coeloms and nephridia in annelids and arthropods

Abstract: The Ecdysozoa hypothesis caused a still ongoing debate on the homology of segmentation in Annelida and Panarthropoda. In this study we reconsider the homology of segmental organs and of the mesoderm based on ultrastructural studies of the anatomy of coeloms and nephridia in the course of their development. Our results challenge main traditional views on coelomogenesis in arthropods, including the previous concept of "coelomic sacs" as mesoblastic organs. Transient occurrence of segmental coeloms is refuted for crustaceans but confirmed for chelicerates and insects. When present, the coeloms proper seem to be largely decoupled from organogenesis. Primarily they only give rise to nephridial anlagen and may also act as pathfinding organs for mesodermal blast cells. The main source for pioneer blast cells is consistently apolar segmental mesoderm in arthropods. Its putative origin from coelomic walls is not corroborated by our results, neither is the assumption that the nephridial organs of arthropods are derived from metanephridial systems. Our data on coelomogenesis in yolk-rich embryos of onychophorans cause further ambiguity on ancestral states of mesoderm development in Panarthropoda. Current insights reveal that the body of evidence supporting a homology of coeloms and nephridia in annelids and panarthropods is remarkably poor. Because the coeloms in arthropods show no obvious functional significance and thus are liable to entire suppression, we nevertheless consider historical reasons (rather than convergence) as the more likely explanation for segmental coeloms and nephridia in annelids and panarthropods, despite marked differences in almost all details of coelomogenesis.

11.1 Introduction

The advancement of the Ecdysozoa hypothesis in the past 20 years demands a reevaluation of the traditional assumption that the metameric body organization of annelids and arthropods is homologous and synapomorphic. Homology of segmentation in these animals is generally defended by its apparent complexity, involving characters of the exoskeleton, nervous system, mesodermal organs, and the formation of limbs (e.g., Scholtz, 2002; 2003; Giribet, 2003). Under the growing body of molecular evidence for Ecdysozoa (see, e.g., Telford et al., 2008; Edgecombe, 2009), reconsiderations of the morphological support culminated in the general conclusion that all characters related to segmentation remain debatable and require a more careful and critical examination (e.g., Zrzavý, 2001b; Giribet, 2003; Schmidt-Rhaesa, 2006). In this regard a plausible approach to testing the homology of complex patterns is to subject the individual substructures of the presumable complex to a homology analysis (cf. 'complexity test' fide Scholtz, 2005) under inclusion of developmental data.

Segmental characteristics of the mesoderm have received little attention in recent anatomical re-examinations despite the contribution of two main arguments in support of the Articulata hypothesis, these being the presence of paired segmental coeloms and nephridia. Comparative morphological studies of these organs indeed suffer from severe difficulties. First, coeloms in arthropods are transient structures and occur only in a brief period during the development of a body segment. In the respective stages of their existence they merely form fluid-filled cavities lined by a simple epithelium and thus are largely character-poor from an anatomical point of view; this implies that comparisons with annelids necessarily need to focus on correspondences in the spatio-temporal development of the coeloms and their fate. Comparative studies into the fate of the coeloms especially include the development of nephridial organs, since their common interpretation as metanephridia in arthropods is largely based on the assumption that the filtration unit of these organs, the sacculus, represents a remnant of the embryonic coelom. Second, unambiguous identification of coelomic cavities requires ultrastructural evidence; precise anatomical descriptions accordingly are methodologically elaborate (see below, *Methodological challenges*). Insights obtained by ultrastructural examinations of the coelomo- and nephridiogenesis exist for a variety of coelomate invertebrates, including annelids (reviewed by Rieger and Purschke, 2005; Bartolomaeus and Quast, 2005) and onychophorans (Bartolomaeus and Ruhberg, 1999; Mayer, Ruhberg, and Bartolomaeus, 2004; Mayer, Bartolomaeus, and Ruhberg, 2005; Mayer and Koch, 2005; Mayer 2006), but are largely missing for arthropods. Until recently Anderson's (1973) survey on the embryology of annelids and arthropods formed the background and main reference for arguments inferred from the mesoderm development in arthropods, but the developmental descriptions and phylogenetic conclusions summarized and discussed therein are based solely on previous light microscopical studies of histological sections.

Within the scope of the *Deep Metazoan Phylogeny* program, we primarily aimed at more precise anatomical descriptions of the mesoderm development in arthropods that can immediately be compared with available insights on annelids. For this purpose, we extended ultrastructural studies into coelomo- and nephridiogenesis onto a variety of arthropod taxa to fill in this obvious gap. Our survey particularly focused on the following questions:

(i) Spatio-temporal development of the coeloms: Do coelomic cavities really occur during the mesoderm development of arthropods? Does any arthropod correspond with annelids in showing an anterior-posterior gradient in the formation of the mesoderm?

(ii) Fate of the coeloms: do the coeloms of arthropods, despite their transient nature, contribute to the same mesodermal organs as in annelids, especially to the somatic and visceral musculature as well as to the circulatory system?

(iii) Function of the coeloms: why does the mesoderm development pass a coelomic stage in arthropods – are there any functional constraints?
(iv) Nephridiogenesis: what is the structural and/or developmental evidence that arthropod nephridial organs are derived from metanephridial systems?

This contribution summarizes our main results and likewise provides an updated overview on the structure and development of coeloms and nephridia in annelids to reevaluate their assumed homology in arthropods. At the same time we provide concise definitions, the lack of which has been considered as part of "*pervasive problems of* [previous] *morphological analyses*" (Giribet, 2003: 306; see also Scholtz, 2010).

11.2 Mesoderm, muscle cells and body cavities – definition of terms

Comparison of mesodermal body cavities and nephridia in annelids and arthropods requires a concise terminological framework, since the term mesoderm is variably used in literature in an ontogenetic, anatomical, and/or evolutionary sense. Inference of morphogenetic changes from anatomical descriptions of successive developmental stages poses the problem that descriptive terms used for a particular stage cannot (and should not) be used in an ontogenetic sense; fixed tissues usually do not indicate their ontogenetic origin, especially in early developmental stages. This problem may be overcome by including insights from 4D-microscopy (see, e.g., Hejnol and Schnabel, 2006; Hunnekuhl and Wolff, 2012), but such information is mostly missing for mesodermal structures. In order to avoid terms implying assumptions on ontogenetic origin, the term mesoderm in anatomical descriptions is purely descriptive only if it refers to topographical relation of tissues. The description of relative positions, however, requires a reference to which structures can be related. In animal anatomy the decisive reference is the extracellular matrix (ECM); anatomically the ECM is the only marker to discriminate mesoderm from other tissues. This implies that the term mesoderm in an anatomical sense can only be applied to animals that have already completed gastrulation, i.e. to advanced embryos, larvae, juveniles, and adults. This restriction nevertheless allows a much more precise identification and description of mesodermal structures, since the more generally a term is defined, the less precise are the criteria to identify structures to which the term can be applied. As the definitions of the terms used in this paper depend on each other, we start hierarchically defining the term *extracellular matrix* followed by the triplet *ectoderm*, *mesoderm* and *entoderm*. These definitions form the backbone for subsequent definition of *mesodermal body cavities*, *nephridia*, and *musculature*. Every definition receives a brief commentary. Stipulative definitions are provided here to keep them free of any homology assumption.

11.2.1 Extracellular matrix

Extracellular matrix (ECM) is the material on which epithelially-organized cells rest or into which non-epithelial cells are embedded. In contrast to Pedersen (1991) this definition excludes extracellular material that faces the outer medium, like glycocalyx or cuticle. The term ECM includes the terms *basal lamina* or *basal membrane* (Laurie and Leblond, 1985, Pedersen, 1991) or *basal matrix* (Fransen, 1982: 178), all of which refer to the densely stainable layer representing the subepidermal part of the ECM, generally composed of a *lamina lucida, lamina densa* and *lamina fibroreticularis*. The latter, however, is continuous with the remaining ECM as already noted by Laurie and Leblond (1985) and Pedersen (1991). There is accordingly no justification to discriminate between basal lamina, basal membrane and basal matrix on the one hand and ECM on the other. According to the above definition the terms *basal lamina, basal membrane* or *basal matrix* describe parts of the ECM, as do their subcomponents (the different laminae noted above). In this sense the ECM contains collagens, laminin, fibrinectin and glycoproteins among other proteins. Of these, collagen type IV, laminin and fibronectin predominate in the *lamina lucida* and *lamina densa* (sensu Pedersen, 1991), while glycoproteins and fibrillar collagen predominate in the *lamina fibroreticularis*.

11.2.2 Ectoderm, entoderm and mesoderm

In comparative zoology these three terms are used either to describe germ layers during ontogenesis (e.g., Siewing, 1969; Brusca, Brusca, and Gilbert, 1997), or as topographical terms which only refer to the relative position of tissues and organs within organisms (Ruppert, 1991b; Bartolomaeus, 1993; Ax, 1995). In this paper they are exclusively used in a topographical sense. Whenever developmental aspects are mentioned in this paper, the presence of an ECM between the three kinds of tissue is used as a marker to distinguish between ectoderm, entoderm, and mesoderm. Thus, application of these terms depends on the distribution of an ECM within the organism, so that these terms can only be applied to organisms and developmental stages which formed an ECM.

Ectoderm is any outer epithelium that faces the surrounding medium. This definition describes all epidermal structures of the Metazoa. Structures sunk below the epidermal surface are referred to as ectodermal if they are continuous with the epidermis and completely surrounded by the subepidermal ECM.

Entoderm is any epithelium lining an extracorporeal cavity that is permanently surrounded by the body and connected to the exterior medium by at least one opening. This definition can only be applied to intestinal structures of the Eumetazoa. Beside the gut, all structures that may be sunken into deeper tissue layer are regarded as entodermal if they are connected to this epithelium and rest on a common ECM. This implies that (1) the entire gut is referred to as entodermal; discriminating certain foregut or hindgut sections as ectodermal is excluded. This restriction results from the observations that the extent of ectodermal sections of the gut varies within the Eumetazoa and that ectodermal parts can only be recognized by analyzing the development or cytological details. Thus, they cannot be considered in a topographical definition of tissues. (2) The definition of entoderm is restricted to the intestinal systems of the Eumetazoa. Thus, the ventral epithelium of the Placozoa, which consists of gland cells and cylinder cells, is not entoderm (since it is not permanently surrounded by the body), although it can plausibly be homologized with the intestinal epithelium of the Eumetazoa.

Mesoderm is any tissue that lies between ectoderm and entoderm and is surrounded by ECM. This definition of mesoderm is nearly identical with the one given by Ruppert (1991b) and characterizes tissues irrespective of their germ layer origin. Any tissue entirely surrounded by ECM between the epidermis and the intestine thus is designated as mesoderm. The mesoderm either lacks or contains fluid-filled spaces. The latter are called mesodermal body cavities. If such cavities are absent, the mesoderm is compact. A single exception to this definition concerns the nephridial organs that require separate definition (see below, *Filtration nephridia*). Segmental cell clusters of the mesoderm are commonly termed "somites" in chordates, and this term has variably been applied to arthropods as well, either to describe segmental mesodermal units (whether compact or coelomate; see Anderson, 1973), or entire body segments. For its inconsistent usage in arthropods (see Scholtz, 2010) and vertebrates (e.g., Piekarski and Olsson, 2007), we entirely avoid the term "somite" in our survey.

11.2.3 Mesodermal body cavities

A mesodermal body cavity is any fluid-filled cavity or space within the mesoderm. Depending on its lining, two types of mesodermal body cavities can be distinguished:

A primary body cavity is a mesodermal body cavity that is lined by ECM. According to this definition the so-called pseudocoel of nemathelminth taxa, the haemocoel or mixocoel in arthropods, and the blood vessels of annelids, among other cavities, are primary body cavities. This definition contrasts against the general belief that the primary body cavity of animals is a remnant of the blastocoel, being called primary body cavity for representing the cavity that occurs first during development (i.e. in an ontogenetic sense). Until today, however, no animal is known in which the primary body cavity of juveniles or adults is derived from the blastocoel.

A secondary body cavity is a mesodermal body cavity that is lined by epithelial cells. According to this definition the coeloms of annelids, the sacculi of nephridial organs in arthropods, or the blood vessels of nemertines and vertebrates are secondary body cavities. The term "secondary body cavity" represents an umbrella term for any fluid-filled mesodermal cavity lined by an epithelium irrespective of homology, and is accordingly not synonymous with "coelom".

A coelom is a secondary body cavity that encloses the intestine. Ruppert (1991b) specified coeloms as "perivisceral coeloms" to express their main characteristic, i.e. the topographical relationship of a coelom to the gut, which is partially or entirely surrounded by a coelom. Since perivisceral coeloms can become "*occluded by specialization of their lining cells*" (Ruppert, 1991b: 9) during development and/or evolution, identification of a secondary body cavity as a coelom requires evidence that this cavity at least transiently fulfills this characteristic.

The definitions of the terms ectoderm and entoderm might lead to uncertainties in the description of structures connecting mesodermal cavities to the exterior. Such ducts are either part of filtration nephridia or are gonoducts that are connected to the exterior by pores and surrounded by an ECM that is continuous with the subepidermal or peri-intestinal ECM. Topographically they are thus difficult to subsume under ectoderm, entoderm, or mesoderm, as they can anatomically be interpreted likewise as ectodermal or entodermal invaginations, or as mesodermal evaginations. To avoid misunderstandings the efferent ducts of mesodermal body cavities are defined separately in the following irrespective of their origin, the way they are formed, and whether they are internally closed or show an open internal connection to secondary body cavities.

11.2.4 Filtration nephridia

Filtration nephridia primarily represent functional units, defined by their function of eliminating nitric wastes from the body in two steps, an initial ultrafiltration followed by modification of the ultrafiltrate (e.g., re-absorption of water and ions). Two kinds of filtration nephridia are known from annelids and arthropods: protonephridia and metanephridial systems.

Protonephridia are filtration nephridia that consist of terminal cells serving in ultrafiltration that are connected to an efferent duct along which the ultrafiltrate is modified. According to this definition, protonephrida are ultimately characterized by the permanent connection between the terminal cells and the duct. Filtration nephridia of this kind accordingly represent not only functional units, but also structural entities.

Metanephridial systems are filtration nephridia that consist of a metanephridium and perivascular podocytes serving in ultrafiltration. In contrast to protonephridia, metanephridial systems do not represent structural entities as the site of ultrafiltration is spatially separate from the efferent duct.

A metanephridium (or gonoduct) is a multicellular, ciliated efferent duct that permanently connects secondary body cavities to the exterior. The terms metanephridia and gonoducts are synonymous in the defined sense, but are generally specified (in their terminology) according to their function. A structural characteristic refers to the ciliation: at least those cells of metanephridia or gonoducts are ciliated that are connected to the lining cells of the secondary body cavity; these cells form a ciliated funnel and its cilia always extend into the secondary body cavity.

11.2.5 Muscular cells and coelomic lining cells

Two basic muscular elements were already recognized in the treatise on coelomic theories by Hertwig and Hertwig (1881): *epithelio-muscle cells* and *fiber muscle cells*. This distinction seemingly got lost through subsequent decades; Rieger (1986) was the first to again focus attention on this central aspect when comparing body cavities.

Epithelio-muscle cells are myofilament containing cells that show a distinct polarity in basally resting on an ECM and apically facing an inner or outer space, and that are integrated into an epithelium by apical adherens junctions. Cells of this kind may be monociliated and basally may attach to ECM via hemidesmosomes and dense plaques. Among eumetazoans they can occur both in the ectoderm and entoderm, and also within mesoderm as lining cells of secondary body cavities, for instance in the coelomic epithelium of annelids.

Fiber muscle cells contain a contractile actin-myosin system, show no specific polarity and are almost completely embedded in an ECM. Two different cell-matrix adhesion structures are known in muscle cells of this kind. Dense plaques connect the intracellular actin-filaments to the ECM, whereas hemidesmosomes link intermediate filaments to the ECM. In contrast to epithelio-muscle cells, they are never interconnected by adherens junctions.

Beside epithelio-muscle cells, secondary body cavities may either be lined by *peritoneal cells* which never contain myofilaments, or by *podocytes* which are peripherally slashed and presumed to be involved in ultrafiltration (see Bartolomaeus, 1994: 174).

11.3 Methodological challenges

Almost all previous descriptions of coelomogenesis in arthropods are based on histological studies, in particular those surveyed by Anderson (1973, 1979). Recent re-investigations by transmission electron microscopy (TEM) (e.g., Bartolomaeus, Quast, and Koch, 2009) demonstrated that insights derived from histology remain ambiguous. This especially concerns species with yolk-rich embryos in which mesodermal body cavities tend to be strongly compressed (Figure 11.1). Since the properties of yolk per se render sectioning difficult (see, e.g., Kimble et al., 2002), previous insights are partially biased in favor of results obtained from less yolky embryos (or body segments, respectively) in which coeloms, as far as they are present, usually have larger dimensions. Independent of yolk content, the epithelial lining of coelomic cavities may furthermore be extremely flat, such that they are hardly recognizable as epithelia in light microscopy (see, e.g., Mayer, Bartolomaeus, and Ruhberg, 2005). Coelomic cavities also seem to underlie osmotic pressure as they tend to collapse or to disrupt depending on the fixative used for histological studies. In comparison to fixatives for ultrastructural studies (usually based on glutaraldehyde, e.g. Karnovsky solution), the popular use of Bouin's fixative (alcoholic solution of picric acid and formaldehyde) in embryonic studies proved in our studies to produce artificial 'body cavities' within factually compact mesoderm (Figure 11.1). The reliability of previous light microscopical studies of organogenesis in arthropods accordingly overall remains dubious.

Unambiguous identification of mesodermal body cavities requires subcellular markers involved in cell adhesion and cell polarity. Classically epithelia are evidenced by apical adherens junctions ("zonula adherens"), linings by ECM instead by hemides-

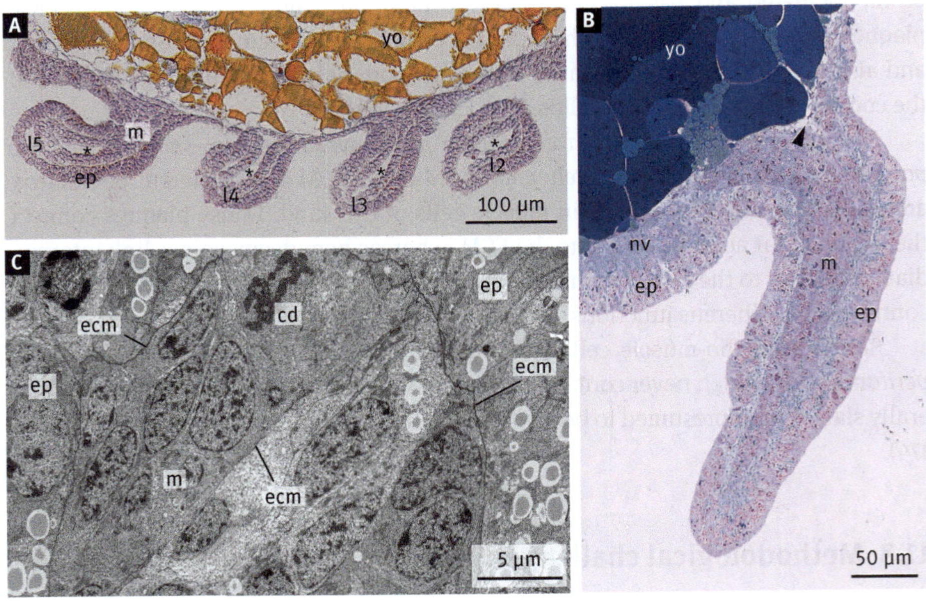

Figure 11.1: *Pholcus phalangioides* (Chelicerata, Araneae), embryonic stage 14. Depiction of fixation artifacts. **A.** Paramedian sagittal section (LM) through Bouin-fixed embryo, anterior is right. Bouin-fixed embryos show limb anlagen with large mesodermal cavities (*asterisks*) seemingly bordered by an epithelium. **B.** Semithin transverse section (LM) through third prosomal limb anlage, embryo fixed with glutaraldehyde and postfixed with osmium tetroxide. The limb anlage is massively filled with mesenchyme (*m*). A small coelomic cavity (*arrowhead*) only occurs above the limb anlage. **C.** Transverse section (TEM) through same limb anlage, revealing its mesoderm as to only comprise apolar mesenchymal cells (*m*) surrounded by extracellular matrix (*ecm*); some cells differ in cytoplasmic staining and underlie cell division (*cd*). *cd* cell division, *ecm* extracellular matrix, *ep* epidermis, *l2–l5* prosomal limbs 2–5, *m* mesodermal tissue, *nv* nerve tissue, *yo* yolk cells of the gut anlage.

mosomes (Figure 11.2). Despite common usage, the definition of an epithelium as a layer of polarized cells (basally resting on ECM, apically interconnected by adherens junctions) basically excludes application of the term epithelium to cellular layers bordering blood vessels (e.g., Pass, 1991) or gonads (follicular "epithelium"; see, e.g., Brown, Sieglaff, and Rees, 2009) in arthropods, since these layers usually do not show the specific polarity of epithelia. Most recent studies on coelomogenesis in invertebrates made use of transmission electron microscopy (TEM) to differentiate the respective subcellular markers. Studies by TEM, however, are highly time consuming. We increased the efficiency of our ultrastructural comparisons via electronic image analyses with the software *Fiji* (open source, see Schindelin et al. 2012). The TrakEM2-module of this software (Cardona et al., 2012) permits aligning micrographs of ultrathin and histological sections of the same specimen (generated during sectioning by regular alternation of the sections' thickness: depending on body size 20–50 semithin sections (0.5–1.0 µm thickness) followed by about 20 ultrathin sections (about 70 nm thickness)) to zoom into apparent body cavities until the nature of their lining becomes evident (Figure 11.3A–D). This approach has several advantages: it enables immediate comparison with previous descriptions of arthropod organogenesis based on light microscopy, it permits restricting ultrastructural studies to a minimum required for unambiguous interpretations of the histological sections, and thereby it offers the possibility to use the histological sections for computer-aided 3D-reconstructions.

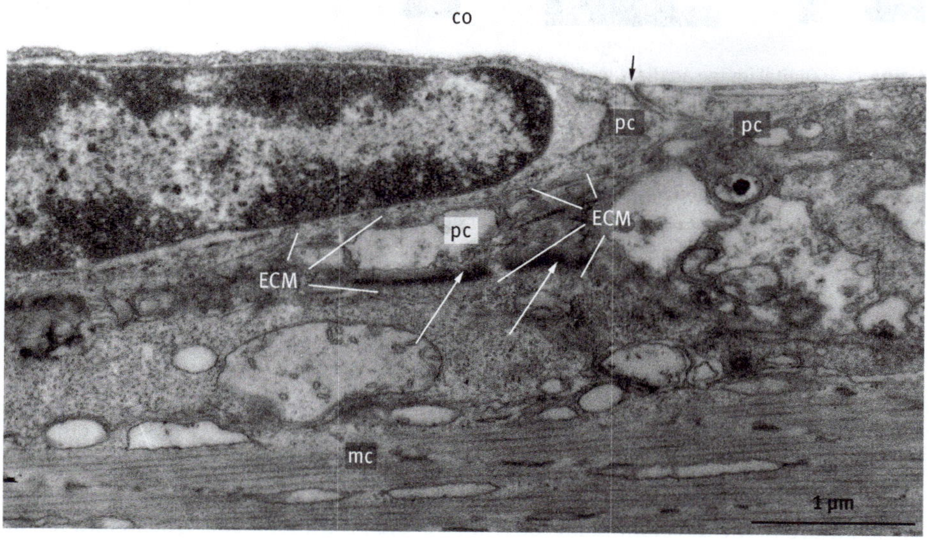

Figure 11.2: *Harmothoe sarsi* (Annelida, Polynoidae), non-fertile adult. Adherens junctions and hemidesmosomes. Somatic coelomic lining consists of a peritoneum whose cells (*pc*) are epithelial and interconnected by adherens junctions (*arrowhead*). Basally the cells adhere to the extracellular matrix (*ECM*) with hemidesmosomes (*arrows*). *co* coelom, *mc* muscle cells.

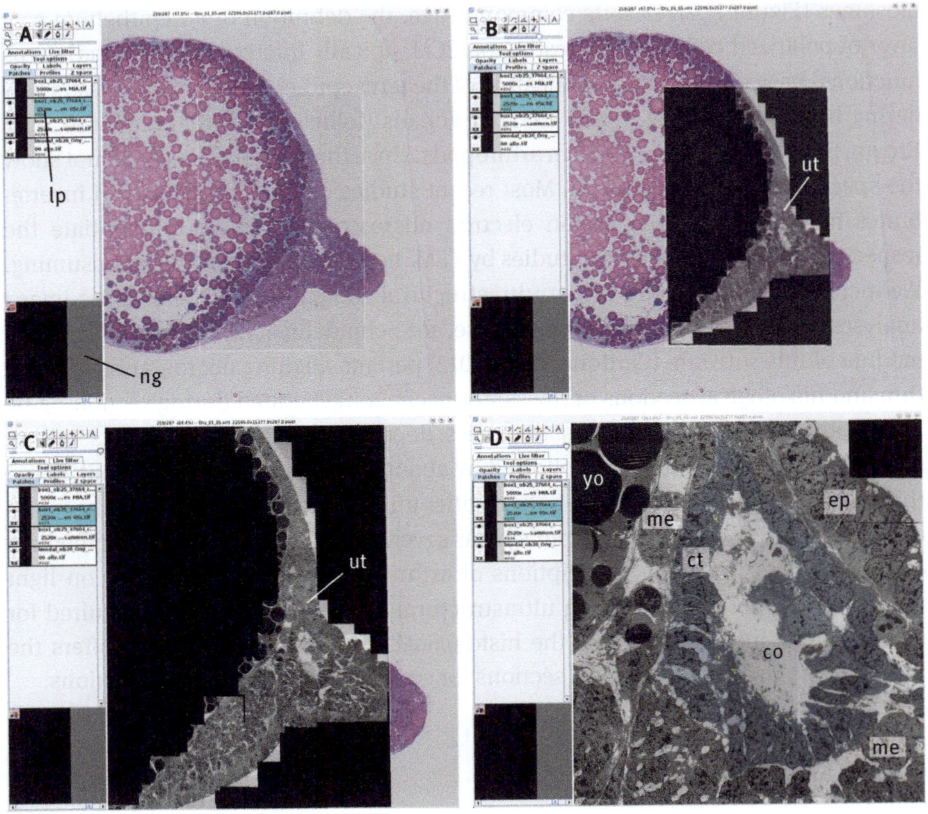

Figure 11.3: Digital image analysis with *Fiji* and the TrakEM2-module. The software combines images of semithin sections with interleaved images of ultrathin sections. **A.** Image stacks of semithin sections are loaded. The software keeps track of the current position within the stack by a patch-list (*lp*) and a canvas navigator (*ng*). **B.** Tiles of ultrastructural images (*ut*) are stitched together and used as overlay on the corresponding semithin section. **C.** After alignment to the semithin section the ultrastructural tiles are made opaque and used for detailed analysis. **D.** The TrakEM2-module allows zooming into the images up to their original resolution; ultrastructural details can be investigated within the overlaid images. Drawing tools are used to label structural features (here: coelothelium labeled transparent blue) at arbitrary levels of resolution. *co* coelomic cavity, *ct* coelothelium, *ep* epidermis, *lp* patch list, *me* apolar mesenchyme, *ng* canvas navigator, *ut* tiles of ultrastructural images, *yo* yolk cells.

We also aimed at including confocal laser scanning microscopy (cLSM) into our studies of coelomogenesis. Phalloidin staining of F-actin particularly facilitated detection of musculature in the developing mesoderm (see, e.g., Bartolomaeus, Quast, and Koch, 2009). In order to detect coelomic cavities, we used various protocols to label specific components of adherens junctions (e.g., cadherins, β-catenin, or Armadillo; see Harris and Tepass, 2010). This technique was successfully applied to embryological studies of *Drosophila melanogaster* (e.g., Sawyer et al., 2009, and references therein) but failed in all our studies of other arthropods. We nevertheless consider this technique as most promising for quantitative approaches, i.e. for a denser taxon sampling.

Current insights summarized in this chapter particularly refer to new ultrastructural data on the following panarthropod species: *Cephalofovea clandestina* Reid, Tait, Briscoe & Rowell, 1995 (Onychophora); *Nymphon gracile* Leach, 1814 and *Achelia vulgaris* (Costa, 1861) (Pycnogonida); *Limulus polyphemus* (Linnaeus, 1758) (Xiphosura); *Pholcus phalangioides* (Fuesslin, 1775) (Araneae); *Artemia salina* (Linnaeus, 1758) (Branchiopoda); *Tisbe* sp. (Copepoda); *Folsomia candida* Willem, 1902 (Collembola); *Ctenolepisma longicaudata* Escherich, 1905 and *Thermobia domestica* (Packard, 1873) (Zygentoma); and *Carausius morosus* (Sinéty, 1901) (Phasmida).

11.4 Body cavities and nephridia in Annelida

Except for certain progenetic species all annelids possess primary and secondary body cavities from the metatrochophore stage onward throughout their entire life. Their primary body cavity is generally called the blood vascular system; their secondary body cavity is the coelom primarily consisting of serially arranged paired cavities that are separated from each other by median mesenteries and anterior and posterior septa (dissepiments) (Ruppert and Carle, 1983) (Figure 11.4A,D). The blood vascular system is traditionally considered a remnant of the blastocoel that remains as a small dorsal and a small ventral blood vessel within the mesenteries between each pair of coeloms and as transverse vessels within the dissepiments between subsequent coeloms (Korschelt, 1936; Brusca, Brusca, and Gilbert, 1997) (Figure 11.4B,C). There is, however, strong evidence from light and electron microscopy that in several species the blood vessels are formed from small interstices within the extracellular matrix (ECM) and are not derived from the blastocoel (Anderson, 1959; Smith, 1986; Bartolomaeus and Ruhberg, 1999). In some annelid subgroups the blood vascular system is reduced, for example in hirudinean and tomopterid species. There is also no blood vascular system in sipunculids, which according to recent molecular studies are members of the Annelida (Struck et al., 2011).

The coelomic cavities are generally assumed to be segmentally arranged in all annelids, but exceptions are found in a large number of subgroups where septa either are lacking throughout the entire body (Tomopteridae, Hirudinea, Echiurida, Sipuncula) or in some body regions (the abdomen of certain species of the Terebellida, the trunk in sternaspidan species, the thorax of arenicolidan species). The septa are incomplete in several annelid species that belong to different subgroups so that incomplete septa allow bypassing the blood vascular system. The coelomic fluid can be transported within and between different coelom compartments by cilia on peritoneal cells (see Figures 11.7D, 11.8B) (Fransen, 1988), or by body wall contraction (Fransen, 1980a), or by both. Moreover, the loss of complete septa is often associated with the reduction of the blood vascular system (Sipuncula, Hirudinea, Tomopteridae)

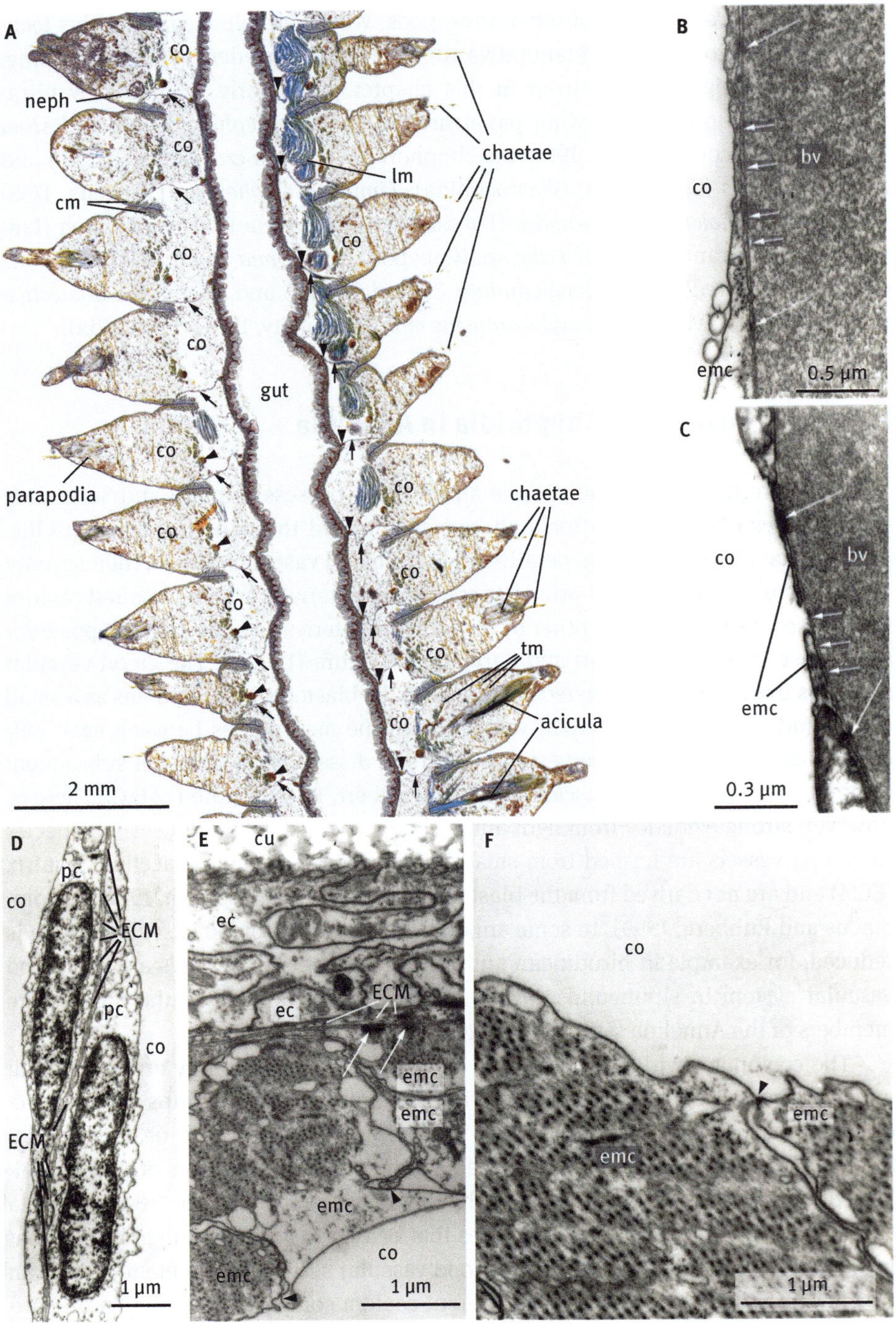
A
co
neph
cm
gut
lm
chaetae
parapodia
tm
acicula
2 mm
B
bv
co
emc
0.5 µm
C
bv
co
emc
0.3 µm
D
pc
co
ECM
1 µm
E
cu
ec
ECM
emc
co
1 µm
F
co
emc
1 µm

◂ **Figure 11.4:** *Nereis diversicolor* (**A**) and *N. pelagica* (**B–F**) (Annelida, Nereididae). Coelom (*co*) and coelomic lining. **A.** Midhorizontal section (light microscopy, Azan staining) to illustrate the general polychaete annelid organization with segmental coeloms that are here filled with developing eggs. Note that septa (*arrows*) are not aligned with the external segmental borders. Longitudinal muscles (*lm*) form strong lateral strands, circular muscles (*cm*) do not form a continuous layer, and transversal muscles (*tm*) adhere to the chaetal sacs and acicula. Blood vessels (*arrowheads*) are segmental. **B, C.** Blood vessels (*bv*) are lined by the lamina densa (*small arrows*) of the ECM. Epithelio-muscle (*emc*) cell with few myofibrils adhere to the lamina densa by hemidesmosomes (*large arrows*) to line the coelomic cavity (*coel*); they may form a fenestrated epithelium (**C**). **D.** Septa are formed by two coelomic linings resting base to base on a common extracellular matrix (*ECM*); its cells are primarily peritoneocytes (*pc*). **E, F.** The somatic coelomic lining of the parapodia consists of epithelio-muscle cells (*ecm*) that apically are interconnected by adherens junctions (*arrowhead* in **F**) and adhere to the extracellular matrix by dense plaques (*white arrows*) *cu* cuticle, *ec* epidermal cell.

as the aseptal coelom adopts the transporting function (oxygen, nitric wastes, nutrients) of the blood vascular system. Ruppert and Carle (1983) therefore introduced the term “coelomic circulatory system” to describe fluid exchange between and within coelomic compartments as an integral functional aspect of coelomic cavities. In addition to transport of coelomic fluid a complete loss of septa in certain body regions generally has functional reasons correlated with specific life habits, for instance with burrowing in arenicolids or with a sedentary life style in a tube or firm burrow in certain terebellidan species. Since reduction of septa evolved secondarily for a variety of functional reasons, the phylogenetic significance of the complete or partial reduction of the septa is most likely restricted to lower hierarchical levels. In no instance did the reduction of septa lead to a change in the lining of the body cavity. It is always an epithelium that lines the body cavity, which characterizes the body cavity as a coelom irrespective of the degree of septa reduction.

In Hirudinea and in several small meiofaunal annelid species the coelomic cavities are partially or completely reduced. Hirudinean phylogeny argues for an evolutionary transformation series from segmental coelomic cavities towards their complete modification into a system of anastomosing, endothelialized vessels (Hammersen and Staudte, 1969; Purschke et al., 1993). The meiofaunal annelids are generally regarded as progenetic species as they show striking similarities with postlarval or juvenile states of macrofaunal annelids (Westheide, 1986). Fransen (1988) classified their different coelomic states in an evolutionary series ranging from epithelia lining a small coelomic cavity to a complete lack of any coelom in *Dinophilus* species. In the latter a typical acoelomate condition with the mesoderm consisting of apolar cells only is found, while in the other progenetic species the epithelial character of the coelomic lining is maintained, irrespective of whether or not a small coelomic space is left (Smith, Lombardi, and Rieger, 1986).

In annelids, fluid-filled coelomic cavities may act as a hydrostatic skeleton (Clark, 1964), in gas exchange (Weber, 1980), in nutrient/waste transport (Eckelbarger and

Grassle, 1982; Zerbst-Boroffka and Haupt, 1975), or as repositories for gametes (Schroeder and Hermans, 1975). The latter two functions explain the tight anatomical correlation between coelomic cavities, nephridia and gonoducts (Bartolomaeus, 1999; Bartolomaeus and Quast, 2005), since the coelomic cavities in annelids are drained by metanephridia that in several annelid species also serve in releasing gametes from the coelomic cavities. These metanephridia are the modifying unit in a functional class of bilaterian excretory organs called filtration nephridia. As such, these organs always consist of an ultrafiltrating unit and a unit that modifies the ultrafiltrate (Ruppert and Smith, 1988; Bartolomaeus and Ax, 1992). Both sites either form a structural unit called a protonephridium or a functional unit called metanephridial system that is spatially separated into metanephridia and podocytes or fenestrated epithelia (Figure 11.14). In annelids with metanephridia, perivascular podocytes serve in filtration, so that ultrafiltered blood fluid is released into the coelom. From here, the ultrafiltrate and, thus, the entire fluid inside the coelomic cavity are modified during its passage through the metanephridia to the exterior. Protonephridia generally do not co-occur with podocytes; their ultrafiltering unit is the terminal cell which is directly connected to the nephridial duct. While nearly all annelid larvae possess protonephridia ("head kidneys"; see, e.g., Bartolomaeus and Quast, 2005; Kato, Lehrke, and Quast, 2011), these organs are sometimes also present in postlarval body segments, so that their segmental nephridia are protonephridia. Within annelids, thus, both kinds of filtration nephridia are present, either as ontogenetically subsequent organs or as permanent segmental nephridia. In annelids the course of nephridiogenesis and its interdependence with coelomogenesis lead to the one or the other kind of segmental nephridia (Bartolomaeus and Quast, 2005). This interdependence will briefly be summarized below (see Section 11.4.3).

11.4.1 Coelomic lining in Annelida

In contrast to the traditional assumption put forth by Hyman (1951) the annelid coelom is not completely lined by a non-muscular peritoneum that covers the intestinal and somatic musculature (Figure 11.5A). Earlier light microscopical studies (Orrhage, 1964) already mentioned that the peritoneum is incomplete in several species; a visceral peritoneum covering the coelomic side of the gut musculature as assumed by Hyman (1951) is lacking. Comparative surveys into the ultrastructure of the coelomic lining by Fransen (1980a; 1980b; 1988) and Bartolomaeus (1994) and additional data on selected species (Chien, Stephans, and Healey, 1972, Gardiner, 1992; Gardiner and Rieger, 1980; Rieger and Lombardi, 1987, Smith, Lombardi, and Rieger, 1986) corroborated this observation and elucidated that a peritoneum is partly or completely lacking in most polychaete species (summarized by Rieger and Purschke, 2005) (Figure 11.5B). Instead, a myoepithelium, the cells of which act as somatic and visceral muscles, lines the coelom in most annelid species (Figure 11.5C). Thus, the terms coelomic lining

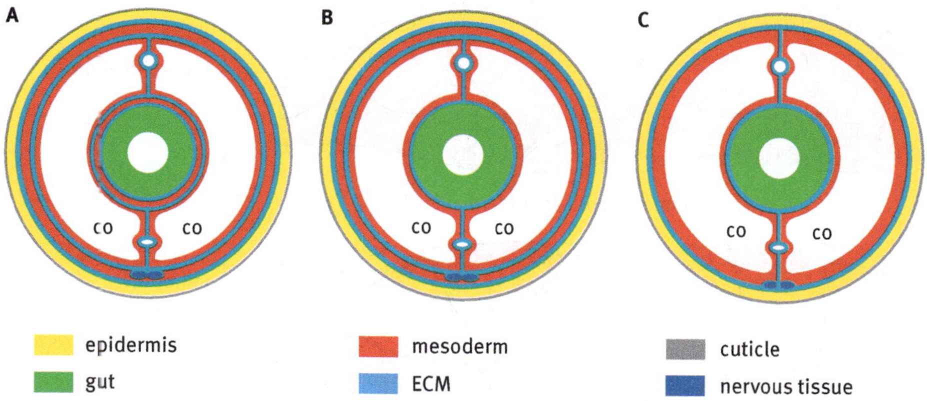

Figure 11.5: Generalized scheme of a transverse section through an annelid body segment. **A.** Hyman's (1951) incorrect representation. Until today a visceral peritoneum has not been described in annelids. **B.** A peritoneum separated from the underlying body wall muscles by ECM has been recorded in some annelid species, while their visceral lining always forms the gut muscles. **C.** In other annelid species a peritoneum is lacking and the coelomic lining forms both the gut and the body wall muscles. *co* coelom.

and peritoneum are not equivalent; a peritoneum is just a special kind of the coelomic lining. Whenever a myoepithelium lines the coelomic cavity its muscle cells form the somatic, visceral, septal, or mesenterial musculature. Rieger and Lombardi (1987) classified the coelomic lining into five types; their classification is adopted here.

Definitions. A coelomic lining is a layer of cells that are situated between a subepithelial basal ECM (consisting of a *lamina densa* plus a *lamina fibroreticularis*) and the coelomic cavity. In annelids it consists of either a **myoepithelium** (Figure 11.6A-D) or a non-muscular **peritoneum** (Figure 11.6E). Myoepithelia are either simple, uniformly pseudostratified, bipartite pseudostratified, or stratified. In a **simple myoepthelium** (Figure 11.6A) all cells are epithelial in a strict sense, i.e. all show apicobasal polarities in always resting on a subepithelial ECM, while their apex faces the coelomic cavity, and all are linked to each other by apical adherens junctions. This is not the case in the remaining types of myoepithelia. In both types of pseudostratified myoepithelia all cells rest on a subepithelial ECM, but not all of them are interconnected by adherens junctions. While in a **uniformly pseudostratified myoepithelium** (Figure 11.6B) epithelio-muscle cells and peritoneal cells face the coelom, only peritoneal cells face the coelom in a **bipartite pseudostratified myoepithelium** (Figure 11.6C). In a **stratified myoepithelium** (Figure 11.6D) those cells that rest on the subepithelial ECM are overlaid by peritoneal cells that do not have contact to any subepithelial ECM. Electron microscopy and serial sectioning are required to unambiguously discriminate between the different types of coelomic linings. Peripherally overlapping epithelio-muscle cells may erroneously cause the impression of a bipar-

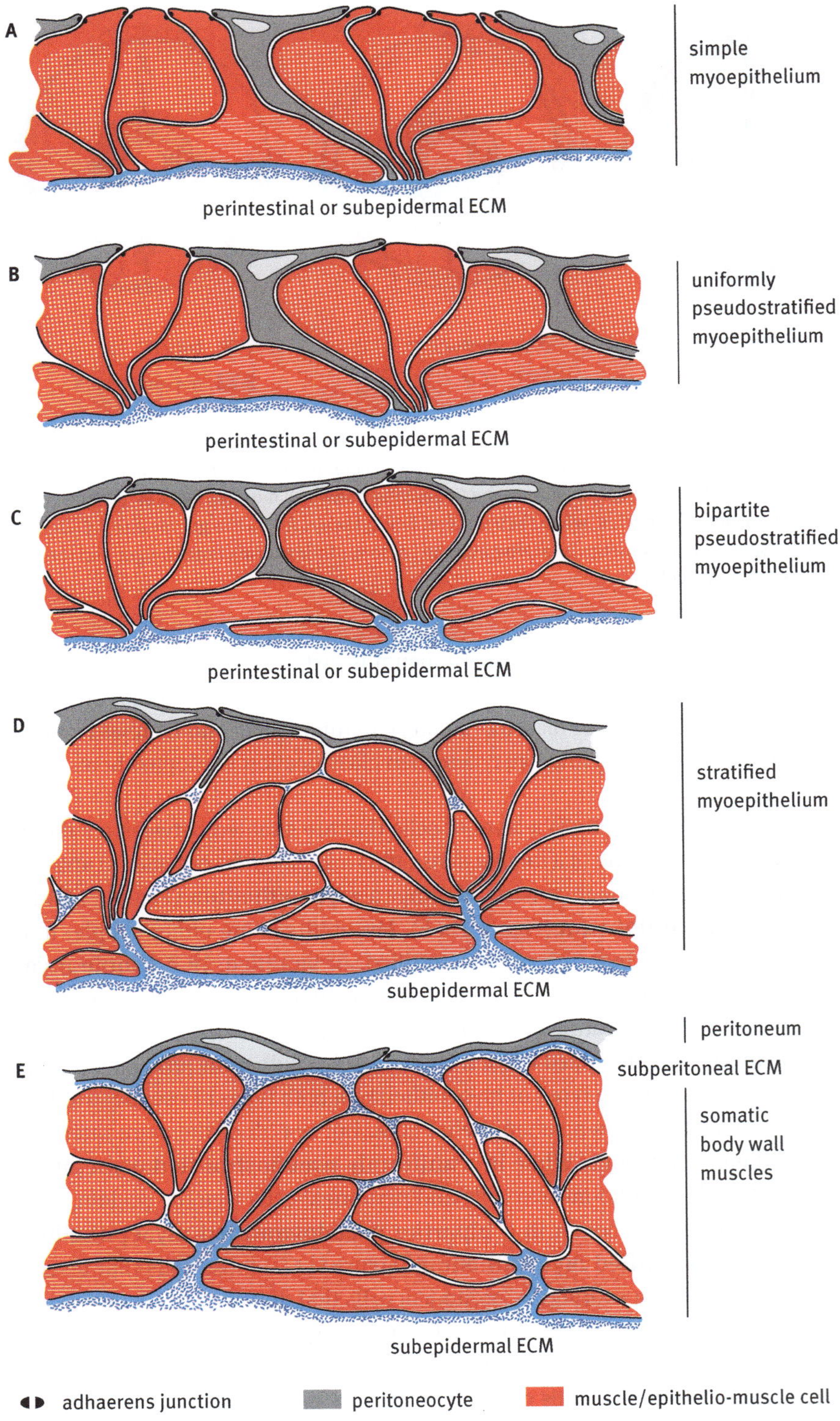
A
simple myoepithelium
perintestinal or subepidermal ECM
B
uniformly pseudostratified myoepithelium
perintestinal or subepidermal ECM
C
bipartite pseudostratified myoepithelium
perintestinal or subepidermal ECM
D
stratified myoepithelium
subepidermal ECM
E
peritoneum
subperitoneal ECM
somatic body wall muscles
subepidermal ECM
adhaerens junction
peritoneocyte
muscle/epithelio-muscle cell

◂ **Figure 11.6:** Scheme of the organization of the different types of coelomic lining. A peritoneum always rests on a subperitoneal ECM, any myoepithelium rests either on the subepidermal, the peri-intestinal, the septal, or the mesenterial ECM. The position of the extracellular matrix (*ECM*) (*lamina densa* blue, *lamina fibroreticularis* blue dots) is decisive for identification of the type of lining.

tite pseudostratified or a stratified myoepithelium (Figure 11.8E). The position of the ECM, however, is the decisive criterion to discriminate a peritoneum from myoepithelia. While a peritoneum always rests on a subperitoneal ECM, a myoepithelium rests either on the subepidermal, the perivascular, the septal, or the mesenterial ECM.

Cells of the coelomic lining. Four different types of cells may contribute to the coelomic lining: epithelio-muscle cells with myofibrils (Figure 11.7A,C), peritoneal cells (peritoneocytes) without any myofibrils (Figure 11.7A,D), podocytes that allow selective fluid transfer between different cavities (Figure 11.7B,E), and, in pseudostratified and stratified myoepithelia, also fiber muscle cells (Figure 11.7A). Each cell of the coelomic lining has a pair of centrioles forming the diplosome, i.e. both centrioles are perpendicular to each other. These centrioles are able to induce cilia, so that some of the lining cells possess a rudimentary cilium (a short cilium without a basal plate and a central pair of microtubules), which either lies within the cell capped by crescent vesicle or arise from the cell surface into the coelom. While in *Owenia fusiformis* ciliated peritoneal cells possess only one cilium (Gardiner, 1992), they are multiciliated in other annelid species (Figures 11.7D, 11.8B) (e.g, *Glycera dibranchiata* (Fransen, 1988; Gardiner, 1992, Fig. 59), *Chaetopterus* sp. (Gardiner, 1992), *Melinna maculata* and *Phragmatopoma lapidosa* (Fransen, 1988), *Harmothoe sarsi* and *Tomopteris helgolandica* (Bartolomaeus, 1994)). Ciliated peritoneocytes are probably involved in transporting coelomic fluid between different segments as outlined above. Since ciliated peritoneocytes have been reported from a few species only, their phylogenetic significance remains unknown. Fransen (1988) noted that the outline of the peritoneocytes varies depending on whether or not the underlying musculature contracts in two directions.

Topography. The annelid coelomic lining can topographically be subdivided into the somatic (or parietal) lining located underneath the epidermis (Figures 11.4E, 11.7A), the visceral (or splanchnic) lining that surrounds the intestine (Figures 11.7C, 11,8C), and the septal and mesenterial lining. The latter two always consist of two monolayers of lining cells that rest on a common ECM. The ECM may partially be widened to form a blood vessel (Figure 11.7E). While mesenteries are formed by the coelomic lining where left and right coeloms meet in the median plane, the septa are formed intersegmentally by the posterior and the anterior lining of two consecutive coelomic cavities. Transverse muscle strands may traverse the coelomic cavity (Figure 11.7F).

– **Mesenterial and septal coelomic lining.** These linings are rather uniform throughout the annelids and generally represent simple myoepithelia. In all

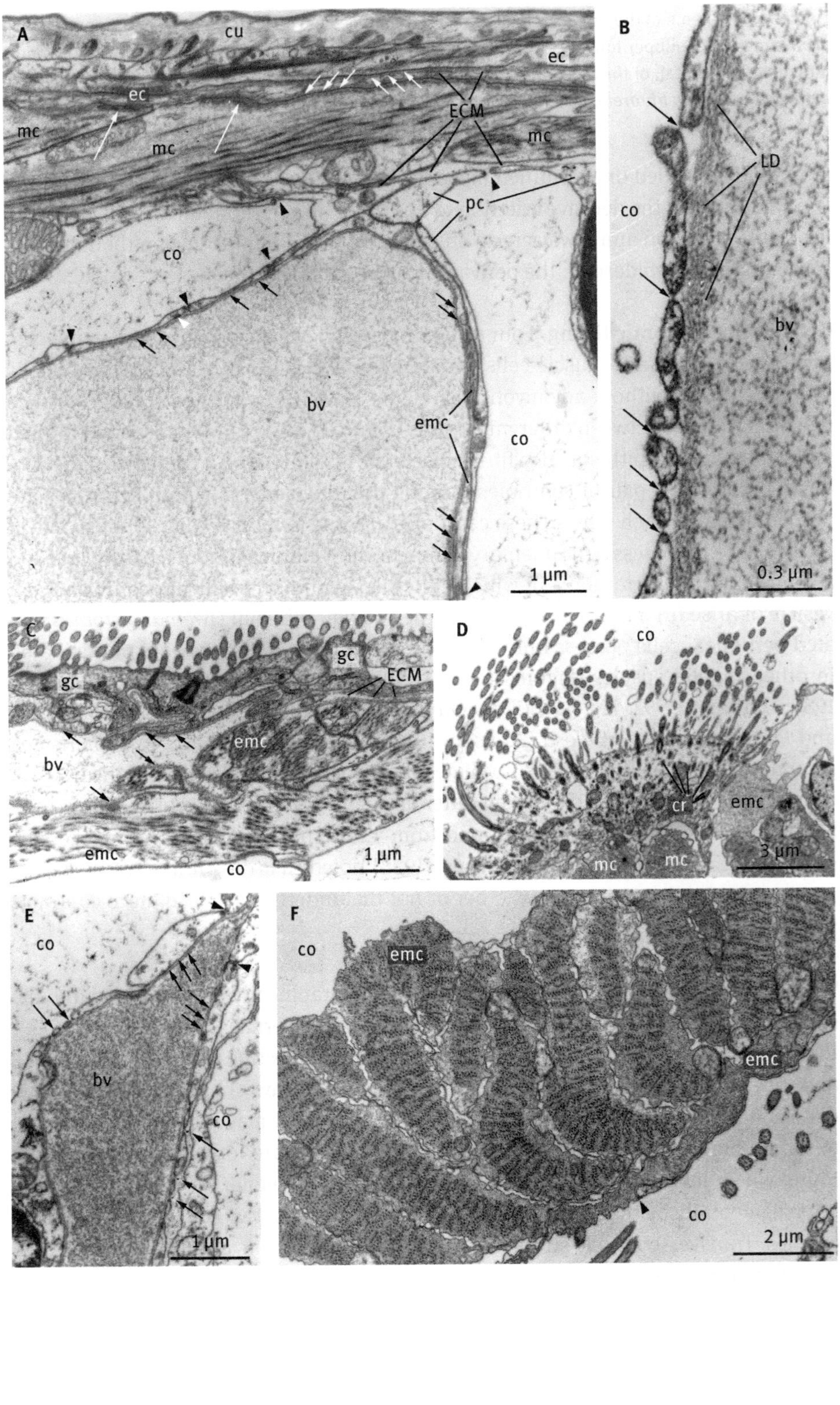
A
cu
ec
ec
mc
mc
mc
ECM
pc
co
bv
emc
co
1 µm
B
LD
co
bv
0.3 µm
C
gc
gc
ECM
emc
bv
emc
co
1 µm
D
co
cr
emc
mc
mc
3 µm
E
co
bv
co
1 µm
F
co
emc
emc
co
2 µm

◂ **Figure 11.7:** Coelomic lining in polychaete annelids. **A–C.** *Pectinaria koreni* (Terebellida), **D.** *Harmothoe sarsi* (Polynoidae), **E.** *Pygospio elegans* (Spionida), and **F.** *Eulalia viridis* (Phyllodocidae). **A.** A peritoneum overlies the somatic muscle. The blood vessel is lined by peritoneocytes (*pc*) and weakly muscularized epithelio-muscle cells (*emc*). Coelomic lining cells basally rest on the lamina densa (*small arrows*) of the ECM, which lines the blood vessel (*bv*). Muscle cells (*mc*) of the subperitoneal body wall musculature adhere to the extracellular matrix (*ECM*) with dense plaques (*white arrows*). The ECM is composed of two laminae densae (*white arrows*) sandwiching the lamina fibroreticularis. **B.** Podocytes rest on the coelomic side of the perivascular lamina densa (*LD*). Diaphragms (*arrows*) bridge the pedicels. **C.** Peri-intestinal blood lacuna, lined by the lamina densa (*arrows*). Gut cells (*gc*) rest on the intestinal side and epithelio-muscle cells (*emc*) on the coelomic side, forming a simple myoepithelium. **D.** Multiciliated peritoneocytes with strong ciliary rootlets (cr) and adjacent epithelio-muscle cells. **E.** The intraseptal matrix is enlarged to form a blood vessel, lined by the lamina densa (*small arrows*). Podocytes rest on its coelomic side and are interconnected by adherens junctions (*arrowhead*). *Large arrows* mark diaphragms. **F.** Transverse muscle. Note apical adherens junctions (*arrowhead*); its epithelio-muscle mimics a peritoneal coverage. *co* coelom, *cu* cuticle, *ec* epidermal cell.

species studied thus far both linings are composed of epithelio-muscle cells, peritoneocytes, and podocytes. The epithelio-muscle cells possess a moderate number of myofibrils (Figure 11.4B,C), except for certain regions where contractile parts of the dorsal blood vessels (heart) are formed by muscle cells derived from the coelothelium. Peritoneocytes and podocytes are always restricted to those regions where the ECM underlying the mesenterial or septal coelomic lining is enlarged to form blood vessels (Figure 11.7B,E). In certain species, like nereid species, the septal lining is fenestrated and allows selective fluid transfer between both compartments, i.e., blood vessels and coeloms (Figure 11.4C) (Nakao, 1974). The lining cells are connected by apical adherens junctions to the layer of cells that form the coelomic face of the somatic and visceral lining.

- **Visceral coelomic lining.** The structure of the visceral coelomic lining ranges from simple myoepithelia to bipartite pseudostratified myoepithelia (Figures 11.7C, 11.8C). Podocytes have thus far never been recorded in the visceral coelomic lining, neither has a peritoneum in the defined sense ever been found. The myoepithelium forms a grid of circular and longitudinal fibers surrounding most parts of the intestine, except for the hindgut that lacks longitudinal fibers in most polychaete species (Fransen, 1988; Bartolomaeus, 1993). In several clitellate species the myoepithelium around the dorsal and dorsolateral section of the gut is transformed into some kind of storage tissue (chloragogen cells).
- **Somatic coelomic lining.** This lining shows the highest structural variation in annelids and comprises simple myoepithelia (Figures 11.4E, 11.8E), uniformly and bipartite pseudostratified myoepthelia, stratified myoepthelia as well as non-muscular peritoneal linings that cover the underlying body wall musculature (Figures 11.7A, 11.8A,B). This variation in the structure of the somatic coelomic lining occurs within individuals and supraspecific levels as well as during ontogenesis (see below) and excludes assigning a special type of lining to a species or

a supraspecific taxon. In *Diplocirrus glaucus*, for instance, the somatic coelomic lining consists of epithelio-muscle cells and clusters of peritoneal cells, which may or may not be separated by an ECM from the underlying muscle cells within the same segment (Figure 11.8D). The somatic coelomic lining of *Fabricia sabella* consists of a peritoneum in the anterior and median segments, whereas this lining is a simple myoepithelium in the caudal segments (Bartolomaeus, 1993). In *Sipunculus nudus* and *Pectinaria koreni* a peritoneum covers the muscle layer of the body wall (Figure 11.7A; 8A). From all information presently available, it must be concluded that the structure of the coelomic lining generally differs not only between species, it also differs with respect to the topographic position within an individual (somatic, visceral, mesenterial), as well as between its different body regions (i.e., thorax, abdomen, etc.). In general it is related to the amount of muscle cells and the degree of muscularization, since a larger amount of body wall muscles is often associated with the presence of a peritoneum.

- **Transverse muscles.** The coelomic lining forms major parts of the musculature. It is complemented by strands of mostly transverse muscles and in some species by a special pharyngeal musculature. The muscle strands cross coelomic cavities and are generally involved in parapodial locomotion or moving the chaetae (Storch, 1968; Tzetlin and Filippova, 2005). These muscle strands are never enwrapped by a peritoneum, not even in those species that possess a somatic peritoneum. They consist of a single epithelio-muscle cell or of a group of such cells (Figure 11.7F) that adhere to the ECM at their origin (generally the subepidermal ECM) on one side and to the ECM of their destination (generally the perifollicular ECM or the subepidermal ECM of the adjacent body side). In any case the transverse muscles are interconnected with the coelothel via adherens junctions, thus providing further evidence for their myoepithelial nature.

Hypotheses on the ancestral structure of the coelomic lining in Annelida. Like the results of earlier investigations (Fransen, 1980a; 1988; Rieger, 1986; Bartolomaeus, 1994; Rieger and Purschke, 2005), our data clearly contradict the traditional view (see

Figure 11.8: Coelomic lining in *Sipunculus nudus* (Sipuncula) (**A–C**), *Diplocirrus glaucus* (Flabelligeridae) (**D**), and *Scolelepis squamata* (Spionida) (**E**). **A.** Body wall muscles are embedded in extracellular matrix (*ECM*) and covered by a peritoneum. **B.** Multiciliated peritoneocytes are connected to adjacent peritoneocytes by adherens junctions (*arrowheads*) and to the ECM by hemidesmosomes (*arrows*). **C.** Visceral lining is a myoepithelium with epithelio-muscle cells interconnected by adherens junctions (*arrowheads*). Dense plaques (*arrows*) adhere them to the ECM. **D.** Transition from a peritoneum to a myoepithelium in the somatic lining. Dense plaques (*arrows*) attach the muscle cells (*mc*) to the ECM and adherens junctions (*arrowheads*) connect peritoneocytes and epithelio-muscle cells. **E.** Epithelio-muscle cells with a large, flat cytoplasmic protrusion that mimics a peritoneal cell. Adherens junctions connect them to adjacent cells (*short arrows*) and dense plaques (*long arrows*) attach them to the ECM. *cm* circular muscles, *co* coelom, *cr* ciliary rootlets, *cu* cuticle, *ec* epidermal cell, *gc* gut cell, *lm* longitudinal muscle.

▸

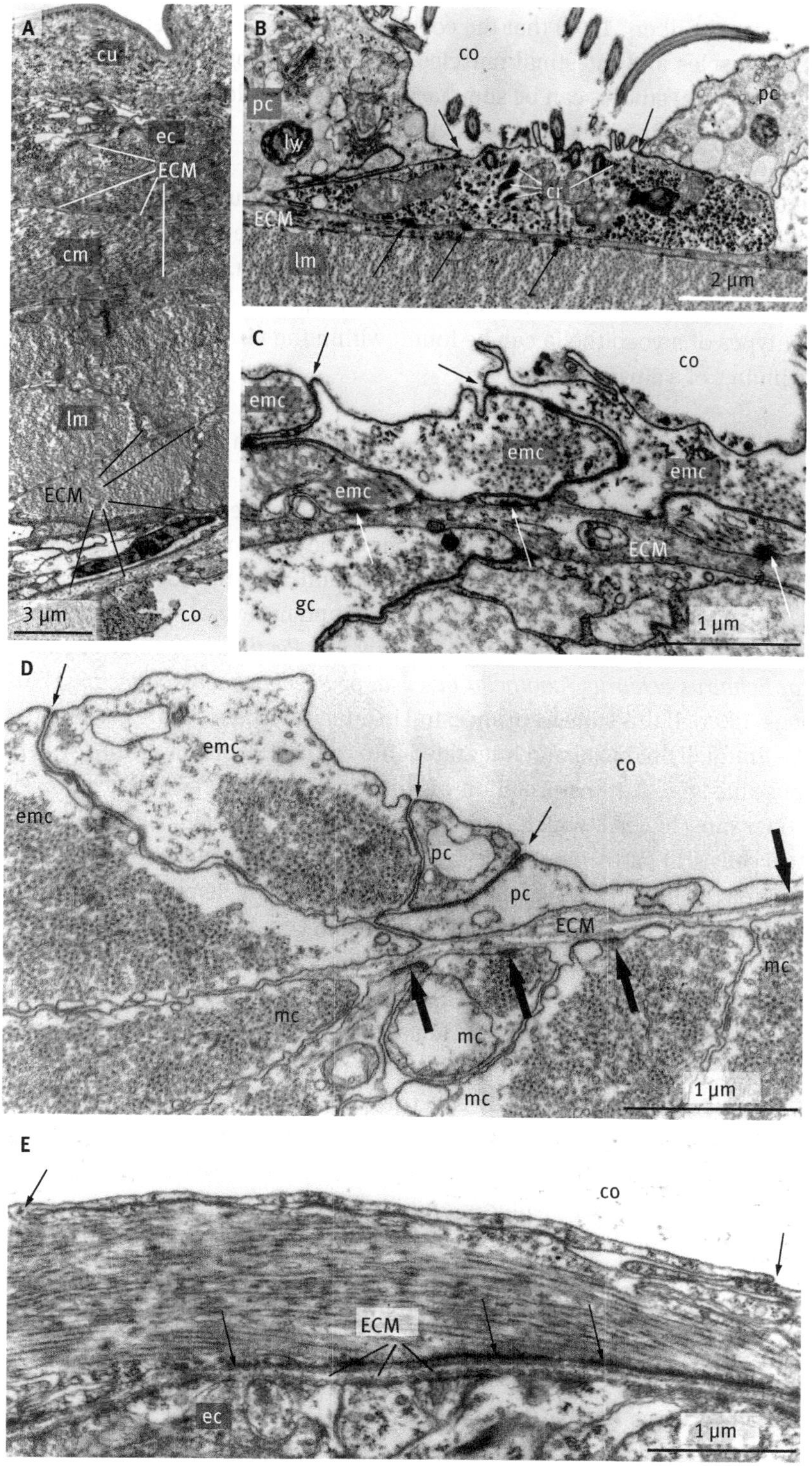
A
cu
ec
ECM
cm
lm
ECM
3 µm
co
B
co
pc
pc
ly
cr
ECM
lm
2 µm
C
co
emc
emc
emc
emc
ECM
gc
1 µm
D
emc
emc
co
pc
pc
ECM
mc
mc
mc
mc
mc
1 µm
E
co
ECM
ec
1 µm

Brusca, Brusca, and Gilbert, 1997) that the coelomic lining is always separated from the body wall muscles and intestinal muscles. The most important results of comparative ultrastructural studies can be summarized with the following statements:

(1) A peritoneum as defined above is restricted to the somatic site of a few annelid species only.
(2) If present, the peritoneum overlies the somatic musculature only. Neither the intestinal visceral muscles, nor the transversal muscles are covered by a peritoneum, and it is also absent along the mesenteries and septa.
(3) Myoepithelia are the common type of coelomic lining.
(4) Different types of myoepithelia can be found within the visceral (intestinal) and somatic linings of a single species.

These results allow two hypotheses on the primary organization and evolution of the coelomic lining in annelids:

(1) In the annelid stem species the coelom was originally lined by a somatic peritoneum that was separated from the underlying body wall musculature by ECM (Figure 11.5B). The mesenterial, dissepimental and visceral coelomic linings consist of a simple myoepithelium. This state is found in several species like *Harmothoe sarsi, Sternaspis scutata, Sipunculus nudus, Pectinaria koreni, Capitella capitata, Echiurus echiurus Axiothella mucosa,* or *Poecilochaetus johnstoni* (Bartolomaeus, 1994). If this state were ancestral in annelids, the separate peritoneum would be gradually reduced and integrated into the underlying muscular system in various lineages. A transformation of this kind involves epithelialization of former fiber muscle cells, which thus become epithelio-muscle cells, while the peritoneal cells still partly overlie the epithelio-muscle cells and extend between them towards the basal ECM, to which they are attached by hemidesmosomes.
(2) In the annelid stem species a simple myoepithelium consisting of epithelio-muscle cells and a few peritoneal cells lined the coelomic cavities (Figure 11.5C). This organization is found in *Owenia* species, some phyllodocidan species, and spionidan annelids. If this organisation were ancestral in annelids, the peritoneal cells of the somatic coelomic lining would increase in number within the annelids to form a peritoneum by completely covering the former epithelio-muscle cells which then transform into fiber muscle cells. Finally, this peritoneum becomes separated from the underlying somatic muscular system by ECM. Like the visceral coelomic lining, the mesenteries and dissepiments keep the primary organization of the coelomic lining, i.e. a simple myoepithelium. According to this interpretation the absence of a peritoneal lining along the transversal muscles represents a primary condition that is maintained irrespective of the evolution of a peritoneum within the annelids.

The latter hypothesis was favored by Fransen (1980a; 1980b; 1988), Gardiner and Rieger (1980), and Rieger (1986) and is in accordance with current phylogenetic hypotheses (Struck et al., Chapter 9), in which Oweniida form the basalmost branching group of annelids. It assumes a uniform kind of lining for the entire coelomic cavity, i.e. a simple myoepithelium as the ancestral state in annelids. Along with the observation that within annelids all other kinds of coelomic linings are restricted to a few regions, mainly to the somatic coelomic lining, this explains why the transversal muscle strands are never covered by a peritoneum, nor contain any peritoneal cells. In addition this hypothesis is in accordance with the molecular annelid tree of Struck et al. (Chapter 9)

Studies into annelid coelomogenesis can be used to test these hypotheses. If the second hypothesis were true, the coelomic cavities should be formed by fluid accumulation (schizocoely) between prospective polarized epithelio-muscle cells, so that polarity of these cells is present prior to schizocoely. Such a mode of coelomogenesis should also be present in those species that possess a somatic peritoneal lining with an underlying ECM and a visceral myoepithelium. The first hypothesis would instead be favored were the peritoneum were formed by epithelialized cells that develop independently from the muscular system. Since the visceral coelothel is never a peritoneum, one would expect that the subperitoneal ECM is reduced during coelomogenesis.

11.4.2 Formation of the coelomic lining and coelomogenesis

Coelomogenesis is the process during which coelomic cavities are formed by fluid accumulation among mesodermally derived epithelia. It has long been known that in polychaete annelids coelomogenesis differs prior to and after metamorphosis. Classical studies report that a single pair of unsegmented coelomic cavities stretches out over the first three to five larval segments, whereby the number of larval segments is species specific. In postmetamorphic stages, in contrast, coelomogenesis gives rise to strictly segmental coelomic cavities. This phenomenon has been termed heteronomy and influenced different theories on the origin of the coelom (Iwanoff, 1928; Wilmer, 1990; Tautz, 2004; Seaver, Thamm, and Hill, 2005).

In order to get deeper insight on the differences underlying heteronomy, the larva of *Spirorbis spirorbis* representing a stage prior to metamorphosis, and juveniles representing postmetamorphic stages were compared on the ultrastructural level. In juveniles coelomogenesis can be analyzed when studying the preterminal (prepygidial) growth zone and its adjacent anterior segments in postlarval stages of annelids. In this region compact cell material between gut and epidermis on either body side (visible in whole-mounts of slightly squeezed juveniles; Figure 11.9) enlarges along the anterior-posterior axis and acquires cavities of increasing dimensions towards the anterior. According to the definitions given initially, this cell material is mesoderm. Due to its development along an anterior-posterior gradient, mesoderm differentiation and formation of body cavities can easily be studied in a single individual. In the following we describe the entire process as a series of steps in time, although we studied the process in a series of segments along the anterior-posterior axis at a single point in time (i.e., any postembryonic stage of the juvenile at which it was fixed for anatomical studies). This way of analyzing postlarval mesoderm differentiation and coelomogenesis has also been used in classical light microscopical studies (see Anderson, 1959), but it has seldom been used in ultrastructural studies (Turbeville, 1986; Bartolomaeus, 1993; Bartolomaeus and Ruhberg, 1999). Anderson (1959) was one of the first authors who showed how the mesoderm anlage gradually transforms into somatic and visceral musculature and into body cavities. In *Scoloplos armiger* he described that both compact and ventrolaterally situated mesodermal cell strands enlarge continuously along an anterior-posterior gradient and develop into the somatic and visceral musculature. During this process the cells of each strand separate into a somatic and a visceral layer, between which fluid accumulates under formation of a cavity on each side of the gut. Since this is an iterative process, segmental coelomic cavities arise. We studied this process in the metatrochophore of *Spirorbis spirorbis* and in juveniles of *Pectinaria koreni*, *Tomopteris helgolandica*, *Ophelia rathkei*, *Glycera alba*, and *Pristina longiseta*.

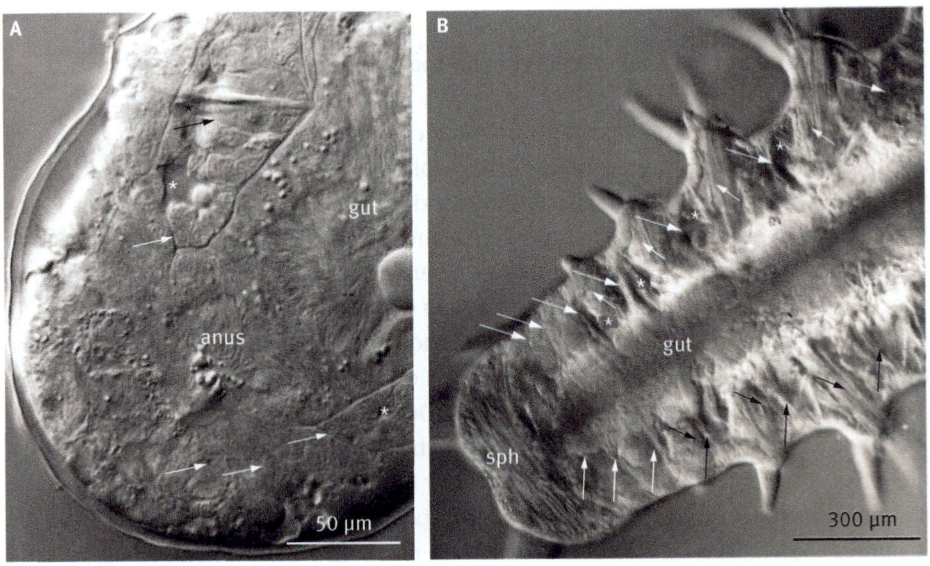

Figure 11.9: Coelomogenesis in a premetamorphic stage of *Ophelia rathkei* (Opheliidae) (**A**) and in a postmetamorphic stage of *Nereis pelagica* (Nereididae) (**B**). **A.** *Arrows* mark early anlagen of the mesoderm, *asterisks* tiny developing coelomic compartments. Note that the earliest anlagen are posterior to the anus. **B.** A strong anal sphincter muscle (*sph*) is located posterior to the growth zone. *Large arrows* mark signs of septation, which initially is incomplete. *Asterisks* mark coelomic cavities. *Small arrows* mark chaetal sacs.

11.4.2.1 Coelomogenesis in premetamorphic stages

Coelomogenesis was studied in a metatrochophore of *Spirorbis spirorbis* with four complete sets of chaetae indicating that the larva is close to metamorphosis (Figure 11.10A). The course of coelomogenesis was inferred from a complete series of sections of the entire larva. According to this method the coelomic lining develops from two mesodermal cell clusters situated caudal to the anus. These cells are polarized in showing an apical pair of diplosomal centrioles and basal hemidesmosomes linking the cells to the ECM that separates the cluster from the surrounding epidermal cells (Figure 11.10B). Groups of dorsally and ventrally located neurites rest on the epidermal side of this ECM. The mesodermal cells contain few short myofibrils in areas where the neurites contact the cells. Both clusters represent the caudal growth zone of premetamorphic stages. We assume that the mesodermal cell clusters in *Spirorbis spirorbis* are derivatives of the 4d cells, which are proposed to undergo a restricted number of iterative, unequal cleavages in polychaete annelids to form a band of small cells (Fischer and Arendt, 2013).

Further frontally both cell clusters medially merge into a U-shaped cluster that surrounds the ventral portion of the gut in front of the anus (Figure 11.10C). The cells show a clear apico-basal polarity; apically they are connected to their adjacent neighbors by adherens junctions, and by desmosomes to those facing their apex (Figure 11.10F). Fluid initially accumulates within spaces between the desmosomes of opposing cells, indicating onset of coelomogenesis. At the same time myofibrils appear inside these cells characterizing them as epithelio-muscle cells. While formation of myofibrils proceeds, the cell size increases. This process occurs along with enlargement and continuous separation of the visceral and the somatic coelomic linings, both of which now form a simple myoepithelium (Figure 11.10G).

The ventral mesenterium is formed where the basal sides of the epithelio-muscle cells of left and right clusters face each other. A small patch of ECM is formed here and expands to gain contact with the peri-intestinal and subepidermal ECM. By this time fluid accumulates within the ECM that enlarges the ventral mesenterium and establishes the anlage of the ventral blood vessel (Figure 11.10D). Differentiation of the coelomic lining on the ventral side precedes its formation on the dorsal side. During this process the epithelio-muscle cells seem to migrate dorsally within the ECM between the gut and the epidermis, as if this ECM were the substrate to guide their expansion. Finally, left and right coeloms completely surround the gut and meet dorsally. An ECM remains between the coeloms that together with their dorsalmost walls will form the dorsal mesenterium (Figure 11.10E). At this time, podocytes with myofibrils appear that rest on the perivascular ECM of the ventral blood vessel and are part of the mesenterial coelomic lining. The peri-intestinal ECM has also been enlarged now to form a blood lacuna.

At the end of coelomogenesis, the epithelio-muscle cells of the visceral lining are relatively flat and contain a few myofibrils only, while the somatic lining consists of tall epithelio-muscle cells with a large amount of myofibrils. Some of the

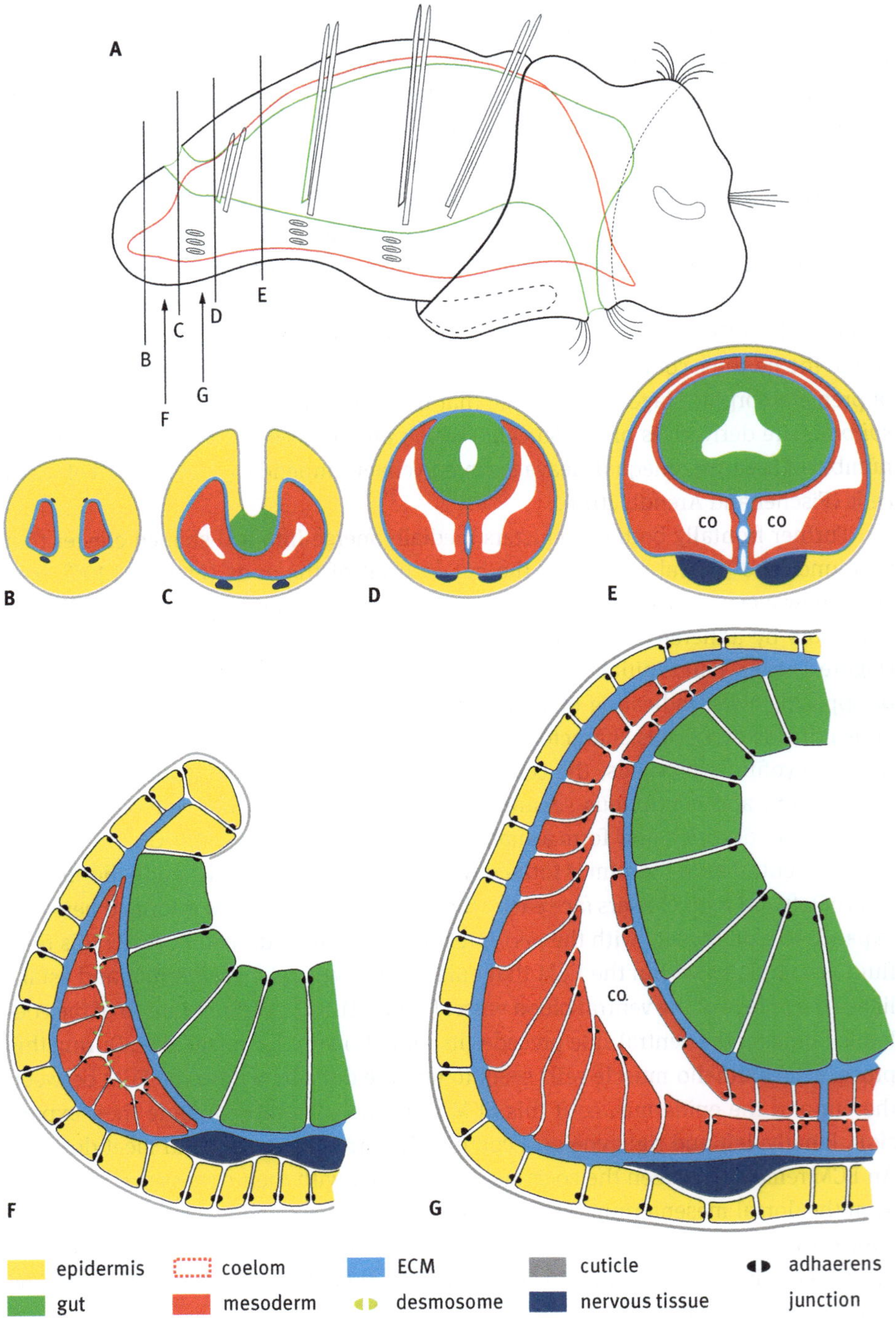
A
B
C
D
E
F
G
co
co
co.
epidermis
gut
coelom
mesoderm
ECM
desmosome
cuticle
nervous tissue
adhaerens junction

◂ **Figure 11.10:** Coelomogenesis in premetamorphic *Spirorbis spirorbis* (Sabellida). **A.** Drawing of the premetamorphic larva with indicated levels of transverse sections shown in **B–E** indicated. **B.** Mesoderm anlage is located posterior to the anus on each side of the larva. **C.** Mesoderm fuses ventrally after coelomogenesis started. **D.** Somatic muscle differentiation is accelerated; double layer of lining cells migrate within the matrix towards the dorsal side. A small blood vessel is formed within the ventral mesenterium. **E.** Mesenterium is formed when left and right coelomic linings meet dorsally. **F.** Visualization of the situation prior to coelom formation at a level between **B** and **C.** Note desmosomes interconnecting opposing coelomic lining cells. **G.** Schizocoely. Situation on a level between **D** and **E.** After the desmosomes disappeared the coelomic cavity increases in size by fluid accumulation. The coelomic lining expands dorsally, while the ventrolateral epithelio-muscle cells increase in size. *co* scoelom.

somatic lining cells contain circularly arranged myofibrils. Ventrolaterally they are located basal to the strong and tall lining cells that contain longitudinally arranged myofibrils. These cells are partially covered by peritoneocytes that adhere to the subepidermal ECM, indicating that the somatic coelomic lining develops from a simple myoepithelium into a uniformly pseudostratified myoepithelium. The visceral lining remains a simple myoepithelium, since it consists of a monolayer of epithelio-muscle cells forming a grid of myofibrils, and of podocytes resting on the ECM surrounding the blood lacunae. All lining cells are apically interconnected by adherens junctions.

No septa are formed during formation of the first four body segments, so that a single large coelomic cavity is located on either side of the larval gut.

11.4.2.2 Coelomogenesis in postmetamorphic stages

Postanal organization and coelomogenesis differ in juvenile annelids from the above given description for premetamorphic stages (Bartolomaeus, 1993). This is exemplified in the following for the tubiculous juvenile of *Pectinaria koreni* in more detail, since this species shows a complete peritoneal lining of the somatic muscular system and thus is of major interest in terms of testing the above two hypotheses on the primary organization of the coelomic lining in annelids. The results for this species are briefly compared to coelomogenesis in other species.

11.4.2.2.1 *Pectinaria koreni* (Terebellida: Pectinariidae)

In the tubiculous juvenile of *Pectinaria koreni* the cuticularized hindgut opens to the exterior in a ventrally located anus. Frontal to the anus the hindgut is surrounded by a partially bilayered ring of fiber muscle cells, the caudal-most of which are mitotically active (Figure 11.11A). The muscle cells comprise the perianal sphincter muscle and are surrounded by the subepidermal and peri-intestinal ECM (Figures 11.11B, 11.13B). The latter is enlarged to form the perianal blood lacuna, which is continuous with the gut lacuna system. Further frontally the cuticular lining of the hindgut disappears and the diameter of the gut lumen decreases. Here, the sphincter muscle is replaced

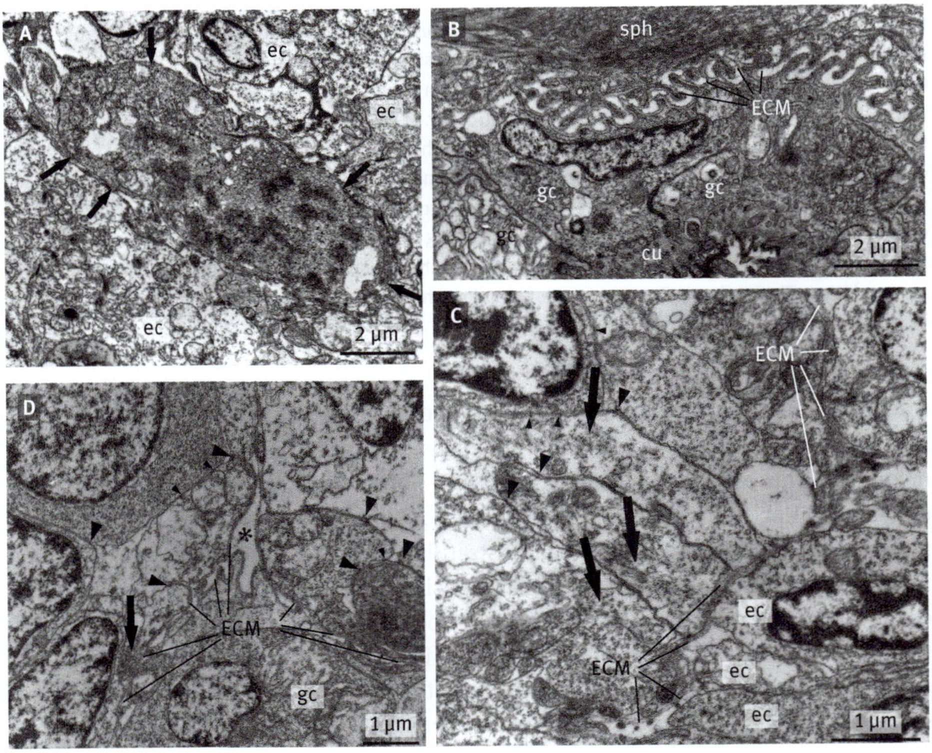

Figure 11.11: Coelomogenesis in *Pectinaria koreni* (Terebellida), transverse sections (TEM). **A.** Posterior-most cell of the anal sphincter is mitotically active. **B.** Sphincter muscle (*sph*) surrounds anus. Note the blood lacuna (lined by extracellular matrix (*ECM*)) surrounding the muscle. **C.** Formation of myofibrils, which initially are exclusively longitudinal (*large arrows*). *Small arrowheads* mark desmosomes, *large arrowheads* adherens junction. **D.** ECM indicates developing transverse septum. Epithelio-muscle cells differ in their number and orientation of myofibrils (*large arrow*); extracellular matrix is enlarged to form a tiny blood vessel (*asterisks*). *Arrowheads* as in C. Distance A–B: 11 µm; B–C: 12 µm, C–D: 12 µm. *cu* cuticle, *ec* epidermis cell, *gc* gut cell.

by a ring of largely undifferentiated cells (posterior growth zone or meristematic ring). It consists of two layers of cells that may overlap each other, but are each attached either to the subepidermal or to the peri-intestinal ECM by hemidesmosomes. Apically, adherens junctions connect adjacent cells and desmosomes opposing cells. Dorsally the peri-intestinal ECM contains fluid and forms a blood space continuous with the perianal blood lacuna (Figure 11.11B).

Onset of myofibril formation turns the undifferentiated cells into epithelio-muscle cells and changes their shape from more or less spherical into asymmetric cells, as they elongate according to the main axis of the myofibrils. An increase in the number of myofibrils moves their nucleus into a lateral position. A bilayered ring of mesodermal cells is now established anterior to the meristematic ring. The cells of the prospective somatic lining form a simple myoepithelium, in which some of them adhere to the subepidermal ECM by a cytoplasmic process only, while others are broadly attached to it (Figure 11.11C). The latter contain myofibrils that are perpendicular to those of the former; these cells will later form the outer circular musculature. Their differentiation clearly precedes that of the inner cells, which will form longitudinal muscles. In the visceral coelomic lining formation of myofibrils is slightly delayed (Figure 11.11C).

Coelomogenesis by schizocoely starts when fluid accumulates between desmosomes that connect opposing epithelio-muscle cells of the prospective somatic and visceral lining (Figure 11.11C,D). At this time, no further cells (such as non-muscular peritoneocytes or podocytes) are found. The gut is surrounded by a simple layer of epithelio-muscle cells with a grid-like arrangement of myofibrils. The content of myofibrils, however, is lower in the visceral lining cells than in the somatic lining cells, due to delay in myofibril formation. The border between the first prepygidial body segments can only be recognized by a transverse ECM interconnecting subepidermal and peri-intestinal ECM. This transverse ECM indicates the prospective septa. Where transversal and peri-intestinal ECM meet, a small fluid-filled space is formed that will become part of the blood vascular system (Figure 11.11C). Up to this stage mesenteries are already formed and continuous with the anterior ones.

Breakdown of the desmosomes between opposing epithelio-muscle cells enlarges the coelomic cavities. Coelomogenesis, however, is also accompanied by with further differentiation of the somatic ECM, which now also can be found between different groups of epithelio-muscle cells and leads to a continuous separation of an outer circular muscle layer from an inner longitudinal layer. During this process some epithelio-muscle cells are transformed into fiber muscle cells, so that finally the somatic circular musculature consists of fiber muscle only, while the prospective longitudinal muscles still maintain their epithelial character. As soon as the ECM between outer circular muscles and inner prospective longitudinal muscles has been established, the inner epithelio-muscle cells start growing and form strong myofibril bundles that are oriented anterior-posteriorly. Again, an ECM seems to be a prerequisite for muscle formation. While the visceral coelomic lining remains a simple myoepithelium, the somatic lining has already transformed from a simple myoepithelium into a stratified myoepithelium.

The first large coelomic cavities formed this way co-occur with chaetal sacs that are formed by epidermal invaginations. At this stage neither the visceral nor the somatic coelomic walls contain peritoneal cells forming a complete or even a partial coelomic lining, nor do podocytes occur within the coelomic lining. Transverse muscles in the anterior body segments are simple strands of epithelio-muscle cell that lack any peritoneal lining.

In later stages, i.e. the juvenile benthic stage, the somatic coelomic lining is made of a peritoneum that rests on a subperitoneal basal lamina and covers the somatic musculature composed of outer circular and inner longitudinal fiber muscle cells, both of which are embedded in an ECM. The transverse muscles and the visceral muscles lack a peritoneal lining. Since during coelomogenesis initially no other cells than epithelio-muscle cells are formed, it must be assumed that peritoneocytes are nothing else but transformed epithelio-muscle cells. During coelomogenesis the somatic coelomic lining thus develops from a simple myoepithelium via a stratified myoepithelium into a peritoneum and an underlying somatic musculature.

11.4.2.2.2 Comparison

Compared to *Pectinaria koreni*, coelomogenesis is almost identical in juveniles of the polychaete species *Ophelia rathkei* (Bartolomaeus and Ruhberg, 1999), *Glycera alba*, *Tomopteris helgolandica*, and the small clitellate species *Pristina longiseta* ("Oligochaeta": Naididae) (Bartolomaeus, 1993) (Figure 11.13) . Differences largely refer to the number of mitosis found in the preanal ring of undifferentiated cells and its extension, as well as to the timing of cell differentiation.

In *Ophelia rathkei* a ring of undifferentiated cells is located on the level of the interface between the hindgut and midgut, anterior to the perianal sphincter (Figure 11.12A). This ring represents the prepygidial meristematic growth zone. Its cells form a bilayer and are either attached to the peri-intestinal or to the subepidermal ECM. The cells of either side are apically interconnected by adherens junctions, those belonging to different layers by desmosomes. Like in *Pectinaria koreni* they differentiate into epithelio-muscle cells along an anterior-posterior gradient, whereby those resting on the peri-intestinal ECM and those resting dorsolaterally on the subepidermal ECM clearly differentiate faster than the ventrolateral cells.

Coelomogenesis by schizocoely starts when fluid accumulates between desmosomes connecting the opposing somatic and visceral epithelio-muscle cells. The tiny spaces fuse as the desmosomes disappear, so that finally a pair of lateral coelomic cavities is formed. At this time their lining is a simple myoepithelium consisting of epithelio-muscle cells only (Bartolomaeus and Ruhberg, 1999). The developmentally delayed clusters of ventrolateral subepidermal cells appear to be rather compact, since their cells are largely undifferentiated. These cells are the precursors of the somatic muscles. Their tight packing causes the epidermis to bulge slightly to the exterior. Since these clusters are serially arranged along the anterior-posterior axis they allow for identifying intersegmental borders. There is, however, no ECM separating these clusters or indicating the position of the prospective septum. Further frontally segmental chaetal sacs are visible that sank in from the epidermis and force the coelomic lining to surround the chaetal sac. Epithelio-muscle cells that rest on the perichaetal ECM are transformed into transverse muscles.

Even further frontally the somatic coelomic lining consists of epithelio-muscle cells with longitudinally arranged myofibrils; between their bases a few circular muscles can be seen. Among the epithelio-muscle cells occasionally multiciliated peritoneocytes can be found. The visceral coelomic lining consists of a monolayer of epithelio-muscle cells with a grid-like arrangement of myofibrils. In adult *Ophelia rathkei* the groups of peritoneocytes cover the strong longitudinal muscles in part. They are interconnected and connected to neighboring epithelio-muscle cells by adherens junctions. The somatic coelomic lining is a pseudostratified myoepithelium.

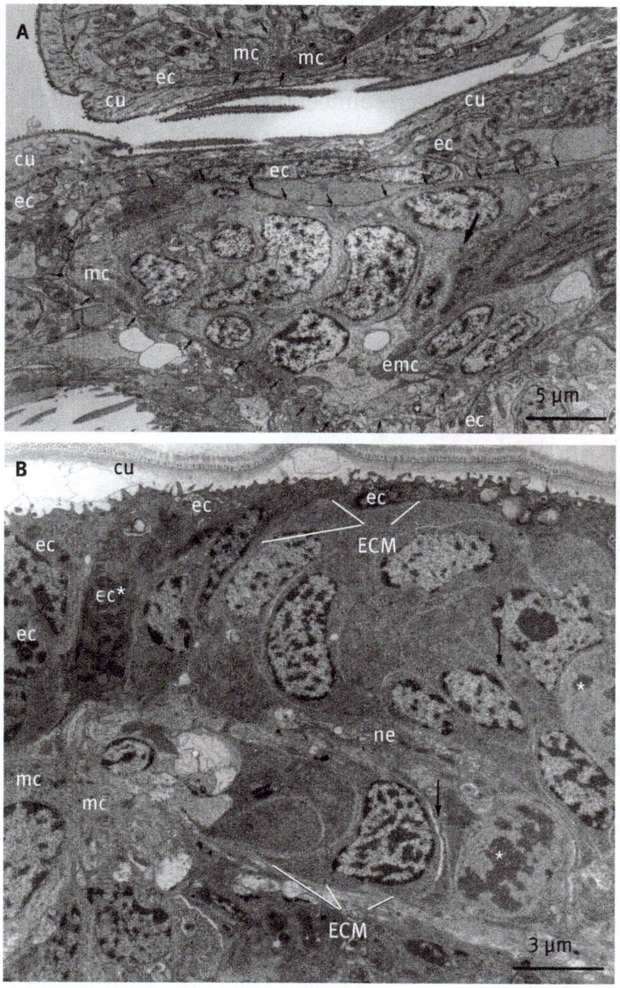

Figure 11.12: Posterior-most region of juvenile *Ophelia rathkei* (Opheliidae) (**A**) and *Pristina longiseta* (Naididae) (**B**). **A.** Sagittal section. An anal sphincter is located posterior to the growth zone consisting of undifferentiated cells. *Small arrows* mark extracellular matrix, *large arrow* a diplosome. **B.** Semisagittal section. Anterior to the anal sphincter the growth zone is marked by undifferentiated cells and mitoses (*asterisks*), which include epidermis cells (*ec**). *Arrow* marks centrioles. *cu* cuticle, *ec* epidermal cells, *ECM* extracellular matrix, *emc* epithelio-muscle cell, *mc* muscle cell, *ne* neurites.

In *Glycera alba* the anal sphincter is located at the transition between the cuticularized hindgut and the midgut (Figure 11.13B). The prepygidial growth zone is represented by a ring of undifferentiated cells that anteriorly continues into cells containing myofibrils; the further anteriorly they are located, the more myofibrils occur within these cells. All cells rest on either the peri-intestinal or the subepidermal ECM and are epithelial (Figure 11.13C). In the first prepygidial segments these cells form a more or less compact cylinder around the gut consisting of a bilayer of prospective epithelio-muscle cells with only tiny interstitial spaces. Like in the other two species, these spaces are found between desmosomes that connect the opposing apices of the prospective visceral and somatic linings, while the neighboring cells of each lining are interconnected by adherens junctions.

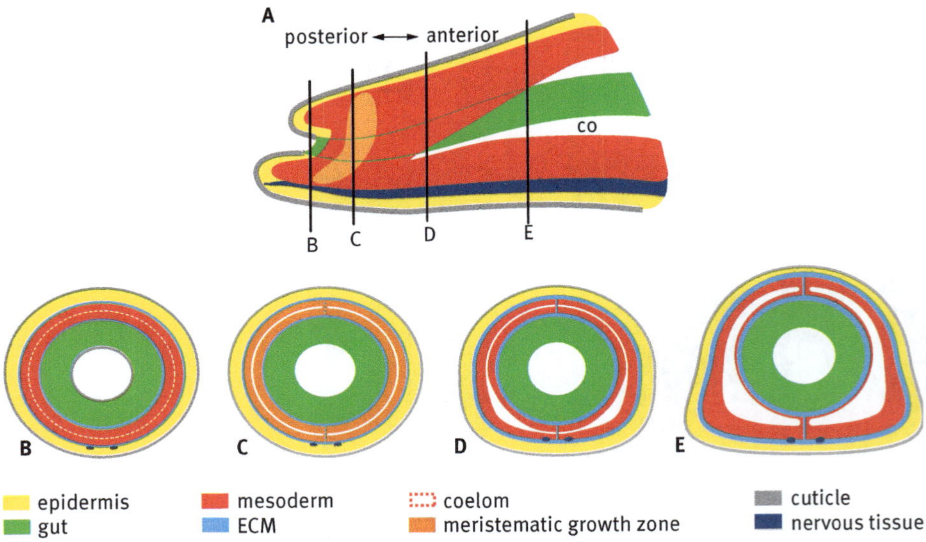

Figure 11.13: Generalized scheme of coelomogenesis in metamorphosed stages. **A.** Sagittal section, septa are not shown, levels of transverse section indicated. **B.** Region of the anal sphincter which is bilayered in some species. **C.** Anterior to the sphincter is a ring of mitotically active, undifferentiated cells that are already polarized. In some species mesenterial extracellular matrix (*ECM*) is continuous with the ECM of the mesenteria of anterior segments. **D.** Coelomogenesis by schizocoely starts after desmosomes connected opposing coelothelial layers disappeared. This process starts generally ventrolaterally. **E.** The coelom expands dorsalward when breakdown of the desmosomal connections proceeds. Note that in contrast to coelomogenesis prior to metamorphosis the entire coelomic lining is already established on either side of the body (compare Figure 11.10D to E). *coel* coelom.

This cylinder is serially subdivided by transverse extensions of ECM originating from the peri-intestinal and subepidermal ECM, respectively. The subdivision is initially incomplete as the transverse extensions of the peri-intestinal and subepidermal ECM are centrally not confluent (Figure 11.18C). With increasing distance from the growth zone the transverse extensions extend deeper into the cylinder and merge to form complete transverse subdivisions from the fifth prepygidial body segment onwards. On each body side, the small interstices between the epithelio-muscle cells gradually fuse into a single coelomic cavity after the septa have been completed. The segmental coeloms thus formed are initially restricted to ventrolateral cavities within the mesodermal cylinder around the gut (Figure 11.13D). During the further development, the coelomic lumen increases in size dorsalward by resolution of the desmosomes interconnecting the somatic and visceral linings to complete the paired coelomic cavities of the body segment (Figure 11.13E).

Coelomogenesis in *T. helgolandica* and *P. longiseta* largely resembles that of *G. alba*. In both species a meristematic ring is located anterior to the perianal sphincter (Figure 11.12B for *P. longiseta*; see also Akesson, 1962 for *T. helgolandica*). Epithelio-muscle cells are formed prior to schizocoely; they form the simple myoepithelium that initially lines the new coelomic cavities. During further development the simple myoepithelium is transformed into pseudostratified or stratified myoepithelia that line the somatic face of the coelom, while the visceral and mesenterial linings remain simple myoepithelia in *T. helgolandica*. In *P. longiseta* the entire coelomic lining remains a simple myoepithelium that is separated from the underlying circular muscle layer by an ECM.

11.4.2.3 Coelomogenesis in clitellate embryos

Mesoderm development in clitellate embryos has not been studied ultrastructurally thus far. Recently, Fischer and Arendt (2013) showed that in the polychaete annelid *Platynereis dumerilii* the two daughters of the 4d cell undergo a restricted number of unequal cleavages that give rise to a band of small cells similar to the mesoteloblast derivatives in Clitellata. While the cycle of iterative cleavages most likely is less than ten in polychaetes, it is increased during clitellate evolution and may comprise up to 35 within this group. Extending the number of iterative, unequal cleavages necessarily must alter the course of coelomogenesis, so that it differs slightly from the beforementioned description of pre- and postmetamorphic stages:

(1) The segmental mesoderm does not originate from a posterior growth zone but from mesoteloblasts (M-cells) that give rise to a linear series of paired, segmentally arranged primary blast cells forming the so-called mesodermal germ band (e.g. Shimizu and Nakamoto, 2001). Labeling and ablation experiments with teloblasts in "oligochaetes" (Tubificidae) and glossiphoniid leeches (Hirudinea) revealed that the entire mesoderm of one hemisegment is derived from a single primary blast cell (Goto et al., 1999). In tubificids about 30 body segments are

formed by the germ bands before the mesoteloblasts become indistinguishable from the surrounding tissue (Shimizu et al., 2001).

(2) Clitellate embryos are characterized by high yolk content, including species with lecithotrophic embryos nourishing on the cocoon liquid (see Hunnekuhl et al., 2009). Body segments and coeloms are formed from paired germ bands that after coalescence in the ventral midline are initially restricted to the ventral side of the embryo and in later stages gradually expand dorsalward to overgrow and to entirely enclose the yolk-sac (Anderson, 1973; Shimizu, 1982). Coelomogenesis in tubificids starts after a stereotyped division of the primary blast cell (M teloblast) in each hemisegment into a cluster of two large, four intermediate and several small mesodermal cells (M-cells). "Tiny cells" (Goto et al., 1999) of this cluster are described as giving rise to the coelomic walls. Whether the entire cell cluster becomes involved in coelom formation remains unclear in previous descriptions, as do the original differentiation of the coelothelial cells and the timing of myogenesis.

(3) Most light microscopical investigations agree that coeloms start to develop before the germ bands expand dorsalward. Initially the entire segmental mesoderm, including the coeloms and their gradually increasing lumen, is restricted to ventrolateral areas of the embryo. Motility of the dorsal coelomic edges is higher than that of the ventral edges. A "jagged" appearance of the dorsal edges (Goto et al., 1999) seems to result from circumferential fibrous extensions containing actin. These apparently contractile fibers are assumed to mediate dorsal expansion, apparently by pulling the dorsal edges of the coelomic wall toward the dorsal side (Goto et al., 1999). Provided that these contractile fibers are inside the coelomic wall, a similar process as described for the dorsal extension of the coelom in larvae of *Spirorbis spirorbis* must be assumed. Dorsal expansion of the coeloms is also maintained in yolk-poor clitellate embryos (e.g., Piscicolidae; Mann, 1962), in which the small gut anlage is primarily surrounded by a large primary body cavity, while the segmental mesoderm is initially restricted, as in yolk-rich embryos, to ventrolateral cell clusters.

The early development of muscle fibers has more recently been studied by phalloidin labeling (Bergter and Paululat, 2007, on *Enchytraeus*; Bergter et al., 2007, on *Limnodrilus* and *Erpobdella*; Hunnekuhl et al., 2009, on *Eisenia*). Unfortunately, these data do not allow distinction of myoepithelia from fiber muscle cells. The sequence of cell differentiation within the coelomic wall and an interdependence of myogenesis and coelomgenesis as observed in pre- and postmetamorphic stages of non-clitellate annelids accordingly still remain unclear.

The resemblance of postembryonic mesoderm development in clitellates, as exemplified above for *Pristina longiseta*, to postmetamorphic stages in non-clitellate annelids supports the idea that the development by germ bands in clitellate embryos is mainly due to the large yolk-sac that needs to be overgrown by the developing body segments (e.g. Anderson, 1973). Embryonic coelomogenesis in clitellates is likely derived from the premetamorphic mode of coelomogenesis as shown by non-clitellate species with a relatively high number of larval body segments, such as *Capitella teleta* (14 larval body segments; Seaver, Thamm, and Hill, 2005). In *Capitella teleta* a synchronous development of the larval body segments has been reported that resembles early stages of the embryonic coelomogenesis in "oligochaetes", in starting from latero-ventral regions of cell proliferation that subsequently move ventrally and then surround the gut anlage by dorsal expansion. The original postmetamorphic mode of body segment formation is primarily maintained in the postembryonic development of clitellates, but became increasingly suppressed during evolution and finally turned into entire suppression of postembryonic body segment formation and, thus, a fixed number of body segments in leeches (Euhirudinea). Ultrastructural details on the transformation of the embryonic coeloms into a vascular system in leeches like *Hirudo medicinalis* are still unknown.

11.4.2.4 Comparative evaluation

Coelomogenesis has been studied thus far at the ultrastructural level in a few polychaete annelid species only (Potswald, 1981; Turbeville, 1986; Bartolomaeus, 1993; Bartolomaeus and Ruhberg, 1999). These descriptions permit the following conclusions:

(1) The mesodermal growth zone is located anterior to the anus in juveniles (Figure 11.10A,B). In larvae (metatrochophore stages) paired clusters of undifferentiated mesoderm cells are located at the level of the anus or caudal to it. This region is characterized by high mitotic activity.
(2) Epithelio-muscle cell differentiation precedes formation of any other cell type that participates in lining the coelomic cavities, and precedes fluid accumulation that indicates onset of coelomogenesis.
(3) Coelomogenesis starts with tiny fluid-filled spaces between desmosomally connected epithelio-muscle cells of opposing layers of the prospective coelomic lining. The cells of each layer (visceral and somatic) are interconnected by apical adherens junctions.
(4) The mesodermal cells of the annelid growth zone are polarized cells from their earliest stage of differentiation onward. They either rest on the peri-intestinal or on the subepidermal ECM. All cells possess a diplosomal pair of centrioles, one of which may induce a rudimentary cilium that is either capped by a crescent vesicle or emanates from the cell apex, underpinning the primary polarity of the cells.
(5) Septa are not formed by repeated formation of paired cell clusters that abut on each other anterior-posteriorly. They are rather formed within a more or less compact strand of mesodermal cells by ECM that is continuously formed between groups of cells to finally connect the peri-intestinal and subepidermal ECM. Formation of the ECM is temporally correlated with the appearance of tiny coelomic

spaces between opposing epithelio-muscle cells, so that the coelom becomes segmentally arranged (Figure 11.17 A–C). Where septa are completely or partially absent in the adult, like in *Tomopteris helgolandica* or in *Pectinaria koreni*, the ECM has not formed in its proper position during early coelomogenesis. Septa thus are not reduced, but never form in species with a large uniform coelom (*Tomopteris helgolandica*) or remain incomplete in species with merely partially subdivided coeloms (*Pectinaria koreni*). This could also explain the occurrence of uniform coelomic cavities in sipunculans and echiurans.

(6) Formation of ventral mesenteries in premetamorphic stages precedes the formation of their dorsal counterparts. The central ECM of ventral mesenteries is formed *de novo*, where the mid-ventral walls of the paired coeloms are opposed. Here the ECM grows from the periphery (subepidermal and peri-intestinal ECM) to the center of the prospective mesenterium, similarly as during septal growth. For forming the dorsal mesenterium, in contrast, lining cells of the dorsal edge of the coelom migrate dorsally within the ECM located between epidermis and gut, thus separating the ECM into a peri-intestinal and a subepidermal layer. Where the edges of left and right coelom meet dorsally, the ECM will become the central ECM of the dorsal mesenterium. In postmetamorphic stages, the dorsal and ventral mesenteries already exist in anterior segments, so that these are used as some kind of template. Generally they are formed by ingrowing subepidermal and peri-intestinal ECM.

(7) Formation of an ECM between different muscle groups or muscle layers is always secondary. Primarily a uniform ECM surrounds both strands of mesodermal cells. All mesodermal cells primarily rest on this matrix and are polarized cells interconnected by apical adherens junctions and cross-linked by desmosomes. At least during coelomogenesis of premetamorphic stages, the ECM seems to guide the growth of the epithelio-muscle cells (i.e. the prospective coelothelia) towards the dorsal side.

(8) Blood vessels are always formed by fluid accumulation between the two *laminae densae* of the ECM that underlies mesenterial and septal coelomic linings. Primary body cavities (i.e., blood vessels) thus never represent persisting spaces of the former blastocoelic cavity, but come into being after ECM has been formed.

According to these results visceral and somatic musculature develop from an epithelialized bilayer of cells. Formation of myofibrils precedes fluid accumulation between the two layers; a simple myoepithelium represents the first coelomic lining formed during coelomogenesis. In this respect all previous ultrastructural studies on coelomogenesis in annelids coincide (Fransen, 1980a; 180b; Gardiner, 1992: fig. 49–51; Turbeville, 1986). Transformation of undifferentiated cells into epithelio-muscle cells may be triggered by neurons, since neurites are found in the subepidermal ECM next to the cells of the early mesoderm anlage.

Any non-muscular peritoneum is always formed rather late. From all descriptions it is evident that peritoneocytes as well as podocytes either differentiate from the same set of undifferentiated cells giving rise to epithelio-muscle cells, or, more evidently, they represent transformed epithelio-muscle cells. Moreover, our data as well as all previous studies provide convincing evidence that fiber muscle cells are transdifferentiated epithelio-muscle cells. Wherever a peri-intestinal bipartite myoepithelium and a somatic peritoneum covering the underlying somatic muscles are present, these linings ontogenetically arose from simple myoepithelia.

Heteronomy. As mentioned above, differences in coelomogenesis prior to and after metamorphosis have long been known and were termed heteronomy by Iwanoff (1928). Recently, it has been shown that in *Nereis virens* these differences correlate with differences in *Hox* gene expression, although a causal relation remains to be shown (Novikova et al., 2013). Morphologically heteronomy is characterized by a relocation of the mesodermal growth zone at the end of the larval phase and prior to metamorphosis. In late larvae the caudal mesoderm tapers into two ventrolateral cones. The posterior end of each cone consists of undifferentiated cells with a distinct polarity. In juveniles the mesoderm is a cylinder between the gut and epidermis. The posterior margin of this cylinder consists of the circum-intestinal anal sphincter. Anterior of the sphincter a ring of undifferentiated cells represents the mesodermal growth zone (see also Akesson, 1962; 1967; Anderson, 1966; 1973). Provided that our and the published results are representative, they indicate that at a certain time during ontogenesis the posterior-most cells stop mitoses and differentiate into muscle cells of the perianal sphincter (compare Figures 11.10 and 11.13). Their anterior neighbors then start mitoses and generate cells that differentiate into epithelio-muscle cells of the coelomic lining. This transition, which has often been said to be correlated with metamorphosis of the annelid larva, was already described in earlier light microscopical studies of mesoderm differentiation and coelomogenesis. Anderson (1973: 38) summarized that at the end of metamorphosis, *"the M-Cells* [teloblasts of the mesoderm strand between gut and epidermis] *of* Scoloplos, Arenicola, Eunice *and* Tomopteris *give rise to a circum-enteric ring of small mesoderm cells around the posterior end of the midgut* [...]. *These cells then continue to proliferate somites after metamorphosis is completed.*" Anderson subsequently critically evaluated Iwanoff's (1928) hypothesis on annelid heteronomy, according to which the mesoderm of all postmetamorphically formed segments ("tritomeres" fide Siewing, 1969) originates from the ectoderm (epidermis) of the growth zone. His critique was later substantiated by Dohle (1979) and is absolutely justified. According to our results all mesoderm that differentiates into the coelomic lining is always surrounded by the subepidermal and the peri-intestinal ECM and never shows any connection to epidermal cells.

Heterochrony. In interstitial annelids except for *Dinophilus* species the coelom is always lined by a simple myoepithelium, even if the coelomic space between the two layers is very small or almost absent (Fransen, 1980b; Smith, Lombardi, and Rieger, 1986; Rieger and Purschke, 2005). Even if a coelomic cavity is completely absent, their coelom-less, and often termed "acoelomate" condition (Smith, Lombardi, and Rieger, 1986) thus differs from that of acoelomate organisms in which ECM is present between true fiber muscle cells. The lack of a coelomic cavity in these organisms can be explained by a progenetic origin of the interstitial annelids (Westheide, 1986). Since in all annelids studied here mesodermal cells are epithelial prior to fluid accumulation, it must be assumed that these organisms retain this early ontogenetic organization; either fluid accumulation simply does not happen or occurs in a very restricted manner only. The states in *Dinophilus* as well as *Diurodrilus* species, however, are presently hard to interpret, because it remains unclear whether their mesoderm is formed from transformed coeloms at all (Worsaae and Rouse, 2008).

11.4.2.5 Conclusions

Our ultrastructural observations on coelomogenesis corroborate the first of the above mentioned hypotheses on the primary organization of the annelid coelomic lining. In accordance with Fransen (1988) and Rieger (1986) our data show that a simple myoepithelium ontogenetically precedes any other lining of the coelom. This observation has also been made for sipunculans (Jägersten, 1963; Rice, 1973; Ruppert and Rice, 1983) and echiurans (own unpl. data). In adults of both the somatic lining is always a peritoneum underlain by a strong body wall musculature composed of fiber muscle cells embedded into ECM (Figure 11.8D), while the visceral lining is a simple or a pseudostratified myoepithelium (Figure 11.8C). Maintenance of a simple myoepithelium in adult annelids most likely represents the primary organization of the somatic and visceral musculature in annelids, so that a peritoneum represents a derived state. Irrespective of the adult state, a simple myoepithelium is the first coelomic lining in both the somatic and visceral layers during the initial steps of coelomogenesis. Thus, the muscular system and the coelomic lining are identical during early coelomogenesis. Coelom formation by schizocoely accordingly does not happen independently of the prospective musculature.

According to current understanding of annelid phylogeny (e.g., Struck et al., 2011; Struck et al., Chapter 9), the embryonic development of clitellates is derived with respect to the formation of the segmental mesoderm and a higher number of embryonic body segments compared to the small number of larval segments (up to 15; see, e.g., Irvine, Chaga, and Martindale, 1999) in polychaete premetamorphic stages. Ultrastructural data are still missing, but we expect to find a similar formation of the coelomic cavities in clitellate embryos and polychaete larvae in terms of ventrodorsal expansion of the coelom during coelomogenesis. We also expect to find an initial myoepithelial lining of the developing coelom with a subsequent transformation of lining cells into fiber muscles, formation of a peritoneum, and in the case of hirudineans a truncated expansion of the coelom, so that merely small coelomic cavities are found in the adults. The fixation of body segment number in hirudineans is most likely correlated with the formation of a posterior sucker (Purschke et al., 1993).

11.4.3 Nephridia and nephridiogenesis

11.4.3.1 Nephridia

The nephridia in annelids are filtration nephridia which eliminate nitric wastes from the body in two subsequent physiological processes, ultrafiltration and modification of the ultrafiltrate (Ruppert and Smith, 1988). As mentioned above, annelids possess two different kinds of filtration nephridia, which differ in the topological relation between those units that mediate ultrafiltration on the one hand and modify the ultrafiltrate on the other hand (Smith and Ruppert, 1988; Bartolomaeus and Quast, 2005).

In protonephridia both units form a structural entity, in which adherens junctions firmly connect the terminal cell that ultrafilters body fluid to the duct that modifies this ultrafiltrate (Figure 11.14A). The terminal cell either bears a perforated cytoplasmic hollow cylinder (filter) or a circle of reinforced microvilli. Both structures surround a small fluid-filled compartment in which one or several cilia (ciliary flame) beat to move the fluid towards the duct. The pressure inside this compartment is thus lower than that of the surrounding medium, which causes an influx of fluid into the compartment. An extracellular matrix that covers the pores and slits of the filter or that covers the microvilli serves as an ultrafilter.

In metanephridial systems ultrafiltration occurs when fluid passes the slit membranes (diaphragms) of podocytes that rest on the coelomic side of the perivascular ECM (Figure 11.14B). Ultrafiltrate is, thus, collected inside the coelomic cavities. The metanephridia consist of a ciliated collecting funnel and a duct that modifies the coelomic fluid (ultrafitrate) by re-uptake (resorption) and secretion to release only nitric wastes and water to the external medium. Because the metanephridia are open connections between the exterior and the coelomic cavities, they are also used during reproduction as gonoducts to release gametes that are produced by proliferating coelomic lining cells and temporarily stored in the coelomic cavity. Gamete release is generally associated with an enlargement of the funnel and an increase in the number of cilia, whose rootlets are enlarged and become interconnected. Gonoducts of this kind accordingly represent transiently modified metanephridia. Due to their structure protonephridia cannot be used for releasing gametes.

In the majority of annelids a pair of protonephridia is the first nephridial organ formed during ontogenesis (*head kidneys* sensu Hatschek, 1878; 1885; see Kato, 2013). All subsequently formed nephridia are segmental protonephridia or segmental metanephridial systems. This ontogenetic sequence is kept in the majority of annelid taxa and it is generally assumed that both, head kidneys and segmental nephridia, result from serial repetition of the same anlage (Smith and Ruppert, 1988; Bartolomaeus, 1989; 1999; Bartolomaeus and Ax, 1992, Bartolomaeus and Quast, 2005; Haase et al.,

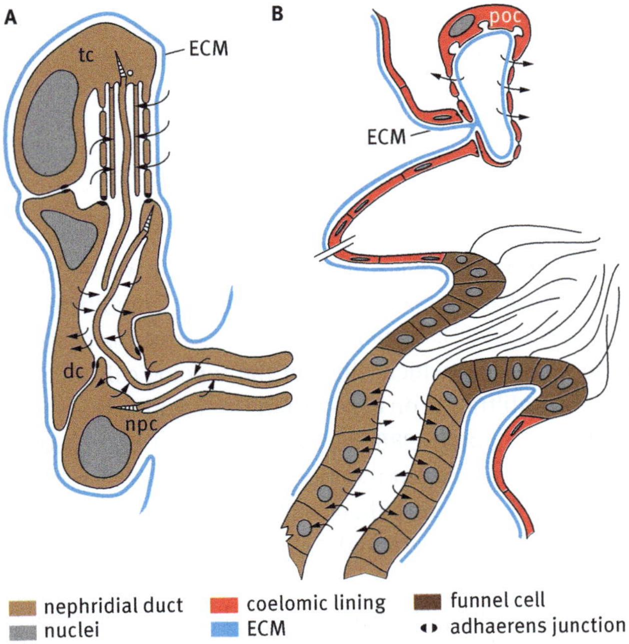

Figure 11.14: Structure of filtration nephridia. **A.** Protonephridium consisting of a terminal cell (*tc*), duct cell (*dc*) and nephropore cell (*npc*). A perforated cytoplasmic hollow cylinder is the supporting structure for exctracellular material (*ECM*) representing the filtration barrier. The ultrafiltrate produced when fluid passes this barrier (*arrows*) is modified by transport processes into and out of the duct and nephridiopore cell (*curved arrows*). The structural integrity of a protonephridium is maintained by adherens junctions that connect adjacent cells. **B.** Metanephridial system. Ultrafiltration occurs while blood fluid (*arrows*) passes the extracellular matrix (*ECM*) that lines the blood vessel; podocytes (*poc*) resting on the coelomic face of blood vessel act as supporting structure, a functional analogue to protonephridial terminal cells. The ultrafiltrate is modified by the metanephridium (*curved arrows*). Cilia originating from its proximal section extend into the coelom, so that this part of the metanephridium is traditionally called ciliated funnel. Site of ultrafiltration (blood vessels) and site of modification (metanephridium) are not directly connected and do not form a structural unit.

2010). Certain meiofaunal annelids, especially myzostomids and certain phyllodocidan annelids, including all species of the Phyllodocidae, Nephthyidae, Glyceridae, and Gonadiidae, possess segmental protonephridia instead of segmental metanephridia. In these species the gametes are either released by body wall rupture, by additional gonoducts that are not derived from nephridia, or by structurally modified protonephridia (Bartolomaeus, 1999). In the latter case the proximal duct cells start to proliferate at the beginning of the reproductive period to form a funnel that opens into the coelom and is connected to the protonephridial duct in lateral opposition to the terminal filtration unit. This funnel is a transitory structure that releases gametes during reproduction and degenerates thereafter (Stecher, 1968).

Historically the transitory funnel of protonephrida and the transient enlargement of the funnel of metanephridia have been used to substantiate the idea that primarily a pair of nephridia and a pair of gonoducts were present in each annelid segment (Goodrich, 1898). Proponents of this idea assumed that during annelid evolution nephridia and gonoducts fused to different degrees and applied different terms for the resulting organ, which still can be found in the literature (Goodrich, 1945; Rouse and Fauchald, 1997). This idea stems from the gonocoel theory (Hatschek, 1878; Meyer, 1890; Lankester, 1900), according to which coeloms are enlarged gonads. The gonocoel theory, however, fails to explain the evolutionary relation between protonephridia and metanephridia in annelids as it does not properly explain the known facts on annelid nephridia and their development (see Bartolomaeus and Quast, 2005). Transitory funnels in protonephridia gave rise to another idea, according to which segmental protonephridia with transitory funnels represent the primary excretory organs in annelids and metanephridia evolved from them by an ontogenetic shift in the formation of the funnel (Bartolomaeus, 1989; Bartolomaeus and Ax, 1992). This hypothesis predicts instances of permanent co-existence of terminal cells and funnels as well as of transitory protonephridial stages during the development of metanephridia. Although such instances were actually described in some species that seem to corroborate this hypothesis (Bartolomaeus, 1993; 1997), none of the annelid phylogenies discussed thus far reveal such an evolutionary sequence. Most meiofaunal annelids possess segmental protonephridia, while their macrofaunal relatives possess segmental metanephridia. The latter observation in particular gave rise to the hypothesis that protonephridia evolved from metanephridia by gradual size reduction of their funnel and transformation of the single funnel cell finally maintained into a terminal cell (Westheide, 1986). Although comparative studies of the nephridia in Syllidae initially seemed to support this assumption (Kuper, 2001), it increasingly turned out to be unlikely with respect to new structural data on the head kidneys, and with respect to recent insights on annelid phylogeny (Bartolomaeus, Purschke, and Hausen, 2005; Struck et al., Chapter 9). The most recent comparative analysis of annelid nephridia (Bartolomaeus and Quast, 2005) explains nephridial diversity in annelids by differing ontogenetic pathways (Figure 11.15). According to Bartolomaeus and Quast (2005) the structure of the nephridia in adults is strongly related to coelomogenesis, as outlined in the following.

11.4.3.2 Nephridiogenesis

As described above, after metamorphosis a meristematic region persists frontal to the anus until the animal reaches its final segment number. This region shows enormous mitotic activity and gives rise to largely identically organized body segments. Nephridial development can easily be studied in a series of sections from the meristematic region to body segments containing large coelomic cavities (see Bartolomaeus, 1993; 1999).

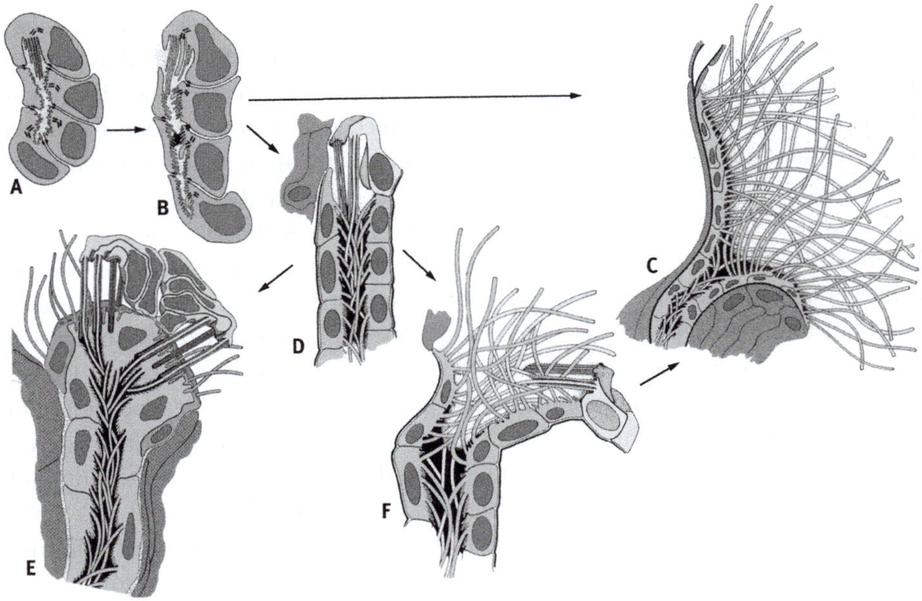

Figure 11.15: General scheme of developmental pathways to different nephridial organs in annelids. Truncation and/or modification of the development causes nephridial diversity in annelids. (**A**) Earliest recognizable anlage consisting of three to four cells. (**B**) Cilia of the proximal cells of the anlage are located laterally to the perikarya, not completely surrounded by the cells of the anlage. These cells either become the ciliated funnels of the metanephridium during coelomogenesis (**C**) or they differentiate into solenocytes (**D**). These either move aside when the proximal section of the duct is passively opened (**F**) or a further cell covers the proximal section of the duct so that a protonephridium with solenocytes is formed (**E**). If the temporarily formed solenocytes (**D**) move aside (**F**), they degenerate during formation of the funnel (**C**).

Figure 11.16: Nephridial anlage in *Pholoe minuta* (Pholoididae) (**A, B**), *Glycera alba* (Glyceridae) (**C**), and *Ophelia rathkei* (Opheliidae) (**D**). **A.** The earliest identifiable nephridial anlage consists of the prospective terminal cell (*ptc*) and two prospective duct cells (*pdc*). They are interconnected by adherens junctions (*aj*) to form a small compartment with microvilli (*mv*). An extracellular matrix only partially surrounds the anlage. **B.** Stage at the onset of schizocoely. The terminal cell (*tc*) possesses one cilium and is connected to an epithelio-muscle cell (*emc*) of the prospective coelomic lining by adherens junctions (*arrowheads*); an extracellular matrix (*ECM*) is restricted to one side of the anlage. Cilia and microvilli extend between epithelio-muscle cells before they enter the duct (*dc*). During schizocoely the terminal cell will passively follow the movements of the adjacent cells (see Figure 11.15C). **C.** The earliest anlage consists of three cells; the prospective terminal cells are monociliated. An ECM only partially surrounds the anlage. **D.** The anlage consists of four cells (*stars*) that are interconnected by apical adherens junctions (*arrowhead*) and surround a small compartment (see also Figure 11.17A). An ECM only partially surrounds the anlage. *Arrow* marks centriole. *ac* accessory centriole, *ce* centriole. ▸

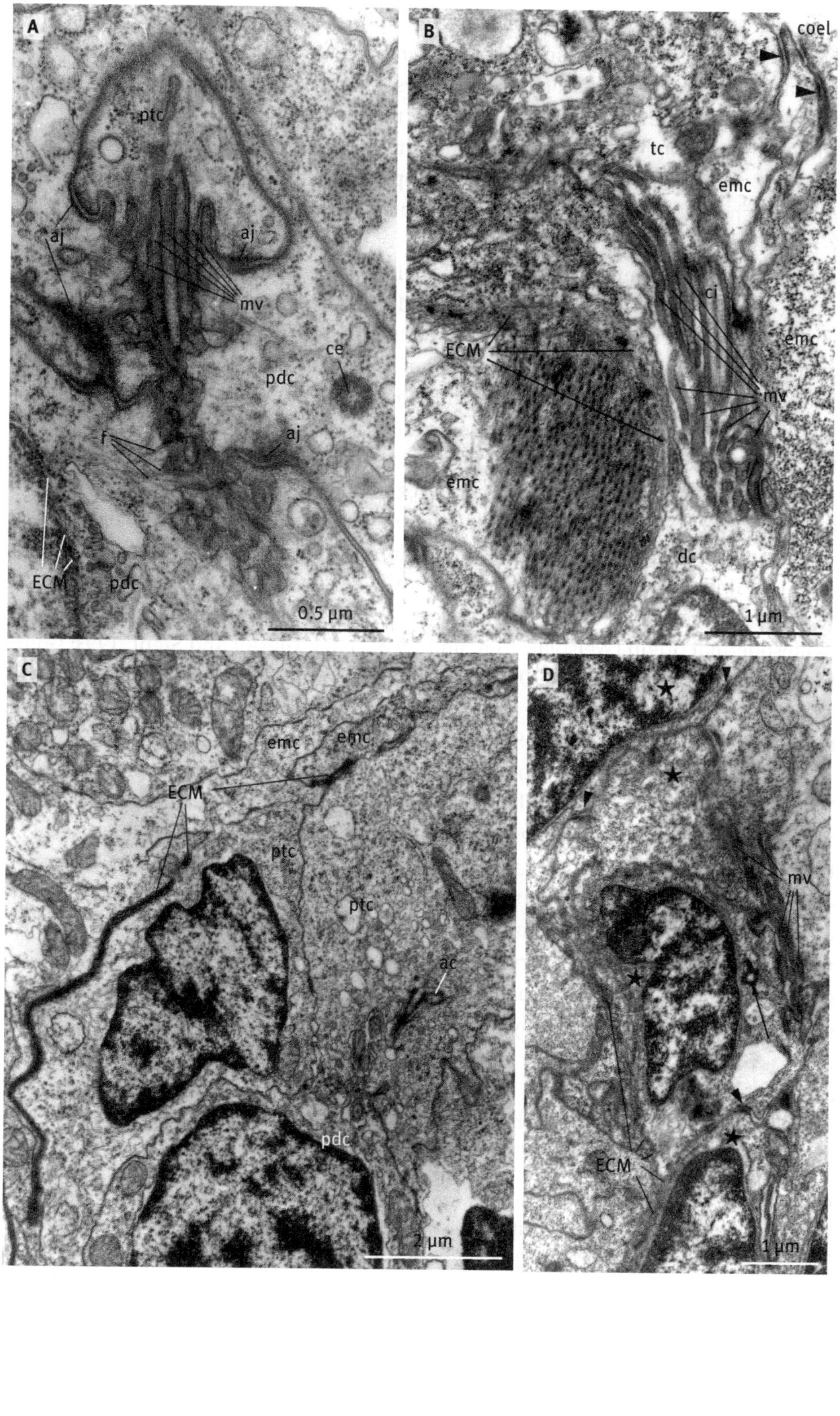
A
ptc
aj
aj
mv
ce
pdc
aj
f
ECM
pdc
0.5 µm
B
coel
tc
emc
cj
ECM
emc
mv
emc
dc
1 µm
C
emc
emc
ECM
ptc
ptc
ac
pdc
2 µm
D
mv
ECM
1 µm

The earliest nephridial anlagen that can structurally be discriminated as such from the surrounding mesodermal tissue consist of three to four cells that border a small compartment, the prospective duct lumen (Figures 11.15A, 11.16. 11.17A). These cells rest on a thin, incomplete ECM and show a distinct polarity with apical adherens junctions and small apical microvilli extending into the compartment (Figure 11.16A, D). This group of cells marks the starting point of nephridial development; any earlier stage does not show structures allowing unambiguous identification as nephridial anlage. While differentiation of the surrounding bilayer of polarized mesodermal cells of the prospective coelom into epithelio-muscle cells proceeds, the cells of the nephridial anlage start to form cilia (Figures 11.16B,C, 11.17B) that extend into the prospective duct lumen bordered by these cells. All cells initially are monociliated; additional cilia are formed with delay. The distal cells of the anlage clearly precede the proximal ones in terms of cell differentiation. While the number of nephridial cells increases, the distal duct cells elongate and the distalmost one pierces the epidermis so that the duct lumen becomes connected to the exterior (Figure 11.17C). The source of the additional nephridial cells is not clear. In regenerating segments they are formed by mitosis of presumptive, but yet weakly differentiated duct cells (Bartolomaeus, 1993). We assume an identical process for the developing nephridia during segment formation. Cell lineage tracing in *Tubifex* species and various Hirudinea revealed that the segmental nephridia develop from descendants of the M-teloblast and therefore are of mesodermal origin (Weisblat, Kim, and Stent, 1984; Gleizer and Stent, 1993; Goto, Kitamura, and Shimizu, 1999). Only the distalmost section of the duct may stem from either the o-bandlet (Martindale and Shankland, 1988 for *Helobdella triserialis*) or the n-bandlet (Goto, Kitamura, and Shimizu, 1999 for *Tubifex hattai*).

Although we were unable to unambiguously identify a nephridial anlage earlier than a 3-cell-stage in the polychaete species we studied, it nevertheless should be noted that the classical studies into clitellate development always mentioned a nephridioblast (Goodrich, 1945 and references therein) that is located retroperitoneally and gives rise to a cell band that anteriorly differentiates a funnel and posteriorly a duct. These studies indicate that nephridial development is decoupled from the formation of coelomic walls at a very early stage, a suggestion that is corroborated by our observations, which reveal a separation of the presumptive coelomic lining and parts of the prospective nephridium by ECM.

Figure 11.17: Development of metanephridia in *Ophelia rathkei* (Opheliidae). **A.** Earliest identifiable anlage consists of four cells surrounding a small compartment with microvilli inside. **B.** Ciliogenesis and enlargement of the anlage, the distal cell pierces the subepidermal extracellular matrix (*ECM*). **C.** Additional cells and stronger ciliation of the proximal cells of the anlage at onset of fluid accumulation in the prospective coelomic cavity. The proximal cell is only partially connected to the adjacent nephridial cell; the cilia run laterally to the proximal cell. **D.** During schizocoely the proximal cells will passively follow the movements of the coelomic lining cells to which they are connected by adherens junctions. The proximal duct cell cilia, thus, finally face the coelom and form the funnel. ▸

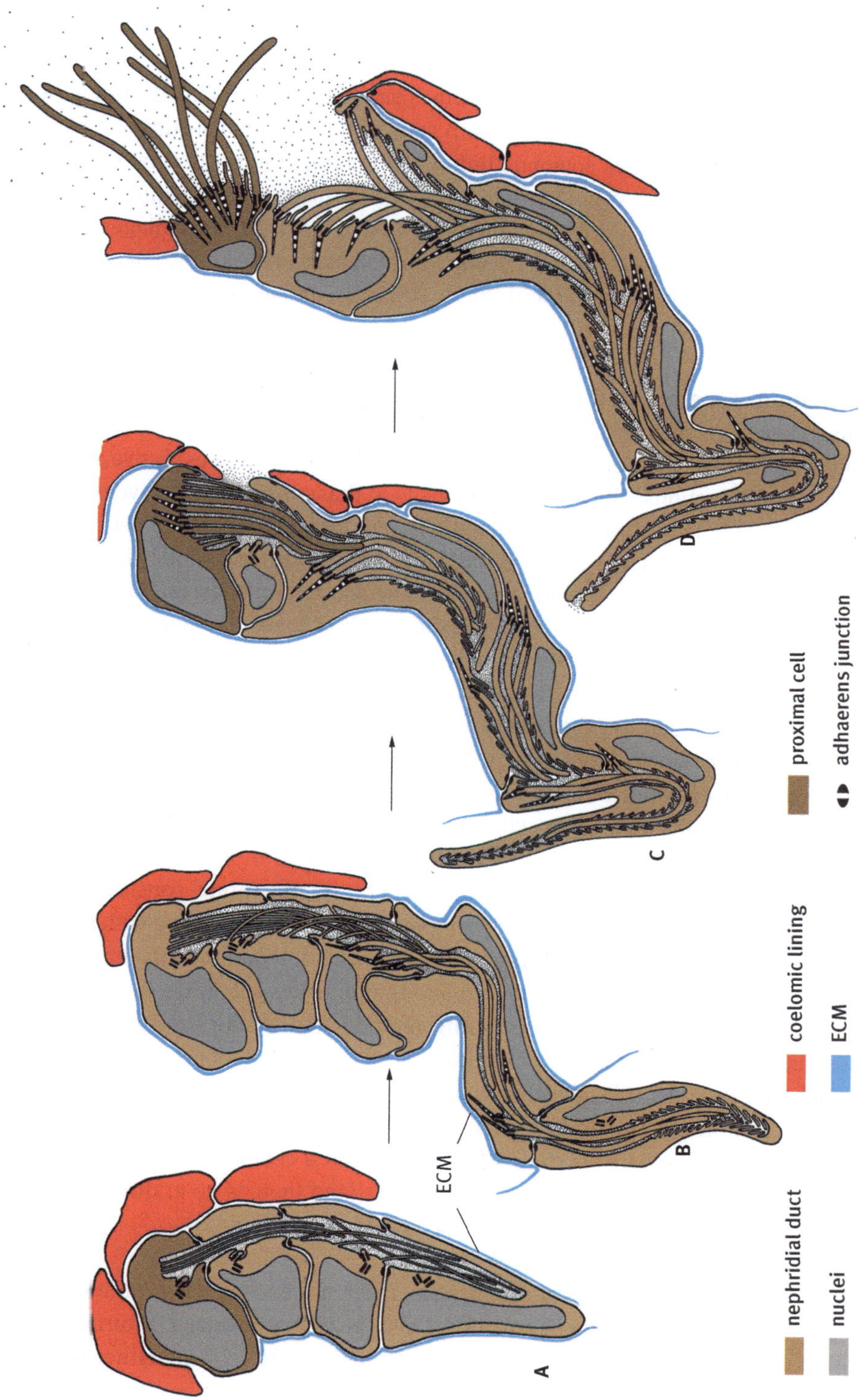
A
B
C
D
ECM
nephridial duct
nuclei
coelomic lining
ECM
proximal cell
adhaerens junction

During development of the nephridia, the ECM only partly surrounds the proximal and distal section of the nephridial anlage (Figure 11.16C,D). There are two cells forming the proximal section. One of them is situated more proximally and is called the innermost cell in the following. Both cells of the proximal section are opposed to cells of the prospective coelomic lining, without any delineation by ECM. These proximal cells may or may not become connected to the prospective lining cells, a difference that determines the further differentiation of the nephridial anlage into either a protonephridium or a metanephridium (Bartolomaeus and Ax, 1992; Bartolomaeus, 1993; 1997; 1999; Bartolomaeus and Quast, 2005) (see in Figure 11.15: D to E versus to F and then C).

Formation of segmental metanephridia. Both proximal cells of the early nephridial anlage become connected to the epithelio-muscle cells of the prospective coelomic lining and to their adjacent cells of the nephridial anlage (the prospective duct cells) by adherens junctions. The latter rest on the ECM that surrounds the developing duct (Figures 11.17B, 11.18A). After additional cilia are formed, the innermost cell partly detaches from the adjacent proximal cell. Thus, a small gap arises and cilia and microvilli of the innermost cell extend lateral to its perikaryon for a few micrometer before they enter the duct (Figure 11.17C). At this time no fluid has accumulated as yet among the epithelio-muscle cells of the prospective coelomic lining. When during schizocoely fluid accumulates among the epithelio-muscle cells to form the coelomic cavity, both proximal cells follow the coincient movement of the epithelio-muscle cells and the gap between them enlarges (Figure 11.17C). Their cilia and microvilli, thus, become exposed to the coelomic fluid and form a tiny funnel (Figure 11.17D). In some naidid clitellates, in aeolosomatids, in syllids and some other polychaetes a single cell will form the funnel nearly throughout the entire life cycle (Bartolomaeus, 1993; Bunke, 1994; 1998; 2000; 2003; Kuper, 2001). In other species, the funnel consists of several cells (Figure 11.15C). During further development and especially along with the reproductive period, the number of funnel cells, their ciliation and the attachment of the cilia by rootlet structures are increased and reinforced to guarantee the mechanical forces needed for gamete release during reproduction. Multiplication of the cells most likely happens by mitoses of the funnel cells. In a metanephridium all cells, i.e. funnel cells and duct cells, develop from a single anlage. In this respect as well as in the course of differentiation formation of the metanephridia is basically identical in polychaete and clitellate species. A fusion between a coelothelially derived funnel and an independently formed duct never occurs, and there is no grafting of ciliated cells onto the duct as proposed by Goodrich (1945).

Formation of segmental protonephridia. There are basically two different types of segmental protonephridia in annelids. One type is characterized by terminal cells with a collar of microvilli surrounding each cilium. These cells are either mono- or multiciliated; a perforated cytoplasmic cylinder (filter) is absent. Such terminal cells are called solenocytes (Goodrich, 1945). The other type is characterized by multiciliated terminal cells with a lateral ciliary flame that is surrounded only peripherally by microvilli. Protonephridia of the latter type are restricted to interstitial annelids (e.g., Westheide, 1985; Clausen, 1986; Bartolomaeus and Quast, 2005), while those of the former type are found in certain phyllodocidan (nephthyids, phyllodocids, glycerids, tomopterids and others) polychaetes only. The development of the segmental protonephridia with solenocytes is quite well understood (Bartolomaeus, 1989; 1993; 1997; 1999; Bartolomaeus and Quast, 2005) (Figure 11.18B). The origin of protonephridia with lateral ciliary flames will be described below. Here, the proximal cells of the anlage are directly opposed, but not connected to the epithelio-muscle cells of the prospective coelomic lining. Adherens junctions connect them to the adjacent cells of the prospective duct, which rest in the ECM that surrounds the developing duct (Figure 11.15). The proximal cells remain monociliated and develop into solenocytes, while the prospective duct cells become multiciliated. In contrast to anlagen of metanephridia the proximal cells, i.e. the prospective solenocytes, largely retain their original position. They extend into the coelom but are not moved apart while the coelomic cavity widens by fluid accumulation. This is because the proximal cells remain interconnected by adherens junctions and because at least one multiciliated coelothelial cell covers the proximal duct cells except for those small areas where the cilium and microvilli of the solenocyte enter the duct (Figure 11.15E) (Goodrich, 1945; Bartolomaeus, 1989).

Metanephridia originating from anlagen of protonephridia. At least two species are known that do not possess such a coelothelial cell covering the proximal duct cells of the nephridial anlage, i.e. *Harmothoe sarsi* and *Pholoe inornata* (Bartolomaeus and Ax, 1991; Bartolomaeus, 1999). In both species early stages of nephridial anlagen are identical to those of Phyllodocidae, Nephtyidae and Glyceridae, consisting of proximal solenocytes and adjacent duct cells (Figure 11.16A, B). However, the most proximal duct cell becomes connected to the adjacent cell of the coelomic lining. When the coelom starts to expand by fluid accumulation the solenocytes accordingly move aside such that the apices of the adjacent duct cells are exposed to the coelomic fluid by forming a funnel. During further development the solenocytes degenerate and a typical metanephridium results from this development (Figures 11.15D to F and C, 11.18C).

Protonephridia originating from anlagen of metanephridia. In interstitial species, presence of segmental protonephridia is correlated with the smallness of the animals and their small coelomic cavities. Based on structural similarities between the protonephridia of interstitial annelids and anlagen of the metanephridium in their macrofaunal relatives, we recently hypothesized that these protonephridia result from a truncated metanephridial development, so that these organs actually represent an early developmental stage of metanephridia that is retained during further development (Bartolomaeus and Quast, 2005). In the stage retained, the proximal cells of the anlage bear a lateral bundle of cilia (ciliary flame) surrounded by peripheral micro-

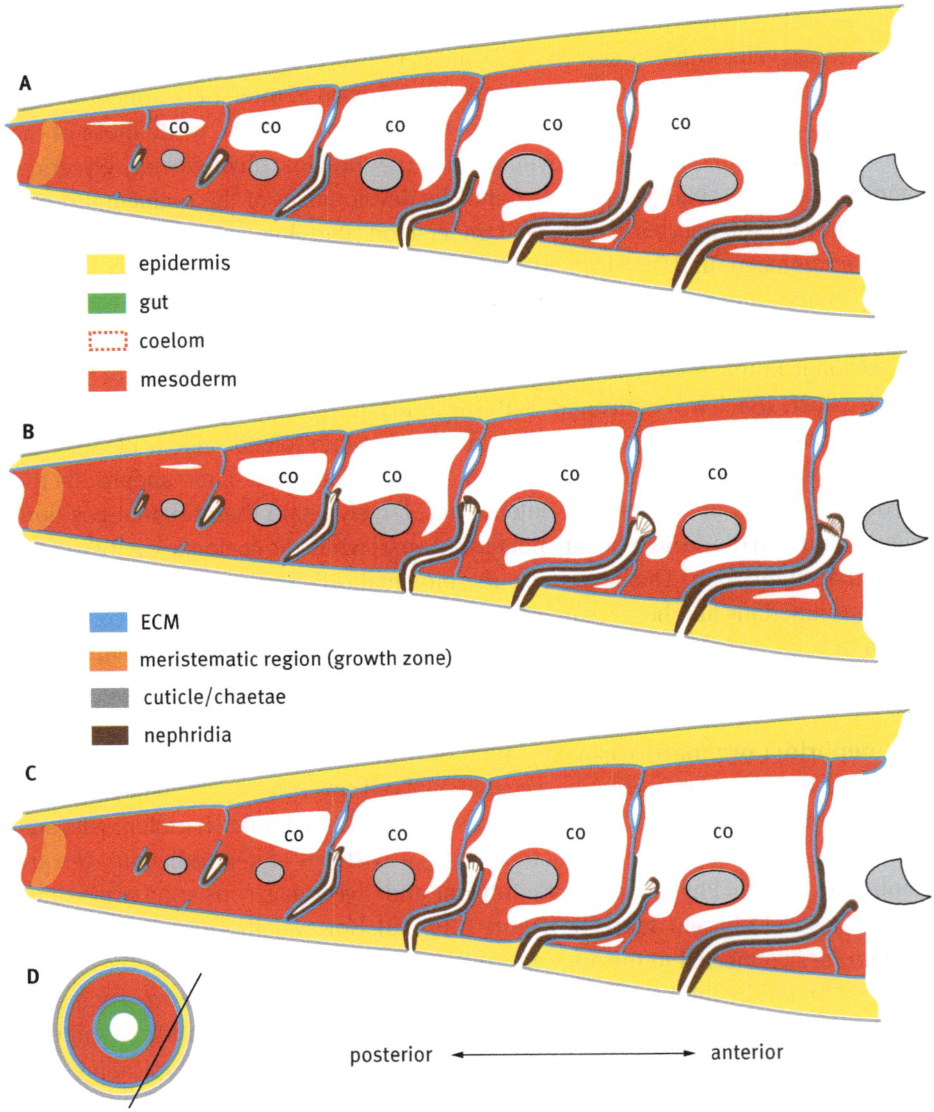

villi. Structurally, this resembles the terminal unit of a protonephridium devoid of a perforated cytoplasmic hollow cylinder (i.e., a filter; see for instance Clausen (1986) for *Microphthalmus ephippiophorus* or Brandenburg (1970) for *Dinophilus gyrociliatus*). Comparative studies, however, reveal that such a protonephridial anlage is identical to a developmental stage of a metanephridium. Kuper (2001) showed the enormous variety of terminal sections in syllid segmental organs ranging from small funnels to slit-like terminal openings and lateral ciliary flames that extend into the duct. Several of them clearly resemble stages of metanephridial development so that the variety of syllid nephridia can be explained by variably truncated nephridial development.

◂ **Figure 11.18:** Schematic semisagittal section through a postmetamorphic stage of *Nereis pelagica* (Nereididae) (**A**) *Phyllodoce mucosa* (Phyllodocidae) (**B**), and *Pholoe minuta* (Pholoididae) (**C**) to illustrate coelomo- and nephridiogenesis. The meristematic growth zone is located anterior to the anal sphincter. The nephridial anlage initially is located between the coleomic lining cell, only partially separated from the latter by an ECM. **A.** During fluid accumulation the proximal (inner) section of the nephridium is passively opened, since these cells are connected to adjacent lining cells. **B.** The proximal cell is a terminal cell that is firmly attached to the adjacent duct cells. The terminal cell does not follow movement of the lining cells during schizocoely. **C.** The proximal cell is a terminal cell that is connected to adjacent lining cells. During schizocoely the terminal cell is moved apart and the proximal section of the duct is exposed to the coelomic fluid. The terminal cell disappears and the proximal cells form the ciliated funnel of the metanephridium. **D.** Sketch of a transverse section to illustrate the position of the schematic semisagittal sections shown in **A–C.** *co* coelom.

11.4.3.3 Conclusion

Head kidneys and segmental nephridia in annelids seem to result from serially repeated information, irrespective of whether the latter are proto- or metanephridia. The actual structure of segmental nephridia depends on different pathways during early nephridial development and its interdependence with coelomogenesis. This especially refers to the early differentiation of the proximal cells of the nephridial anlage (Figure 11.15) (Bartolomaeus and Quast, 2005).

Two observations from our study are important in this context:

(1) Smith and Ruppert (1988) found evidence for protonephridial terminal cells being transformed into podocytes of the metanephridial system in *Sabellaria cementarium*, suggesting interconvertibility between both cell types. One therefore would expect that those proximal cells that become solenocytes in annelids with segmental protonephridia develop into podocytes in annelids with metanephridia. This idea of interconversion of terminal cells and podocytes was proposed by Ruppert and Smith (1988) based on their ultrastructural study into the fate of the head kidney (larval protonephridium) in *Sabellaria cementarium*. Such interconversion, however, was never observed in any species we studied. Instead we found that formation of podocytes is completely decoupled from nephridiogenesis. Our studies into coelomogenesis in annelids clearly show that podocytes are transformed coelomic lining cells that develop independently from the nephridia. One therefore would expect that certain annelids possess metanephridia although podocytes are absent and that others possess podocytes in addition to protonephridia. Both instances actually have been found in annelids: *Protodrilus rubropharygeus* has typical protonephridia, while podocytes rest on the wall of the interseptal blood vessels (von Nordheim and Schrader, 1994), other annelid species (e.g. nereidid species) possess metanephridia while podocytes have never been recorded. In some of the latter instances blood fluid instead is ultrafiltered through fenestrated epithelia (see Figure 11.4C).

(2) The first matrix components formed inside the segmental mesodermal cell cluster that gives rise to the nephridia and the coelomic lining is the ECM separating the distal cells of the nephridial anlage from the surrounding prospective coelomic lining cells (Figure 11.18). When the distal cells of the nephridial anlage pierce the subepidermal basal lamina, this ECM becomes connected to the subepidermal basal lamina. In a similar way the peri-nephridial ECM of the proximal cells expands towards the peri-intestinal ECM when prospective coelomic lining cells next to the proximal cells start forming an ECM. During coelomogenesis this matrix will become the interseptal ECM, with coelomic lining cells resting on both the anterior and posterior face of the septum. Since the proximal cells of the nephridial anlage are formerly asymmetric with an anterior-laterally located bundle of cilia and a posteriorly located ECM, the funnel these cells give rise to will open into the coelomic cavity anterior to the nephridial anlage. Because the distal cells of the nephridial anlage are completely surrounded by ECM, prospective coelomic lining cells will adhere to this matrix. The entire metanephridial duct will, thus, be located posterior to the prospective septum. This transsegmental position of the nephridia is characteristic of annelids and results from early asymmetries in matrix formation of the early nephridial anlage. Our data seem to indicate that in phyllodocidan polychaetes formation of septa is closely related to nephridial development – an idea that has to be tested in further studies.

11.4.4 Summary

Annelids possess paired, segmentally arranged coelomic cavities that are lined by myoepithelia that also represent the body wall (somatic) musculature and the intestinal (visceral) musculature. A distinct peritoneal lining, if present, is restricted to the somatic wall of the coelomic cavities. There is strong evidence from comparative and developmental data that such a peritoneum represents a derived condition. Coelomic cavities in annelids arise from solid clusters of epithelialized mesodermal cells situated between the epidermis and intestine. These bands are subdivided by ECM into blocks of epithelialized cells that are primarily arranged as a single layer of epithelio-muscle cells containing a very low content of myofilaments. The epithelio-muscle cells become separated by fluid accumulation (schizocoely) to form coelomic cavities. Later during development these lining cells develop into peritoneal and muscular cells that may become separated by ECM. Most likely epithelio-muscle cells lining the coelomic cavities represent the primary state in annelids. Such a lining is consistently the first that occurs during coelomogenesis and is retained in the organization of the mesenteria and the dissepiments as well as in the visceral muscular system of most species. These findings finally explain why transverse muscles that cross the coelom are never lined by a peritoneum. Nephridia and coelomic linings develop from polarized mesodermal cells that take different pathways early during development.

The earliest nephridial anlage that can be identified by specific structural features is already separated from the adjacent prospective lining cells by a thin ECM, so that formation of the nephridia in terms of increasing cell number is independent from the remaining mesoderm. Whether the nephridial anlage develops into a protonephridium or metanephridium depends on the structure of the proximal cells of the anlage and coelomogenesis. Podocytes are never formed from cells of the nephridial anlage.

11.5 Arthropoda

According to previous histological studies (surveyed by Anderson, 1973; 1979) the extent to which coeloms contribute to organogenesis in arthropods is quite variable. Size and extension of coeloms seem to vary among taxa to the same degree as among body segments of an individual embryo (or larva, respectively), ranging from large, trilobed shapes at the peak of their development over globular or diminutive forms to partial absence in some body segments, or entire absence in all body segments as in *Drosophila melanogaster*. When present, descriptions of the coelomogenesis in arthropods generally agree on the following aspects: (i) the primordial mesoderm only partially differentiates into "coelomic sacs", with a variable number of mesodermal cells not participating in their formation; (ii) during coelomogenesis the "coelomic sacs" form various diverticula (or lobes) differentiating into mesodermal organs (especially musculature, fat body, hemocytes, and coelomoducts) before the coeloms disintegrate; (iii) disintegration of the coeloms coincides with 'mixocoely' (i.e., fusion of the coeloms' lumen with the hemocoel); (iv) the walls of the "coelomic sacs" form the main source for somatic, visceral, and appendicular muscles, as well as for the fat body, circulatory system (including pericardial septum), and the anlagen of gonads; (v) the filtration unit (sacculus) of nephridial organs represent remnants of the original coeloms. The main problem with all previous histological studies is the lack of precision in the documentation of the results; especially in illustrations of histological sections cell borders within "coelomic sacs" and their diverticula are usually omitted. The limited resolution of light microscopy is particularly evident in published light micrographs of histological sections, the thickness of which ranged from 5 up to 40 µm (see, e.g, Weygoldt, 1975; Scholl, 1977). Skepticism is particularly caused by claims that diverticula of the (original) coeloms form 'secondary coeloms' before disintegrating into single or clusters of mesenchymal cells (e.g., Siewing, 1969), implying repeated dedifferentiation and re-epithelialization of the original coelom and/or its derivatives. Apparently, an "ordered" arrangement of mesodermal cells in layers was considered as a primary indication for epithelialization. Doubts are also raised by illustrations showing an anterior-posterior sequence of alleged coeloms that are intersegmentally separated only by a monolayer of mesodermal cells or just by trabecula-like branches of them (e.g., Weygoldt, 1964: figures 50, 51), whereas true dissepiments between consecutive coeloms consist of two cell layers with ECM in between. The insufficient resolution of the composition of the "coelomic sacs" and their derivatives in previous histological studies indeed proved in electron microscopy to be the main source of basic misconceptions.

11.5.1 Occurrence and fate of transient coeloms

Our taxon sampling for ultrastructural studies primarily focused on basal lineages of the high-ranking arthropod subgroups (e.g., Pycnogonida among Chelicerata, Xiphosura among Euchelicerata, Collembola among Hexapoda, Zygentoma among insects), which have the strongest impact on ancestral character state reconstructions. To this choice we added taxa that were (or still are) considered as representative for a particular clade, e.g. *Artemia salina* for (entomostracan) crustaceans, web spiders (sampled by the haplogyne species *Pholcus phalangioides*) for arachnids, and the phasmid *Carausius morosus* for hemimetabolous pterygote insects. Among these taxa, complete absence of coeloms proved to be characteristic for pycnogonids and collembolans. Insights obtained for *Artemia salina* indicate that coeloms are also generally absent in crustaceans. Transient formation of segmental coeloms is nevertheless confirmed for horseshoe crabs (Xiphosura), web spiders (Araneae), and insects. An anterior-posterior gradient in the development of the coeloms, however, has never been observed; nor did coeloms prove to occur in all body segments. The following descriptions especially detail our results for chelicerates, which seem to be most informative for comparisons with onychophorans and annelids; for other arthropod subgroups the main deviations and new insights are highlighted.

11.5.1.1 Pycnogonida

Achelia vulgaris and *Nymphon gracile* represent species with a free-living, ectoparasitic protonymphon larva (Figure 11.19A), which is traditionally considered as the ancestral hatching stage in pycnogonids (for discussion see Brenneis, Arango, and Scholtz, 2011a; Burris, 2011). They show a massive embryonic mesoderm that entirely fills the space between the gut and epidermis in all investigated stages (Figure 11.19B,C). The mesodermal cells are apolar and surrounded by extracellular matrix, to which they are connected by hemidesmosomes and dense plaques (Figure 11.19D). Intercellular connections between mesodermal cells are restricted to spot desmosomes; adherens junctions that might point to rudiments or remnants of epithelial organization within the mesoderm have neither been detected at any embryonic stage nor in the protonymphon larva. In the late protonymphon stage the musculature of the limbs is well developed, as are the muscles of the pharynx. The pharyngeal muscles proved not to form myoepithelium but fiber muscle cells beneath the wall of the pharynx. The pharyngeal wall is made up of a thin monolayer of epidermal cells and a thin cuticle. Attachment of the muscle cells at the pharyngeal cuticle is indirectly achieved via

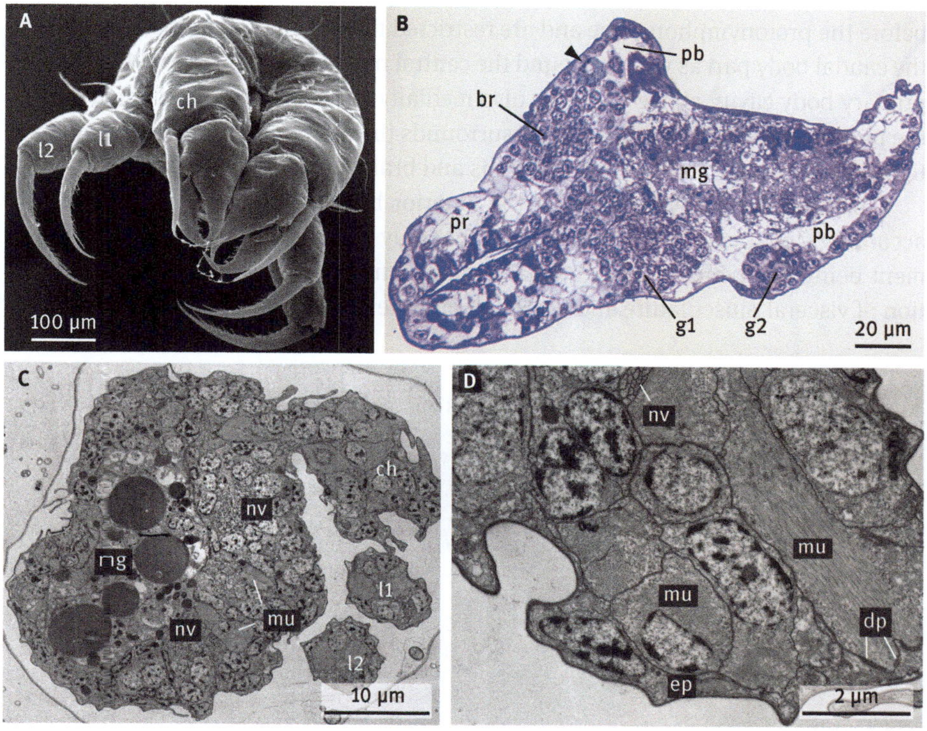

Figure 11.19: Protonymphon stage of (A,B) *Nymphon gracile* and (C,D) *Achelia vulgaris*. **A.** Frontal view (SEM) on chelifores (*ch*) and subsequent legs (*l1, l2*). **B.** Semithin sagittal section (LM), anterior to the left. The central nervous system consists of the brain (*br*) and two paired ganglia (*g1, g2*). The median eyes (*arrowhead*) are situated dorsal to the brain (*br*). Larger primary body cavities (*pb*) occur dorsally around the brain and in the ventrocaudal part of the body. **C.** Paramedian sagittal section (TEM) of protonymphon shortly before hatching, showing entire absence of any body cavities. Ventrally to the midgut anlage (*mg*) the apolar mesoderm comprises muscle cells (*mu*) and forms a massive tissue into which the nervous system (*nv*) is embedded. **D.** Transverse section of leg 1-anlage. Muscle cells (*mu*) and nerve strands (*nv*) massively fill the limb. Dense plaques (*dp*) connect the muscle cells to the epidermis (*ep*). *dp* dense plaques, *pr* proboscis.

microtubules passing through the epidermal cells in the same direction as the myofibrils of the muscle cells. Visceral musculature does not develop before hatching and is still absent in the protonymphon stage. In this stage, the sac-shaped midgut anlage consists of a massive aggregation of cells moderately to sparsely filled with yolk droplets; a midgut lumen is still lacking, as is a distinct border to adjacent mesodermal cells apart from ECM (Figure 11.19D). In accordance with previous descriptions (summarized by Arnaud and Bamber, 1987; see also Brenneis, Arango, and Scholtz, 2011b), embryos and protonymphon larva show neither any traces of the proctodaeum, nor of nephridial organs, or of primordial germ cells. Mesodermal body cavities do not occur before the protonymphon stage and are restricted to marked primary body cavities in the caudal body part as well as around the central nervous system (Figure 11.19B). The primary body cavity above the brain also medially separates the two median eyes of the protonymphon larva and partially surrounds them, but most parts of the eyes are tightly squeezed in between the epidermis and brain (Figure 11.19B).

The mesoderm development of the anterior body segments of pycnogonids is accordingly characterized by the formation of only somatic musculature, its development being independent of segmental coeloms and obviously preceding the formation of visceral musculature and any other mesodermal organs.

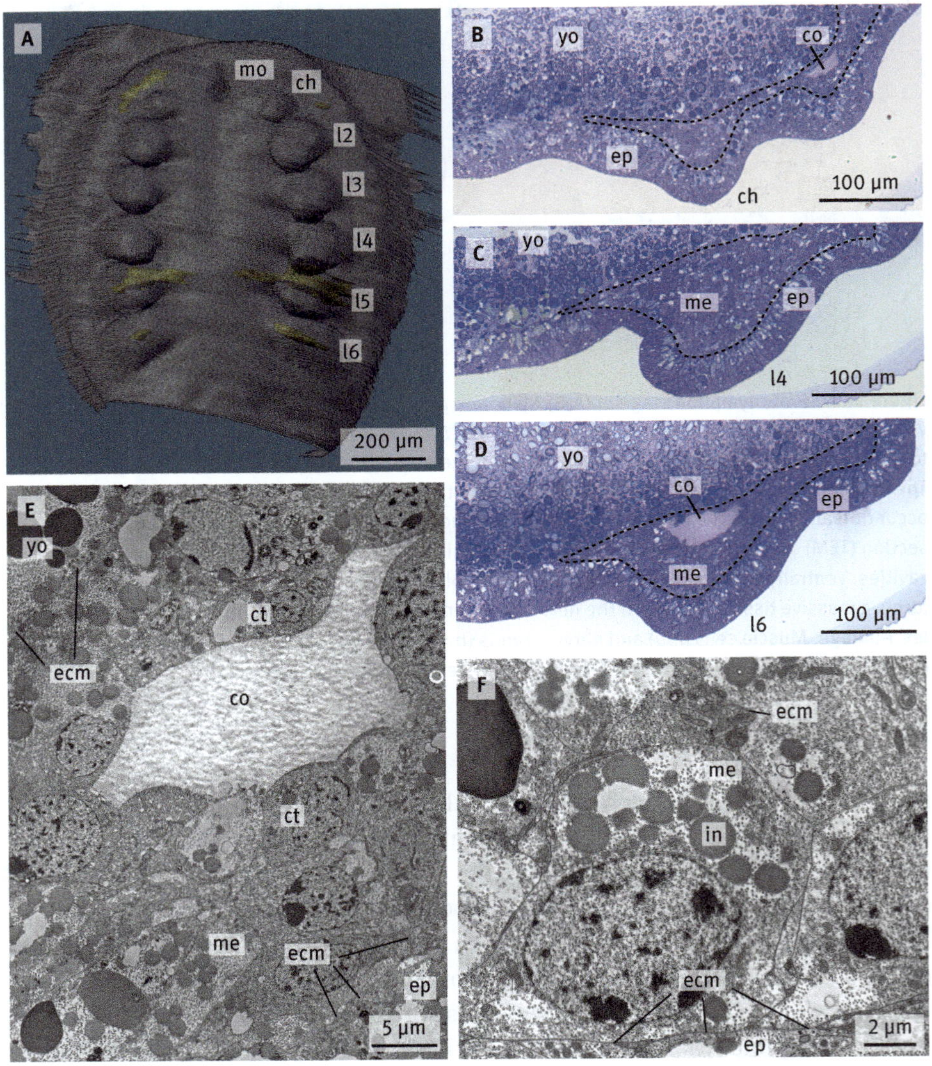

11.5.1.2 (Eu-) Chelicerata

Large segmental coeloms transiently extending from the ventrolateral base of the limb anlagen towards the dorsal midline have been considered as characteristic for arachnids (Weygoldt, 1975). Our studies by TEM confirm that the coeloms of some body segments attain such dimensions not only in *Pholcus phalangioides* (Araneae) but also in *Limulus polyphemus* (Xiphosura).

11.5.1.2.1 *Limulus polyphemus* (Xiphosura)

Coelomogenesis in *Limulus polyphemus* has been studied ultrastructurally at embryonic stages 15 to 20–2 fide Sekiguchi, Yamamichi and Costlow, (1982; reviewed by Shuster and Sekiguchi, 2003). The embryonic mesoderm primarily consists of two massive, longitudinal bands of apolar cells squeezed in-between the epidermis and yolk cells of the midgut anlage. The bands are medially clearly separated from each other and laterally extend into the massive early limb buds. The apolar cells of these bands do not show any particular order but are largely arranged irregularly, with two to three cells superimposed upon each other along the longitudinal bands. Epithelialization within the mesodermal bands is first recognizable by adherens junctions interconnecting some cells. The interconnected cells primarily maintain the shape of adjacent apolar cells and are accordingly hardly distinguishable from them. Shortly after simultaneous formation of the prosomal limb buds (embryonic stage 15) the first body cavities arise above the rudiments of the chelicerae as minute slits within the compact mesoderm (Figure 11.20A,B); with the occurrence of the first opisthosomal

◂ **Figure 11.20:** *Limulus polyphemus*, embryonic stage 15. **A.** Reconstruction of the germinal plate and underlying coelomic cavities *(yellow)*, anterior is top. Coelomic cavities occur laterally to the cheliceral limb buds (*ch*) and underneath the limb buds of the fifth and sixth leg-bearing prosomal segments (*l5*, *l6*). Leg-bearing prosomal segments 2–4 (*l2*–*l4*) lack coeloms at this and any other stage. **B–D.** Series of semithin transverse sections (LM) of the germinal plate, from anterior to posterior. *Dotted lines* demarcate the extension of the mesoderm. **B.** Section through the left cheliceral anlage (*ch*). A small coelomic cavity (*co*) occurs in the mesoderm latero-dorsally to the limb bud. **C.** Section through left limb bud of fourth prosomal segment (*l4*). The limb bud is filled with massive apolar mesoderm (*me*). **D.** Section through left limb bud of sixth prosomal segment (*l6*). A large coelomic cavity occurs centrally above the limb bud. The massive apolar mesoderm (*me*) within the limb bud consists of several cell layers. **E.** Transverse section (TEM) through the ultimate prosomal segment. The coelomic cavity (*co*) is lined by cuboidal coelothelial cells (*ct*). Thin extracellular matrix (*ecm*) occurs between the visceral coelothelium and the gut anlage (*yo*). The somatic coelothelium is underlaid by apolar mesenchymal cells (*me*). Very thin extracellular matrix (*ecm*) is detectable between the apolar mesenchymal cells and the epidermis (*ep*). **F.** Transverse section (TEM) through apolar mesenchymal cells (*me*) in left limb bud of fourth prosomal segment (*l4*). Some mesodermal cells contain numerous inclusions (*in*). Extracellular matrix (*ecm*) occurs at the border to the epidermis (*ep*) and in areas where several apolar cells meet. *mo* mouth opening, *yo* yolk cells of gut anlage.

limb buds further slit-like cavities occur above the rudiments of the penultimate and ultimate prosomal segment (Figure 11.20D). These slits proved to represent coelomic cavities as they are lined by an epithelium, the cells of which basally adhere to ECM via hemidesmosomes and are apically equipped with rudimentary cilia (Figures 11.20E, 11.21B,F). The flat visceral wall of the coelothel comes to lie immediately adjacent to the central yolk cells, whereas the slightly thicker somatic wall largely remains underlain by apolar mesodermal cells (Figure 11.20E,F). Extracellular matrix separates the entire coelothel from the yolk cells (Figure 11.21F), from the epidermis, and from apolar mesodermal cells at its frontal, lateral, and caudal border. Among the apolar cells some interstitial regions of the ECM are widened into minute lacunae (Figures 11.20F; 21D,E). Apart from these lacunae (invisible in light microscopy), neither any distinct primary body cavities, nor muscles, nor nephridial anlagen are formed at the earliest stage of coelomogenesis. In subsequent stages, the prosomal coeloms increasingly encompass the yolk cells into dorsal, medioventral, and longitudinal directions, partly by cell divisions within the coelothel, but mostly by stretching and coincident flattening of the coelothelial cells. Paired coelomic cavities also develop simultaneously in the first three opisthosomal segments (embryonic stage 19) and attain similar dimensions as in the last two prosomal segments (Figure 11.22C,D,E). At the peak of their development the coeloms are flat and crescent-shaped in transverse section, and triangular in horizontal section as the somatic wall anteriorly slightly protrudes towards the ectoderm. Most parts of the coelomic cavity, however, form an irregular network of minute interstitial compartments (Figure 11.22D,E,G). This is due to repeated close approximations of the somatic and visceral walls along the dorsolateral extension of the coeloms, along which the walls are often separated only by rudimentary cilia protruding into the coelomic cavity (Figure 11.22H). At the same time large primary body cavities and single or clusters of apolar mesodermal cells adhering to either the ectoderm or to the coelothel (Figure 11.22F) delimit the somatic wall of the coelom from the ectoderm. Dissepiments do not form between the coeloms of successive body segments, since consecutive coeloms intersegmentally approximate only along a very limited dorsolateral area, while being otherwise separated by primary body cavities and apolar cells. All cavities arising within the originally massive limb buds proved to be primary body cavities that are lined by ECM and irregularly traversed by strands of apolar cells (Figures 11.21A,D,E, 11.22B,I); a coelomic lobe extending into the limb buds does not exist at any stage.

The complex system of cavities arising during coelomogenesis renders distinction of coeloms from primary body cavities in histological sections factually impossible (Figures 11.21A, 11.22B–E). We furthermore found no traces of coeloms in the mesoderm of limb-bearing prosomal segments 2–4, which corroborates previous claims (Kishinouye, 1891; Scholl, 1977) that they are absent in these body segments. The mesoderm of segments devoid of coeloms gives rise to strands of apolar cells with thin protrusions forming an irregular network within primary body cavities. In embryos of the viviparous scorpion *Heterometrus scaber*, Mathew (1956) described

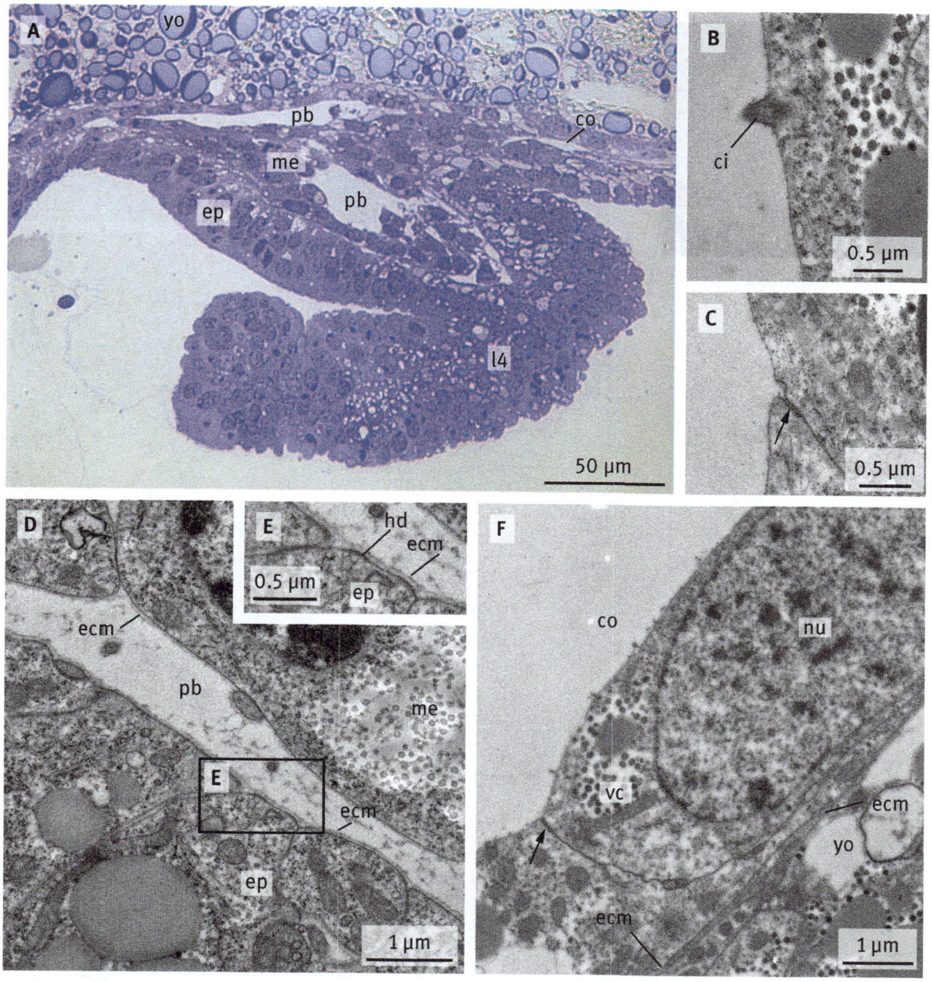

Figure 11.21: *Limulus polyphemus*, discrimination of primary body cavities and coeloms. **A.** Semithin transverse section (LM) through anlage of third walking leg (*l4*) at embryonic stage 17, median is left. Primary body cavities (*pb*) and coelomic cavities (*co*) cannot be distinguished by light microscopy, since mesenchymal cells (*me*) enclose these cavities similarly. **B–F.** Ultrastructural characteristics (TEM) of the lining of primary body cavities and coeloms. **B.** Coelothelial cell with apical rudimentary cilium (*ci*). **C.** Adherens junctions (*arrow*) interconnect the coelothelial cells apically. **D.** Primary body cavity lined by extracellular matrix (*ecm*) between epidermal (*ep*) and mesenchymal (*me*) cells. **E.** Detail of D in higher magnification, showing hemidesmosomes (*hd*) connecting the extracellular matrix (*ecm*) to the basal cell membrane of the epidermal cells (*ep*). **F.** Embryonic stage 15, visceral coelothelium (*vc*) in the cheliceral body segment. Coelothelial cells are apically interconnected by adherens junctions (*arrow*). Extracellular matrix (*ecm*) is only occasionally detectable between the basal side of the coelothelial cells and the yolk cells (*yo*) of the gut anlage. *hd* hemidesmosome, *nu* nucleus, *yc* yolk cells of the gut anlage.

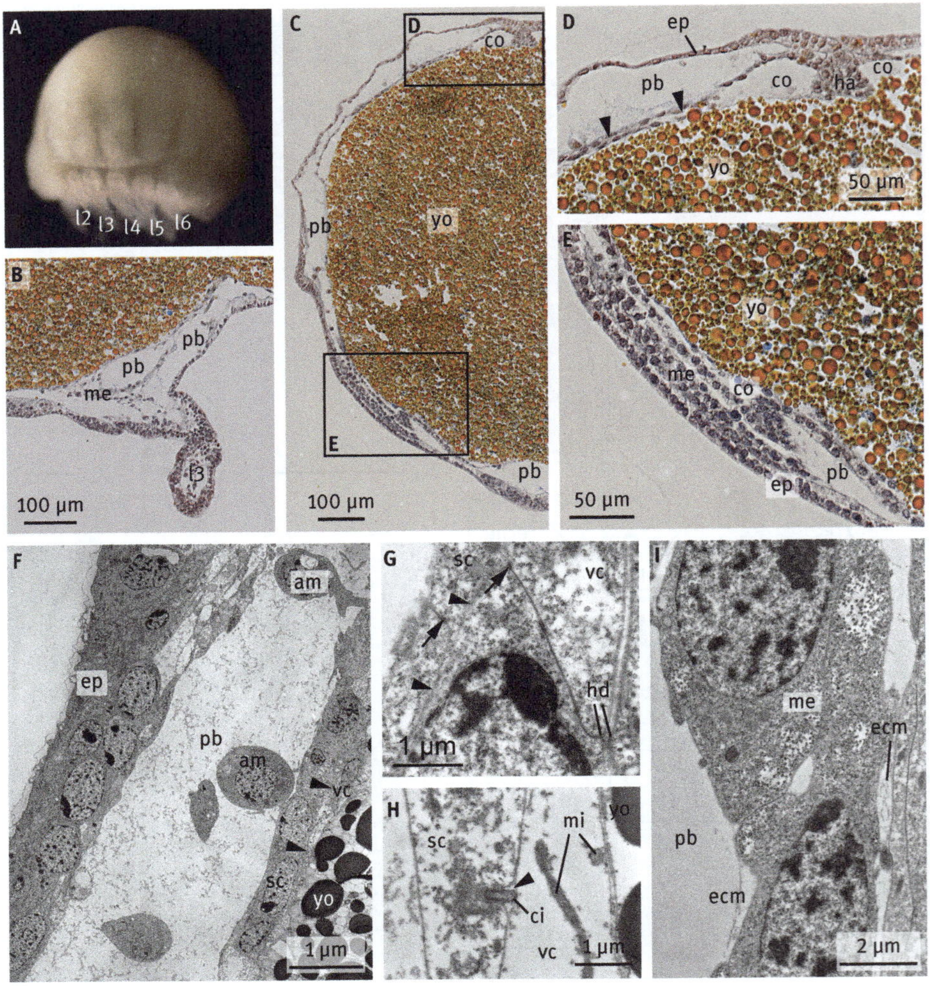

similar mesodermal networks as "trabeculae" in mesosomal segments that he supposed to lack coelomic cavities.

Origin of muscles. The main source for the formation of somatic muscles in *Limulus polyphemus* is the early cluster of apolar mesodermal cells that do not participate in coelom formation. This cluster showed steady mitosis, whereas cell divisions within coelothelia were rarely observed. The distribution of the growing mass of apolar cells within a body segment primarily happens along the somatic wall of the extending lobes of the coeloms, partly along the lateral ectoderm (Figures 11.22E,F, 11.23A,B,E); single or loosely interconnected groups of apolar cells thereby seem to be shifted towards the dorsal and ventral midline and particularly below the dorsolateral ectoderm of a body segment. The first cells with muscle fibers become evident in embryonic stage 18 when all coeloms are still complete and in the process of exten-

◂ **Figure 11.22:** *Limulus polyphemus*, embryonic stages 19–1 and 19–2. **A.** Lateral view of embryo after removal of chorion and vitelline membrane. **B–E.** Transverse sections (LM) through embryo at stage 19–2. **B.** Third prosomal limb anlage (*l3*), strands of mesodermal tissue (*me*) extend through large primary body cavities (*pb*) into the limb base. **C.** Second opisthosomal segment, right half, showing large primary body cavities laterally bordering against the coelom. *Boxes* mark areas shown in D and E. **D.** Dorsal portions of the coeloms (*co*). Their lumen is dorsolaterally compressed (*arrowheads*), such that the somatic lining lies adjacent to the visceral lining. Towards the dorsal midline the lumen widens. The dorsal coelomic tips border against the compact mesenchymal heart anlage (*ha*). **E.** Ventrolateral part of right coelom (*co*). Its compressed cavity slightly widens ventromedially. Laterally primary body cavity (*pb*) and presumptive muscle cells (*me*) border against the somatic coelothelium. **F–H.** Transverse sections (TEM) through second opisthosomal segment at stage 19–1. **F.** Dorsolateral primary body cavity (*pb*) with "amoeboid" mesenchymal cells (*am*). Medially the primary body cavity borders against the somatic wall (*sc*) of the coelom. The coelomic cavity (*arrowheads*) forms small slits between the somatic (*sc*) and visceral (*vc*) coelothelia. **G, H.** Extremely flattened areas of the coelomic cavity; the apical cell membranes of somatic and visceral coelothelia contact each other (*arrowheads*). **G.** Adherens junctions (*arrows*) interconnect the cells of the coelothelia. **H.** Rudimentary cilium (*ci*) projecting into the slit-like coelomic space (*arrowhead*). **I.** Transverse section (TEM) through third leg-bearing prosomal segment at stage 19–1. Mesodermal strand (*me*) of apolar cells extending through primary body cavity into the limb anlage. The strand is lined by extracellular matrix (*ecm*). *ep* epidermis, *hd* hemidesmosome, *l2–l6* prosomal limb anlagen 2–6, *mi* mitochondrion, *yo* yolk cells of the gut anlage.

sion (Figure 11.24C,D). Below the dorsolateral ectoderm, apolar cells differentiate into prospective oblique muscles (fide Shultz, 2001) and into extrinsic appendicular muscles (Figures 11.24D and 11.25A–C). Dorsoventral muscles develop out of strands of apolar cells that pass into the yolk and regularly give rise to intersegmental grooves within the yolk (Figures 11.22A, 11.24A,B). At the same time apolar cells within the limb anlagen start differentiating into intrinsic musculature. The epithelial cells lining the coeloms never showed muscle fibers at the stages studied, nor did we find any indication that prospective muscle cells proliferated from the coelothel. This also seems to be the case for the heart musculature and the anterior aorta. Until the last stage studied by us neither the lumen of the heart and aorta, nor lateral vessels were formed yet, but within the opisthosoma a massive cluster of apolar cells accumulated in the dorsal midline bridging the former gap between the tips of the dorsal coelomic lobes (Figures 11.21B,C, 11.23A–D, 11.25B,C). These apolar cells likely give rise to the heart wall; their presence renders the possibility unlikely that coelothelial cells themselves transform into the myocardium.

Nephridiogenesis. In *Limulus polyphemus* loose aggregations of apolar cells (Figure 11.26B,D) laterally above the limb anlagen of prosomal segments 2–5 give rise to podocytic organs, the prospective sacculi ("nephridial lobes" fide Briggs and Moss, 1997) of the so-called coxal glands. The early anlagen of the podocytic organs have no clearly defined outline; in the prosomal segments devoid of coeloms they

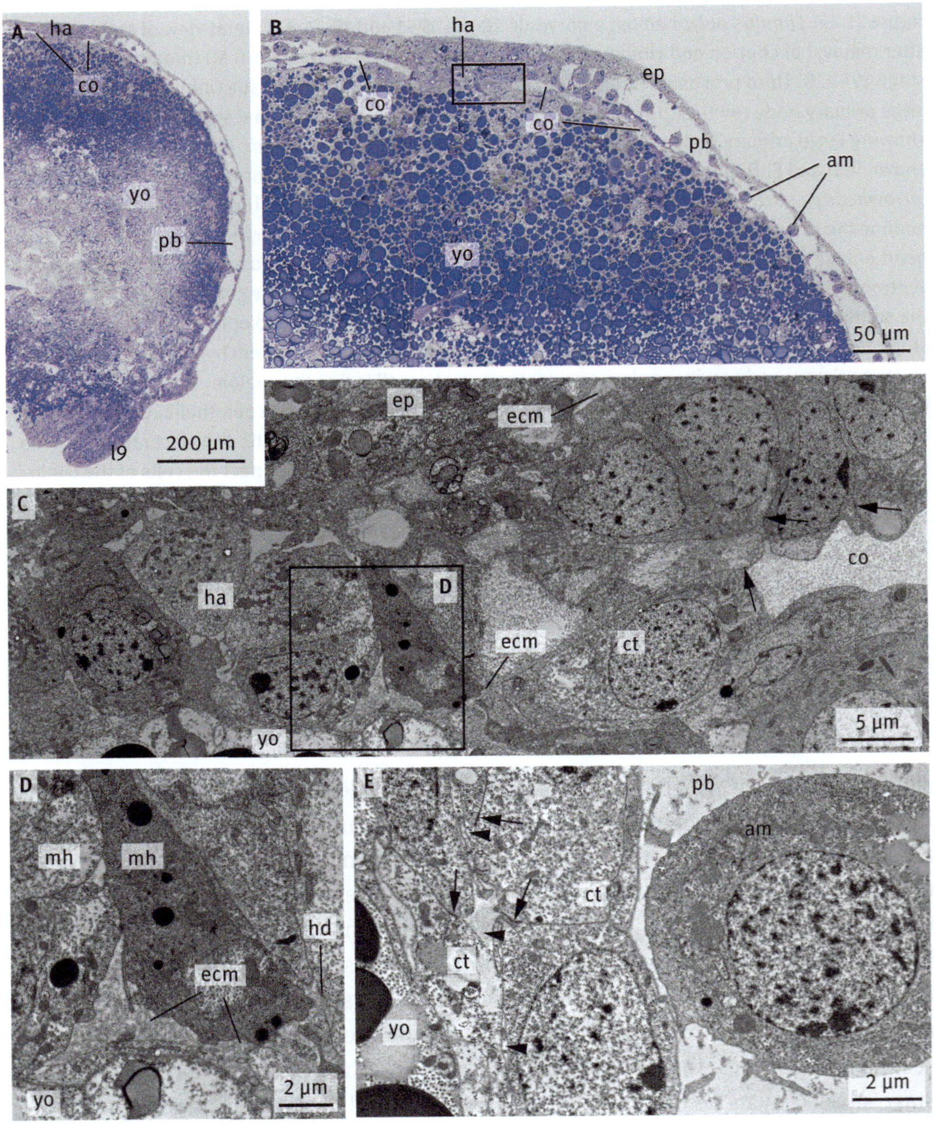

form part of the irregular network of apolar cells (Figure 11.26B). Our ultrastructural studies revealed the first podocytes within this network in embryonic stage 17, in which small groups of 2–3 apolar cells with pedicels (linked by diaphragmata) (Figure 11.26D,E) are interconnected by apical adherens junctions, each group jointly delimiting a slit-like cavity. Precursors of these podocytic organs still lack pedicels, but already show intercellular junctions as well as distinctive aggregations of vesicles (Figure 11.26C,D). In histological sections they did not become distinguishable from adjacent apolar cells before embryonic stage 19, by different staining properties

◄ **Figure 11.23:** *Limulus polyphemus*, embryonic stage 17. **A.** Semithin transverse section (LM) through third opisthosomal segment. The coelomic cavities (*co*) of both sides extend dorsally up to the heart anlage (*ha*). **B.** Detail of A, showing the dorsal part in higher magnification. The coelomic cavities (*co*) are visible aside the heart anlage (*ha*); laterally they are flattened to slit-like lumina. A large primary body cavity (*pb*) with single "amoeboid" mesenchymal cells (*am*) occurs underneath the epidermis (*ep*) on each side of the embryo. **C.** Transverse section (TEM) through heart anlage (*ha*) and adjacent coelothelium as indicated in B. The columnar coelothelial cells (*ct*) are apically interconnected by adherens junctions (*arrows*). The heart anlage is formed by a cluster of apolar mesodermal cells (*ha*) filling the dorsomedian space between epidermis (*ep*), yolk cells (*yo*), and the dorsal tips of the coeloms. **D.** Detail of C at higher magnification. The apolar mesenchymal cells (*mh*) of the heart anlage are surrounded by extracellular matrix (*ecm*). Hemidesmosomes (*hd*) connect the coelothelial cells to their basal lamina. **E.** Transverse section (TEM) through dorsolateral part of the coelom. Its lumen is restricted to slit-like lacunae (*arrowheads*). The coelothelial cells (*ct*) are slightly flattened. Apolar "amoeboid" mesenchymal cells (*am*) within the primary body cavity (*pb*) seem to migrate along the somatic coelothelium towards their target position. *l9* third opisthosomal limb, *mh* mesenchymal cell of heart anlage, *yo* yolk cells of gut anlage.

of their cytoplasm (Figure 11.26B). Patten and Hazen's (1900) description is refuted by our studies, according to which "nephridial cells" are already recognizable in the earliest stages of the segmental mesoderm development, including the cheliceral and ultimate prosomal body segments, in which we never found nephrocytes (or podocytic organs). Our results also contradict their statement that the podocytic organs in the penultimate prosomal segment develop much later than in the preceding body segments. All podocytic organs indeed proved to develop simultaneously and – as demonstrated by the penultimate prosomal segment – independent of coeloms. In the penultimate prosomal segment, the cluster of presumptive podocytes formerly lies immediately lateral to the somatic wall of the coelom; in later stages (late embryonic stage 18) the podocytes are partly integrated into the coelothel such that the podocytic organ seems to form a lateral lobe of the original coelom. At this stage, the somatic wall of the coelom is medioventrally, almost opposite to the lateral "nephridial lobe" anlage, already connected to the ventral epidermis caudal to the base of the limb anlagen by an efferent duct ("nephric duct" fide Briggs and Moss, 1997) (Figure 11.26A). The anlage of this duct is primarily devoid of both a continuous lumen and ectodermal opening but opens into the coelom via a funnel-shaped expansion of the duct anlage. The cuboidal epithelium of the funnel and duct anlage lacks cilia and is formerly clearly distinguished from the flat coelothel. However, as soon as the duct shows a continuous lumen and is equipped with an ectodermal opening, the cells of the funnel adopt the shape of the coelethelial cells, and the funnel has shifted its orientation from originally towards the yolk to the ventrolateral ectoderm. While our results accordingly corroborate the view that at least the proximal part of the "nephric duct" likely represents an outgrowth of the coelothel, they remain ambiguous about whether the coelothel also gives rise to the so-called end

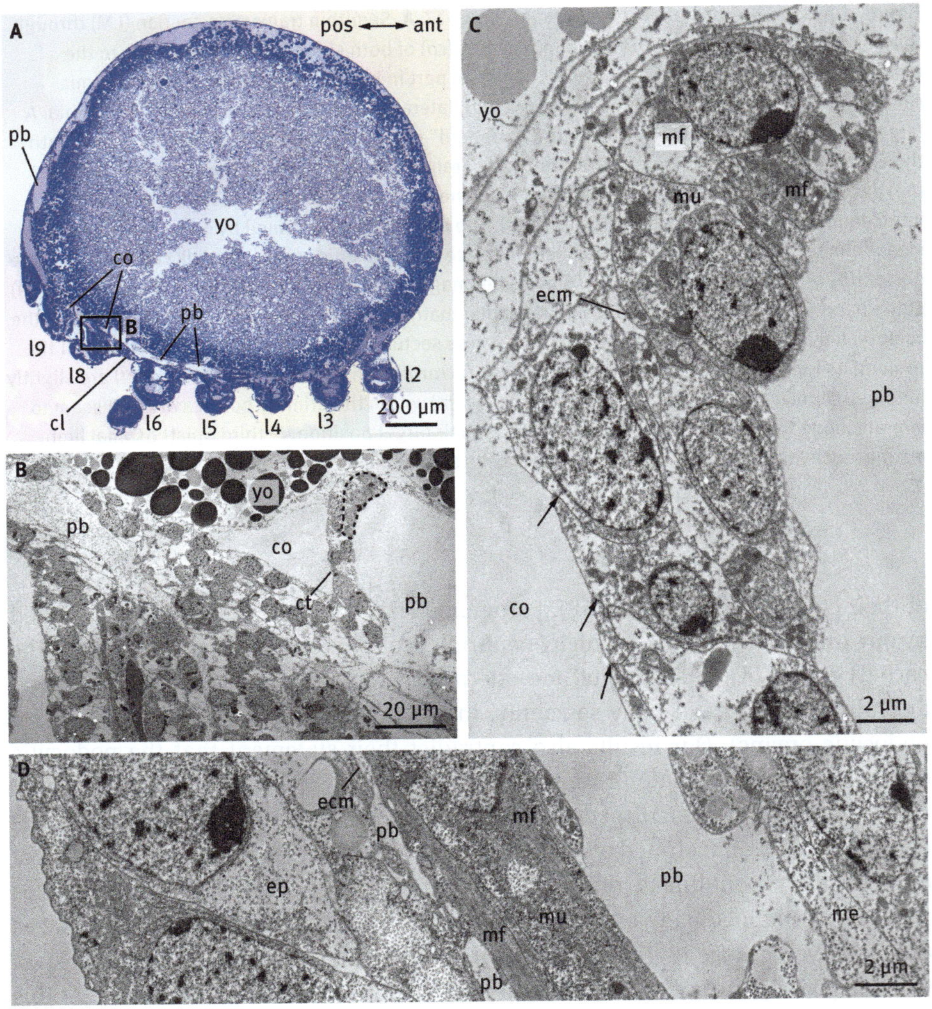

Figure 11.24: *Limulus polyphemus*, embryonic stages 19–1 (A–C) and 17 (D). **A.** Paramedian sagittal semithin section (LM), anterior is right, showing triangular coelomic cavities (*co*) above the anlagen of the opercula (*l8*) and first pair of book gills (*l9*). *Box* demarcates area depicted in B. **B.** Coelomic cavity (*co*) of second opisthosomal segment in higher magnification (TEM). The coelothelium (*ct*) anteriorly borders against a muscle strand (*dotted outline*) that extends along a furrow of the gut anlage (*yo*) dorsalward. **C.** Sagittal section of coelomic cavity (*co*) of third opisthosomal segment in higher magnification (TEM). Minute cavities lined by extracellular matrix (*ecm*) occur between the coelothelial cells (*ct*) and muscle cells (*mu*). The coelothelial cells are interconnected by apical *zonula adherens* (*arrows*). **D.** Transverse section (TEM) through penultimate prosomal segment. A somatic muscle strand (*mu*) extends through the primary body cavity (*pb*) from the base of the limb anlage dorsolaterally. *cl* chilarium anlage, *ep* epidermis, *l2–l6* prosomal limb anlagen 2–6, *l8* operculum anlage, *l9* anlage of first book gill, *me* mesenchymal cell, *mf* muscle filaments, *yo* yolk cells of the gut anlage.

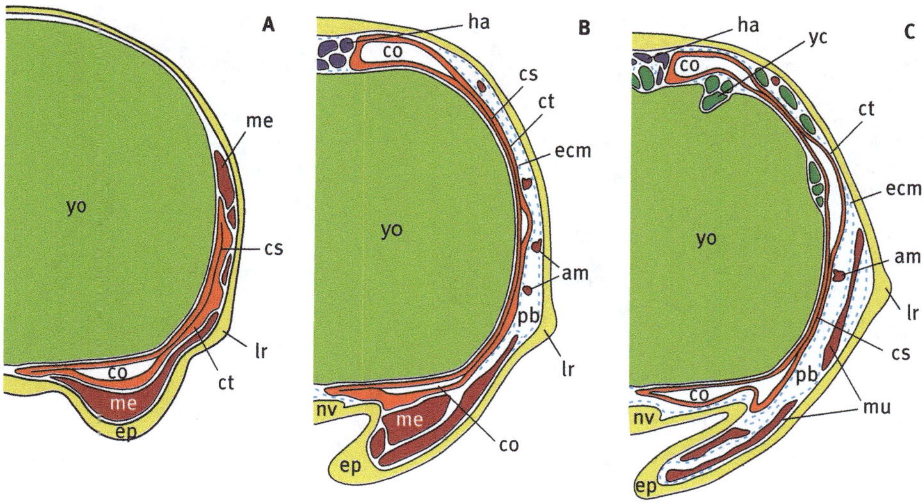

Figure 11.25: *Limulus polyphemus*, schematic illustration of coelomogenesis in a hemisegment. **A.** Embryonic stage 15, transverse section through ultimate prosomal segment. The coelom (*co*) extends from near the ventral midline across the lateral rim (*lr*) towards the lateral midline of the embryo. The limb bud is filled by apolar mesenchymal cells (*me*); additional apolar mesoderm occurs laterally and dorsally to the coelom. Only a very thin layer of extracellular matrix (not shown) can only occasionally be detected between the mesoderm and the yolk cells (*yo*) or the epidermis (*ep*). **B.** Stage 17, transverse section through second opisthosomal segment. The coelom extends to the dorsal heart anlage (*ha*) that consists of apolar mesenchymal cells. Most areas of the coelomic cavity are compressed and only detectable as intercellular slits (*cs*) between the apical surfaces of the somatic and visceral coelothelia (*ct*). The visceral coelomic wall borders immediately against the yolk cells (*yo*). Large primary body cavities (*pb*) lined by extracellular matrix (*ecm, blue dotted lines*) occur between the somatic coelomic wall and the epidermis. Apolar mesoderm fills the anlage of the limb bud. Single "amoeboid" mesenchymal cells (*am*) are distributed within the large lateral primary body cavity. **C.** Embryonic stage 19–1, transverse section through second opisthosomal segment. The apolar mesoderm within the limb anlage is split into strands of musculature (*mu*) separated by extracellular matrix (*ecm*) and primary body cavities (*pb*). "Free yolk cells" (*yc*) are distributed within dorsal primary body cavities around the coeloms and between the cells of the heart anlage (*ha*). *ep* epidermis, *nv* ventral nerve strand, *yo* central mass of yolk cells.

sac[1] of the coxal gland that interconnects the efferent duct with the sacculi in adults. Any duct interconnecting the podocytic organs in the preceding prosomal segments ("stolon" fide Briggs and Moss, 1997) were not formed until the last stage studied, nor did we observe any other coeloms to form coelomoducts.

1 The "end sac" of crustacean nephridial organs refers to a podocytic organ (the sacculus), whereas Briggs and Moss (1997) use this term for a compartment devoid of podocytes into which the sacculi jointly open (directly in the penultimate segment and via a "stolon" in preceding body segments).

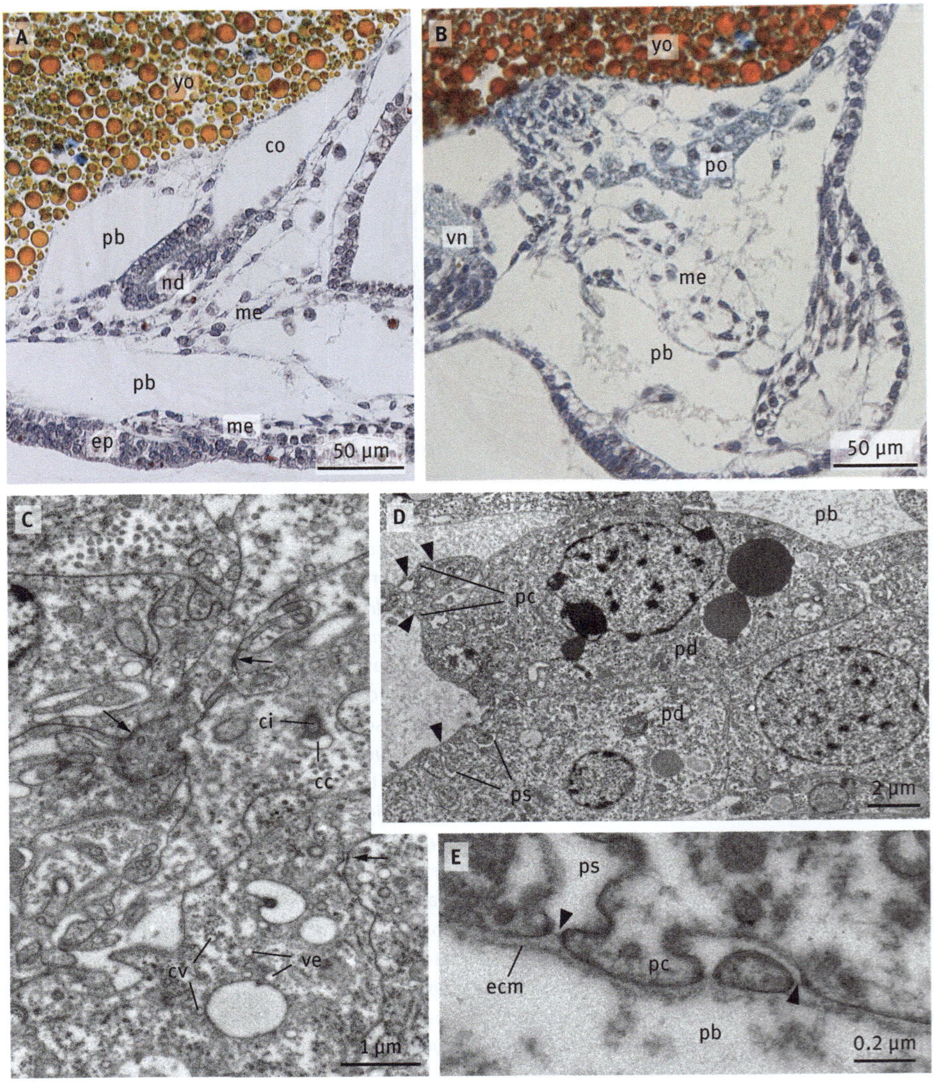

11.5.1.2.2 *Pholcus phalangioides* (Araneae)

Coelomogenesis in *P. phalangioides* was studied from shortly before inversion up to ventral closure, covering embryonic stages 11–21 in *Cupiennius salei* fide Wolff and Hilbrant (2011) and stages 10–14 in *Parasteatoda tepidariorum* fide Mittmann and Wolff (2012). Both staging schemes refer to the development of entelegyne spiders but can also be applied to haplogyne spiders (like *P. phalangioides*) despite differences in the flexure of the opisthosoma. Embryos of *P. phalangioides* show a ventrally (or forwardly) flexed opisthosoma, which seems to be the ancestral flexure type in Araneae (e.g., Yoshikura, 1975).

◂ **Figure 11.26:** *Limulus polyphemus*, embryonic stage 19–1, nephridial anlage. **A.** Transverse sections (LM) through penultimate prosomal segment, left side. The coelom (*co*) is continuous with the nephridial duct anlage (*nd*) that passes to the posterior end of the limb base in this segment. Coelom and nephridial duct anlage are surrounded by apolar mesenchymal cells (*me*) and primary body cavities (*pb*) that are confluent with primary body cavities beneath the epidermis (*ep*) and the gut anlage (*yo*). **B.** Transverse section of third limb-bearing prosomal segment. Podocytic organs (*po*) are characterized by the distinct staining properties and large cell bodies of the podocytes. The podocytes are arranged in an irregularly shaped cluster underneath the gut anlage (*yo*). Apolar mesenchymal cells (*me*) and primary body cavities (*pb*) surround the podocytic organ. **C–E.** Transverse sections (TEM) through podocytic organ. **C.** Central area of the podocytic organ. The podocytes are interconnected by adherens junctions (*arrows*); the lumen between the cells is very narrow. Occasionally, rudimentary cilia (*ci*) project into crescent shaped vesicles (*cc*). Numerous coated (*cv*) and smooth (*ve*) vesicles occur within the cytoplasm of the podocytes. **D.** Peripheral area of the podocytic organ. Towards the surrounding primary body cavity (*pb*) the podocytes (*pd*) form peripheral lacunae (*ps*) and pedicels (*pc*). The pedicels are interconnected by slit membranes (*arrowheads*). **E.** Pedicels (*pc*) and peripheral lacunae (*ps*) at higher magnification. Slit membranes (*arrowheads*) bridge the gaps between the pedicels. An extracellular matrix (*ecm*) covers the pedicels and the slit membranes. *ve* smooth vesicle, *vn* ventral nerve chord, *yo* yolk cells of the gut anlage.

The spatial development of the segmental mesoderm in *Pholcus phalangioides* corresponds in general outline to the descriptions for *Limulus polyphemus*. The main differences from the mesoderm development in *Limulus polyphemus* concern the occurrence of paired coelomic cavities in all prosomal segments, their asynchronous formation (Figure 11.27A–D), and the absence of rudimentary cilia in coethelial cells. Coelomic cavities start to develop as small, ovoid spaces in formerly massive mesodermal tissue above the bases of the limb anlagen (Figure 11.1B). The coelothel primarily consists of few cuboidal cells interconnected by apical adherens junctions. Subsequent coelomic growth is mainly based on stretching and flattening of the coelothelial cells; mitotic activity was rarely observed within the coelothel. The first coeloms develop in the cheliceral-precheliceral region (Figures 11.27A, 11.28A,B), followed by the third limb-bearing segment of the prosoma (i.e., the first ambulatory segment). At the beginning of inversion large, complete coeloms spread out from the cheliceral into the precheliceral mesoderm (in outline similar to Pross's illustration (1977: fig.1) of the cheliceral and precheliceral coeloms, but in *P. phalangioides* the respective coelomic compartments are interconnected), as well as in the second opisthosomal segment (Figure 11.27A). In the latter the coelomic cavity is very flat, encompassing the yolk from lateroventral to mediodorsal, its lumen being recognizable only at its dorsal apex as well as above the limb buds; in remaining areas, the somatic and visceral walls immediately face each other, with few irregular minute spaces (invisible in light microscopy) in between the walls' apical sides. Other coeloms in this stage are still small and ovoid in limb-bearing prosomal segments 4–6, slightly enlarged at the limb base and tapering towards the lateral ectoderm in the third prosomal segment

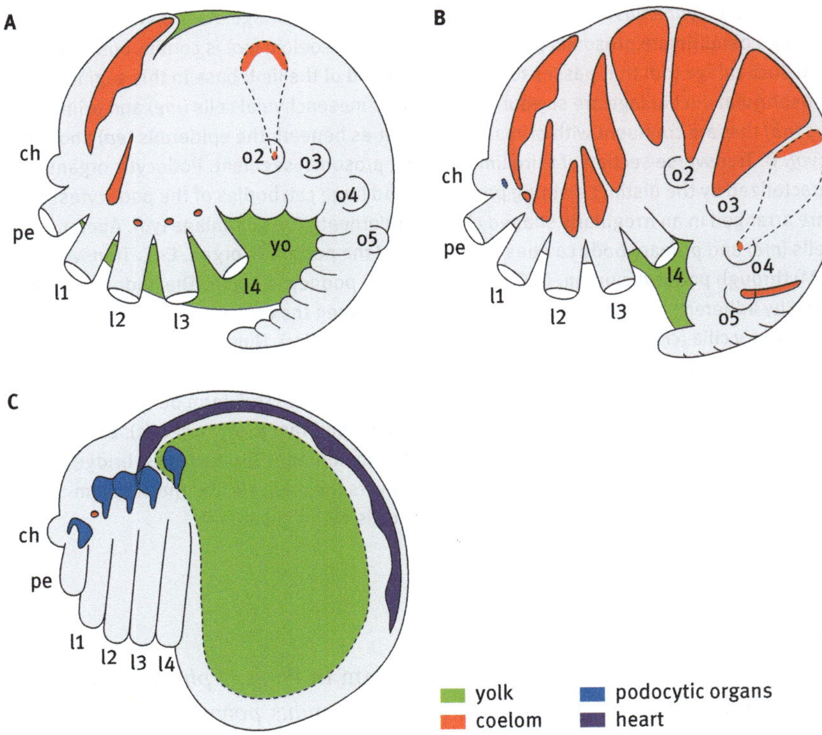

(Figure 11.29A,B), and still absent in the pedipalpal segment as well as in the remaining opisthosomal segments (Figure 11.27A). During inversion the coeloms show individual, partly divergent fates (Figure 11.27B): the anterior-most coeloms further enlarge posterior-dorsally but at the same time disintegrate ventrally in the cheliceral segment; small, ovoid coeloms arise in the pedipalpal segment above the limb buds but do not enlarge at all in subsequent stages (Figure 11.27B,C); the paired coeloms in the ambulatory prosomal segments form narrow dorsal extensions, among which only the last pair approach in the dorsal midline; in the opisthosoma the coeloms attain their largest size in the first three anterior body segments, each with a flat, continuous, dorsally broad and ventrally narrow lumen (Figure 11.28D–F); the coelomic lumen in the fourth opisthosomal segment becomes recognizable only in the tips of the dorsal extensions as well as above the limb buds, and is restricted to a uniform narrow stripe above the limb buds in the fifth opisthosomal segment (Figure 11.28G). No traces of coeloms were detected in remaining opisthosomal segments. After ventral closure, when intersegmental borders are externally no longer delimited within the opisthosoma, only the small coeloms of the pedipalpal segment are still complete. In all other body segments up to the fifth opisthosomal segment, the former coeloms are

◂ **Figure 11.27:** *Pholcus phalangioides*, schematic illustration (lateral view) of succeeding developmental stages of coeloms (*orange*) and podocytic organs (*blue*) during and shortly after completion of inversion. **A.** Embryonic stage 16. Dorsal extension of coeloms precedes development of podocytic organs. The coeloms of the cheliceral (*ch*) and second opisthosomal segment (*o2*) extend into the dorsal region of the germ bands, while the coeloms of prosomal segments 3 to 6 (corresponding to walking limbs 2 to 4 *l2–l4*) are small and restricted to ventrolateral areas above the limb buds. In the pedipalpal segment (*pe*) coelomic cavities are still absent; in the first ambulatory segment (l1) they slightly extend dorsalward. The coelomic cavities of the second opisthosomal segment are largely compressed; their lumen (*orange*) is only detectable near the limb bud and in dorsalmost areas (complete extension of the coelom marked by stippled line). **B.** Embryonic stage 17, the first podocytic organs (*blue*) are detectable in the cheliceral segment ventrolateral to the cheliceral coelom. The coelomic cavities of the ambulatory segments (*l1–l4*) variably extend further dorsalward; in the pedipalpal segment (*pe*) a small ventral coelom is detectable. Large coelomic cavities occur in opisthosomal segments 1 to 3, while in opisthosomal segments 4 and 5 the coelomic cavities are either compressed (*stippled outline*) or restricted to a small lateral stripe. **C.** Embryonic stage 19. Most coelothelia are disintegrated except for the pedipalpal coeloms; remnants persist in prosomal segments as nephridial duct anlagen. Podocytic organs are present in all prosomal segments; those of the cheliceral and pedipalpal segments as well as those of the first to third ambulatory segments are dorsally confluent. Podocytic organs of the fourth ambulatory segment remain separate and are located slightly more dorsally than the others. All podocytic organs form a medioventral lobe towards the limb buds, but they do not extend into them. The heart anlage (*purple*) extends from the penultimate prosomal segment up to the fifth opisthosomal segment. *yo* yolk.

only recognizable by coelothelial fragments of the former somatic and visceral walls, if at all (Figures 11.27C, 11.29C).

Thus, despite their occurrence in all body segments (at least up to the fifth opisthosomal segment), the entire series of coeloms from anterior to posterior is difficult to trace at any particular stage because of their variable size and development in individual segments (while some are either still absent or in the process of growth, others already disintegrate (Figure 11.27A–C); the temporal sequence of coelomogenesis indeed shows no obvious pattern. As in *Limulus polyphemus*, neither dissepiments, nor mesenteries are transiently formed. Intersegmental areas where coelomic walls of successive segments locally meet are formerly filled with apolar mesodermal cells and later by extensive primary body cavities. Left and right coeloms dorsally meet only in opisthosomal segments (Figure 11.30A,B). In these segments their dorsalmost edges locally become confluent medially by a flat bridge (devoid of a lumen) above the yolk (Figures 11.30C,F, 11.31A–C). Medioventral extensions of the coeloms arise only in anterior prosomal segments beneath the prospective stomach, but these extensions never meet medially.

The canonical view that the coeloms form “appendicular lobes” deeply extending into the outgrowing limb anlagen in arachnids proved in our studies as a misinterpretation based on a fixation artifact (Figure 11.1A,B). Glutaraldehyde-fixed specimens demonstrate that the mesoderm within the limb buds remains massive long after coeloms are formed (Figure 11.1B,C). At the peak of coelomic extension

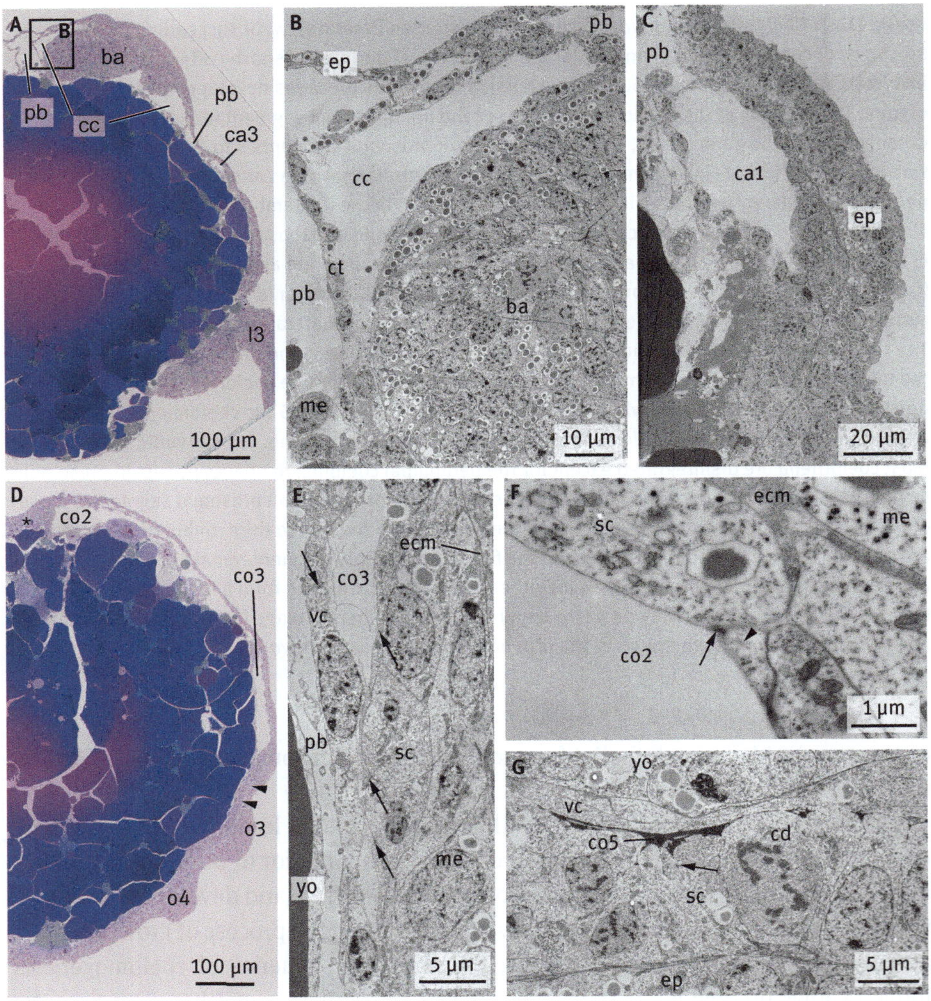

they may slightly protrude into the limb base, but at much later stages and never to such a strong extent as depicted in Wallstabe's (1908) classical illustrations of anterior-posterior sequences of putative coeloms in embryos of *Agelena labyrinthica* (Clerck, 1757) (Araneae). His illustrations were often reproduced in surveys of arachnid embryogenesis (e.g., Gilbert, 1997; Foelix, 2011), but our studies support the view that Wallstabe's "coeloms" represent artificial primary body cavities. Wallstabe used hot Zenker's fixative, which has a similarly high content of substances with strong osmotic activity as Bouin's fluid. His notion that Zenker's fixative caused the chorion to explode corroborates our observation that fixatives of this kind effect strong swellings of the ECM, locally dividing the rather loose mass of mesodermal cells into single strands. This behavior of the early mesoderm was also found in Bouin-fixed embryos of *Pholcus phalangioides* (Figure 11.1A). In embryos fixed with Karnovsky solution,

◂ **Figure 11.28:** *Pholcus phalangioides*, embryonic stages 15 (A–C) and 17 (D–G). **A.** Transverse section (LM) through third ambulatory prosomal segment. Dorsal lobes of the cheliceral coeloms (*cc*) partly surround the brain anlage (*ba*) and medially deliminate a large primary body cavity (*pb*), in which clusters of apolar cells will give rise to the anlage of the aorta. Dorsolaterally, primary body cavities (*pb*) and the dorsal lobe of the coelom (*ca3*) of the third ambulatory segment underlie the epidermis. **B.** Transverse section (TEM) of left cheliceral coelom (*cc*), depicting area marked in A in higher magnification. Medially the coelothelium (*ct*) borders against a primary body cavity, in which apolar mesenchymal cells will form the aorta. This cavity is dorsally bordered by the epidermis (*ep*). **C.** Transverse section (TEM) through coelomic cavity (*ca1*) of the first ambulatory segment. Dorsally, the coelom borders a primary body cavity (*pb*). **D.** Transverse section (LM) through opisthosoma. The coelomic cavities (*co2*) of the second opisthosomal segment extend almost to the dorsal midline; apolar mesenchymal cells are situated between the dorsal lobes of the coeloms (*asterisk*). The coelom (*co3*) of the third opisthosomal segment show partly widened lumen, partly a slit like space (*arrowhead*). **E.** Transverse section (TEM) of coelomic cavity (*co3*) in third opisthosomal segment. A small primary body cavity (*pb*) occurs between the gut anlage (*yo*) and the coelom. Apolar mesodermal cells (*me*) laterally border against the somatic coelothelium (*sc*). **F.** Detail (TEM) of flattened somatic coelothelium (*sc*) in second opisthosomal segment, transverse section. Actin filaments (*arrowhead*) are connected to apical adherens junctions (*arrow*). **G.** Transverse section (TEM) through slit-like coelomic cavity (*co5*) of fifth opisthosomal segment, containing condensed extracellular material. The somatic coelothelium (*sc*) consists of cuboidal cells that undergo mitoses (*cd*); the visceral coelothelium (*vc*) is flattened. *ca1*, *ca3* coeloms of ambulatory segments 1, 3, *cc* coelom of chericeral segment, *cd* cell division, *co2–co5* coeloms of opisthosomal segments 2–5, *ecm* extracellular matrix, *l3* limb anlage of third ambulatory segment, *o3*, *o4* limb buds of opisthosomal segments 3, 4, *yo* yolk cell.

in contrast, the first mesodermal cavities within limb anlagen occurred not before ventral closure; all of them proved to represent primary body cavities.

Origin of muscles. In *Pholcus phalangioides* the first muscular fibers (still invisible in light microscopy) were detected in opisthosomal segments during inversion, in apolar cells between yolk and the visceral wall of the coeloms. Up to ventral closure, well-developed muscles comprise the dilator muscles of the prospective stomach, extrinsic muscles of the prosomal legs, and the myocardium (Figure 11.30F,G), whereas intrinsic muscles of the prosomal legs just start to develop (Figure 11.29C,D). The wall of the heart and anterior aorta differentiates from posterior to anterior out of apolar cells that accumulated in the dorsal midline during dorsal growth of the coeloms (Figures 11.30A,B,D, 11.31A–C). The cluster of mesodermal cells between the dorsal lobes of the (pre-)cheliceral coeloms give rise to the anterior aorta; single cardioblasts between the dorsal coelomic lobes of the ultimate prosomal segment up to the fifth opisthosomal segment form the wall of the heart (Figure 11.30A). The prospective heart lumen is formerly bordered ventrolaterally by the tips of the dorsal coelomic lobes and dorsally by the ectoderm (Figure 11.30F,G). Muscle fibers develop in a very thin (in histological section invisible) monolayer of apolar cells gradually attaching from dorsal to ventral at the tips of the dorsal coelomic lobes; the forma-

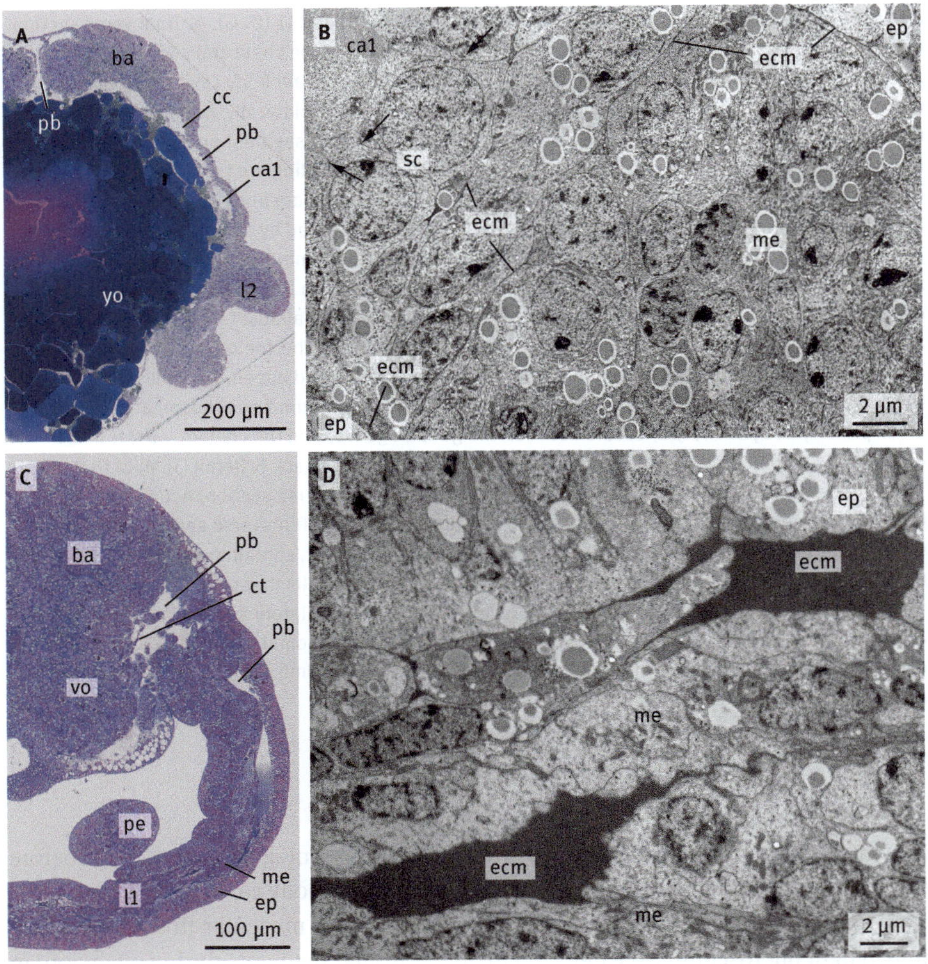

tion of the heart accordingly passes stages where its definitive wall is dorsolaterally present but ventrolaterally still missing. Shortly after ventral closure, the heart still extends into the ultimate prosomal segment, its transition into the anterior aorta being clearly indicated by a marked constriction between the penultimate and ultimate prosomal segments (Figure 11.30C). At this stage, three pairs of intersegmental ostia are already present in the opisthosomal part of the heart (Figure 11.30E), each of which is equipped with muscular strands extending to the lateral ectoderm. These strands likely represent the anlagen of the prospective lateral ligaments (fide Huckstorf et al., 2013). In the coelothelial cells we did not observe muscle fibers at any stage.

Nephridiogenesis. The sacculi of the coxal glands of *Pholcus phalangioides* are formed out of paired aggregations of apolar, globular 'nephrocytes' that differentiate into podocytic organs. Nephrocytes become distinguishable from other apolar

◂ **Figure 11.29:** *Pholcus phalangioides*, leg development at embryonic stages 15 (A–B) and 20 (C–D). **A.** Transverse section (LM) through prosoma at the level of the second walking legs (*l2*). The limb anlagen are massively filled with mesoderm. **B.** Transverse section (TEM) through base of left walking leg 1, proximal is left. The somatic coelomic wall (*sc*) of the coelom of the ambulatory segment 1 (*ca1*) borders against apolar mesenchymal cells (*me*) filling the limb anlage. The coelothelial cells are interconnected by zonula adherens (*arrows*). Thin layers of extracellular matrix (*ecm*) demarcate the border between epidermis (*ep*) and mesoderm and are occasionally also detectable between apolar mesenchymal cells. **C.** Transverse section (LM) through prosoma at the level of the anlagen of the first walking legs (*l1*). Large primary body cavities (*pb*) together with apolar mesenchymal tissue (*me*) occur within the limb anlage. Only few remnants of the coelothelia (*ct*) are detectable. The brain anlage (*ba*) and ventral organs (*vo*) fill the central space of the prosoma; the yolk-rich gut anlage is displaced into the opisthosoma. **D.** Transverse section (TEM) through anlage of first walking limb. Apolar mesenchymal cells (*me*) border against small primary body cavities filled with condensed extracellular matrix (*ecm*). *cc* cheliceral coelom, *l1*, *l2* walking leg 1, 2, *pe* pedipalpus, *yo* yolk cells of the gut anlage.

mesodermal cells during inversion. They are characterized histologically by distinctive staining properties of the cytoplasm, and ultrastructurally by irregular infoldings of the cell membrane and high density of endoplasmic reticula and cytoplasmic vesicles (Figure 11.32A,D,E). The first paired aggregations of nephrocytes occur in the cheliceral segment, soon after the postero-ventral part of the precheliceral-cheliceral coeloms disintegrates (Figure 11.27B,C). The nephrocytes formerly cluster posteriorly at the base of the still massive anlagen of the chelicerae, partly projecting into primary body cavities (that at this stage already represent spaces of the 'mixocoel'). In subsequent stages the paired clusters extend from the cheliceral segment into the base of the pedipalpal anlagen, forming the so-called pedipalpal gland; according to Yoshikura (1955: 49) this gland "*is recognized in most true spiders, but it has been overlooked by many embryologists.*" Since the first components of this gland occur in the cheliceral segment, it likely comprises podocytic organs of both the cheliceral and pedipalpal segments in later stages. In accordance with Yoshikura's (1955) descriptions for the liphistiid spider *Heptathela kimurai*, further nephrocytic organs develop in the first three ambulatory segments. These nephrocytic organs become dorsally confluent, forming a single pair of massive, cord-like cell strands laterally (and separate) from the coeloms (or their remnants, respectively) in these body segments; ventrally they keep segmental extensions towards the bases of the limb anlagen, similarly as in the cheliceral and pedipalpal segments (Figure 11.27C). Another cluster of nephrocytes also occurs in the ultimate prosomal segment; this pair remains separate from the nephrocytes in preceding segments and shows a more dorsal extension. Hence, according to our results all limb-bearing prosomal segments transiently show anlagen of podocytic organs, but podocytes developed only in the anterior three pairs up to the last stage studied by us. At this last stage all coeloms of the prosomal segments except for those of the pedipalpal segment have already disintegrated. In the ambula-

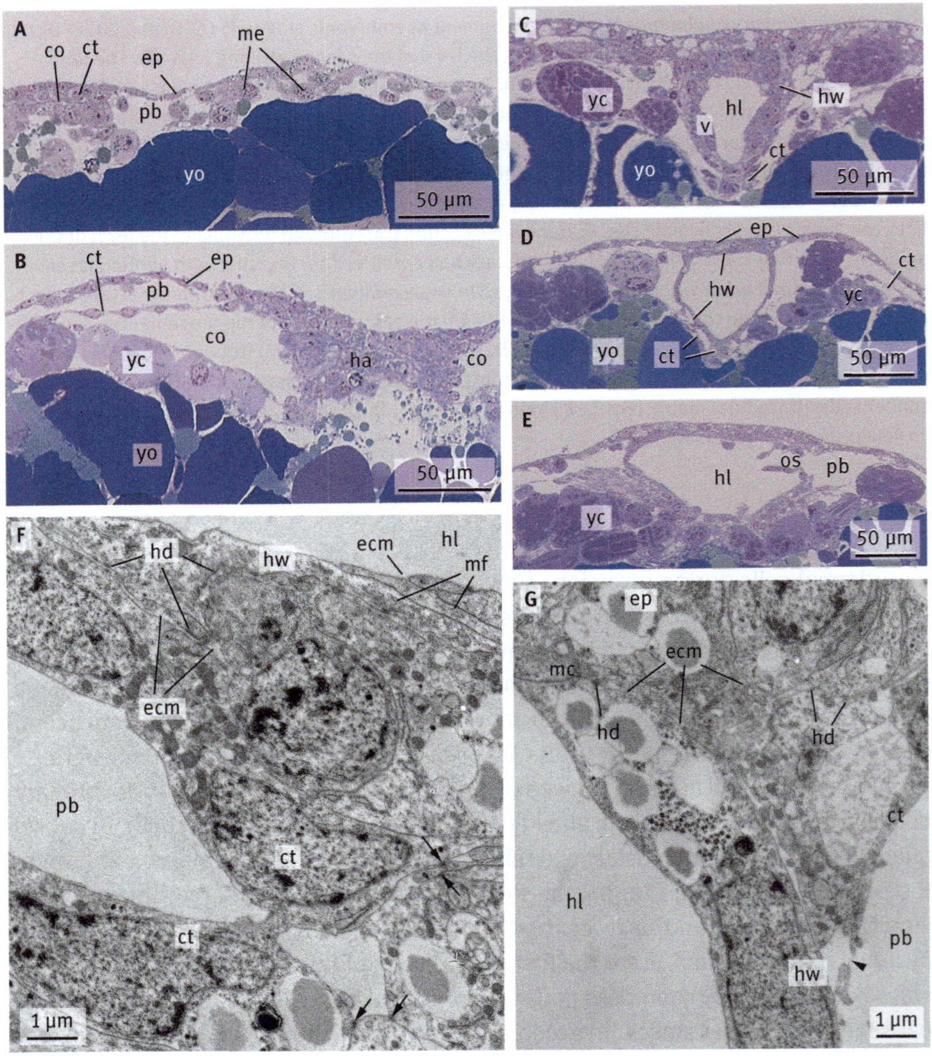

tory segments, remnants of the coelothel form a funnel immediately below the podocytic organs, most conspicuously in the first ambulatory segment (Figure 11.32A–C). Remaining parts of the nephridial ducts (whether mesodermal or ectodermal) are not recognizable up to this stage (and do not develop up to the last stage studied by us).

In sum, our ultrastructural results on *Limulus polyphemus* and *Pholcus phalangioides* refute the traditional view that the embryonic segmental coeloms form the major source for the musculature, sacculi, or the circulatory system in chelicerates. According to our results the most accurate characterization of organogenesis in chelicerates is Weygoldt's statement (2000: 17) that "*the mesoderm surrounding the coelomic cavities gives rise to the dorsal (pericardial septum) and ventral diaphragm, heart,*

◂ **Figure 11.30:** *Pholcus phalangioides*, development of the heart. **A–E.** Heart anlage at embryonic stages 15, 17, and 20. **A.** Embryonic stage 15, transverse section (LM) through third opisthosomal segment. The tips of the dorsal coelomic lobes (only right side shown) are formed by cuboidal coelothelial cells (*ct*) lining a slit-like lumen (*cs*). The coeloms are medially separated by a primary body cavity (*pb*) that fills the space between epidermis (*ep*) and gut anlage (*yo*) and contains numerous mesenchymal cells (*me*). **B.** Embryonic stage 17, transverse section (LM) through second opisthosomal segment. The dorsal tips of the coeloms (*co*) are markedly widened and lined by flattened coelothelia (*ct*). The heart anlage (*ha*) is formed by apolar mesenchymal cells. "Free yolk cells" (*yc*) border against the visceral coelomic walls. A primary body cavity (*pb*) occurs dorsolaterally between the dorsal coelomic lobes and the epidermis (*ep*). **C.** Embryonic stage 20, transverse section (LM) through the transitional area between fifth and sixth prosomal segments. The heart anlage forms a valve delimiting the frontal aorta. The valve flaps (*v*) are parts of the heart wall (*hw*) extending towards the heart lumen (*hl*). Remnants of the coelothelia (*ct*) occur between the ventral heart wall and the gut anlage (*yo*). The gut anlage forms a deep groove underneath the heart. **D.** Embryonic stage 20, transverse section (LM) through second opisthosomal segment. The heart wall (*hw*) consist of a single layer of muscle cells that also form the roof of the heart underneath the epidermis (*ep*). Remnants of the coelothelia (*ct*) occur underneath the heart and laterally overlay "free yolk cells" (*yc*). **E.** Embryonic stage 20, transverse section (LM) through heart anlage at the border of third and fourth opisthosomal segments. The heart lumen (*hl*) communicates with the hemocoel (*pb*) via funnel-like ostia (*os*). **F–G.** Embryonic stage 20, transverse sections (TEM) through heart anlage in second opisthosomal segment. **F.** Extracellular matrix (*ecm*) separates remnants of the coelomic walls (*ct*) from the muscular cells of the heart wall (*hw*). Both cell types are connected to ECM by hemidesmosomes (*hd*). ECM also borders the cells of the heart wall against the heart lumen (*hl*). Myofilaments (*mf*) occur only in cells of the heart wall. Some coelothelial cells are still interconnected by apical zonula adherens (*arrows*). **G.** The dorsal epidermis (*ep*) and the muscular cells (*mc*) of the heart wall (*hw*) are separated by extracellular matrix (*ecm*), to which both layers are connected via hemidesmosomes (*hd*). Remnants of the coelomic walls (*ct*) occur at the dorsolateral edge of the heart wall; they have lost their apical adherens junctions, and extracellular matrix occurs on their apical surface (*arrowhead*). *mc* muscle cell, *me* apolar mesenchymal cell, *yo* yolk cell of the gut anlage.

arteries, endosternite and muscles." In contrast to his (and common) belief, the apolar mesoderm surrounding the coeloms does not seem to originate from the embryonic coelothel. The function of the coeloms seems to be limited to pioneering spaces between ectoderm and yolk, along which apolar mesodermal cells can move toward their target position, including intersegmental and segmental areas between the walls of adjacent coeloms. The muscular heart wall proved to be formed not by the coelomic wall itself, but by apolar cardioblasts that abut retroperitoneally on the dorsomedial wall of the coelothel. Since the opisthosomal coeloms are still complete when the cardioblasts form the definitive heart wall, we can exclude that the cardioblasts originate from "*decomposition of the coelomic sacs into autonomous cell groups*" (Janssen and Damen, 2008: 163). Mesenteries and dissepiments are not formed. Intersegmentally, consecutive coeloms may locally overlap but to the most extent remain separated by apolar mesodermal cells and primary body cavities; within a body segment,

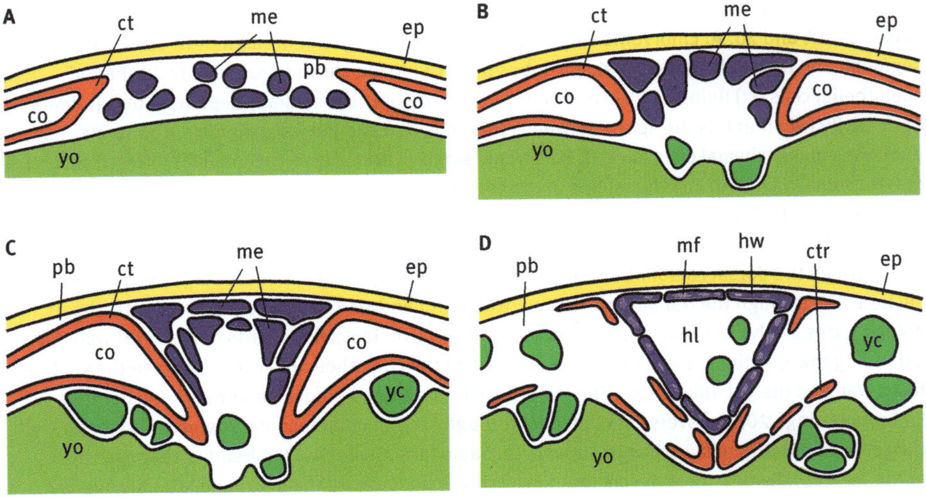

Figure 11.31: *Pholcus phalangioides*, schematic illustration of heart development through embryonic stages 15–20, depicted by transverse sections. **A.** Embryonic stage 15, third opisthosomal segment. A large dorsomedian primary body cavity (*pb*) loosely filled with apolar mesenchymal cells (*me*) spans between the dorsal edges of the coeloms (*ct*). **B.** Embryonic stage 17, first opisthosomal segment. The dorsal tips of the coeloms (*co*) approached further medially and extend towards the gut anlage (*yo*). Apolar mesodermal cells (*me*) accumulate dorsomedially between the dorsal edges of the coeloms (*co*). "Free yolk cells" (*yc*) medially aggregate above the gut anlage. **C.** Embryonic stage 17, third opisthosomal segment. The dorsal tips of the coeloms (*ct*) extend into a groove formed along the dorsal midline of the gut anlage (*yo*). Dorsomedian parts of the coelothelia form the lateral walls of a trough that is ventromedially bordered by "free yolk cells". The epidermis (*ep*) and flattened apolar mesenchymal cells (*me*) form the roof of the trough. Some apolar mesenchymal cells arrange retroperitoneally along the dorsomedian walls of the coeloms. **D.** Embryonic stage 20, third opisthosomal segment. All apolar mesodermal cells are transformed into a layer of muscle cells forming the heart wall (*hw*). Remnants of the coelomic walls (*ctr*) still surround the heart laterally and ventrally. *hl* heart lumen, *mf* myofilaments, *yo* yolk of the gut anlage.

left and right coeloms form medioventral and mediodorsal extensions devoid of a cavity that above may become confluent the yolk. Coelomogenesis does not include formation of appendicular lobes, nor of sacculi. Remnants of the coelothel are only involved in interconnecting the prospective sacculi with the exterior by contributing a funnel and (at least) the proximal part of the nephridial duct. The previous assumption that the nephrocytes forming the sacculus anlage (i.e., the podocytic organs) are also derived from the coelothel (e.g., Moritz, 1959, for Opiliones; Weygoldt, 1975, for Amblypygi) is not corroborated by our results. The canonical view that extant horseshoe crabs (Xiphosura) show the most ancestral states within Chelicerata (e.g., Mittmann and Wolff, 2012) is also rejected. Under the assumption that segmental formation of coeloms and sacculi is plesiomorphic in arthropods, horseshoe crabs are more derived than web spiders (Araneae) in entirely suppressing coeloms and podocytic organs in some prosomal segments, while the entire set is maintained in at least *Pholcus phalangioides*. Any contribution of embryonic coeloms to the development of the gonads is also questioned by our studies, since primordial germ cells were not recognized within the mesoderm up to the last stages studied by us.

11.5.1.3 Crustacea

Anderson (1973) considered the development of the segmental mesoderm in crustaceans as "*uniquely different from that of other arthropods*" (p. 336). Differences include the suppression of segmental coeloms that "*are never more than vestigial*" (p. 333). Genuine observations of previous embryologists surveyed by Anderson up to 1982 indeed vary remarkably in the spatial and temporal occurrence of "vestigial coeloms". Spatial variation concerns apparent differences in whether coeloms occur in (almost) all body segments, in postnaupliar segments only, or in some thoracic body segments only, if at all, and in which compartments of the segmental mesoderm they occur. Temporal differences relate to whether coeloms occur already in the early segmental mesoderm (naupliar segments only) or not before subdivision of the early segmental mesoderm into separate and/or interconnected "blocks" of mesodermal cells (naupliar and/or postnaupliar segments); whether they are transient or persisting in adults; and whether the paired coeloms occur simultaneously in one body segment or are delayed in time (in one hemisegment sooner than in the other). One is indeed tempted to conclude from previous histological studies that any minute space at any stage of the mesoderm development was described as a coelom, as long as the spaces appeared entirely surrounded by mesodermal cells.

Weygoldt (1958) based his conclusions on coelomogenesis in crustaceans on comparisons of the last thoracic segments only. He (like Manton, e.g. 1928) considered other segments to strongly deviate because of higher yolk pressure; in these segments coeloms were believed not to be absent but to be undetectable for being compressed by the yolk. Based on this assumption Weygoldt mainly revealed dorsal spaces within the presumptive pericardial floor (i.e., an initially bilayered pericardial septum) as a common pattern of crustacean coelomogenesis and considered apparent deviations from this pattern (e.g. Nair, 1956; Weygoldt, 1961) as exceptions. Anderson (1979; 1982) finally restricted his interpretation of vestigial coeloms in crustaceans to these dorsal spaces within the pericardial floor, as well as to the sacculi of the antennal and maxillary nephridial organs. The main justification for the focus on the pericardial floor is its association with primordial germ cells, as well as the observation that the gonoducts "usually" arise from the coelom within the pericardial floor of the genital segment. Primordial germ cells were described as either aggregating on its lower (ventral) wall, or arising in the wall of the vestigial coelom within the pericardial floor. The putative coeloms, however, apparently do not contribute to gonad development any further – there is common agreement that the lumen (or lining, respectively) of the gonads in

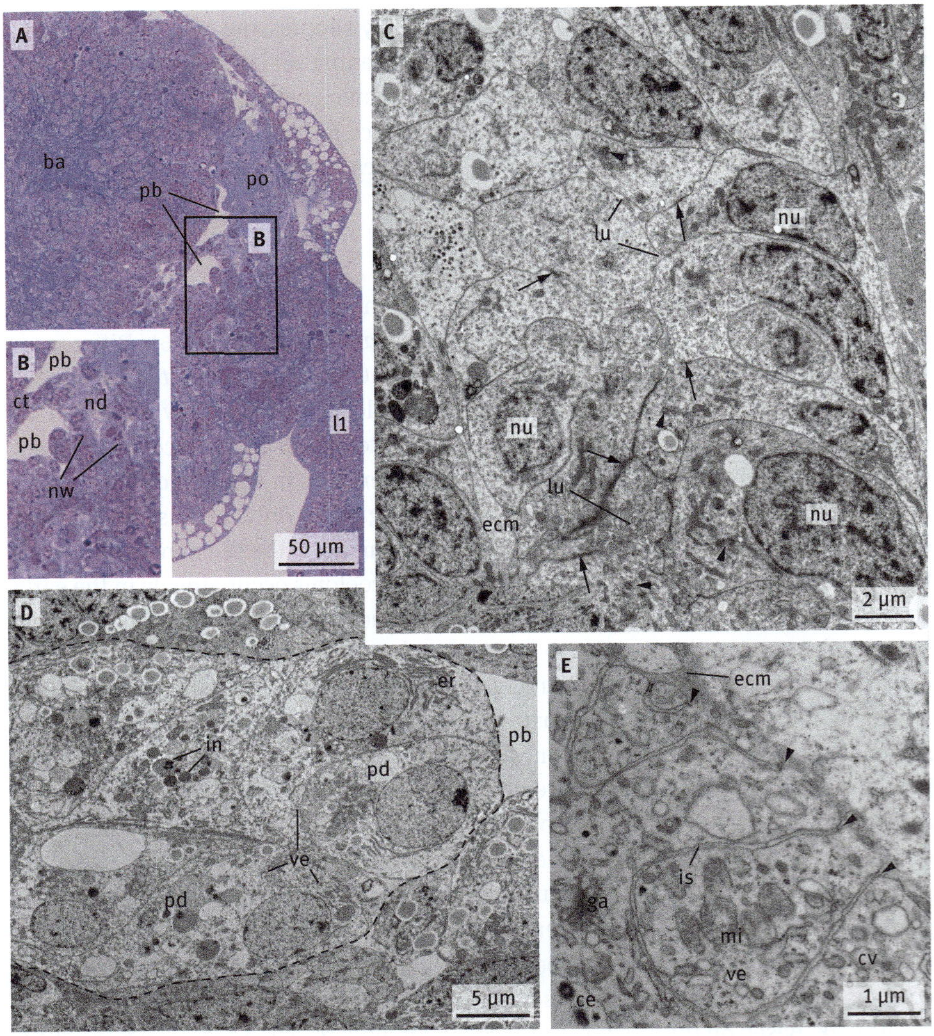

crustaceans forms de novo out of apolar mesoderm surrounding the gonad rudiment. Neither do the transient spaces considered as coeloms within the presumptive pericardial floor undergo mixocoely; according to Mayrat, McMahon, and Tanaka (2006: 54) "*there is no dissociation of coelomic cavities as has often been described by various authors*"; instead the "*small coelomic cavity of ephemeral existence* (...) *later will close on itself.*" The so-called coelom in the anlagen of the sacculi of the cephalic nephridial organs likewise does not undergo mixocoely but persists as the lumen of the presumptive "end-sacs". Thus, irrespective of the lack of evidence for epithelialization within the embryonic mesoderm, Weygoldt's and Anderson's conclusions basically raise doubts on the existence of coeloms in crustaceans by revealing two main differences to the coeloms in non-crustacean arthropods: in crustaceans they develop in later

◂ **Figure 11.32:** *Pholcus phalangioides*, embryonic stage 20. Anatomy of nephridial anlage and podocytic organs. **A.** Transverse section (LM) through first ambulatory prosomal segment, showing remnants of the coelothelium with a nephridial duct anlage (*box*) that opens into the 'mixocoel' (i.e., primary body cavity, *pb*). A podocytic organ (*po*) lies dorsolateral to the duct anlage (*nd*). **B.** Detail of A in higher magnification. The wall of the nephridial duct (*nw*) is continuous with remnants of the coelothelium (*ct*) extending medially. **C.** Transverse section (TEM) through nephridial duct in first ambulatory prosomal segment. Distally the lumen (*lu*) of the duct anlage is narrow, but all duct cells are interconnected by apical adherens junctions (*arrows*). The duct cells contain diplosomes (*arrowheads*) and their cytoplasm is less densely stained than in neighboring cells. Nuclei (*nu*) are situated basally within the duct cells. Extracellular matrix (*ecm*) borders the nephridial anlage against adjacent tissues. **D.** Transverse section (TEM) through podocytic organ (*outline stippled*) in cheliceral segment. The podocytes (*pd*) form cell clusters within the primary body cavity (*pb*) and are characterized by a high number of cell inclusions/lysosomes (*in*), vesicles (*ve*), and large areas of rough endoplasmic reticulum (*er*). **E.** Detail of podocytes at higher magnification, showing characteristic intercellular slits (*is*) that are apically bridged by slit membranes (*arrowheads*). Coated (*cv*) and smooth (*ve*) vesicles are numerous within the apical cytoplasm. The cells are partly covered by extracellular matrix (*ecm*). *ba* brain anlage, *ce* centriole of a diplosme, *ga* Golgi stack, *l1* walking leg 1, *mi* mitochondrion, *nd* nephridial duct anlage, *ve* vesicle.

stages of mesoderm development (after the initial segmental mesoderm subdivided or compartmentalized into various "blocks" of mesodermal cells), and they show a different fate (no mixocoely). These differences together with the minute size of the putative "coeloms" and various exceptions from an apparent pattern per se question homologization with segmental perivisceral coeloms (fide Ruppert, 1991b), even if the respective cavities in crustaceans proved to represent secondary body cavities. Previous conclusions in addition seem to be biased towards observations for malacostracan crustaceans, in which the segmental mesoderm develops from teloblasts (see, e.g., Scholtz and Wolff, 2013). During teloblastic growth daughter cells regularly form chains that in light microscopy resemble epithelia in their ordered arrangement. We suspect that chains of daughter cells growing in the same direction can transiently adopt a similar shape as coeloms, especially when they become locally compressed by yolk or other tissues during dorsal growth.

In our ultrastructural studies on *Artemia salina* (Branchiopoda) (Bartolomaeus, Quast, and Koch, 2009) and *Tisbe* sp. (Copepoda) (Figure 11.33C,D) we never found coelomic cavities within the segmental mesoderm – neither in naupliar body segments, nor in postnaupliar body segments. The mesoderm of the yolkless nauplius larvae mainly consists of well-developed appendicular muscles and a fine network of intestinal muscles, both surrounded by a large primary body cavity (Figure 11.33A,B). In *A. salina* mesoderm of postnaupliar segments arise from a compact posterior growth zone entirely filling the space between epidermis and hindgut in the caudalmost region of the nauplius. In early and late nauplius stages of *Tisbe* sp., in contrast, the primary body cavity also involves the caudal region; a posterior growth

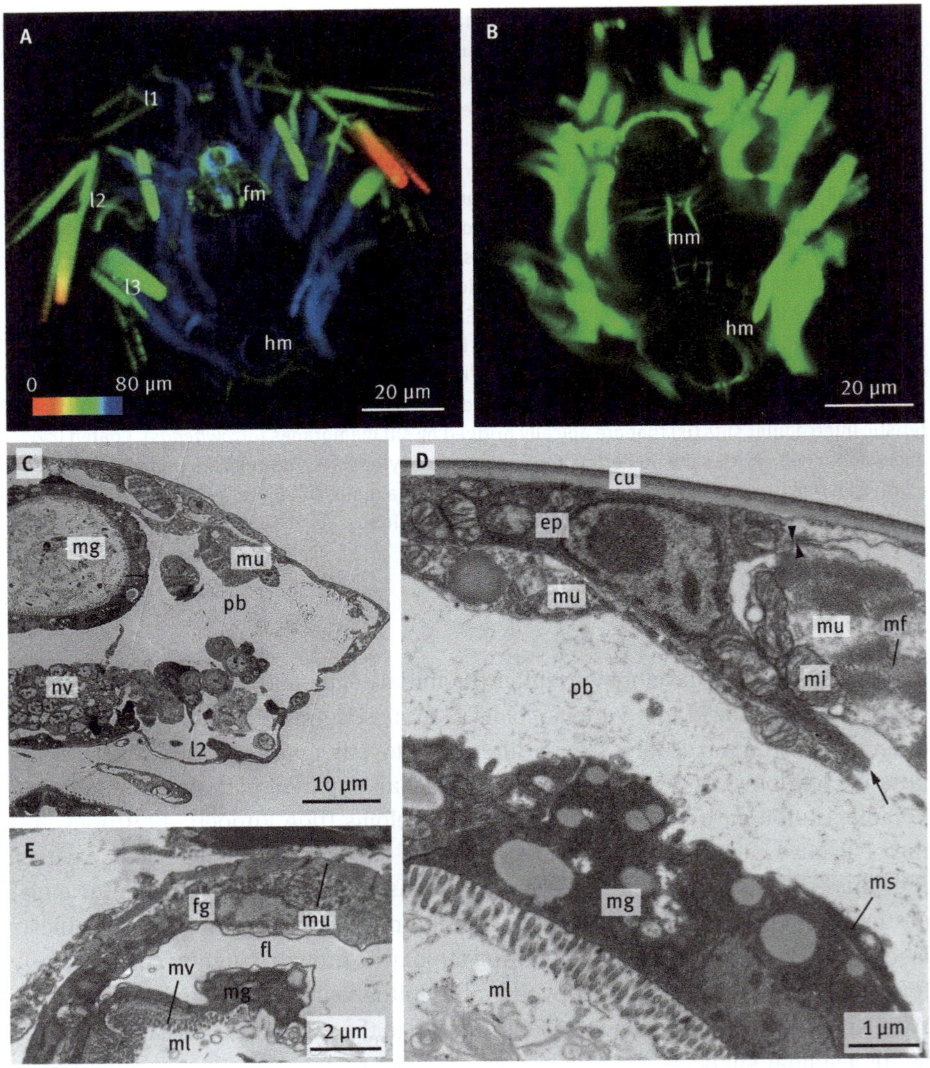

zone does not exist for the mesoderm. In this species, single apolar mesodermal cells and small clusters of them are irregularly arranged along the somatic body wall of the nauplius larva; in the postmandibular region they seem to form the only source for the segmental mesoderm of postnaupliar body segments. We have not found any evidence for enhanced proliferation of the posterior-most mesodermal cells and are accordingly reluctant to confirm the presence of a posterior mesodermal proliferation zone made of few "marginal cells" or "*Urmesodermzellen*" (surveyed for Copepoda by Kohler, 1976). Neither could we unambiguously identify anlagen of nephridial organs in nauplius stages of *Tisbe* sp., whereas they are clearly recognizable in the antennal segment of the early nauplius of *A. salina*. Nauplius stages of *Tisbe* sp. also differ

◂ **Figure 11.33:** *Tisbe* sp., nauplius stage shortly before and after hatching. **A–B.** Phalloidine labeling of musculature, projections of complete image stacks. **A.** Depth coded projection from ventral; color bar indicates relative depth of the labeled structures. The musculature of the larval limbs (*l1–l3*) and of the foregut (*fm*) is most prominent. The limb musculature consists of striated muscles passing straightly into the limbs. The foregut musculature (*fm*) comprises several circular and few longitudinal strands, which also show striated muscle fibers. **B.** Overexposed image stacks allow detection of the fine smooth muscles of the midgut (*mm*) and hindgut (*hm*). Ring muscles and longitudinal muscles are arranged in regular distances of several micrometers. **C–E.** Transverse sections (TEM) through hatchling. **C.** Section at the level of the base of the second larval limbs (*l2*). A large primary body cavity (*pb*) surrounds musculature (*mu*), the midgut (*mg*), and the ventral nervous system (*nv*). **D.** Detail of primary body cavity (*pb*), dorsolateral to midgut (*mg*). Epidermal cells (*ep*) are connected to muscle cells (*mu*) by dense plaques (*arrowheads*). The cells of the epidermis (*ep*) are flattened in areas where they interconnect muscle cells with the cuticle (*cu*), such that their perikarya and cytoplasmic protrusions extend deeply into the primary body cavity (*arrow*). Single smooth muscle cells (*ms*) form a loose network of circular and longitudinal muscles around the midgut (*mg*). **E.** Transitional area between foregut (*fg*) and midgut (*mg*). Musculature (*mu*) around the foregut show striated Myofilaments with regularly arranged z-discs. The midgut (*mg*) is distinguishable by numerous microvilli (*mv*) projecting into its lumen (*ml*). *fl* foregut lumen, *l1–l3* larval limbs 1–3, *mf* myofilaments, *mi* mitochondrion, *nv* ventral nervous system.

from the nauplius of *A. salina* in the presence of large mesodermal cells with almost translucent cytoplasm that are associated with the ventral midgut wall in the postmandibular region. These cells partly seem to suspend the gut onto the somatic body wall, thereby subdividing the primary body cavity into ventral and dorsal spaces, but their fate remained unclear.

In *A. salina* vestigial coeloms in postnaupliar body segments were stated to occur either in dorsal and ventral compartments of the mesoderm (Weisz, 1947), or in so-called segmental blood glands (Benesch, 1969). In our studies, all transient spaces within dorsal and ventral compartments of the mesoderm proved to be primary body cavities that arise by mere disaggregation of the formerly compact clusters of apolar cells (Bartolomaeus, Quast, and Koch, 2009). The blood glands, in contrast, maintain a compact organization throughout their development. The few cells forming a blood gland transiently show cytoplasmic areas that appear translucent in light microscopy, and we accordingly assume that these translucent areas have been mistaken for body cavities. The pericardial floor is at no stage bilayered but composed of thin strands of interconnected, overlapping apolar cells with local thickenings in areas containing muscle fibers. Similarly to blood glands, translucent cytoplasmic areas within the pericardial floor can easily be misidentified as minute body cavities in light microscopical studies. The formation of gonads could not be traced in our studies; up to the last metanaupliar stages studied by us we did not observe primordial germ cells to cluster at or below the pericardial floor. The gut musculature is already well developed in hatchlings and all subsequent stages. We therefore consider the early, still undifferentiated segmental mesoderm in postnaupliar segments to comprise prospective somatic tissue only, based on an early separation of somatic and visceral mesoderm in the caudal region of the body. The same may be true for *Tisbe* sp. as the gut muscles are similarly well developed from hatching onwards while the undifferentiated mesodermal cells in the postnaupliar body region mainly cluster along the somatic body wall.

Nephridiogenesis. The nephridial organs in the antennal and maxillary segments of *A. salina* start to develop each from a single, formerly compact cluster of few mesodermal cells (Bartolomaeus, Quast, and Koch, 2009). The anlagen become recognizable as nephridial organs first by the formation of an intercellular duct. A cap of presumptive podocytes connected to the duct cells by adherens junctions subsequently gives rise to the sacculus ("end-sac" in the terminology of carcinologists). Based on electron microscopical studies a similar formation of the antennal gland from a tubular primordium was also described for the decapod *Astacus leptodactylus* by Khodabandeh et al. (2005). Other studies based on histology (e.g. Strömberg, 1965) remained ambiguous as to whether the cavity first recognized within the antennal gland anlage represents the lumen of the end-sac, or of the nephridial duct. The frequent assumption that the duct and sacculus of the cephalic nephridial organs initially arise from two separate blocks of mesodermal cells that coalesce in subsequent stages (reviewed by Mayer, 2006) are refuted by our results. In *A. salina* the presumptive duct and sacculus are already interconnected by adherens junctions when the first cavity forms within the former. The interpretation of the end-sac's lumen as a vestige of a coelom is accordingly questioned, since its formation out of a tubular precursor is reminiscent of the development of protonephridia.

Our insights on the nephridiogenesis in chelicerates furthermore raise the expectation that only the nephridial duct is a derivative of a coelom, whereas the podocytic component (the presumptive sacculus) develops independently adjacent to segmental coeloms. In this regard Cannon's description (1924) of the maxillary gland development in the branchiopod *Estheria* sp. is most remarkable as it seems to match this expectation. He described the sacculus ("end-sac") as to form at the junction between "dorsal coelom" and nephridial duct, rather than as a remnant of the "dorsal coelom". His text-fig. 3 depicts the sacculus (and its rudiment) as a lateral lobe of the "dorsal coelom" in a manner immediately comparable to a state of the coxal gland development that we observed in the penultimate prosomal body segment of *Limulus polyphemus* after its podocytic organ coalesced with the coelom. Presence of "dorsal coeloms" in *Estheria*, however, still requires unambiguous confirmation, as does the possibility that the end-sac anlage fuses to the "dorsal coelom" rather than being formed by its coelothel.

The assumption that segmental organs in the trunk of various crustaceans, like the blood glands of *A. salina*, are homologous to the sacculi of the cephalic nephridial organs (see, e.g., Hessler and Elofsson, 1995) is still contentious. Here we can only state that the mesoderm development in crustaceans, as far as known, does not provide anatomical or developmental support for the view that their cephalic nephridial organs or the respective segmental organs in the trunk are derived from segmental coeloms. This is because no unambiguous evidence for the existence of transient segmental coeloms has been provided for crustaceans to date, while our results corroborate doubts on their existence raised by the available data.

11.5.1.4 Hexapoda

Organogenesis in insects is traditionally considered as extraordinarily uniform (e.g., Johannsen and Butt, 1941; Siewing, 1969), irrespective of whether embryos develop from short germs, long germs, or any intermediate types of germ bands. Wiesmann's (1926) survey of the formation of segmental coeloms in insects, however, followed by subsequent studies surveyed by Anderson (1973) highlighted some remarkable variation in whether and how segmental "coelomic sacs" are formed. This seems to include (i) multiple instances of convergent suppression of coeloms under either loss of their "*typical trilobation*" (Anderson, 1973: 197), or entire loss and "*direct development*" (Anderson, 1973: 242) of the segmental mesoderm; (ii) several instances of apparent abandonment of schizocoely in favor of coelom formation by either 'infolding' (or 'curling') of the lateral parts of an initially transverse monolayer of mesodermal cells, or by withdrawing of the lateral parts into the outgrowing limb buds, the latter two modes coinciding with the transient occurrence of 'open coeloms' that are initially confluent with the primary body cavity before they enclose by fusion; and (iii) the frequent absence of an anterior-posterior gradient in coelomogenesis due to simultaneous development of coeloms in some or all trunk segments. It is also noteworthy that marked differences in the development of "larval" (premaxillary) and postmandibular mesoderm have never been observed in insects (Larink, 1969), whereas in crustaceans the basic distinction of naupliar and postnaupliar mesoderm is strongly conserved even in species with embryonized naupliar stages. The classic distinction of the segmental mesoderm into somitic, subsomitic, and median parts, with only the former part being involved in coelomogenesis, was introduced by Eastham (1931) for endopterygote insects. In accordance with Wiesmann's (1926) interpretation, Eastham considered the coelomic part of the segmental mesoderm in endopterygote insects only as a (dorsolateral) remnant of the trilobed coeloms of apterygote and hemimetabolous insects. Recent developmental studies (e.g., de Velasco et al., 2006) use the histological description of the mesoderm development in the beetle *Tenebrio molitor* by Ullmann (1964) as a principal reference on coelomogenesis in insects.

Our studies focused on the springtail *Folsomia candida* (Collembola), the silverfish species *Thermobia domestica* and *Ctenolepisma longicaudata* (Zygentoma), as well as the hemimetabolous stick insect *Carausius morosus* (Phasmatodea). While *Folsomia candida* provides another example of entire suppression of segmental coeloms, their transient occurrence is confirmed for the insects studied by us.

In *Folsomia candida*, the embryonic mesoderm remains largely compact until shortly before hatching (Figure 11.34A). Up to this stage, larger body cavities within the mesoderm only form around the hindgut; elsewhere mesodermal cells only slightly disaggregate, if at all, and give rise to an irregular network of minute lacunae (Figure 11.34B–D). All cavities within the mesoderm proved to be lined by ECM and accordingly represent primary body cavities. Segmental coeloms do not occur, as was already confirmed for *Tetrodontophora bielanensis* by Klag and Ksiazkiewicz-Kapralska (1989) based on electron microscopical studies, in contrast to previous histological studies. In *F. candida*, the only secondary body cavities formed within the embryonic mesoderm are the sacculi of the paired nephridial organs in the labial (= second maxillary) head segment (Figure 11.35A,B). We could not determine whether the sacculus and its nephridial duct arise from two separate anlagen. Their anlagen within

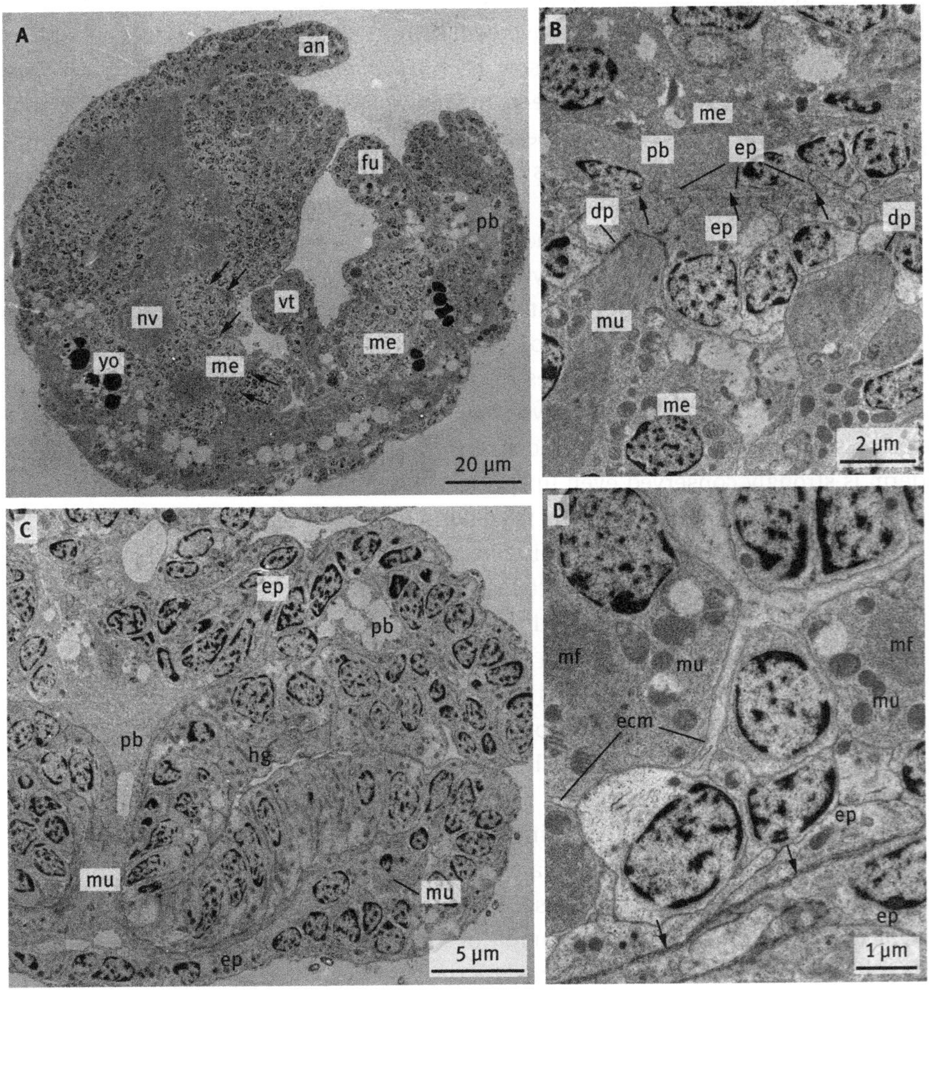

the embryonic mesoderm do not become recognizable before differentiation of podocytes and formation of the duct's lumen, respectively. As soon as they become recognizable, the proximal duct cells and podocytes are interconnected by apical adherens junctions. Shortly before hatching, the sacculus is made of 5–8 podocytes and still is largely compact, its lumen being confined to peripheral slits and minute lacunae below the podocytes and their pedicels (Figure 11.35B,D). The nephridial duct is mesodermal except for its distalmost, ectodermal part that becomes lined with cuticle. Further nephridial organs made of podocytes but devoid of an efferent duct ("*rudimentäre Nephridien*" fide Feustel, 1958) were reported for postembryonic stages of various species and are positioned in the head close to the antennal bases as well as in the first and third abdominal segments (e.g., Humbert, 1975). We could not identify their anlagen in embryonic stages. The somatic musculature, especially longitudinal muscles of the trunk and appendicular muscles, is largely differentiated until hatching (Figure 11.34B,D), whereas visceral muscles only become recognizable as a single layer around the hindgut, but are still missing around the midgut (Figure 11.34A,C). The midgut is composed of yolk cells that become interconnected by adherens junctions but still is devoid of a lumen until hatching.

Coelomogenesis. Insights obtained from embryos of silverfish and the stick insect *Carausius morosus* are largely congruent. In the respective stage studied by us (stage S3 fide Mittmann, 2000; stage IV fide Scholl, 1969) segmental coeloms are present in all trunk segments except the ultimate abdominal segment (Figure 11.36A–D). The coelomic cavities are roundish to oval or crescent-shaped both in transverse and horizontal section and, irrespective of their variable size, show no areas of compression along their extension between the ectoderm and the yolk-sac. The coeloms coexist with large primary body cavities that segmentally occupy major spaces dorsolateral and medioventral to the coeloms (Figure 11.36B,D,E,F). Intersegmentally the primary body cavities are medially confluent and laterally separate consecutive coeloms. An overlap of consecutive coeloms exists dorsolaterally. In histological sections this

◂ **Figure 11.34:** *Folsomia candida*, embryo prior to hatching (TEM). **A.** Paramedian sagittal section. The entire mesoderm (*me*) is apolar and almost compact. A larger primary body cavity (*pb*) only occurs around the hindgut. The midgut anlage consists of yolk-rich cells (*yo*). Due to the bending of the embryo, the ventral epidermis forms deep infoldings between the anterior trunk segments (indicated by *arrows*). **B.** Mesodermal tissue frontally and caudally to a ventral epidermal infolding (indicated by *arrows*). Muscle cells (*mu*) are connected to the epidermis (*ep*) by dense plaques (*dp*). Minute primary body cavities (*pb*) sporadically occur between epidermis and mesodermal cells (*me*). **C.** Sagittal section through hindgut (*hg*). The hindgut is lined by a single layer of muscle cells (*mu*). A primary body cavity (*pb*) occurs between this muscular layer and the epidermis (*ep*). **D.** Muscle cells (*mu*) underlying the epidermis (*ep*) near an epidermal infolding (*arrows*). The muscle cells are surrounded by extracellular matrix (*ecm*). *an* antenna, *fu* furca, *me* apolar mesenchyme, *mf* myofilaments, *nv* nervous system, *vt* ventral tube, *yo* yolk cells of midgut anlage.

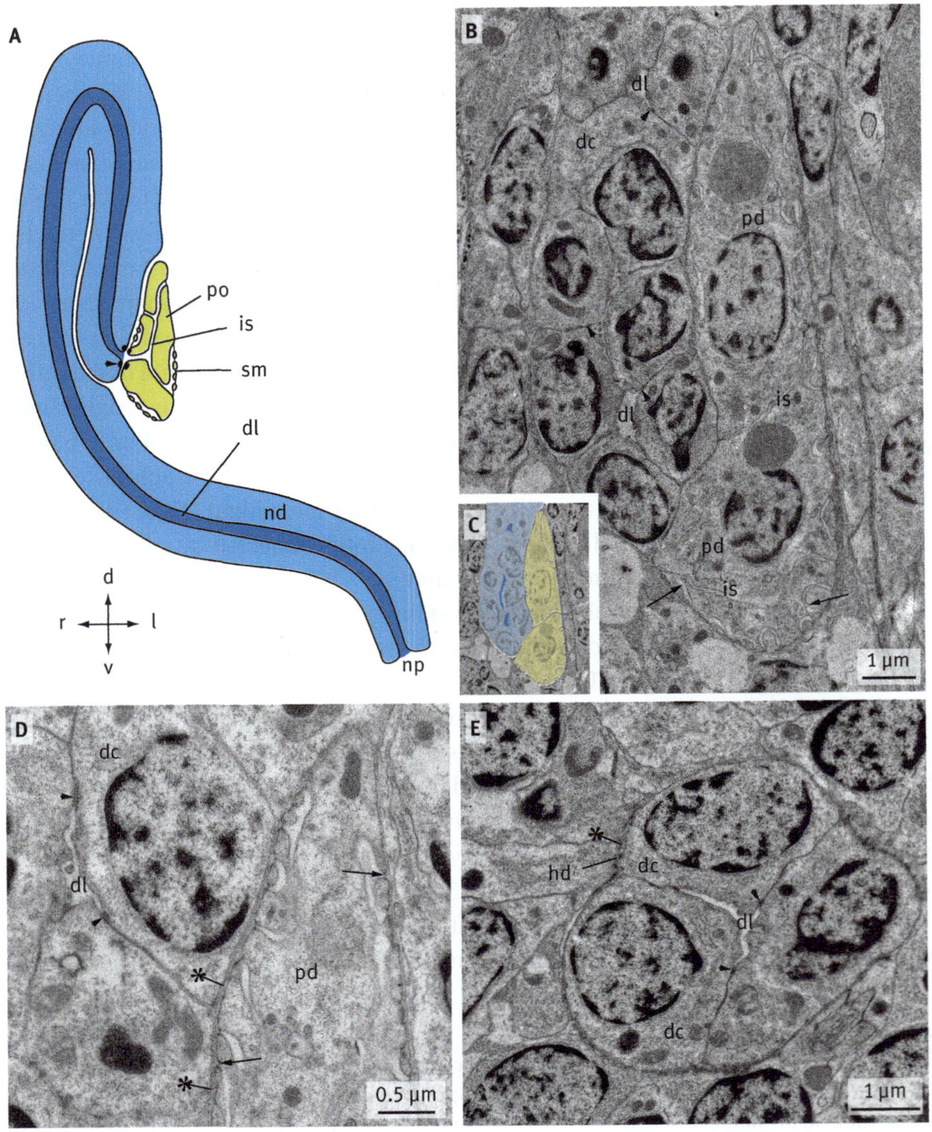

overlap raises the false impression that coeloms particularly of abdominal segments are intersegmentally confluent in the respective area. An anterior-posterior gradient in coelomogenesis is not apparent; the coeloms of the thoracic segments are larger (i.e. more advanced) than in abdominal segments, but within each tagma the coeloms are at the same developmental stage, thus indicating simultaneous development of the segmental mesoderm in thorax and abdomen. In the silverfish embryos, size and arrangement of the coeloms in both the first abdominal segments deviate from those

◂ **Figure 11.35:** *Folsomia candida*, anlage of the nephridial organ of the labial (second maxillary) head segment. **A.** Schematic reconstruction of the right nephridial anlage in embryo prior to hatching. The nephridial duct (*nd*) extends from the nephridiopore (*np*) dorsally and forms a tightly bent loop. At the proximal end a cluster of cells form a podocytic organ (*po*). The cells of the cluster show peripheral lacunae and pedicels. Slit membranes (*sm*) bridge the gaps between the pedicels. Narrow intercellular slits (*is*) occur between the podocytes. The system of intercellular slits and peripheral lacunae is continuous with the lumen of the duct. Apical adherens junctions (*arrowhead*) connect the podocytes to the duct cells. **B–D.** Transverse sections (TEM) through the nephridial anlage shortly before hatching. **B.** Proximal part of the anlage. The cluster of podocytes (*pd*) lies adjacent to the dorsal loop of the nephridial duct. The duct cells are adluminally interconnected by adherens junctions (*arrowheads*). A network of peripheral lacunae (*arrows*) demarcates the podocytes (*pd*). These lacunae are continuous with the intercellular slits (*is*) between the podocytes. **C.** Labeled areas of proximal duct (*blue*) and podocytic organ (*yellow*) shown in B. **D.** Detail of transitional area between podocytic organ and nephridial duct. A thin layer of extracellular matrix (*asterisks*) surrounds the podocytic organ except for areas connected to the duct. The lacunae between the pedicels are bridged by slit membranes (*arrows*). The duct cells (*dc*) are apically connected by adherens junctions (*arrowheads*), the duct lumen (*dl*) is narrow and slit-like. **E.** Distal part of the nephridial duct. Its lumen (*dl*) is intercellular, bordered by cuboidal cells (*dc*). Adluminal adherens junctions (*arrowheads*) interconnect the duct cells. Hemidesmosomes (*hd*) connect the duct cells to the basally underlying extracellular matrix (*asterisk*). *d* dorsal, *l* left, *po* podocytic organ, *r* right, *v* ventral.

in thoracic and other abdominal segments due to the presence of large glandular structures (Figure 11.37A). These glands likely represent the so-called pleuropodia that at stage IV of *C. morosus* still protrude outwards like limb buds but at later stages are withdrawn into the body before they degenerate (Woodland, 1957; Louvet, 1976). Our studies confirm that within the head the formation of coeloms in the mandibular and maxillary segments is retarded (Figure 11.37A).

All trunk coeloms anteriorly show extensions towards the limb buds (Figures 11.36D, 11.37B), but they never bulge into them. Apical adherens junctions as well as the absence of extracellular material on the apical side of the coelothel provide the main evidence for epithelia lining these cavities (Figures 11.36G, 11.37E). All cells of the coelothel show a diplosomal pair of centrioles close to the apical cell membrane (Figure 11.36G). Due to their position the paired centrioles likely represent vestiges of cilia, rather than indicating mitotic activity. We found only few instances of mitosis within the coelothel, and no evidence for cells delaminating from the coelothel into the primary body cavity. More numerous cell proliferations occur within the apolar mesoderm. A large cluster of apolar mesodermal cells is situated immediately below the somatic wall of the coeloms (Figures 11.36C,E, 11.37C,E). In histological section this cluster of apolar cells raises the impression of being part of the somatic wall of the “coelomic sacs”, which accordingly appear to be thicker than their visceral wall. Our ultrastructural studies, however, revealed that these apolar cells do not show a firm connection to the coelothel. Apart from few desmosomes sporadically interconnecting coelothelial cells with adjacent apolar cells, most of the apolar cell cluster are separated from the somatic wall of the coeloms by its basal lamina (i.e., ECM),

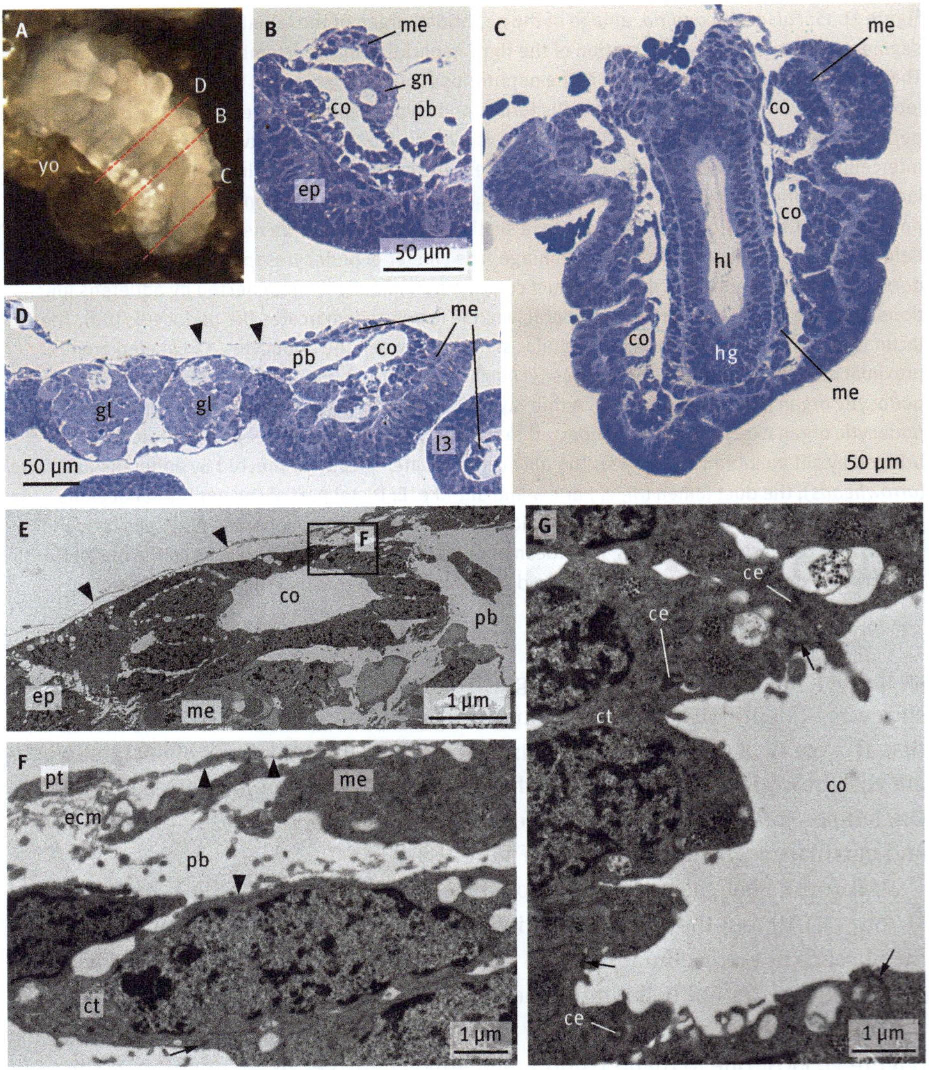

particularly at the base of limb buds. Somatic and visceral walls of the coeloms thus do not differ in thickness, except for areas of the visceral wall into which primordial germ cells are integrated (Figures 11.36B, 11.37C,E). Additional loose aggregations of apolar mesodermal cells occur medioventrally to the segmental coeloms as well as intersegmentally between consecutive coeloms. In histological sections these aggregations appear to form massive 'diverticles' of the coeloms, but our ultrastructural studies again revealed no cellular connections between these apolar cell aggregations and the coeloms. Single apolar cells sporadically occur between the visceral wall of the coeloms and the yolk-sac. The yolk-sac is made of massive aggregations

◂ **Figure 11.36:** *Carausius morosus*, embryonic stage IV fide Scholl (1969). **A.** Light micrograph of embryo dissected from chorion, ventral view. The embryo is dorsally attached to the yolk mass (*yo*), but during dehydration (after fixation) almost all yolk detached from the embryo. Stippled red lines indicate the level of semithin sections shown in **B–D. B.** Transverse section (LM) of coelom (*co*) of right abdominal hemisegment. Apolar mesenchymal cells (*me*) and a large primary body cavity (*pb*) surround the coelom. The gonad rudiment (*gn*) is partly integrated into the coelothelium. **C.** Longitudinal section (LM) through posterior abdomen. Apolar mesenchymal cell layers (*me*) surround the hindgut (*hg*) and the somatic wall of the coeloms (*co*). **D.** Transverse section (LM) through metathoracic segment (yolk cells detached). Somatic and visceral wall of the coelom to different degrees border against apolar mesenchyme (*me*) that also fills the limb bud (*l3*). A thin layer of extracellular matrix (*arrowheads*) demarcates the border of the embryo to the (detached) yolk. A large primary body cavity (*pb*) separates this layer from the coeloms and the ventral ganglia (*gl*). **E–G.** Transverse ultrathin sections (TEM). **E.** Right coelomic cavity (*co*) in metathoracic segment. The coelothelial cells are more electron dense than apolar mesenchymal cells (*me*) adjacent to the somatic coelomic wall. *Arrowheads* mark extracellular matrix bordering the (detached) yolk. Medially the coelom borders against a large primary body cavity (*pb*). **F.** Detail of E at higher magnification. The primary body cavity (*pb*) is lined by extracellular matrix (*arrowheads*) that basally underlies the coelothelial cells (*ct*). Cell protrusions (*pt*) from yolk cells are attached to the extracellular matrix (*ecm*). **G.** Detail of the coelomic wall (*ct*) at lateral edge of the coelom (*co*). Adherens junctions (*arrows*) apically interconnect the coelothelial cells. Centrioles (*ce*) frequently occur within the apical cytoplasm of the coelothelial cells. *ep* epidermis, *hl* hindgut lumen, *l3* walking leg 3, *yo* yolk cell.

of yolk cells bordered against the primary body cavity by a thin extracellular lamina (Figure 11.36D–F). This lamina proved to be independent from the basal lamina of the visceral coelothel, even in areas where the visceral walls of the coeloms lie immediately adjacent to the yolk-sac (Figure 11.36E), and accordingly seems to arise from the yolk cells. In *C. morosus*, the ECM bordering the yolk-sac is enhanced in some areas by minute cell protrusions (Figure 11.36F), whose origin remains ambiguous (either from adjacent apolar mesodermal cells, or from the yolk cells, or both); we can exclude though that these cell protrusions arise from the coelothel. Around foregut and hindgut, apolar cells form a thick mesodermal layer (still devoid of muscle cells) that is clearly separated from the coeloms (as far as present in the head) by the primary body cavity (Figure 11.37B,D).

Musculature. Only few myofilaments are differentiated at stage S3 of silverfish embryos and stage IV of *C. morosus*, respectively; they are restricted to apolar mesodermal cells at the base of the thoracic limb buds as well as to elongate apolar cells traversing the body segments from medioventral to ventrolateral below the coeloms and posterior to the limb buds. In the limb anlagen, intrinsic musculature is still absent. Limb buds are initially massively filled with apolar mesoderm, as demonstrated by the early state of the mouthpart anlagen and the abdominal limb rudiments. At later stages, as shown by the thoracic limb anlagen, the massive apolar mesoderm irregularly disaggregates into loose cell aggregations under formation of primary body cavities (Figure 11.37B). The segmental coeloms do not form appendicular lobes in any body segment, nor do the coelothelial cells contain myofilaments.

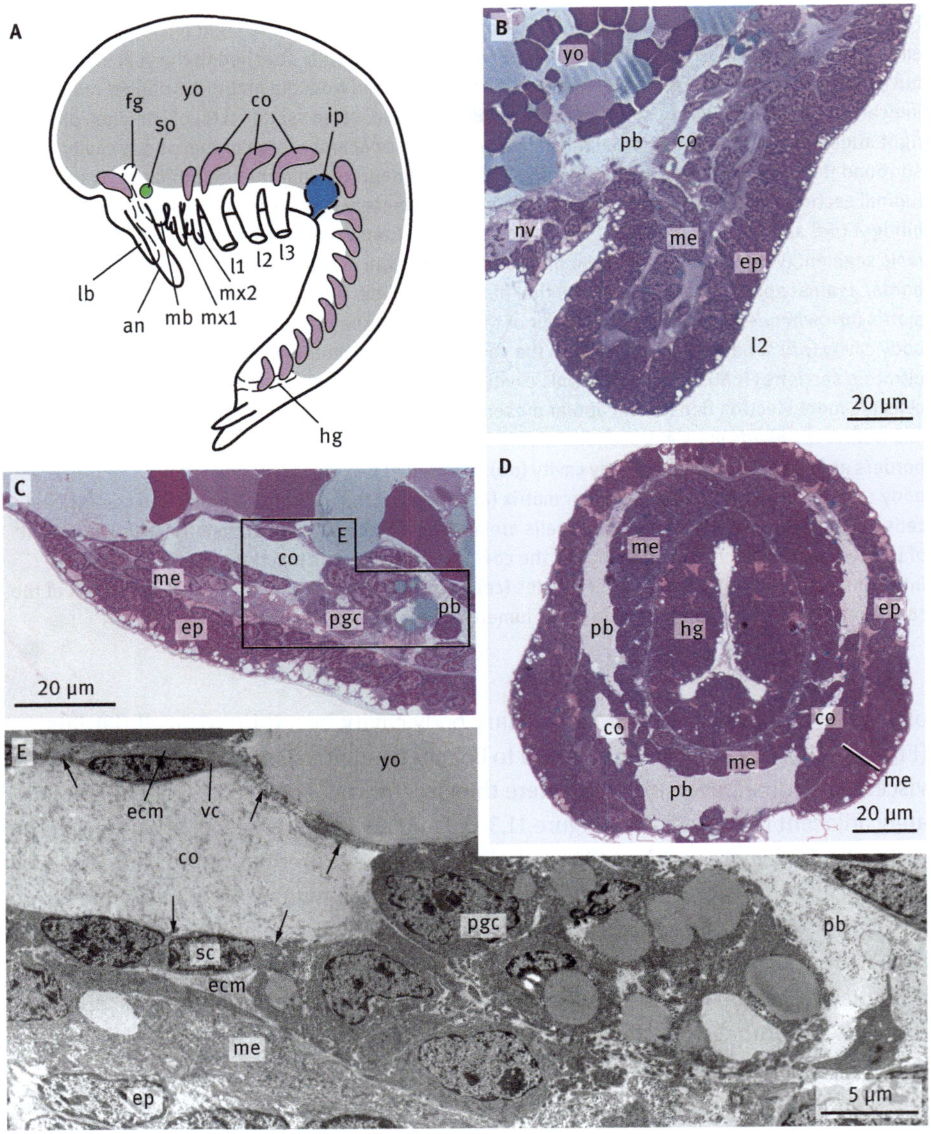

Gonad rudiment. Among the various descriptions of how primordial germ cells (PGCs) associate with segmental coeloms in insects (reviewed by Ewen-Campen et al., 2013), our studies confirm that the PGCs in abdominal segments transiently form small segmental clusters, the most lateral cells of which become medioventrally integrated into the visceral wall of the segmental coeloms (Figure 11.37C,E). The integrated PGCs interconnect to adjacent coelothelial cells by apical adherens junctions, while such junctions do not exist to remaining PGCs of the cluster. In *Thermobia domestica*, PGCs

◂ **Figure 11.37:** *Thermobia domestica*, embryonic stage S3 *fide* Mittmann (2000). **A.** Schematic illustration of coeloms (*co*), left body side. In most segments the coeloms are situated ventrolateral to the yolk cells (*yo*) of the midgut anlage. In the mandibular (*mb*) and maxillary head segments (*mx1*) coeloms are still absent. Paired subesophageal bodies (*so*) occur in the intercalary head segment. The coeloms of the two posterior-most segments are located ventrolateral to the hindgut (*hg*). Invaginated pleuropodia (*ip*) occur ventrolaterally in the first abdominal segment. Only the basal parts of the thoracic leg anlagen (*l1–l3*) are shown. **B–D.** Semithin transverse sections (LM). **B.** Base of left mesothoracic limb anlage (*l2*). The coelom (*co*) lies above the limb base and lacks an appendicular lobe. A primary body cavity between yolk cells (*yo*) and the coelom slightly extends into the limb anlage, which otherwise is filled with apolar mesenchymal cells (*me*). **C.** Right coelom (*co*) in metathoracic segment. The visceral coelomic wall immediately borders against yolk cells, while apolar mesenchymal cells (*me*) separate the somatic wall from the epidermis (*ep*). Ventromedially a cluster of primordial germ cells (*pgc*) is partly integrated into the coelomic lining. **D.** Penultimate abdominal segment. The hindgut (*hg*) is surrounded by a thick layer of apolar mesenchymal cells (*me*). Large primary body cavities (*pb*) occur medioventral and dorsal to the coeloms (*co*). **E.** Transverse ultrathin section (TEM) through coelom shown in **C.** The coelothelial cells are interconnected by apical adherens junctions (*arrows*). Primordial germ cells (*pgc*) line the coelomic cavity ventromedially. Extracellular matrix (*ecm*) borders the visceral coelothelium (*vc*) against yolk cells (*yo*) and the somatic coelothelium (*sc*) against adjacent apolar mesenchymal cells (*me*). *an* antenna, *ep* epidermis, *fg* foregut, *l1–l3* thoracic limb anlagen 1–3, *lb* labrum, *mb* mandible, *mx1*, *mx2* first, second maxilla, *nv* nervous system, *yo* yolk cells of gut anlage.

are not restricted to segmental clusters within abdominal segments at stage S3, but also extend as a pair of large, continuous strands ("genital ridges") from the first abdominal segments anteriorly up to the posterior border of the mandibular head segment. In these body segments, the genital ridges also are segmentally integrated into the visceral walls of the segmental coeloms (as far as present). In *C. morosus*, the PGCs form continuous genital ridges from the second to seventh abdominal segment at stage IV. Our results corroborate the view that PGCs in insects arise from a subset of segmental mesoderm, and that this subset, as inferred from the still acoelomate head segments in silverfish embryos, differentiate into PGCs prior to formation of segmental coeloms (Ewen-Campen et al., 2013). The PGCs initially form segmental clusters adjacent to and partly integrated into the coeloms but in later stages merge intersegmentally under formation of paired genital ridges that become restricted to some trunk segments.

Nephridiogenesis. Among insects nephridial organs are only maintained in primarily wingless species (Archaeognatha and Zygentoma) and are restricted, as far as known, to the labial (second maxillary) head segment (e.g., Seifert, 1979). In the silverfish embryos we could not recognize their anlagen at the stage studied by us. In this stage, however, seemingly glandular organs well-known from winged insects as so-called subesophageal bodies are present in the mesoderm of the intercalary head segment (Figure 11.38A–C). These organs are consistently devoid of an efferent duct and are considered, based on their anatomy, as nephrocytic organs (e.g., Harrat,

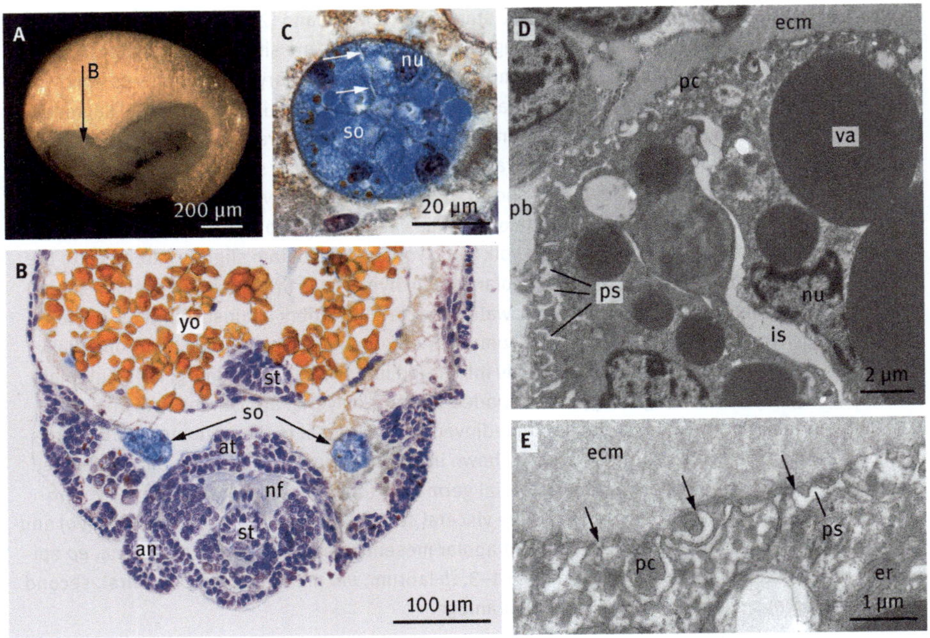

Figure 11.38: *Ctenolepisma longicaudata*, embryonic stage S3 *fide* Mittmann (2000). **A.** Micrograph of embryo within chorion, lateral view. *Arrow* depicts the position of subesophageal bodies. **B–C.** Transverse sections (LM) through embryo as indicated in A. **B.** The subesophageal bodies (*so*) lie lateral to the anlagen of the anterior tentoria (*at*) and slightly posterior above the antennal anlagen (*an*). Distal and proximal sections of the stomodaeum (*st*) are cut at this level as the stomodaeum extends in a frontally bended loop from the mouth opening to the midgut anlage. Paired ventral nerve strands (*nf*) border ventrally against the anterior tentorial anlagen. **C.** Detail of the left subesophageal body (*so*). Small intercellular slits (*arrows*) occur between the cells. **D–E.** Transverse sections (TEM) through subesophageal body **D.** The subesophageal bodies lie within a primary body cavity (*pb*) and are surrounded by extracellular matrix (*ecm*). Beneath the ECM a peripheral system of slits and lacunae (*ps*) occurs in each cell of the subesophageal body. Interdigitating cytoplasmic protrusions form pedicels (*pc*) that cover the peripheral lacunae. Widened intercellular slits (*is*) separate the cells from each other. The cells of the subesophageal body contain several large vacuoles (*va*). **E.** Detail of peripheral slits (*ps*) and pedicels (*pc*). The pedicels are bridged by slit membranes (*arrows*). *er* endoplasmic reticulum, *nu* nucleus, *yo* yolk cells.

Ihsan, and Schoeller-Raccaud, 1999) and/or endocrine organs (e.g., Tsuzuki et al., 2005). Our ultrastructural studies revealed that the subesophageal bodies in silverfish embryos are composed of podocytes that delimit a secondary body cavity, made of minute peripheral lacunae and intercellular clefts, against the primary body cavity (Figure 11.38D,E). Since coeloms are not as yet formed in the mandibular (and maxillary) segment at stage S3, we can exclude that the subesophageal bodies in silverfish embryos arise from the mandibular coeloms, as has been claimed for orthopteran insects (e.g., Harrat, Ihsan, and Schoeller-Raccaud, 1999). The differentiation of the subesophageal bodies in silverfish embryos as podocytic organs within the intercalary head segment corroborates previous assumptions (e.g., Wheeler, 1893) that the subesophageal bodies of insects are homologous to the sacculi of the antennal nephridial organs ("green gland") in crustaceans.

The assumption that the subesophageal bodies may show variable origins in insects (from either the intercalary mesoderm, or the mandibular coeloms; Harrat, Ihsan, and Schoeller-Raccaud, 1999) is based on histological studies that seemingly reveal appendicular coelomic lobes in the early anlagen of the mouthparts. Our ultrastructural studies, however, refute the presence (or even origin) of coeloms in any limb bud. According to our results, the development of the mesoderm in the limb anlagen is indeed identical in the insects and chelicerates studied by us, starting from massive aggregations of apolar mesodermal cells that subsequently disaggregate under formation of primary body cavities. We accordingly assume that the impression of presence of appendicular coeloms deeply extending into the early limb anlagen in previous histological studies of insect embryos is caused by the same fixation artifacts as in chelicerates. This implies that previous observations of coelom formation by 'infolding' and subsequent enclosure by fusion of the infolded parts likewise are misinterpretations.

Our main insight on coelomogenesis in insects is that the segmental mesoderm passes stages (the respective stages studied by us) that closely resemble the early stages of coelomogenesis in *Limulus polyphemus* and *Pholcus phalangioides*, including lack of a trilobation of the coeloms. As in chelicerates, we found no evidence for diverticles arising from the embryonic coeloms in insects, nor for any cells delaminating from the coelothel into the primary body cavity. We accordingly hypothesize that the diverticles of the "coelomic sacs" in insects actually represent parts of the early segmental mesoderm that do not participate in coelomogenesis but remain apolar. The generalized description of the mesoderm differentiation in a "typical" insect by Johannsen and Butt (1941) determines this apolar mesoderm adjacent to the coeloms as to form splanchnic and somatic layers. The mesodermal cells of these apolar layers give rise to the fat body, body musculature, and cardioblasts, primarily under maintenance of intact coeloms. The previous view that the splanchnic and somatic walls of the coeloms give rise to the apolar splanchnic and somatic layers is seriously questioned by our results. Although we could not trace the fate of the coelomic lining cells with the stages studied thus far, our results cause us to doubt that the coelothelial cells show any major contribution to organogenesis in insects, and thus became gradually or entirely suppressed in many insect lineages. According to this hypothesis, the main difference in the coelomogenesis of chelicerates and insects only refers to the suppression of pronounced dorsal and ventromedial extension of the coeloms, which in insects have never been described as approaching above or below the yolk-sac. This difference may correlate with different modes of how dorsal closure and incorporation of the yolk-sac in the midgut anlage is achieved.

11.5.1.5 Myriapoda

Current knowledge of organogenesis in myriapods still is mainly based on the classic embryonic studies by Heymons (1901, on Chilopoda), Tiegs (1940 on Symphyla; 1947 on Pauropoda), and Dohle (1964, on Diplopoda), and has been summarized by Anderson (1973). Ultrastructural studies on the mesoderm development in myriapods are still missing. Based on our previous results, however, we expect that apparent differences in coelomogenesis among myriapods as described thus far are largely comparable to the differences recorded by us for chelicerates, insects, and crustaceans, respectively.

Transient occurrence of segmental coeloms in myriapods is supported by the previous descriptions of the earliest stage of coelomogenesis. The descriptions of this stage basically seem to be congruent with our results on the relative timing of the occurrence of coeloms, their formation by schizocoely as well as their original shape and position within the segmental mesoderm in chelicerate embryos. It is peculiar though that according to Heymons (1901) prospective somatic and visceral cells of the "coelomic sacs" in *Scolopendra* embryos are already distinguishable within the early segmental mesoderm before they interconnect to form somatic and visceral walls, and that these walls are already distinguishable by the different shape of their cells before a coelomic lumen arises between the walls. Subsequent stages, in contrast, remain dubious in all previous descriptions of the mesoderm development in myriapods with respect to the fate of the coeloms. Heymons's (1901) observations on coelomic growth in *Scolopendra* embryos recall, in general outline, the expansion of the coeloms in yolk-rich chelicerate embryos. However, given that the embryonic body cavities in arthropods can tremendously be effected in terms of expansion and integrity by the way the embryos were fixed, some apparent differences in Heymons's descriptions can be pointed out that in our view require thorough reconsideration. This especially concerns his claims that the coeloms in *Scolopendra* embryos disintegrate – apparently under mixocoely – before the yolk is entirely encompassed by expanding dorsal and medioventral coelomic lobes, hence far earlier than in chelicerate embryos; that these lobes transiently form 'open coeloms' that become connected with their 'open' ends to the ectoderm and yolk-sac, respectively, while continuing to expand dorsally and medioventrally; that the tip of the dorsal lobes with their ventrally 'open' coeloms enclose their "original coelomic lumen" against the primary body cavity to persist as the lining (and lumen) of the paired gonad anlage; and that the heart anlage and pericardial septum arise above the gonad anlagen from blast cells delaminating from the persisting tip and somatic wall of the dorsal coelomic lobes, respectively.

Tiegs's (1940; 1947) and Dohle's (1964) descriptions overall raise doubts that the segmental coeloms in progoneate myriapods show a marked expansion at all. This is because of their claim that the early segmental coeloms, before they start to form dorsal lobes, coalesce both segmentally and intersegmentally by narrow, medioventral "coelomic bridges". A fusion of the segmental coeloms appears highly unlikely in the light of our ultrastructural results on chelicerate and insect embryos; claims of this kind rather indicate that Tiegs and Dohle could not properly discriminate coeloms from primary body cavities and apolar mesoderm. This is particularly evident in Tiegs's (1940) documentation of his results, according to which a "trilobation" of the coeloms in *Hanseniella* (Symphyla) seems not to be based on an expansion of the coeloms, but to be merely due to a compression of the early, primarily ovoid coelomic lumen by neighbouring mesodermal and ectodermal tissue. It is remarkable (and in our view unlikely) that according to Tiegs's (1940) description the coeloms in *Hanseniella* embryos do not undergo mixocoely, but "enclose on itself" in the course of their compression. The alleged origin of the neighbouring mesodermal tissue (i.e. mesoblasts) from the coelothel still requires confirmation. In accordance with our results, an independent origin of blast cells and coeloms (and their lining, respectively) from the early segmental mesoderm was pointed out by Tiegs (1947) for *Pauropus* embryos. His results on the development of pauropods furthermore indicate noteworthy convergences with some crustaceans: similarly as in *Artemia salina*, the segmental mesoderm apparently gives rise to somatic tissue only and suppresses coelom formation entirely in body segments formed postembryonically in the course of anamorphosis.

The previous descriptions of coelomogenesis in myriapods accordingly indicate variable steps of progressive decoupling of organogenesis from transient coeloms, details of which still remain to be clarified due to apparent difficulties to discriminate the coeloms proper within the developing mesoderm. Tiegs's studies (1940, 1947) indicate that the decoupling of organogenesis from transient coeloms also concern the development of the segmental nephridial organs. Their development in head segments as described by Tiegs is in accordance with our current insights on nephridiogenesis in that they develop out of tubular precursors, the blind end of which forms *de novo* a podocytic 'end-sac' (i.e. sacculus) primarily devoid of a lumen; and in that only the nephridial duct anlage is derived from the coelom, whereas the 'end-sac' – as pointed out by Tiegs for the premandibular head segment of *Hanseniella* embryos – forms independent of the "coelomic sacs" out of a cluster of mesenchymal cells that later become attached to the expanding nephridial duct anlage. Generalization of this insight for myriapods must await ultrastructural studies on nephridiogenesis in centipedes, for which Heymons (1901) recognized only ectodermal head glands.

11.5.2 Function of embryonic coeloms in arthropods

The traditional view that the embryonic coeloms of arthropods primarily form the main source for mesodermal blast cells (cardioblasts, myoblasts, etc.) is seriously questioned by our results. We accordingly focused on anatomical evidence indicating any alternative function of the embryonic coeloms. In previous studies presence and size of coeloms were often assumed to correlate with either the amount of yolk (e.g., Weygoldt, 1979), or with variable participation of individual body segments in the formation of the circulatory system, the gut musculature, and the segmental set of somatic muscles (Scholl, 1969: 43). Absence of particular somatic muscles in a body segment, however, is a weak argument for smaller sizes of coeloms in the respective body segment as the "diverticles" of segmental mesoderm giving rise to the somatic muscles also occur in body segments that lack the respective muscles postembryonically (Scholl, 1969: 43).

As for the amount of yolk, we particularly wondered whether yolk-rich embryos require segmental "bladders" (analogous to the embryonic allantois of amniote vertebrates) for transient storage of metabolic waste products arising from degradation of yolk. This possibility, however, basically contrasts with previous observations that embryonic coeloms are much larger in yolkless embryos (or yolkless body segments, respectively). Transient storage of waste products furthermore seems to be mainly assumed by single cells or clusters of a few cells ('hemocytes' that act as phagocytes; Seifert and Rosenberg, 1976) floating in the primary body cavity. Such cells are particularly conspicuous in embryos of chelicerates, in which large cells with rod-shaped inclusions transiently accumulate laterally in septa-like incisions of the yolk-sac as well as dorsally adjacent to the heart. The origin of these cells is still unclear. In previous studies they were called "free yolk cells" as they are assumed to serve for distribution of nutrients within the hemocoel (e.g., Yoshikura, 1955), or they were denominated as "*Streifenzellen*" or "*Stäbchenzellen*" (e.g., Scholl, 1977) for their striated appearance in histological sections. Because the intracellular rods are crystalline in *Limulus polyphemus* and *Pholcus phalangiodes* (Figure 11.30C,E), the inclusions may not represent nutrients but waste products (e.g., uric acid). This view is supported by the timing of the disappearance of the striated cells which coincides with the formation of segmental podocytic organs. Further reasons against the view that segmental coeloms serve for storage of nitrogen waste consider the lack of structural evidence for transportive activity within the cells lining the coeloms, as well as the early formation of primary body cavities. Transient storage of urea in the coeloms would necessarily increase their osmolarity and would accordingly hinder accumulation of water – another copious byproduct of yolk degradation – within the ECM. Soaking by water accumulation is a basic requirement for the formation of primary body cavities within the primarily compact mesoderm; their early occurrence, large extent, and co-existence with segmental coeloms inevitably argues against the presence of an osmotic gradient between primary body cavities and the coeloms. Anatomically, maintenance of an osmotic gradient requires tight intercellular enclosure of the coelomic lumen, which in invertebrates is usually achieved by septate junctions (e.g., Gilula, Branton, and Satir, 1970) that we, however, never found among the coelomic lining cells.

A correlation between segmental coeloms and limb bud formation is indicated in some body segments by the relative position of the early coelomic anlagen close to the area where limb buds arise, as well as by the almost coincident first occurrence of both coeloms and limb buds (see, e.g., Figure 11.25A). In the respective body segments the small, fluid-filled and thus incompressible coelomic cavities must necessarily impose increased local pressure against the ectoderm (via the apolar mesoderm below the coeloms), since they cannot outweigh the pressure of the yolk-sac against the coeloms. We accordingly assumed that this pressure by the early coeloms may represent a mechanical induction for the ectoderm to form limb buds at the respective range of pressure. Our results, however, overall question an effect of this kind. In *Limulus polpyhemus*, some limb buds arise irrespective of size and relative position of the coeloms, which may be minute and located far laterally to the early limb bud as exemplified by the cheliceral segment (Figure 11.20B), and simultaneously also occur in body segments lacking coeloms entirely (Figure 11.20C). In *Pholcus phalangioides* the prosomal limb anlagen are already in an advanced developmental state and massively filled with apolar mesoderm, while the coelomic cavities in the respective body areas are still minute and almost devoid of a lumen (Figure 11.1B), whose pressure against the ectoderm is likely insignificant. Whether the segmental mesoderm in insects, irrespective of its organization, contributes to "limb field initiation" likewise remains unclear (Weihe, Milan, and Cohen, 2009: 77). A mechanical induction of limb bud formation by segmental coeloms in any case cannot be generalized for arthropods.

Dorsal expansion of the coeloms in later stages has traditionally been considered to correlate with the formation of the circulatory system, on the one hand by providing cardioblasts that form the heart tube, and on the other hand by the formation of dissepiments between consecutive coeloms that in at least chelicerates were considered to provide the cellular source for the formation of the lateral arteries of the heart and aorta (see, e.g., Weygoldt, 1975 versus Scholl, 1977). Although we refute both the origin of cardioblasts from coeloms and formation of dissepiments (septa), our results favor the view that the mesenchymal cells giving rise to the circulatory system use the expanding coeloms for being directed towards their target position. Based on this observation and the fact that coeloms show expansions in various directions (dorsally, ventromedially, and antero-posteriorly) we hypothesize that the segmental coeloms may act more generally as pathfinding organs for the entire apolar part of the segmental mesoderm that does not participate in coelomogenesis and that we consider as the main source for organogenesis in arthropods. Targeted migration of apolar mesenchymal cells along the coeloms would also explain why the coeloms in yolkless embryos (or yolkless body segments), as described in previous histological studies, become vastly larger than in yolk-rich embryos as the target areas need to be delimited accordingly.

We found no evidence for any other functional correlation or constraints for coelomogenesis, like the size of the eggs, anatomical characteristics of postembryonic or adult stages, or terrestrial life habits. Also, a potential correlation with pathfinding for apolar mesenchymal cells cannot be generalized for arthropods, but primarily considers chelicerates and thus maybe ancestral functions of the embryonic coeloms in arthropods. At present we can only conclude that the functional significance of the embryonic coeloms in arthropods seems to be much more limited than previously assumed, and that the segmental coeloms possibly for this reason were liable to multiple, convergent losses.

11.5.3 Summary and conclusions on the evolution of arthropod coeloms

Histological studies are insufficient for analyzing the development of the mesoderm in arthropods. The traditional and still used concept of "coelomic sacs" in particular proved to be misleading. Any description of the embryonic segmental mesoderm in arthropods requires precise distinction of the segmental coeloms proper from apolar mesoderm surrounding the coeloms. The coeloms proper, when present, consistently form by schizocoely within primarily apolar mesoderm and represent the first organs to be formed within the early segmental mesoderm. With our taxon sampling, however, we found no instance showing the entire segmental mesoderm to participate in coelomogenesis. Mitotic growth of the segmental mesoderm mainly involves the apolar mesoderm, whereas coelomic growth is mostly based on extension and coincident flattening of the coelothelial cells. The previous assumptions that the coelomic lining in arthropods "usually" bear no flattened cells and that the somatic wall of the coeloms becomes thicker (i.e. pseudostratified or stratified) than the visceral wall (Mayer, Bartolomaeus, and Ruhberg, 2005: 69) are refuted by our results.

The coeloms in chelicerates and insects are closely associated with their visceral walls to the gut anlage (or yolk-sac, respectively), whereas the somatic walls are mostly separated from the epidermis by apolar mesoderm and/or primary body cavities. The coeloms thus undoubtedly fulfill the main characteristic of perivisceral coeloms fide Ruppert (1991b) in their topographical relation to the gut (anlage), in contrast to the sacculus anlagen of the nephridial organs. Coelomic growth occurs in dorsal, medial, and longitudinal directions, but appendicular lobes do not form. A trilobation of the embryonic coeloms hitherto considered 'typical' for arthropods is accordingly refuted by our results. Anlagen of limbs are primarily massively filled with apolar mesoderm and subsequently develop primary body cavities only. Our studies support the view that previous observations of alleged coeloms within limb buds are based on fixation artifacts. We assume that the same artifacts caused previous authors to claim that coeloms sometimes form by 'infolding' instead of schizocoely and transiently pass a stage of 'open coeloms' that are confluent with the primary body cavity.

Along the trunk segments an anterior-posterior gradient in the development of the segmental coeloms was not observed in any species studied. Instead, coelomogenesis either appears to be synchronized (i.e., simultaneous) in a body tagma, or is asynchronous in a seemingly unpatterned sequence. Consecutive coeloms dorsally approach, but they never form dissepiments. Delamination of mesodermal cells from the coelothel and/or epithelial-mesenchymal transitions was never observed in our studies. The traditional assumption that the segmental coeloms are involved in organogenesis by contributing mesodermal blast cells is accordingly refuted by our results. The fate of the coelothel after mixocoely, however, still remains unclear. Myofilaments do not occur within the coelothelial cells at any stage. The main components of the presumptive visceral and somatic musculature as well as of the circulatory system are already formed before the coelothel disaggregates. Participation in gonad formation is indicated in insects by the transient integration of primordial germ cells into the coelomic lining, but is not corroborated by our results for chelicerates and crustaceans. Unambiguous evidence that the coelomic walls in insects give rise to the wall of the presumptive gonads is still lacking. We accordingly presume that after mixocoely remnants of the coelothel at most contribute to the formation of the fat body, presumptive cells of which were not recognized in our studies, or simply undergo apoptosis.

In chelicerates the only organs formed by the embryonic coeloms are the ducts of nephridial organs, but in crustaceans and hexapods the nephridioducts develop independently of coeloms. The formation of sacculi in arthropods generally seems to be independent of coeloms. Proof for this view is provided by *Limulus polyphemus*, in which the sacculi arise from separate anlagen within the apolar segmental mesoderm adjacent to coeloms and also occur in body segments that entirely suppress formation of coeloms. Presence of sacculi accordingly cannot be considered as proof for (transient) formation of coeloms in the respective body segment. The independence of coeloms and sacculi in arthropods also implies that segmentally arranged podocytic organs in the trunk of various crustaceans (Cephalocarida, Copepoda, Decapoda, Syncarida, Mysidacea, and Isopoda; reviewed by Hessler and Elofsson, 1995) and collembolans might well represent homologues to sacculi of nephridial organs ("*rudiments of former nephridial organs*"; Hosfeld and Schminke, 1997). This may also concern the segmental cell clusters called blood glands in branchiopod crustaceans. However, as the respective cells do not differentiate into podocytes (or 'nephrocytes' fide Seifert and Rosenberg, 1976), the segmental arrangement of the cell clusters in similar positions within the respective body segments as the segmental podocytic organs remains the only evidence in support of homology. Previous assumptions that the sacculus and duct of the cephalic nephridial organs have separate anlagen in crustaceans that subsequently coalesce into a single organ are not supported by our results (Bartolomaeus, Quast, and Koch, 2009), although we expected separate anlagen for the two components based on our results on nephridiogenesis in chelicerates. In crustaceans and hexapods the cephalic nephridial organs become recognizable ultrastructurally not before the presumptive sacculus and duct are interconnected. We accordingly doubt that these two components can be recognized as separate anlagen in histological studies. However, the observation of sacculi or segmental podocytic organs devoid of a duct may still be considered as evidence for the origin of nephridial organs from two separate anlagen, one of which – the duct anlage – can be entirely suppressed independently from the sacculus anlage. Our results accordingly corroborate the view that the subesophageal bodies of insects are homologous to the "green glands" (sacculi of the antennal nephridial organs) in crustaceans. Available data on the development of the nephridial organs in arthropods, however, provide no unambiguous support for the view that these organs are derived from metanephridial systems.

Based on our results we hypothesize that the variation in the occurrence and development of coeloms in arthropods is mainly based on a decrease in the number of cells of the early apolar segmental mesoderm participating in coelomogenesis (Figure 11.39A–C). Outgroup comparison with Onychophora (e.g., Mayer, 2005) favors the view that arthropods primarily maintained segmental coeloms. However, pioneer blast cells, especially those giving rise to musculature as well as to podocytic organs (sacculi), originate in arthropods no longer from the coelomic lining cells, but from mesodermal cells that remain apolar within the early segmental mesoderm

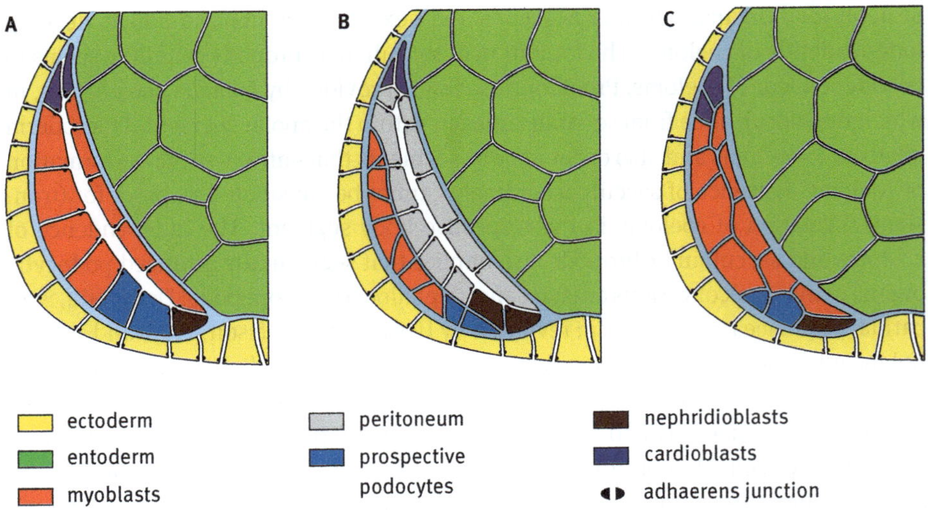

Figure 11.39: Hypothetical schematic illustration of the origin of mesodermal blast cells in panarthropods, as inferred from current ultrastructural studies. **A.** Putatively ancestral state in Panarthropoda, as shown by (some) Onychophora. The entire cell cluster of the early segmental mesoderm epithelializes; all mesodermal blast cells derive from the coelomic walls. **B.** Putatively ancestral state in Arthropoda, as shown by chelicerates. Only portions of the early segmental mesoderm epithelialize, other portions remain apolar. Mesodermal blast cells mainly directly derive from the apolar segmental mesoderm, including the prospective podocytic organ that gives rise to the sacculus of the nephridial organ. Only the nephridial duct anlage (i.e., nephridioblasts) still arises from the coelothelium. The same state may apply (primarily) to insects, but the origin of the nephridial anlage has not as yet been clarified. The fate of the coelothelial cells after mixocoely remains unclear. **C.** "Direct development" of the mesoderm in arthropod body segments lacking transient coeloms. Epithelialization of the early segmental mesoderm is omitted; all mesodermal blast cells directly derive from apolar mesoderm. This state is inferred from the mesoderm development in pycnogonids and collembolans, and may also apply to (some) crustaceans. In the latter, however, the segmental mesoderm may lack blast cells for visceral tissues due to early separation of somatic and visceral mesoderm in the posterior growth zone, as shown by *Artemia salina* (Branchiopoda).

(Figure 11.39B). As inferred from our results on chelicerates, the function of the embryonic coeloms in arthropods is primarily restricted to the formation of nephridioducts as well as to pathfinding for mesodermal blast cells, the latter based on maintenance of a marked (especially dorsal) extension of the coeloms. Based on previous descriptions of coelomogenesis in centipedes (e.g. Heymons, 1901; Knoll, 1974; Hertzel, 1984) we expect that this state is primarily also maintained in myriapods. More advanced states of coelomogenesis include a decoupling of nephridiogenesis from coelomogenesis, possibly primarily under maintenance of embryonic coeloms (although we did not observe this state in the studied species), and loss of the pathfinding function by suppression of marked coelomic growth, as indicated by insects. According to this scenario, the evolution of the mesoderm development in arthropods is characterized by a progressive decoupling of organogenesis from embryonic coeloms, which already started in the arthropod stem species and in various lineages, including crustaceans, pycnogonids, and collembolans, convergently yielding entire suppression of coelom formation in favor of "direct development" of the early apolar segmental mesoderm. This hypothesis, however, conflicts with the current phylogenetic view that crustaceans are paraphyletic with respect to hexapods (see, e.g., Giribet and Edgecombe, 2012) as it implies, under parsimony, a regain of coeloms in insects. Our hypothesis furthermore relies on the view that the participation of originally the entire segmental mesoderm in coelomogenesis (Figure 11.39A), as has been described for *Epiperipatus biolleyi* (Mayer, Ruhberg, and Bartolomaeus, 2004; Mayer, 2005), represents the ancestral mode of the early mesoderm development in Onychophora (and Panarthropoda), a view that with respect to the considerable variation of developmental modes in velvet worms presently remains questionable.

11.6 Onychophora and the problem of polarizing ancestral developmental modes in Panarthropoda

Recent ultrastructural studies of the mesoderm development in *Epiperipatus biolleyi* confirmed that coelomogenesis starts in front of a posterior growth zone within paired strands of mesodermal cells on either side of the presumptive gut (Mayer, 2005). The ultrastructure of the posterior growth zone and the initial mesodermal strands have not yet been described in detail. Coelomic cavities seem to consistently arise within the mesodermal strands by a schizocoelic mode under epithelialization of apparently the entire segmental mesoderm. In *E. biolleyi* the coeloms occur before limb buds arise and along the trunk show an anterior-posterior gradient in their development. During coelomic growth the coeloms of velvet worms generally seem to pass a trilobed stage consisting of a dorsal lobe, a lateral lobe that may extend into the limb buds, and a medioventral lobe. This trilobation may be acquired either by transformation of a formerly crescent-shaped coelom that almost entirely fills the dorsoventral space between entoderm and ectoderm (e.g. *Epiperipatus biolleyi*; Mayer, Ruhberg, and Bartolomaeus, 2004), or by gradual extension of formerly small ventral coeloms (e.g. *Peripatopsis capensis*; e.g. Sedgwick, 1888) or of small dorsal coeloms (e.g. *Opisthopatus roseus*; Mayer, Bartolomaeus, and Ruhberg, 2005) at dorsal or ventral direction, respectively. In the latter case the lobes thus formed may pinch off under subdivision of the coelom within a hemisegment into separate dorsal and ventral coelomic compartments. This variation in coelomic growth apparently coincides with differences in the relative timing of the formation of primary body cavities, apolar mesenchymal cells, limb buds, as well as of mixocoely and nephridiogenesis. In the only two species in which coelomic growth has been studied ultrastructurally thus far, i.e. the yolk-free embryos of *Epiperipatus biolleyi* and *Opisthopatus roseus*, the relative timing seems to show two divergent extremes. In *E. biolleyi*, mixocoely already occurs before limb buds arise and the main sinuses (i.e., primary body cavities) and muscle cells form no earlier than shortly before and after degradation of the coeloms. In *O. roseus*, in contrast, coeloms are initially restricted to small dorsal cavities and primarily coexist with large sinuses that are separated by muscular septa. In later stages, the dorsal coeloms seem to pinch off ventral coeloms that pass into the early limb anlagen. Whether both dorsal and ventral coeloms in later stages undergo mixocoely, or the dorsal coeloms only, has not been clarified sufficiently thus far. Irrespective of an eventual subdivision of the segmental coeloms, nephridial anlagen generally arise from either the ventromedian or the ventrolateral wall of the coeloms. Mayer, Ruhberg, and Bartolomaeus (2004) and Mayer (2006) demonstrated that in *E. biolleyi* only the nephridial duct is a derivative of the coelom. The duct anlage encloses against the mixocoel and in later stages forms the sacculus de novo out of the enclosed tip of the nephridial duct anlage. In this species the sacculus accordingly arises from a tubular precursor in an apparently similar mode as observed by us in the branchiopod *A. salina* (Bartolomaeus, Quast, and Koch, 2009). In onychophoran species with subdivided coeloms, in contrast, the ventral compartments were never explicitly stated to undergo mixocoely but instead seem to be transformed as such into the nephridial organ, with parts of the former coelomic lumen persisting within the prospective sacculus (Sedgwick, 1888).

Sedgwick's (1895: 23) view that "*the mesoderm seems to be formed entirely by the proliferation of the cells of the mesoblastic somites*" was confirmed by Mayer's (2006) ultrastructural studies on embryos of *E. biolleyi*. In this species apolar mesenchymal cells are stated to arise from coelothelial cells both before disintegration of the coelothel by detachment and migration below the coelomic lining, and after disintegration of the coeloms by epithelial-mesenchymal transformation of remnants of the coelothel. An (apparently) exceptional deviation from this mode was described by Pflugfelder (1948) for yolk-free embryos of *Paraperipatus amboinensis*. In this species, the early apolar segmental mesoderm is considered to partition into a median cluster that forms the coelom and a lateral cluster that remains apolar and directly gives rise to mesodermal blast cells. This observation is remarkable with respect to our results for arthropods, but it still requires confirmation.

Given this variation in mesoderm development among velvet worms current developmental research focuses on ovoviviparous species with yolky eggs (e.g. Eriksson and Tait, 2012) as ovoviviparity with yolky, rather than non-yolky eggs is considered the ancestral reproductive mode in Onychophora (Reid, 1996). Walker and Tait (2004) showed that the body segments of yolk-rich embryos in early stages are split into halves by dorsal and ventral extra-embryonic ectoderm, which is reminiscent of the median segment split of spider and centipede embryos during "inversion" or "lateral migration/spreading", respectively (see, e.g., Chipman, Arthur, and Akam, 2004; Brena and Akam, 2012; Mittmann and Wolff, 2012). Eriksson and Tait (2012) revealed for *Euperipatoides kanangrensis* that the hemisegments are already separate in the earliest embryonic stage of the "split germ band". The mesoderm is stated to primarily form one to two cell layered, "epithelia-like" strands that are surrounded by ECM ("basal lamina" fide Eriksson and Tait, 2012) and are apparently bordered against the ectoderm and entoderm by primary body cavities (see Eriksson and Tait, 2012, their Fig. 6). First coeloms seem to "bud" off the mesodermal strands as monolayerd vesicles. Presence or absence of additional apolar mesoderm in the first body segment with coelomic vesicles is not expressly stated by Eriksson and Tait (2012).

For immediate comparisons with our results on yolk-rich arthropod embryos we added yolk-rich embryos of *Cephalofovea clandestina* at stage IV fide Walker and Tait (2004) to our ultrastructural studies. At this stage the hemisegments are still separated by extra-embryonic ectoderm, dorsally to a greater extent than ventrally (Figures 11.40A, 11.41A). The segmental mesoderm at this stage is composed of coeloms, primary body cavities, and large masses of apolar mesenchymal cells (Figures 11.40A,C, 11.41A,B). Coeloms occur in all body segments but in at least the oral papilla segment and the anterior-most three trunk segments show no marked gradient in their development. In these segments the coeloms are mostly crescent-shaped (in transverse view) along their anterior-posterior extension (Figure 11.41A). Dorsoventrally they extend from above the level of the limb buds towards the dorsal transition of embryonic into extra-embryonic ectoderm. The somatic wall of the coeloms lies immediately beneath the epidermis, whereas the visceral wall is separated from the compact yolk-sac by minute primary body cavities and elongate visceral mesenchymal cells, some of which contain myofilaments. Visceral muscles also occur between the ventral neuroectoderm and the yolk-sac (Figures 11.40C, 11.41B). A ventromedial coelomic lobe does not exist. Dorsal to the coeloms, primary body cavities become quite large at the level of the limb buds and contain single, roundish apolar cells that may represent cardioblasts (Figure 11.40A, *asterisk*). Primary body cavities also occur intersegmentally and border consecutive coeloms. Somatic and visceral walls of the coeloms do not differ in thickness (Figures 11.40B, 11.41B). The coelothel forms a monolayer of cuboidal cells that along its dorsal extension slightly flattens and elongates. The coelomic lumen is largely compressed. In histological sections the lumen is recognizable only in the posterior ventralmost part of the coeloms where it enlarges within the limb buds (Figure 11.40B), but is otherwise either indistinguishable from

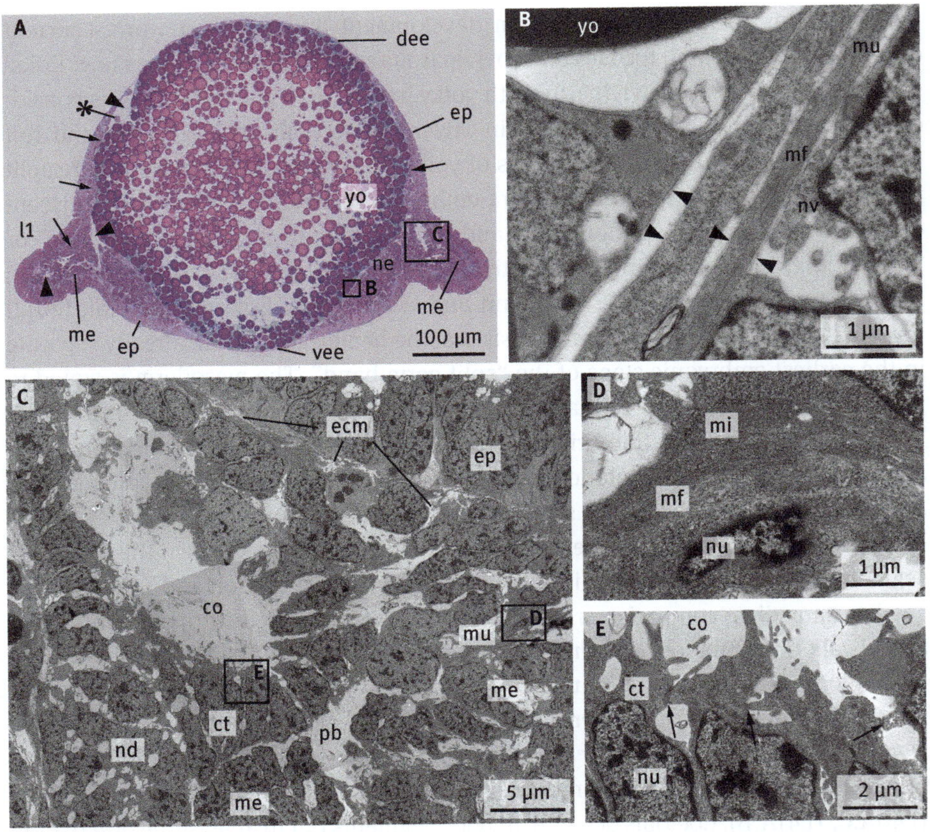

Figure 11.40: *Cephalofovea clandestina*, stage IV embryo *fide* Walker and Tait (2004). **A.** Transverse section (LM) through segment of first walking legs (*l1*). The coeloms (*arrows*) have a narrow lumen that widens only at the limb base. The flat dorsal coelomic lobe is sandwiched between epidermis (*ep*) and the yolk cells of the gut anlage (*yo*). Another flattened coelomic lobe extends into the limb buds below their dorsal wall. Their ventral wall is covered by apolar mesenchymal cells (*me*) extending into the tip of the limb buds as well as to their dorsal wall behind the appendicular lobe of the coelom. Small primary body cavities (*arrowheads*) border the coelomic walls against apolar mesoderm and the gut anlage. Note single mesodermal cells within dorsal primary body cavities (*asterisk*). Ventrolaterally, the neuroectoderm (*ne*) forms most of the tissue aside the gut anlage. **B–E.** Transverse sections (TEM) at the level of the limb base of the prospective first walking legs. **B.** Thin circular muscle strands (*mu*) extend between gut anlage (*yo*) and neurites (*nv*) of the neuroectoderm. Extracellular matrix (*arrowheads*) borders the yolk cells and entirely surrounds the muscle cells. **C.** Detail of limb base shown in A. The coelomic cavity (*co*) is lined by cuboidal or columnar coelothelial cells (*ct*). Ventrolaterally apolar mesenchymal cells (*me*) border against the somatic coelomic wall. Ventromedially the coelothelium continues into the nephridial duct (*nd*). **D–E.** Details of **C**, showing (**D**) apolar mesenchymal cells with myofilaments (*mf*) in the tip of the limb anlage, and (**E**) coelothelial cells (*ct*) apically interconnected by adherens junctions (*arrows*). *dee* dorsal extra-embryonic ectoderm, *ecm* extracellular matrix, *mi* mitochondrion, *nu* nucleus, *nv* nerve cell, *pb* primary body cavity, *vee* ventral extra-embryonic ectoderm, *yo* yolk cells of gut anlage.

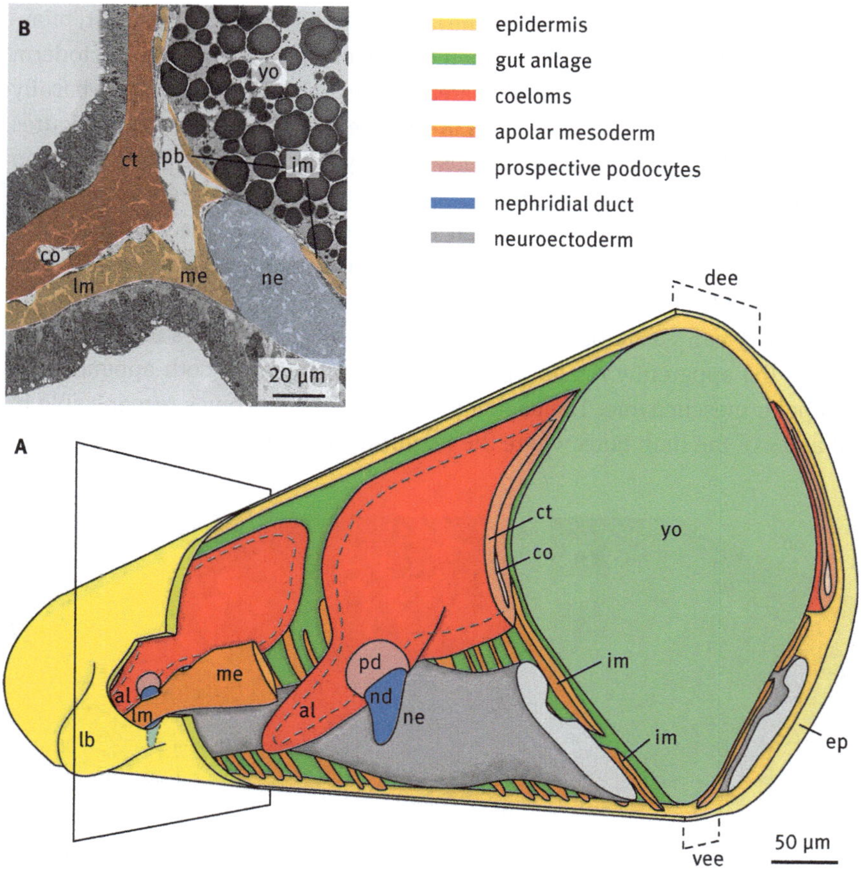

Figure 11.41: *Cephalofovea clandestina*, stage IV embryo fide Walker and Tait (2004). **A.** Schematic reconstruction of anterior trunk segments. Epidermis (*ep*) and parts of the apolar mesenchyme removed on right half of the body, frontal is right. Coeloms (*co*) border dorsolaterally against the yolky gut anlage (*yo*). A ventrolateral apendicular lobe (*al*) of the coeloms fills the dorsal part of the corresponding limb bud (*lb*). The coelomic cavity (outline indicated by *dotted lines*) widens only at the limb base, but is flattened or completely compressed in remaining areas. A longitudinal strand of apolar mesenchymal cells (*me*) (cut off in anterior segment) lies underneath the coeloms and also extends into the limb bud where it fills its ventral part (*lm*). Muscle cells are most prominent in the appendicular extension of this strand of apolar mesenchymal cells. Thin strands of transverse muscles (*im*) extend ventrolaterally between the neuroectoderm (*ne*) and the yolky gut anlage. Dorsolaterally they contact longitudinal strands of apolar mesoderm devoid of myofilaments. Nephridial duct anlagen (*nd*) originate from the frontal side of the ventrolateral lobe of each coelom and end blindly within the ventral epidermis adjacent to the limb base (*arrow*). The coelothelium shows specialized cells (*pd*) around the opening of the nephridial duct that may represent prospective podocytes. **B.** Transverse section (TEM) through limb base at the level indicated in A. Mesodermal tissues and neuroectoderm transparently colored according to the reconstruction shown in A. *al* apendicular lobe of coelom, *ct* coelothelium, *dee* dorsal extra-embryonic ectoderm, *im* transverse muscles at gut anlage, *lm* mesenchyme of limb bud, *me* longitudinal strand of mesenchyme, *vee* ventral extra-embryonic ectoderm, *yo* yolk cells of gut anlage.

primary body cavities or invisible. Within the limb buds the coeloms form a flat, bilayered appendicular lobe with a tubular lumen that extends along the dorsal ectoderm of the limb buds up to their tip. Remaining areas of the limb bud epidermis are basally underlaid by apolar mesenchymal cells that include somatic muscle cells (Figures 11.40B, 11.41A,B). Thin strands of apolar cells with myofilaments also occur between the appendicular coelomic lobe and the dorsal limb bud epidermis. A primary body cavity centrally separates the appendicular coelomic lobe from the apolar mesenchyme (Figures 11.40B, 11.41B). In the oral papilla anlagen the prospective ectodermal slime gland protrudes into this central cavity (Figure 11.42A). In the trunk segments the central cavity can easily be misidentified in histological studies as a coelomic cavity due to the apparently regular, peripheral arrangement of both appendicular lobe and apolar mesenchyme. The coelomic lumen proper becomes recognizable in proximal areas of the limb buds where the nephridial anlage opens into the coelom

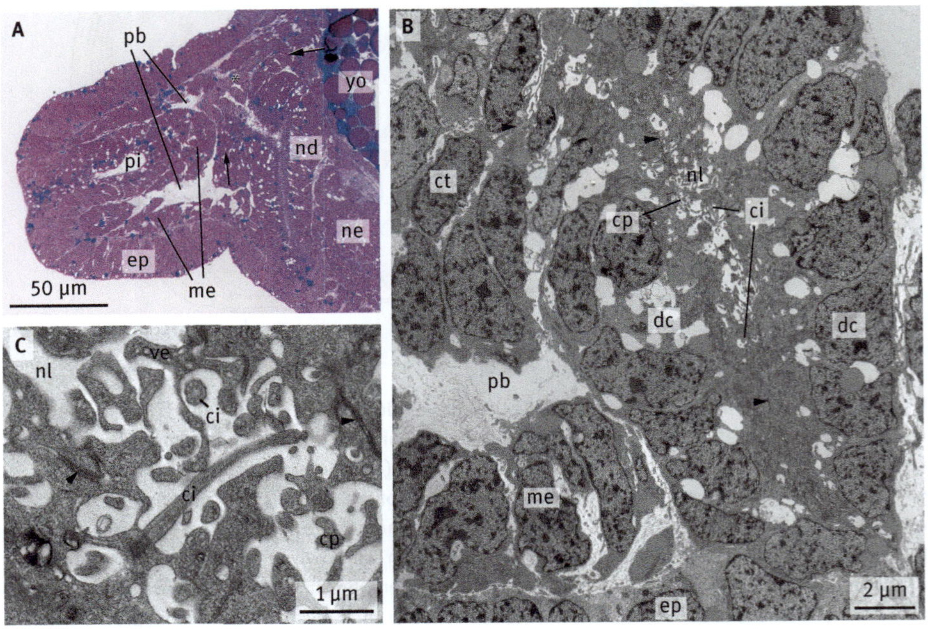

Figure 11.42: *Cephalofovea clandestina*, stage IV embryo fide Walker and Tait (2004), nephridial anlage. **A.** Transverse section (LM) through anlage of right oral papilla. The nephridial duct anlage (*nd*) is medioventral to the limb base. Around the duct's opening into the coelomic cavity (*asterisk*) the coelothelium remarkably differs in cell size and lighter staining from the otherwise thick lining cells of the coelom (*arrows*). **B.** Transverse section (TEM) through nephridial duct anlage in the prospective first walking legs, right side. The duct cells (*dc*) are interconnected by adluminal adherens junctions (*arrowheads*). **C.** Detail of the lumen of the nephridial duct anlage shown in B. Cilia (*ci*) and cell protrusions (*cp*) project into the lumen (*nl*), while adluminal adherens junctions (*arrowheads*) interconnect the duct cells. *ep* epidermis, *me* apolar mesenchyme, *ne* neuroectoderm, *pb* primary body cavity, *pi* epidermal invagination of oral papilla, *ve* vesicles, *yo* yolk cells of the gut anlage.

(Figure 11.42A). The nephridial anlage forms a short, ciliated straight duct extending towards the ventral epidermis (Figure 11.42A,B). In the oral papilla segment the duct already has an external opening in the ventral epidermis, whereas in subsequent body segments the nephridial ducts end blindly within the epidermis. Around the opening of the nephridial duct anlage into the coelomic lumen the coelothel shows specialized cells that differ from remaining coelothelial cells in their shape and staining properties (Figure 11.42A). We doubt that these specialized cells represent the presumptive funnel of the nephridial duct as they do not show the dense ciliation of the duct cells. The specialized cells of the coelothel instead are reminiscent of the presumptive podocytes in the chelicerate embryos studied by us, in being less electron-dense and equipped with several vesicles and finger-like cell protrusions.

The coeloms of the jaw segment are more advanced and they are remarkably large in the antennal segment. In these head segments the coelomic lining is extremely flattened, such that perikaryeal regions form marked cell protrusions. In the jaw segment the coelomic lumen is slightly larger within the limb base than in subsequent body segments, but is similarly compressed within the appendicular and dorsal lobes. The antennal coeloms show no areas of compression but form large cavities at the base of the antennae, with deep extension into the antennal anlagen and caudal extension along the pharynx anlage. A large primary body cavity with single apolar cells separates left and right coeloms along the dorsal midline. Apolar mesenchyme basally surrounds the epidermis of the pharynx anlage, which ends blindly against the yolk-sac within the jaw segment. Nephridial anlagen are absent in the antennal segment. In the jaw segment the nephridial anlagen are less advanced than in subsequent body segments.

The coexistence of coeloms, large primary body cavities, and apolar mesenchyme (including muscle cells) in early embryonic stages of *Cephalofovea clandestina* is immediately comparable to the mesoderm development in *Limulus polyphemus* and *Pholcus phalangioides*. The main differences of the coeloms in embryos of *C. fovea* concern the formation of appendicular lobes, absence of ventromedian lobes (at least before 'ventral closure'), and presence of specialized cells around the opening of the nephridial anlage that resemble presumptive podocytes. The source of the early apolar mesenchyme in *C. fovea* could not be clarified by our studies, as in the arthropod embryos studied by us we found no anatomical evidence that the apolar mesenchyme is derived from the coelothel. The overall resemblance of the segmental mesoderm during coelomogenesis in *C. fovea* and chelicerate embryos is nevertheless remarkable and calls into question whether previous insights inferred from yolk-free embryos of *Epiperipatus biolleyi* are representative for Onychophora. Current insights indicate that the mesoderm development in *E. biolleyi* underlies accelerated mixocoely (with respect to limb bud formation), but we cannot exclude that correspondences observed in yolk-rich embryos of onychophorans and arthropods are merely caused by the high yolk content. Shifts in timing of developmental events (i.e. heterochrony; Mayer, 2005) may explain the developmental plasticity of the mesoderm development in velvet worms, but the direction of evolutionary change still remains to be resolved.

11.7 Conclusions

The theory of evolution predicts graded similarity between organisms due to shared information. Similarity, however, has two possible explanations in evolutionary biology, transformation or convergence. To substantiate homology hypotheses, additional assumptions on the transformation are required that explain structural differences. As a general rule complexity of structure and number of identical substructures are used to estimate the support for transformation, i.e. for homology. Primary homology hypotheses for similar structures are accordingly the weaker supported the less complex the structures are.

Coeloms and nephridia in annelids and panarthropods are such less complex structures, since at the comparable developmental stages structural diversity of the coelomic lining and details in the cellular composition of nephridia are low. Our studies revealed that the anatomical body of evidence in support of homology is indeed poor and presently remains restricted to three basic correspondences:
(i) the (at least primarily) segmental arrangement of coeloms and nephridia,
(ii) their origin each from one pair of anlagen per body segment, and
(iii) the formation of coelomic cavities by schizocoely after epithelialization of the segmental mesoderm.

In any other details of their anatomy in the course of their development and spatio-temporal relationships these organs show marked differences:
(1) *Earliest step in coelomogenesis: epithelialization of the mesoderm*. In pre- and postmetamorphic stages of annelids the mesoderm is already epithelialized in the posterior growth zone, whereas in arthropods the segmental mesoderm is initially apolar. The state in onychophorans remains unclear, as does the anatomy of the initial embryonic segmental mesoderm in clitellates among Annelida.
(2) *Initial differentiation of the coelomic walls*. In annelids the prospective coelothel is already differentiated into a myoepithelium before the coelomic cavity develops; only in later stages, if at all, may the initial myoepithelium partially transform into a peritoneum. In onychophorans and arthropods, in contrast, the coelomic walls are peritoneal (i.e. devoid of myofilaments) throughout their existence. While remnants of the coelothel seem to undergo epithelio-mesenchymal transition in (some) onychophorans, the fate of the coelomic lining after mixocoely remains unclear in arthropods.
(3) *Extent of mesodermal epithelialization*. In both pre- and postmetamorphic stages of annelids all mesodermal cells within the cluster of a hemisegment (maybe to the exception of the nephridioblasts, the origin of which still is unclear) partici-

pate in the formation of the coelomic wall. Whether this also holds true for the embryonic development of coeloms in clitellates requires confirmation by ultrastructural studies. Among panarthropods, only yolkless onychophoran embryos seem to correspond in the apparently entire epithelialization of the segmental mesoderm. In arthropod embryos with transient coeloms, the segmental mesoderm partially remains apolar, with only a variable subset of the initially compact cluster of apolar cells per hemisegment participating in coelom formation. The same state as in arthropod embryos with transient coeloms may occur in yolk-rich onychophoran embryos, but the origin of the early apolar segmental mesenchyme (including musculature) coexistent with the early coelom within a hemisegment still remains unclear. Absence of segmental coeloms in arthropods such as pycnogonids, collembolans, or crustaceans, is due to entire omission of epithelialization of the segmental mesoderm.

(4) *Segmentation of the early mesoderm.* In pre- and postmetamorphic stages of annelids intersegmental subdivision of the mesoderm, i.e. formation of septa, is triggered by the subepidermal and peri-intestinal ECM transversely expanding into and coalescing within the myoepithelial cell strands. Omission of ECM formation leads to unsegmented coeloms. In arthropods the ECM plays no obvious role in segmenting the early mesodermal cell strands. Whether ECM triggers mesoderm segmentation in clitellate embryos and onychophorans remains unclear.

(5) *Formation of mesenteries.* Neither intersegmental septa nor mesenteries form in panarthropods, whereas both structures can be assumed to represent ancestral features of annelids. Formation of mesenteries in postmetamorphic stages of annelids is triggered by ECM that forms de novo between basally opposing myoepithelial cells of the prospective coelomic lining and becomes continuous with the peri-intestinal ECM and the subepidermal ECM. Omission of ECM formation accordingly prevents medial separation of left and right coeloms (i.e. mesenteries are incomplete or entirely absent). The same concerns the ventral mesenterium in premetamorphic stages, whereas the dorsal mesenterium, due to the dorsal expansion of the coeloms, forms from the initial (subepidermal and peri-intestinal) ECM between epidermis and gut that becomes squeezed in between the adjoining dorsal coelomic lobes.

(6) *Formation of the coelomic cavity.* In annelids the coelomic cavity enlarges during fluid accumulation by gradual degradation of desmosomes interconnecting the apical cell membranes of opposing epithelio-muscle cells. In panarthropods such desmosomes do not exist between opposing cells of the prospective somatic and visceral walls of the coeloms, which likely correlates with the lack of intermediate filaments in panarthropods. In yolk-rich embryos of onychophorans and chelicerates the coelomic cavities widens only partially, apparently with a preference for areas where the pressure from yolk is less strong. In the remaining areas, the somatic and visceral walls of the coeloms remain in close opposition but do not

show apical, intercellular junctions. In insects, the pressure from yolk does not seem to influence the dimension of the coelomic cavity.

(7) *Anterior-posterior gradient in coelomogenesis*. A gradual development of the coeloms along the anterior-posterior body axis is strictly conserved in segmented annelids, which allows examination of the entire sequence of coelomogenesis with a single specimen at any larval or juvenile stage. Among panarthropods, a comparable anterior-posterior gradient in the development of the coeloms, including their earliest state up to their degradation (in terms of mixocoely), is restricted to some embryonic stages in onychophorans before segment formation in the posterior growth zone ceases. In arthropods, an anterior-posterior gradient in the development of the segmental mesoderm as such may be present but is not reflected in coelomogenesis. The development of the coeloms of arthropods is either synchronized in body tagmata, or variably retarded (or accelerated, respectively) or entirely suppressed in some body segments such that the sequence of coelom formation from anterior to posterior is asynchronous or (apparently) random. Examination of coelomogenesis in arthropods is accordingly much more elaborate than in annelids as the entire developmental sequence can never be inferred from a single embryo (or juvenile stage, respectively).

(8) *Dorsal expansion of coeloms*. Dorsalward coelomic growth is restricted in annelids to premetamorphic and embryonic stages, in which the early segmental mesodermal cell clusters initially encompass the gut only ventrolaterally. In postmetamorphic stages, the initial, still "acoelomate" segmental mesoderm encompasses the gut entirely and coelomic "growth" along the dorsoventral axis is restricted to the coelomic cavity that gradually widens from ventral to dorsal (by degradation of the desmososmes interconnecting opposing coelothelial cells). Yolkless embryos of onychophorans resemble the postmetamorphic mode of coelomogenesis in annelids in that the initial segmental mesoderm already encompasses the majority of the dorsoventral space between gut and epidermis, such that only the dorsomedian and ventromedian areas of the gut are not (and never become) surrounded by the coeloms. In yolk-rich embryos of onychophorans and arthropods the initial segmental mesoderm is restricted to ventrolateral areas. When the germ band(s) starts to overgrow the yolk-sac the coeloms follow this movement by expanding dorsally, similarly as in clitellate embryos. Dorsal expansion of the coeloms has accordingly been considered to correlate with high yolk content (e.g. Dohle, 1999). A correlation of this kind, however, is questioned by yolkless clitellate embryos, in which dorsal expansion of the coeloms is maintained despite absence of a yolk-sac, and by insects in which a marked dorsal expansion of the coeloms does not occur despite presence of a bulky yolk-sac.

(9) *Formation of the circulatory system*. In annelids all blood vessels arise from fluid accumulation within ECM under formation of primary body cavities. Blood vessels can accordingly occur in any spaces of the ECM, irrespective of the coelomic lining. In arthropods, in contrast, the circulatory system (including the heart)

generally seems to be formed by apolar mesodermal blast cells migrating in the primary body cavity. The traditional assumption that the cardioblasts of panarthropods are derived from the coelomic walls still lacks unambiguous evidence. Details on heart formation in onychophorans are still missing.

(10) *Myogenesis*. In annelids the entire segmental musculature is derived from the coelomic walls, which initially form monolayered myoepithelia and variably become pseudostratified or stratified. Among panarthropods, transient pseudostratification or stratification of the coelomic lining has, as yet, only been evidenced for the somatic walls in yolkless onychophoran embryos. The stratification of the coelomic wall may be considered as evidence for the assumption that in onychophorans myoblasts derive from the coelomic lining by delamination. This assumption is otherwise only inferred from the fact that the entire segmental mesoderm initially seems to become integrated into the coelomic lining, i.e. for the apparent lack of any other source for myoblasts. Before the coeloms disintegrate in onychophorans, the coelothel itself does not transform into a myoepithelium. In arthropods, the coelomic walls also form a monolayered peritoneum throughout their existence. The traditional view that in arthropods myoblasts originate from the coelomic walls still lacks unambiguous evidence.

(11) *Origin and position of the nephridial anlage*. The anlagen of segmental nephridia develop in prospective intersegmental position in annelids and are separate from the prospective coelomic lining in their earliest recognizable stage. In panarthropods the nephridial anlagen have a segmental position in their earliest recognizable stage and at least primarily (as shown by onychophorans and chelicerates) are immediately derived from the coelomic lining. The intersegmental position within the prospective septa and the specific manner in which ECM surrounds the nephridial anlage causes the anlage in annelids to become transsegmental, i.e. to involve (at least) two different body segments in their development, such that the (eventual) opening of the nephridial duct into a coelom lies in a different (preceding) body segment than its external opening. In panarthropods, the external opening of the nephridial duct develops in the same body segment as the segmental mesoderm that gives rise to the nephridial anlage, irrespective of whether coeloms are formed within this mesoderm. Among panarthropods, no instance of body segments with transient coeloms is known in which the nephridial anlage and the coeloms arise from separate anlagen.

(12) *Timing of nephridiogenesis*. In annelids the anlagen of segmental nephridia are already recognizable before the onset of schizocoely in the mesoderm surrounding the nephridial anlage. In panarthropods the nephridial anlagen become recognizable at considerably later stages in the development of the segmental mesoderm, in species (or body segments) with transient coeloms shortly before mixocoely (i.e. degradation of the coelothel).

(13) *Podocytic organs*. In annelids podocytes are formed by the coelomic lining by transformation of epithelio-muscle cells and never get in contact with the

nephridial anlage. Previous assumptions that protonephridia in annelids may also give rise to podocytes (by conversion of solenocytes) are refuted by our results. In panarthropods, the sacculus of the nephridial organs in some instances, including yolkless onychophoran embryos and crustaceans, develops in a similar manner as segmental protonephridia of annelids in that the most proximal cells of the nephridial anlage transform into a filtering unit. In contrast to annelids, however, the filtering unit of panarthropods consistently develops into a podocytic organ. In chelicerates the anlagen of the podocytic organs are initially separate and independent of the nephridial anlage and of the coeloms that give rise to the nephridial anlage. Yolk-rich embryos of onychophorans presently provide the only evidence that podocytic organs (i.e. the sacculus of the nephridial organs) in panarthropods may originally arise from coelomic linings. The supposed prospective podocytes within the coelomic lining of yolk-rich onychophoran embryos, however, differ from those in annelids in their relative position and immediate connection to the prospective nephridial funnel, as well as in their independence from blood vessels.

(14) *Gonad formation.* In annelids germ cells are usually released to the exterior via the coeloms, whether or not discrete gonads occur, whose location and number are highly variable (see, e.g., Eckelbarger, 2005). At the ultrastructural level, primordial germ cells (PGCs) are not unambiguously recognizable in the early segmental mesoderm. In onychophorans and insects, PGCs proved to become transiently associated with the segmental coeloms but they are never released into the coeloms. In yolkless embryos of onychophorans, PGCs arise at the caudal end of the embryo and become attached retroperitoneally to the visceral coelomic walls, portions of which in the genital and few pregenital segments later (apparently after disintegration of the coeloms) contribute the gonadal wall (Mayer and Tait, 2009). In the insects studied by us, putative PGCs arise within the early segmental mesoderm of many body segments (in silverfish embryos including the posterior head segments) and initially become transiently integrated into the visceral coelomic linings. Later the PGCs aggregate into paired genital ridges, which become restricted to some abdominal segments; whether the coelomic lining contributes to gonad formation any further remains unclear. A general transient association of PGCs to the coeloms in panarthropods, however, is not corroborated by our results for chelicerates and crustaceans, which instead corroborate previous assumptions on a late mesodermal origin of PGCs (see, e.g., Extavour and Akam, 2003), i.e. in chelicerates not before degradation of the segmental coeloms.

Homology of segmental coeloms and nephridia in annelids and panarthropods can accordingly be hardly substantiated from the viewpoint of comparative anatomy, nor from insights on their development inferred from comparative anatomy. Under either the Articulata or the Ecdysozoa hypotheses homology of these organs implies considerable anatomical and developmental transformations in either the annelid ancestor, or in the panarthropod and the arthropod ancestors. One may concede that recent arthropods in comparison to onychophorans are too strongly derived and too variable in in the mesodermal organization as to shed light on this question. This particularly concerns current doubts raised by our studies that the coeloms in arthropods (still) represent major mesoblastic organs. The remarkable variation of organogenesis in onychophorans, however, presently hinders reconstruction of putatively ancestral states. Some differences in yolk-rich embryos of onychophorans and arthropods superficially resemble similar changes of the mesoderm development in clitellate embryos. Although this resemblance is doubtless due to convergence and apparently correlated with high yolk content, a more profound anatomical understanding of the mesoderm development in clitellate embryos may reveal correspondences that render convergent transformation in panarthropods more likely.

The tremendous anatomical and developmental differences between annelids and arthropods seem to be due to the apparent dispensability of transient coelom formation in the latter, as corrobated by the multiple instances of entire suppression of coeloms among arthropods. We accordingly suppose that the involvement of coeloms in the development of the mesoderm in panarthropods is primarily correlated with a lobopodian body organization, whereas arthropodization and/or arthrodization (see, e.g., Liu et al., 2011) coincided with a loss of the functional significance of the embryonic coeloms. This hypothesis centers the ongoing discussion of the evolution of coeloms in the context of body segmentation basically on the question on why the segmental mesoderm has to to pass a coelomic stage in lobopodians at all.

The derivation of the lobopodian body organization from a coelomate ancestor still provides the most reliable explanation, i.e. inheritance. The Ecdysozoa-hypothesis, in contrast, favours the more parsimonious scenario that the panarthropods evolved from an acoelomate ancestor. Our findings provide additional support for the Ecdysozoa-hypothesis as they demonstrate low degree of correspondence in characters traditionally considered as synapomorphic of annelids and panarthropods, which questions the homology of these characters. Epistemologically, however, we cannot conclude from differences on convergence – such conclusions can ultimately be drawn only from a congruence test. Since molecular analyses congruently support Ecdysozoa, the interpretation of the differences in the organization, formation and fate of the coeloms and nephridia between annelids and panarthropods as a result of convergent evolution is more likely at present. A caveat against this scenario, however, still is that our current insights provide no plausible explanation for an independent gain of transient embryonic coeloms in Panarthropoda.

Acknowledgments

We thank the German Research Foundation for continuous support within the scope of the *Deep Metazoan Phylogeny* priority program (DFG-projects BA 1520/9-1, BA

1520/9-2, BA 1520/9-3). Many thanks are due to Chiharu Kato, Ed Enos and Alison Leschen for sampling and rearing *Limulus polyphemus* at the Marine Biological Laboratory Woods Hole. Noel Tait is greatly acknowledged for his help in collecting gravid females of *Clandestina cephalofovea* in the Blue Mountains of New South Wales, and Jenny Norman for providing lab space and support in the fixation of the onychophoran embryos at the Electron Microscope Unit of The University of New South Wales. We thank Georg Mayer (Univ. Leipzig, Germany) for discussion of our results for *C. cephalofovea*. Robert Woltmann contributed studies by TEM on *Nymphon gracile* within the scope of his Diploma thesis, and Liesa Rütjes for *Pholcus phalangioides* within the scope of her Diploma thesis. Technical support for our studies on *Tisbe* sp. by Tatjana Bartz are likewise acknowledged. The reviewers are thanked for valuable suggestions on our manuscript.

J. Wolfgang Wägele and Patrick Kück

12 Arthropod phylogeny and the origin of Tracheata (= Atelocerata) from Remipedia-like ancestors

Abstract: This review summarizes some major events in the evolution of body plans along the backbone of the arthropod tree, with a special focus on the origin of insects. The incompatibility among recent molecular phylogenies motivates a discussion about possible causes for failures: there is a worrisome lack of information in alignments, which can be visualized with spectra of split-supporting positions, and there are systematic errors occurring even when using correct models in maximum likelihood methods (Kück et al., this book). Currently, these problems cannot be avoided. Combining information from the fossil record and from extant arthropods, the morphology-based evolutionary scenario leads from worm-like stem-lineage arthropods via first euarthropods to the crown group of Mandibulata. The evolution of the mandibulate head is well documented in the Cambrian Orsten fossils. The evolution within crustaceans is also the evolution that leads to characters of the bauplan of myriapods and insects. It is argued that morphologically myriapods do not fit to the base of the mandibulatan tree and that this placement is also not plausible from a paleontological point of view. Available morphological evidence suggests that myriapods are the sister-group to Hexapoda and that tracheates evolved from a marine ancestor that was similar in many ways to Remipedia. In the extant fauna, the Remipedia are the sister-group of Tracheata.

12.1 Introduction

It is beyond the scope of this chapter to summarize the fossil record and to review the literature published on the phylogeny within different arthropod taxa. While the following chapters discuss important aspects of the morphological evolution and molecular phylogenies inferred for subgroups of Arthropoda, this overview deals mainly with the relationship between the large and well-discernible monophyla Chelicerata, Myriapoda, Insecta, and groups of crustaceans.

A major concern is the conflict between published hypotheses on animal evolution. There are still strong contradictions between the available (and frequently ignored) morphological evidence and molecular tree topologies. This conflict cannot be disregarded. Some important sources of error still remain undetected and there are currently too few attempts to discover the mechanisms that mislead our analyses. We therefore discuss briefly some aspects of the theory of phylogenetics.

To highlight the nature of contradictions, the case of the Tracheata hypothesis and the question of the origin of insects are discussed in greater detail.

12.2 Avoidance of misconceptions

Before we discuss the integration of available observations into an evolutionary scenario, we have to point out that several misunderstandings have been obfuscating the view on the larger phylogeny of arthropods. These include an overvaluation of molecular data, arguments against the homology of varying or adaptive characters, and the role of complexity in homologization of similarities.

12.2.1 Phylogenies obtained from different genes are not independent evidence

At this point it is remarkable that authors tend to believe that phylogenies obtained from different genes are independent evidence, for example, when the basal placement of myriapods within Mandibulata is found in different analyses. It is a fact that genes selected from the same taxa are **samples from the same genome** and the same phylogeny. All these genomes went through the same historical processes and are imprinted by the same rapid or slow evolution, by population bottlenecks and rapid radiations. Therefore, systematic errors caused by branch length ratios (see Kück et al., 2012) should be found independently of gene selection. They occur due to critical branch length relationships in the true history of lineages. When Kusche et al. (2003) described that hemocyanin genes are evidence for a closer relationship between crustaceans and insects, excluding myriapods, they sampled the same genome patterns as e.g. Regier, Shultz, and Kamble (2005) or Dunn et al. (2008).

12.2.2 Adaptation is no argument against homology

"Most of the presumed synapomorphies (...) [of insects and myriapods] are clearly adaptations to terrestrial life and, therefore, the possibility of them arising by convergence cannot be ruled out" (Averof and Akam, 1995a: 299). This argument is not relevant, even though it has been repeated many times. The probability of homology does not depend on the adaptive value of a character: probably most phylogenetically important characters are adaptations (e.g. feathers and wings of birds, the compact skull and beak of turtles, the suckers of leeches, book lungs in Arachnida).

Adaptation of an organ means that it evolved for a specific function. Function is no argument in favor of or against hypotheses of homology. Homologous organs frequently change function. For example, mandibles can be used for chewing, piercing, digging, or are exclusively used as defense organs (many ants, stag beetles). To refute a homology hypothesis it must be shown that the structural similarity is only superficial, that there is no shared identity of details supporting the homology hypothesis, or that the structures have different genetic or phylogenetic origins (an a posteriori argument independent of character quality). Such arguments support the convergent evolution of body shape in dolphins and sharks. On the other hand, to substantiate homology it must be shown that shared structural or genetic complexity cannot be explained as chance similarity.

12.2.3 Co-occurrence of characters increases the probability of homology

In discussions of characters of the Tracheata by proponents of the Pancrustacea hypothesis usually only few anatomical features are mentioned (e.g. tracheal system, Malpighian tubules, absence of second antennae). The estimation of the probability of homology is then restricted to the argument that each character could be an adaptation, and it is said that adaptations are unreliable characters.

We want to point out that the probability of homology increases with the number of details shared in two body plans (Wägele, 2005). If in a pure stochastic world a character X has the probability $\mathbf{P_x}$ to evolve along a lineage, the probability that it is found in two lineages by chance is $\mathbf{P_x}^2$. If two lineages share six different characters A–F, the total probability P is much lower than for each single character, namely:

total probability of a pattern A–F: $\mathbf{P = P_A{}^2 * P_B{}^2 * P_C{}^2 * P_D{}^2 * P_E{}^2 * P_F{}^2}$

Even if the probability for the convergent evolution of a single adaptation is estimated to be high, the fact that such a character occurs simultaneously with many other characters in two different body plans increases the probability of homology drastically for both, the single detail and the complete body plans. This is also true when we compare body plans of insects and myriapods.

12.2.4 Variation is no argument against homology

Homologous characters can vary. It is generally accepted that the various shapes of insect mandibles or of tetrapod appendages do not contradict homology of mandibles or of tetrapod limbs. In other cases, variability has been used as an argument. For example, a movable tooth-like process on mandibles, the *lacinia mobilis*, can be found below the incisor process in several crustaceans, including Remipedia, and in Symphyla and several hexapods. Because there are variations in shape and size, Richter, Edgecombe, and Wilson (2002) propose that this structure evolved five times convergently. In a similar way, the variation in tracheal systems has been used to argue against a common origin of the respiratory system of Tracheata (Kraus and Kraus, 1994; Dohle, 1997; Hilken, 1998; Kraus, 2001, see discussion below). These arguments are only selectively applied and have no logical basis. The case of the tracheae is similar to a putative discussion of non-homology of insect mandibles. Evidence for homology cannot be refuted by pointing only to variations. As already mentioned, to argue against homology other reasons are needed.

Of course, a hypothesis of homology has no basis when no complex similarities exist or when a well-founded tree-topology indicates parallel evolution in different lineages (a posteriori argument not based on character quality, Wägele, 2005). The same arguments are valid for the tracheal system. In view of the structural details shared by insects and myriapods (see below), the primary assumption is the existence of shared genetic information inherited from a common ancestor.

12.3 Early arthropod evolution

12.3.1 What are arthropods?

Most arthropods are easily identified. They have a rigid exoskeleton composed of extracellular material (alpha-chitin, proteins such as resilin, sometimes carbonates) which is unique among living organisms. Arthropods have many appendages arranged in segmental pairs, with articles separated by elastic joints. This basic equipment makes it possible to build a large variety of mouthparts, legs, paddles, gills, different types of tools (e.g. scissors, forceps, pliers, daggers, fans, palps, brushes, sieves), and even wings. This variability explains why arthropods have the largest number of species among living animals. Internally, arthropods have the same basic anatomy as annelids: a ventral nervous system with segmental ganglia, a brain with mushroom bodies located above the esophagus region (see Loesel in this book), dorsally a longitudinal heart, paired segmental nephridia, a development that is originally anamorphic with new segments added in a preanal region. The earliest nauplius-like larvae are composed only of anterior head segments and the last segment carrying the anus, as many annelid larvae.

Some taxa are highly derived and cannot easily be identified as being arthropods. Adult parasitic Rhizocephala, for example, which live on or in other crustaceans, have no segmentation and no appendages, and only their larvae show that they belong to the Thecostraca (e.g. Høeg et al., 2009a). Another group of parasites occurring in vertebrates, the **Pentastomida**, lack such larvae and are difficult to place in the tree of life. They were thought to be a link between Cycloneuralia and Arthropoda (de Oliveira Almeida, Christoffersen, de Sousa Amorim, 2008), possible stem-lineage representatives of Euarthropoda (Waloszek, Repetski, and Maas, 2005), or – according to their sperm ultrastructure and placement in molecular phylogenies – they could be within crustaceans the sister-group of the ectoparasitic **Branchiura** (e.g. Møller et al., 2008).

Whether **Tardigrada** and **Onychophora** should be included in a taxon Arthropoda is a matter of definition and tradition. If they are included, they can be separated as “prot-“ or “pararthropods” from Euarthropoda. Waloszek, Maas, Chen et al. (2007)

call them with good reasons "stem arthropods" together with a series of lobopodian fossils. Both more worm-like taxa had instable positions in molecular phylogenies. Tardigrades appeared close to nematodes in some molecular analyses (e.g. Dunn et al., 2008, probably an artifact, there exist no substantial morphological arguments for a placement of lobopods as sister-group to Cycloneuralia) or close to euarthropods (e.g. Campbell et al., 2011). Onychophorans and Tardigrada are more often placed as sister-taxa to euarthropods based on morphology and molecular data (e.g. Campbell et al., 2011; Haug, Rota-Stabelli et al., 2011). We find in both taxa internalized mouthparts that seem to be derived from a pair of appendages. With their elastic cuticle and short unsegmented legs (lobopods with claws) both taxa bridge the gap between annelids and arthropods. The dwarfish tardigrades have a simplified anatomy, but onychophorans show some typical arthropod characters, such as the dorsal ostiate heart with a pericardial space separated from the body cavity by a transverse membrane or the segmental nephridia with sacculus. Both protarthropods fit well to a series of marine fossils known as paraphyletic "Lobopodia", which show a stepwise transition from soft-bodied animals with lobopods to armored arthropods. Many authors count tardigrades and onychophorans as extant lobopods (e.g. Waloszek et al., 2007; Haug et al., 2012a and further references therein).

In the following we use the terms crown-group Arthropoda or Euarthropoda that exclude tardigrades and onychophorans.

12.3.2 Early steps in arthropod evolution

Since the origin of arthropods is still hotly debated (Ecdysozoa versus Articulata hypotheses), the assumed number of steps required to build a first arthropod are very different. Starting from annelids the basic anatomy is already there, namely the coelomic segmentation, the anameric development with preanal segment addition, nephridial organs, a complex circulatory system with a dorsal heart, a complex brain with mushroom bodies, ventral segmental ganglia, segmental appendages with innervation and musculature. Starting with a cycloneuralian, all these structure have to evolve convergently to annelids, or we must assume that the first bilaterians were already highly complex animals that evolved from scratch (i.e. from a cnidarian-like anatomy, which is equivalent to first building a Porsche to later invent the donkey cart).

Taking the more parsimonious solution we start with an annelid-like ancestor. The novelties we need in this case to build a first **stem-lineage arthropod** or protarthropod are (see also Waloszek et al., 2007):

Character 1: Appendages uniramous, unjointed, tubular, ventrolaterally directed, possibly with a pair of terminal claws (in contrast to parapodia).

Character 2: Open circulatory system with a dorsal ostiate heart pumping hemolymph frontally (plesiomorphic state: a closed circulatory system, heart without ostiae).

Character 3: As a consequence of Character 2: closed nephridial sacs (instead of open nephridial funnels).

Assuming that these animals already had a first antenna or an equivalent appendage in preoral position, they also would have had a brain composed of proto- and deutocerebrun. This is the level of organization of the lobopods, including tardigrades and onychophorans.

12.3.3 Evolution of the first euarthropod

Waloszek et al. (2007) call this the "second phase in arthropod evolution". New features are:

Character 4: A pair of compound eyes in addition to single median ommatidia.

Character 5: A uniramous first limb (called antennula in mandibulates) used for food gathering.

Character 6: A large tergite on the second body segment that serves as a shield.

Character 7: A strongly sclerotized cuticle and as a consequence arthrodial membranes between segments and limb articles, as well as inner apodemes and other endoskeletal elements and a sclerotized pygidium (named telson in euarthropods).

Character 8: Biramous limbs with multisegmented endopod and a flattened exopod (Figure 12.1: *Shankouia*) stemming from a single first article (often called protopod or basipod).

Figure 12.1: Evolution of mandibulate head appendages started from distant ancestors like *Shankouia*, which possessed neither second antennae nor mouthparts. *Skara* is an example of the Cambrian Orsten mandibulates which already show some specializations: note endites (orange) of the second and third head appendage and the similarity of the following two limbs which resemble thoracopods. In Mystacocarida and Cephalocarida the adult second antenna has no endites; the last pair of maxillae (maxilla 2) is similar to a walking leg, in Mystacocarida also the maxilla 1. Cephalocarida possess epipods (blue), which also occur on the second maxilla. Note that stem-lineage mandibulates and many lower crustaceans possess a primary abdomen (green). (*Shankouia* after Waloszek et al., 2005; *Skara* after Müller and Waloszek, 1985; Mystacocarida after Hessler, 1969 and Hessler and Sanders, 1965; Cephalocarida after Gooding, 1963) ▸

Character 9: Mouth located ventrally, anterior esophagus bent ventrally and posteriorly towards mouth.

A good example for this level of organization are the fuxianhuiids, early Cambrian arthropods with a head shield, a pair of antennae, a short tritocerebral appendage

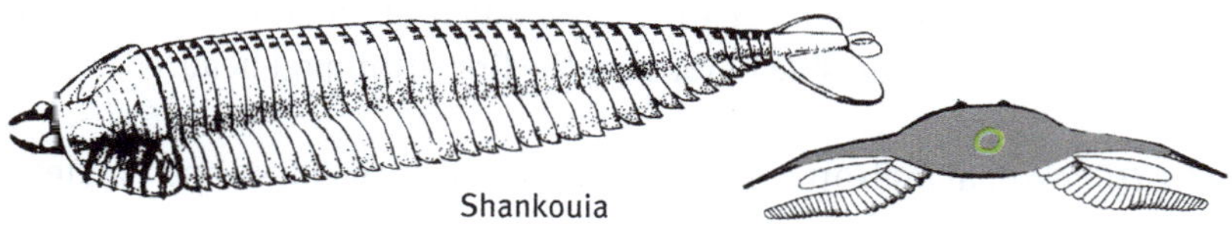

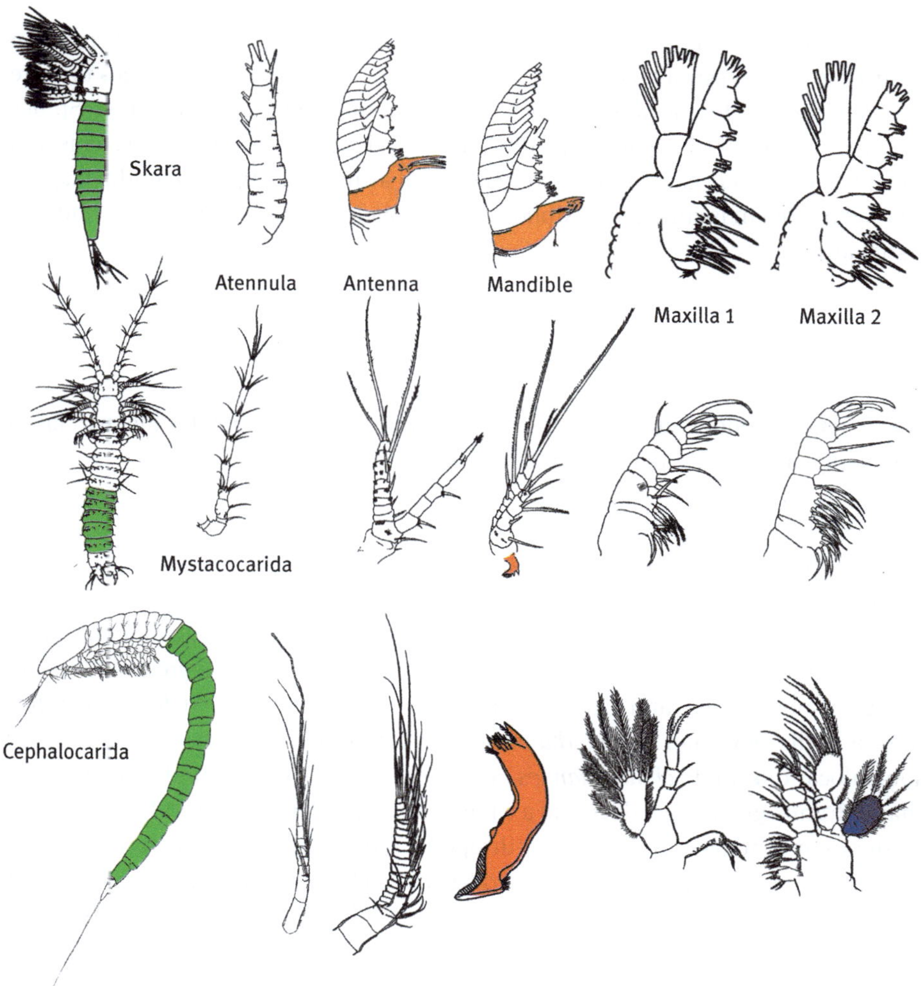

probably used for sweep feeding, biramous limbs lacking endites, a narrow abdomen without appendages (Yang et al., 2013).

The next phase is the development of a rigid head and the specialization of trunk appendages:

Character 10: Anterior segments bearing eyes, a pair of antennules and three more limbs dorsally fused to head shield, forming the first arthropod head.

Character 11: Proximal limb articles with medially directed spines and endites or precursors of endites, limbs therefore involved in locomotion *and* food gathering. As a consequence, the first appendage can evolve to shorter chelicerae or to a sensory organ, as in Chelicerata or Mandibulata.

These animals probably developed via a head larva as seen in extant crustaceans. In contrast to the ancestral "Lobopodia", euarthropods have trunk appendages that are not only essential for locomotion but also for gathering and transport of food. Spines and setae are directed towards a medioventral food path formed between stems and endopods (e.g. Haug et al., 2012b). Among extant arthropods, this longitudinal space for food treatment is still seen in xiphosurans, branchiopods, cephalocarids, and leptostracans.

12.3.4 Chelicerata and Mandibulata

The first major split in the extant crown group arthropod tree separates the two clades Chelicerata and Mandibulata, taxa that are easily identified.

12.3.4.1 The origin of Chelicerata

Extant chelicerates probably evolved from "great appendage arthropods". These are a paraphyletic assemblage of basal arthropods that have instead of an antenna a pair of large uniramous limbs probably used to capture prey. Typical representatives are the Anomalocarididae and Cambrian arthropods like *Yohoia*, *Leanchoilia*, *Fortiforceps*, and the Devonian *Schinderhannes*. According to Kühl, Briggs and Rust (2009) anomalocaridids and *Schinderhannes* are taxa at the base of the euarthropods. They share a frontal great appendage and a circular mouth. *Schinderhannes* already has the biramous appendages typical for euarthropods. *Yohoia* and *Branchiocaris* are seen as stem-lineage representatives of chelicerates s.str., characterized by a shorter "great appendage" that is homologous with chelicerae (Kühl, Briggs and Rust, 2009 and further references therein).

Chelicerates have no head separated from a trunk. There are two tagmata with a unique composition:

Character 12: The **prosoma** with dorsally fused segments includes the head region and trunk segments that bear walking appendages.

Character 13: The **opisthosoma** looks like an abdomen and carries gills (in primarily aquatic species) or contains lungs and/or tracheae.

Gills and lungs are derived from paired appendages, implying that the opisthosoma is not an appendage-free abdomen like the one found in basal Mandibulata. The number of segments is constant (prosoma: acron plus six somites, opisthosoma 12 somites plus telson), the first pair of appendages are short chelicerae, always followed by another five pair of appendages. The phylogeny within Chelicerata is discussed by Dunlop, Borner and Burmester (this book).

Character 14: First pair of limbs transformed to short **chelicerae**, homologous to the first antenna of mandibulates.

Character 15: Five pairs of walking legs (the first of these secondarily transformed to pedipalps in arachnids).

12.3.4.2 The origin of Mandibulata

Mandibulates have in comparison with chelicerates a very different and much more variable construction. Ancestors of mandibulates possibly were elongated, flattened animals which did not possess a distinct abdomen, as seen e.g. in *Tanazios* (Siveter et al., 2007). Note that in the publications of the Waloszek group all stem-lineage mandibulates are called crustaceans, because it is thought that insects and myriapods evolved independently and that Crustacea are monophyletic (see e.g. Haug et al., 2012b). Their crustacean ground pattern is equivalent to our ground pattern of Mandibulata.

A constant feature of mandibulates is

Character 16: The structure and composition of the **head**, which includes a minimum of five appendage-bearing segments. The biramous second antenna and the third head appendage (which evolves later into a mandible) are subsimilar. The following two pairs of head appendages resemble thoracopods. In extant taxa these appendages are differentiated into two pairs of antennae, one pair of mandibles, and two pairs of maxillae. Early arthropods in the stem lineage of mandibulates probably had only four head appendages (as in *Agnostus*) before a further trunk segment fused with the head (Haug et al., 2012b).

Character 17: A fleshy outgrow at the rear of the hypostome, the **labrum**, helps to retain chewed particles.

Character 18: Second antenna and mandible basally subdivided into coxa and basis.

Character 19: The first article of postmandibular limbs (the sympod or basis) is equipped with a **proximal endite** (see papers on the Orsten fauna by Waloszek and his team), the basis and articles of the endopods also have enditic lobes with setae and spines. The proximal endite evolves later progressively into a rigid basal limb article, the **coxa**, however not in all appendages of all taxa (see Waloszek, 2003a, b).

The second antenna and mouthparts are originally less specialized than in insects or higher crustaceans. Mandible and antennae are quite similar, with an enlarged endite used to stuff food into the mouth, while the following head appendages are less differentiated and not "real mouthparts" (see *Skara* and Mystacocarida in Figure 12.1). Therefore, all limbs behind the third head appendage are more similar to each other than to antennae and mandibles (see also review in Haug et al., 2012b).

Arthropods with this level of organization have been named **Labrophora** (see Waloszek, 2003a, 2003b; Siveter, Waloszek and Williams, 2003). This is a subtaxon of Mandibulata and includes all species with a proximal endite enlarged to form a coxa and showing a well-differentiated labrum. They include Phosphatocopida, crustaceans, insects and myriapods. In their first representatives the postantennular head appendages were similar. A functional differentiation into second antenna, mandible and maxillae does not exist.

Character 20: The trunk is originally divided into a **thorax** with legs, and an appendage-free **primary abdomen** (Figures 12.1, 12.2). The latter is sometimes reduced (as in Remipedia) or replaced by a *secondary abdomen*, which is a thorax region with leg rudiments (as in insects). It is a characteristic feature of the Orsten stem-lineage mandibulates and of crustacean taxa.

Character 21: Second and third head appendage (the future second antenna and mandible) with enlarged endites, differing from the following head appendages.

A typical organism with this level of organization is the Cambrian fossil *Skara* (Figure 12.1). A feature that is typical for these early mandibulates is that appendages are directed ventrally, in contrast to most other Cambrian euarthropods (see e.g. cross section of *Shankoui*a in Figure 12.1), also in contrast to chelicerates.

Further steps towards crown-group mandibulates (the first real "crustaceans"):

Character 22: Second head appendage loses in the adult its function as mouthpart and is transformed to a **second antenna** (see Mystacocarida and Cephalocarida in Figure 12.1).

Character 23: The **mandible** becomes the major masticatory appendage; however, it is originally still a biramous limb (Figure 12.1). The maxillae may still look like trunk appendages and probably are still used for walking, as seen in Mystacocarida.

Character 24: Ommatidia in compound eye with crystalline cone. This character has not been (could not be) studied in stem-lineage fossils of Mandibulata.

There are more details that could be discussed. However, in the following we focus briefly on some major events in the evolution within crown-group Mandibulata and especially on the placement of Myriapoda either as sister-group to Chelicerata or as taxon of the Mandibulata. Molecular phylogenies are presented by von Reumont and Wägele (for crustaceans, this book) and in chapters by Simon, Hadrys, Meusemann et al. (for insects, this book).

12.3.5 Phylogeny within primarily marine Mandibulata (crustaceans)

Crustaceans are paraphyletic with respect to hexapods. Characters of the more derived higher crustaceans, of myriapods and insects evolved in the stem lineage of Mandibulata and in ancestral crustacean lineages. An exemplary study of the stepwise evolution of endoskeletal elements of the head and other characters was published by Fanenbruck (2009, unfortunately in German) and is used here as backbone for the tree topology in Figure 12.2. We will not enumerate all characters discussed in the literature.

Node 1 represents all the previously discussed novelties in the ground pattern of crown-group mandibulates. The most conspicuous evolutionary steps along the following backbone tree within Mandibulata are:

Character 25: Cephalic endoskeleton with fourth transverse (intermaxillary) tendon connected to anterior endoskeletal complex (Fanenbruck, 2009).

This character evolved after the branching of the Mystacocarida. Possibly the specialization of the first maxilla (see Figure 12.1) is also a character that evolved later.

Character 26: Addition of a parlabral connection of the endoskeleton (lacking in Mystacocarida and Copepoda).

Character 27: First maxilla with less than four endites (usually only two, on coxa and basis; four occur in Mystacocarida).

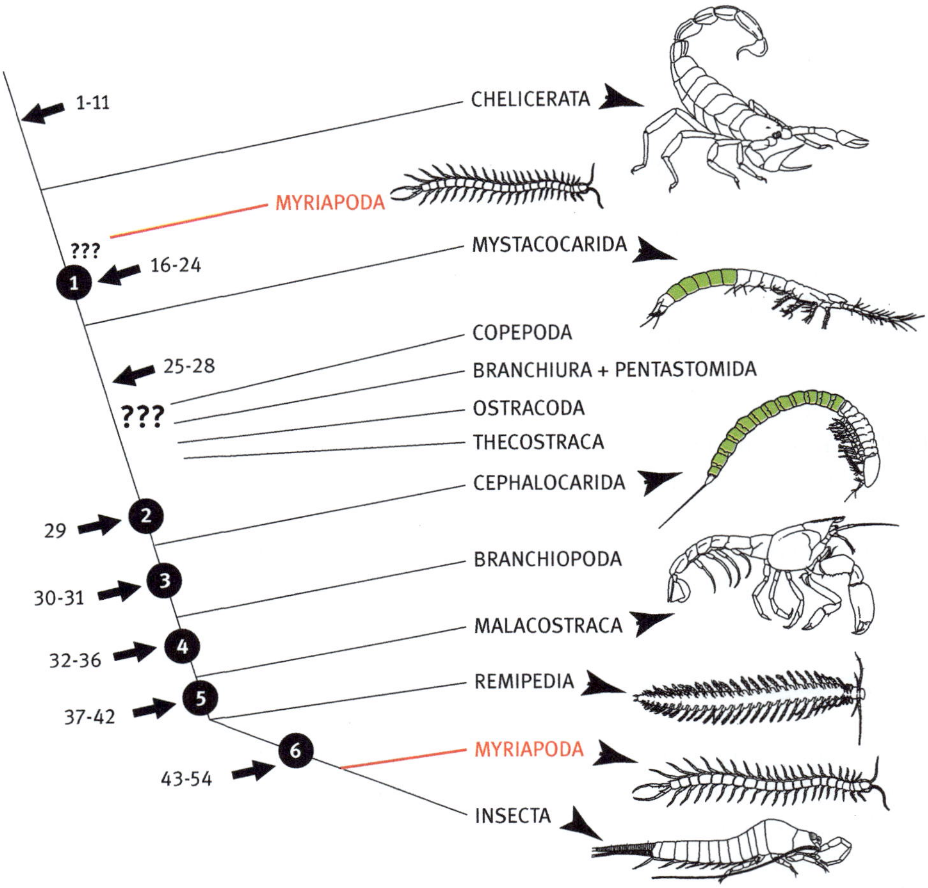

Figure 12.2: Phylogeny and evolution of morphology within Mandibulata as discussed by Fanenbruck (2009). The placement of taxa differs from various molecular phylogenies, especially the placement of myriapods, which from a morphological point of view are related to insects. Taxon names: node 1: Mandibulata; node 2: Thoracopodomorpha; node 3: Rotignatha; node 4: Caudoabdicata; node 5: Archilabiata; node 6: Tracheata. Arrows indicate where apomorphic characters discussed in the text appear for the first time. The primary abdomen is shown in green color.

Character 28: Palp of mandible uniramous (plesiomorphic state: two rami occur in Mystacocarida, Ostracoda and Copepoda).

Character 29: Thoracic appendages and second maxilla with a plate-like lateral outgrowth (epipod) primarily used for osmoregulation and respiration. Such outgrowths are also known from other arthropods and evolved several times convergently. However, these typical “crustacean gills” are absent from lower crustaceans and appear only in node 2 (Figure 12.2). The gills move during the further evolution from the basis of the exopod (in Cephalocarida) to the coxal region (Malacostraca) and increase their surface by subdivision and branching.

Hessler (1992) pointed out that the epipods are an important character and named the taxon in node 2 the "Thoracopoda". To avoid confusion with the similar appendage name Fanenbruck (2009) proposed the name **Thoracopodomorpha**. A further conspicuous feature is the larger number of trunk segments in comparison with the lower crustaceans. This is, however, a variable character. We must also consider that dwarfish taxa like Mystacocarida, Copepoda and Ostracoda show clear signs of secondary size reduction and anatomical simplification.

Character 30: Presence of a transverse mandibular tendon connected to a mandibular adductor muscle, which is enforced and has a radial arrangement of its parts, allowing rotating movements of the mandibles (a difference to ostracods with transverse tendons).

Character 31: Mandible with broad grinding *pars molaris*, possibly a consequence of the new mobility of the mandible.

These characters exist in all taxa following node 3 (Figure 12.2). Fanenbruck (2009) named this group the **Rotignatha**. These also seem to have a shorter anterior range of the ventral longitudinal muscles, which end in the intermaxillary region or more posterior (in contrast to e.g. Cephalocarida).

New characters in node 4 (**Caudoabdicata** sensu Fanenbruck, 2009):

Character 32: Reduction of the primary abdomen (see Figure 12.2). A rudiment of the abdomen is possibly the last pleon segment of Leptostraca, which lacks appendages. In fossil Phyllocarida this abdomen rudiment is clearly visible (e.g. Bergmann and Rust, 2013). Character 32 is an especially important character in node 4.

Character 33: Mandible with *lacinia mobilis* between *incisor* and *pars molaris*. This structure is present in all taxa connected to node 4 except myriapods. Its absence in most (but not all) insects can be a secondary loss, which (as an evolutionary process and gain of genetic information) is easier to achieve than multiple acquisitions.

Character 34: Nauplius not feeding, lecitotrophic, with reduced mouth and anus (in all aquatic Caudoabdicata). The fact that tracheates do not have aquatic head larvae is no contradiction to the assumption that this type of nauplius appeared for the first time in node 4 of Figure 12.2 (see discussion of the Tracheata hypothesis).

Character 35: First antenna with a second flagellum (absence in Tracheates requires the assumption of a secondary loss in terrestrial Caudoabdicata; the alternative is parallel evolution in Remipedia and Malacostraca).

Character 36: Exopod of second antenna scale like (present in Remipedia and Malacostraca).

New characters in node 5 (**Archilabiata**: Remipedia and Tracheata) that will be discussed further on in greater detail:

Character 37: Mandible without palp.

Character 38: Second antenna reduced (very small in Remipedia, absent in Tracheates).

Character 39: Pair of second maxillae basally fused, forming a labium.

Character 40: Coxa of thoracopods immobilized, fused to pleural region.

Character 41: Erected brain (Protocerebrum placed dorsally of the other ganglia).

Character 42: Reduction of gills (epipods), assuming Character 29 is a homology.

Characters of node 6 (Tracheata) will be discussed in the following.

12.4 The Tracheata hypothesis

The Tracheata (= Atelocerata) hypotheses, i.e. the assumption that insects and myriapods are sister-taxa, has been challenged by a few morphological (e.g. Edgecombe, 2004, contra Bitsch and Bitsch, 2004) and all hitherto published molecular analyses, except when molecules and morphology are combined (e.g. Wheeler, Cartwright, and Hayashi, 1993; Edgecombe et al., 2000; Edgecombe, 2010). Even though currently the Pancrustacea hypothesis is widely accepted based on multigene and phylogenomic analyses, a strong contradiction between molecular and morphological data (see below) remains, and until now no explanation for this contradiction has been offered.

The clade Tracheata is compatible with the Mandibulata hypothesis (e.g. Snodgrass, 1938a, 1950, 1951) that assumes that the mandibulate head with its typical appendages evolved only once. However, the currently popular Pancrustacea (= Tetraconata) hypothesis places myriapods outside or at the base of the Mandibulata (e.g. Dohle, 1997; Dohle, 2001; Giribet, Edgecombe, and Wheeler, 2001; Shultz and Regier,

2000; Regier and Shultz, 2001; Richter, 2002; Regier, Shultz, and Kamble, 2005; Ungerer and Scholtz, 2008; Aleshin et al., 2009) as a parallel lineage to Crustacea that is much older than insects, making it necessary to assume that a large set of unique characters shared by insects and myriapods evolved twice (see below).

In the following we explain that the morphological evidence supporting the traditional Tracheata is clearly more numerous and less fuzzy than evidence for the Pancrustacea (see below). We propose a new scenario for the origin of insects, where myriapod-like ancestors are a link between Remipedia and Hexapoda. In contrast to the Pancrustacea concept, this scenario is compatible with paleontological data and it explains why Remipedia have a myriapod-like body and share so many characters with insects. We also discuss some causes for the mutual incompatibility of molecular phylogenies and for the consistent failure to recover the clade Tracheata in sequence analyses.

12.4.1 Molecular evidence for the placement of myriapods

Early analyses of DNA sequences (at the beginning often restricted to fragments of single nuclear rDNA genes and a few species), placed myriapods as sister-group of Chelicerata (**"Myriochelata"**), leaving the remaining euarthropods in a clade composed of Hexapoda and Crustacea. First, it was thought that hexapods are the sister-group of crustaceans (Field et al., 1988; Friedrich and Tautz, 1995; Turbeville et al., 1991; Ballard et al., 1992). Later analyses suggested a variety of combinations, some still supporting the Myriochelata clade (e.g. Min, Kim, and Kim, 1998; Anderson, Córdoba, and Thollesson, 2004; Mallatt, Garey, and Shultz, 2004; Pisani et al., 2004; Petrov and Vladychenskaya, 2005; Hassanin, 2006; Mallatt and Giribet, 2006; Gerlach et al., 2007; Mallatt, Waggoner Crag, Yoder, 2010), sometimes with myriapods as first lineage of euarthropods (Regier, Shultz, Kamble, 2005), or placing chelicerates within Myriapoda (Negrisolo, Minelli, and Valle, 2004a), while others recovered myriapods as first lineage of Mandibulata (**"Pancrustacea"** hypothesis: e.g. Giribet, Edgecombe, Wheeler, 2001). Also larger phylogenomic data sets did not allow inference of stable phylogenies: myriapods appear partly outside, partly inside Mandibulata (e.g. Reumont et al., 2009; Roeding et al., 2009; Regier et al., 2010). The results of Koeneman et al. (2010) suggest that the position of myriapods cannot be clarified with phylogenomic sequence analyses, while Giribet, Richter, Edgecombe et al. (2005) believed this is only a problem of placing the root correctly.

From a morphological point of view it seems that myriapods slip down the tree and end up where they share similarities either with chelicerates or with basal crustaceans. Until now nobody tried to check the quality of those molecular characters that attach myriapods to basal edges of the arthropod tree (support values are no indication for data quality!).

12.4.2 Molecular evidence for the placement of Hexapoda

There exists a large variety of topologies (a selection of mutually incompatible clades is depicted in Figure 12.3). The most important ones are:

- Hexapods are not monophyletic, with entognathous taxa spread in various ways among crustaceans (Giribet and Ribera, 2000; Giribet, Edgecombe, and Wheeler, 2001; Cook, Yue, and Akam, 2005; Hassanin, 2006; Carapelli et al., 2007). Sometimes, these topologies have been published by authors who had stated in earlier papers that hexapods are monophyletic (e.g. Cook et al., 2001; Cook, Yue, and Akam, 2005)
- Regier, Shultz, and Kamble (2005) proposed the grouping {Hexapoda, Branchiopoda}
- Copepoda as sister-taxon of hexapods (Mallatt, Waggoner Crag, and Yoder, 2010: "undet. Cyclopidae", Reumont, Meusemann, Szucsich et al., 2009)
- Thoracopodomorpha (= Malacostraca + Cephalocarida + Branchiopoda) as sister-group of hexapods (Carapelliet al., 2007)

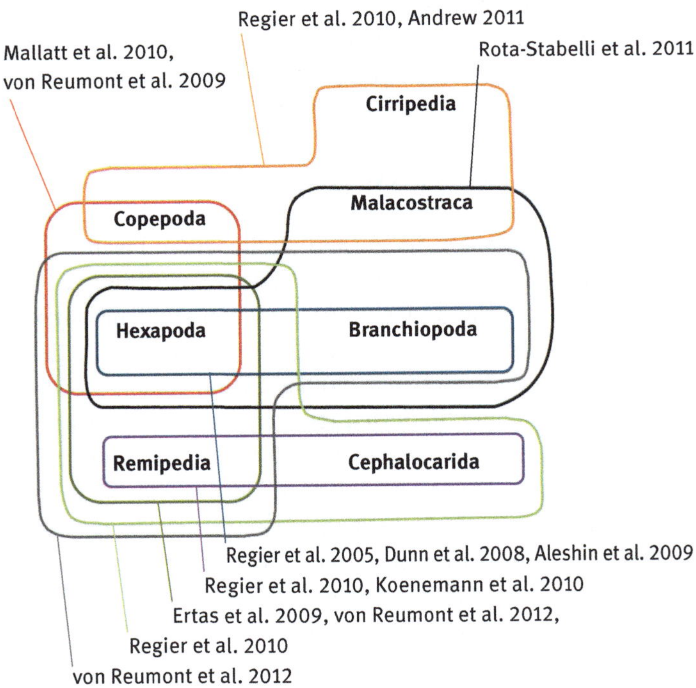

Figure 12.3: A selection of groupings proposed in recent molecular analyses of large data sets (multigenic or transcriptomic data) with the corresponding references. The overlapping lines visualize contradictions.

- The clade {Remipedia + Cephalocarida} is the sistergroup of Hexapoda (Regier et al., 2010)
- Hexapods are paraphyletic with respect to Remipedia, Cephalocarida and Malacostraca (Koenemann et al., 2010)
- The sister-group to Hexapoda are the Remipedia (Ertas et al., 2009, von Reumont et al., 2012). The next larger clade is {Branchiopoda (Remipedia, Hexapoda)}, excluding Malacostraca (von Reumont et al., 2012)
- More incomplete data sets (e.g. lacking Remipedia, Copepoda, Cirripedia or other crustaceans) also show branchiopods close to Hexapoda (e.g. Gaunt and Miles, 2002; Dunn et al., 2008; Aleshin et al., 2009; Rota-Stabelli et al., 2011)

None of the trees recovered the Tracheata. Publications often contain several topologies that are evidence for how sensitive the results are to variations of taxon sampling, alignment, gene selection, and substitution modeling (e.g. Regier et al., 2008; Koenemann et al., 2010). The reader then usually has the difficulty that no hard criteria for the selection of the "best" topology exist. All in all, the comparison of published results suggests that molecular phylogenetic analyses of the deep phylogeny of Arthropoda do not produce reliable results. There are too many contradictions and until now no criteria exist to discern between qualities of data sets and to assess qualities of analyses. Different authors of mutually incompatible results usually state that their data are excellent and the analyses adequate. If authors contradict their own earlier work they fail to explain the mechanisms that produce errors and usually only tell that there are differences in data sets and substitution models.

Figure 12.4 illustrates with the example of the data of Regier et al. (2010) the typical structure of the information content of deep phylogeny alignments. Using the software SAMS (see Wägele and Mayer, 2009) it is possible to select conserved split-supporting positions to demonstrate with a "spectrum of split support" how many mutually compatible and incompatible splits are represented in a data set. In Figure 12.4 all columns shaded in grey are incompatible with the best supported splits and represent the noise in the data. It is typical that only those taxa that are relatively young or separated by long branches are also well supported by conserved sequence positions, which is equivalent to a strong phylogenetic signal in the data. None of the deeper nodes relevant for arthropod phylogeny are found among the 150 best splits. The weak phylogenetic signal explains why phylogenetic analyses of these data produce so many incompatible results (Figure 12.3).

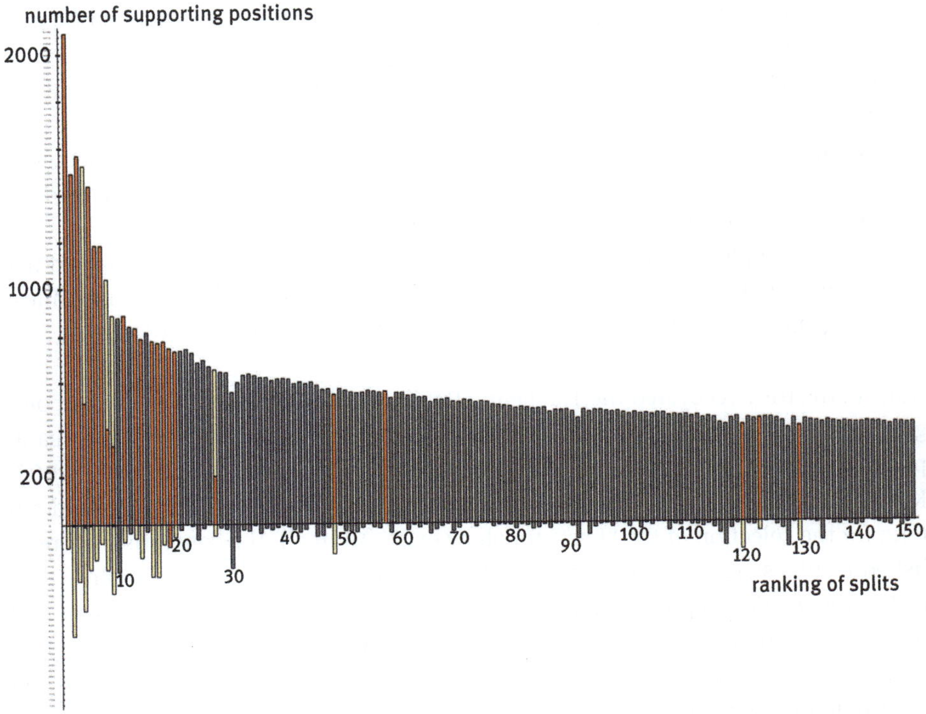

Figure 12.4: Spectrum of conserved split-supporting ingroup positions (**CIPs**) for the data set of Regier, Shultz, Zwick et al. (2010) drawn with SAMS (see Wägele and Mayer, 2007). The vertical axis indicates the number of alignment positions that fit to a split. Each bar represents a bipartition (split) in the complete set of taxa, with the group that contains the majority of the conserved positions above the horizontal axis. The best supported splits that fit on a single binary tree are shown in orange and yellow (orange: more conserved positions, yellow: noisy positions); grey splits are incompatible with the best supported tree. Note that deep nodes are not among the best 150 splits. CIPs selected with SAMS include binary positions (black), asymmetric positions (conserved only in the ingroup: orange) and noisy positions (with some substitutions in the ingroup: yellow). The mutually compatible taxa separated by the best supported splits are: 1: Archaeognatha; 2: Tardigrada; 3: Branchiopoda; 4: Malacostraca; 5: Odonata; 6: Ephemeroptera; 7: Symphyla; 8: Pycnogonida (partim) ; 9: Thecostraca; 11: Onychophora (partim) ; 13: Lepidoptera (partim); 14: Xiphosura; 16: Lepidoptera; 17: Copepoda; 18: Onychophora (partim); 19: Branchiura + Pentastomida; 27: Pycnogonida; 48: Collembola. Split 57 is a remiped and an arachnid, 120 are Cirripedia, 123 Balanidae, 130 Scorpiones.

12.4.3 Taxon-slippage: Evolutionary processes can produce sequence patterns that break up the clade Tracheata

Due to the systematic errors caused by differences in branch lengths (class I effect (symplesiomorphies) and class II effect (signal erosion), Wägele and Mayer, 2007), we expect to see in special cases **taxon-slippage** in molecular phylogenies estimated from real data (Figures 12.5 and 12.6). The occurrence of these errors has been detected and explained with simulations (Kück et al., 2012). Since these errors are not caused by noise, increases of taxon-sampling and of alignment lengths *will not necessarily cure the problem* but *increase the statistical support for the wrong tree.*

According to morphological evidence, there are two mandibulatan taxa that are major candidates for taxon-slippage: The Myriapoda, morphologically best placed as the sister-group of Hexapoda (see below), and the Malacostraca, that clearly are highly derived crustaceans with characters shared with insects, Remipedia, and Branchiopoda (e.g. Harzsch, 2002a; Fanenbruck, Harzsch, and Wägele, 2004; Grimaldi, 2010; Strausfeld, 2011), but which in molecular trees group with lower crustaceans (e.g. close to Cirripedia and Copepoda: von Reumont et al., 2012, however not in Rota-Stabelli et al., 2011).

The placement of myriapods at the base of Mandibulata or even as earliest lineage of Euarthropoda (e.g. in von Reumont et al., 2012) is implausible from morphological and paleontological points of view (see below). In the following we focus on artifacts that might cause a wrong arthropod tree.

The simulation studies of Kück et al. (2012) have shown that **class II long-branch artifacts** where one single long branch slips down the tree due to signal erosion along this branch (Figure 12.5), are expected to be rare. This happens only when the internal branch supporting the correct clade (the stem lineage) is very short in relation to

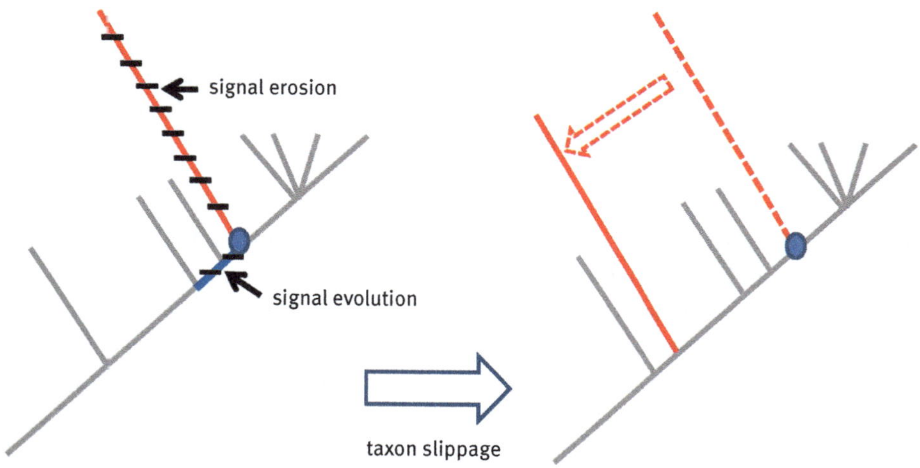

Figure 12.5: Cartoon illustrating the mechanism that can produce the class II long-branch artifacts (systematic errors) in molecular phylogenies (see text). Whenever the stem lineage of a clade (blue line) is short, there will be little phylogenetic signal available to infer the monophyly of this clade. A long-branch taxon within this clade (red line) can lose most of this signal by multiple substitutions. The consequence is taxon-slippage. Where this taxon attaches to the tree depends on the number of characters (plesiomorphies and chance similarities) shared with other lineages. This error cannot be avoided with currently available tree-inference methods (Kück et al., 2012).

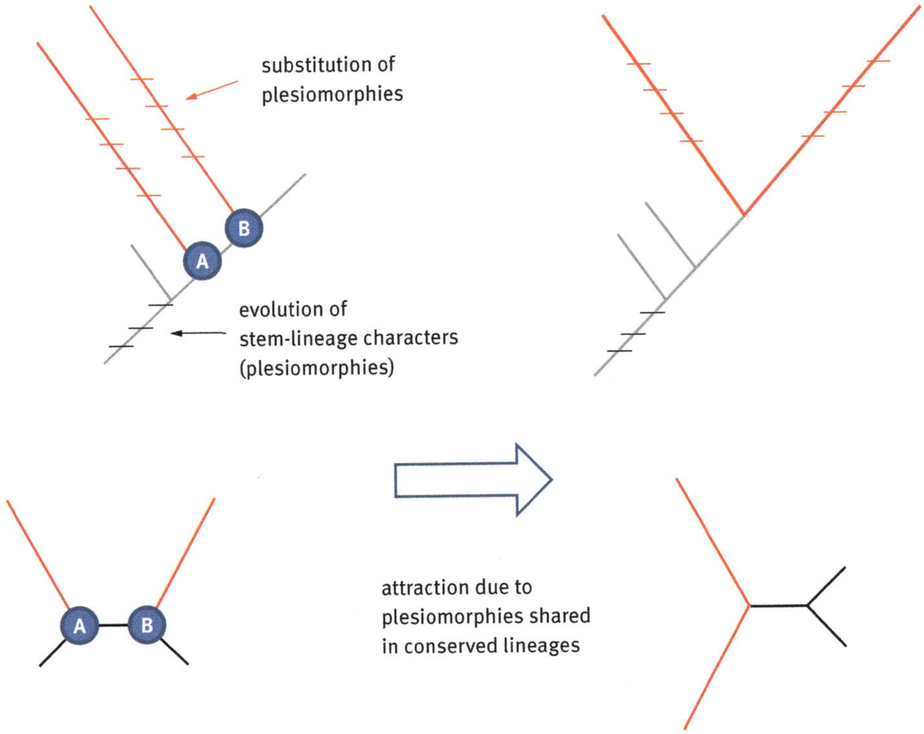

Figure 12.6: Cartoon illustrating the effects of plesiomorphies in a four-taxon tree (shown above as rooted, below as unrooted topology). The Felsenstein-effect is not only the attraction of long branches due to shared chance similarities evolving along the long branches. The accumulation of shared plesiomorphies in short branches is faster and attracts short branches. This effect can also be seen in multi-taxon trees (unpublished simulations).

the length of the slipping branch. In the mentioned simulations, the ratio of stem-lineage length to long-branch length has to be 1 : 70 or more to produce the artifact. However, nearly all topologies for deep phylogenies show the critical situation: short inner branches in combination with long terminal ones. Therefore, at least some of the misleading signal erosion will take place. The lack of a distinct signal for deeper nodes has already been discussed (Figure 12.4).

The second artifact (**class I effect**, caused by symplesiomorphies), the attraction of unrelated short branches, is a systematic error caused when other taxa evolved faster and accumulate more derived characters than the short branches. This is the typical Felsenstein situation (Felsenstein, 1978) which is usually interpreted as "long-branch attraction" (Figure 12.6). While it has become a tradition to assume that in the Felsenstein case an accumulation of chance similarities along long branches are causing this attraction, our own simulations show that symplesiomorphies accumulate much faster, not only in four-taxon topologies. The class I effect cannot be avoided even when the correct substitution model is used for the ML tree inference (see Kück et al., 2012, also Kück, Misof and Wägele, this book).

Currently there exist no methods to identify the footprints left by evolutionary processes in the form of specific site patterns of alignments. We are still trying to understand in simulations how the situations that produce systematic errors are reflected in site patterns. However, there are promising first observations. Using the data from Regier et al. (2010) (see split spectrum in Figure 12.4) we searched for conserved ingroup positions (**CIPs**) in splits relevant for deeper nodes. CIPs are defined as alignment positions with a conserved character state in a functional ingroup of a split that differs from character states of taxa of the corresponding functional outgroup. CIPs are putative synapomorphies for monophyletic functional ingroups. Our alternative hypotheses are:

- To confirm that myriapods are the sister-group to Pancrustacea, there should be distinct conserved evidence in the form of CIPs for the group {Myriapoda + Crustacea + Hexapoda} in comparison with the remaining taxa.
- To confirm that myriapods are the sister-group to hexapods, there should be conserved evidence in the form of CIPs for the group {Myriapoda + Hexapoda}.

What we did find was a surprise (see also Figure 12.7): myriapods share more invariant characters (CIPs) with protarthropods (onychophorans and tardigrades) than with Pancrustacea, and more with hexapods or with chelicerates than with Pancrustacea. The ranking order of splits according to the occurrence CIPs (as percentage of the whole alignment) is:

(1) {(Hexapoda + Myriapoda), remaining taxa} with 2.29 %,
(2) {(Remipedia + Myriapoda + Hexapoda), remaining taxa} with 2.28 %,
(3) {(Chelicerata + Myriapoda), remaining taxa} with 2.26 %,
(4) {(Hexapoda + Remipedia), remaining taxa} with 1.83 %,
(5a) {(Crustacea + Hexapoda + Myriapoda), remaining taxa} with 0.01 %,
(5b) {(Crustacea + Hexapoda), remaining taxa} with 0.01 %.

There are very few conserved characters for the taxa Mandibulata (0.01 %) and Pancrustacea (0.01 %) in comparison to the distinct number of conserved characters for Myriochelata (2.26 %), Tracheata (2.29 %), and for the Archilabiata, the combination of Tracheata and Remipedia (2.28 %) (Figure 12.7).

Crustaceans as a group are more derived than the other clades. This is obvious when we count the CIPs conserved within groups: 4.37 % for Myriapoda, 4 % for Chelicerata, 1 97 % for Hexapoda, only 0.07 % for Crustacea (a paraphyletic group!).

Comparing the taxon-specific CIPs *shared between arthropod taxa and protarthropods* the difference is also obvious: We find e.g. for Myriapoda 3.70 %, for Tracheata 1.37 %, for {Tracheata + Remipedia} 1.36 %, for Hexapoda 0.99, for Chelicerata 0.48, for crustaceans 0.01.

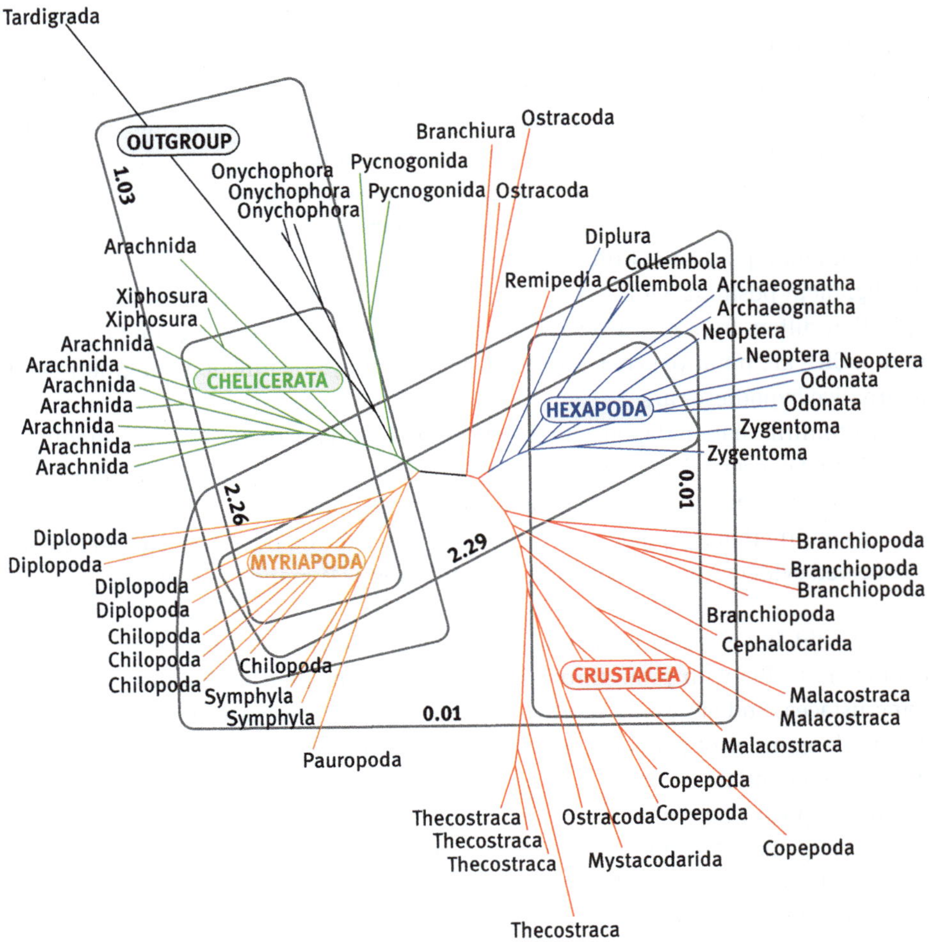

Figure 12.7: Distribution of split-supporting conserved ingroup positions (CIPs) in an arthropod tree. ML-tree estimated for the data of Regier et al. (2010) (RAxMLHPC-PTHREADS 7.2.6 GTR+Gamma+I for 60 taxa). Numbers indicate the percentage of split-supporting conserved ingroup positions (CIPs) for selected groups of taxa (delimited with boxes). For further details see text.

The number of conserved protarthropod characters of a taxon depends on the substitution rate in the taxon's stem lineage and on the diversity within the taxon. Our interpretation is that due to the faster evolution of lineages of crustaceans with their very diverse body plans (compare e.g. copepods, barnacles, crabs), other taxa like myriapods and chelicerates retained in comparison substantially more plesiomorphic characters. Remipedia are a conserved relict taxon, Hexapoda sequences share more CIPs with protarthropods than with crustaceans. This explains the higher number of CIPs for the clades Archilabiata (= Tracheata + Remipedia) and Tracheata.

More ancient protarthropod characters are conserved in hexapods (0.99) than in crustaceans (0.01 %), and there are more in myriapods (3.70) than in chelicerates (0.43). The ratio of numbers of shared old character states could explain why myriapods are usually placed in molecular phylogenies close to the outgroup taxa of mandibulates or even as sister-taxon of chelicerates. These plesiomorphies are no evidence for monophyly, but they form a strong signal that distorts phylogenies (Kück et al., 2012).

To explain the distinct signal for Tracheata there are three different interpretations: (a) these could be plesiomorphies retained in myriapods and hexapods, or (b) new shared character states (synapomorphies) of Tracheata, or (c) a combination of both. Comparison with protarthropods allows us to discern these cases: There are 1.36 % CIPs shared with protarthropods: these are candidates for symplesiomorphies. There are in addition 2.28 % CIPs unique for Tracheata: these are candidates for synapomorphies.

In the light of the morphological evidence (see below) and the observed site patterns in alignments we can postulate that in molecular ML analyses myriapods slip down the tree due to a systematic error of the class I type (symplesiomorphy effect). Our observations suggest that the Tracheata hypothesis can be correct despite the fact that Tracheata is not recovered in phylogenetic analyses of sequence data.

12.4.4 Are there morphological apomorphies of Pancrustacea (=Tetraconata) primarily absent in Myriapoda?

If we assume that myriapods branched off very early within Mandibulata and are the sister-lineage to Pancrustacea, we should not only see derived characters apomorphic for Pancrustacea, but these should clearly have a corresponding plesiomorphic state in Myriapoda. Characters that are only "different" in myriapods could have evolved from a state seen in insects and are no evidence against the Tracheata hypothesis.

The following characters have been proposed as evidence for the Pancrustacea – Myriapoda dichotomy or for a placement of Myriapoda as sister-group to Chelicerata:

(1) Eye structure (e.g. Nilsson and Osorio, 1998; Paulus, 2000; Dohle, 2001; Richter, 2002; Harzsch, Melzer, and Müller, 2006; 2007): In chelicerates and trilobites ommatidia of compound eyes have cuticular lenses that focus incident light on rhabdomeres. In trilobites, eye lenses are exoskeletal material and consist of packed lenses. At each ecdysis a new lens is produced from the apical part of epidermal cells (review in Clarkson, 1979). In horseshoe crabs, the only extant Chelicerata with facetted lateral eyes, the lenses of ommatidia are also formed by the exoskeleton, namely by internal projections of the transparent cuticle (Fahrenbach, 1975). In arachnids, lateral eyes are highly modified, obviously by fusion of groups of ommatidia. The light is also focused by cuticular lenses (Paulus, 1979; Weygoldt and Paulus, 1979a).

The well-known ommatidia of crustaceans and insects share a new character, namely a large lens not formed by the cuticle but by vitreous bodies, the so-called crystalline cone, which is usually produced by four cone cells (Semper cells) (e.g. Debasieux, 1944; Paulus, 1979; Land, 1981; Cronin, 1986; Klass and Kristensen, 2001; Richter, 2002). The crystalline cone can be covered externally by a cuticular lens, however, there is no cuticular cone as in other arthropods. The crystalline cone has been considered to be a synapomorphy either of Mandibulata (e.g. in Paulus, 1979; Wägele, 1993; Klass and Kristensen, 2001; Fanenbruck, 2009) or of Pancrustacea (e.g. Dohle, 1997a; Harzsch and Waloszek, 2001; Richter, 2002; Strausfeld and Andrew, 2011), depending on the placement of myriapods in the arthropod tree.

Myriapod eyes have rarely been studied in detail. Müller, Sombke, and Rosenberg (2007) summarized new data: among myriapods, the dwarfish Symphyla, Pauropoda, and the Polydesmida lack eyes, also all Geophilomorpha. Even though myriapod eyes are variable and often deviate from the conserved composition known from many lateral eyes of adult crustaceans and insects, a common pattern is seen in some species of chilopods and diplopods. The presence of a multipartite crystalline cone has been documented for *Scutigera* by Müller, Rosenberg, Richter et al. (2003) and for Penicillata (Diplopoda) (Müller, Sombke, and Rosenberg, 2007). Spies (1981) had already noted that in the eyes of *Polyxenus* (Penicillata) each ocellus can be derived from a single insect-type ommatidium. Therefore Müller, Sombke, and Rosenberg (2007) consider the myriapod ommatidium to be derived from the mandibulatan eye and they define the typical character of mandibulatan ommatidia as "common possession of crystalline cone cells and a bilayered dual type retinula". Another pattern conserved in many (but not all!) insects and crustaceans but absent in myriapods is the restriction of cell numbers in the retina (8 cells) and cornea (2 cells).

The argument that because of the lack of the crystalline cone the eye of myriapods is more plesiomorphic than that of insects and crustaceans can be refuted with reference to the presence of rudiments of the crystalline cone in *Scutigera* and Penicillata, but also pointing out that myriapod-like eyes occur within insects. The larval eyes (stemmata) of tiger beetles, for example, have a single corneal lens and an underlying rhabdom layer with a rhabdomeric pattern similar to that seen in chilopods (Toh and Mizutani, 1994). Larval stemmata of holometabolous insects are possibly derived from the most posterior ommatidia of the complex eye seen in hemimetabolous insects (Sbita, Morgan, and Buschbeck, 2007). Absence of crystalline cones and four Semper cells in stemmata is a secondary modification, because primarily they are present (e.g. Melzer and Paulus, 1994; Briscoe and White, 2005).

Modifications of eyes also occur in adult insects. During metamorphosis of Chaoboridae (Diptera), for example, the structure of larval ommatidia changes profoundly: the larval cornea transforms into strongly curved lenses and the originally present crystalline cone is completely reduced (Melzer and Paulus, 1994). In Strepsiptera, the compound eye is replaced by a few eyelets which consist of a biconvex thick cuticular lens, corneal cells, a cup-shaped retina, but there are no crystalline cones (Buschbeck, 2005). Irrespective of the homology between single stemmata of insects and lateral ommatidia or accessory eyes of other arthropods, we must acknowledge that there occur structurally similar eyes in insects and in myriapods. If modifications of lateral eyes happened within insects, they could as well have occurred in lineages of the Myriapoda. The alternative, formulated by Paulus (2000) under the impression made by molecular phylogenies, is that the eye of *Scutigera* is the more plesiomorphic precursor of the mandibulatan eye. Unfortunately, there is no further morphological evidence for the placement of *Scutigera* as a representative of basal, pre-crustacean mandibulatan lineage.

The most parsimonious assumption is that the crystalline cone did not evolve several times independently but is a character of Mandibulata, often reduced in modified eyes like larval stemmata of insect or in many myriapods, as already explained in great detail by Paulus (1979). Therefore, the occurrence of variations in eye structure is not an argument against the Tracheata hypothesis.

(2) Eye development (Harzsch, Melzer, and Müller, 2006; 2007): When eyes of myriapods grow, new ommatidia are added along the anterior border of the eye (between the eye and the insertion of the antenna). The authors claim that this pattern is the same as in chelicerates and therefore an argument for the basal position of myriapods in the arthropod phylogeny. This character has rarely been studied and there are no detailed comparisons across arthropod taxa. However, eye growth by addition of ommatidia on the anterior eye margin has also in principle been observed in some crustaceans (Wägele, 1987), while other crustaceans add ommatidia all around the edge of the eye (Keskinen et al., 2002). It is not clear which condition is derived and which variations exist. Therefore, this character has currently no value to clarify the origin of myriapods.

(3) Presence of a third optic lobe neuropil, the lobula or medulla interna (Osorio, Averof, and Bacon, 1995; Melzer, Petyko, and Smola, 1997; Strausfeld, 1998; 2005; Harzsch, 2006; Harzsch and Hafner, 2006; Strausfeld and Andrew, 2011): Eumalacostraca and Hexapoda share specific brain structures, among these a third neuropil in the optical lobes, the medulla interna (e.g. Harzsch, 2002a). Other Mandibulata usually have only two neuropils connected by parallel bundles. The argument pro Pancrustacea is that the third neuropil occurs in Pancrustacea and is absent in myriapods.

In his phylogenetic analysis of arthropod brain characters, Strausfeld (1998) lists as a character supporting Pancrustacea "lamina and medulla, shared by non-malacostracans and apterygotes" (character 2 in his Fig. 8). This argument can only support the Pancrustacea hypothesis if it can be shown that the situation found in Myriapoda (medulla absent in chilopods) is plesiomorphic. Difficulties in the interpretation of the homology of brain characters may be the methodological cause for monophyly of Crustacea and polyphyly of Myriapoda in the cladistic analysis by Strausfeld (1998: Fig. 2).

It has been assumed that within Malacostraca the brain of the phylogenetically old Leptostraca lacks the third neuropil (Elofsson and Dahl, 1970) and is more similar in the anatomy of optic lobe neuropils to apterygotes than to pterygotes. More recent studies revised this view, Leptostraca have four neuropils (Kenning et al., 2013). Obviously, brain anatomy has to be studied with modern tools in more taxa before conclusions about brain evolution are possible. New analyses of excellently preserved Cambrian early euarthropods (*Fuxianhuia*) indicate that optic neuropils evolved much earlier than hitherto thought (Ma et al., 2012).

Absence (Remipedia) or strong modifications of eyes (Myriapoda) may be the cause for the reduction of neuropils. Myriapods and remipedes could easily have had an ancestor that possessed a third neuropil. Lobula and second chiasma are also absent in the apterygote *Lepisma*, which indicates secondary loss within Hexapoda (Strausfeld, 2005). Therefore, presence or absence of this morphological detail is a questionable character. Other cases of parallel acquisition or secondary loss of brain characters are discussed in Klass and Kristensen (2001).

(4) Brain complexity has been suggested to be a character supporting the monophyly of Pancrustacea (Fanenbruck, Harzsch, and Wägele, 2004; Fanenbruck and Harzsch, 2005). However, in their comparison of new findings in Remipedia with other available data, Fanenbruck, Harzsch, and Wägele (2004) point out that data are still missing for many crustaceans and for myriapods.

The central complex (midline neuropils) is known for chilopods, hexapods, several crustaceans (Loesel, Nässel, and Strausfeld, 2002). In chelicerates it has a different shape (arcuate body). Myriapods have the same central complex as other mandibulates (see Loesel in this book; contra: Strausfeld and Andrew, 2011). The fact that this complex is absent in diplopods proves that brain anatomy can vary profoundly in derived taxa. The absence of the central complex does not necessarily imply that Diplopoda do *not* belong to Euarthropoda or to Myriapoda, it can be explained as a secondary reduction. Other differences seen only in diplopods are protocerebral neuropils that are not lateralized but extended bilaterally across the brain (Strausfeld, 1995). Loesel, Nässel, and Strausfeld (2002) include in the list of characters supporting Pancrustacea (= Tetraconata) the lateral protocerebral neuropil and the protocerebral chiasma. Strausfeld and Andrew (2011) state that "a synapomorphy of Tetraconata is the presence of midline neuropil complexes that *includes a neuropil protocerebral bridge*", but also admitting that this character is not present in Branchiopoda, which implies secondary loss.

The structure of mushroom bodies (innervated mainly by olfactory interneurons) is the same in myriapods and other mandibulates, while chelicerates have a structure more similar to that of onychophorans (an argument against the Myriochelata hypothesis, see Loesel in this book). Also, the structure of the deutocerebrum with separate neuropils for processing of chemo- and mechanosensory information originating from the (first) antennae is a homology of Mandibulata absent in Onychophora and Chelicerata (Sombke et al., 2012).

The supporting evidence for Pancrustacea is not convincing. Differences between myriapods and insects may have evolved due to eye and neuropil reductions and due to internalized frontal eyes in myriapods. However, to support the Pancrustacea it is important to show that myriapods have mandibulatan plesiomorphies (*with substantiated homology statements*), while the corresponding derived state should be found in clades that include hexapods and exclude myriapods.

(5) Similarities in neurogenesis have been said to support the Pancrustacea hypothesis (Osorio, Averof, and Bacon, 1995; Whitington, Meier, and King, 1991; Whitington, Leach, and Sandeman, 1993; Whitington, 1995; Stollewerk, Tautz, and Weller, 2003; Chipmann and Stollwerk, 2006; Pioro and Stollewerk, 2006; Ungerer and Scholtz, 2008; Mayer and Whitington, 2009a).

In a spider embryo, ventral groups of ectodermal cells form neural precursors that do not divide further (e.g. Stollewerk, Tautz, and Weller, 2003). They form about 30 invagination groups per hemisegment. Invaginating cells are replaced by divisions of cells at the surface, providing material for later invagination of further neural precursors. Single neuroblasts are not present. Mittmann (2002) found in *Limulus* a mechanism similar to that described in spiders.

Whitington, Meier, and King (1991) discovered that in a chilopod the developing ventral nerve cord also has no neuroblasts. Kadner and Stollewerk (2004) proposed that myriapods have a neurogenesis more similar to Chelicerata. They found that in the *Lithobius* embryo evenly spaced groups of bottle-like cells invaginate ventrally, as in spiders. However, there is some variation in this character: diplopods differ from spiders and chilopods because cell groups do lie over and above each other, the neuroectoderm is multilayered. Invaginating cells form stacks and are not all basal as in spiders (Dove and Stollewerk, 2003). Also, in the chilopod, cell groups detach from the apical surface sequentially, starting anteriorly, while in the examined spider and the diplopod there are four waves of invagination. Stollewerk and Simpson (2005) noted that neural precursor formation is correlated with cell proliferation in myriapods, but not in spiders. In myriapods and spiders there are about 30 spots of invaginating cell groups per hemisegment.

In *Drosophila*, there is per hemisegment a group of initially equivalent cells in the ventral neuroepithelium, from which one is selected that delaminates into the embryo. In insects, there are about 25 to 30 such neuroblasts in each hemisegment. The number is essentially the same as the invagination sites of other arthropods. In *Leptodora* (Cladocera) there are neuroblasts that produce ganglion mother cells "by highly unequal division perpendicular to the surface" (Gerberding, 1997). Neuroblasts also occur in Malacostraca (e.g. Harzsch and Dawirs, 1995; Harzsch et al., 1998; Stollewerk, 2005), but their origin is different. Malacostraca have specialized stem cells, the so-called ectoteloblasts that generate epithelial cells from which, after some rounds of divisions, neuroblasts are formed by perpendicular asymmetrical divisions. Ganglion mother cells of Malacostraca differ from those of insects in that they do not delaminate from the surface neuroectoderm and they are not associated with specialized sheath cells (Whitington and Bacon, 1998). However, more detailed examinations suggested homology of neuroblasts (Mittmann, 2002; Ungerer and Scholtz, 2008). It was also observed that in some non-malacostracan crustaceans there seem to be no neuroblasts and "neurons arise by inwards proliferation of ectodermal cells" (Whitington and Bacon, 1998). More observations are needed to understand how mechanisms of neurogenesis evolved.

It seems that myriapods do not differ greatly from insects in their gene expression patterns during neurogenesis. Pioro and Stollewerk (2006) could show that "the expression pattern of homologs of the *Drosophila* proneural genes *daughterless*, *atonal*, and *SoxB1* are partially conserved in *Glomeris marginata*". This is probably a feature of most arthropods.

The major difference in all these variations seems to be the timing of cell proliferation and the number of invaginating cells as a result of the site where cell proliferation occurs (at the surface, or after delamination). The formation of grooves is correlated with invagination of cell groups. It is not clear how to homologize these patterns. In many chelicerates and myriapods most cell divisions take place in the apical layer of the neuroectoderm, while in crustaceans and insects cell divisions of single neuroectodermal cells give rise to smaller cells that are pushed into the embryo. This is the most conspicuous difference. But, there is no specific and complex pattern that substantiates homology of the situation seen in chelicerates and myriapods, and there are differences. It cannot be excluded that a change in cell division timing could transform the mechanism seen in insects into the myriapod neurogenesis. As stressed by Harzsch (2003a), homologous neurons could well arise through divergent developmental pathways.

(6) Similar axonogenesis: There are also differences between insects and myriapods in the formation of the first longitudinal axonal pathways (Whitington, Meier, and King, 1991). In a studied centipede, the first axon pathways arise from neurons located in the brain, while axonogenesis by segmental neurons begins later. It must be noted that these first axons are not connected to segmental ganglia and therefore are not homologous to axons arising from segmental somata. In insects, the latter are the first neurons producing longitudinal axons. The difference between insects and myriapods is therefore a difference in timing of axon growth for the first longitudinal axonal pathways. However, it was also shown later by Whitington, Leach, and Sandeman (1993) that in Malacostraca first axonal pathways can also arise from brain neurons or from the mandibular segment. Also, comparing insects, the pathways of segmental axons can differ (Whitington, 1995, Whitington, Harris, and Leach, 1996). Early axonogenesis is less conserved in arthropods than frequently circulated by proponents of the Pancrustacea hypothesis. The situation seen in myriapods can either be a derived character or a general plesiomorphic pattern shared with some crustaceans and chelicerates.

(7) Expression of segmentation genes (Patel, 1994; Averof and Akam, 1995b; Popadic et al., 1996; Dohle, 1997): Dohle (1997) stressed that in hitherto examined insects and crustaceans *engrailed* is expressed during embryogenesis in a transversal row of cells at the anterior part of a parasegment, while the engrailed antibody did not bind in the myriapod *Glomeris* (which might be a derived state). Furthermore, before visible segments form, in *Glomeris* there are many tightly packed small ectodermal cells. These data only show that myriapods are different, but there is no evidence for a derived or plesiomorphic state in comparison with insects.

Similarities in stomach (= proventriculus) morphology have been noted between some Decapoda and some insects (Klass, 1998). The similarity is essentially the presence of teeth. One must keep in mind that Decapoda are highly evolved Malacostraca, while more plesiomorphic characters can be found in Leptostraca, which lack such teeth. The location and number of teeth is different in insects and decapods.

The major feature in Malacostraca are not the teeth but ventromedian longitudinal folds that serve as filter channels. These are not only present in Decapoda, but have been described in detail in peracarids and leptostracans, where teeth are not developed. The channels are covered with fine setae that retain larger particles from the fluid that flows through the channels into the digestive glands (Haffer, 1965; Scheloske, 1976; Storch, 1987; Wägele, 1992). In insects and myriapods a filtering stomach and the digestive glands are absent. It seems that the teeth of Decapoda evolved only within Malacostraca and it is more probable that the teeth in some insect taxa evolved convergently. Remipedia do not have a complex gizzard (Felgenhauer, Abele, and Felder, 1992). If present in a common ancestor, the reduction of filter channels could have happened in the lineage leading to insects and myriapods or earlier, and this could easily be explained because the lack of midgut tubules makes filtering of the stomach chyme superfluous. However, currently there is no evidence for the existence of such filters outside Malacostraca. In summary, there is no evidence for homology of stomach characters present in a common ancestor of higher crustaceans and Hexapoda.

12.4.5 Putative derived homologies occurring in insects and myriapods (Tracheata)

The list of shared derived characters that occur in insects and myriapods is much longer than usually discussed. There are more similarities than just tracheal systems and Malpighian tubules. An earlier review (Klass and Kristensen, 2001) already illustrated several of these characters which are not compatible with the Pancrustacea hypothesis. Some are derived characters of Tracheata, others also occur in higher crustaceans. The following list is probably not complete, but it summarizes better known characters assumed to belong to the ground pattern of Tracheata:

(1) mandible without palp (*Character 37*)
(2) second antenna reduced, however, the tritocerebrum is present (*Character 38*)

(3) maxillae with two terminal, frontally directed endites, long protopod and long uniramous palps (*Character 43*)
(4) maxillae 2 basally fused (forming a labium, *Character 39*)
(5) cephalic endoskeleton with anterior tentorial arms (*Character 44*)
(6) first embryonic appendage article develops into pleural sclerites ("subcoxa") (*Character 45*)
(7) midgut glands reduced (*Character 46*)
(8) ectodermal Malpighian tubules present (*Character 47*)
(9) tracheal system with paired segmental spiracles, spiracles originally located on pleurae dorsally or dorsocaudally near leg insertion (*Character 48*)
(10) thoracic limbs are uniramous stenopodia, first free appendage article of thoracopod with stylus (*Character 49*)
(11) coxal eversible vesicles (*Character 50*
(12) indirect sperm transfer (*Character 51*)
(13) primary abdomen absent (*Character 32*)
(14) Tömösváry organ?
(15) Dorsally smooth head, head shield includes ocular and antennular segment (*Character 52*)
(16) erected brain (*Character 41*)
(17) gonoducts opening terminally (*Character 53*)
(18) praetarsus with single flexor muscle (*Character 54*)
(19) similar mushroom bodies?

The fact that these characters are not present in all myriapod or insect species is not worrying: plesiomorphic characters are often substituted or reduced in derived taxa, as in the case of the five toes of the ancestral tetrapod or the egg shell of amniotes. We typically expect to see ground pattern characters in less derived taxa.

In the following we discuss briefly the homology of these characters and the placement of homologies in nodes of a most parsimonious tree topology compatible with these characters (Figure 12.2).

(1) mandible without palp (*Character 37*)
The absence of the mandibular palp is not unique for insects and myriapods, it is also lacking in many crustaceans (e.g. Branchiopoda, Cirripedia, Oniscidea, Valvifera). Primary homology of the character cannot be substantiated. It is a ground pattern character of Tracheata and Remipedia (node 5 in Figure 12.2), but not of Crustacea.

(2) second antenna reduced, however the tritocerebrum is present as in crustaceans (*Character 38*)
The tritocerebrum belongs to the second limb-bearing head segment of Mandibulata. The segment is present in myriapods and insects, but the appendage is completely reduced (Heymons, 1901). This absence in all life stages of myriapods and insects

is unique among Mandibulata. Size reductions of antennae or absence in life stages also occur elsewhere, but not the complete absence in all stages. For example, both pairs of antennae are absent in adult Cirripedia, but present in larval stages. This is a character of Tracheata (node 6).

The reduction of the second antenna in myriapods and insects is – taken alone – a weak character. However, there is also a positive character associated with it, namely the unique pattern of expression of the *collier* gene (Janssen, Damen, and Budd, 2011). And, the fact that in crustaceans the second antenna is always present, at least in larval stages, increases the probability of homology of the reduction for myriapods and insects (as a rare event).

(3) maxillae with two terminal, frontally directed endites, long protopod and long uniramous palps (Figure 12.8) (*Character 43*)
In hexapods, the endites of the first maxilla (maxillula) are called lacinia and galea, those of the second maxilla (the labium) glossa and paraglossa. In myriapods, these endites are reduced in size, but they are present in most taxa, while in chilopods the maxillula protopod bears only a single apical lobe and the labium has reduced endites and short, fused protopods. A remarkable aspect is the terminal position of the endites on a comparatively long protopod (composed of cardo and stipes (maxillula) or mentum and praementum (maxilla)), and that the endites are **directed frontally**. This feature is not seen in crustaceans. In myriapods, both pairs of maxillae form a functional unit. In chilopods the long palps we see in insects are only present on the second maxilla. The latter functionally covers the lateral parts of the preoral space, while maxilla 1 fills the median area between maxilla 2.

Both pairs of maxillae of Progoneata form a lower lip without palps. The components are still separated in Symphyla (two pairs of maxillae), while in the other taxa they are fused to a gnathochilarium (see discussion of this character in Kraus, 2001). The small terminal endites are seen in all taxa, though often in reduced numbers. In diplopods the lateral areas corresponding to the first maxilla bear two pairs of lobes (reduced endites) on the stipes, the area of maxilla 2 bears only 1 pair of apical lobes. Comparison of Symphyla and basal insects suggests that the second maxilla might originally have had more than two endites, as in crustaceans.

The maxillae of crustaceans vary, but typically endites are located on the medial margin of the appendage and **directed medially** (Waloszek and Müller, 1998a, 1998b; Boxshall, 1998), also in Remipedia. The endopod is often absent and usually not longer than the protopod of the appendage. The maxillae of Remipedia are strongly modified, the palp is an exceptionally large raptorial endopod. The first maxilla has a protopod with two medially directed endites, and a third one on the first palpal article. The second maxilla has three endites. This number is also seen in Malacostraca.

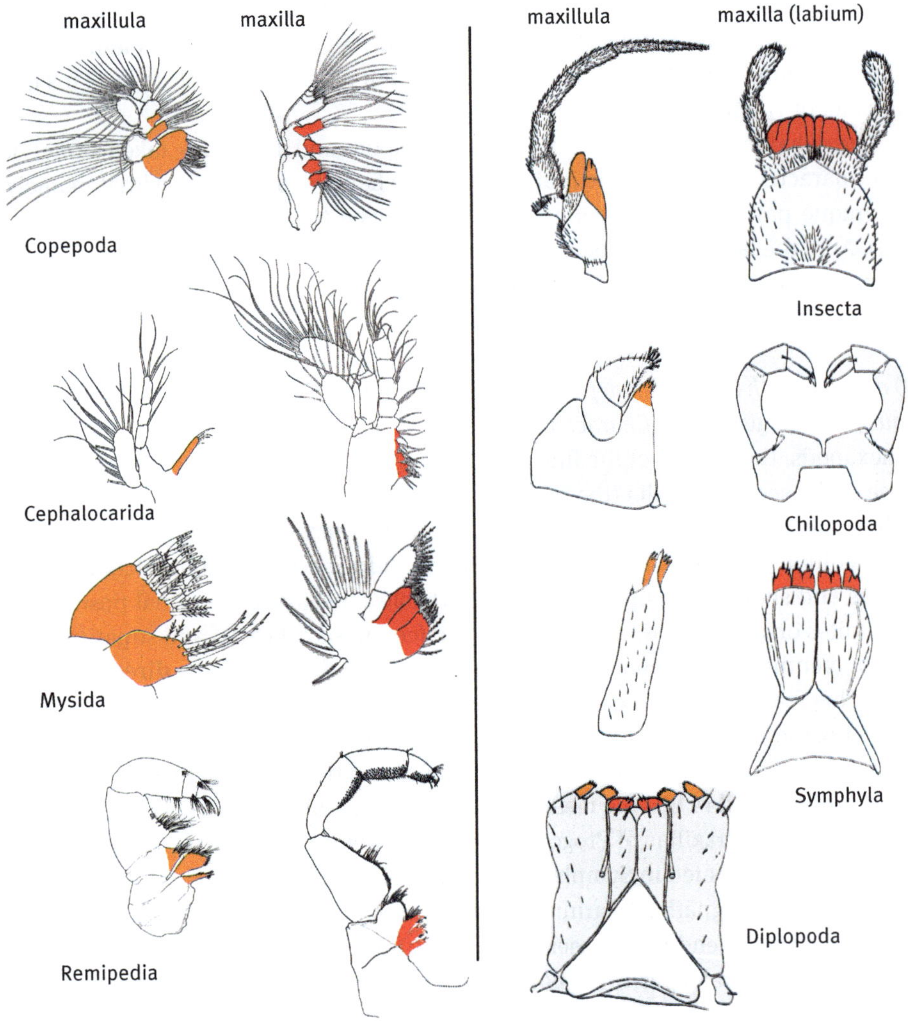

Figure 12.8: Comparison of maxillula (first maxilla) and maxilla (second maxilla) of crustaceans (left) and tracheates (right). In tracheates, the endites are directed frontally (and not medially as in crustaceans), both pairs of maxillae have stout protopods in relation to endite and endopod size (larger than in most crustaceans). In the second maxilla, protopods are fused. In Remipedia, the second maxilla is fused basally in the coxal region, which can be seen in undissected specimens (Fanenbruck, 2009). Copepoda: generalized copepod mouthparts (after Boxshall, 1991); Cephalocarida: *Sandersiella bathyalis* (after Hessler and Sanders, 1973); Mysida: *Haplostylus australiensis* (after Woolridge, Greenwood, and Greenwood, 1992); Remipedia: *Kaloketos pilosus* (from Koenemann, Iliffe, and Yager, 2004); Chilopoda: *Lithobius forficatus* (Mx1) and *Scolopendra cingulata* (Mx2) (after Attems, 1926); Symphyla: *Scutigerella immaculate* (after Attems, 1926); Diplopoda: *Polydesmus collaris* (after Attems, 1926).

The large basal articles (protopod) of the maxillae (cardo and stipes of maxilla 1 in insects) and the comparatively small terminal endites inserted terminally on the protopods of both pairs of maxillae are a potential synapomorphy of Tracheata (node 6 in Figure 12.2) absent in crustaceans. The elongated palps are also present in Remipedia and could – in view of the other evidence (see below) – have been a character already present in the last common ancestor of Remipedia and Tracheata (node 5 in Figure 12.2).

(4) maxillae 2 basally fused (forming a labium) (Figure 12.8) *(Character 39)*
The labium of insects consists of a pair of second maxillae with medially fused basal articles. This basal fusion is also typical for the myriapod maxilla, i.e. myriapods also have a labium. This medial region of the labium is small in chilopods, while in symphylans palps are missing and the medial parts of the labium are large and close the preoral cavity ventrally, as in insects. This is also similar in Diplopoda, where in addition the labium is laterally fused with the first maxilla to form the gnathochilarium.

The labium is a putative homology, but not a synapomorphy of tracheates, because the basal fusion is also seen in Remipedia if the two appendages of the labium are not dissected separately (Fanenbruck, 2009). In addition, in tracheates and in remipedes the maxilla has, where present, a relatively long palp (see also Character 2). The evolution of the labium is therefore compatible with a clade {Remipedia, Tracheata} (node 5 in Figure 12.2).

(5) cephalic endoskeleton with anterior tentorial arms (*Character 44*)
Anterior tentorial arms are a feature that cannot be explained as adaptation to terrestrial life. These hollow apodemes begin laterally of the labral insertion. Lateral tracts of the endoskeleton and their paralabral roots found in crustaceans in this part of the head are missing (Fanenbruck, 2009). The tentorial arms occur in ectognathous insects, symphylans, chilopods and myriapods (Koch, 2001; Snodgrass, 1935) and are a character of high probability of homology due to the complexity of the cephalic endoskeleton: "... the anterior tentorium itself is among the most noteworthy potential myriapod/hexapod synapomorphies" (Klass and Kristensen, 2001) (node 6 in Figure 12.2).

(6) first embryonic thoracopod article develops into pleural sclerites ("subcoxa") (Figure 12.9) (*Character 45*)
This character was well known to entomologists a hundred years ago (e.g. Heymons 1899; Verhoeff, 1902; 1906; Snodgrass, 1909) and was later forgotten (with rare exceptions: Manton, 1979; Deuve, 1994). Heymons (1899) was probably the first who homologized pleural sclerites of insects with a subcoxal article. In adult specimens of insects and chilopods, pleural sclerites surround the insertion of the "coxa", forming semilunar rings. In winged insects the sclerites are larger, reinforcing the pleural area. Similar sclerites do not exist in chelicerates and crustaceans. Since these sclerites evolve during ontogenesis from the limb base (see also Roonwal, 1936; Ibrahim,

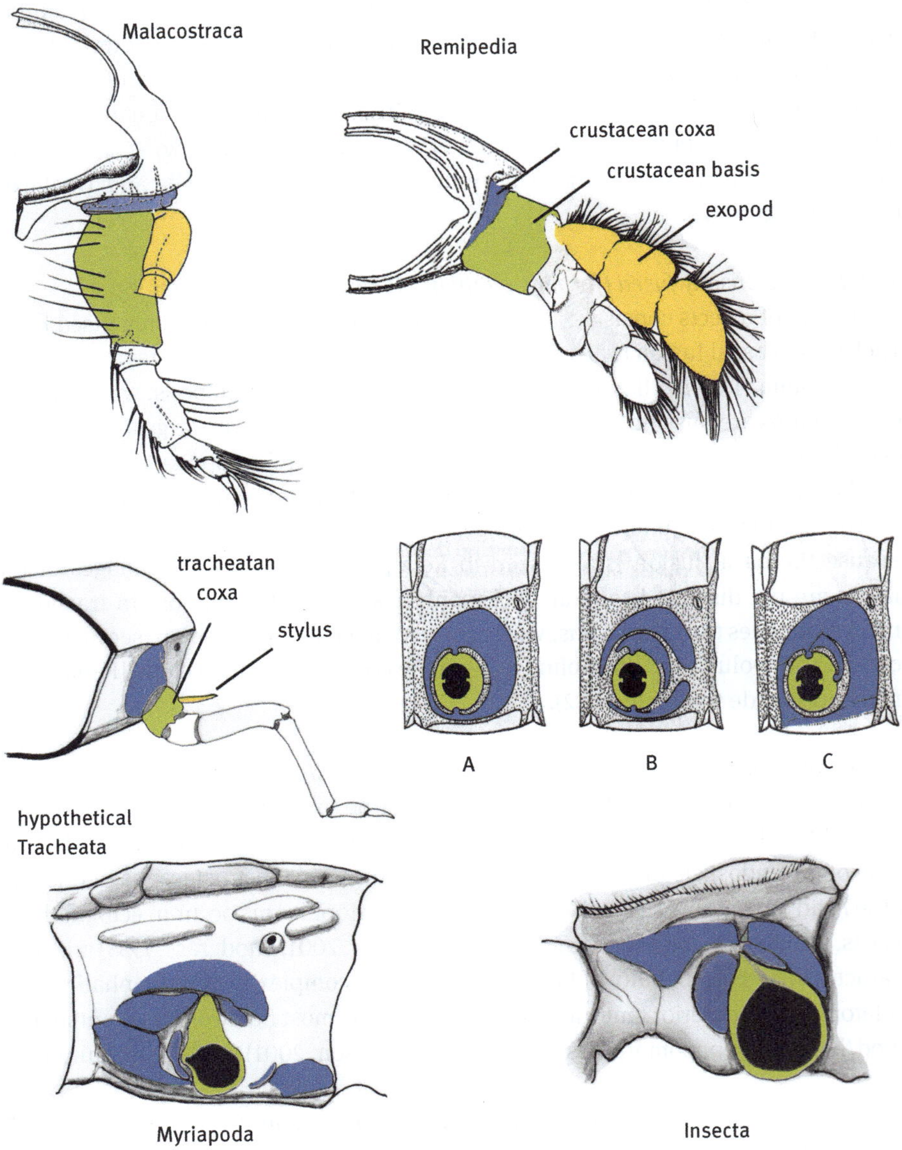

1958; Bretfeld, 1963) and seem to be derived from an appendage article, Heymons (1899) named these sclerites "subcoxa", a view shared by Snodgrass (1927), Ewing (1928) and Weber (1928; 1933). Interestingly, an intermediate state is seen in Remipedia (Hessler and Yager, 1998), where an immobile semilunar coxa (this is the crustacean coxa, not the insect coxa!) is fused to the pleural area. This observation is highly relevant, because according to recent molecular analyses Remipedia are placed close to insects (Reumont, Jenner, Wills et al., 2012).

◂ **Figure 12.9:** The subcoxa theory explains the transformation of the crustacean leg into the thoracopod seen in insects and myriapods (see Bäcker, Fanenbruck, and Wägele, 2008). Note that the number of appendage articles is the same in Malacostraca and basic Tracheata, if we assume that the crustacean coxa (blue) is transformed into subcoxal (pleural) sclerites and the crustacean basis (green) is the coxa of Tracheata. The styli are possibly exopod rudiments. **A.** pleural area of a hypothetical ancestral tracheatan; **B.** further evolution of pleural area with separation of the outer ring of eupleurites; **C.** subcoxa of a wing-bearing hexapod segment (after Snodgrass, 1935). Malacostraca: *Diastylis rathkei* (after Hessler, 1982); Remipedia: combined after Hessler and Yager (1998) and Koenemann, Iliffe, and Yager (2004); hypothetical Tracheata and evolution of hexapod subcoxa after Eidmann and Kühlhorn (1970); Myriapoda: *Cryptops hortensis* (after Bäcker, Fanenbruck, and Wägele, 2008); Insecta: *Neomura cinerea* (Bäcker, Fanenbruck, and Wägele, 2008). The outer ring of subcoxal sclerites are the eupleurites, the inner ring the trochantinopleurites. Fragmentation of the pleurites varies in Myriapoda and basal Hexapoda.

Imms (1938: p. 30) writes that in insects "the coxa [= crustacean basis] has replaced the subcoxa as the functional base of the leg". A secondary effect of the transformation of the crustacean coxa into subcoxal sclerites of tracheates would be the transformation of the **crustacean basipodite** (the second article), which carries exo- and endopod, into the **tracheate coxa** (the first movable article) (character of node 6 in Figure 12.2) And indeed, the latter often bears two branches in less derived hexapods and myriapods, namely the fully developed stenopodial endopod and in addition the enigmatic stylus in the place of the exopod (see Character 10).

Assuming that remipedes are a sister-lineage of tracheates, the scenario for the evolution of subcoxal sclerites starting with the intermediate condition of an immobile coxa fused to the pleuron as seen in Remipedia is plausible and – as opposed to the assumption of a parallel evolution of pleural sclerites, styli and coxal vesicles in myriapods and hexapods – is more parsimonious. The immobilized coxa is a character of node 5 in Figure 12.2.

Recent studies on the development of the insect leg have focused on the influence of gene expression on article formation, neglecting the area of leg attachment, where the subcoxa (= crustacean coxa) should be seen. However, cross sections of imaginal discs of *Drosophila* which develop to adult legs show in the disc epithelium outside the specific ring that develops to the insect coxa an additional ring that could be the anlage of the subcoxa (e.g. Fig. 1 in Kojima, 2004). Heymons (1899) described the subcoxa in a water bug (Naucoridae) as an embryonic article located between the future coxa and the pleuron.

The subcoxa is not a single sclerite. The breaking up of the cuticle of an article into several sclerites is not unique for the subcoxa; it can also occur in other leg articles, as in coxa and trochanter of *Scutigera* (e.g. Becker, 1923).

Again, the fact that variations of pleural sclerites exist (extensively discussed in Bäcker, Fanenbruck, and Wägele, 2008) are no argument against homology. The arguments pro homology are: origin from embryonic limb base, sclerites or their fragments forming two (usually fragmented) concentric semilunar rings (named e.g. anapleurite and coxopleurite: Snodgrass, 1935). It is possible to derive the different shapes from a hypothetical ancestral state (see Bäcker, Fanenbruck, and Wägele, 2008).

Similar structures do not occur in other arthropods. Therefore, even if the homology of the subcoxa of tracheates with the crustacean coxa is rejected, the fact remains that myriapods and insects share the same peculiar pleural sclerite system.

This allows a new interpretation of the homology of trunk appendages in higher crustaceans and tracheates (Table 12.1).

Table 12.1: A proposed homology of trunk limb podomeres of higher crustaceans and Tracheata. The dactylus is missing in paddle-shaped appendages of Remipedia, their coxa is fused to the pleural region. The dotted line indicates where a pronounced knee is formed in Tracheata.

	Malacostraca	Insecta	Myriapoda	
1	coxa	subcoxa	subcoxa	
2	basis with exopod	coxa with stylus	coxa with stylus	
3	ischium	trochanter	trochanter	endopod
4	merus	femur	prefemur + femur	endopod
5	carpus	tibia	tibia	endopod
6	propodus	tarsus	tarsus	endopod
7	dactylus with claws	praetarsus with claws	praetarsus with claws	endopod

(7) midgut glands reduced (*Character 46*)
Most crustaceans and chelicerates have tubular, often branched digestive glands that open into the anterior midgut. These glands are important organs for the production of digestive enzymes, for resorption of food, for storage of glycogen and lipids, detoxification, and for synthesis of blood pigments (e.g. Picaud, Souty-Grosset, and Martin, 1989; Hennecke, Gellissen, and Spindler, 1991; Lovett and Felder, 1990; Lallier and Walsh, 1991; Brunet, Arnaud, and Mazza, 1994). Crustaceans that are closer to insects in molecular phylogenies like larger Branchiopoda (e.g. *Triops*), Cephalocarida and Malacostraca all have (often voluminous) tubular digestive glands. It is therefore surprising to see that these organs are absent in Remipedia, myriapods and insects. This is not merely a reduction, but it means that other tissues must take over the function of digestive glands.

The digestive tract of insects and myriapods is essentially a straight tube, sometimes coiled when the posterior gut is longer than the body length. The foregut is of ectodermal origin, the midgut bears no cuticle and is entodermal, the hindgut is again lined by a thin cuticle, as in other arthropods. The foregut can have a crop and a differentiated gizzard. The transition from foregut to midgut often forms a valve, as in other arthropods. The midgut epithelium contains cell groups clustered in crypts or between furrows. Production of digestive enzymes is restricted to the midgut epithelium. The well-developed fat body is the main storage tissue for glycogen, lipids, vitellogenins etc. in insects and myriapods (e.g. Seifert, 1979). The peritrophic membrane is well developed, as in other arthropods (e.g. Fidalgo, 1990; Martin, 1992; Brunet, Arnaud, and Mazza, 1994; Halcrow, 2001). It seems that in tracheates the function of the midgut glands was taken over by the midgut (digestion) and by the fat body (storage, syntheses).

Reductions of the midgut glands are rare in crustaceans and typically occur in dwarfish species with little space in their body (e.g. Cladocera, Mystacocarida). It is therefore remarkable that the comparatively large Remipedia as well as myriapods and insects do not possess these glands. The character supports the clade {Remipedia, Tracheata} (node 5 in Figure 12.2).

(8) ectodermal Malpighian tubules present, originating at the junction between midgut and hindgut (*Character 47*)
This character is frequently cited as part of the ground pattern of Tracheata. The excretory tubules of insects and chilopods have fundamentally the same ultrastructure (e.g. Füller, 1963; Seifert, 1979) and position. They are formed during embryogenesis from the anterior part of the proctodaeum. The main problem discussed in literature is that similar tubes are known from Arachnida, however, these tubules have an entodermal origin (see Seifert, 1979) and evolved convergently when chelicerates adapted to terrestrial life. Obviously, the posterior gut of euarthropods has the potential to form such tubes. It is little known that posterior tubules (of unknown function) are also present in amphipod crustaceans (Schmitz, 1992). A hypothesis of convergent evolution within Tracheata cannot be rejected, but for a common ancestor of insects and myriapods a single origin is the more parsimonious hypothesis (character of node 6 in Figure 12.2).

(9) tracheal system with paired segmental spiracles originally on pleurae, spiracles located dorsally or dorsocaudally near leg insertion (Figure 12.10) (*Character 48*)
Because respiratory systems have to be adapted when aquatic animals evolve to terrestrial life forms, a frequently cited argument is that the tracheal systems of insects and myriapods could have evolved convergently (e.g. Dohle, 1997; 1998; Koch, 2001). And indeed, tracheal systems of insects and myriapods are not identical in every detail. Some authors think that these variations are indications for non-homology (discussed in Hilken, 1998). However, this argument has no logical basis (Klass and Kristensen, 2001), variation is no evidence against homology (see above). An important observation is that there exist no stem-group myriapods or insects with a "primitive" bauplan that lack tracheae, as expected in a scenario where respiratory systems evolved several times after colonization of terrestrial habitats, as seen in terrestrial Isopoda (Schmidt and Wägele, 2001).

"Most centipedes possess a respiratory system comparable with that of insects ..." (Minelli, 1993). There is no other arthropod group that has a segmental arrangement of paired pleural spiracles and ectodermal tubules. The spiracles are pleural open-

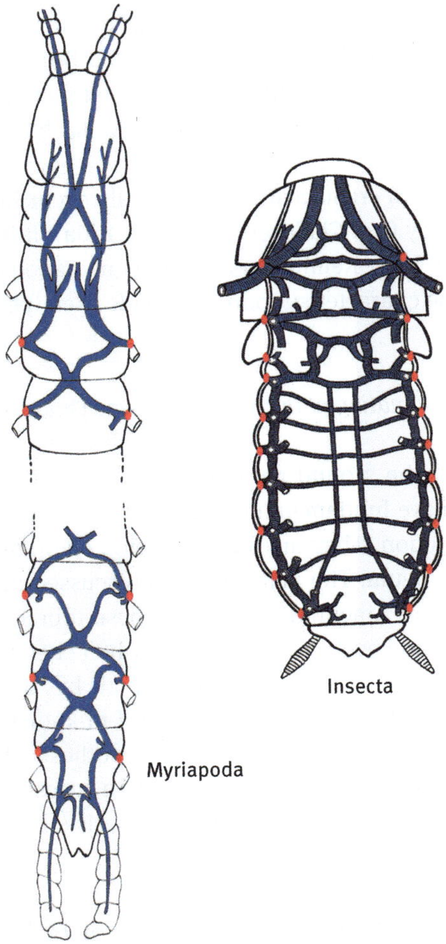

Figure 12.10: Number of spiracles, ramifications and anastomoses of the tracheal system of Myriapoda and Hexapoda vary. However, the principal pattern of segmental spiracles (red) located on the pleurae is the same and does not occur in other arthropods. Myriapoda: principal tracheae of *Geophilus carpophagus* (after Dubuisson, 1928). Insecta: principal ventral tracheae of *Periplaneta* (after Imms, 1938).

ings usually placed close to the leg insertion in similar spatial relations to pleural sclerites when insects and Pleurostigmophora are compared (Klass, 2000). Of course, there are some variations, as the migration of spiracles into a dorsal position (in Scutigeromorpha), or the ventral "sternal" position in Dignatha that is easily explained by the large expansion of tergites that in Diplopoda cover diplosegments laterally, which causes a ventral position of *all* pleural structures. Snodgrass (1958: 20) already argued that the lateral spiracle plates occurring in some diplopods are pleurites. This implies that the spiracles of diplopods are topologically essentially in the same position as in the ground pattern of chilopods and not a new "sternal" structure.

The reduction of the respiratory system in dwarfs (Pauropoda) is typical for secondarily miniaturized animals (see discussion in Klass and Kristensen, 2001). Internally, the tracheal system consists of segmental ducts that open into the spiracles, and longitudinal tubes that usually are connected in various ways, with modifications even among closely related taxa (see e.g. Fig. 10 in Minelli, 1993). Nevertheless, all these variations repeat the main pattern. The most parsimonious assumption is that these variations are derived from a common pattern of segmentally arranged pleural spiracles and tracheal tubules. This assumption requires a common ancestry of insects and myriapods (a character of node 6 in Figure 12.2).

(10) thoracopods are uniramous stenopodia, first free appendage article of adult thoracopod with stylus (Figure 12.9) (*Character 49*)
In crustaceans, exopod and endopod insert on the second thoracopod article, the basipodite. In insects and myriapods, the homologous article is named coxa by entomologists, while the crustacean coxa is fused with the pleural area (see Character 6). Therefore, in tracheates we find styli on the article homologous to the one that in crustaceans bears exopods (a character of node 6 in Figure 12.2).

This opens the possibility that styli are rudimentary exopods (Sharov, 1966; Klass and Kristensen, 2001, see also Table 12.1). Since their function is not clear and more derived hexapods do not possess them, the interpretation as rudiments is a plausible explanation. Styli are small and unsegmented, intrinsic musculature is lacking in myriapods, but exists in insects. Styli occur in Symphyla, Entognatha and basal Ectognatha and are "a potentially important synapomorphy of Tracheata ..." (Grimaldi, 2010). There has been some discussion about the endopod nature of abdominal styli in insects (Klass and Kristensen, 2001), but in combination with the presence of coxal vesicles, the arrangement of peculiar coxal outgrowths is the same in symphylans and hexapods. Styli are absent in more derived Hexapoda and many Myriapoda, they are obviously not a necessity for terrestrial life. However, their presence in different lineages of Tracheata is best explained by common ancestry, they are a rudiment of the biramous appendage of crustaceans.

Another important detail is the stenopodium. This is an endopod with stout, essentially cylindrical articles that can bear the body weight. Among mandibulates, stenopodia evolved several times convergently. It seems that many crustacean lineages started with swimming, epibenthic lifestyles, as seen in Branchiopoda, in basal Malacostraca, Remipedia, Cephalocarida. These animals have weaker and flattened endopods not suitable for walking. They can use their limbs to rest on the sediment or to stir up food particles, but they move by swimming. Animals that walk on the ground supporting the body weight with stout stenopodia are the more derived and benthic decapods and peracarids among Malacostraca. The transformation of a thoracopod into a stenopodium can be studied in the evolution of Decapoda, from swimming shrimps to heavy crabs. A parallel transformation is seen in peracarids, from fairy shrimps to isopods. We must assume that a similar evolution took place in the stem lineage of Tracheata. The shared similar musculature (see Character 18) and the formation of a pronounced knee between femur and tibia (Table 12.1) are additional arguments for the homology of the stenopodia of myriapods and insects.

(11) coxal eversible vesicles (*Character 50*)
These membranous sacs (some with extrinsic muscles) occur ventrally on coxae or between sternites in coxal areas in basal hexapods and in myriapods (not in Chilopoda) (e.g. Tiegs, 1945; Drummond, 1953; Weyda, 1974; Eisenbeis, 1982). They are an adaptation to absorb water from thin films. Since such structures are absent in crustaceans, they probably evolved in the stem lineage of Tracheata as adaptation to terrestrial life. In Chilopoda, different coxal organs occur on posterior legs, where under several cuticular pores a transport epithelium suitable for water uptake occurs (e.g. Rosenberg, 1983). It is not clear if these are modified coxal vesicles. Klass and Kristensen (2001) conclude after a discussion of the homology of vesicle retractor muscles that at least the vesicles of Insecta, Diplura and Progoneata "may indeed be homologous". In contrast to parallel evolution it is more parsimonious to assume that the common ancestor of Myriapoda and Hexapoda possessed coxal vesicles (node 6 in Figure 12.2).

(12) indirect sperm transfer (*Character 51*)
In crustaceans, fertilization usually occurs by transfer of spermatophores or sperm masses which are attached to the female (as in Cirripedia or Copepoda) or transferred into the female genital system (many Malacostraca), from where eggs may be fertilized within the female or externally during spawning. Free spawning into the water and *indirect* insemination has not been documented for crustaceans (Subramonian, 1993). This is different in tracheates. Indirect sperm transfer is typical for basal tracheates: in chilopods, males spin a web and deposit on it a stalked spermatophore. In Diplopoda there is no male-female contact in Penicillata. All apterygote insects also use – as far as is known – either sperm droplets (as in Machilidae) or stalked spermatophores. It is interesting that spinning of threads by males is observed in Chilopoda, Machilidae, Lepismatidae, Lepidothrichidae. Collembola produce stalked droplets placed on the ground. *Direct* sperm transfer is obviously a derived state that evolved parallel in Diplopoda and in the stem lineage of Pterygota. The most parsimonious assumption is that the combination of spinning of webs or threads by males and secretion of a stalked spermatophore for indirect insemination evolved in the stem lineage of Tracheata as an adaptation to terrestrial life (summary in Bitsch and Bitsch, 1998) (a character of node 6 in Figure 12.2)

(13) primary abdomen absent (Figure 12.2) (*Character 32*)
We define here "primary abdomen" as a posterior trunk region of Mandibulata with primarily absent limbs. It is relevant to consider in this context the fossil record. It is clear that stem-lineage arthropods, tardigrades and onychophorans, and also many Cambrian arthropods like trilobites have no subdivision of the trunk into thorax and limb-free abdomen. However, all stem-lineage Mandibulata have such an abdomen (see e.g. Orsten fauna in Waloszek, 1995; Waloszek and Müller, 1998a, 1998b; Waloszek, 2003a; 2003b; comments in Fanenbruck, 2009). Its reduction is rare and clearly

derived within crustaceans, as in dwarfish Branchiopoda, or in all Malacostraca. The same applies to insects and myriapods. It seems that absence of the primary abdomen is caused by a loss of posterior segments. Averof and Akam (1995b) have shown that Hox gene (AbdA, AbdB) expression in the branchiopod thorax is the same as in the entire insect pregenital trunk, suggesting that the crustacean multisegmented abdomen (as in *Artemia*) is absent in insects, but present in branchiopods. And indeed, the insect "abdomen" can form larval legs and often carries leg rudiments (styli) or genital appendages in adults. It is a "secondary abdomen", derived from a trunk that had limbs on all segments and that still conserves the ability to express genes for the formation of paired appendages.

A primary multisegmented abdomen is also absent in Malacostraca (their pleon bears swimming appendages), Remipedia, and Myriapoda. Morphologically, Malacostraca are not primitive crustaceans and often were thought to be close to insects (Harzsch, 2002a; Fanenbruck, Harzsch, and Wägele, 2004; Grimaldi, 2010; Strausfeld, 2011). Tagmosis and head details of Remipedia resemble myriapods and in more recent molecular analyses Remipedia appear close to insects. Remipedia, Malacostraca and insects share a complex derived brain anatomy (Fanenbruck, Harzsch, and Wägele, 2004): it is highly probable that they share a last common ancestor with this type of brain.

The resulting scenario: we agree with Moura and Christoffersen (1996) and Fanenbruck (2009) that the primary abdomen was reduced in a lineage of crustaceans that gave rise to Malacostraca, Remipedia, and Tracheata (the Caudoabdicata of Fanenbruck, 2009) (a character of node 4 in Figure 12.2). The character shared by insects and myriapods is a plesiomorphic homology and not compatible with the Pancrustacea hypothesis.

(14) Tömösváry organ

This organ is always located in the region between the eye and the insertion of the antenna. It consists of a pit covered by a cuticular plate which may have pores or slits (Haupt, 1979). The pit contains sensory cells innervated by the protocerebrum. The sensory cells have branched or unbranched dendrites projecting into the pit, partly extending into the pores. In Chilopoda, this structure is found only in anamorph centipedes (Tichy, 1973; Minelli, 1993). The innervation has been studied in *Lithobius* (Petyko et al., 1996): the organ is connected to neuropil areas proximal to the second optic neuropil, i.e. in the dorsolateral protocerebrum. The proturan *Eosentomon transitorium* has only one pore in the endocuticle (Haupt, 1972). Collembolans, symphylans and pauropods have very similar pore fields (Haupt, 1972; 1973).

As pointed out earlier (e.g. Wägele, 1993; Klass and Kristensen, 2001), a similar organ is known from crustaceans. It was described first for isopods (e.g. Bellonci 1881; Chaigneau, 1971; 1976) and was later found in many crustacean taxa, including other Malacostraca, Copepoda, Mystacocarida, it possibly also exists in Branchiopoda (Renaud-Mornant, Pochon-Masson, and Chaigneau, 1977; Martin, 1992; Boxshall,

1992; Hosfeld, 1995). Therefore, this may be a mandibulatan character not compatible with the Myriochelata hypothesis. To discover characteristics of this organ exclusive for tracheates more detailed comparative analyses are required.

(15) Dorsally smooth head, head shield includes ocular and antennular segment (Figure 12.11) (*Character 52*)
This character was studied in detail by Haug (2011). In Tracheata, the head has no freely outgrowing shield margins, and the compound eyes and antennulae are situated dorsally on the "head capsule". Looking at the frontal area of a head of myriapods or insects one sees a smooth surface. In contrast to crustacean heads, there is no rostrum, rim, fold or suture at the boundary between the areas of eyes/antennae and the remaining dorsal areas of the head. Pleurostigmophora possess a flat "head plate", which however can be derived from the head shield shape of other myriapods.

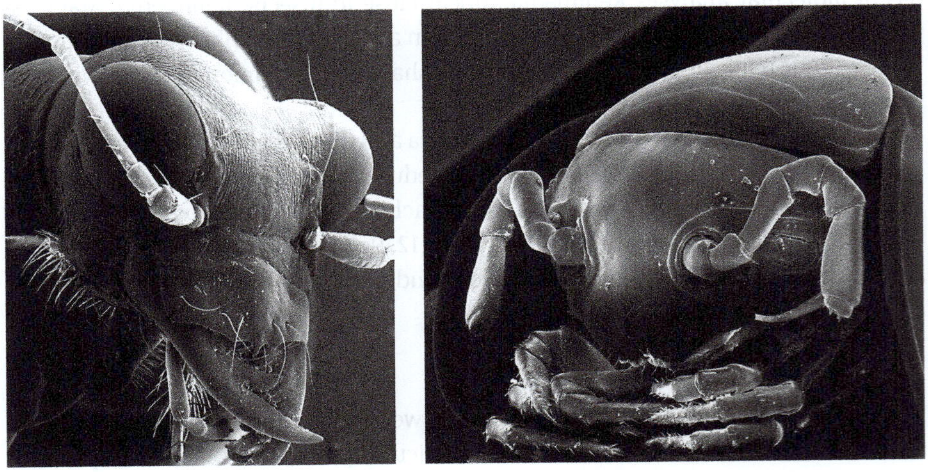

Figure 12.11: Tracheate heads are smooth, have no carapace and no protruding rims or folds of the cephalic shield. The head shield encloses eyes and antennae. (Insect: a grasshopper; myriapod: a diplopod).

In crustaceans, the dorsal parts of the mandibulatan head and often additional thorax segments are dorsally fused with the head shield. In addition, shield margins are protruding to different degrees, often covering dorsally and laterally additional segments and appendages. In some taxa, as in isopods (highly derived Malacostraca), the head shield is strongly reduced and free lateral extensions are lacking. Therefore the head resembles that of Tracheata. However, the anterior margin that separates the antennal segments from the remaining head is often visible and can possess a rostral point. Following Haug (2011) we propose that the smooth head of myriapods and insects with the dorsalized eyes and antennae evolved in the stem lineage of Tracheata (node 6 in Figure 12.2).

(16) Erected brain (*Character 41*)
A peculiar feature seen in myriapods and insects is the spatial arrangement of the anterior brain in comparison with crustaceans. It has essentially a vertical position in relation to the longitudinal axis of the animals. Opening the head capsule dorsally, one sees the protocerebrum covering the other parts of the brain (e.g. Minelli, 1993). This position of the protocerebrum correlates with the dorsalized position of eyes and antennae seen in myriapods and insects (Figure 12.11). The erection of the brain was also described for Remipedia, where even in dorsal view the protocerebrum is located posteriorly to the deutocerebrum (Fanenbruck, Harzsch, and Wägele, 2004). This character supports the clade {Remipedia, Tracheata} (node 5 in Figure 12.2).

(17) Gonoducts opening terminally (*Character 53*)
Sexes are separated in myriapods and insects, as in nearly all arthropods. Reproductive organs basically consist of paired organs ending terminally. Reproductive organs in crustaceans may open in different parts of the body. The most frequent position of gonopores is the region between thorax and primary abdomen (Schram, 1986). In malacostracans, where the primary abdomen is missing, gonopores are found in posterior segments of the anterior thorax before the pleon (in males eighth, in females sixth thoracic segment). Chilopods have two preanal genital segments, the penultimate segment bearing the genital orifice. In the monophyletic progoneate lineage (Symphyla, Pauropoda, Diplopoda: Shear and Edgecombe, 2010) the genital opening is relocated to anterior body segments, a mutation that must have occurred in the stem lineage of Progoneata. Since the genital opening of insects is also located terminally, it is most parsimonious to assume that the last common ancestor of myriapods and insects also possessed this character (node 6 in Figure 12.2).

(18) praetarsus with single flexor muscle (*Character 54*)
As discussed by Bitsch and Bitsch (2004), the last article of thoracopods is movable by two antagonistic muscles in Chelicerata and Crustacea, while in Myriapoda and Hexapoda only a single muscle exists, a flexor which inserts in the tibia (sometimes additionally in the femur) and has a remarkably long tendon. This is certainly a shared derived state with two modifications: the reduction of the extensor and the elongation of the muscle-tendon assemblage. The most parsimonious assumption is that this character evolved only once in the stem lineage of Tracheata (node 6 in Figure 12.2).

(19) similar mushroom bodies
Loesel and Heuer (2010) compared mushroom bodies (MBs) in the brain of arthropods and annelids. They conclude that these specific neuroanatomical structures are homologous in annelids and arthropods and that there are further characteristics in

some clades. They summarize patterns that indicate a close relationship of hexapods and myriapods. Myriapods have "... clusters of small-diameter globuli cells ... that supply ramifications to MBs which comprise a pedunculus and lobes which are connected to the antennal lobes via a tract of interneurons .... In *Lithobius variegatus* the lobes have been described to represent spherical outswellings, a motif similar to the MB organization of the apterygote hexapod *Lepisma saccharina* ...". This could also be a character of node 6 in Figure 12.2. Even though taxon sampling is still poor because of the technical difficulties of the reconstructions, hitherto published observations indicate that mushroom bodies evolve slowly and are a good phylogenetic marker.

There is more evidence. However, often taxon sampling is poor and more comparisons are necessary.

- Hennig (1969) assumed that paired tarsal claws belong to the ground pattern of Tracheata. They are known in all Ectognatha and in Diplura, and in some Myriapoda (Symphyla, Pauropoda). However, crustaceans also possess claws, usually in the form of a single terminal article, sometimes with accessory claws, and a detailed comparison is lacking.
- Wägele (1993) reviewed aspects of the structure and function of neurohemal organs of arthropods, which during recent years have not been studied any more. It seems that in insects and myriapods the neurohemal projections of protocerebral neurosecretory cells (such as the *corpus allatum* and *corpus cardiacum* of insects) are placed more posteriorly and more separated from the brain than in crustaceans.
- Innervation patterns of dorsal longitudinal muscles are the same in insects and chilopods (Heckmann and Kutsch, 1990). The exploration of this character requires more comparisons with other arthropods.
- Jannsen and Budd (2010) discuss that there is possibly a conserved mechanism of the regulation of the Hox gene Ubx in myriapods and *Drosophila*.
- Myriapods and insects have broad sternites with lateral endoskeletal furcal rami (insects) or apophyses (in myriapods). Unfortunately, the evolution of sternal sclerites in crustaceans remains unstudied, i.e. the outgroup character state is unknown. (Note that sternites are also primarily present in Chilognatha: Kraus & Brauckmann, 2003)
- The mandibles of Malacostraca, Remipedia, some Myriapoda and some insects have a distal incisor part, a proximal molar process, and between these processes some spines and in addition a movable tooth or movable stout spine underneath the *pars incisiva*, coined *lacinia mobilis* for crustacean mandibles. Though Richter, Edgecombe, and Wilson (2002) argue that differences in shape, asymmetry and position are evidence against homology of the *lacinia*, suspiciously similar structures are nevertheless present. The mandible of an immature ephemeropteran, for example, can be confounded with that of peracarid crustaceans. It is more parsimonious to assume that the genetic information for this structure evolved

only once in a common ancestor of Malacostraca, Remipedia and Tracheata. As already mentioned, variation is no argument against a homology hypothesis.

Placement of Myriapoda as sister-group to Pancrustacea (= Tetraconata) would make all these characters autapomorphies of Myriapoda that evolved convergently later in time in the stem lineage of Hexapoda.

12.4.6 Taxonomic consequences: Caudoabdicata and Archilabiata

Fanenbruck (2009) already introduced the names Caudoabdicata for the clade {Malacostraca, Remipedia, Tracheata} and Archilabiata for {Remipedia, Tracheata}. The first name refers to the reduced primary abdomen, the second to the basal fusion of maxilla 2. Fanenbruck also listed and discussed the derived character states that support the monophyly of these clades.

12.4.7 Fossil record and the implausibility of a Cambrian origin of Myriapoda

If, as seen in molecular phylogenies, the myriapod lineage branches off as earliest mandibulatan clade, direct ancestors of modern myriapods should have existed together with early crustaceans. Several fossils of the mandibulatan or crustacean stem lineage appear during the Cambrian, as demonstrated with the studies of the Orsten fauna (Müller, 1983; Müller and Waloszek, 1986; Waloszek, 1999; Siveter, Williams, and Waloszek, 2001; Siveter, Waloszek, and Williams, 2003; Waloszek, 2003a, 2003b). These early fossils have – in contrast to myriapods – no well-developed specialized mouthparts and the species had a long primary abdomen. A summary of the fossil record for Myriapoda was published by Shear and Edgecombe (2010). There are no Cambrian or Ordovician myriapods or insects. Scutigeromorphs are known from the Late Silurian (418 m.y.a.) and thus are the oldest known Chilopoda, of the Chilognatha earliest fossils are also of Silurian age (Cowiedesmida, Zosterogrammida, Eoarthropleurida). This coincides with the first expansion of vegetation on land (Kenrick and Crane, 1997) and precedes or coincides with the early evolution of insects (Labandeira, Beall, and Hueber, 1988; Gaunt and Miles, 2002).

Under the "early Mandibulata" scenario, the myriapod lineage must have evolved since the Lower Cambrian in the ocean until these animals conquered land in the Silurian. This, however, is mere speculation (a "ghost lineage"), because corresponding fossils have never been found (Edgecombe, 2004; 2010). The Early Cambrian fossil *Ercaia minuscula* could in theory be such a fossil (Chen, Vannier and Haug, 2001). It is elongated, with legs on all segments, i.e. it has no primary abdomen. However, mouthparts are not known, it may not have mandibles, and there is no synapomorphy of the myriapod lineage. Such fossils could also belong to the stem-group of Remipedia. Under the Tracheata scenario such speculations are not required.

What we know about the evolution of habitats on earth is only compatible with the hypothesis that the first terrestrial mandibulatans, including myriapods, evolved at a time when many lineages of crustaceans already existed. The first geological period where a transition to terrestrial life was possible for an arthropod fauna was probably the Silurian, when the first land plants started to spread (Kenrick and Crane, 1997). This is also the period when chelicerates conquered land and the first arachnids evolved (Dunlop and Selden, 2009), while the radiation of spiders and their prey, the insects, begins in the Devonian (Penney, 2004). Though it is conceivable that a proto-myriapod already lived earlier on shores, feeding on algae, the evolution of terrestrial arthropods that abandon amphibic life cycles requires a well-developed terrestrial vegetation as a source for food, moisture and shade. This implies that a derivation of a myriapod lineage at the base of the mandibulatan tree, i.e. earlier than all known crustacean lineages and much earlier than insects (and arachnids) is not plausible, because at the beginning of the Cambrian (or even earlier) there was no suitable terrestrial habitat. There is also not a single marine Cambrian fossil that might count as marine representative of the myriapod lineage.

12.5 A plausible scenario: Remipedia as last living marine relatives of Tracheata

(a) Conquering land: the morphological scenario (Figure 12.12)

The morphology of Remipedia is a puzzling mixture of characters. They have no primary abdomen, but biramous antennules and a second antenna with a scale-like exopod, just as malacostracan crustaceans, while the posterior gonopores, the pleural "subcoxal" ring, the fused second maxillae, and the absence of a hepatopancreas (midgut glands) remind of myriapods and insects. Moura and Christoffersen (1996) already assumed that these are homologies and proposed the clade {Malacostraca, Remipedia, Tracheata}. This is the clade Caudoabdicata of Fanenbruck (2009). Important arguments are:

(1) The trunk of Remipedia has a myriapod-like appearance with many similar and homonymous appendage-bearing segments, just as myriapods. A pleon with pleopods (present in Malacostraca) or a primary abdomen (present in many arthropods, but not in Malacostraca and Hexapoda) are missing. The absence of the primary abdomen is a derived character. No other crustacean group has such a myriapod-like body.

(2) There are several derived characters of the mouthparts: The mandibular palp is lacking in Remipedia, as in myriapods and insects. The mandibles of Remipedia bear a lacinia mobilis, i.e. a mobile tooth located below the cutting edge of the pars incisiva. A very similar lacinia is found on the same place in several

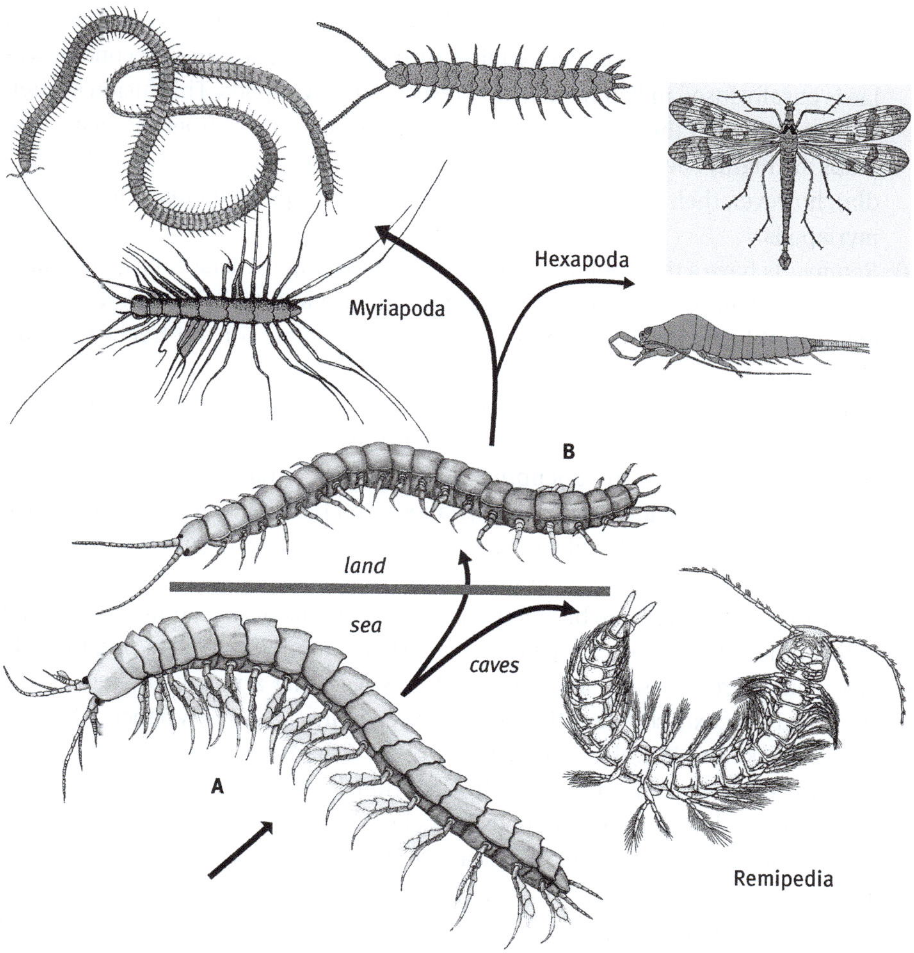

Figure 12.12: Cartoon illustrating the Archilabiata scenario: the marine ancestors of Tracheata (A) must have been similar in many respects to extant Remipedia, however, lacking the special adaptations to life in caves (e.g., eyes are not reduced). Since these creatures conquered land, their endopods must have been suitable for walking. The first tracheate (B) is morphologically a link between myriapods and the remiped-like aquatic ancestors. Head and thoracopods have the adaptations seen in myriapods and insects (e.g. absent second antenna, subcoxal sclerites, exopod reduced to a stylus). Note that basal insects still possess coxae and styli on the segments of their abdomen (which is a modified posterior trunk).

Tracheata (Symphyla, Diplura, larvae of Ephemeroptera and some Coleoptera) and in Malacostraca, indicating that the same genetic information was inherited from a shared common ancestor. The first maxilla of Remipedia has endites of the basipodite that do not protrude into the mouth (as in Malacostraca) but cover the lateral gap between paragnaths and labrum, concealing part of the mandible (described in Fanenbruck, 2009). This situation is also seen in insects with chewing mouthparts (Archaeognatha, Blattoidea, Mantoidea, Phasmatoidea) and possibly also in Symphlya (requires further study). The pair of second maxillae is basally fused in Remipedia, as in insects and myriapods. This fusion is often overlooked when the appendages are separated (dissected) from the head. The palps are highly specialized predatory appendages (an autapomorphy of Remipedia), however, their elongation reminds of the long palps occurring in insects and myriapods.

(3) Remipedia have a tiny cephalic shield, while many other crustaceans have a large carapace. Remipedia do not have a large carapace and their head construction could in this respect well represent the ancestral state for the tracheate head (Figure 12.12).

(4) The cephalic endoskeleton of Remipedia has a derived structure within Mandibulata. A derived feature among crustaceans is the transversal tendon between the mandibles (described in Fanenbruck, 2009). This character however is shared with Branchiopoda, Malacostraca, Myriapoda and Hexapoda. It supports the idea that Remipedia are not basal crustaceans and that myriapods not an early lineage of Mandibulata (see Figure 12.2).

(5) The trunk appendages are biramous, as typical for crustaceans, however, the first article, the coxa, is not a fully functional article but an incomplete ring fused to the pleural area (Hessler and Yager, 1998), suggesting a precursor stage of the so-called subcoxal sclerites, the situation seen in insects and myriapods (Bäcker, Fanenbruck, and Wägele, 2008, Figure 12.9). Furthermore, the dactylus and the terminal claws are only present on the first trunk limb, which is a maxilliped. The remaining paddle-shaped endopods lack claws and have only four instead of the five podomeres present in Malacostraca and insects, probably an adaptation to permanent swimming in Remipedia.

(6) The second antenna is absent in tracheates, in Remipedia it is strongly reduced in comparison with the first antenna. This suggests that already in Remipedia the second antenna lost its functional importance in comparison with e.g. Malacostraca.

(7) The nauplius of Remipedia is a modified lecitotrophic larva that does not feed. This derived type of nauplius is also seen in Malacostraca (in taxa where the nauplius still hatches) and indicates common ancestry (Koenemann, 2007; Koenemann, Olsen, Alwes et al., 2009). The absence of larvae in tracheates is the result of an adaptation to terrestrial life and does not allow statements about the ancestral type of larval stage, but there is no contradiction to the assumption that the marine ancestors had aquatic larvae. The character distribution is the same as for the derived characters "biramous antennule" and "exopod of antenna scale-like". A non-feeding larva is a good preadaptation for the transition to terrestrial life.

(8) The structure of the brain does not support the idea that remipedes are primitive within crustaceans, as suggested by Schram (1983) and Wills (1997), but its

complexity is matched only in Malacostraca and Hexapoda, taxa that share many neuroanatomical details (Fanenbruck, Harzsch, and Wägele, 2004).

(9) Midgut tubules are absent. Therefore, midgut and fat body probably take over functions of the digestive glands. Among Mandibulata of normal size (excluding dwarfs), this is a peculiar novelty seen only in Remipedia and Tracheata.

(10) Phylogenetic analysis of the hemocyanin gene of Remipedia suggests a sister-group relationship to insects (Ertas et al., 2009), and together, remiped and hexapod hemocyanins are in the sister-group position to the hemocyanins of malacostracan crustaceans, however, excluding myriapods. In multigene analyses, Remipedia also appear as sister-taxon to insects (von Reumont et al., 2012), however, Malacostraca are placed among lower crustaceans. We caution to take molecular phylogenies from the first as reliable evidence. It has not been considered until now that taxon-slippage may occur. Also, it is not surprising that single gene analyses give similar results as multigene studies, because the data are samples from the same source (the genome). The artifacts in phylogeny inference will have the same causes.

Shared common ancestry for Remipedia, Myriapoda and Hexapoda was already proposed by Fanenbruck (2009). He explained how body plans could have evolved and assumed that a common marine ancestor with homonymous body segmentation must have existed.

Characters of the last terrestrial ancestors in the lineage of Tracheata evolved from a bauplan related to that of aquatic Remipedia. Figure 12.12 depicts a possible evolution of body plans, starting with a common ancestor of Remipedia and Tracheata (hypothetical ancestor A, i.e. of the clade Archilabiata). The common marine ancestors must have looked like modern Remipedia, however lacking the unique adaptations of Remipedia to a predatory life in caves (lack of eyes, raptorial maxillae and maxillipeds, paddle-like endopods adapted to permanent swimming). The eyes must have been well developed and the thoracic appendages probably allowed walking on endopods, in analogy to the more derived Malacostraca. This is a prerequisite for an aquatic myriapod-like animal to be able to crawl out of the water. It is very probable that the fusion of the first free thoracic segment with the head seen in extant Remipedia happened only in the stem lineage of Remipedia. The first tracheate must have had a free first segment as in extant insects and myriapods. The mouthparts must have evolved from a state more similar to that seen today in Malacostraca to the typical tracheatan mouthparts. In the first myriapod-like tracheatan they were composed of strong mandibles without palp and connected by a transverse tendon, a first maxilla with a long basipod, the two terminal endites covering the gap between paragnath and labrum, a basally fused second maxilla, also with terminal endites, elongated palps on both pairs of maxillae. The second antenna possibly was small in the aquatic species and possessed a tiny scale-like exopod as in Remipedia, the first antenna possibly was still biramous (as in Malacostraca), but in the last, already terrestrial common ancestor of extant Tracheata (hypothetical ancestor B in Figure 12.12) the second antenna and the second flagellum of the first antenna were completely reduced. These reductions are probably adaptations to terrestrial life, where functions of the accessory flagellum (e.g. to fan water towards the aesthetascs) and of the second antenna (e.g. sensing the surroundings in murky water) became obsolete. The thoracic appendages had no functional coxal article, as in Remipedia. Its remains are fused to the pleural region, which is reinforced with the new "subcoxal" pleural sclerites (absent in crustaceans). This is an advantage when legs and pleurae have to bear the full body weight (negligible buoyancy in air!). A parallel case is the evolution of coxal plates in isopods, which strengthen the area of the leg insertion by transformation of the coxa into a rigid plate. As in myriapods and basal insects, the first moveable leg article (the former crustacean basipodite) must have had two rami (exo- and endopod in remipedes, stylus and endopod in myriapods and insects). The first thoracic appendage was a normal leg (not a maxilliped). The myriapod-like body ended with a shortened telson (bearing the anus) and furcal rami, an abdomen was absent. The gut system lacked digestive midgut glands (as in Remipedia).

The nauplius of the marine species was originally lecitotrophic (as in Malacostraca and Remipedia), lacking mouth, anus and gut. Larval gnathobases of second antennae and mandibles are missing, the head appendages lack a proper articulation (see Koenemann, 2009). A nauplius that does not feed on plankton is a first step towards a complete reduction of this stage and direct development, a further requirement to conquer land. It could be that – as in terrestrial crabs – for some time eggs were spawned in the sea, but it is also possible that larval stages were already reduced in the aquatic phase, as in crayfish.

The comparison of insects and myriapods allows us to infer some ground pattern characters. The *first* tracheate arthropod living on land possessed some novelties absent in the remiped-like aquatic ancestor. It probably evolved further and acquired adaptations like tracheal respiration. The *last* common ancestor of all extant myriapods and insects (ancestor B in Figure 12.12) must have had the following new characters:

- complete reduction of the second antenna and of the second flagellum of the first antenna
- complete reduction of rims of the cephalic shield
- complete reduction of aquatic larval stages
- reduction of exopods (formerly used for swimming) and transformation to small styli
- transformation of the crustacean coxa into sclerites supporting the soft pleural region
- transformation of endopods into stout stenopodia suitable for walking
- evolution of "coxal vesicles" (used to take up films of moisture)
- segmental spiracles leading into segmentally arranged tracheal tubules
- Malpighian tubules.

Other features, like the absence of the primary abdomen and the absence of a hepatopancreas were already present in the remiped-like marine ancestors (ancestor A in Figure 12.12).

For the later transformation of a myriapod-like animal into a hexapod we must assume that a suppression of leg development in the trunk posterior to the fourth body segment and shortening of the posterior thorax led to the hexapod body plan. The evolution of a shorter body and of longer anterior legs had the advantage to increase mobility (not speed), to hide in crevices, while speed for escape reactions was acquired by jumping as in Archaeognatha or Collembola (Manton, 1979).

Interestingly, many Myriapoda have an anamorphic development. In Pauropoda, Penicillata, and many Chilognatha development includes a first juvenile stage with only three pairs of legs (corresponding to thoracic appendages 2–4). Therefore, in the development of early tracheates there probably were life stages that could have been "experimental substrate" for the evolution of paedomorphic body plans with fewer legs, which resulted in the hexapod construction. The genetic mechanism that leads to a suppression of leg development by mutations in the Ubx/AbdA pathways is partly understood and explains this aspect of the insect body plan (Averof and Akam, 1995b; Ronshaugen, McGinnis, and McGinnis, 2002; Angelini and Kaufman, 2005). In any case it has to be assumed that hexapods are derived from ancestors that had a posterior trunk with serial pairs of legs. Hexapods still retain the ability to develop legs in their posterior trunk (the "insect abdomen"), as seen in the larvae of Symphyta, Lepidoptera, in embryos of e.g. *Sialis*, and also in form of genital appendages of Ectognatha. Pleural filamentous gills like those of Megaloptera may also be modified segmental appendages.

These are the only major changes required to transform a myriapod-like tracheate into a hexapod. The internal anatomy, the structure of mouthparts and antennae, and even the articulation and structure of the legs remains the same.

Modern myriapods are by no means "primitive" in the sense that they conserve a Palaeozoic bauplan. Their body shape with thoracic legs on all trunk segments is a plesiomorphic condition, not different from the condition seen in Remipedia. However, many characters are clearly derived and support monophyly of extant Myriapoda, as discussed by Edgecombe (2004). Myriapods possibly went through a phase of rapid changes during their early evolution, a process that affected both their genome and their morphology. Evolution of new terrestrial predators may have produced the selection pressure that triggered rapid changes in the myriapod stem lineage. While hexapods developed a shorter body, escape reactions (like jumping) and the ability to crawl in small crevices (see lifestyle of apterous primitive hexapods), myriapods adapted to a cryptic, edaphic lifestyle. A consequence is probably the simplification and partial reduction of their complex eyes with correlated changes in brain anatomy seen in all Myriapoda. They acquired flattened bodies (e.g. in Chilopoda), or became grubbing worm-like diplopods, or dwarfs (Pauropoda, Symphyla, Polyxenida). They are thus able to hide in sediment and beneath stones to avoid predators like arachnids and vertebrates.

(b) Conquering land: ecology and more adaptations

As already discussed above, the evolution of terrestrial arthropods could only have taken place after the establishment of terrestrial vegetation, probably in the Silurian. A Cambrian evolution of myriapods is not plausible.

We argue that morphological evidence suggests that in this palaeozoic period tracheates evolved from a marine crustacean with a body shape similar to that of Remipedia. Such an animal must have crawled on beaches and adapted to terrestrial life. The transitional stage (before the evolution of the tracheal system) must have been an amphibic animal that already moved like a myriapod on beaches but still had to keep its surface moist for respiration. A model for such a mode of life are the primitive terrestrial isopods of the family Ligiidae (Edney and Spencer, 1955; Warburg, 1993; Schmidt and Wägele, 2001) which can live many hours submerged in seawater as well as in air and still lack specialized organs for respiration on land. They use the integument of their pleopods for gas exchange in water and in air.

The physiology of respiration in Remipedia is still unknown, but it is clear that they have no gills. Therefore, other body surfaces must be used for gas exchange. Amphibic myriapod-like animals probably used thin integument areas for respiration, and these had to be kept moist (as the pleopods in Ligiidae). Thin cuticle areas are found on the pleurae, the areas where later the tracheal system evolved. Only after evolution of tracheal systems could the proto-myriapod leave the humid habitats of marine shores. The same process is seen in oniscid isopods: starting with smooth respiratory surfaces on pleopods (as in *Ligia*), first internalized pockets and – later – tracheal systems evolved from these same areas (Ferrara, Paoli, and Taiti, 1996; Schmidt and Wägele, 2000). Equipped with internalized respiratory surfaces, modern woodlice are able to live even in deserts (Coenen-Stass, 1989; Ferrara, Paoli and Taiti, 1996; Baker, Shachak, and Brand, 1998).

Remipedes have unique predatory mouthparts that do not occur elsewhere. Since the mouthparts of other non-parasitic or otherwise specialized crustaceans as well as those of basal insects and myriapods are of the chewing type, we can assume that the first amphibic myriapod-like animals also had chewing mouthparts. Again, the supralittoral sea slater *Ligia* is a good analogy: these animals have chewing mouthparts and they can feed on algae, mosses, and all sorts of plant and animal wastes (Jöns, 1965; Carefoot, 1984; Pennings et al., 2000). *Ligia* has omnivorous and scavenging habits and can also be cannibalistic. For an amphibic arthropod, this type of food is available on most shores. There is no reason to assume that the ancestors of myriapods and/or hexapods had specialized mouthparts or feeding habits.

A prerequisite for terrestrial life is the direct development of early stages within the egg, if the aquatic development is abandoned. Since remipedes (and other crustaceans) have swimming nauplii, the hatching of the nauplius stage must have been suppressed in the stem lineage of tracheates. This process probably was facilitated by the already lecitotrophic larvae without functional mouth and gut, as seen in remipedes and malacostracans. An analogous evolution of direct development also happened in many crustacean lineages (e.g. Cladocera, Leptostraca, all Peracarida, some shrimps, freshwater crayfish).

Sperm transfer in crustaceans is usually direct via spermatophores or sperm bundles and leads to fertilization in or on the body of the females. It is therefore remarkable that we find in insects and myriapods a different mechanism: sperm droplets or spermatophores are placed on the ground and taken up by females in Symphyla, Chilopoda, Diplopoda (polyxenids), Diplura, Collembola, Archaeognatha, Zygentoma, i.e. in myriapods and basal insect groups (both Ectognatha and Entognatha) (e.g. Klingel, 1960; Proctor, 1998; Gols, Ernsting, and van Straalen, 2004). It is also interesting that in basal Ectognatha and in myriapods with indirect sperm transfer, silk threads are used to guide the female to the spermatophore. This suggests that sperm transfer in the first myriapod-like tracheates was indirect via sperm attached to the ground. Interestingly, the same mechanism evolved in terrestrial Chelicerata (Arachnida). It is typical for the less derived taxa (e.g. scorpions, whip spiders and whip scorpions) and probably a character of the first arachnid (Scholtz and Kamenz, 2006).

In this scenario, a further adaptation in the stem lineage of Tracheata must have been the evolution of stenopodia, as already mentioned above (Character 49). Most crustacean taxa move by swimming, and this is also true for Remipedia, basal Malacostraca and Branchiopoda. To conquer land, the animals had to reduce the now superfluous exopods and stabilize the endopods and the pleural area to bear the weight of the body.

It thus seems that myriapod-like tracheates are the ideal link between Remipedia and Hexapoda.

12.6 Discussion

12.6.1 Molecules

During the past 20 years the results of molecular phylogenetics have had an enormous influence on the interpretation of arthropod evolution. The fact that sequence data can be analyzed numerically and that powerful computers are required has led many biologists to believe in molecular phylogenies and to adapt explanations for character evolution to the new tree topologies. The Pancrustacea hypothesis is a typical example: the assumption that all the characters shared by myriapods and insects evolved independently is not really parsimonious, also in view of the inconsistency with paleontological evidence, but it is necessary if the molecular phylogenies are correct.

This strong reliance on molecular systematics is astonishing because published tree topologies obviously contradict each other and document that they are not trustworthy; only one tree can be correct. The usual attitude is to believe in the latest version, if this is based on more data and more sophisticated analysis methods than in previous publications. This attitude is based on the assumption that "more data" implies less background noise, "more taxa" means fewer long-branch artifacts, and complex models allow more realistic inferences of sequence evolution. However, this methodical repertoire cannot cope with systematic errors such as those described in Kück et al. (2012, and also this book). A problem is that addition of taxa is not possible for many lineages, because there are no surviving species (e.g. stem lineages of annelids, mollusks, lobopods, copepods, etc.).

A widespread assumption is that topologies are reliable when they are well resolved and when nodes have a high "statistical support" (e.g. boostrap values). Scientists interested in mathematics have been stressing that support values give no hint for possible systematic errors (e.g. Lento et al., 1995; Wägele and Mayer, 2007; Penny et al., 2008). However, this fact has persistently been ignored by biologists. Because there are no established theories and tools to deal with systematic errors, their detection is very difficult.

Our analysis of systematic errors in ML analyses of sequence data (Wägele and Mayer, 2007; Kück et al., 2012) can explain contradictions between molecular phylogenies and morphological or paleontological evidence. We have shown in simulation studies that these artifacts occur even when the correct model (the one used to evolve simulated sequences) is used for ML-tree inference.

In view of the morphological data, we have to assume that historical branch length ratios are the cause for systematic errors in molecular phylogenies. Differences in branch lengths at the scale examined here can partly be the result of relatively fast radiations followed by many millions of years of evolution of separated lineages. In addition, there are probably lineage-specific rate accelerations. These have been documented for many (more recent) cases among animals and plants (e.g. Hafner, Sudman, Villablanca et al., 1994; Hoeh, Steward, Sutherland et al., 1996; Bromham, Rambaut and Harvey, 1996; Darling, Wade, Kroon et al., 1997; Friedrich and Tautz, 1997; Pawlowski, Bolivar, Fahrni et al., 1997; Schubart, Diesel and Hedges, 1998; Andreasen and Baldwin, 2001; Omilian and Taylor, 2001; Castro, Austin and Dowton, 2002; Hebert, Remigio, Colbourne et al., 2002; Wilcox, de León, Hendrickson et al., 2004), as expected from theoretical considerations (Ohta, 1997). A typical case is that of planktonic Foraminifera (Pawlowski et al., 1997): because these species go through repeated population breakdowns, the evolutionary rate is 50 to 100 times higher than in benthic species.

The analysis of site patterns (Figures 12.6 and 12.7) we used to search for footprints left in genes by historical evolutionary processes requires more research. The aim of this approach is not to build fully resolved trees, but to understand under which conditions it is possible to use data for ML analyses, and when we have to expect the occurrence of systematic errors.

12.6.2 Morphology

It is interesting to compare the morphological evidence supporting Pancrustacea versus Tracheata. Most characters said to be evidence for the Pancrustacea clade are fuzzy or faulty. Several characters refer to developmental processes for which homology of a plesiomorphic state is difficult to ascertain or has not been discussed, and in addition empirical data are sparse (eye development; early neurogenesis; first axonogenesis; expression of segmentation genes). The discussion about brain anatomy and the number of neuropils shows that there are differences between myriapods and some insects and crustaceans, but again it is not clear how to distinguish secondary modifications and reductions from plesiomorphic states. The crystalline cone (Character 24, see above) turned out to exist in myriapods as well as in other Mandibulata and cannot be used as argument.

Most of the characters supporting the Tracheata have a different quality and they are more numerous. Most refer to structures and not to processes, and these structures are composed of details (e.g. head appendages, styli and vesicles on the tracheate coxa, praetarsus muscles) that allow the precise formulation of homology statements. Therefore, from our point of view the balance is much heavier on the side of the Tracheata hypothesis.

12.6.3 Evolutionary scenarios

The scenario described here, with a myriapod-like first tracheate as a link between Remipedia and Hexapoda, is not an original idea. It was first published by Moura and Christoffersen (1996). Since then, more evidence accumulated especially for characters shared between Remipedia and Hexapoda (Fanenbruck, Harzsch, and Wägele, 2004; Ertas et al., 2010, von Reumont et al., 2012). We now have explanations for the repeated failure of molecular analyses (Figure 12.3, see other examples with contradictions in this book). Systematic errors (see Kück, Misof and Wägele, this book) could be the reason why the clade Tracheata is not recovered in molecular phylogenies. And we can describe an alternative scenario for a transition from marine mandibulates to the first tracheates which does not require the assumption of multiple parallelisms in myriapods and insects and which is compatible with the fossil record.

The idea that myriapods and insects evolved directly from Onychophora-like ancestors (the Uniramia hypothesis: Manton, 1973) will not be discussed further. The most recent scenario discussed in literature is the origin of hexapods from branchiopod-like ancestors, formulated under the impression made by some molecular phylogenies (Jenner, 2006).

The branchiopod scenario: In some molecular phylogenies, the sister-group of Hexapoda are the Branchiopoda, usually when Remipedia are missing in the data set (e.g. Andrew, 2011). The idea formulated by Jenner (2006) is that the ancestors of insects were crustaceans living in freshwater around 410 million years ago. This idea has not been elaborated further. There is no explanation for the obvious differences in lifestyle and morphology which at first sight do not suggest that branchiopods could be the stem group of insects.

Branchiopods are – in comparison with higher Malacostraca – defenseless crustaceans. They are epibenthic or planktonic and their appendages are not suitable for walking. Their cuticle is weak, the appendages are soft and have no chelae, strong spines or acute claws. Branchiopods do not have escape reactions as seen in shrimps and are a preferred food of fish. Marine branchiopod-like crustaceans, which once must have existed, disappeared from marine habitats (except planktonic dwarfish Onychopoda secondarily derived from similar freshwater species). The larger branchiopod species survive only because they can colonize ephemeral continental waters where fish cannot live.

A key adaptation of branchiopods is the production of cysts (or "eggs") that survive unfavorable conditions in the sediment, often including desiccation. Further morphological apomorphies of branchiopods (delicate, leaf-like legs with partly fused articles, first antenna tiny, reduced first and second maxilla) do not occur in the ground pattern of Hexapoda. Other derived characters shared exclusively by Hexapoda and Branchiopda are not known. In addition, branchiopods show plesiomorphic conditions absent in Malacostraca, Remipedia and Tracheata, like the presence of a primary abdomen, many endites on thoracic legs, a feeding nauplius (when present). The branchiopod brain is simpler than that of Malacostraca, Hexapoda or Remipedia (Fanenbruck and Harzsch, 2005; Strausfeld and Andrew, 2011), wherefore Andrew (2011) had to assume that branchiopods have a secondarily reduced brain complexity ("adaptive simplification"). The more parsimonious explanation is that Branchiopoda are more plesiomorphic in this respect (and others) and the brain of Hexapoda is derived from a complex brain as seen in Malacostraca and Remipedia.

There is no morphological evidence that supports the "branchiopod-like ancestor" scenario, and the lifestyle of branchiopods (planktonic or epibenthic swimming, filtering, planktotrophic larvae) is not the ideal foundation for the conquest of land.

The *marine hexapod* scenario: Haas, Waloszek, and Hartenberger (2002) described a Devonian fossil from the Hunsrück slates (Germany) which was interpreted as a marine representative of Hexapoda. This finding would imply that adaptations to terrestrial life evolved convergently in myriapods and insects. However, a re-examination of the holotype and of further similar specimens demonstrated that this enigmatic species (synonymized with *Wingertshellicus backesi*, an arthropod with unknown phylogenetic position) is neither a tracheate nor a crown-group mandibulatan (Kühl and Rust, 2009).

If the first tracheate was myriapod-like, it might have been several cm long, not different from extant Remipedia. Kraus and Kraus (1994) published a different view, with first tracheates that were tiny animals feeding on algae and fungi. This assumption relies mainly on the fact that several Hexapoda (mainly Entognatha) and some myriapods (Pauropoda, Symphyla) are dwarfish animals. There is also a tendency in these taxa to entognathy, to suck out cells and fungal hyphae. Since the dwarfish species show signs of secondary reductions (tracheal system, eyes, circulatory system) it is more plausible to assume that they evolved from larger ancestors with a more complete anatomy. Analogous cases of evolution of dwarfs occur in many other animal groups (e.g. Syncarida, Isopoda, Acari, Palpigradi, Polychaeta).

Wherever we find a strong contradiction between substantial morphological and molecular evidence, we should hesitate to accept the molecular trees. The probability that phylogenies based on molecular data are wrong because of a lack of phylogenetic signal is especially high for early (palaeozoic) phases of metazoan evolution. Future theoretical research is needed for the development of tools for the detection of systematic errors in molecular phylogenies and for the distinction between distinct phylogenetic signal and misleading patterns.

Acknowledgments

The authors want to thank the arthropod team of the Museum Koenig in Bonn for many years of cooperation in the "Deep Metazoan Phylogeny" project, especially Prof. B. Misof, Dr. C. Mayer, Dr. B. von Reumont, and Dr. K. Meusemann. Dr. C. Mayer prepared the SAMS graph for Figure 12.4.

Sabrina Simon and Heike Hadrys

13 Phylogeny of the most species-rich group on Earth, the Pterygota: Ancient problems, living hypotheses and bridging gaps

Abstract: Insects are by far the most successful animal group on Earth. Radiations of bauplan modifications and ecological adaptations have reached an unchallenged diversity in insects. The most influential insect bauplan modifications took place during and after the transition from apterygote (wingless) to pterygote (winged) insects. Key questions (and transitions) relate to: (i) the origin of Pterygota (invention of wings), (ii) the origin of Neoptera (insects with wing flexion) and (iii) the origin of Holometabola (insects with complete metamorphosis). Unfortunately, directions of evolutionary change and locations of important diversification shifts remain unresolved since deep phylogenetic relationships between insect orders are not yet resolved. In this chapter, (i) molecular genetic data at different levels of genetic and genomic organization and (ii) new bioinformatic tools and algorithms to improve data analyses and data quality assessment are introduced for inferring winged insect relationships. A representative sampling of all pterygote infraclasses were analyzed with a special focus on the most basal pterygote orders. The studies presented contribute several comprehensive molecular phylogenetic studies to resolve the insect evolutionary tree and in a grander scheme the arthropod and metazoan evolutionary tree. In addition, the improved phylogenetic analyses of several molecular marker systems at different genetic levels demonstrated that data quality assessment prior to tree reconstruction yielded better resolution in several places. However, the still observed incongruence between the inferred topologies highlights the challenges in interpreting the strength and weakness of the underlying data and of the analyses conducted.

13.1 Introduction

Insects compose the bulk of arthropod species and are amongst the most successful animal groups of our planet, representing approximately 60 % of the named diversity of life on our planet (Grimaldi and Engel, 2005). This enormous diversity is linked to the key bauplan innovations that occurred during insect evolution. At least four events are commonly cited as being of major macro-evolutionary importance: (i) the evolution of the insect bauplan, (ii) the acquisition of wings (Pterygota), (iii) the evolution of wing flexion (Neoptera) and (iv) the evolution of a complete metamorphosis (Holometabola), (Carpenter, 1992; Carpenter and Burnham, 1985; Evans, 1984; Gullan and Cranston, 2000; Hutchinson, 1959; Mayhew, 2003; Yang, 2001). These unique features are major contributors to the vast diversification of insects, since they allowed the accessibility of ecological niches (insect bauplan), dispersal in the air (wings), exploration of concealed places (folding of wings) and specialization of juvenile and adult forms (complete metamorphosis) (Mayhew, 2007). Unfortunately, directions of evolutionary change and locations of important diversification shifts are still elusive since deep phylogenetic relationships between insect orders are not yet resolved (Mayhew, 2002; Mayhew, 2003). One potential explanation for this situation is that the divergences took place within a limited time frame (Whitfield and Kjer, 2008). This circumstance is accompanied by the major challenge in finding useful molecular markers accurately tracking these short ancient internodes.

Below we review existing hypotheses and old problems in pterygote evolution and introduce new molecular approaches (i) to compile genetic data at different levels of genetic and genomic organization, (ii) to develop and apply new bioinformatic tools and algorithms to improve data analyses and – most important – to create a strategy to assess data quality instead of quantity and finally (iii) to bridge the gap between morphological and genetical pathways to key bauplan innovations. A representative sampling of all pterygote infraclasses has been analyzed within this subproject of the "Deep Metazoan Phylogeny" (DMP) priority program. The sampling comprises 181 newly generated sequences for single genes across all pterygote orders, two new mitochondrial genomes and five EST projects with a special focus on the most basal orders and the bauplan transitions.

13.1.1 Pterygote phylogeny: Ancient problems, living hypotheses

13.1.1.1 The basal pterygote divergence or the never-ending "Palaeoptera problem"?

Undoubtedly, the invention of wings is one of the most important key innovations in insect evolution since wings allowed the conquest of a large number of new ecological niches and improved the capability of dispersal and evasion of predators (Evans, 1984; Grimaldi and Engel, 2005). However, the interrelationships of the earliest pterygote lineages – the "Palaeoptera" ("old winged insects" without wing flexion), and therefore the question of the earliest diverging lineage of Pterygota are still hotly debated. Here the three branches Ephemeroptera (mayflies), Odonata (dragonflies and damselflies) and Neoptera (all other extant insects with wing flexion) compete in three different scenarios: (i) the traditional scenario – the Palaeoptera hypothesis – suggests a monophyletic Palaeoptera clade, including Ephemeroptera and Odonata (Hennig, 1969); (ii) the Metapterygota hypothesis – a clade formed by Odonata and Neoptera – places the Ephemeroptera as the earliest diverging lineage of Pterygota (Kristensen, 1991b) and (iii) the Chiastomyaria hypothesis – a clade formed by Ephemeroptera and Neoptera – places Odonata as the earliest diverging lineage of Pterygota (Boudreaux, 1979). Each hypothesis is still considered viable and supported by morphological (but see Blanke et al., 2012) as well as molecular data; see also Table 13.1 and Figure 13.1A.

Molecular data support all three hypotheses depending on the analyses applied (e.g. Hovmöller, Pape, and Källersjö, 2002; Mallatt and Giribet, 2006; Ogden and Whiting, 2003). Even on a larger scale, phylogenomic studies have not yet achieved a robust phylogenetic hypothesis regarding the evolution of these ancient lineages. In recent studies, using the same EST libraries for Odonata and Ephemeroptera and the same approach for orthology prediction (HaMStR) (Ebersberger, Strauss, and von Haeseler, 2009), all three hypotheses have been supported: Palaeoptera (Meusemann et al., 2010: BI analyses; von Reumont et al., 2012); Metapterygota (Simon et al., 2012) and Chiastomyaria (Simon et al., 2009). In the most recent study of Simon et al. (2012) it has been shown that for the ancient and rapidly diverged lineages, like the basal pterygote lineages, not only the taxon sampling but also the function of the genes have an enormous impact on the inferred relationships, claiming the "Palaeoptera problem" more enigmatic than before.

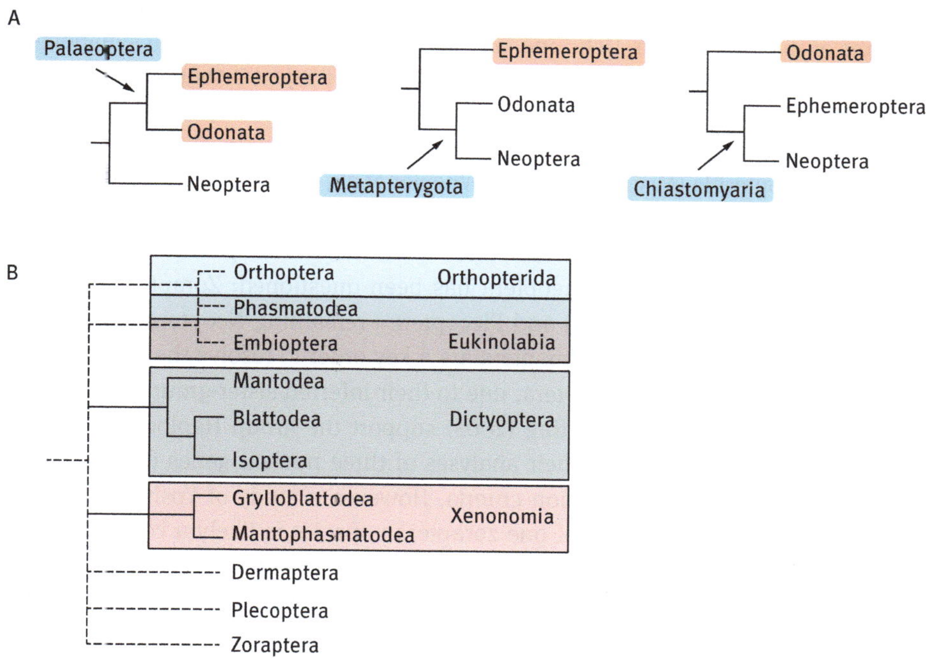

Figure 13.1: A. Proposed phylogenetic relationships for the earliest branching lineages of Pterygota: Palaeoptera, Metapterygota and Chiastomyaria hypothesis. Figure modified after Simon et al., 2009. **B.** The interrelationships of the polyneopteran orders. Dashes lines indicate not widely accepted placements due to lack of convincing morphological synapomorphies and contradictory or only poorly resolved relationships based on molecular data sets.

13.1.1.2 The polyneopteran relationships

The neopteran infraclass Polyneoptera includes Plecoptera, Zoraptera, Dermaptera, 'Blattodea', Isoptera, Mantodea, Mantophasmatodea, Phasmatodea, Grylloblattodea, Embioptera and Orthoptera. Although there is increasing support for a probably monophyletic origin of Polyneoptera (Ishiwata et al., 2010; Letsch et al., 2012; Simon et al., 2012; Yoshizawa, 2011), the interrelationships of these 11 polyneopteran orders are far from being resolved and the consensus topology of the published studies still resembles a nearly unresolved "comb" (Kristensen, 1991b; Trautwein et al., 2012) (Figure 13.1B). Only two clades, Dictyoptera (= 'Blattodea' + Isoptera + Mantodea) and Xenonomia (= Grylloblattodea + Mantophasmatodea) (Terry and Whiting, 2005) have become better understood. Other proposed groups within Polyneoptera are not widely accepted due to a lack of convincing morphological synapomorphies and contradicting or only poorly resolved relationships based on molecular data sets (Table 13.1). One of these groups is Orthopterida (= Orthoptera + Phasmatodea); here the wing base structure (Yoshizawa, 2011) and the increased number of cervical sclerites and the shortened and simplified cercus favor this group (Beutel and Gorb, 2006). In contrast, head morphology (Friedemann et al., 2012; Wipfler et al., 2011) and structural features of the eggs (Jintsu, Uchifune, and Machida, 2010) are potential synapomorphies for Eukinolabia (= Phasmatodea+Embioptera) and reject Orthopterida. Molecular support also exists for both groups: Eukinolabia (Ishiwata et al., 2010; Kjer et al., 2006; Letsch et al., 2012; Terry and Whiting, 2005; Whiting, Bradler, and Maxwell, 2003) or Orthopterida (Misof et al., 2007; Wheeler et al., 2001).

The phylogenetic position of the three remaining polyneopteran orders (Dermaptera, Plecoptera and Zoraptera) is even more unclear. Even the placement of Plecoptera and Zoraptera within Polyneoptera has been questioned; Zoraptera + Paraneoptera (Beutel and Weide, 2005) and Plecoptera + remaining Neoptera (Beutel and Gorb, 2006). In this context, Dermaptera are a key order to resolve the phylogenetic position of Plecoptera and Zoraptera, due to their inferred sister-group relationship to both of them. Terry and Whiting (2005) support the group Haplocerata (= Dermaptera + Zoraptera) based on their analyses of three nuclear genes (18S, 28S and Histone3) using direct optimization criteria. However, a study of Yoshizawa (2010) showed that the 18S sequence of one zorapteran species is likely a contamination which is a serious problem especially if using the direct optimization method. In addition, Terry and Whiting (2005) also state that all morphological characters for the group Haplocerata might be homoplasious based on their matrix due to the highly derived status of Zoraptera. The study of Jarvis, Haas, and Whiting (2005) shows that morphological characters supporting this group are highly sensitive to different analytical approaches. In contrast, the grouping of Dermaptera + Plecoptera has in the past received increased support based on molecular data (Ishiwata et al., 2010; Kjer et al., 2006; Misof et al., 2007; Simon et al., 2012; Simon, Schierwater, and Hadrys, 2010; von Reumont et al., 2009) while possible morphological synapomorphies of these orders are still rare (see discussion in Yoshizawa, 2011).

The Zoraptera belong to the most enigmatic insect lineages with more than ten discussed positions within Polyneoptera as well as Paraneoptera (Yoshizawa, 2007). The term "Zoraptera problem" (Beutel and Weide, 2005) is nowadays even more deserved than the "Strepsiptera problem" (Kristensen, 1981) (see also Section 13.1.1.3). Here, the sequence information available for Strepsiptera, including several nuclear coding genes, EST projects, a complete mitochondrial genome and a whole genome project greatly improved the resolution of the phylogenetic position of Strepsiptera (Longhorn, Pohl, and Vogler, 2010; McKenna and Farrell, 2010; McMahon, Hayward, and Kathirithamby, 2009; Niehuis et al., 2012; Talavera and Vila, 2011; Wiegmann et al., 2009). In contrast, molecular sequence data for Zoraptera are still rare. Based on these data and/or morphological characters four of the ten discussed phylogenetic positions of Zoraptera have received increased support: (i) Zoraptera + Dictyoptera (Boudreaux, 1979; Ishiwata et al., 2010; Wheeler et al., 2001; Yoshizawa and Johnson, 2005), (ii) Zoraptera + Dermaptera (= Haplocerata) (Jarvis, Haas, and Whiting, 2005; Terry and Whiting, 2005; Xie et al., 2009), (iii) Zoraptera + Paraneoptera (Beutel and Weide, 2005; Beutel and Gorb, 2001; Beutel and Gorb, 2006; Hennig, 1969) and (iv) Zoraptera + Embioptera (= Mystroptera) (Engel and Grimaldi, 2000; Grimaldi and Engel, 2005; Rafael and Engel, 2006; Yoshizawa, 2007; Yoshizawa, 2011). Indeed, in a recent study of Simon et al. (2012) – including the first transcriptomic data for the three ambiguous polyneopteran orders (Dermaptera, Plecoptera and Zoraptera) – the hypotheses of Zoraptera + Paraneoptera and Plecoptera + remaining Neoptera were contested and conclusive support for monophyletic Polyneoptera was provided.

Table 13.1: Existing phylogenetic hypotheses within the palaeopterous and polyneopteran lineages.

Relationship	Clade	Criterion	References
Ephemeroptera + Odonata	Palaeoptera	morphology	(Bechly et al., 2001; Blanke et al., 2012; Haas and Kukalova-Peck, 2001; Kukalova-Peck, 1991)
		molecular	(Hovmöller, Pape, and Källersjö, 2002; Meusemann et al., 2010)
Odonata + Neoptera	Metapterygota	morphology	(Beutel and Gorb, 2006; Kristensen, 1981; Kristensen, 1991b; Kristensen, 1999; Staniczek, 2000; Willkommen and Hornschemeyer, 2007)
		molecular	(Ogden and Whiting, 2003; Simon et al., 2012; Wheeler et al., 2001; Whiting et al., 1997; Zhang et al., 2008)
Ephemeroptera + Neoptera	Chiastomyaria	morphology	(Boudreaux, 1979; Matsuda, 1970; Staniczek, 2000)
		molecular	(Kjer, 2004; Kjer et al., 2006; Misof et al., 2007; Simon et al., 2009; von Reumont et al., 2009; Yoshizawa and Johnson, 2005)

Relationship	Clade	Criterion	References
Blattodea + Isoptera + Mantodea	Dictyoptera	morphology	(Inward, Beccaloni, and Eggleton, 2007; Yoshizawa, 2011)
		molecular	(Kjer et al., 2006; Terry and Whiting, 2005; von Reumont et al., 2009; Zhang et al., 2010b)
Grylloblattodae + Mantophasmatodea	Xenonomia	morphology	(Friedemann et al., 2012; Grimaldi and Engel, 2005; Klass et al., 2002; Wipfler et al., 2011)
		molecular	(Ishiwata et al., 2010; Kjer et al., 2006; Terry and Whiting, 2005)
Orthoptera + Phasmatodea	Orthopterida	morphology	(Beutel and Gorb, 2006; Yoshizawa, 2011)
		molecular	(Misof et al., 2007; Wheeler et al., 2001)
Phasmatodea + Embioptera	Eukinolabia	morphology	(Friedemann et al., 2012; Jintsu, Uchifune, and Machida, 2010; Wipfler et al., 2011)
		molecular	(Ishiwata et al., 2010; Kjer et al., 2006; Terry and Whiting, 2005; Whiting, Bradler, and Maxwell, 2003)
Zoraptera + Paraneoptera	Parametabola	morphology	(Beutel and Weide, 2005)
Plecoptera + remaining Neoptera	(paraphyly Polyneoptera)	morphology	(Beutel and Gorb, 2006)
Dermaptera + Zoraptera	Haplocerata	morphology	(Jarvis, Haas, and Whiting, 2005)
		molecular	(Jarvis, Haas, and Whiting, 2005; Terry and Whiting, 2005; Xie et al., 2009)
Dermaptera + Plecoptera	–	molecular	(Ishiwata et al., 2010; Kjer et al., 2006; Misof et al., 2007; Simon, Schierwater, and Hadrys, 2010; von Reumont et al., 2009)
Zoraptera + Dictyoptera	–	morphology	(Boudreaux, 1979; Wheeler et al., 2001)
		molecular	(Ishiwata et al., 2010; Wheeler et al., 2001; Yoshizawa and Johnson, 2005)
Zoraptera + Embioptera	Mystroptera	morphology	(Engel and Grimaldi, 2000; Grimaldi and Engel, 2005; Rafael and Engel, 2006; Yoshizawa, 2007; Yoshizawa, 2011)

13.1.1.3 Paraneoptera and Holometabola

Paraneoptera has been traditionally considered as the sister-group to Holometabola – Eumetabola hypothesis. In fact, this group shares several synapomorphies (see also Beutel and Pohl, 2006) but most topologies derived from molecular sequence data alone did not recover this clade at all (Ishiwata et al., 2010; Meusemann et al., 2010; Misof et al., 2007; von Reumont et al., 2009; Wheeler et al., 2001; Whiting et al., 1997) or only with low support (Kjer, 2004; Simon et al., 2012; Simon, Schierwater, and Hadrys, 2010). Indeed, even the monophyly of Paraneoptera (Hemiptera, Thysanoptera, Psocodea) remains debatable. Several morphological synapomorphies provide evidence for this group, e.g. cerci lost, labial pals reduced or lost, asymmetrical mandibles and other modifications of attachment structures (Beutel and Gorb, 2001; Beutel and Gorb, 2006; Grimaldi and Engel, 2005). However, most characters supporting their monophyly are based on shared losses (Kristensen, 1981) and also molecular data sets are inconsistent: they support monophyly (Kjer et al., 2006; Misof et al., 2007; Wheeler et al., 2001) or contest it (Ishiwata et al., 2010; Letsch et al., 2010; Yoshizawa and Johnson, 2005).

In contrast, recent progress has been made in the resolution of the interrelationships of Holometabola. Particular attention deserves the resolution of the earliest-branching lineage within Holometabola, Hymenoptera, and the inferred sister-group relationship of Coleoptera and Strepsiptera. The support for Hymenoptera as sister to all remaining holometabolan lineages provided by a phylogenomic study of Savard et al. (2006) was subsequently confirmed by other phylogenomic studies (Meusemann et al., 2010; Simon et al., 2012; Simon et al., 2009; von Reumont et al., 2012), rRNA studies (Misof et al., 2007; von Reumont et al., 2009), protein-coding gene analyses (Ishiwata et al., 2010; Longhorn, Pohl, and Vogler, 2010; Wiegmann et al., 2009) and also recently by morphological analyses (Beutel et al., 2011). The other mentionable progress in inferring Holometabola evolution involves the resolution of the "Strepsiptera problem". The placement of the enigmatic, parasitic Strepsiptera (twisted-winged parasites) within Holometabola has been one of the major obstacles in insect systematics. Nowadays a sister-group relationship of Strepsiptera and Coleoptera (named Coleopterida) is well supported by independent molecular data sets (Ishiwata et al., 2010; Longhorn, Pohl, and Vogler, 2010; Niehuis et al., 2012; Talavera and Vila, 2011; Wiegmann et al., 2009) and comprehensive morphological analyses (Beutel et al., 2011; Friedrich and Beutel, 2010; Lawrence et al., 2011).

13.1.2 Systematic studies in the era of phylogenomics

In insect systematics, whether based on morphological or molecular data, numerous characters have been assembled and their states have been interpreted to unravel the evolutionary history of these arthropods. But at present neither the evolutionary origin of the insect wings (the divergence of Pterygota from "apterygote" insects) nor their radiation into a broad spectrum of sophisticated flight organs are well understood. One potential explanation for this situation is that the divergences of winged insects took place within a limited evolutionary time frame. Whitfield and Kjer (2008) pointed out that the "ancient rapid radiation" causes the main problem to resolve insect relationships with molecular data. Due to the short ancient internodes connecting the taxonomic groups, inadequate (e.g. saturated, homoplastic) molecular data sets and conflicting results within or among data sets have been observed in many phylogenetic studies (Kjer et al., 2006; Misof et al., 2007; Ogden and Whiting, 2003; von Reumont et al., 2009; Wheeler et al., 2001).

The way in which molecular systematic studies are conducted has significantly changed during the last few years due to the new sequencing technologies as outlined below. Instead of using only one gene that might be a useful molecular marker to infer a species tree (see Sections 13.2.1 and 13.2.2) or complete mitochondrial genomes (see Section 13.2.3), phylogenomic (multi-gene) approaches derived from whole genome or EST (Expressed Sequence Tags) projects have become the state-of-the-art in molecular systematic studies (see Section 13.2.4). Several studies have shown that phylogenomic approaches are promising to infer unresolved issues in deep metazoan relationships (Dunn et al., 2008; Philippe, Lartillot, and Brinkmann, 2005; Roeding et al., 2007; Savard et al., 2006) and also to calculate divergence times (Rehm et al., 2011). In contrast, it has also been shown that a phylogenomic approach is not necessarily a panacea for the accurate resolution of a species tree (Gadagkar, Rosenberg, and Kumar, 2005; Philippe et al., 2001a) and that in some problematic cases resolving deeper phylogenetic relationships by simply increasing the number of nuclear sequence markers seems to fail (see also Section 13.2.4). In this context, comparative morphological studies provide in the genomic era crucial independent data sets and also allow the placement of fossils. With the application of new technologies, e.g. computer-based 3D reconstruction or synchrotron micro-computed tomography (SR-microCT), and new approaches to analyze possible homoplasy and concerted convergence (e.g. Blanke et al., 2013), morphology plays an important part in assessing the accuracy of the hypotheses of insect evolution derived from large gene analyses (Beutel and Kristensen, 2012).

In the following sections, focusing mainly on the molecular studies conducted within the project of the "Deep Metazoan Phylogeny" priority program to address pterygote evolution, it is demonstrated that for each approach from single-gene to multi-gene analyses the assessment of the phylogenetic utility of the data and the evaluation of certain systematic bias remains the biggest challenge in molecular studies. Especially inferring pterygote evolution – the ancient rapid radiation into the enormous species diversity – has turned out to be far more complex than can be resolved by only increasing data quantity (see also Section 13.2.4). Thus, an increasing number of systematists believe that in many of those cases the inclusion of not only transcriptomic or whole genome data but also data on gene organization and complexity as well as new target genes will be essential (Boero, Schierwater, and Piraino, 2007). Eventually, comparative research on regulatory genes may become helpful for deep phylogenetic studies and might bridge some gaps between description and causal explanations (Hadrys et al., 2012) (see also Section 13.2.5).

13.2 Molecular systematic studies to infer pterygote evolution

13.2.1 Single-gene analyses

The most commonly used molecular markers to infer relationships between pterygote orders have been single mitochondrial genes and nuclear rRNA gene fragments. However, both have shown severe limitations for phylogenetic analyses in Pterygota (Kjer, 2004; Ogden and Whiting, 2003; Whiting, 2001; 2002a) (see 2.2 and 2.3). Consequently, the evaluation of new markers, e.g. nuclear protein-coding genes, has become a necessity in pterygote phylogeny. Apart from Histone 3 (H3 hereafter) – a nuclear protein-coding gene frequently used in insect systematics – some additional nuclear protein-coding genes have increasingly been used, including the Elongation factor-1α (EF-1α hereafter). EF-1α has been applied in some terrestrial arthropods (Caterino, Cho, and Sperling, 2000; Hughes and Vogler, 2004; Zakharov, Caterino, and Sperling, 2004). The variation in silent nucleotide sites makes this marker useful for phylogenetic analyses among species groups and genera (Belshaw and Quicke, 1997; Cho et al., 1995; Mitchell et al., 1997; Reed and Sperling, 1999), while amino acid replacements provide phylogenetic information for deeper divergences (Regier and Shultz, 1997; Regier and Shultz, 1998).

In order to evaluate the phylogenetic value of both nuclear protein-coding genes, EF-1α and H3, these genes have been isolated and characterized for a representative sampling of pterygote orders (Simon, Schierwater, and Hadrys, 2010). The comparative phylogenetic and statistical analyses of the EF-1α and H3 in this study highlighted the superiority of EF-1α over H3. The results indicate that H3 should be replaced and that EF-1α should be used as an additional informative molecular marker in pterygote systematics, as previously suggested by Kjer at al. (2006). The statistical analyses revealed that H3 shows extreme conservation in the second position and that substitutions occur mainly at silent sites, whereas in EF-1α substitutions occur more frequently. The phylogenetic analyses showed that a small exon fragment of EF-1α provides a topology separating the infraclasses that corresponds to evidence from other molecular markers and morphology, while the H3 topology has no resolution at these deep nodes.

In addition, analyses of intron positions of the amplified EF-1α fragment revealed the same potential "structural" characters. Other studies demonstrated that these "structural" characters may mirror relationships within insect orders and that recent intron insertions can characterize monophyletic groups (Brady and Danforth, 2004; Goetze, 2006). The identification of an intron that is present in almost all pterygote orders but is absent in the derived holometabolous orders indicates an intron loss in the evolution of the Holometabola.

In a recent study of Ishiwata et al. (2010) three other nuclear protein-coding genes (DNA polymerase delta and the two largest subunits of RNA polymerase II) have been evaluated on the sequence level to infer insect evolution also with promising results.

13.2.2 Nuclear rRNA genes

The rRNA genes for some time played a dominant role in inferring metazoan intra- and interrelationships due to their availability for several representatives of phyla, classes, orders and even families. However, the structural complexity of rRNA molecules – highly variable expansion segments (length and sequence heterogeneity) between conserved and slowly evolving core regions – challenges the assignment of homologous positions within alignments. The development of new semi-automated tools like MXSCARNA (Tabei et al., 2008) or RNAsalsa (Stocsits et al., 2009) provides here a consistent algorithmic framework for the consideration of secondary structures as alignment guide. The incorporation of existing knowledge on secondary structure has been proven to be a promising source for increasing alignment accuracy (Hickson, Simon, and Perrey, 2000; Kjer, 1995; Misof et al., 2007). Moreover, the inferred consensus secondary structure and the individual structure predictions can be used (i) to incorporate the site covariation in the phylogenetic analyses and (ii) to evaluate the phylogenetic signal in the structure of rRNA molecules.

Results of a cooperative work addressing arthropod relationships have shown that the implementation of biologically realistic model parameters, such as site interaction (mixed DNA/RNA models) and compositional heterogeneity of base frequency (time-heterogeneous approach), is fundamental to robustly reconstructing phylogenies from rRNA alignments (von Reumont et al., 2009). Based on this study, both tree topologies using time-heterogeneous or time-homogeneous substitution models support the monophyly of Pterygota and the Chiastomyaria hypothesis (Ephemeroptera + Neoptera), but the major clades, Hexapoda, Entognatha and Ectognatha are only recovered in the time-heterogeneous approach. However, the improved topology estimates based on non-stationarity models still show some weakly supported ancient splits within Pterygota, especially within the basal Neoptera. In addition, Polyneoptera appears polyphyletic in both topology estimates, which contradicts the monophyletic origin of this clade supported by large gene analyses (Ishiwata et al., 2010; Letsch et al., 2012; Simon et al., 2012). Plecoptera and Dermaptera cluster together in both topologies as frequently suggested by recent rRNA and phylogenomic analyses (Kjer, 2004; Letsch et al., 2010; Simon et al., 2012; Yoshizawa and Johnson, 2005). However, the placement of this clade within Neoptera varies depending on the reconstruction method: sister-group to Dictyoptera in the time-heterogeneous approach, sister-group to Holometabola in the time-homogeneous approach. Paraneoptera and Holometabola receive high support in both tree topologies for their monophyly and Hymenoptera form the sister-group to remaining Holometabola in both topologies. In contrast, Antliophora (Siphonaptera + Mecoptera, Diptera) is not recovered in both topologies, which is incongruent to the current morphological hypotheses of holometabolan evolution (Beutel, Kristensen, and Pohl, 2009). In this context, two other studies have demonstrated that especially for the hexapod rRNA data set a standard DNA model setup might perform better in some instances than a mixed DNA/RNA model setup due to homoplasy occurring in the loop regions (Letsch and Kjer, 2011; Letsch et al., 2010). Using a standard DNA model setup Antliophora were recovered in the study of Letsch et al. (2010) because mixed DNA/RNA models might down weight the phylogenetic signal within the stem regions leading to increased noisy phylogenetic signal stemming from the saturated loop regions and consequently inferred erroneous phylogenetic relationships.

13.2.3 Mitogenomics

Complete mitochondrial genomes of 230 pterygote species are available from the GenBank nonredundant database (GenBank May 16, 2011), but with a strong bias to derived orders. Especially some of the most enigmatic insect orders are underrepresented (e g. Odonata (2) and Ephemeroptera (5)) or even missing (Zoraptera, Embioptera, Siphonaptera and Trichoptera).

Until now, no mitogenomic analysis has been conducted on the complete data set. Although Cameron et al. (2004; 2006a) reviewed the existing problems in mitogenomic phylogenies inferring deep inter-ordinal relationships among insects and more general among arthropods, they did not present the inferred insect relationships based on mitochondrial genomes comprising 29 insect orders (Cameron et al., 2006a) or had only a limited taxon sampling comprising only 7 insect orders (Cameron et al., 2004). On the other hand, several studies had limited scope inferring only intra-ordinal relationships or inter-ordinal relationships in the pterygote infraclasses ('Palaeoptera', Polyneoptera, Paraneoptera and Holometabola). For example Zhang et al. (2008) published the first mitogenomic study to resolve the "Palaeoptera problem" and supported the Metapterygota hypothesis (Ephemeroptera, Odonata + Neoptera), but their data set is characterized by sparse taxon sampling. In contrast, Lin et al. (2010) used a denser taxon sampling within the basal hexapod lineages, including representatives of all sequenced polyneopteran lineages, but excluding all paraneopterans and holometabolan species. Based on this data set they supported Odonata as the earliest branching pterygote lineage but also the clade Ephemeroptera + Plecoptera, challenging the monophyly of Neoptera.

To further improve the taxon sampling of the basal pterygote orders and to resolve the "Palaeoptera problem", two mitochondrial genomes of palaeopterous orders, of the dragonfly *Boyeria irene* and the mayfly *Baetis sp.*, were isolated and characterized (unpubl. data). Using a data set comprising 31 taxa including two apterygote outgroup orders (Archaeognatha and Zygentoma), four odonates, four mayflies, eleven polyneopteran taxa, two paraneopterous taxa and five holometabolus taxa, the sequence analyses clearly demonstrate the main problem of mitogenomic data – the lack of phylogenetic signal for the deep nodes. Although monophyletic Pterygota was well recovered in the phylogenetic analyses, there was no significant support for the earliest branching lineage of Pterygota but the sister-group relationship of Ephemeroptera + Plecoptera was rejected, thus confirming monophyletic Neoptera. In general, significant support was only recovered for the intra-ordinal relationships (unpubl. data).

The limitations of mt sequence data for the resolution of inter-ordinal relationships in insects were also displayed in a collaborative mt-network within the "Deep Metazoan Phylogeny" priority program (Simon and Hadrys, 2013). Using dense data sets across Metazoa (648-metazoan-taxa, 300-metazaon-taxa, 100-metazoan-taxa) (Bernt et al., 2013a) and a hexapod data set (209 taxa; 4 crustacean and 205 hexapod taxa) all phylogenetic analyses resulted in overall poor statistical support for the deep nodes among Hexapoda and strong incongruence with morphological and molecular nuclear data (Simon and Hadrys, 2013). Features of hexapod mitochondrial genomes like accelerated evolutionary rates likely related to phenotypic changes in body size or to parasitic lifestyle (Castro, Austin, and Dowton, 2002) and the mutational saturation of mt genomes are both known to cause biases in the data resulting in systematic errors.

Using the largest taxon sample of hexapods and metazoans available to date, it was illustrated that no support could be achieved for the deep branches connecting the hexapod orders. This is in strong contrast to previously published mt-genome based phylogenies using a reduced data set and strongly supported relationships, e.g. for the paraphyly of Hexapoda (Carapelli et al., 2007; Cook, Yue, and Akam, 2005; Nardi et al., 2003). Consequently, it is highly recommended interpreting mt-genome based phylogenies with reduced data sets particularly for hexapods with caution although the inferred relationships are highly supported.

13.2.4 EST projects of five enigmatic taxa – the phylogenomic approach

Comparative phylogenomic analyses for higher-insect systematics are still rare and genome and/or EST data from several pterygote orders are non-existent (Delsuc, Brinkmann, and Philippe, 2005; DeSalle, 2005; Eisen and Fraser, 2003) (see also Figure 13.2). Focusing on the lack of information at the base of the Pterygota, EST projects were conducted for two representatives of the palaeopterous Odonata (*Ischnura elegans*) and Ephemeroptera (*Baetis* sp.) in order to (i) address the "Palaeoptera problem" (Simon et al. 2009) and (ii) to reconstruct a backbone tree for arthropods (Meusemann et al. 2010). In addition, EST projects were performed on three enigmatic polyneopteran orders, Dermaptera (*Forficula auricularia*), Plecoptera (*Nemurella pictetii*) and Zoraptera (*Zorotypus cf. gurneyi*) (Simon et al., 2012).

Besides the primary aim of all three studies to reconstruct a robust phylogeny, the data were also tested for their quality and their phylogenetic signal in different approaches. In the study of Meusemann et al. (2010) a data set was selected by the application of a newly developed bioinformatic tools (MARE: Meyer, Meusemann, and Misof, April 2011) to increase the number of taxa with potentially informative genes and excluding poorly represented taxa and uninformative genes. This optimized data set included 117 taxa and 129 genes and robustly resolved major arthropod

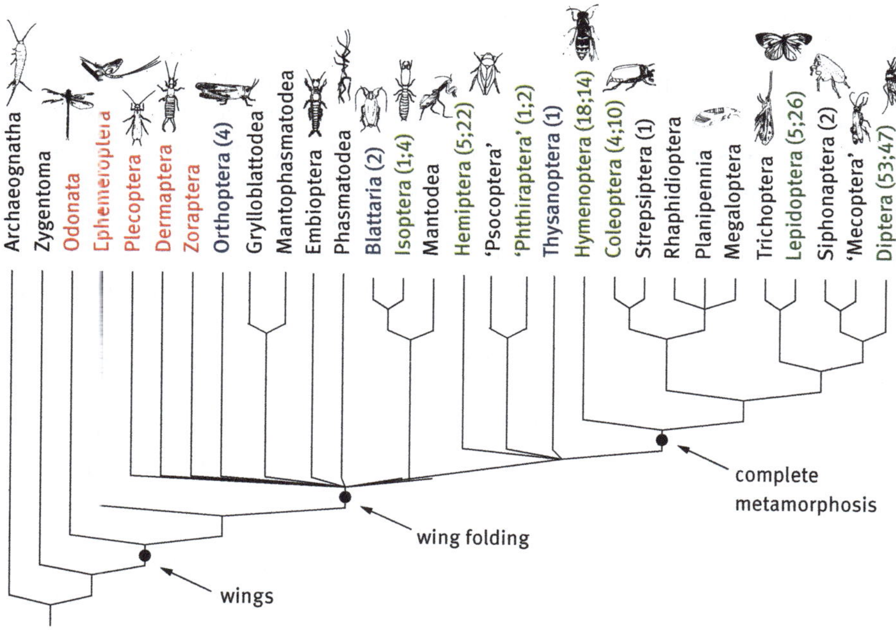

Figure 13.2: Phylogeny of insect orders, modified from Hadrys et al. (2012) The phylogeny is based on information and figures of the studies conducted within the DMP (but see also Section 13.2.4 and Simon et al., 2012). Macro-evolutionary events in insect evolution, which are cited as being major bauplan transitions, are mapped on the phylogeny. Numbers in parenthesis indicate numbers of whole genome and/or EST (> 1,000 contigs) projects conducted or announced as of September 2011. Color code: green, species only with EST projects; blue, species with genome and EST projects; red, pterygote species for which EST projects were conducted within the DMP. Note: 'Pthiraptera' and 'Psocoptera' are polyphyletic to each other and are joined in the clade Psocodea.

relationships with strong support, in contrast to the initial complete data set, which is in many respects unresolved, or shows low support values. However, the analyses on the optimized data set also show for some deep splits among arthropods a remarkable sensitivity to methods of analysis, again for example with respect to the resolution of the basal pterygote divergence.

The study of Simon et al. (2009) used another approach to evaluate the phylogenetic content of EST data – with the focus on the basal pterygote relationships. In this context the lack of resolution still observed in phylogenomic approaches is frequently interpreted as a "molecular signature of radiations compressed in time" (but see also Baurain, Brinkmann, and Philippe, 2007; Rokas, Kruger, and Carroll, 2005). Indeed, it is obvious that ancient rapid radiations like the divergence of pterygotes that lead to short internodes connecting the clades are the greatest challenge for phylogenetic studies. It has been suggested that a polytomy can be resolved by using genes that evolve at the optimal rate in the relevant timescale (Townsend, 2007; Townsend, Lopez-Giraldez, and Friedman, 2008) and that the evolutionary history of the genes that compose a data set has a direct impact on the reconstructed phylogeny (Aguileta et al., 2008; Comas, Moya, and Gonzalez-Candelas, 2007b). To address this issue and considering that the functional role of genes is important for the phylogenetic signal they carry (Graur and Li, 2000), the biological function of the genes in the data set was assessed in the study of Simon et al. (2009). The phylogenetic and statistical analyses based on two data sets, differing in their represented taxa and genes, their content of missing data and consequently the overall number of characters, rejected the traditional Palaeoptera scenario as well as the Metapterygota scenario and unambiguously supported the Chiastomyaria hypothesis (Odonata, Ephemeroptera + Neoptera). In addition, statistical tests of both and on the subdivided data set – according to their biological function – have shown that only certain proteins provide the phylogenetic signal to support the Chiastomyaria hypothesis.

Following the assumption of Simon et al. (2009) that the biological function of the genes might have an impact on the phylogenetic utility of the data, this approach was extended by using a dense taxon sampling across the diverse insect lineages (Simon et al., 2012). Several phylogenetic hypotheses within the Insect Tree were addressed and how the support for these hypotheses might be influenced by the quality of the data including missing data, matrix selection, gene and taxa number/choice and the biological function of the genes. To address these issues, comparative phylogenomic analyses comprising the three new polyneopteran EST data sets were performed. In addition, the strength of support for the concatenated phylogenetic hypotheses along the concatenation process was assessed using a new bioinformatic approach, RADICAL (Narechania et al., 2012). RADICAL generates a library of trees by sequentially concatenating randomly chosen gene partition, starting from one gene to whole-matrix concatenation. The dynamics of concatenation is than monitored by calculating support statistics for the whole-matrix concatenation topology against the library of trees. Based on these analyses conclusive support for monophyletic Neoptera and monophyletic Polyneoptera was provided. In addition, although the interaction of gene choice and taxon sampling still remains unknown, no influence of selecting different compositions of taxa and genes based on an original supermatrix could be identified in the resulting inferred insect relationships. In contrast, additional analyses highlighted that especially for the ancient rapid radiation of the

Figure 13.3: RADICAL analysis of nodal support. Shown are the RADICAL curves for the three conflicting hypotheses of basal pterygote divergence (at the top) of a phylogenomic data set (P_matrix_g) and of the four functional subgroups based on this data set; cell = cellular processes and signaling (85 genes), info = information storage and processing (80 genes), meta = metabolism (78 genes) and poorly = poorly recognized (42 genes). The y-axis shows the CFI (Consensus Fork Index) and the x-axis the concatenation size (number of genes). Note that depending on the biological function of the genes different evolutionary hypotheses or pterygote divergence are supported, e.g. genes involved in cellular processes and signaling (cell) and information storage and processing (info) support the Metapterygota hypothesis while genes involved in metabolism (meta) or poorly recognized genes (poorly) support the alternative Palaeoptera hypothesis. Figure modified after Simon et al., 2012. ►

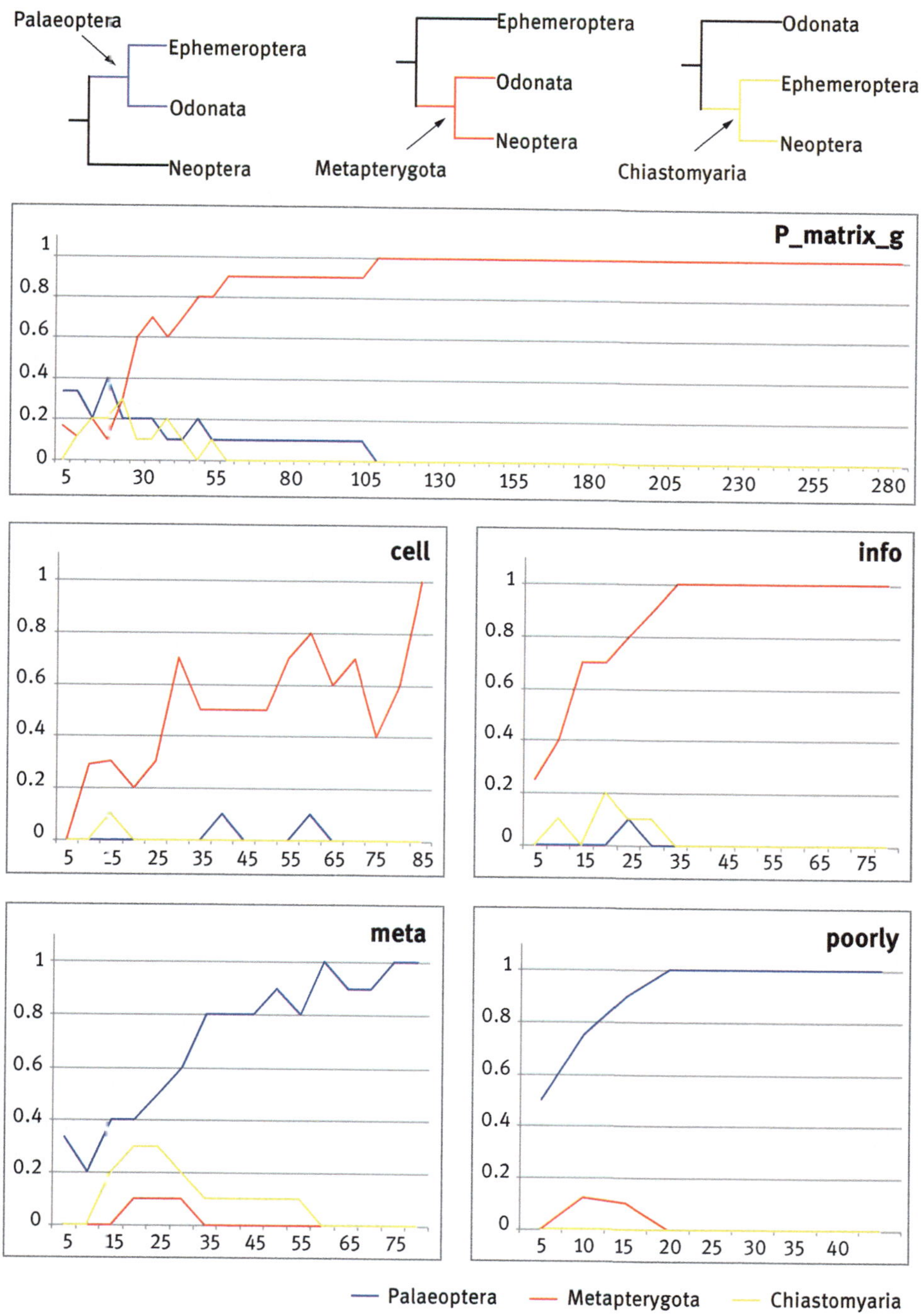
Palaeoptera
Ephemeroptera
Odonata
Neoptera
Ephemeroptera
Odonata
Neoptera
Metapterygota
Odonata
Ephemeroptera
Neoptera
Chiastomyaria
P_matrix_g
cell
info
meta
poorly
Palaeoptera
Metapterygota
Chiastomyaria

insects, e.g. basal pterygote divergence or split of neopteran infraclasses, the function of the genes and the taxon sampling may have the most influential impact on the inferred relationships (see also Figure 13.3).

13.2.5 Bridging gaps: New target genes for the analysis of gene organization, function and morphology

A promising new approach to use target genes in deep phylogeny is a combined sequence and functional analysis of regulatory genes – a direct bridge to morphology. Regulatory genes have often been discussed in a phylogenetic context and when studying the complicated evolutionary history of insects. The explanatory power of these genes in developmental research is without question and their genomic organization has also supported important clades in metazoan phylogeny (Brooke, Garcia-Fernandez, and Holland, 1998; Kamm et al., 2006). At the sequence level, these genes may also harbor some phylogenetic information. There is – for example – strong evidence that changes in the sequence and expression (function) of the homeotic genes *ultrabithorax* (*Ubx)* and *sex combs reduced (Scr*) in basal pterygote orders relate to various forms of abdominal appendages (Galant, Walsh, and Carroll, 2002; Marden and Thompson, 2003). In the derived holometabolous insects, changes in Ubx gene expression patterns relate directly to the evolution of morphological diversity in hindwings (Weatherbee et al., 1998; Weatherbee et al., 1999). Comparative sequence and functional studies of these genes may shed some light on the early evolution of insect flight, the most important key innovation in hexapod diversification. However, no comparative data exist for basal insects/hexapods. Indeed, no Hox genes at all have been isolated from 20 out of 34 hexapod orders yet, and the existing sequences mainly derive from two holometabolous orders only. In a current study, 35 new partial homeobox sequences of *Antp*-class genes (*lab, pb, Dfd, Scr, ftz, Antp, Ubx, abd-A* and *Abd-B*) for basal Hexapoda, "apterygote" and pterygote orders were isolated (Hadrys et al., 2012). By comparing rates of molecular evolution of insect Hox genes with the divergence rates of other arthropod classes and Mammalia these analyses have shown that an increased radiation rate correlates with an increased rate of sequence evolution in the regulatory Hox genes. Further studies of temporal and spatial Hox gene expression and their function in different pterygote orders by comparative in situ hybridization studies as well as comparative RNAi studies might bridge the gap between gene sequence, function, body architecture and finally evolutionary pathways of key bauplan innovations.

13.3 Conclusion

The studies conducted within the DMP contribute several comprehensive molecular phylogenetic studies to resolve the insect evolutionary tree and in a grander scheme the arthropod and metazoan evolutionary tree. Several molecular marker systems at different genetic levels from crucial insect orders were isolated and characterized providing the first step to get a deeper insight into the macro-evolutionary events among insects. The improved phylogenetic analyses of all molecular markers demonstrated that data quality assessment prior to tree reconstruction yielded better resolution in several places. The incongruence between the inferred topologies mirrors the challenges in interpreting the strengths and weaknesses of the underlying data and of the analyses conducted. Indeed, the phylogenomic analyses have shown that the large data matrices have a complex phylogenetic signal and that a greater number of independent loci recover different scenarios of insect evolution.

Currently, it seems that the available molecular data sets and/or bioinformatic approaches are still insufficient to recover the most debated ancient splits within insects. As long as we do not have independent alternative characters, e.g. genetic characters such as gene order, genome rearrangements, intron and transposon positions, as well as information on comparative gene function and their impact on morphology (which might provide a greater understanding of insect evolution in a grander scheme), we can only suggest which "markers" might reflect the "true evolutionary history" of insects. In sum, exploring the extent and source of biases in phylogenetic studies remains the biggest challenge. Here, inferring insect relationships offers a great opportunity to assess the data quality and how the recovery of ancient rapid radiations might be influenced by the choice of taxa and genes.

Acknowledgments

We would like to express our gratitude to the German Research Foundation (Deutsche Forschungsgemeinschaft [DFG]) which supported this project, grant given to H.H. (DFG HA 1947/5), within the special priority program "Deep Metazoan Phylogeny" SPP1174. We thank Jennifer Angermann, Rolf G. Beutel, Christian Epe, Horst Hertel, Ann Kathrin Ketelsen, Sara Khadjeh, Klaus Klass, Boris C. Kondratieff, Ryuichiro Machida, Albert Melber, Karen Meusemann, Leonardo Calderon Obaldia, Mike D. Picker, Reinhard Predel, Sven Sagasser, Valentina Teslenko and Peter Zwick for help in collecting specimens, for providing tissue or other DNA samples or for laboratory help. We would also like to thank all collaborators within the DMP for this fruitful cooperation, especially: Matthias Bernt, Thorsten Burmester, Alexander Donath, Ingo Ebersberger, Patrick Kück, Karen Meusemann, Bernhard Misof, Lars Podsialowski, Peter Rehm, Falko Roeding, Peter Stadler, Sascha Strauss, Arndt von Haeseler, Björn M. von Reumont and J. Wolfgang Wägele. Special thanks go also to Stephen Dellaporta, Robert DeSalle, Sven Sagasser and Bernd Schierwater for providing constructive comments, which greatly improved the project.

Martin E.J. Stegner, Martin Fritsch, and Stefan Richter

14 The central complex in Crustacea

Abstract: The central complex is an intricate cluster of three interconnected neuropils (i.e. the unpaired central body and protocerebral bridge, and the paired lateral accessory lobes) situated in the brain of insects and crustaceans, which has been interpreted as a synapomorphy of all Tetraconata. However, the neuropils have been reported to be missing in several other crustacean taxa such as Cephalocarida and Mystacocarida, allowing for the alternative hypothesis that the central complex is a synapomorphy of only some tetraconate taxa. In order to better evaluate the disparity from a phylogenetic perspective, our contribution is to establish a well-founded comparison of the central complex across Crustacea, including its morphological structures as well as immunoreactive domains of neuro-active substances such as serotonin, RFamide, or histamine. In extension of the glossary of neuroanatomical terms developed within the Deep Metazoan Phylogeny framework, we define and clarify morphological terms related to the central complex. On the basis of improved comparability, we describe the central complex for all crustacean taxa, presenting selected data from our own immunohistochemical and histological investigations into Cephalocarida, Mystacocarida, Branchiopoda, Malacostraca, Copepoda, Ostracoda, and giving an updated review of the relevant literature. Our results and literature review add further support that the central complex is part of the tetraconate ground pattern. We discuss that in the light of all current phylogenetic hypotheses, the central complex would have been reduced several times independently during crustacean evolution. Central-body-like immunoreactive domains such as in Cephalocarida or Mystacocarida are here interpreted as plesiomorphic rudiments. Whether the central complex is indeed a tetraconate or rather a mandibulate apomorphy remains unclear, unless Myriapoda are studied in higher detail.

14.1 Introduction

One of the most interesting topics in arthropod neuroanatomy over the past decades has been the central complex, an intricate cluster of neuropils and tracts that occurs in the brain of various arthropods (reviewed by Harzsch, 2006; 2007; Homberg, 2008; Strausfeld, 2012). Following the definition proposed by Richter et al. (2010), the central complex consists of three interconnected neuropils: (1) the unpaired central body, (2) the unpaired protocerebral bridge, and (3) a pair of lateral accessory lobes. Although descriptions of these three neuropils date back to early neuroanatomists (e.g., Holmgren, 1916; Hanström, 1928; 1940; 1947), it was Williams (1975) who first recognized that, in the locust *Schistocerca gregaria*, the neuropils are interconnected by tracts, together forming one functional unit for which he then suggested the term 'central complex'.

The finding that the shape and arrangement of unpaired midline neuropils in Decapoda widely correspond to those in Hexapoda led Hanström (1928) to propose a sister-group relationship between Hexapoda and Crustacea – long before a monophyletic 'Tetraconata' composed of Hexapoda and Crustacea found support in molecular data. Detailed studies, especially of the hexapod (e.g., Boyan and Williams, 2011) and decapod central complex (e.g., Utting et al., 2000) have revealed further correspondences in the pattern of interconnecting tracts. A central complex has also been documented in three other crustacean taxa, namely anostracan Branchiopoda (Harzsch and Glötzner, 2002), Remipedia (Fanenbruck, Harzsch, and Wägele, 2004; Fanenbruck and Harzsch, 2005; Stemme et al., 2012) and Copepoda (Andrew, Brown, and Strausfeld, 2012). Although earlier investigators have described both a protocerebral bridge and a central body in the myriapod brain (Holmgren, 1916; Hanström, 1928; Joly and Descamps, 1987), this view has been questioned repeatedly by recent authors assuming that only one unpaired midline neuropil termed either 'central body' or 'central neuropil' occurs in Myriapoda (Lösel, Nässel, and Strausfeld, 2002; Loesel, 2004; Harzsch, 2006; Homberg, 2008; Sombke, Rosenberg, and Hilken, 2011). Unlike the central body, which is embedded within the protocerebrum, the single unpaired midline neuropil in Onychophora and Chelicerata is situated in the periphery of the brain and termed the 'arcuate body' (Strausfeld et al., 2006a; Loesel et al., 2011).

Besides these obvious differences between the major arthropod taxa, morphological disparity is also found within the taxa. Within Tetraconata, for example, the central body of Hexapoda is split into two distinct subunits, the ellipsoid body and the fan-shaped body (Strausfeld, 1976; Hanström, 1940; Kollmann, Huetteroth, and Schachtner, 2011), while the central body of Decapoda (e.g., Harzsch and Hanson, 2008; Krieger et al., 2010; 2012) appears as a horizontally layered but structurally coherent unit. Several other crustacean taxa, e.g. Mystacocarida (Brenneis and Richter, 2010) and Cephalocarida (Stegner and Richter, 2011) have even been reported to lack the neuropils of a central complex completely, or only to exhibit certain components of a central complex, e.g., phyllopod Branchiopoda (Fritsch and Richter, 2010).

Various studies have been carried out in an attempt to shed light on the morphological disparity in unpaired midline neuropils across Arthropoda and unravel their evolutionary origin. Approaches have taken in the innervation pattern, internal structure, development during ontogeny and functional aspects of these neuropils (reviewed by Homberg, 2008, see also Strausfeld, 2012). The association between both the arcuate body (e.g. Strausfeld, Weltzen, and Barth, 1993; Loesel, 2004) and the central complex (Liu et al., 2006) on the one hand and the optic pathway on the other initially led to the assumption that unpaired midline neuropils might constitute second order visual centers (reviewed by Homberg, 2008). However, a well-developed central body also occurs in blind taxa such as Symphyla (Joly and Descamps, 1987), Remipedia (Fanenbruck, Harzsch, and Wägele, 2004) and spelaeogriphacean and cumacean Malacostraca (this study). As convincingly demonstrated by Strauss (2002) and Bender, Pollack, and Ritzmann (2010), all the evidence points towards the central complex playing a cardinal role in controlling motor activity, irrespective of whether in response to visual stimuli (Triphan et al., 2010) or tactile stimuli (Harley and Ritzmann, 2010).

Strausfeld et al. (2006a) suggested an evolutionary scenario in which the onychophoran/chelicerate arcuate body, the myriapod central neuropil and the tetraconate central complex all come from a hypothetical central neuropil in an ancestral urarthropod. Broadly in line with this hypothesis, Harzsch (2006) suggested that the central complex – consisting of a central body, a protocerebral bridge and lateral accessory lobes – constitutes an apomorphy of Tetraconata. The morphological correspondences between Decapoda, Hexapoda and Remipedia certainly suggest homology (Fanenbruck and Harzsch, 2005; Harzsch, 2006), but what about the "missing" central complex neuropils reported in other crustacean taxa? Since the phylogenetic relationships within Tetraconata are unclear (see for example Richter, Møller, and Wirkner, 2009; Jenner, 2010; Regier et al., 2010; von Reumont et al., 2012; Oakley et al., 2013), an explanation alternative to Harzsch (2006) would be that a central complex only evolved within Tetraconata as a synapomorphy of a limited group of tetraconate subtaxa (including Hexapoda, Malacostraca and Remipedia).

Methods of neuroanatomical investigation have advanced considerably over the past two decades thanks to progress in immunolabeling, confocal laser scanning microscopy and computerized 3D modeling. Given the improved morphological detail which present-day techniques make it possible to obtain, neuroanatomical structures such as the central complex can be used increasingly to address phylogenetic questions. This promising discipline has been coined 'neurophylogeny' (e.g., Harzsch, 2006), and the 'Invertebrate Neurophylogeny' section was a major constituent of the Deep Metazoan Phylogeny (DMP) program. Modern neurophylogenetic studies permit a reevaluation of earlier morphological descriptions and character definitions such as the presence of a central body. Using immunolabeling and confocal microscopy, immunoreactive neurons can be described individually right down to their neurites (e.g., Harzsch and Waloszek, 2000), bringing a wealth of new characters to phylogenetics. The serotonin-like immunoreactive pattern has proven especially useful in this respect (Harzsch, 2002b; 2003b; 2004; Harzsch, Müller, and Wolf, 2005; Brenneis and Richter, 2010; Stegner and Richter, 2011; Fritsch and Richter, 2012).

Since early neurophylogenetic studies focused on the long-debated relationships between the major arthropod taxa (Chelicerata, Myriapoda, Hexapoda, Crustacea), investigators of the central complex have tended to concentrate on hexapod or malacostracan representatives of Tetraconata – i.e. those species which are larger and more easily available – neglecting the less common non-malacostracan crustacean taxa.

One of our aims within the framework of DMP was to unravel phylogenetic relationships between the individual tetraconate subtaxa and, ultimately, to trace the evolution of the central complex within Tetraconata. Given the sparse data available on non-decapod Crustacea, it seemed the natural step to reinvestigate representatives of those taxa for which neuroanatomical data was missing or incomplete. We are aware of the recent approach by Strausfeld and Hirth (2013) who identified potential "deep homology" between the hexapod central complex and the vertebrate basal ganglia based on common gene expression. However, we focus on structural correspondences existing between arthropods only.

In this DMP project, we studied Mystacocarida and Cephalocarida, two taxa whose nervous system had long been neglected and which it was finally possible to re-investigate using present-day methods. We revealed that the organization of the mystacocarid brain is comparatively simple, while the ventral nerve cord and peripheral nervous system broadly correspond to those in other crustaceans (Brenneis and Richter, 2010). In contrast, the cephalocarid nervous system contains several conspicuously complex features, including the olfactory system (Stegner and Richter, 2011), the thoracic commissures and the pattern of serotonin-like immunoreactive neurons (Stegner, Brenneis, and Richter, in press).

To establish a well-founded comparison of central complex components – and potentially homologous structures – across Crustacea, we present results from the DMP program pertaining to the mystacocarid and cephalocarid brain along with the results of independent investigations pertaining to Branchiopoda, Ostracoda, Copepoda and peracarid Malacostraca, complemented by data from the literature.

14.2 Definitions

In our view, morphological descriptions and terminology should be free of homology assumptions of any kind (see Vogt, Bartolomaeus, and Giribet, 2010). This is not to deny that homology exists, but simply to acknowledge the distinction between the various steps which need to be followed in (evolutionary) morphology, starting with pure description and only then going on to conceptualize evolutionary characters (Wirkner and Richter, 2010). In this context it is important to point out that not only must the terms used to denote concepts be free of homology assumptions, the concepts themselves should also be based on 'pure' description alone.

To establish a well-founded comparison of morphological structures and immunoreactive patterns across the different taxa, we use a standardized nomenclature based on the suggestions for a neuroanatomical glossary developed within the 'Invertebrate Neurophylogeny' section of the DMP program (Richter et al., 2010).

In continuation and elaboration of the glossary, our definitions are as follows:

14.2.1 Protocerebral bridge

The protocerebral bridge is an unpaired or paired midline neuropil which is situated in the anterior periphery of the protocerebrum, directly adjacent to the anteromedial protocerebral somata. If the protocerebral bridge is paired, the neuropils are transversely connected via a tract.

14.2.2 Central body

The central body is an unpaired midline neuropil. It is situated within and surrounded by the protocerebral neuropil. The central body may be composed of columnar neurons and/or tangential neurons. The terms 'columnar neurons' and 'tangential neurons' refer to the orientation of their respective columnar or tangential neurite. If the central body occurs as a part of the central complex, it is connected to the lateral accessory lobes and the protocerebral bridge.

14.2.3 Lateral accessory lobes

The lateral accessory lobes are a pair of neuropils located laterally in the protocerebrum. The lateral accessory lobes are connected with the protocerebral bridge and the central body via tracts.

14.2.4 PB-CB tracts

A PB-CB tract is a tract which connects the protocerebral bridge to the central body. Where the specific arrangement of four PB-CB tracts per body side is present, the tracts are termed the W, X, Y, Z tracts.

14.2.5 Immunoreactive domains

An immunoreactive domain is a distinct cluster of immunoreactive neurites. It may overlap with all or part of a distinct neuropil.

Background: The term 'immunoreactive domain' refers to a category which needs to be specified by naming which neuro-active substance is labeled to become a structural term. A number of neurons have been found to synthesize neuro-active substances such as serotonin, tachykinin, RFamide and histamine. The immunoreactive neurites of these neurons extend into the protocerebral neuropil where they arborize to form chemical synapses with other neurons, thus contributing to immunoreactive domains. We have avoided calling these immunoreactive domains 'neuropils'. An immunoreactive domain may overlap with part but not all of a neuropil, and immunoreactive domains may even be present where distinct neuropils cannot be identified. Although they may correlate with each other, neuropils and domains and their respective shape, position and innervation pattern should be described separately. While neuropils are recognizable in semithin sections and histological and immunohistochemical stains by their density and their spatial and structural separation from the surrounding brain neuropil, immunoreactive domains stand out from the surrounding neuropil solely because of their pattern of immunoreactivity.

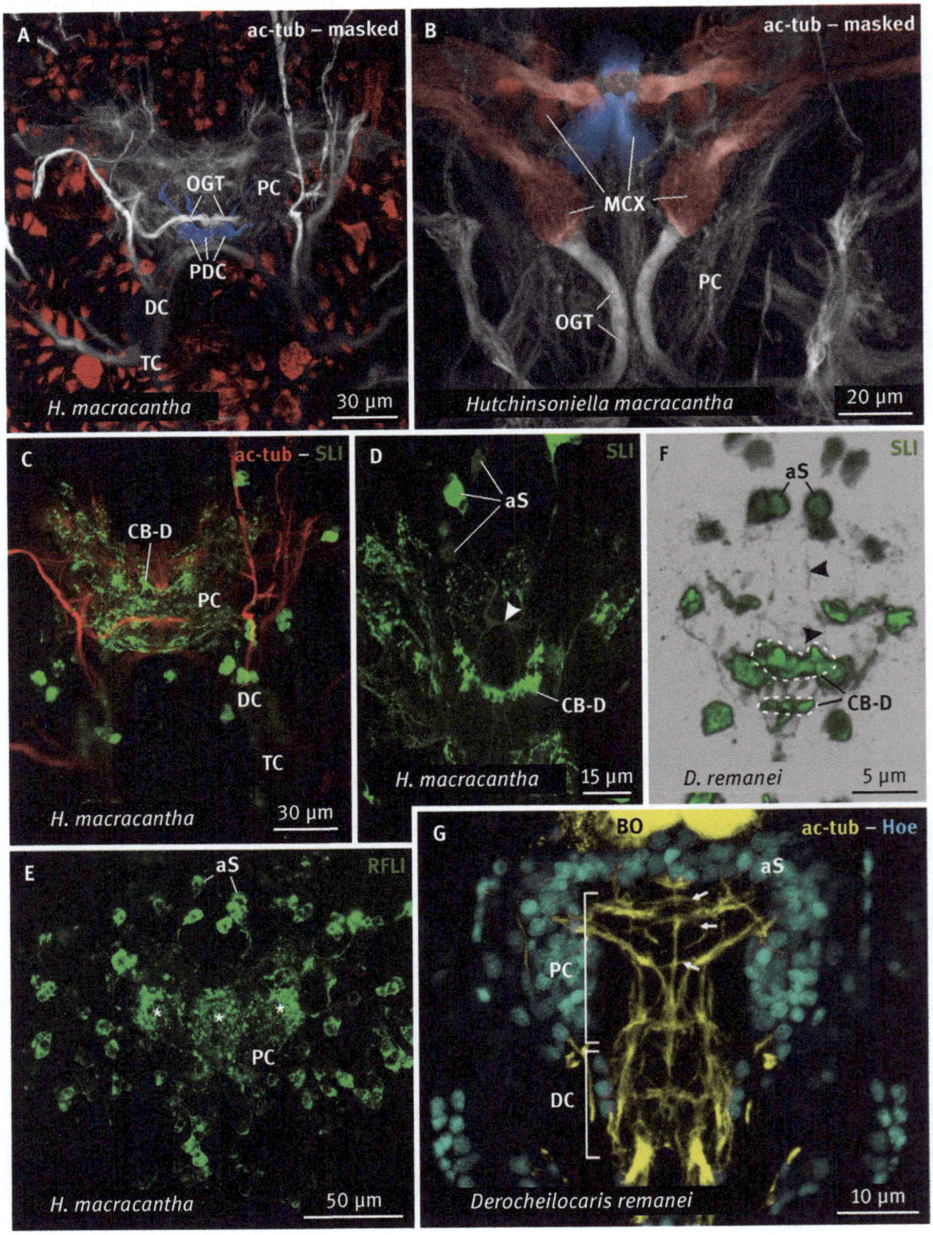

The brain may contain domains in different regions. For the purposes of this publication, three types are of special interest:

- the anterior immunoreactive domain, here termed the 'protocerebral bridge-like immunoreactive domain' (PB-like domain)
- the central immunoreactive domain, here termed the 'central body-like immunoreactive domain' (CB-like domain)
- a pair of lateral immunoreactive domains, here termed 'lateral accessory lobe-like immunoreactive domains' (LAL-like domain)

Those immunoreactive domains which are not associated with the central complex in any way are ignored in this publication.

We use the following abbreviations for the various types of immunoreactivity: serotonin-like immunoreactivity/immunoreactive (SLI/SL-ir); histamine-like immunoreactivity/immunoreactive (HLI/HL-ir); RFamide-like immunoreactivity/immunoreactive (RFLI/RFL-ir).

14.3 Results

14.3.1 Cephalocarida

The cephalocarid species *Hutchinsoniella macracantha* lacks distinct neuropils such as a central body, a protocerebral bridge or lateral accessory lobes (Elofsson and Hessler, 1990; Stegner and Richter, 2011). The only distinct unpaired midline neuropils described in the cephalocarid brain are the median neuropil of the posterodorsal neuropil cluster (Figure 14.1A) and three neuropils of the multi-lobed complex (Figure 14.1B), none of which correspond to the central complex neuropils

◄ **Figure 14.1:** Confocal horizontal micrographs of the protocerebrum in Cephalocarida **(A–D)** and Mystacocarida **(F,G)**. **A.** Dorsal view of a whole-mount. The 'posterodorsal neuropil cluster' (*blue*) includes an unpaired midline neuropil but cannot be related to the central complex of other arthropods. **B.** Transverse vibratome section. The multi-lobed complex includes both paired (*red*) and unpaired (*blue*) components which do not correspond to central complex neuropils. **C.** Dorsal view of a whole-mount. The unpaired sickle-shaped SL-ir domain (*CB-D*) lies within the protocerebrum. **D.** Horizontal vibratome section. The SL-ir domain receives decussating SL-ir axons (*arrowhead*) from anterior somata. **E.** Dorsal view of the brain. None of the cephalocarid RFL-ir domains (*asterisks*) can be related to the central complex neuropils of other crustaceans. **F.** Ventral view of the mystacocarid brain showing two CB-like SL-ir domains receiving ipsilateral neurites from anterior somata. **G.** Ventral view of a whole-mount. *Arrows* point at various commissures, but midline neuropils are not present. *ac-tub* acetylated α-tubulin, *aS* anterior somata, *BO* Bellonci organ, *CB-D* CB-like domain, *DC* deutocerebrum, *Hoe* Hoechst nuclear counterstain, *MCX* multi-lobed complex, *OGT* olfactory globular tract, *PC* protocerebrum, *PDC* posterodorsal neuropil cluster, *RFLI* RFamide-like immunoreactivity, *SLI* serotonin-like immunoreactivity, *TC* tritocerebrum. Originals by Stegner and Richter (A–E), Brenneis and Richter (F,G).

in other arthropods (see Stegner and Richter, 2011). There is a CB-like SL-ir domain (Figure 14.1C,D) in *Hutchinsoniella macracantha* which corresponds to the central body of other arthropods in four respects: (1) its elongated shape, (2) its central position within the protocerebral neuropil, (3) the fact that SL-ir neurites project into it from anteromedial somata, (4) the fact that these neurites cross the midline (Stegner and Richter, 2011). In contrast, none of the RFL-ir and HL-ir unpaired immunoreactive domains in *H. macracantha* (labeled rd1, hd2, hd3 by Stegner and Richter, 2011) correspond to neuropils of the central complex (see, for example, the middle asterisk in Figure 14.1E).

14.3.2 Mystacocarida

Mystacocarida do not exhibit any distinct neuropils in the protocerebrum, and also lack a central body, a protocerebral bridge and lateral accessory lobes (Baccari and Renaud-Mornant, 1974; Elofsson and Hessler, 2005; Brenneis and Richter, 2010). There are two SL-ir CB-like domains in the mystacocarid species *Derocheilocaris remanei* (Figure 14.1F) which correspond to the central body of other arthropods in three respects: (1) their elongated shape, (2) their central position within the protocerebral neuropil, (3) the fact that SL-ir neurites project into them from anteromedial somata (Brenneis and Richter, 2010). Unlike in Cephalocarida, these neurites do not cross the midline before entering the SL-ir CB-like domains (Brenneis and Richter, 2010). Although embedded centrally within the protocerebral neuropil, the two mystacocarid SL-ir CB-like domains lie more dorsally than the cephalocarid SL-ir CB-like domain. Protocerebral RFL-ir domains in Mystacocarida do not correspond to neuropils of the central complex (Brenneis and Richter, 2010).

Figure 14.2: Central complex neuropils in Isopoda **(A)**, Amphipoda (*Parhyale hawaiensis*, **B**), Cumacea **(C)**, Spelaeogriphacea **(D)**, Copepoda **(E)** and Ostracoda **(F)**. **A.** 3D reconstruction of the brain based on semithin sections in posterior view. Somata in light blue, neuropil semi-transparent. The W tract lies posterior to the decussating olfactory globular tracts, and the X, Y, Z tracts anterior to them. **B, C.** Vibratome section in posterior view. **D.** Transverse semithin section through the protocerebrum. The protocerebral bridge is split, and the Z and Y tracts cannot be distinguished. The W tract, which passes posterior to the olfactory globular tract, is situated in a more posterior section (not shown). **E.** Ventral view of a whole-mount. Within the protocerebrum no distinct central complex neuropils are identifiable. **F.** Ventral view of a whole-mount. Within the protocerebrum prominent tracts (arrows) but no distinct central complex neuropils are identifiable. *ac-tub* acetylated α-tubulin, *aS* anterior somata, *CB* central body, *DC* deutocerebrum, *LAL* lateral accessory lobes, *OGT* olfactory globular tract, *OL* olfactory lobe, *PC* protocerebrum, *Syt* Sytox nuclear counterstain, *W/X/Y/Z* PB-CB tracts. Originals by Stegner, Wirkner, Richter (A,D), C. Wittfoth (B,E), C. Döring (C), M. Dreiling (F). ▸

14.3.3 Malacostraca

All the major malacostracan taxa exhibit a well-developed central complex which includes a central body, a protocerebral bridge and lateral accessory lobes (Hanström, 1928; 1934; 1947; Sandeman et al., 1992; Sandeman, Scholtz, and Sandeman, 1993; Sandeman and Scholtz, 1995; Utting et al., 2000; Krieger et al., 2010; 2012; Strausfeld, 2012). This is also true for Leptostraca, which is the sister-group to all remaining Mala-

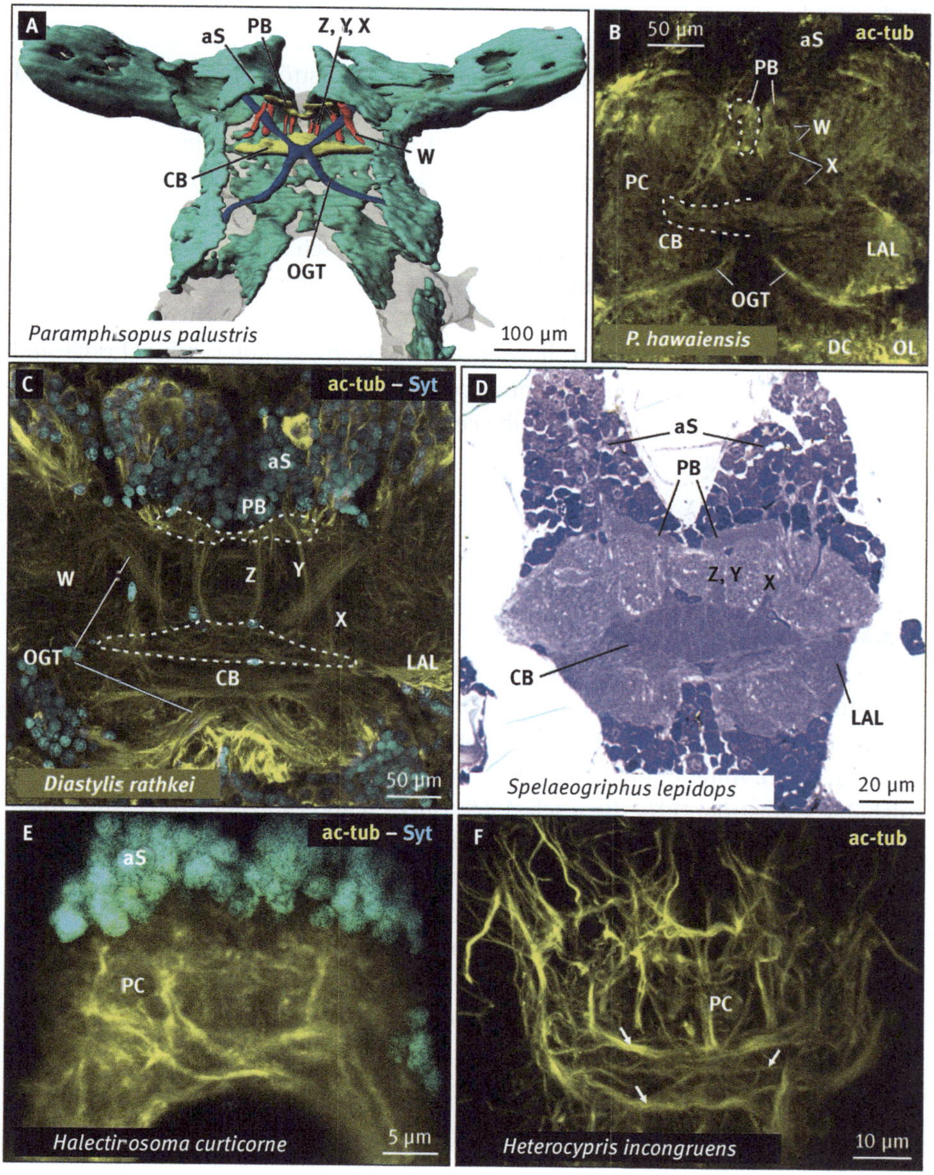

costraca (Kenning et al., 2013). Utting et al. (2000) studied the central complex of the lobster *Cherax destructor* in great detail, tracing the anatomy of individual neurons and showing that the lobster's PB-CB tracts occur in a specific arrangement of four major tracts on each side of the body: these are termed the W, X, Y and Z tracts. A corresponding pattern is found in the leptostracan *Nebalia herbstii* (Kenning et al., 2013), the isopod *Paramphisopus palustris* (Figure 14.2A), the amphipod *Parhyale hawaiensis* (Figure 14.2B), the cumacean *Diastylis rathkei* (Figure 14.2C), while the spelaeogriphacean *Spelaeogriphus lepidops* displays three tracts: W, X, Y/Z (Figure 14.2D). This implies that a central complex featuring W, X, Y and Z tracts can be assumed to be part of the ground pattern of Malacostraca (Kenning et al., 2013). No distinct W, X, Y and Z tracts have been described in Stomatopoda (Hanström, 1947; Sandeman and Scholtz, 1995; Strausfeld, 2012).

There are several SL-ir domains in Malacostraca: unpaired domains which overlap with the central body and the protocerebral bridge respectively, and paired domains which overlap with the lateral accessory lobes (Sandeman, Sandeman, and Aitken, 1988; Langworthy et al., 1997; Utting et al., 2000; Harzsch and Hansson, 2008). The central body and lateral accessory lobes comprise RFL-ir domains, while the protocerebral bridge is not RFL-ir (Harzsch and Hansson, 2008). Synapsin-like immunoreactive (SYNL-ir) domains have been shown to overlap with the protocerebral bridge and lateral accessory lobes, while the central body is only weakly SYNL-ir (Harzsch and Hansson, 2008). All the neuropils of the central complex in Decapoda exhibit HL-ir domains (Langworthy et al., 1997).

14.3.4 Branchiopoda

14.3.4.1 Anostraca

In the anostracan *Artemia salina* a well-developed central body is present. The unstratified and evenly-shaped mesh of the spindle-shaped central body is clearly identifiable via histological sectioning (Hanström, 1924; Benesch, 1969; Harzsch and Glötzner, 2002). Anterior to the central body, a slender protocerebral bridge is distinguishable (Hanström, 1924; Benesch, 1969; Harzsch and Glötzner, 2002). Posterolaterally on both sides of the central body, lateral accessory lobes have been described (Benesch, 1969; Harzsch and Glötzner, 2002).

An HL-ir and SL-ir domain which broadly overlaps with the central body is also present (Harzsch and Glötzner, 2002). Anterior to this domain, Harzsch and Glötzner (2002) discovered a tripartite HL-ir and SL-ir domain. In our view, only the median component of this HL-ir and SL-ir domain corresponds to the unpaired and slender protocerebral bridge identified histologically by Hanström (1924) and Benesch (1969). HL-ir and SL-ir domains overlap with the lateral accessory lobes (Harzsch and Glötzner, 2002).

Recent immunohistochemical investigations into the larvae of the anostracan *Branchinella* sp. have revealed a central body and a slender protocerebral bridge (Figure 14.3A), but lateral accessory lobes were not distinguished using nervous system markers (Frase and Richter, unpublished). Frase and Richter (unpublished) were also able to reveal two potential PB-CB tracts in the developing protocerebral architecture of *Branchinella* sp. larvae. Each tract emanates anterolaterally in the protocerebral superficial somata cluster, runs posteriorly, enters the central body anterolaterally and interconnects with the internal mesh. However, no true connection between these tracts and the protocerebral bridge was found.

14.3.4.2 Notostraca

Both a central body and a slender protocerebral bridge have been identified in the Notostraca (Hanström, 1924; Fritsch and Richter, 2010; Strausfeld, 2012). As in the anostracans, the central body in the Notostraca consists of a mesh of evenly distributed neurites. No lateral accessory lobes have yet been found. Two potential PB-CB tracts are discernible in the protocerebrum of *Triops cancriformis* larvae. Each tract emanates from the superficial anterolateral layer of protocerebral somata and enters the central body anterolaterally, but no true connection has been found between these tracts and the protocerebral bridge. Using neuro-active substance markers, both a CB-like domain (tachykinin: Loesel, 2004 in *Triops longicaudatus*, histamine and serotonin: Fritsch and Richter, 2010 in *Triops cancriformis*) and a PB-like weakly HL-ir domain (Fritsch and Richter, 2010) are detectable in the protocerebrum (Figure 14.3B). No reports of the presence of LAL-like domains exist.

14.3.4.3 Laevicaudata, Spinicaudata and Cyclestherida ("Conchostraca")

A central body and a slender protocerebral bridge are also identifiable in the protocerebrum of the Laevicaudata (Hanström, 1934), Spinicaudata (Hanström, 1934) and Cyclestherida (Figure 14.3C–E). The unstratified and even mesh of the central body is shaped like a spindle. The inner scaffold of the protocerebral bridge forms a horizontal network connecting the two lateral protocerebral hemispheres. Two pairs of potential PB-CB tracts are identifiable in the adult protocerebrum of *Lynceus brachyurus* (Laevicaudata) and *Leptestheria dahalacensis* (Spinicaudata) (arrows: Figure 14.3C,D). Both pairs of tracts emanate from the anterolateral protocerebral somata layer, run posteriorly to the anterolateral margins of the central body and interconnect with the internal mesh of the central body. No distinct connection to the protocerebral bridge has yet been found.

HL-ir and SL-ir domains overlapping with the protocerebral bridge and central body are present in the protocerebrum of the spinicaudatan *L. dahalacensis*, but neither histological sectioning nor neuro-active substance markers reveal the presence of lateral accessory lobes or LAL-like domains.

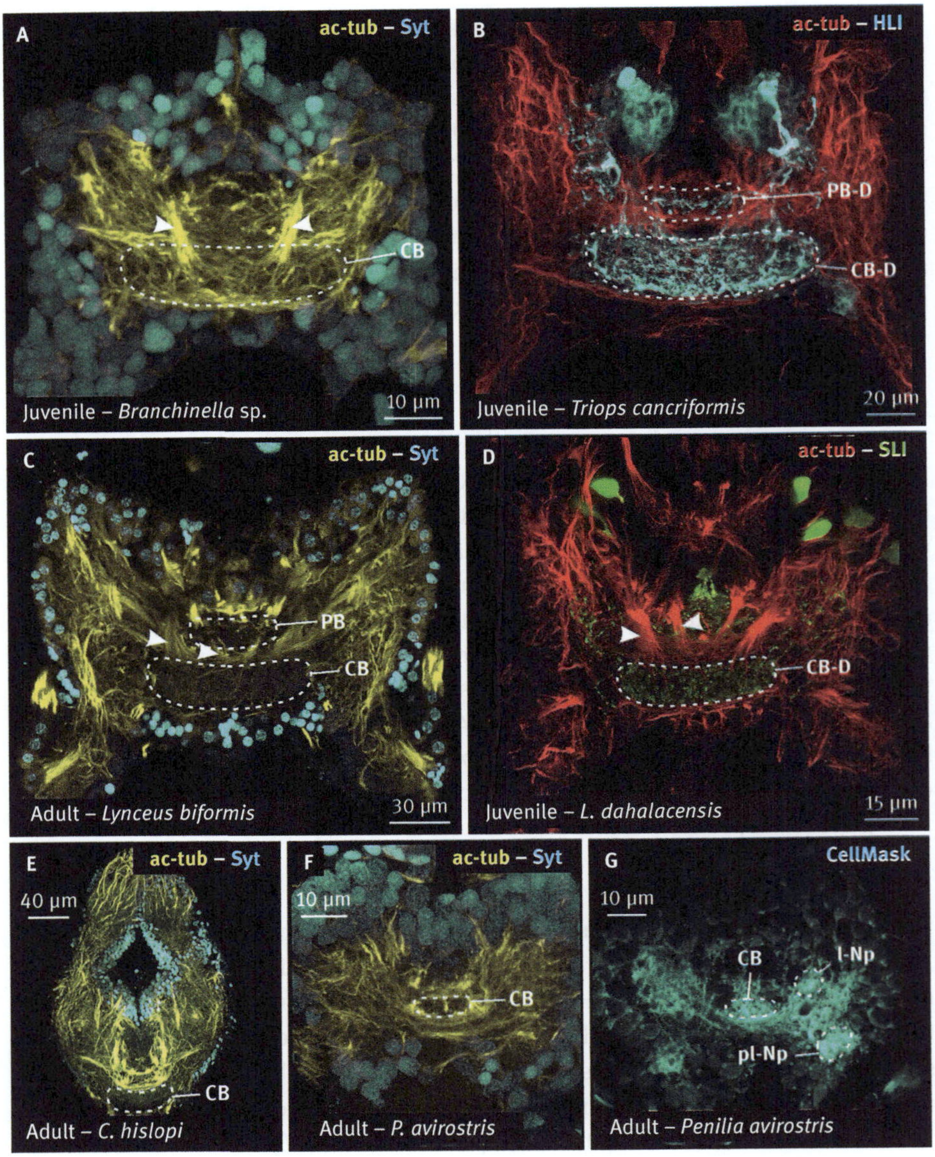

14.3.4.4 Cladocera

Within the protocerebrum of the cladocerans, a well-formed central body is present (Claus, 1876; Leder, 1915; Hanström, 1931; 1934; Bullock and Horridge, 1965; Kirsch and Richter, 2007; Strauß et al., 2011; Weiss et al., 2012; see also Fig. 14.3F,G). A protocerebral bridge has been documented in the haplopod *Leptodora kindtii* (Kirsch and Richter, 2007) and, less distinctive, in the ctenopod *Penilia avirostris* (this study, not depicted). Lateral accessory lobes have been described for the protocerebrum of *L.*

◂ **Figure 14.3:** Ventral view of the branchiopod protocerebrum. Anostraca **(A)**, Notostraca **(B)**, Laevicaudata **(C)**, Spinicaudata (*Leptestheria dahalacensis*, **D**), Cyclestherida (*Cyclestheria hislopi*, **E**) and Cladocera, Ctenopoda **(F, G)**. **A–F.** Immunohistochemical stainings. **A.** A well-developed central body spans the protocerebrum. Two PB-CB tracts (*arrowheads*) enter the central body anteriorly. **B.** The PB-like weakly HL-ir domain is situated anterior to the prominent central body-like HL-ir domain. **C.** Altogether four PB-CB tracts (*arrowheads*) enter the prominent central body at the anterior margin. **D.** Using serotonin markers only, a distinct CB-like SL-ir domain is identifiable. Four PB-CB tracts (*arrowheads*) are present altogether. **E.** Only a central body can clearly be identified. **F.** Using immunohistochemical markers only, the central body can be distinguished in the protocerebrum. **G.** In histological CellMask stainings additional anterior and lateral neuropils are visible next to the central body. *ac-tub* acetylated α-tubulin, *CB* central body, *CB-D* CB-like domain, *HLI* histamine-like immunoreactivity, *l-Np* lateral neuropil, *pl-Np* posterolateral neuropil, *PB* protocerebral bridge, *PB-D* PB-like domain, *PC* protocerebrum, *SLI* serotonin-like immunoreactivity, *Syt* Sytox nuclear counterstain. Originals by T. Frase (A), M. Fritsch (B–E), M. Becker (F,G).

kindtii (Kirsch and Richter, 2007), but while lateral neuropils are detectable anteriorly and posteriorly on each side of the central body in *Daphnia pulex* (Anomopoda) (Strauß et al., 2011; Weiss et al., 2012), and *P. avirostris* (Ctenopoda) (Fig. 14.3G), they do not necessarily correspond to the lateral accessory lobes. In fact we think that the posterolateral neuropils actually constitute the deutocerebral lobes in which the nerve of the antennula ends (see Figure 14.3G). Using neuro-active substance markers (monoaminergic amines: Aramont and Elofsson, 1976; pigment disperse hormone: Strauß et al., 2011), a CB-like domain is detectable in the protocerebrum of *D. pulex*.

14.3.4.5 Branchiopod central complex

A well-formed central body is present in all recent branchiopod taxa, as is, anteriorly, a slender protocerebral bridge. These neuropils are assumed to be part of the branchiopod ground pattern. Lateral accessory lobes are only described in the anostracan *Artemia salina* and the haplopod *Leptodora kindtii*, with none identifiable in the anostracan *Branchinella* sp. or the Notostraca, Laevicaudata, Spinicaudata and Cyclestherida. In Anomopoda, Ctenopoda and Onychopoda, paired lateral neuropils have been identified, but their homology to lateral accessory lobes is questionable. Thus, whether lateral accessory lobes are part of the branchiopod ground pattern remains unclear. We suggest that a single potential PB-CB tract in Anostraca and Notostraca represents the branchiopod ground pattern. The pattern of two tracts per side as found in "conchostracans" would have evolved within Branchiopoda.

14.3.5 Copepoda

The literature on the copepod brain does not paint a consistent picture of the central complex. As reviewed by Andrew, Brown, and Strausfeld (2012), a central body had been described in some calanoid (Lowe, 1935) and cyclopoid Copepoda (Aramant and Elofsson, 1976), while the absence of a central body and protocerebral bridge had been reported for many harpactoid Copepoda (Lang, 1948; Fahrenbach, 1962). Most of the studies concerned were based on traditional histological methods. In a detailed study combining immunolabeling with histological methods, Andrew, Brown, and Strausfeld (2012) described a central body in the protocerebrum of the harpacticoid species *Tigriopus californicus*, and the decussation of a pair of associated tracts which the authors interpreted as PB-CB tracts indicative of a protocerebral bridge. They did, however, mention the difficulties involved in identifying this neuropil on the basis of semithin sections. A lobed lateral protrusion on each side of the protocerebrum was suggested to correspond to the lateral accessory lobes (Andrew, Brown, and Strausfeld, 2012). Immunolabeling and methylene-staining of the harpactoid species *Halectinosoma curticorne* (Wittfoth and Richter unpublished, see Figure 14.2E) did not reveal distinct neuropils in the protocerebrum. Two SL-ir domains were found in both the protocerebrum of *T. californicus* (Andrew, Brown, and Strausfeld, 2012) and that of *H. curticorne* (Wittfoth and Richter unpublished), but were not related to neuropils of the central complex. Although domains of varying immunoreactivity have been found in calanoid Copepoda, none of them correlated with the neuropils of a central complex (Benzid, Morris, and Barthéléma, 2005; Barthélémy et al., 2006; Hartline and Christie, 2010: serotonin; Hartline and Christie, 2010: histamine, dopamine; Wilson and Christie, 2010: allatostatin; Sousa et al., 2008: tachykinin; but see Aramant and Elofsson, 1976 who noted a central body on the basis of an unspecified "monoamine").

14.3.6 Ostracoda

In Myodocopida the protocerebrum innervates a frontal filament organ, a nauplius eye and the compound eyes, while in Podocopida it only innervates a nauplius eye (see Hartmann, 1967). Within the protocerebrum, histological sectioning has revealed a well-defined central body and, anteriorly, a protocerebral bridge in the myodocopids and podocopids (Hanström, 1924; 1928; Cannon, 1931; see Hartmann, 1967). None of the studies mentioned report PB–CB tracts in Ostracoda. Aramant and Elofsson (1976: their Fig. 25a) suggested that a spheroidal domain of unspecified "monoaminergic" immunoreactivity may constitute the central body in a podocopid species. In a preliminary immunohistochemical study in podocopids, neither an unpaired immunoreactive domain nor an unpaired midline neuropil was found.

14.3.7 Branchiura

All the studies into the branchiuran brain which exist to date have been based on traditional histological methods (Leydig, 1889; Martin, 1932; Zaćwilichowski, 1935; Zaćwilichowska, 1948; Overstreet et al., 1992). Putative neuropils of a central complex have not been either described or depicted. Nothing is known about the immunoreactive pattern in the branchiuran brain.

14.3.8 Cirripedia

The reduced brain of adult Thoracica consists of two supraesophageal, transversely connected ganglia (Bullock and Horridge, 1965; Gwilliam and Cole, 1979; Walker, 1992; Webster, 1998; Callaway and Stuart, 1999; Harrison and Sandeman, 1999). Division into a proto-, deuto- and tritocerebrum is vague to non-existent in the adult brain but clearly distinguishable in thoracican cypris larvae (Semmler et al., 2008). In the protocerebrum of larval *Balanus improvisus* a CB-like SL-ir domain (labeled pcn) exists which corresponds to the central body of other arthropods in (1) its central position within the protocerebral architecture and (2) its reception of SL-ir neurites from lateral and posterolateral somata (Semmler et al., 2008). Neuro-active substances fail to reveal any other reactive domains associated with the central complex in the larval protocerebrum.

The nervous system of the Acrothoracica and the Rhizocephala is even more concentrated and reduced than that of the Thoracica (Bullock and Horridge, 1965). No detailed information is available on the protocerebral architecture of either the adult or the larval brain. In his general morphological description of the rhizocephalan cypris larva, Glenner (2001) described the nervous system but made no mention of an unpaired midline neuropil.

14.3.9 Remipedia

Remipedia are equipped with a central complex made up of a small central body, a large protocerebral bridge and lateral accessory lobes (Fanenbruck, Harzsch, and Wägele, 2004; Fanenbruck and Harzsch, 2005). In fact, Remipedia exhibit four main lateral protrusions from each side of the protocerebral neuropil which Fanenbruck and Harzsch (2005) termed 'lateral protocerebral sublobes a–d'. The same authors demonstrated that only 'sublobes d' (which are further divided into the four subunits d1-d4) are associated with the central body, suggesting that only they correspond to the lateral accessory lobes in other arthropods.

In a detailed investigation of the remipede brain based on immunolabeling methods, Stemme et al. (2012) recently showed an unpaired SL-ir domain (labeled sd2) overlapping with the central body and several paired SL-ir domains overlapping with the protocerebral bridge (sd4–6) and lateral accessory lobes (sd9). The SL-ir domains of the central body and protocerebral bridge were found to be connected by four well-separated SL-ir neurites on each side of the body, providing the first indication of W, X, Y and Z tracts in Remipedia. It remains unclear from which somata these neurites originate. Neither the SL-ir domains (sd4–6) of the protocerebral bridge nor that of the central body (sd2) are innervated directly by an "anterior" assemblage of protocerebral somata (Stemme et al., 2012, "anterior" here refers to the neuraxis, because the remipede brain bends posteriorly). However, a group of SL-ir "anterior" somata (sA1) projects SL-ir neurites passing the protocerebral bridge to a convergence point with other SL-ir neurites which extend medially and cross the midline to innervate the SL-ir domain (sd2) of the central body (Stemme et al., 2012, see especially their Figure 14.7). Another "anterior" group of SL-ir somata (sA2) projects SL-ir neurites across the midline of the protocerebral neuropil to supply the SL-ir domain (sd9) of the lateral accessory lobes (Stemme et al., 2012).

14.4 Discussion

The results of our investigation and short literature review are summarized in Table 14.1, based on the definitions given above. Neuropils and specific immunoreactive domains of the central complex are treated separately. It has long been known that the neuropils of the central complex are innervated by neurons containing stainable neuro-active substances (e.g., Aramant and Elofsson, 1976). In other words, the neuropils overlap, to some extent, with immunoreactive domains. Interestingly, our investigations into Mystacocarida (Brenneis and Richter, 2010) and Cephalocarida (Stegner and Richter, 2011) revealed immunoreactive domains in the protocerebrum even where complementary staining methods failed to detect correlated neuropils (i.e. condensed and distinct compartments of protocerebral neuropil). Vice versa, in taxa whose central body occurs as a distinct neuropil, those domains immunoreactive to a certain neuro-active substance were found to overlap with part but not all of the neuropil in question (e.g., Homberg and Hildebrand, 1991; Homberg, 2002; Loesel, 2004; Strausfeld, 2012).

14.4.1 Structural comparison

Given the correspondences in the overall shape and position of central complex neuropils in different crustaceans (Table 14.1), it would be interesting to know whether these correspondences extend to internal neuropil anatomy too. Moreover, how are immunoreactive domains in taxa without distinctive neuropils – e.g. Mystacocarida and Cephalocarida – related to immunoreactive domains and neuropils in other taxa?

The internal anatomy of central complex neuropils, especially in Hexapoda and Malacostraca, has been investigated in impressive detail, as recent reviews make clear (Homberg, 2008; Strausfeld, 2012). We concentrate here on five features of particular relevance for the interpretation of our new data.

14.4.1.1 Columnar neurons

The central body is generally supplied by neurites of 'columnar neurons' whose somata lie anteriorly in the brain. In taxa with a protocerebral bridge, columnar neurites may connect the bridge with the central body ('PB-CB-tracts'). In both Hexapoda (e.g. Kollmann, Huetteroth, and Schachtner, 2011; Boyan, Williams, and Herbert, 2008) and Malacostraca (Utting et al., 2000), the columnar neurites have been known to be arranged into four distinct bundles on each side, respectively termed the 'W, X, Y and Z tracts'. Serotonin-labeling recently revealed corresponding tracts in Remipedia (Stemme et al., 2012). The W, X, Y and Z tracts in decapod (Utting et al., 2000), various peracarid (this study) and leptostracan Malacostraca imply that the tracts are part of the ground pattern of Malacostraca (see Kenning et al., 2013). The central body of "conchostracan" Branchiopoda receives two PB-CB-tracts of columnar neurites per body side, and that of Copepoda (Andrew, Brown, and Strausfeld, 2012) and anostracan and notostracan Branchiopoda only receives one tract (Fritsch and Richter, 2010; 2012; this publication). In a conspicuous correspondence to those taxa, the SL-ir CB-like domain of Mystacocarida and Cephalocarida is also supplied by columnar neurites which project from anterior somata. In contrast, columnar neurites have never been found in Ostracoda, despite this taxon reportedly being equipped with a protocerebral bridge and central body (Hanström, 1924; 1928; Cannon, 1931; Hartmann, 1967).

14.4.1.2 Decussation of columnar neurites

After originating from anterior somata, columnar neurites may either remain ipsilateral or eventually traverse the midline in one of two ways. *Outside the midline neuropil:* Neurites from both sides of the body form a visible decussation before reaching the central body, as shown in Pterygota (Boayn, Williams, and Herbert, 2008; Boyan and Williams, 2011) and Copepoda (Andrew, Brown, and Strausfeld, 2012). The columnar neurites of Cephalocarida decussate in this way before reaching the CB-like domain. *Inside the midline neuropil:* Columnar neurites initially remain ipsilateral and only cross the midline within the central body itself. This pattern is clearly followed by the W, X, Y and Z tracts in Malacostraca (Utting et al., 2000; Kenning et al., 2013; this study) and Remipedia (Stemme et al., 2012), and by the columnar neurites in Branchiopoda (this study). Columnar neurites visibly decussate within the arcuate body of Onychophora and Chelicerata (e.g. Lösel, Nässel, and Strausfeld, 2002; Strausfeld et al., 2006a; Loesel et al., 2011), and within the central body of Chilopoda (Loesel, Nässel, and Strausfeld, 2002). Unlike in Cephalocarida, the CB-like SL-ir domains of

Table 14.1: Central complex in Mandibulata. The distinctive neuropils of the central complex (protocerebral bridge, central body, lateral accessory lobes) are here set against the co-localized immunoreactive domains with which they overlap, and against the arrangements of columnar and tangential neurites in the different taxa. As seen in Mystacocarida, Cephalocarida and Cirripedia, even taxa that lack the distinctive neuropils of a central complex may exhibit CB-like immunoreactive domains or columnar and tangential neurites. Abbreviations: absent – ; present +. CB, central body; ...-D, immunoreactive domain; LAL, lateral accessory lobes; PB, protocerebral bridge; X, decussation of columnar neurites outside midline neuropil.

	Structures and patterns	distinctive neuropil			serotonin-like immunoreactive domains			histamine-like immunoreactive domains			columnar neurites	tangential neurites	Selected literature
Taxa		PB	CB	LAL	PB-D	CB-D	LAL-D	PB-D	CB-D	LAL-D	decussation (X)		
Cephalocarida		–	–	–	–	+	–	–	–	–	1 bundle (X)	–	Elofsson and Hessler, 1990; Stegner and Richter, 2011
Maxillopoda	Mystacocarida	–	–	–	–	++	–	?	?	?	1 bundle	–	Baccari and Renaud-Mornant, 1974; Elofsson and Hessler, 2005; Brenneis and Richter, 2010
	Ostracoda	(+)	(+)	–	?	?	?	?	?	?	–	–	Hanström, 1924; 1928; Cannon, 1931; Hartmann, 1967; Aramant and Elofsson, 1976
	Copepoda	+	+	(+)	–	–	–	–	–	–	1 PB-CB tract (X)	+	Lowe, 1935; Lang, 1948; Fahrenbach, 1962; Aramant and Elofsson, 1976; Hartline and Christie, 2010; Andrew, Brown, and Strausfeld, 2012
	Cirripedia	–	–	–	–	+	–	?	?	?	–	+	Bullock and Horridge, 1965; Callaway and Stuart, 1999; Harrison and Sandeman, 1999; Semmler et al., 2008
	Branchiura	–	–	–	?	?	?	?	?	?	?	?	Leydig, 1889; Martin, 1932; Zaćwilichowski, 1935; Zaćwilichowska, 1948; Overstreet et al., 1992

Branchiopoda	Anostraca	+	+	+	+	+	+	+	+	+	1 PB-CB tract	+	Hanström, 1924; Benesch, 1969; Harzsch and Glötzner, 2002; this study
	Notostraca	+	+	–	–	+	–	+	+		1 PB-CB tract	–	Hanström, 1924; Loesel, 2004; Fritsch and Richter, 2010; Strausfeld, 2012
	"Conchostraca"	+	+	–	–	+	–	+	+	–	2 PB-CB tracts	–	Hanström, 1934; this study
	Cladocera	+	+	(+)	?	?	?	?	?	?	?	?	Hanström, 1931; 1934; Bullock and Horridge, 1965; Aramont and Elofsson, 1976; Kirsch and Richter, 2007; Strauß et al. 2011; Weiss et al., 2012; this study
Malacostraca	Leptostraca	+	+	+	+	+	+	?	?	?	W, X, Y, Z tracts	+	Hanström 1928; Kenning et al. 2013
	Stomatopoda	+	+	+	?	?	?	?	?	?	?	?	Hanström, 1928; 1947; Sandeman and Scholtz, 1995; Strausfeld, 2012
	Caridoidea	+	+	+	+	+	+	+	+	+	W, X, Y, Z tracts	+	Hanström, 1928; 1934; Utting et al., 2000; Krieger et al., 2010; 2012; this study
Remipedia		+	+	+	+	+	+	?	?	?	W, X, Y, Z tracts	+	Fanenbruck, Harzsch, and Wägele, 2004; Fanenbruck and Harzsch, 2005; Stemme et al., 2012
Hexapoda		+	+	+	+	+	+	–	+	?	W, X, Y, Z tracts (X)	+	Williams, 1975; Strausfeld, 1976; Homberg, 2002; Boyan, Williams, and Herbert, 2008; Kollmann et al., 2011; Strausfeld, 2012
Chilopoda		?	+	?	?	?	?	?	?	?	several neurites	+	Loesel, Nässel, and Strausfeld, 2002; Loesel, 2004; Sombke, Harzsch, and Hansson, 2011

Mystacocarida receive non-decussating SL-ir 'columnar neurites' which may cross the midline within the domain (Brenneis and Richter, 2010).

14.4.1.3 Tangential neurons

Additionally to the columnar neurites, lateral somata may also send out 'tangential neurites' (Strausfeld et al., 2006a). Columnar and tangential neurites have been shown to form a more or less rectangular network in the arcuate body of onychophorans (e.g. Strausfeld et al., 2006) and chelicerates (Loesel et al., 2011). Although the neurite arrangement in the tetraconate central body was initially suggested to be fundamentally similar between all subtaxa (Strausfeld et al., 2006a), it appears to be less geometrical than in other arthropods (see, e.g., Harzsch and Hanson, 2008; Krieger et al., 2010; Strausfeld, 2012; and this study). Our study revealed that tangential neurites supply the central body in various peracarid Malacostraca but only some Branchiopoda (Harzsch and Glötzner, 2002; this study). The SL-ir domains in Cephalocarida and Mystacocarida do not receive SL-ir tangential neurites laterally, and in fact it is questionable whether tangential neurites exist at all in these two taxa since the lateral side of the protocerebrum is soma-free in both cases.

14.4.1.4 Innervation of SL-ir domains

In what is a conspicuous correspondence, unpaired SL-ir immunoreactive domains are found in Mystacocarida, Cephalocarida (e.g., this study) and Cirripedia (Semmler et al., 2008), but also in association with the unpaired neuropils in Hexapoda (Homberg, 2002), Malacostraca (Utting et al., 2000; Harzsch and Hansson, 2008), Branchiopoda (Harzsch and Glötzner, 2002; Fritsch and Richter, 2010; this study) and Remipedia (Stemme et al., 2012). Are there further correspondences in these domains with regard to their innervation by SL-ir neurons? In the three taxa mentioned which do not have a central complex, SL-ir neurites originate from different assemblages of somata, anteriorly in the protocerebrum of Mystacocarida and Cephalocarida and laterally in the protocerebrum of Cirripedia (Semmler et al., 2008). Both patterns are found in other Tetraconata (in taxa where the brain bends upwards, we translate the position of somata to an imagined "straight neuraxis"). *Innervation by anterior somata (columnar neurons):* The anterior SL-ir somata of Anostraca (Harzsch and Glötzner, 2002) and Malacostraca (Utting et al., 2000: their Fig. 9A; Harzsch and Hansson, 2008) only innervate the protocerebral bridge, with the result that the PB-CB tracts in these taxa are not SL-ir. The anterior SL-ir somata (labeled sA1) in Remipedia only connect indirectly to the central body (Stemme et al., 2012: sd2). (Remember that the W, X, Y and Z tracts in Remipedia are SL-ir but do not originate from anterior somata.) Only in Hexapoda do anterior SL-ir somata innervate not only the protocerebral bridge but also the fan-shaped body (Homberg and Hildebrand, 1991: somata S1, S2, S4, S5). This pattern is reminiscent of that of the CB-like domain found in Mystacocarida and Cephalocarida. *Innervation by lateral somata (tangential neurons):* Lateral SL-ir somata innervate the central body and lateral accessory lobes in Malacostraca (Utting et al., 2000) and Anostraca (Harzsch and Glötzner, 2002). In Hexapoda, lateral SL-ir somata (Homberg and Hildebrand, 1991: S3) innervate the fan-shaped body of the central body only. This lateral innervation pattern is reminiscent of that of the SL-ir domain in Cirripedia (Semmler et al., 2008). In contrast, lateral SL-ir somata in Remipedia only innervate the lateral accessory lobes (Stemme et al., 2012).

14.4.1.5 Horizontal layers of the central body

The clear differentiation of the central body into an 'ellipsoid body' and a 'fan-shaped body' is unique to Hexapoda (Williams, 1975; Strausfeld, 1976). Interestingly, however, both the hexapod ellipsoid body (Müller, Homberg, and Kühn, 1997; Loesel and Homberg, 1999; Homberg, 2002) and the malacostracan central body (Strausfeld, 2009; Krieger et al., 2012) display "horizontal layers" which are innervated by neurites from well-separated groups of somata that may differ in their immunoreactivity. Layers of varying immunoreactivity were not found in the central body of notostracan Branchiopoda (Fritsch and Richter, 2010). Nothing indicates that the CB-like SL-ir domains in Mystacocarida and Cephalocarida constitute layers of a larger neuropil. Domains of other immunoreactivity (RFLI, HLI) were not found to correlate to any putative central complex neuropil (Brenneis and Richter, 2010; Stegner and Richter, 2011).

14.4.1.6 Conclusion of structural comparison

Since we have seen that both the protocerebral bridge and the central body may be innervated by columnar neurons (anterior SL-ir somata) in other Tetraconata, a comparison of innervation patterns alone delivers no clear indication of how to classify the SL-ir domains in Mystacocarida and Cephalocarida. However, given their elongated overall appearance, their central position and their innervation by columnar neurites, we suggest that the SL-ir domains in Mystacocarida and Cephalocarida correspond to the SL-ir domain of the central body in other tetraconates, rather than to that of the protocerebral bridge.

14.4.2 'Neurophylogeny' of Tetraconata

What do the distribution of central complex neuropils in general and their internal anatomy in detail contribute to the phylogenetic debate? The 'Invertebrate Neurophylogeny' section of the Deep Metazoan Phylogeny program is currently developing a comprehensive morphology-based phylogenetic matrix into which neuroanatomical data including our own on the central complex and many other structures and patterns will be integrated. Since this work is still in progress, our data on the central complex of Crustacea is discussed here on the basis of some of the phylogenetic hypotheses currently suggested in the literature. Our aim is to determine which of the evolutionary scenarios proposed is the most plausible (requiring the lowest number of transformational steps from one morphological pattern to the other).

Within Mandibulata, an overwhelming number of molecular studies and a growing number of morphological studies point towards a sister-group relationship between Myriapoda and Tetraconata, the latter of which comprise by definition all hexapods and crustaceans. The relationships between crustaceans have been the subject of numerous morphology- and molecular-based studies and under controversial phylogenetic debate for decades. This debate continues for Tetraconata (reviewed by Richter, Møller, and Wirkner, 2009; Jenner, 2010). While the monophyly of Hexapoda, Branchiopoda, Malacostraca, Cephalocarida, Remipedia and the maxillopod subtaxa is widely accepted, their interrelationships remain unclear.

While earlier hypotheses based on external morphology assigned Cephalocarida (Sanders, 1957), Remipedia (Schram et al., 1986) and Malacostraca (Walossek, 1993, without Remipedia) to the base of Crustacea, neuroanatomical data suggest alternative hypotheses. Correspondences in the optic neuropils (Strausfeld, 2005) support a close relationship between Hexapoda and Malacostraca, while some correspondences in the olfactory system (hemiellipsoid bodies, chiasma of olfactory globular tracts) and larval features (Koenemann et al., 2007; 2009) support a sister-group relationship between Remipedia and Malacostraca (Fanenbruck and Harzsch, 2005). Higher-level correspondences in the olfactory systems (olfactory lobes, olfactory globular tracts, second-order olfactory centers) of Cephalocarida, Remipedia, Hexapoda and Malacostraca were initially interpreted as synapomorphies (Fanenbruck, Harzsch, and Wägele, 2004; Fanenbruck and Harzsch, 2005; Harzsch, 2006). Today, however, in the light of a number of recent investigations into Cephalocarida (Stegner and Richter, 2011), Myriapoda (Sombke, Harzsch, and Hansson, 2011; Sombke, Rosenberg, and Hilken, 2011; Sombke et al., 2012), Hexapoda (Strausfeld, 2009; Loesel and Heuer, 2010) and Malacostraca (Wolff et al., 2012), they must be considered symplesiomorphies dating back to the mandibulate ground pattern. Our reinvestigation of the cephalocarid nervous system found neither brain (Stegner and Richter, 2011) nor ventral nerve cord features (Stegner, Brenneis, and Richter, in press) supportive of a close relationship, i.e. synapomorphies, between Cephalocarida and Hexapoda, Malacostraca or Remipedia (but see Stemme et al., 2013). Unlike the cephalocarid nervous system with its plesiomorphic complexity, the mystacocarid nervous system is considerably reduced (Brenneis and Richter, 2010), rendering potential synapomorphies with other tetraconate taxa elusive. The hypothesis that Mystacocarida and Cephalocarida occupy a "basal" position within Tetraconata is not contradicted by neuroanatomical data. A recent phylogenetic analysis considering neuroanatomical (Strausfeld and Andrew, 2011) and EST data (Andrew, 2011) did not include Mystacocarida and Cephalocarida.

Where do molecular hypotheses place the tetraconates lacking a central complex? Giribet et al. (2005: their Fig. 5), combining molecular with morphological data, suggested that Branchiura form the sister-group to all other tetraconates. Within the latter, taxa with a reportedly well-developed central complex (Malacostraca, Hexapoda, Remipedia, Branchiopoda, Copepoda) are not closely related. Regier et al. (2010: their Fig. 1) suggested that Tetraconata consist of two sister-groups, one termed 'Oligostraca', the other 'Altocrustacea'. Conspicuously, Oligostraca are mainly comprised of crustacean taxa which have been found to lack a distinct central complex – Mystacocarida, Branchiura and Pentastomida, which together form the monophyletic unit 'Ichthyostraca' – and Ostracoda, which do exhibit a protocerebral bridge and central body. The Ichthyostraca hypothesis was also supported by Koenemann et al. (2010). Altocrustacea (Regier et al., 2010) comprise all other tetraconates, including the groups mentioned which have a well-developed central complex and the remaining groups which do not possess distinctive neuropils, namely Cephalocarida and Cirripedia. Although the monophylum 'Xenocarida' composed of Cephalocarida and Remipedia found support in various studies (Regier et al., 2010; Spears and Abele, 1998; Giribet, Edgecombe, and Wheeler, 2001; Koenemann et al., 2010), it may be an artifact. Koenemann et al. (2010), for example, mentioned possible long-branch attraction between Remipedia and Cephalocarida, and von Reumont et al. (2012) demonstrated that the support for Xenocarida has, in fact, often been weak. According to a recent phylogenomic analysis (von Reumont et al., 2012), Ostracoda occupy a basal position within Tetraconata. Within the remaining tetraconates, Remipedia, Hexapoda and Branchiopoda together form one monophylum, and the remaining crustaceans form another (von Reumont et al., 2012). These authors did not include Cephalocarida.

On the basis of all the phylogenetic analyses mentioned, at least the protocerebral bridge and central body, potentially also the lateral accessory lobes, can be traced back to the tetraconate ground pattern, thus in line with Harzsch's (2006) hypothesis that the central complex is an autapomorphy of Tetraconata. As a consequence, in every evolutionary scenario that is strictly based on the mentioned current analyses (Giribet et al., 2005; Regier et al., 2010; Koenemann et al., 2010; von Reumont et al., 2012), the lack of all central complex neuropils in Cephalocarida, Mystacocarida, Cirripedia and Branchiura, and the lack of lateral accessory lobes in some phyllopod Branchiopoda, can only be explained if we accept that the respective neuropils underwent an independent reduction. The columnar neurons and CB-like SL-ir domains in the mystacocarid and cephalocarid brain may be rudiments of the columnar neurons and central body found in other tetraconates.

While our knowledge on the different tetraconate taxa is successively improving, data on their outgroup Myriapoda has remained sparse. The contemporary assumption of a single central neuropil in the myriapod ground pattern (Loesel, 2004; Harzsch, 2006; Homberg, 2008) is in fact based on representatives of only one genus (Loesel, Nässel, and Strausfeld, 2002: *Scolopendra*, but see Holmgren, 1916; Hanström, 1928;

Joly and Descamps, 1987). Sombke, Harzsch, and Hansson (2011: Fig. 2C) labeled a protocerebral bridge in the myriapod *Scutigera coleoptrata*, but did not comment on this interesting finding, because they focused on the deutocerebrum. A thorough re-investigation of the protocerebrum in different myriapod taxa is of great interest to reconstruct the ground pattern of Myriapoda and finally Mandibulata. The debate on the origin of the central complex will go on.

Acknowledgments

We thank Marlen Becker, Georg Brenneis, Clemens Döring, Maria Dreiling, Thomas Frase, Christian Wirkner and Christin Wittfoth for original images and valuable information on the brain in different crustaceans. We are grateful to Matthes Kenning for his expertise in leptostracan brain anatomy. We thank Steffen Harzsch, Rudi Loesel, Andy Sombke, and two anonymous reviewers for critical reading of our manuscript and constructive comments, and Lucy Cathrow for improving the English. Finally, we thank the Deutsche Forschungsgemeinschaft (DFG) for funding (RI 837/9-1,2 and RI 837/10-1,2). This project is part of the DFG priority program "Deep Metazoan Phylogeny" (SPP 1174).

Björn Marcus von Reumont and J. Wolfgang Wägele

15 Advances in molecular phylogeny of crustaceans in the light of phylogenomic data

Abstract: Phylogenetic analyses based on large molecular data sets are reviewed, including results of the "Deep Metazoan Phylogeny" project. The phylogeny of Crustacea is still little understood and tree topologies based on molecular data are not congruent with many other morphological data especially for deeper nodes. Earlier analyses using DNA data were not mutually consistent, but more recent analyses have achieved some stability due to better taxon sampling, longer alignments and more sophisticated data selection and modeling. Orthology prediction and gene selection strongly influence results. The current phylogenies place Remipedia and not Branchiopoda close to insects, usually Malacostraca appear closer to Cirripedia and Copepoda, monophyletic Maxillopoda were never recovered.

15.1 The diverse and difficult crustaceans

The genealogical relationships within arthropods are still partly unresolved, and currently hypotheses explaining the phylogeny within major arthropod groups are still contradictory. Most scientists today favor the hypothesis of a **monophyletic Euarthropoda**, based on morphological and molecular data (Wägele, 1993; Rota-Stabelli et al., 2011), while the precise evolutionary history and phylogeny of the four major euarthropod clades partly remains to be solved (Budd and Telford, 2009; Edgecombe, 2010; Giribet and Edgecombe, 2012). For further details of arthropod evolution see the chapter by Wägele and Kück in this book (Chapter 12). In this chapter we focus on the phylogeny of crustaceans.

Crustaceans represent probably the euarthropod group with the largest diversity in body shapes and evolutionary adaptions that evolved in more than 500 Mio. years, starting in marine habitats of Cambrian oceans (Waloszek and Müller, 1998a; Waloszek, 2003a). Today, probably only a fraction of this diversity has survived, which is still astonishing. Extant crustaceans can be as small as the Mystacocarida (*Derocheilocaris typicus*) with only 0.3 mm average body size, animals with the syndromes of dwarfish arthropods and whose mouthparts are little differentiated, others are highly evolved and as large as the Japanese giant crab (*Macrocheira kaempferi)* with a span width of almost 4 m. Several highly transformed parasitic forms evolved in different crustacean lineages and coevolved with their hosts. The pentastomids or tongue worms that parasitize in the respiratory tract of terrestrial vertebrates are so aberrant that it is still discussed if they represent either crustaceans (Møller et al., 2008) or, based on the interpretation of fossils, an even more ancient lineage derived from lobopod-like arthropods (Waloszek, Repetski and Maas, 2006).

It is challenging that, after the first phylogenetic tree of crustaceans published by Haeckel in the year 1866 nearly 150 years ago, the internal evolution of crustaceans is still unresolved in many aspects, which is probably a result of the enormous diversity of this group and the scarcity of conserved characters (see Figure 15.1).

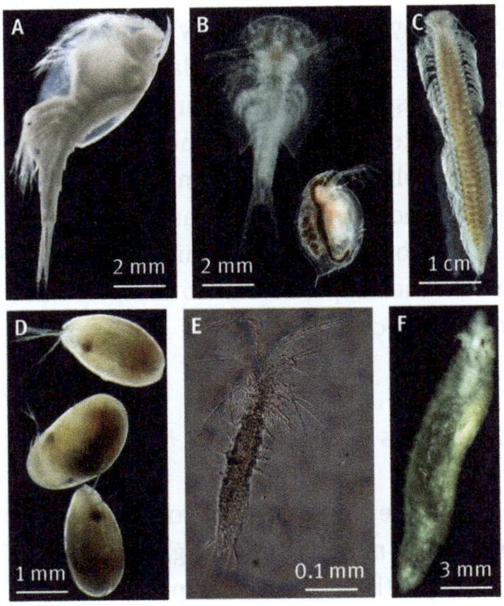

Figure 15.1: Representatives of major crustacean groups: **A.** Malacostraca (*Nebalia antarctica*, Leptostraca), **B.** Branchiopoda (*Triops cancriformis*, left and *Daphnia magna*, right), **C.** Remipedia (*Speleonectes tulumensis*), **D.** Ostracoda (*Myodocopinae* sp.), **E.** Mystacocarida (*Derocheilocaris typicus*), **F.** Pentastomida (*Raillietiella* sp.).

Generally, five major crustacean taxa are widely accepted as being monophyletic (Martin and Davis, 2001): the Malacostraca (Latreille, 1802), Branchiopoda (Latreille, 1817), Remipedia (Yager, 1981), Cephalocarida (Sanders, 1955) and Ostracoda (Latreille 1802). Maxillopoda (Dahl, 1956) are to date still disputed and mounting evidence, from different data sources, indicates that this clade probably does not represent a monophyletic taxon (Martin and Davis, 2001). Smaller taxa not included in this list are Mystacocarida, Branchiura, Pentastomida. Monophyly of some crustacean groups like the Malacostraca, Branchiopoda or Thecostraca (the latter within “Maxillopoda”) is now generally accepted and highly supported, but for many crustacean clades their internal phylogenetic relationships are still unclear, see reviews Martin and Davis (Martin and Davis, 2001) and more recently Richter et al. (2009). This chapter aims to present the status quo and advance of crustacean phylogeny based on a molecular perspective.

15.2 Are crustaceans monophyletic?

Addressing crustacean phylogeny one central question concerns the possible paraphyly of this group. Until the end of the 20th century the monophyly of crustaceans was generally accepted in many textbooks. However, typical diagnostic characters that distinguish crustaceans from other euarthropod groups, like the presence of two pairs of antennae (Schram, 1986) are just diagnostic, but not synapomorphic, as critically discussed by Lauterbach (1983). Most of the diagnostic characters found in crustaceans are just plesiomorphic character states within Mandibulata (Lauterbach, 1983), especially when crustaceans are being compared with hexapods. In a critical assessment only two characters are thought to remain as possible autapomorphies of crustaceans: (1) the nauplius eye, and (2) the reduction of the segmental nephridial organs to two pairs in the segments of the second antennae and maxilla. However, Lauterbach concluded a monophyly of crustaceans based on his studies despite the lack of a complex apomorphic homology that serves as evidence for the monophyly of Crustacea (Lauterbach, 1986). Walossek and Müller (1998) distinguish additional characters that constitute the monophyly of crustaceans and separate stem-lineage and crown group crustaceans. Some of these characters are, however, critically discussed (Schram and Koenemann, 2004). The fact that the nauplius is absent in insects and myriapods, but present in many Crustacea, can be explained by a reduction of this stage and evolution of direct development in terrestrial Mandibulata. Therefore, the nauplius might not serve as a strong apomorphy for a clade Crustacea.

Some cladistic analyses based on morphological characters from the late 1990s to the early 21st century corroborate monophyly of crustaceans (Wheeler, 1998; Giribet et al., 2005). The relevance of such studies can only be understood discussing every character state of the data matrix, which is beyond the scope of our review focusing on molecular data. The majority of currently available data (especially molecular data sets) suggest that "Crustacea" are a collection of primarily marine Mandibulata, with different stages of evolution body complexity. In this scenario they can be regarded as the stem-group of terrestrial Mandibulata, recent molecular and morphological data support a grouping with Hexapoda, see also Figure 15.5. The first morphological study that proposed paraphyletic crustaceans was published by Moura and Christoffersen (1996). These authors suggested also a rather derived position of Remipedia within Mandibulata as sister-group to Tracheata (Moura and Christoffersen, 1996).

Starting in the late 1990s, the first molecular analyses of arthropods and crustaceans were published, which draw a different picture compared to the textbook hypotheses. The traditional **Atelocerata** (= Tracheata) hypothesis, which assumes a sister-group relationship of hexapods and myriapods (Pocock, 1893a; Heymons, 1901), is to date not recovered in any molecular study (see Figure 15.2). Since first analyses of small molecular data sets e.g. Friedrich and Tautz (Friedrich and Tautz,

1995) molecular analyses usually support an origin of hexapods from within "Crustacea". A monophyletic clade composed of crustaceans and hexapods is recovered by *most molecular* studies (Meusemann et al., 2010; Regier et al., 2010; Friedrich and Tautz, 1995; von Reumont et al., 2012). This clade was coined **Pancrustacea** by Zrzavý and Stys (1997) based on molecular, developmental and ultra-structural data.

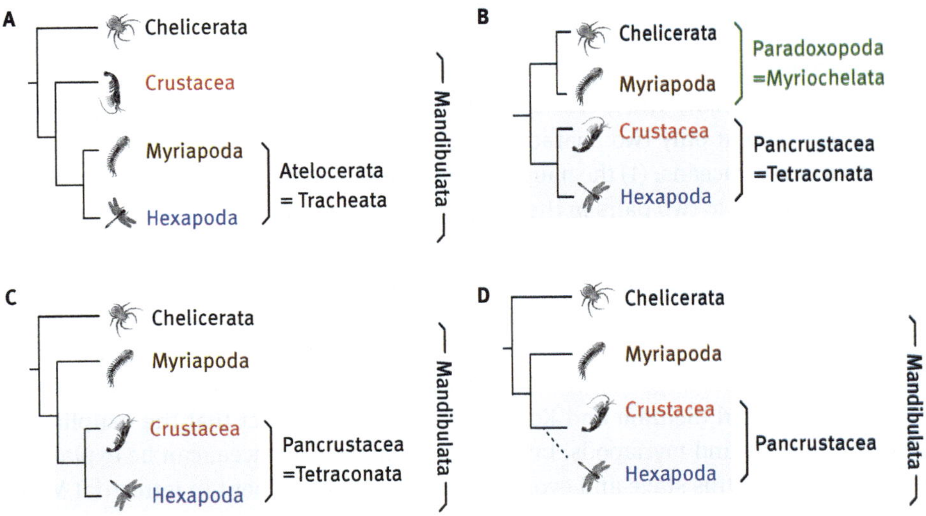

Figure 15.2: Different phylogenetic hypotheses of euarthropod groups. **A.** Traditional Mandibulata sensu Crustacea + Tracheata. **B.** The Myriochelata sensu Pisani, Poling, Lyons-Weiler et al. (2004) (= Parodoxopoda *sensu* Mallat et al. (2004)) that are exclusively reconstructed in molecular trees as sister-group to Tracheata represent an artifact induced by molecular methods as shown in Stabelli et al. (2011). **C.** The Manidbulata sensu Myriapoda + Tetraconata. **D.** Mandibulata sensu Myriapoda + Tetraconata based on paraphyletic crustaceans as to date corroborated by morphological and molecular data.

First morphological observations that were interpreted as evidence for a relationship of crustaceans and hexapods were described early in the 1920s (e.g. Hanström, 1924). An important review of available arguments was published by Paulus (1979). Later in 2001, this taxon was named **Tetraconata** by Dohle (2001) (see Figure 15.2) based on the shared presence of crystalline cones in ommatidia in the compound eyes. Tetraconata and Pancrustacea are synonyms, the older name being Pancrustacea, however, both originally discuss potentially paraphyletic crustaceans. The name Pancrustacea (predominately used in molecular studies) has been critically discussed because its use could induce confusion with the concept "Pan-crustacea" (little known in the English literature), which addresses monophyletic stem-lineage and crown group crustaceans, see Richter (2002). Many authors published after Dohle (2001) further evidence from different sources corroborating the clade Tetraconcata/Pancrustacea in morphological (Richter, 2002; Strausfeld, 2009) and molecular studies including phylogenomic data (Meusemann et al., 2010; Rota-Stabelli et al., 2011; von Reumont et al., 2009; Reumont et al., 2012; Ertas et al., 2009). That clade seems to be widely accepted, see also Richter et al. (2009). But see also a contradicting scenario based on paleontological evidence and a large set of morphological data (Wägele and Kück, this book).

15.3 Which is the crustacean sister-group to Hexapoda?

Since all molecular studies indicate paraphyly of crustaceans with respect to Hexapoda, this raises the question of the hexapod sister-group. Unfortunately, topologies reconstructed from molecular data are ambiguous and do not offer a consistent answer (Jenner, 2010; von Reumont, 2010). There are various effects on the shape of topologies that are caused by the selection of genes, completeness of data matrices, alignment and modeling methods, and taxon sampling. The latter point might sound trivial, but sampling of many non-malacostracan taxa is indeed not that simple and even a challenge if specimens have to be collected and conserved in remote geographic regions, e.g. for phylogenomic studies (von Reumont et al., 2012).

Analyses of mitochondrial data produced the most ambiguous topologies, even suggesting polyphyletic crustaceans and non-monophyletic hexapods (Nardi, 2003; Carapelli et al., 2006; Cook, Yue and Akam, 2005). More recent studies (Hassanin, 2006) conclude that often inappropriate methods have been used and it seems that with mitochondrial data a sufficient resolution to infer pancrustacean phylogeny cannot be obtained.

First molecular studies that used single nuclear genes focused on general arthropod evolution included only few crustacean taxa (Friedrich and Tautz, 1995; Wheeler, Cartwright, and Hayashi, 1993). The first substantial “crustacean” study with larger taxon sampling used 18S rDNA fragments and was published by Spears and Abele (1998). Many of the 18S rDNA sequences from Spears and Abele were later used as a base for further studies. That was problematic, because the first published 18S rDNA of Mystacocarida (*Derocheilocaris typicus*) turned out to be a contamination (von Reumont, 2010; Koenemann et al., 2010).

However, Spears and Abele critically discussed difficulties that still challenge molecular phylogenetics today. Their topologies showed e.g. the grouping of Remipedia + Cephalocarida, a result that Spears and Abele discussed critically as an artifact induced by heterogeneous substitution processes and/or pseudogenes for 18S rDNA in Remipedia. However, 18S and 28S rDNA have since then been one of the standard housekeeping genes to infer crustacean phylogeny. Based on these markers, Copepoda were inferred as crustacean sister-group to Hexapoda in three large studies (Mallatt, Garey and Shultz, 2004; Mallatt and Giribet, 2006; Reumont et al., 2009). Yet, Mallat and Giribet (2006) critically discussed their own results and proposed Branchiopoda as sister-group to Hexapoda, despite the placement of copepods as sister-group to Hexapoda. Later on, the study of von Reumont et al. (2009) was the first rDNA study that covered representatives of all crustacean groups including Remipedia, Pentastomida, Mystacocarida, and Cephalocarida. It was shown that application of a secondary structure-based alignment optimization and the use of substitution models that consider covariation of paired nucleotides and account for inhomogeneous base composition improves tree reconstruction. Some of the crustacean groups are clearly affected by heterogeneous substitution processes, which is the case for the Cephalocarida, as assumed by Spears and Abele over 10 years earlier. A second outcome was that the placement of Copepoda is not well supported in topologies based on the more realistic heterogeneous substitution models, However, despite the complex modeling and more complete data (completely sequenced 18S and 28S rDNA, in total over 8,000 base positions) and with a more complete taxon sampling it could be demonstrated that the resolution of the 18S and 28S genes is obviously not sufficient for many aspects of internal crustacean phylogeny. The study by Koenemann et al. (2010) is important to underpin the limitations of the frequently used "housekeeping genes" 16S rDNA, 18S rDNA and COI by all meanings if used to infer crustacean or even arthropod phylogeny. Respectively the uninformativeness of alignments and resulting ambiguous topologies from sequenced fragments of these genes (respectively if used with standard homogeneous modeling) is critically presented here. Yet, some publications refer, rather unaware of this critical discussion of general shortcomings, to some topologies in Koenemann et al., e.g. generally non-monophyletic hexapods and partly not supported Pancrustacea.

Interestingly, Branchiopoda were found in several studies as sister-group to Hexapoda (Schram and Koenemann, 2004; Shultz and Regier, 2000; Regier and Shultz, 1997; Timmermans et al., 2008) or at least closer to Hexapoda (Mallatt, Garey and Shultz, 2004; Mallatt and Giribet, 2006; von Reumont et al., 2009). Based on this, Glenner et al. (2006) proposed in 2006 an evolutionary scenario according to which Hexapoda evolved from freshwater crustaceans (Figure 15.3C, Figure 15.5). For some authors, the Branchiopoda became top candidates as possible sister-group to Hexapoda. This result was also supported by new large-scale data sets obtained by EST sequencing (Meusemann, 2010; Dunn et al., 2008; Philippe et al., 2008). However, it must be commented that this scenario did not include a discussion of the evolution of body plans. Branchiopods are highly specialized crustaceans with reduced maxillae and specialized, mostly unsegmented legs, and with a special life cycle that includes the formation of highly resistant cysts. This is an adaptation to a planktonic life in ephemeral continental waters, where fish usually cannot survive. Yet most importantly, there is no complex morphological synapomorphy shared by insects and branchiopods. For an alternative scenario see Wägele and Kück (this book).

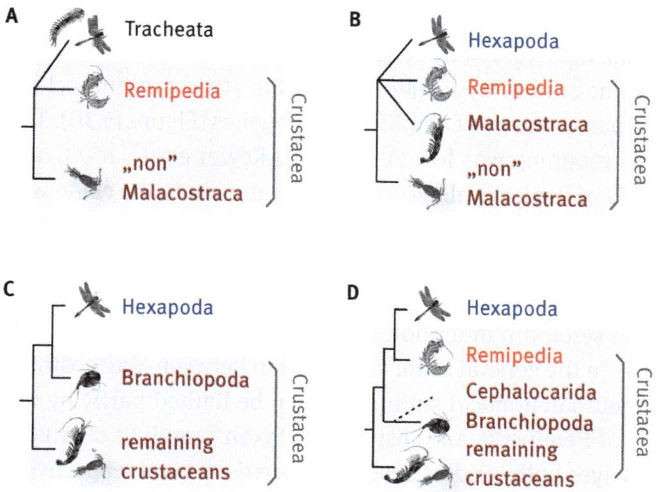

Figure 15.3: Different phylogenetic hypotheses on the possible crustacean sister-group to Hexapoda. **A.** Remipedia as sister-group to Tracheata (see chapter 12). **B.** Polytomic clade of Remipedia + Malacostraca + Hexapoda supported by neuroanatomical and larval development studies (Fanenbruck, Harzsch, and Wägele, 2004; Fanenbruck and Harzsch, 2005; Koenemann et al., 2009). **C.** Branchiopoda as sister-group to Hexapoda as proposed by Glenner et al. (2006) **D.** The present hypothesis corroborated by recent molecular data (von Reumont et al., 2009; 2012; Oakley et al., 2013). Phylogenomic data for Cephalocarida is, however, still missing.

The newest progress is rapid sequence generation using "Next Generation Sequencing" technology. These data sets may suffer from the same difficulties as the early single-gene studies, including especially incomplete taxon sampling, a problem that can be overcome with time. Fresh specimens of species that often are difficult to find have to be collected and preserved appropriately.

Notably one important and difficult to collect taxon, the Remipedia, was missing in most earlier studies, including so far all EST-based analyses. As mentioned before, the Remipedia are especially interesting because of their supposed closer relationship to Tracheata (Moura and Christoffersen, 1996), see Figure 15.3A. Interestingly, this result was corroborated later in a neuroanatomical study by Fanenbruck, Harzsch, and Wägele, 2004. The older hypothesis that Remipedia may represent a rather ancient split within crustaceans (Ax, 1999; Schram, 1986; Emerson and Schram, 1990) was replaced by the assumption that this group could possibly be closely related to hexapods within a monophyletic Pancrustacea. It thus would be a rather derived lineage within crustaceans (Fanenbruck, Harzsch, and Wägele, 2004; Fanenbruck and Harzsch, 2005; Koenemann et al., 2009), but data are not sufficient to resolve the polytomy of Malacostraca + Remipedia + Hexapoda (see Figure 15.3B).

The first molecular study that corroborates a close relation between remipedes and Hexapoda is based on the discovery of three similar hemocyanin subunits (Ertas et al., 2009). Larger multi-gene studies by Regier et al. (2008; 2010) present a clade Remipedia + Cephalocarida based on sets of amplified target genes (Figure 15.3D). The "statistical" support for this grouping was low or absent in Regier et al. (2008) and moderate to high in the study of Regier et al. (2010). In the latter study the clade was named Xenocarida. However, in 2011, Stegner and Richter (2011) present new results of cephalocarid neuroanatomy, which indicate that a close relationship of Remipedia + Cephalocarida and a derived position of Cephalocarida is not supported by neuroanatomical data. Based on serotonin immunoreactive interneurons, Stemme et al. find (2012) shared characters in the general brain organization between Malacostraca and Remipedia. However, neuroanatomical studies seem to be limited partly by the highly derived morphology of Remipedia and insufficient taxon sampling of crustacean and early hexapod lineages both hinders specific comparisons and respectively interpretation of the direction of character evolution. See the critical discussion of difficulties interpreting the evolution of the euarthropod central complex in Stegner and Richter (2011) and in Chapter 14 (Stegner, Fritsch, and Richter, this book).

The most recent study of pancrustacean phylogeny based on phylogenomic data (von Reumont et al., 2012) critically discusses the gene choice of Regier et al.(2010) and eventual difficulties concerning the orthology of some genes. The Remipedia are positioned as sister-group to Hexapoda in the analysis of von Reumont et al. (2012), but, taxon sampling is still not sufficient (e.g. missing Cephalocarida). Anyhow, it was demonstrated that orthology prediction of genes and gene selection strongly influence the reconstructed topologies. Stabelli et al. (2011) re-analyzed the data of Regier et al. (Regier et al., 2010) revealing that tree reconstructions in their study were biased caused by serine codon usage. Interestingly, one major (but often forgotten) conclusion in the first molecular study by Spears and Abele in 1998 was that respectively the clade Cephalocarida + Remipedia (now named "Xenocarida") could likely present an analysis artifact by long-branch effects.

From today's perspective, based on data from different sources the Remipedia are highly supported of being close relatives and possibly the sister-group to Hexapoda (see Figure 15.3D). Yet, some aspects still have to be addressed. (1) Phylogenetic analyses of molecular data strongly contradict combined morphological evidence in respect to the placement of Malacostraca. A vicinity of Malacostraca to Hexapoda and Remipedia was so far never found with molecular data. (2) The grouping of Remipedia with Cephalocarida found in earlier studies requires a revision using phylogenomic data that include cephalocarids and, of course, further missing crustacean taxa. (3) Generally, an improved taxon sampling of early hexapod lineages is mandatory to test which crustacean group resembles the possible closest relatives to Hexapoda. The progress in the 1KITE consortium (http://www.1kite.org) sequencing 1000 hexapod transcriptomes representing all major groups is promising regarding this last point.

15.4 Internal crustacean phylogeny and monophyly of higher crustacean taxa

Crustaceans were and still are enormously controversial concerning their internal phylogenetic relationships (Martin and Davis, 2001; Richter, Møller and Wirkner, 2009) (see Figure 15.4). In their "updated classification of the recent crustaceans", Martin and Davis discuss five major crustacean classes that are generally accepted, the Remipedia, Cephalocarida, Malacostraca, Branchiopoda, and Ostracoda. The Maxillopoda are presented as a sixth class, but described as "continuously terribly controversial assemblage".

For some crustacean clades monophyly is consistently reconstructed and internal relationships are mostly congruent in analyses of morphological and molecular data. The best example is **Branchiopoda**. Morphological (Olesen, 2007) and recent molecular studies including phylogenomic data (Regier et al., 2010; von Reumont et al., 2009; 2012; Stenderup, Olesen, and Glenner, 2006) with representatives of all branchiopod clades support monophyly of this group. Interestingly, the phylogeny obtained from phylogenomic data (von Reumont et al., 2012) is identical with the phylogeny presented in Olesen (Olesen, 2007) and Regier et al. (2010). Phylogenomic data maximally support Notostraca as second lineage within Branchiopoda with paraphy-

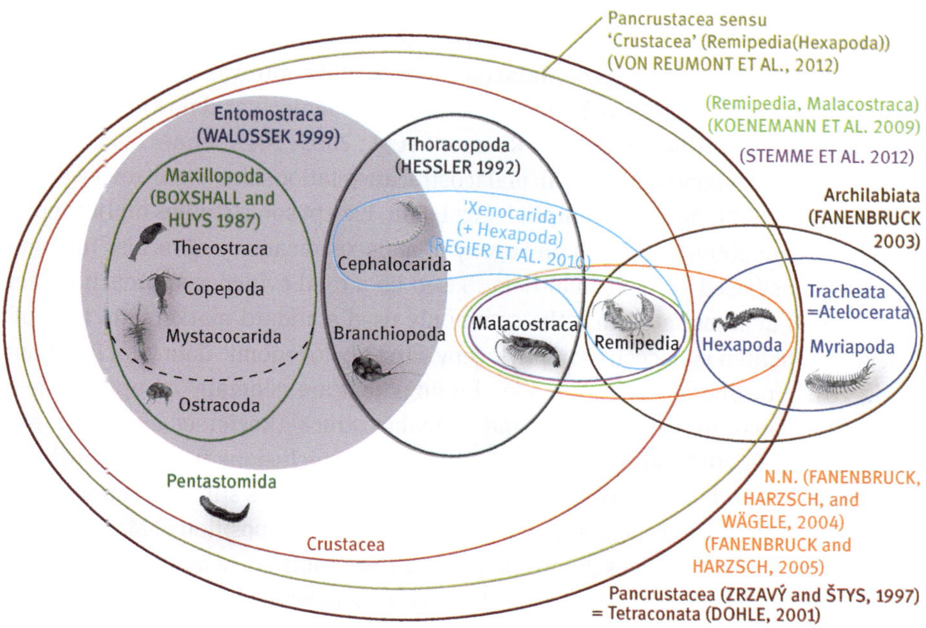

Figure 15.4: Ambiguous phylogeny hypotheses that address crustacean evolution based on morphological data. The most recent molecular results, e.g. (Regier et al., 2010; von Reumont et al., 2012) are integrated.

letic diplostracan taxa (Laevicaudata and Spinicaudata) sensu: Anostraca, (Notostraca, (Laevicautdata, (Spinicaudata, (Cladoceropmorpha)))). This result differs from the analyses in Stenderup et al. (2006), in which each of the different positions of Notostraca were only moderately supported.

Earlier molecular studies, often restricted to rDNA genes, suffered from methodological problems and poor phylogenetic signal. Von Reumont et al. (2009) could demonstrate that combining 18S and 28S rDNA data and including a complex analysis setting results in high support values, but is not sufficient to resolve plausibly the internal phylogeny of Branchiopoda. Cephalocarida are obviously affected by inhomogeneous base compositions and not correctly positioned in rDNA analyses using standard modeling. However, with time-heterogeneous substitution models the Cephalocarida were positioned for the first time based on molecular data as sister-group to the Branchiopoda, which is congruent with several morphological studies (Waloszek, 1993; Waloszek and Müller, 1998a). Hessler (1992) stated in an essay about the phylogenetic position of cephalocarids: "[...] If cephalocarids have any special affinity with other living groups, it is with the branchiopods and malacostracans". A recent study by Oakley et al. (2013) including new ostracod transcriptomes combined with morphological characters corroborate the Cephalocarida as sister-group to Branchiopoda based on molecular and morphological data. A closer relationship of Branchiopoda to Hexapoda, Remipedia, Malacostraca is indicated by some molecular (Regier et al., 2010; von Reumont, et al., 2012; Oakley et al., 2013) and neuroanatomical data (Harzsch, 2006; Fanenbruck, Harzsch, and Wägele, 2004).

The internal relationships of **Malacostraca** are still controversial although their monophyly is well supported by morphological and molecular data (Regier et al., 2010; Mallatt and Giribet, 2006, von Reumont et al., 2009; 2012; Jenner et al., 2009). Most recently, ambiguous internal malacostracan relationships were reviewed by Richter et al. (2009) based on morphological data. Except some studies on Decapoda (Bracken et al., 2009; 2010), the phylogeny of malacostracans remains also ambiguous in studies based on molecular data. Based on a total evidence approach Jenner et al. (2009) demonstrated that the commonly used standard markers are not sufficient to enlighten malacostracan phylogeny. First phylogenomic data (von Reumont et al., 2012) corroborate some previous findings that are generally accepted, like a basal split separating Leptostraca and Eumalacostraca (Regier et al., 2010; von Reumont, 2009; 2012; Jenner et al., 2009). Further conclusions based on phylogenomic data are to date not possible because taxon sampling is still too scattered and biased towards decapod species. To complete confusion, the position of Malacostraca among early derived lineages in most molecular trees contradicts mounting morphological and neuroanatomical data (see also Figure 15.5), which indicate that Malacostraca are eventually highly evolved crustaceans close to Remipedia and Hexapoda (Fanenbruck and Harzsch, 2005; Koenemann et al., 2009; Stemme et al., 2012).

Monophyletic **"Maxillopoda"** were never found in molecular analyses. A critical discussion of this taxon was presented in Martin and Davis (2001). For some clades that were included in the Maxillopoda sensu Dahl (e.g. within Thecostraca), monophyly is corroborated by recent molecular data.

The **Thecostraca** include Cirripedia, Ascothoracica, Acrothoracia and Facetotecta, Their phylogeny was enlightened in many aspects by Høeg et al. (2009b) and Pérez-Losada, Høeg, and Crandall (2009). Open questions concern the position and internal phylogeny of the parasitic Rhizocephala (Glenner and Hebsgaard, 2006; Pérez-Losada, Høeg, and Crandall, 2009), see also discussion in Richter, Møller and Wirkner (2009). However, data are insufficient to fix the position of Thecostraca within crustaceans. Most molecular data support Thecostraca as sister-group to Malacostraca (Regier et al., 2010), which morphologically is not supported by shared synapomorphies. Analyses of phylogenomic data (Meusemann et al., 2010; von Reumont et al., 2012) demonstrate that there is conflicting signal. A placement of Thecostraca as sister-group to Malacostraca can be strongly supported in one analysis, but the same is true for a grouping with Copepoda in another. Morphologically more plausible seems to be the latter case (Figure 15.5).

Monophyly of **Copepoda** has never been seriously questioned. However, internal relationships are still unclear and so is also the position within crustaceans. Interestingly, a supposed close relationship between Copepoda and Mystacocarida (Waloszek and Müller, 1998a) is contradicted by new phylogenomic data (von Reumont et al. in prep.), in which **Mystacocarida** are close related to **Ostracoda**. This is somehow in congruence with a clade "Oligostraca" (Regier et al., 2010; Oakley et al., 2013) that includes (monophyletic) Ostracoda, Mystacocarida, Branchiura and Pentastomida. Yet the Ostracoda are in this clade the sister-group to (Mystacocarida (Branchiura, Pentastomida)). For the latter two taxa, however, phylogenomic EST data is still missing.

15.5 Promises and pitfalls of analyses of phylogenomic data

A promising perspective is that detailed workflows were designed and presented to process phylogenomic data (Meusemann et al., 2010; von Reumont et al., 2012). The difficult steps and issues that demand attention are elaborately addressed and yet partly tested and tackled. Important points are:

(1) Gene selection and orthology prediction. Orthology assessment is a crucial step that is normally conducted before data quality is generally assessed (von Reumont, Meid, and Misof, 2011). As shown in von Reumont et al. (2012), different sets of putative orthologous genes might result in different topologies. In general, it seems that more attention should be paid to precise orthology prediction. Today, the HaMStR software (Ebersberger, Strauss, and von Haeseler, 2009) is the only method recommendable to infer orthologous gene sets. However, several points need improvement, like the simultaneous generation of nucleotide data that correspond to aligned protein sequences.

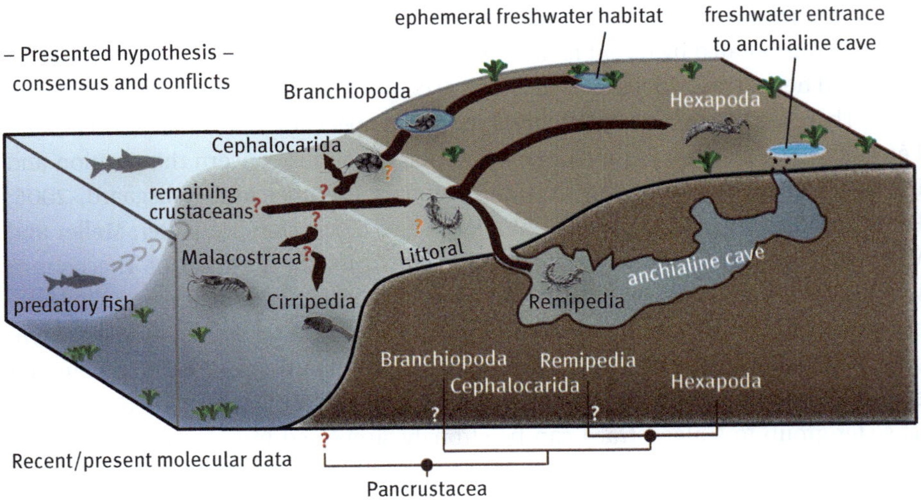

Figure 15.5: Supposed evolutionary scenario that summarizes new insights and open issues concerning crustacean phylogeny based on molecular and morphological data. Brown arrows and lines represent evolutionary lineages. Circles represent nodes that are strongly supported by morphological and molecular data. Dashed lines indicate more weakly supported relationships. Question marks indicate branches whose position is uncertain: variously because of ambiguity in the molecular data (*red*), conflict with morphological data (*red*), or a large gap in the fossil record (*orange*). Molecular analyses generally place Branchiopoda closer to Hexapoda, while selected morphological, neuroanatomical, and larval development data suggest a closer relationship of Malacostraca to Remipedia and Hexapoda. A similar pattern is found for the Cephalocarida, their close relationship to Remipedia in molecular analyses is not supported by recent neuroanatomical data (Stegner and Richter, 2011). Interestingly, a previous comprehensive rDNA study on euarthropods (von Reumont et al., 2009) reconstructed an SG relationship of Cephalocarida and Branchiopoda, which was recently supported by Oakley and colleagues in a combined transcriptomic analysis including morphological and molecular data (Oakley et al., 2013). Paleontological data (e.g. Waloszek, 1993) would support this scenario. Thecostracan position is ambiguously resolved in molecular data, morphologically the closer relationship to Malacostraca that is reconstructed in most trees is less plausible than the alternative position as sister-group to Copepoda (Meusemann et al., 2010; von Reumont et al., 2012).

The handling of phylogenomic data sets is computationally extremely expensive. Notably, the selection of the correct substitution model and subsequently the final phylogenetic reconstruction are hindered by large data set sizes (e.g. beyond 100 MB). From this perspective, tools like MARE (Meusemann et al., 2010; von Reumont, 2011) that select informative gene subsets are important. But MARE is even more essential because it reduces systematic errors caused by missing data and it excludes genes that have little signal-like information. Careful selection of subsets of genes clearly improves tree inference (Meusemann et al., 2010; von Reumont et al., 2012).

(2) Assessing data quality. Data quality is usually estimated a posteriori, i.e. judging the quality of a resulting tree and its fit to data and model assumptions. One aspect of a priori homology assessment is an evaluation of alignments. Several studies show that the search for low-quality regions in alignments, where aligned positions are only randomly similar, and exclusion of such regions can improve data quality and sequence analyses (Reumont et al., 2009; von Reumont, 2011; Misof and Misof, 2009; Kück et al., 2010), see also chapter of Misof and coworkers in this book. Furthermore, it is important to evaluate eventual conflicts in the data based on network reconstructions or comparison of different gene sets (Meusemann et al., 2010; von Reumont et al., 2012; Koenemann et al., 2010). Important tools were designed to accomplish these issues, however, still missing are automatic procedures that allow an easy inspection of conflicts or ambiguous signals for separate genes. Difficult are the sheer number of genes and sizes of phylogenomic data sets.

(3) Methodologically promising is, further, that a comparison between nucleotide and protein data might help to identify conflicts in the data and to resolve some nodes. However, some aspects like the dramatically increased data size create new computational difficulties to handle the data. Persistent are also open questions that address in which amount phylogenetic reconstructions are biased by heterogeneous substitution processes in EST data and by long-branch effects.

(4) New software tools are needed to avoid the systematic errors described by Kück and Wägele (this book).

A present pitfall is a still limited taxon sampling regarding important pancrustacean taxa. Taxon gaps cannot be completely filled with data from multi-gene studies (e.g. Regier et al., 2010) because the number of shared genes in sets for different species is partly too small and orthology prediction might be problematic (von Reumont et al., 2012). Phylogenetic analyses of transcriptomic data require a specific tissue preservation and therefore collection of fresh specimens. The present overrepresentation of Decapoda and of a few parasitic forms must be overcome. Data for many phylogenetically important crustacean clades are still missing. The completion of transcriptomic and complementary genomic data requires the close cooperation of the scientific community.

Acknowledgments

We would like to thank all those collaborators who assisted in fieldwork and helped us to collect species for the molecular work. BMvR thanks generally all colleagues for fruitful discussions on morphological and general aspects, but specially: Steffen Harzsch, Carsten Müller, Joachim Haug (Universität Greifswald), Martin Stegner (Universität Rostock) and Dieter Waloszek (Universität Ulm). Stefan Koenemann was extremely helpful to discuss Remipedia evolution. Karl Kjer (Rutgers University, New Jersey) is to mention for great discussions on rDNA, Ronald Jenner (NHM London) on general evolutionary aspects. Bernhard Misof and Karen Meusemann (ZFMK, Bonn) were close cooperation and discussion partners, without these two colleagues and their input our data and results would have remained incomplete. This work was funded by the DFG within the Spp-1174 Deep Metazoan Phylogeny priority program, BMvR was funded by DFG grants WA-530/34 and RE-3454/1.

Jason Dunlop, Janus Borner, and Thorsten Burmester

16 Phylogeny of the Chelicerates: Morphological and molecular evidence

Abstract: The arthropod subphylum Chelicerata encompasses ~110,000 described living species. Here we review the present state of knowledge on chelicerate phylogeny, thereby including morphological, paleontological and molecular evidence. We must conclude that chelicerate relationships are largely unresolved. The minimal consensus tree supported by various methods is (Pycnogonida, (Xiphosura, (Scorpiones, ((Amblypygi, (Thelyphonida, Schizomida)), Araneae)))). In addition, the Acari – combining Acariformes and Parasitiformes – are probably a valid taxon. The positions of Opiliones, Palpigradi, Pseudoscorpiones, Ricinulei, and Solifugae are, however, essentially unresolved. Even a novel large multi-gene data set derived from expressed sequence tags from various chelicerate taxa did not help improving the tree. The lack of phylogenetic signal may be explained by rapid adaptive radiation associated with the terrestrialization of the arachnids.

16.1 Introduction

Chelicerata is one of the major branches of the arthropods, encompassing arachnids and their relatives. With nearly 110,000 described living species, they are second only to insects in diversity. Arachnids obviously dominate, with spiders (more than 43,000 species) and mites (more than 55,000) representing the 'megadiverse' orders. Spiders are ubiquitous, and a key group of predators in most terrestrial ecosystems. Mites are ecologically more diverse and range from free-living predators to parasites, to detritivores and even herbivores. Some are aquatic and others are of economic significance as crop pests or disease vectors. Further species-rich arachnid groups include the harvestmen (ca. 6,500 species), pseudoscorpions (3,400), scorpions (2,000), and camel spiders (1,100). The remaining arachnids are largely restricted to the tropics and are known from, at most, only a few hundred species. In addition to the arachnids, Chelicerata also includes the marine xiphosurans (horseshoe crabs) known only from four living species. Finally pycnogonids (sea spiders) are another marine group, comprising nearly 1,500 species.

Relationships between these animals have long been debated (e.g. Pocock, 1893b; Börner, 1904). A significant step forward was the introduction of cladistic methods in a seminal paper by Weygoldt and Paulus (1979b), followed by the first computer-assisted cladograms (Shultz, 1990) and the first applications of molecular data (Wheeler and Hayashi, 1998) (Figure 16.1). In recent years, additional molecular phylogenetic studies using a variety of markers and more sophisticated techniques have attempted to enhance our understanding of chelicerate relationships (Shultz and

Regier, 2000; Hassanin, 2006; Podsiadlowski and Braband, 2006; Jones et al., 2007; Shultz, 2007; Dunn et al., 2008; Jeyaprakash and Hoy, 2009; Roeding et al., 2009; Pepato, Rocha, and Dunlop, 2010; Regier et al., 2010; Rehm et al., 2011; Rota-Stabelli et al., 2011; Rehm, 2012). Yet despite this wealth of modern data there is, at present, no single accepted phylogeny for Chelicerata and discrepancies remain between trees derived from morphology and molecules. This striking lack of resolution – in spite of the range of analytical techniques applied – is surely the key issue for contemporary work on chelicerate relationships.

16.2 Chelicerate origins: Mandibulata or Myriochelata?

Traditionally, chelicerates were placed within the Euarthropoda as the sister-group of the Mandibulata (Brusca and Brusca, 2003), a taxon that comprises the subphyla Myriapoda, Crustacea, and Hexapoda. Synapomorphies that unite the Mandibulata are the mandibles, which are mouthparts used for biting and chewing, the possession of antennae and the division of the body into three tagmata: head, thorax and abdomen. The monophyly of Mandibulata receives support from some molecular phylogenetic studies, which include analyses of selected single genes (Giribet and Ribera, 2000; Kusche et al., 2003), concatenated alignments (Regier et al., 2010), total evidence (Giribet, Edgecombe, and Wheeler, 2001), and mircoRNA (Rota-Stabelli et al., 2011).

However, other molecular studies have challenged the monophyly of Mandibulata and suggested a common clade of chelicerates and myriapods (the "Myriochelata" or "Paradoxopoda" hypothesis; e.g., Friedrich and Tautz, 1995; Hwang et al., 2001; Mallatt, Garey, and Shultz, 2004; Pisani et al., 2004; Dunn et al., 2008). Morphological evidence in favor of Myriochelata is poor, and essentially restricted to some similarities of neurogenesis (Kadner and Stollewerk, 2004) and embryonic development (Mayer and Whitington, 2009b). These characters, however, may also be plesiomorphic and reflect the ancient state in stem-line arthropods. An analogous explanation may explain similarities in the organization of the onychophoran and chelicerate brain, which even led to the proposal of a sister-group relationship between these two taxa (Strausfeld et al., 2006a). According to Rota-Stabelli and Telford (2008), the choice of outgroup in analyses of mitochondrial sequences influences whether Myriochelata is recovered.

16.2.1 Evidence from the fossil record of chelicerates

Chelicerate origins in deep time are uncertain. Historically they were often linked to the extinct Trilobita; a hypothesis largely based on superficial similarities between trilobites and horseshoe crabs (Xiphosura). Newly hatched xiphosurans retain vestiges of segmentation and are still commonly referred to as 'trilobite larvae'. The names

Arachnomorpha or Arachnata can often be found in the paleontological literature and encompass a broad, but not necessarily well-defined, group of chelicerates, trilobites and various early Paleozoic arthropods. The concept of Arachnomorpha was explicitly challenged by Scholtz and Edgecombe (2005), who pointed out that most of the potential synapomorphies shared by trilobites and xiphosurans – like the broad head shield – are not seen in pycnogonids or arachnids. They preferred to group trilobites with other antennae-bearing arthropods, i.e. the Mandibulata.

Where does this leave Chelicerata? One emerging hypothesis (e.g., Chen, Waloszek, and Maas, 2004) places a number of Cambrian fossils – sometimes called 'great appendage' arthropods or megacherians – on the chelicerate stem-lineage. These extinct arthropods share with chelicerates a modification of the first pair of head appendages into increasingly raptorial and claw-like structures (see also Haug, Briggs, and Haug, 2012). In this scenario, these raptorial head limbs would represent precursors of the chelate chelicerae (see below). Yet claws are functional adaptations, easy to evolve, and thus potentially prone to homoplasy. Homologizing head appendages between early fossil arthropods and their living relatives remains controversial and other interpretations of the 'great appendage' (e.g., Budd, 2002) have been published. Resolving the chelicerate stem-lineage is nevertheless important with a view towards selecting appropriate outgroup taxa for polarizing morphological character states.

16.3 Chelicerate phylogeny

The name Chelicerata literally means 'claw-bearer' and was coined by Heymons (1901) for arthropods in which the first pair of head appendages – the chelicerae – are used for grasping and/or tearing up prey. These chelicerae are either claw-like or else modified, in groups like spiders into fangs which take the form of a pocket knife. In its original conception, Chelicerata comprised arachnids and xiphosurans only.

16.3.1 Position of the sea spiders (Pycnogonida)

Sea spiders (Pycnogonida) were often added to the chelicerates, although this placement does not have universal support. It is fair to say that pycnogonids are strange-looking creatures with many unusual features. For example, they have a large proboscis for sucking up prey and displace many organ systems from their remarkably narrow body into the legs. Some authors simply accumulated these autapomorphies as evidence for the 'uniqueness' of pycnogonids, and suggested that they evolved independently from the other arthropods (for a review, see Dunlop and Arango, 2005). The name Cormogonida (Zrzavý, Hypsa, and Vlaskova, 1998) has been suggested for a clade of arthropods excluding the pycnogonids. In support of this hypothesis is the fact that in cormogonid arthropods (i.e. euchelicerates plus mandibulates) the genital opening is on the body, while in pycnogonids the genital openings are on the proximal articles of the legs. Additional support for the Cormogonida hypothesis comes from the combined analysis of molecular sequences and morphological characters (Giribet, Edgecombe, and Wheeler, 2001) and from comparative neuroanatomy, which suggests a unique innervation of the chelifores (Maxmen et al., 2005) (but see below).

Morphologically, the key issue for a sister-group relationship of Pycnogonida and Euchelicerata is equating the pycnogonid chelifores (or cheliphores) with the euchelicerate chelicerae. Both structures are fundamentally chelate, but as noted above claws are functional elements which could potentially evolve in parallel. As mentioned, there have also been proposals that the chelifores and chelicerae are not serially homologous elements, i.e. they are not innervated from the same part of the brain (Maxmen et al., 2005). However, Hox gene (Jager et al., 2006) and neuroanatomical studies (Brenneis, Ungerer, and Scholtz, 2008) have argued that this is the same appendage – and in this context chelate chelifores/chelicerae remain the best autapomorphy of a monophyletic Chelicerata.

Most of the modern literature favors the traditional textbook concept of Chelicerata, namely a sister-group relationship of Pycnogonida and Euchelicerata. Molecular phylogenetic studies using hemocyanin sequence and structure (Rehm et al., 2012), selected genes (Regier et al., 2010; Sanders and Lee, 2010) or expressed sequence tags (ESTs) (Dunn et al., 2008; Meusemann et al., 2010) strongly support this topology. An ingroup position of the Pycnogonida with the Arachnida – as deduced from mitochondrial DNA sequences (Hassanin, 2006; Podsiadlowski and Braband, 2006; Jones et al., 2007; Park, Lee, and Hwang, 2007; Jeyaprakash and Hoy, 2009) – is not supported by other data and is best considered an artifact.

16.3.2 Euchelicerata

The term Euchelicerata was introduced by Weygoldt and Paulus (1979b) for all chelicerates, excluding pycnogonids. In their original definition it included an extinct fossil group called Aglaspidida, but these animals – which superficially resemble xiphosurans – are now usually placed on the mandibulate stem-lineage instead; see Ortega Hernandez, Legg, and Braddy (2013; and references therein). Euchelicerata is an uncontroversial group, easily defined by the presence of plate-like appendages on the underside of the opisthosoma. These are clearly visible in xiphosurans as the movable, flap-like opercula covering the paired genital openings and the five subsequent pairs of gills. The presence of lamellate respiratory organs (i.e. book gills or book lungs) is also characteristic for euchelicerates; although numerous arachnids have lost the lungs and replaced them either partially or wholly with trachea. In general, it can also be argued that Euchelicerata have the typical prosoma–opisthosoma division of the body; the front half focused on feeding, locomotion and sensory systems and the back half focused on digestion, respiration and reproduction. A caveat to this is that in xiphosurans the dividing line is less clear cut. Comparative morphology and embryology reveals that parts of the anterior dorsal opisthosoma contribute to the prosomal dorsal shield, while the first opisthosmal appendages (the small, stubby chilaria) are functionally integrated into the ventral prosoma (Shultz, 2001). The textbook prosoma–opisthosoma division is also questionable for acariform mites, in which the principal division of the body actually runs between the second and third pair of legs (Dunlop, Kruger, and Alberti, 2012), rather than behind the fourth pair as in, e.g., spiders.

Within euchelicerates, two classes were traditionally recognized. Merostomata include xiphosurans and the extinct Paleozoic Eurypterida (sea scorpions). The other class is the (largely) terrestrial Arachnida. As critiqued by Kraus (1976), this is primarily an ecological division rather than an explicitly phylogenetic one. Note that Lamsdell (2013) recently challenged the monophyly of Xiphosura, arguing that certain Paleozoic fossils traditionally placed here are not stem-group horseshoe crabs, but represent stem-group euchelicerates instead.

Weygoldt and Paulus (1979b) placed eurypterids closer to arachnids in a clade they called Metastomata. This was ostensibly defined by a plate-like metastoma immediately behind the leg coxae. This metastoma is a fundamental part of the eurypterid body plan, but it could represent the modified first opisthosomal appendages; thus making them potentially homologous to the chilaria of xiphosurans. A direct equivalent in Arachnida is harder to demonstrate, and in general arachnids are thought to lack appendages on the first opisthosomal segment (Shultz, 1990). As an alternative diagnostic character, in exceptionally preserved eurypterids, Kamenz, Staude, and Dunlop (2011) identified precursor sclerites within internal parts of the genitalia which go on, eventually, to form spermatophores. Since xiphosurans simply release their sperm during mating, spermatophore-mediated sperm transfer offers a good potential synapomorphy for (Eurypterida + Arachnida); a clade for which the new name Sclerophorata was proposed.

16.4 Arachnids: Conquerors of the land

The name Arachnida can be traced back to Lamarck (1801) and emerged, historically, from among the 'wingless insects' as recognized by eighteenth century zoologists. Current schemes usually accept sixteen arachnid orders: twelve extant (Acariformes [mites], Amblypygi [whip spiders], Araneae [true spiders], Opiliones [harvestmen], Palpigradi [micro-whip scorpions], Parasitiformes [predatory mites and ticks], Pseudoscorpiones [false scorpions], Ricinulei [hooded tickspiders], Schizomida [schizomids], Scorpiones [scorpions], Solifugae [camel spiders], and Thelyphonida [whip scorpions]), and four extinct (Haptopoda, Phalangiotarbida, Trigonotarbida, Uraraneida). Cladistic studies employing morphological characters have almost invariably recovered Arachnida as a monophyletic group (Shultz, 1990; Wheeler and Hayashi, 1998) (Figure 16.1). Putative synapomorphies supporting Arachnida include a reduction in the width of the prosomal dorsal shield – usually just called the carapace in the taxonomic literature – and the possession of "terrestrial" respiratory organs like book lungs or trachea. Other proposed arachnid characters include the presence of tiny 'strain gauges' in the cuticle called slit sense organs, although these are not seen in all arachnid orders. Another potential synapomorphy is the loss of appendages from the first opisthosomal segment (see above). This is problematic as scorpions retain limb buds here, at least during their embryological development.

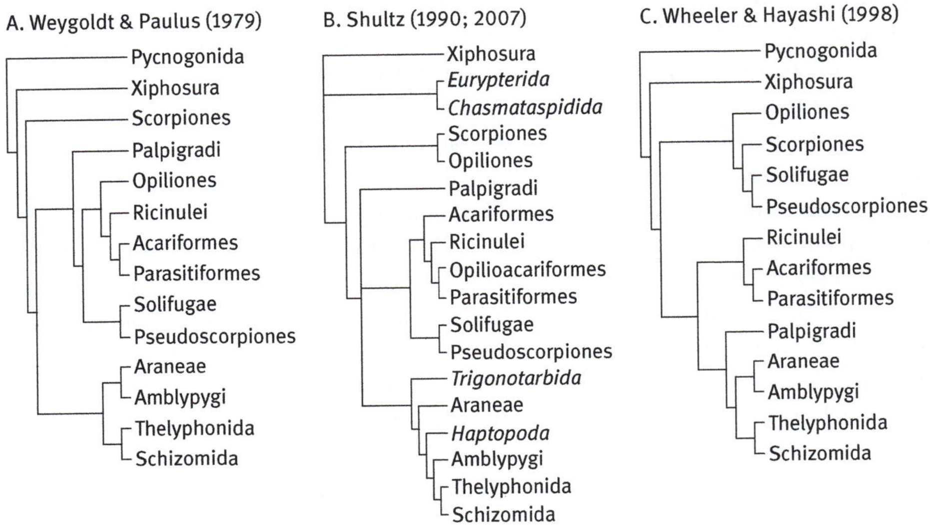

Figure 16.1: Chelicerate phylogeny by **A.** Weygoldt and Paulus (1979b), **B.** Shultz (1990; 2007) and **C.** Wheeler and Hayashi (1998). The names in italics in (B) denote extinct taxa.

16.4.1 Are arachnids monophyletic?

Some paleontological studies have challenged arachnid monophyly, specifically with respect to scorpions and eurypterids (reviewed in Dunlop and Braddy, 2001). Both express a similar body plan in which the last five segments of the opisthosoma are somewhat narrow and ring-like. The question is whether this is synapomorphic or homoplastic? Although it is tempting to envisage scorpions evolving directly from 'sea scorpions', it has to be conceded that the most scorpion-like eurypterids – the mixopteroids, complete with a telson shaped like a sting – actually resolve in quite a derived position among the eurypterids (e.g. Tetlie, 2007).

Several molecular phylogenetic studies also failed to recover arachnids as a monophyletic clade (Roeding et al., 2009; Meusemann et al., 2010; Sanders and Lee, 2010). Sanders and Lee (2010) used concatenated sequences of 18S rRNA, 28S rRNA, elongation factors 1α and 2, and RNA polymerase II subunit in Bayesian analyses and obtained Xiphosura nested within the arachnids. Phylogenomic studies employing large multi-gene alignments of 11,168 (Roeding et al., 2009) or 37,476 (Meusemann et al., 2010) amino acid positions, which had been inferred from EST-derived orthologous genes, consistently obtained the Acari as the sister-group of all other euchelicerates. However, these studies only included representatives of the taxa Xiphosura, Araneae, Scorpiones and Acari.

In a more recent approach, additional taxa, namely Amblypygi, Opiliones, Pseudoscorpiones, Solifugae, and Thelyphonida, were included (Borner, Rehm and Burmester, unpublished). 454 pyrosequencing was used to obtain ESTs, resulting in a concatenated multi-gene alignment consisting of 197 orthologous genes (18,163 aa positions, 32.4 % missing data). However, employing Maximum Likelihood and Bayesian methods for tree reconstructions (similar to the methods described in Meusemann et al., 2010), even this impressive data set was not sufficient to identify the relative relationships among the arachnid taxa, with the exception of support for Euchelicerata, Tetrapulmonata and Pedipalpi (Figure 16.2).

Connected to this debate is the question of whether arachnids all share a common, terrestrial ancestor, or whether multiple lineages moved onto land independently. Terrestrialization was achieved in some (if not all) groups by at least the late Silurian period (Jeram et al., 1990). Note that some Paleozoic fossil scorpions have been interpreted as aquatic animals which, if correct, would imply at least two separate terrestrialization events: one for the scorpions themselves and at least one for the remaining arachnids. However, aquatic scorpions have not been universally accepted (for a recent critique see Kühl et al., 2012) and the trend is now to see most, if not all, fossil scorpions as terrestrial. Scholtz and Kamenz (2006) compared the book lungs of scorpions with those of tetrapulmonate arachnids, identifying numerous common features in their fine structure which imply a single origin for the book lungs in a terrestrial arachnid ancestor.

16.4.2 Tangled relationships: The arachnid groups

The monophyly of each of the twelve arachnid orders is undisputed. However, the relationships between them remain controversial. Traditionally, scorpions were often envisaged as 'primitive' arachnids – perhaps again influenced by their resemblance to eurypterids – and placed basal to the remaining Arachnida, a clade for which Pocock's (1893b) name Lipoctena has been adopted. In Weygoldt and Paulus (1979b), lipoctenic arachnids were characterized by, for example, the fine structure of both their sperm cells and the lateral eyes, and a reduction in the number of respiratory openings. Alternatively, Shultz (2007, and references therein) developed a novel hypothesis, largely based on skeletomuscular characters, in which scorpions are the sister-group of harvestmen (Opiliones). The name Stomothecata was proposed for this

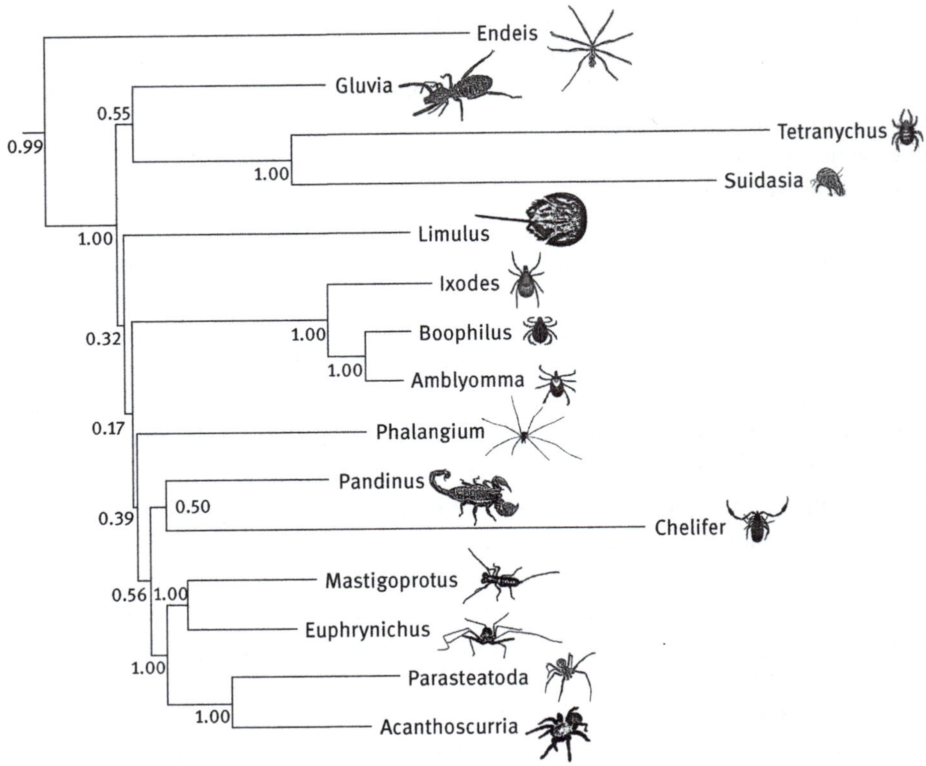

Figure 16.2: Bayesian phylogenetic analysis of 15 chelicerate taxa based on a multi-gene alignment selected from EST data sets (Borner, Rehm and Burmester, unpublished). 197 orthologous genes (88,044 amino acid positions) were selected by the aid of HaMStR (Ebersberger, Strauss, and Haeseler, 2009) and aligned with MAFFT (Katoh and Toh, 2008a). Poorly aligned sections were removed by Gblocks employing the "less stringent" criteria (Castresana, 2000; http://molevol.cmima.csic.es/castresana/Gblocks_server.html). The final alignment covered 18,163 positions with 32.4 % missing data. Tree reconstructions were carried out with PhyloBayes3.3e (Lartillot, Blanquart, and Lepage, 2008) assuming the CAT mixture model with discrete gamma model (four categories) and the Dirichlet process. 20 chains were run for 20,000 cycles each. The numbers at the nodes represent the Bayesian posterior probabilities.

clade (Shultz, 2007), and refers to the shared presence of a distinct preoral chamber – the stomotheca – which in scorpions and harvestmen is explicitly constructed from projections (apophyses) derived from the first two pairs of leg coxae. A critique of this hypothesis would be that some early fossil scorpions have a much simpler coxosternal arrangement and appear to lack these so-called coxapophyses. With regard to harvestman feeding ecology, a further interesting point to note is that they have an omnivorous diet and are one of the few arachnid orders which are (still) capable of ingesting solid food. This is presumably the plesiomorphic condition, and is seen in xiphosurans (as an outgroup) and also some mites (Walter and Proctor, 1998). Most arachnids liquefy their food preorally.

Börner (1904) introduced the name Haplocnemata for a group comprising camel spiders (Solifugae) and pseudoscorpions (Pseudoscorpiones). Apart from their large pedipalpal claws, basal pseudoscorpions do indeed resemble camel spiders. Haplocnemata can be defined on a range of morphological characters, such as two pairs of tracheal openings on the third and fourth opisthosomal segments and a very short femur such that the leg bends principally from the patella–tibia articulation (hence the confusion about the 'missing' patella). In fact, cladistic studies (Weygoldt and Paulus, 1979b; Shultz, 1990) based on morphological characters have recovered Haplocnemata, although we should note that male genital characters do not support this relationship (Alberti and Peretti, 2002).

16.4.3 Are Acari monophyletic and arachnids at all?

Mites and ticks were traditionally grouped together as a single order named Acari. Despite this, acarologists have long recognized (e.g., Grandjean 1935) that there are at least two fundamental lineages, which can be differentiated on whether or not the cuticle of their setae shows birefringence; as well as a swathe of other internal and external characters (reviewed by Dunlop and Alberti, 2008). The parasitiform (or anactinotrichid) mites include opilioacarids, holothyrids, mesostigmatids and ticks. The acariform (or actinotrichid) mites include all the rest. The possibility that Acari is not monophyletic has long been discussed; championed primarily in the non-cladistic work of Zachvatkin (1952) and van der Hammen (1989). All mites share one very good synapomorphy: the gnathosoma. This is a specific functional unit at the front of the body encompassing the chelicerae, mouth lips and pedipalps, which can move independently of the rest of the body (or idiosoma). Any scenario in which mites are not monophyletic must contend with the idea that the gnathosoma is homoplastic. For a further discussion of this key character see Alberti et al. (2011). Some authors have speculated that opilioacariform mites are – as their name implies – related to harvestmen, or have compared palpigrades to the acariform mites (van der Hammen, 1989). However, these hypotheses tend to rely on superficial resemblances and are rarely backed by robust sets of apomorphies.

Perhaps the strongest recurring model is that all mites (or at least the Parasitiformes) are related to the rare ricinuleids. This is the Acaromorpha concept. Acari and Ricinulei both uniquely have hexapodal larvae, i.e. the hatching instar has only three pairs of legs and acquires the fourth pair later in ontogeny. Other authors have argued that ricinuleids also have a gnathosoma, or at least the fused pedipalpal coxae which contribute towards a gnathosoma (Shultz, 1990). However, this interpretation is open to question and other authors treat the gnathosoma as a mite feature only. A further challenge to Acaromorpha is the fact that ricinuleids also share a number of putative synapomorphies (reviewed by Dunlop, Kamenz, and Talarico, 2009) with the extinct arachnid order Trigonotarbida; a lung-bearing taxon which is evidently close to the Tetrapulmonata.

Alberti and Peretti (2002) identified potential synapomorphies in sperm and testis structure shared between camel spiders (Solifugae) and the acariform branch of the mites only. Both groups also share a so-called sejugal furrow, a sulcus dividing the body between the second and third pair of legs (e.g. Dunlop, Kruger, and Alberti, 2012). Taking up this theme, the molecular study of Dabert et al. (2010) formally recognized (Solifugae + Acariformes), and the same result was found independently in a combined molecular and morphological analysis published shortly afterwards by Pepato, Rocha, and Dunlop (2010); who proposed using the historical name Poecilophysidea for this clade. Where, in this scenario, the pseudoscorpions and the other mite clade (Parasitiformes) belong is uncertain, although Dabert et al. (2010) recovered these two groups together.

Other molecular phylogenetic studies also gave conflicting results on both the monophyly and the position of the Acari, respectively. Wheeler and Hayashi (1998) treated the Acari as monophyletic. By using total evidence, combining rRNA and morphological characters, they recovered Acari as sister-group of Ricinulei within the monophyletic Arachnida. A large body of literature exists in which mitochondrial amino acid or nucleotide sequence data have been employed to resolve the phylogenetic position of Acari. Fahrein et al. (2007) derived a paraphyletic Acari, with Acariformes + Ricinulei being sister-group of the Araneae. Diphyletic Acari were also found by Ovchinnikov and Masta (2012), who recovered the clades Acariformes + Pseudoscorpiones and Parasitiformes + Ricinulei (cf. the Acaromorpha concept, as discussed above). By contrast Park, Lee, and Hwang (2007) derived a monophyletic Acari as sister-group of the Araneae. In analyses employing orthologous genes selected from the ESTs, we either found the Acari as sister-group to the other Euchelicerates (Roeding et al., 2007; Roeding et al., 2009; Meusemann et al., 2010) or, upon the inclusion of additional arachnid taxa, Parasitiformes and Acariformes in an unresolved position within the euchelicerates (Figure 16.2) (Borner, Rehm and Burmester, unpublished).

16.4.4 Tetrapulmonata

Perhaps the least controversial higher arachnid taxon in terms of morphology and molecular data is Tetrapulmonata. This group has been consistently recovered – in one form or another – since the early work of Pocock (1893b). It essentially comprises the spiders (Araneae) and their closest relatives: namely whip spiders (Amblypygi), whip scorpions (Thelyphonida), schizomids (Schizomida) plus three extinct orders: Trigonotarbida, Uraraneida and Haptopoda. What these arachnids all share in common is a ground pattern of two pairs of book lungs, opening on the second and third opisthosomal segments. They also have similar chelicerae, which are less chelate and shaped instead more like a pocket knife with a fang articulating against a basal region. Another typical feature is a degree of constriction between the prosoma and opisthosoma, although this is also seen in other arachnids such as palpigrades and camel spiders. The name Tetrapulmonata was introduced by Shultz (1990) for spiders, whip spiders, whip scorpions and schizomids (Figure 16.1B). Study of well-preserved trigonotarbid fossils indicates that they belong to this assemblage too (Shear et al., 1987). Tetrapulmonata was thus subsequently expanded to a clade Pantetrapulmonata by Shultz (2007) specifically to encompass this extinct order. The monophyly of extant Tetrapulmonata has been consistently recovered by the molecular phylogenetic studies using various sets of genes (Wheeler and Hayashi, 1998; Shultz and Regier, 2000; Jones et al., 2007; Pepato, Rocha, and Dunlop, 2010; Rehm et al., 2012), as well as by a large concatenated multi-gene alignment (Borner, Rehm and Burmester, unpublished).

Within the pantetrapulmonates there remains some debate about the precise phylogenetic relationships of the individual orders. Integrating the recently identified (and probably quite spider-like, Selden, Shear, and Sutton, 2008) fossil uraranenids into the scheme of Shultz (2007), we would recover a hypothesis along the lines of (Trigonotarbida, ((Uraranenida, Araneae), (Haptopoda, (Amblypygi, (Thelyphonida, Schizomida)))))). There seems little doubt that the similar-looking whip scorpions and schizomids are closely related – together forming the Uropygi in some nomenclature schemes – and historically they were further combined with the whip spiders (Amblypygi) into a single arachnid order: Pedipalpi. The Pedipalpi concept remains a robust hypothesis on morphological grounds (Shear et al., 1987; Shultz, 1990, 2007) and is justified by synapomorphies such as modification of the pedipalps into subchelate limbs for prey capture, and the first pair of legs becoming long and slender such that they are used more like tactile organs. Further skeletomuscular details can be added to this list (Shultz, 1999).

The alternative hypothesis, the Labellata concept (Petrunkevitch, 1949; Weygoldt and Paulus, 1979b) groups spiders (Araneae) with whip spiders (Amblypygi) and is justified morphologically by the presence of a muscular sucking stomach within the prosoma which aids the ingestion process, and an especially narrow junction between the prosoma and opisthosoma which is often referred to explicitly here as a pedicel or petiolous (Figure 16.1A). Wheeler and Hayashi (1998) recovered Labellata based on the phylogenetic analysis of ribosomal RNA sequences and total evidence, and more recently, Ovchinnikov and Masta (2012) found this topology with mitochondrial amino acid sequences. The unusual topology of a clade comprising Araneae + Uropygi was identified by analyses of 18S + 28S ribosomal RNA (Pepato, Rocha, and Dunlop, 2010). Most other molecular analyses, however, recover monophyletic Pedipalpi (Amblypygi and Uropygi) as one of the best-supported taxa within the arachnids. The data sets applied in these studies included two nuclear genes (elongation factor 1-a and the large subunit of RNA polymerase II) (Shultz and Regier, 2000), selected mitochondrial data (Jones et al., 2007), single and concatenated hemocyanin sequences (Rehm et al., 2012), and the above-mentioned multi-gene alignment derived from ESTs (Borner, Rehm and Burmester, unpublished).

16.4.5 Araneae: The true spiders

Spiders are the most prominent representatives of the chelicerates, and are characterized by the possession of opisthosomal silk glands opening via spinnerets (Selden, Shear, and Sutton, 2008), cheliceral venom glands and male pedipalps modified as a copulatory organ. The monophyly of Araneae has never been disputed among morphologists and has also been consistently recovered in molecular studies (Wheeler and Hayashi, 1998; Roeding et al., 2009; Rehm et al., 2012). Two suborders of spiders are recognized; the Mesothelae are considered as the most basal clade, being sister-group to the Opisthothele. The latter is divided into the infraorders Mygalomorphae (tarantulas and relatives) and Araneomorphae (orb weaving spiders, wandering spiders, and others).

16.5 Dating chelicerate evolution

The fossil record of chelicerates is fragmentary, but still helps to date their evolutionary history (Dunlop and Selden, 2009; Dunlop, 2010). The origin of the subphylum Chelicerata dates back at least to the Cambrian period. The earliest fossil evidence for the split of Pycnogonida and Euchelicerata is a larval sea spider from the upper Cambrian ~ 500 million years ago (MYA) (Waloszek and Dunlop, 2002). A molecular clock approach, which compares the DNA or amino acid sequences, is an additional and alternative approach to date chelicerate origin and evolution. Initial clock analyses suggested that the earliest divergence within the chelicerates took place already 813–632 MYA (Regier, Shultz, and Kambic, 2005). This estimate is most likely too old and predates the first unequivocal evidence for metazoan life. We employed ESTs and calculated that the split between Pycnogonida and Euchelicerata occurred ~ 546 MYA (Rehm et al., 2011). Other calculations employing hemocyanin sequences came to a similar conclusion and suggest a divergence of 543 MYA (Rehm et al., 2012). A similar date was found in the molecular clock study of Rota-Stabelli, Daley, and Pisani (2013). These estimates only slightly predate the earliest putative (stem) chelicerate fossils, which derive from the Lower Cambrian ~ 530 MYA (Chen, 2009).

Likewise, the fossil and molecular clock dates of euchelicerate evolution essentially agree. The earliest true euchelicerates are xiphosuran fossils from the Early Ordovician ~ 480 MYA of Morocco (Van Roy et al., 2010). Studies employing hemocyanin sequences derived ~ 463 MYA (444 to 489 MYA) as the time of divergence of arachnids and xiphosurans (Rehm et al., 2012). The first unambiguous arachnid is a ~ 428 MYA old Silurian scorpion (Dunlop and Selden, 2009; Dunlop, 2010); we calculated that scorpions and Tetrapulmonata split ~ 419 MYA (405 to 440 MYA). The Tetrapulmonata likely emerged in the Devonian period, but the first fossils (i.e. recognizable mesothele spiders, whip spiders and whip scorpions) are from the Carboniferous (Dunlop, 2010). An early (i.e. Devonian) origin was supported by the molecular clock calculations (~ 369 MYA). According to the hemocyanin data set Araneae and Pedipalpi diverged ~ 369 MYA (357 to 414 MYA), Amblypygi and Uropygi 334 MYA (316 to 344 MYA). The oldest known non-mesothele spider is a mygalomorph dating 240 MYA (Triassic), while Araneomorphae are also Triassic in age (Selden et al., 1999). From the hemocyanin data set we derived that Mygalomorphae and Araneomorphae diverged ~ 271 MYA (254 to 288 MYA), which is somewhat older and implies a Permian origin of today's two principal spider clades (Rehm et al., 2012).

The most successful subgroup within the Araneomorphae are the Entelegynae. These spiders are characterized by their complex genitalia, and are further subdivided into the Orbicularidae (i.e. orb weavers and their relatives) and the spiders of the RTA clade which includes both web-builders and hunting or ambushing spiders which have secondarily given up their web-building behavior. A recent fossil suggests that the Orbicularidae already occurred in the Jurassic 165 MYA (Selden, Shih, and Ren, 2011). Our calculations derived a date ~ 239 MYA for the divergence of Orbicularidae and the RTA clade (Rehm et al., 2012), which suggests that the origin of orb-weaving spiders may even be earlier.

16.6 Perspectives: Resolving the chelicerate tree

Despite more than 100 years of morphological research and the recent advance of molecular techniques, relationships among the chelicerate orders are still largely unresolved. The minimal consensus, which is well supported by various methods, would be Pycnogonida, (Xiphosura, (Scorpiones, ((Amblypygi, (Thelyphonida, Schizomida)), Araneae)))). There is also evidence, albeit disputed, that the Acari – combining the Acariformes and Parasitiformes – are a valid taxon. However, the positions of the remaining arachnid orders remain elusive and different results have emerged from different data sets. Even our very large multi-gene data set that also includes Opiliones, Pseudoscorpiones and Solifugae does not improve the tree. Alternative data sets in which either slow or fast evolving genes, or both, or long branching taxa have been excluded, did not result in any improvement of the tree (not shown). Likewise, alternative methods, such as tree reconstructions using rare genomic changes such as indels failed at the same nodes.

This observation is remarkable, and in sharp contrast to the other arthropod subphyla, in which the generation of large sequence alignments and the application of similar methods led to a significant advance in the resolution of the trees (Roeding et al., 2007; Roeding et al., 2009; Meusemann et al., 2010; Regier et al., 2010; von Reumont et al., 2012). A possible explanation for the lack of phylogenetic signal within the arachnid data sets may be a rapid diversification in the early Paleozoic, which may in turn be evidence of an adaptive radiation associated with the terrestrialization of these animals (see also: Pisani et al., 2005; Rota-Stabelli, Daley, and Pisani, 2013). Such an interpretation is tentatively supported by the very short internal branches at the relevant nodes (cf. Figure 16.2) and is in line with the notorious difficulties encountered when trying to resolve the chelicerate tree with the help of morphological characters.

Acknowledgments

We thank Peter Rehm for his help with data acquisition, Paul Selden and Gerd Alberti for helpful discussions, and the comments of a reviewer.

Martin Schlegel, Michael Weidhase, and Peter F. Stadler

17 Deuterostome phylogeny – a molecular perspective

Abstract: Molecular data improved our knowledge on deuterostome phylogeny and challenged traditional hypotheses. Basically, deuterostomes split in monophyletic Ambulacaria, Xenacoelomorpha, and Chordata. Within Ambulacraria, the sister-taxa Echinodermata and Hemichordata are supported in numerous studies. Mitogenomic analyses support the echinoderm monophyly as well as the monophyly of the five major clades Asteroidea, Crinoidea, Echinoidea, Holothuridea, and Ophiuroidea. However, their mutual relationships remain ambiguous. Likewise, the placement of hemichordate subtaxa is still fluent, most analyses showing Pterobranchia to be nested in paraphyletic enteropneusts. Chordata are subdivided in Cephalochordata and Olfactores, in which the latter includes the two sister-taxa Tunicata and Vertebrata. Inside Tunicata three clades are distinguishable: (Phlebobranchia + Thaliacea + Aplousobranchia), Appendicularia, and Stolidobranchia. Finally, Vertebrata comprise the sister-taxa Gnathostomata and Cyclostomata, the latter containing Hyperoartia and Hyperotreti. Within Gnathostomata sequencing of the coelacanth genome and lungfish transcriptome has resolved a long standing issue by supporting the lungfish-tetrapod dichotomy.

17.1 Introduction

Based on morphological data, deuterostome phylogeny was commonly regarded as stable. Neither the monophyly of this taxon, nor the branching order was disputed. Traditionally, three monophyla were recognized: the Echinodermata, Hemichordata (Pterobranchia + Enteropneusta), and Chordata (Tunicata + Cephalochordata + Vertebrata). Name-giving and also regarded as a strong synapomorphy uniting these taxa was the fate of the blastopore, which develops into the anus in deuterostome embryos (Ax, 2001). This systematization has been challenged in recent years. Not only is deuterostomy now regarded as a plesiomorphic character (Martín-Durán et al., 2012), but also further taxa have been included within Deuterostomia, namely the enigmatic *Xenoturbella* and Acoelomorpha (Acoela + Neodermatida) (Philippe et al., 2011b). Major rearrangements have been largely propelled due to the increasing availability of molecular data for systematic analyses. In a landmark study solely based on the 18S rRNA gene, Halanych (1995) investigated the position of pterobranches, finding them nested within Enteropneusta (basically synonymizing this taxon with Hemichordata), and found those as sister-taxon to the Echinodermata. This idea goes back to Metschnikoff (1881), who called this clade Ambulacraria. Subsequently, additional single gene analyses dealing with deuterostome relationships or relationships of deuterostome subtaxa have been published (e.g. Cameron, Garey and Swalla, 2000; Kuraku et al., 1999; Littlewood et al., 1997; Mallatt and Sullivan, 1998; Stach and Turbeville, 2002; Swalla et al., 2000). Analyses relying on one or a few genes, however, frequently suffer systematic errors such as long-branch attraction, stochastic errors and a limited taxon sampling (Jeffroy et al., 2006). As a result, the use of different genetic markers led to incongruent topologies (Rokas et al., 2003a). Subsequently, technical advances in both, sequencing and analysis techniques, triggered a revolution of molecular systematics leading this field into a post-genomic era. So-called phylogenomic analyses (Philippe and Telford, 2006) are now state of the art to investigate deep animal relationships when using molecules. Genome scale approaches bear the advantage that not only the raw sequence data became available for analyses, but also analyses using rare genomic changes (Rokas and Holland, 2000) as phylogenetic markers led to promising results. The latter marker systems are of special interest, as it has been shown that even massive data sets can be positively misled due to systematic biases within the data, e.g. as demonstrated for long-branch attraction (Wägele and Mayer, 2007). Therefore, rare genomic changes such as gene order (Perseke et al., 2008), codon reassignment (Perseke et al., 2011), or microRNA-content (Heimberg et al., 2010) have been explored for their utility in phylogenetic analyses of deuterostome relationships. In the following section we aim to give an overview of the phylogeny of Deuterostomia and its major groups from a molecular systematic point of view.

17.2 Deuterostome phylogeny

Referring to recent phylogenomic analyses, deuterostomes split in two sister-taxa: the Chordata (Cephalochordata + Olfactores) and a clade uniting Ambulacraria plus Xenacoelomorpha (Figure 17.1) (Philippe et al., 2011b; Swalla and Smith, 2008).

This view is completely at odds with the traditional picture. In particular, the position of the Xenacoelomorpha clade is highly disputed. This taxon includes some simple organized animals such as *Xenoturbella*, Acoela, and Neodermatida. The latter two groups were originally classified as flatworms (Ehlers, 1995), or have been regarded as the sister-taxon of all other Bilateria by others (e.g. Ruiz-Trillo et al., 1999). A position at the base of the bilaterian tree was further corroborated by the Hox gene content of these animals, which resembles those of cnidarians (Cook et al., 2004). *Xenoturbella* exhibits a similar organized (reduced?) Hox cluster (Fritzsch et al., 2008) and its phylogenetic position is a hotly debated topic since its first description in 1949 (Westblad, 1949). Originally described as acoel flatworms, later authors linked them with deuterostomes (e.g. Reisinger, 1960). Surprisingly, a first molecular study found strong support that these animals were molluskan relatives (Norén and Jondelius, 1997), a result which seemed to receive further corroboration due to the presence of eggs and larvae resembling those of bivalves (Israelsson, 1997).

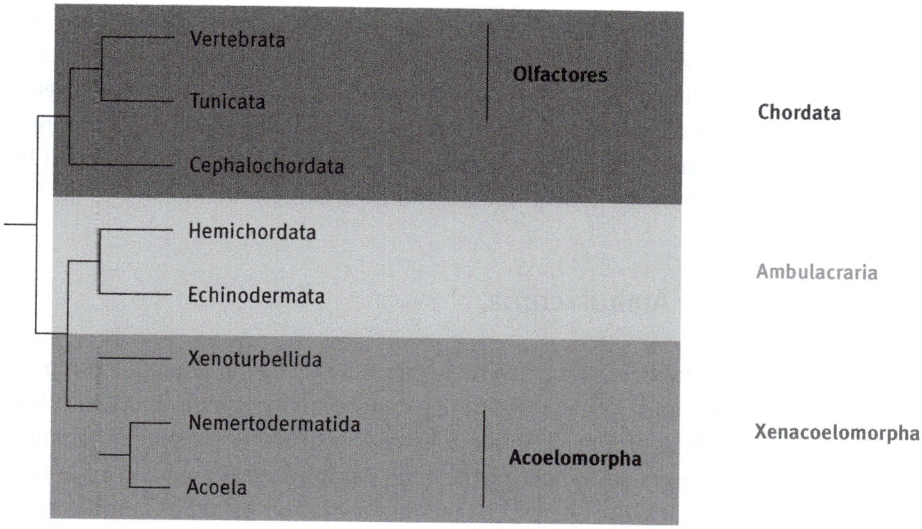

Figure 17.1 Deuterostome phylogeny according to Philippe et al. (2011b).

As intriguing as this result was, it was later disenchanted as an artifact, demonstrating that the analyzed sequence data were bivalve DNA from food ingested by *Xenoturbella* (Bourlat et al., 2003). Instead, a nested position within deuterostomes, as sister-taxon to the Ambulacraria was favored by these authors, a result also found in an analysis based on expressed sequence tags (ESTs) by Dunn et al. (2008). Analyses of mitochondrial genomes remained inconclusive regarding its phylogenetic position. Whereas one analysis supported the sister-group relationship to Ambulacraria (Bourlat et al., 2006), another analysis based on the CAT model recovered a sister-group relationship of *Xenoturbella* and the remaining Deuterostomia (Perseke et al., 2007). The placement of *Xenoturbella* is especially challenging due to the long branch this taxon exhibits in sequence based analyses, and consequently the CAT model, which has been shown to be robust against long-branch attraction artifacts (Lartillot, Brinkmann and Philippe, 2007), was chosen by the latter authors. Interestingly, Perseke et al. (2007) found in one of their analyses *Xenoturbella* clustering with the acoel *Paratomella* in the basal part of the bilaterian tree. A close relationship of acoels and *Xenoturbella* was later on suggested by several phylogenomic analyses based on ESTs. Most notably, an analysis by Hejnol et al. (2009) based on 844 genes supported just this scenario. A sister-group relationship of these morphologically simple organized animals to all other bilaterians appeared tantalizing in regard to evolutionary scenarios explaining the evolution of animal body plans (e.g. Bourlat and Hejnol, 2009; Hejnol and Martindale, 2008b). However, this view was challenged by another EST-based phylogenomic analysis, which also included an analysis of the microRNA content (Philippe et al., 2011b). In this analysis, the authors used three independent phylogenetic marker sets that support a nested position of Xenacoelomorpha within

Deuterostomia as sister-taxon to the Ambulacraria. Consequently, the simple organized body architecture (absence of deuterostome characters like coelomic cavities, a through gut and gill slits) as well as the Hox cluster organization had to be interpreted as evolutionary losses for this taxon. Given that more recent analyses favor a different placement for Acoelomorpha and *Xenoturbella* (e.g. Mwinyi et al., 2010), it must be summarized that this phylogenetic problem is far from being settled.

17.3 Phylogeny of Ambulacraria

Support for a taxon Ambulacraria came first from 18S rRNA gene sequence comparisons (Cameron, Garey and Swalla, 2000). An extended analysis including 18S rRNA, mitochondrial 16S rRNA and cytochrome *b* genes suggested a placement of pterobranchs within enteropneusts, rendering the latter paraphyletic (Cannon et al., 2009). In a comprehensive analysis of deuterostome mitochondrial genomes Perseke et al. (2011) found prevailing evidence for Ambulacraria however with Hemichordata only being represented by enteropneusts due to the unusual genome of the pterobranch *Rhabdopleura compacta* (Perseke et al., 2011). Because being so highly derived, it had to be excluded from the study (Figure 17.2).

However, weighted as two autapomorphies in the mitochondrial genetic code, the assignment of "ATA" to isoleucine (Castresana, Feldmaier-Fuchs and Pääbo, 1998; Perseke et al., 2011) and the unique anticodon sequence CUU for the tRNALys support Ambulacraria including Pterobranchia (Perseke et al., 2011). Several histone variants from *Balanoglossus clavigerus* could be determined from polyadenylated EST data comprising sequences with similarities to cleavage specific (CS) histone variants of H2B and H3. The cleavage specific histone genes have been described only from sea urchins (Echinodermata). The presence of histone variants in Hemichordata orthologous to the CS histones in sea urchins constitutes an additional strong autapomorphy for the taxon Ambulacraria (Perseke et al., unpubl.).

17.3.1 Echinodermata

The Echinodermata are a large invertebrate taxon with more than 7,000 described recent species and twice as many fossil ones (Pawson, 2007). Easily recognizable due to their pentaradial symmetry and the possession of a calcareous mesodermal skeleton it is not surprising that the monophyly of this group is undisputed (Ax, 2001). Sequence analyses of mitochondrial genomes and the assignment of the codon "AGA" to asparagine further support echinoderm monophyly (Perseke et al., 2011). Within extant echinoderms, five major taxa are distinguished: Crinoidea, Echinoidea, Ophiuroidea, Asteroidea, and Holothuroidea. The monophyly of each of these taxa is well supported and also substantiated by both mitochondrial genome sequences and gene

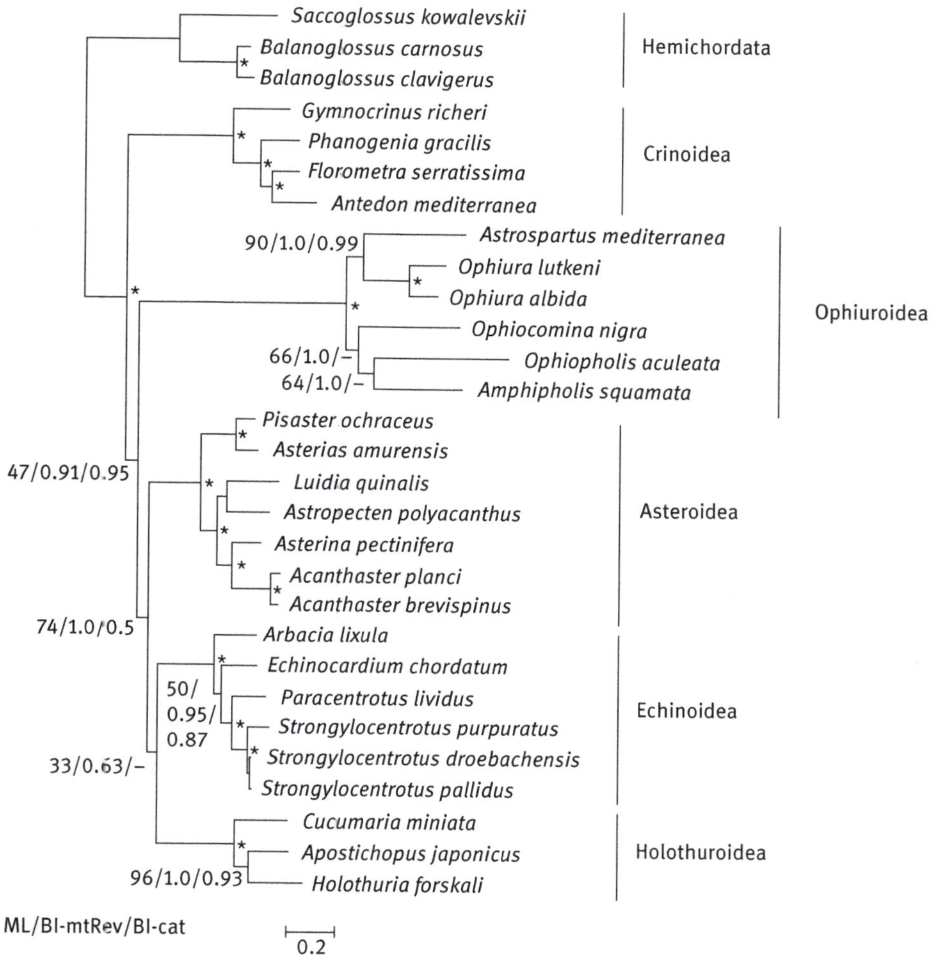

Figure 17.2: Maximum-likelihood tree of the amino acid sequences of all thirteen protein coding genes using the taxon set "Hemichordata". The numbers on the nodes show the bootstrap percentages (BP) for ML, the posterior probabilities (PP) of BI-mtRev and BI-cat in this order. Asterisks indicate highest support values obtained with all methods (Perseke et al., 2010).

order evolution (Perseke et al., 2010; 2011). However, the relationships between them are far from being settled. From a morphological point of view there is strong support for a sister-group relationship of crinoids representing the taxon Pelmatozoa, and the other four groups forming the Eleutherozoa (Nielsen, 2012). The results of molecular analyses however remained ambiguous. Maximum Likelihood and maximum parsimony analyses of 28S and 18S rRNA gene sequences (ca. 4,000 nucleotides) recovered the morphology-based division into Crinoidea and Eleutherozoa, and of the latter into Asterozoa (Asteroidea and Ophiuroidea) and Echinozoa (Echinoidea and Holothuroidea) (Littlewood et al., 1997; Mallatt and Winchell, 2007). In mitochondrial genome comparisons the classical phylogeny was recovered in Maximum Likelihood and Bayesian analyses when hemichordates were used as outgroup representatives. However, an inconsistent picture arose when vertebrates were used as outgroups in maximum parsimony and neighbor joining reconstructions (Perseke et al., 2010). In the currently most comprehensive data set, the Ophiuroidea branched off first in a well-supported clade (Perseke et al., 2011). Within the remaining four taxa most analyses suggested a split between Crinoidea and the Asteroidea, Echinoidea, and Holothuroidea. The resolution of that group remained unstable, depending on the different data sets used, i.e. from the taxon sampling (Perseke et al., 2011), although earlier studies on mitochondrial gene order evolution had supported the clades Asterozoa and Echinozoa (Littlewood et al., 1997; Smith et al., 1993). The analyses of Perseke et al. (2010; 2011) detected the highest mtDNA substitution rates in Ophiuroidea. The largest number of gene order rearrangements and large amounts of unassigned sequences were also detected in brittle stars, suggesting a higher evolutionary rate compared to all other echinoderm taxa. Fast evolving lineages are difficult to handle in phylogenetic analyses and tend to a basal branching even with high support (Scouras et al., 2004). A different feature in the Crinoidea mitochondrial genomes may also complicate phylogenetic analyses. A gene inversion including the origin of replication lead to opposed directed strand specific nucleotide pressure (Scouras and Smith, 2006) and adaptation of the protein genes to this bias with subsequent loss of phylogenetic signal.

17.3.2 Hemichordata

The Hemichordata are a small taxon including approximately 130 described species, containing about 100 solitary-living enteropneusts and around 30 colonial-living pterobranchs (Cameron, Garey and Swalla, 2000; Zhang, 2011). Despite their key role for understanding deuterostome evolution, hemichordate phylogenetic relationships and evolution are poorly understood and molecular data covering both subgroups were until now limited to rRNA sequence comparisons. Previous studies based on conserved 28S rRNA regions suggested that enteropneusts are monophyletic (Winchell et al., 2002), whereas 18S rRNA sequence comparisons suggested paraphyletic enteropneusts (e.g. Bourlat et al., 2003). An increased number of hemichordate taxa used in phylogenetic analyses of 18S rRNA gene sequences divided enteropneusts into two distinct clades, with pterobranchs grouping with Harrimaniidae (Figure 17.3) (Cannon et al., 2009). Following Cannon et al. (2009) this indicates that colonial pterobranchs possibly evolved from a solitary acorn worm-like ancestor. Consequently pterobranchs cannot constitute an ancestral form of deuterostomes like often supposed (Cannon et al., 2009).

This issue is far from being solved, however, since hemichordate phylogeny still relies solely on rRNA data and additional information from other molecular markers

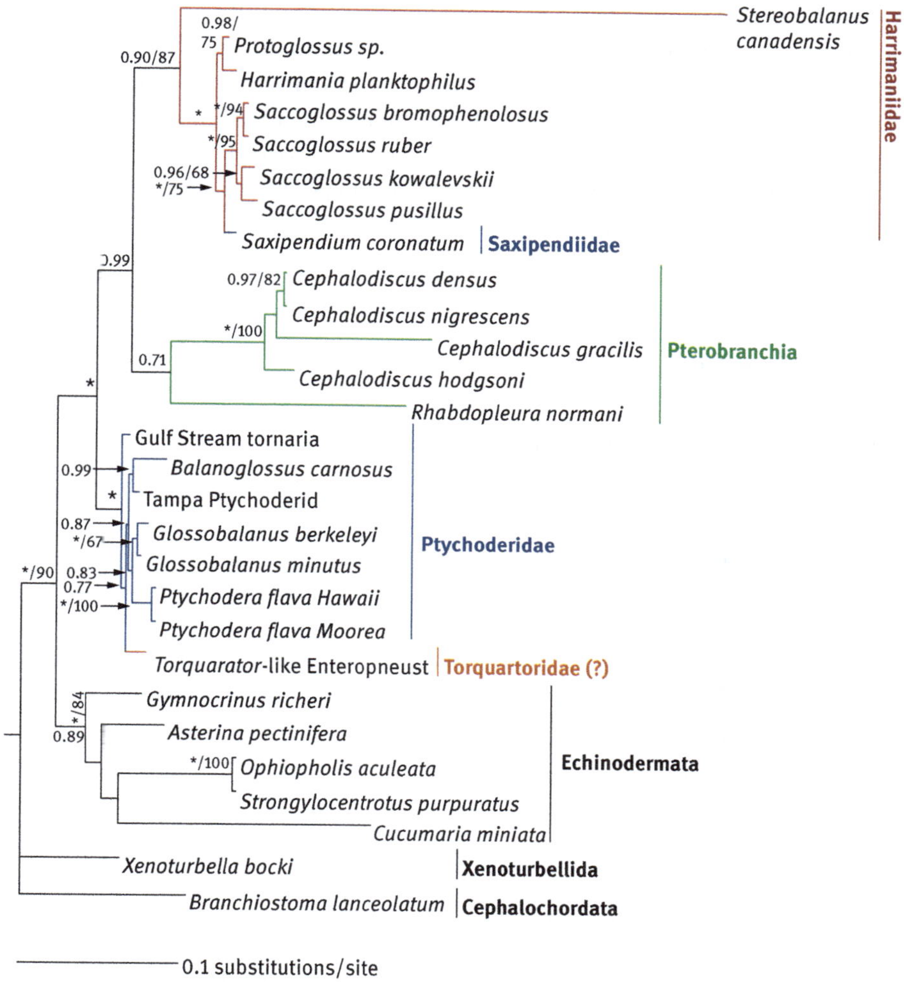

Figure 17.3: 18S rDNA tree of the Hemichordata. Bayesian tree is shown with posterior probabilities greater than 0.60 and Maximum Likelihood bootstrap values above 60 indicated above the nodes. Bayesian posterior probabilities of 1.00 are indicated by asterisks. Methodological details are given in the text (Cannon et al., 2009).

is scarce. Three complete mitochondrial genomes are available from enteropneusts, and only one was determined for the pterobranch *Rhabdopleura compacta* (Perseke et al., 2011). Although corresponding in size and gene content to typical metazoan mitochondrial genomes, it showed the strongest strand-specific mutational bias in the nucleotide composition among deuterostomes with a very GT-rich coding strand. The protein coding genes have been highly affected by a strand-specific mutational pressure showing unusual codon frequency and amino acid composition, causing extremely long branches in phylogenetic analyses. Similar to the case of crinoids, this strong reversed asymmetrical mutational constraint may have originated from an inversion of a genomic region including the replication origin and adaptation of the protein genes to this bias, hampering phylogenetic analyses. As a prime example for misplacement in a phylogenetic tree due to long-branch attraction, *R. compacta* groups within Echinodermata to the Ophiuroidea (Figure 17.4) (Perseke et al., 2011). However, as already pointed out, features of the tRNA system of *R. compacta* support monophyly of Ambulacraria (Perseke et al., 2011).

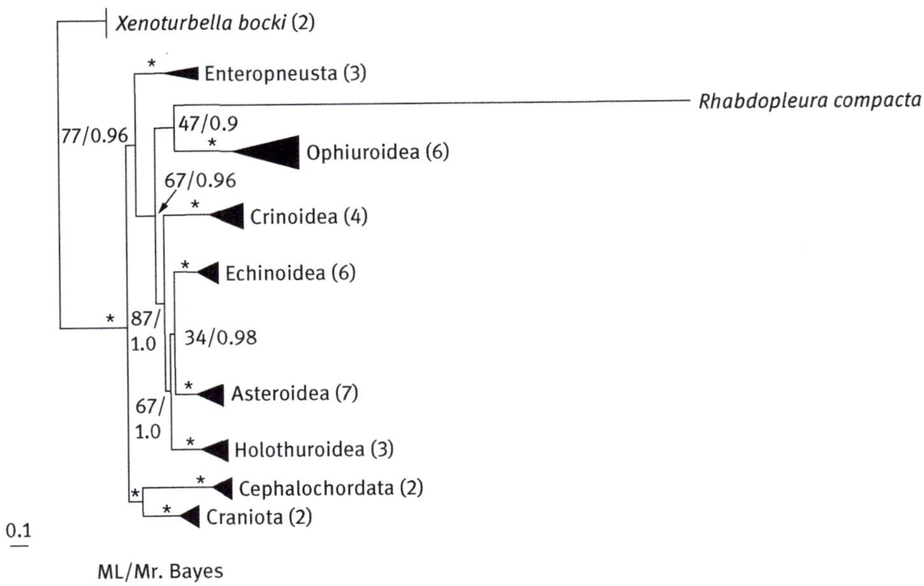

Figure 17.4: Maximum-likelihood tree of the major deuterostome groups excluding tunicates using the amino acid sequences of all thirteen mitochondrial protein-coding genes. The numbers on the nodes show the bootstrap percentages (BP) for ML and the posterior probabilities (PP) of the Bayesian analyses in this order. Asterisks indicate highest support values. The numbers in parenthesis behind the deuterostome groups show the number of analyzed species. The length of the triangles for each group indicates the variability within the group. The extreme long branch leading to the pterobranch *R. compacta* is based on its highly different protein-composition compared to other deuterostome sequences (Perseke et al., 2011).

17.4 Phylogeny of Chordata

The relationships within Chordata seemed almost to be written in stone. Most morphological as well as early molecular phylogenetic analyses converged to the hypothesis that Tunicata are sister-taxon of the Cephalochordata + Vertebrata (e.g. Cameron, Garey and Swalla, 2000; Turbeville, Schulz and Raff, 1994). This view was dramatically challenged by phylogenomic analyses based on large EST data sets, where strong support was given for a sister-group relationship of tunicates and vertebrates (Bourlat et al., 2006; Delsuc et al., 2006; 2008, Dunn et al., 2008; Hejnol et al., 2009). Additional support for this hypothesis came also from a data set based on mitochondrial genomes (Singh et al., 2009). The term Olfactores for this clade was proposed by Jefferies (1991) based on fossils exhibiting a putative homologous olfactory apparatus.

The assumption of a sister-group relation of Cephalochordata and Vertebrata also rendered the evolution of the Hox clusters problematic: While all anterior and middle group genes are clearly orthologous in Vertebrata, Tunicata, Cephalochordata, and Ambulacraria, this is not the case of the posterior group. However, the Olfactores hypothesis accommodates the orthology of vertebrate and tunicate posterior Hox genes and indicates an independent expansion of the HOX clusters in Olfactores, Cephalochordata, and Ambulacraria (Feiner et al., 2011; Freeman et al., 2012; Fritzsch et al., 2008; 2010; Ikuta, 2011; Thomas-Chollier et al., 2010). The inclusion of the Xenacoelomorpha, on the other hand, postulates a drastic reduction in the Hox gene system relative to the protostome-deuterostome ancestor, since the orthology of the anterior paralog groups 1–5 between protostomes and deuterostomes is well established (Figure 17.5) (de Rosa et al., 1999; Thomas-Chollier et al., 2010).

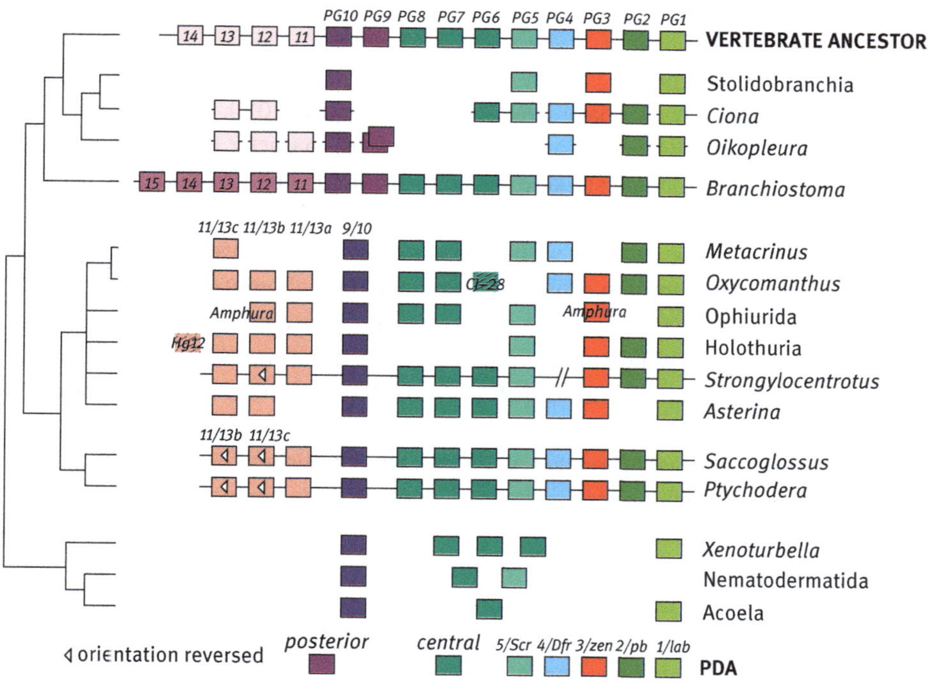

Figure 17.5: Evolution of Hox gene clusters in deuterostomes, updated from Fritzsch et al. (2008) by including the hemichordate Hox clusters (Freeman et al., 2012) and the Hox genes of acoels (Jimenez-Guri et al., 2006, Moreno, Permanyer and Martinez, 2011).

17.4.1 Cephalochordata

This taxon comprises some 30 described species (Ax, 2001). Cephalochordata have long been known as "living fossils" representing many anatomical features (autapomorphies) of the last common ancestor of deuterostomes, such as the dorsal neural tube, mesodermal notochord, anterior gut with gill slits, somites and a post-anal tail. Within cephalochordate, phylogeny has received little attention from the point of view of molecular data beyond a few limited studies using mitochondrial genomes (Kon et al., 2007; Xiao et al., 2008). A comprehensive analysis using different molecular markers seems to be missing. On the other hand, the nuclear genome of *Branchiostoma floridae* has been sequenced (Putnam et al., 2008). This study also addressed the phylogenetic position of cephalochordates within chordates. In both, Bayesian and Maximum Likelihood analyses of 1090 orthologous genes *B. floridae* branched off before tunicates and vertebrates. An analysis of intron gain and loss patterns provided independent evidence for a sister-group relationship of tunicates and chordates to the exclusion of cephalochordates (Putnam et al., 2008).

17.4.2 Tunicata

Tunicata (a.k.a. Urochordata) comprise a group of peculiar marine invertebrates and nearly 2,800 extant species have been described (Zhang, 2011). The monophyly of tunicates is well established. A particularly intriguing character is the ability of cellulose biosynthesis which most likely has been acquired via horizontal gene transfer in the last common ancestor of extant tunicates (Matthysse et al., 2004). Due to their important phylogenetic position within deuterostomes they have been regarded as a key to understand chordate evolution ever since their chordate ancestry was proposed by Kowalewsky (1866). Traditionally, Tunicata were classified into three major clades: Ascidiacea, Appendicularia, and Thaliacea (Nielsen, 2012). However, phylogenetic analyses of molecular as well as combined molecular and morphological data sets rejected this hypothesis (Stach and Turbeville, 2002; Tsagkogeorga et al., 2009). According to Tsagkogeorga et al. (2009), who published the most comprehensive tunicate phylogeny, tunicates are divided into a clade composed of members of the Phlebobranchia + Thaliacea + Aplousobranchia, an Appendicularia clade, and a Stolidobranchia clade. Consistent with these findings, analyses of tunicate mitochondrial genomes show a similar picture (Stach, Braband and Podsialowski, 2010). However, many tunicate taxa have an elevated rate of sequence evolution and long branches make molecular systematic analyses of this taxon difficult. As such, phylogeny of higher ranked tunicate taxa, especially those of Appendicularia, should be treated with caution.

17.4.3 Vertebrata

While it is largely undisputed that gnathostomes (jawed vertebrates) form a monophyletic clade, two of the three possible relationships among the lamprey (Hyperoartia), hagfish (Hyperotreti) and gnathostome lineages have been the subject of intense debate. Comparative morphology tends to support a sister-group relationship of lampreys and gnathostomes, while both conventional molecular phylogenies using protein coding sequences (Delarbre et al., 2002; Furlong and Holland, 2002; Kuraku et al., 1999) and the existence of exclusively shared microRNAs (Heimberg et al., 2010) unites lampreys and hagfishes in the clade Cyclostomata. New insights into their development revealed that hagfishes secondarily lost many vertebrate characteristics including vertebrae, thus undermining the morphological support for a basal position of hagfishes and the existence of a lamprey-gnathostome clade (Ota and Kuratani, 2007; Shimeld and Donoghue, 2012). The sister-group relation of cyclostomes and gnathostomes, and hence the synonymity of Vertebrata and Craniata, appears to be widely accepted now.

The genomes of extant gnathostomes share two rounds of whole genome duplications, the first of which can be traced to the stem lineage of the Vertebrata (Dehal and Boore, 2005; Putnam et al., 2008). The timing of the second round has long been the topic of intense debate (Kuraku, Meyer and Kuratani, 2009; Kuraku, 2010). Sequencing the genome of lamprey *Petromyzon marinus* has provided strong evidence that the lamprey and vertebrate lineages share both the 1st and the 2nd whole genome duplication event, although less parsimonious alternatives cannot be ruled out completely (Smith et al. 2013). Decisive conclusions are complicated by the massive reductions of genomic DNA in lamprey and hagfish development (Smith et al., 2009, 2012; Smith, Saha and Amemiya, 2010).

Within the gnathostomes, the phylogenetic relations of most of the major groups are firmly established and largely undisputed (Chen et al., 2012). A major unresolved question concerns basal structure of the sarcopterygian branch, i.e., the relationship of the tetrapod, lungfish, and coelacanth lineages. Analyses of small amounts of sequence data favor a lungfish+tetrapod clade but failed to convincingly exclude alternatives (Brinkmann et al., 2004; Shan and Gras, 2011; Takezaki et al., 2004).

The sequencing of the coelacanth genome and of the transcriptome of the lungfish *Protopterus annectens* has provided unambiguous support for the lungfish-tetrapod dichotomy and firmly places the coelacanth as the earliest branch in sarcopterygian lineage (Amemiya et al., 2013).

17.5 Outlook

Monophyly of deuterostomes and their major subgroups were confirmed by molecular data, and plausible evidence for the decision of current issues emerged, such as the monophyly of Ambulacraria. Some new and challenging results concern the inclusion of a taxon Xenacoelomorpha as sister-group to Ambulacraria or the basal split of Cephalochordata in the chordate lineage. Other issues remain open, such as the relationships between echinoderm clades or the phylogenetic position of pterobranchs within Ambulacraria. Despite the enormous increase of molecular genomic data sets, some problems in phylogeny reconstruction have not been solved, such as fast rates in different lineages as shown for the crinoids in echinoderms, or for tunicates within chordates. Mechanisms for these different evolutionary rates at least could be discussed for mitochondrial genomes. In the future, not only more evidence from improved taxon sampling, e.g. in pterobranchs, but also from genomic analyses and structural features of genomes, such as gain and loss patterns of introns or microRNAs will help to shed further light on deuterostome phylogeny and evolution.

Acknowledgments

This work was supported by the German Research Foundation (DFG) through the project "Deep Metazoan Phylogeny" within SPP 1174 (projects STA 850/3-2, STA 850/2, SCHL 229/14-1), and with extra funds from the rectorate of the University of Leipzig.

Thomas Stach

18 Deuterostome phylogeny – a morphological perspective

Abstract: Deuterostomia is one of the major monophyletic groups in the animal kingdom. Interest in deuterostome evolution stems from the diversity of life cycles and anatomies present in this group and moreover from the fact that vertebrates are part of Deuterostomia. Phylogeny is a prerequisite for the understanding of evolution. In the present review I critically assess our current understanding of deuterostome phylogeny based on molecular and morphological data. Conflict in phylogenetic signal in molecular phylogenies is revealed utilizing data display networks and split analyses. Phylogenetic argumentation supports the hypothesis that the fossil *Pikaia gracilens* is a member of the stem lineage of Craniata. A first experimental data matrix based on "evo-devo" characters is presented and it is demonstrated that such a cladistic approach holds great potential for phylogenetic analysis of molecular developmental studies. The phylogenetic analysis of morphological characters is in conflict with several aspects of molecular phylogenies but more consistently explains the evolution of phenotypes based on a paradigm of descent with modification. An evolutionary scenario based on the phylogeny derived from the morphological data is presented.

18.1 Introduction

18.1.1 Diversity fascinates, similarities inform

As soon as one examines the diversity of metazoans, one simultaneously marvels at the extraordinary and exquisite disparity of animals and cannot but realize patterns of graded similarities between species. These patterns of graded similarities become obvious, independent of the biological level of comparison. The pattern may appear easy to describe in numerical terms if only four different character types are compared as in the case of nucleotides, the pattern can be technically difficult to reveal as in the case of gene regulatory networks during ontogenies, or the pattern might be found in complex, intricate, often complicated structures composed of billions of substructures like in many morphological characters. Morphological characters are essential in functional considerations concerning tissues and organs, in identifications of fossilized structures, or in elucidating evolutionary character transformations.

Because of its long scientific tradition, morphology has accumulated knowledge about numerous characters that are in addition potentially phylogenetically informative, although the use of morphological characters in clarifying phylogenetic patterns is currently controversial (Scotland, Olmstead and Bennett, 2003; Tautz, 2006;

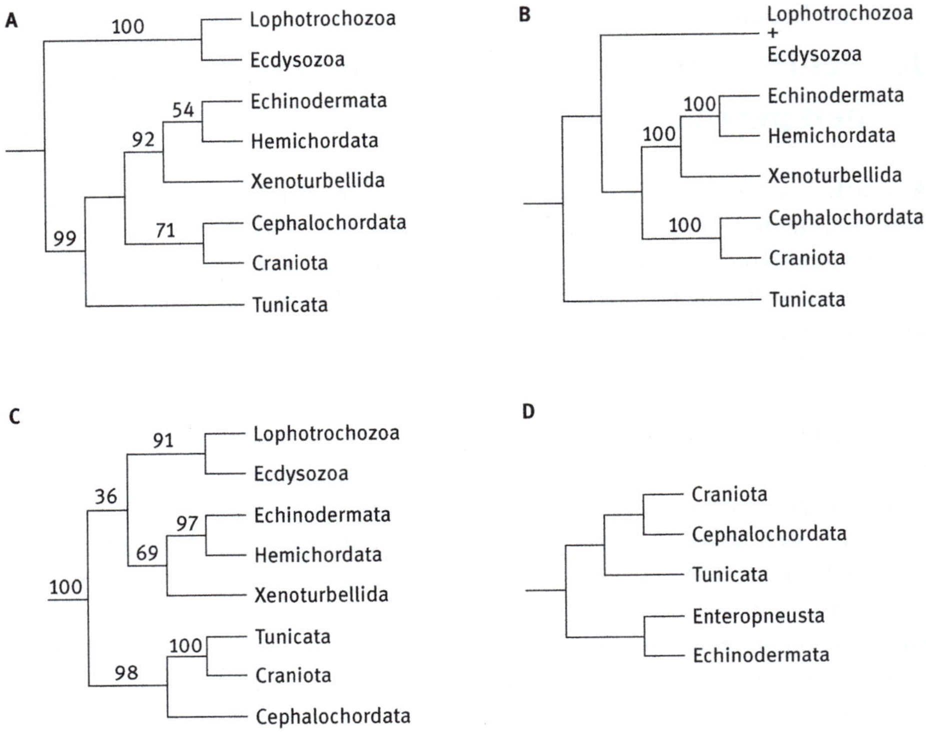

Figure 18.1: Four differing molecular phylogenies of deuterostomes. **A.** Phylogenetic relationship of higher metazoan taxa after Bourlat et al. 2003, Figure 1. Values are bootstrap percentages. **B.** Phylogenetic relationship of higher metazoan taxa after Bourlat et al. 2006, Figure 1C. Values are bootstrap percentages. **C.** Phylogenetic relationship of higher metazoan taxa after Delsuc et al. 2008, Figure 2B. Values are bootstrap percentages. **D.** Phylogenetic relationship of higher metazoan taxa after Tsagkogeorga et al. 2009, Figure 4. Statistical support values were only given for internal tunicate relationships.

Giribet, 2010; Marlétaz and Le Parco, 2010; Ragsdale and Baldwin, 2010). It is often claimed that phylogenetic analysis of morphological characters is subjective and that mapping morphological characters on "objective" molecular phylogenies explains character evolution in a better way (Scotland, Olmstead and Bennett, 2003; Bleidorn, 2007; Kocot et al., 2011). However, in my opinion this view suffers from at least two major erroneous conceptions. First, it assumes, often tacitly, that molecular phylogenetic hypotheses are superior to morphology-based phylogenies. This assumption is difficult to substantiate if one recalls the differing hypotheses resulting from different molecular phylogenetic studies (see Figure 18.1 for an example concerning deuterostomes; Turbeville, Schulz and Raff, 1994; Bourlat, et al., 2003; 2006; Delsuc et al., 2006; 2008; Goloboff et al., 2009; Tsagkogeorga et al., 2009). Despite the seeming simplicity of nucleotide or amino acid data, the reasons for discrepancies are manifold and therefore difficult to isolate. Reasons for divergent phylogenies comprise different alignment algorithms, including various reasoning to exclude "ambiguously aligned" sections of the data (Ogden and Rosenberg, 2006; Morrison, 2009; Blair and Murphy, 2011). Other reasons might encompass usage of different models for phylogeny reconstruction. It has been shown for example that the choice of evolutionary model crucially influences the outcome of molecular phylogenetic studies (Kolaczkowski and Thornton, 2004; Kumar et al., 2012). Of course, the fit of the model used in the analysis and the real, unique, and idiosyncratic evolutionary process remains principally elusive. Another issue that becomes increasingly scrutinized is the quality of the molecular data (Wägele, 2001; Wägele and Mayer, 2007). Not the quality of the practical research results is meant here. This is usually exquisite. The quality of the data in the present context means their intrinsic quality of phylogenetic informativeness. During the last couple of years several software utilities have been developed (Wägele and Mayer, 2007; Morrison, 2010; Kück, 2012; Mao et al., 2012) with the explicit goal to reveal specifically this potentially confounding problem and the results are sobering. Especially the large "genomic" data sets used for phylogenetic inferences seem to be relatively noisy and inferences become challenging, but – what is worrisome – at the same time highly supported by statistical values (Kück et al., 2012).

Secondly, mapping is not using morphological data to their full extent (Assis and Rieppel, 2011) while simultaneously retaining the supposedly subjective element of morphological analyses. Morphological structures integrate numerous ontogenetic interactions at different biological levels and each recognizable and heritable character contains potentially phylogenetic information (Hennig, 1950; Kitching et al., 1998; Wiley and Lieberman, 2011). This information is used if it is coded in the form of a cladistic data matrix and analyzed under an optimality criterion to result in a cladogram. The coding of characters is the crucial step in such an analysis (Pleijel, 1995; Winther, 2009). The information coded in a data matrix is equal to the pattern of character distribution across the investigated taxa. And the resulting cladogram is merely a different representation of the same information content. The coding in essence is the suggestion of homology of character states compared between taxa. Thus, suggesting specific hypotheses of homology is the crucial central step in a cladistic analysis (de Pinna, 1991; Wägele, 2001; Wiley and Lieberman, 2011). And it is exactly this key step of a cladistic phylogenetic analysis that is retained in character mapping: characters of different taxa mapped on a molecular phylogeny are supposed to be homologous. Yet, in case the mapped character does not fit the molecular phylogeny, this contradictory evidence is not considered as a potential falsifier of the molecular phylogenetic hypothesis. Or in other words: relevant information for a scientific hypothesis is more than neglected, it is dismissed.

In the following, I will argue that considering morphology not only visualizes evolutionary transformations based on molecular data but that morphology also improves phylogenetic analyses.

18.1.2 Morphology improves phylogenetic studies

In the preceding introduction I have alluded to the discrepancies found in different molecular phylogenetic studies. Figure 18.1 shows a compilation of such divergent results concerning large-scale deuterostome evolution published recently. Examining these cladograms reveals that there are repeatedly difficulties in recovering Chordata monophyletic or Deuterostomia monophyletic, despite attempts to present current molecular phylogenies as more or less consensual (Delsuc et al., 2006; Swalla and Smith, 2008). Several lines of evidence show that Tunicata (often called Urochordata) is a rogue taxon that appears at different positions in the cladogram (Zhong et al., 2009; Tsagkogeorga et al., 2010), disrupting even widely-believed true monophyletic groups such as Chordata or Deuterostomia: Tunicata groups together with nematodes (Goloboff et al., 2009), as sister-taxon to the remaining Bilateria (Bourlat et al., 2006), as sister-taxon to Craniota (Delsuc et al., 2006; 2008; in the former study Chordata is not monophyletic, in the latter study Deuterostomia is not monophyletic), or, as traditionally held, as sister-taxon to Notochordata (Tsagkogeorga et al., 2009; Tatián et al., 2011).

The accumulation of enormous amounts of molecular data in recent years has prompted attempts to evaluate data quality when such large amounts of data are used for certain goals. For example, software packages have been designed to evaluate phylogenetic information content or amount of conflicting signal in data utilized for phylogenetic analyses (Wägele and Mayer, 2007; Morrison, 2010; Kück, 2012). As an example, Figure 18.2 shows a network analysis and a split decomposition analysis of one of the so-called phylogenomic data sets used for large-scale animal phylogeny with a focus on deuterostome relationships, published by Philippe and co-authors (Philippe et al., 2011b). These analyses demonstrate that conflicting information is a

Figure 18.2: Exploratory data analysis of phylogenomic data from Philippe et al. (2011b) utilizing Splitstree 4 (Huson and Bryant, 2008) and SAMS (Wägele and Mayer, 2007). **A.** Network analysis. Numbers close to edges correspond to bars in the SAMS-Analysis depicted in B. Colors highlight deuterostome taxa (according to Philippe et al. 2011b). **B.** Split composition. Numbers refer to the following monophyletic groupings supported and present in the preferred phylogenetic hypothesis and marked in the network analysis depicted in A. 2: *Flaccisagitta enflata* + *Spadella cephaloptera*; 3: *Amphimedon* queenslandica + *Suberites domuncula*; 4: *Hydra magnipapillata* + *Hydractinia echinata*; 6: *Brachionus plicatilis* + *Philodina roseola*; 7: *Convolutriloba longifissura* + *Symsagittifera roscoffensis*, 11: *Danio rerio* + *Eptatretus burgeri* + *Gallus gallus* + *Leucoraja erinacea* + *Petromyzon marinus* + *Xenopus tropicalis, 26: Flaccisagitta enflata* + *Isodiametra pulchra* + *Spadella cephaloptera* + *Symsagittifera roscoffensis*; 13: *Meara stichopi* + *Nemertoderma westbladi*, 17: *Macrostomum lignano* + *Paraplanocera* sp.; 20: *Ptychodera flava* + *Saccoglossus kowalevskii*; 23: *Bugula neritina* + *Cristatella mucedo*; 24: *Pedicellina cernua* + *Symbion pandora*; 30: *Convolutriloba longifissura* + *Isodiametra pulchra* + *Philodina roseola* + *Symsagittifera roscoffensis*; 32: *Helobdella robusta* + *Tubifex tubifex*; 35: *Convolutriloba longifissura* + *Isodiametra pulchra* + *Meara stichopi* + *Symsagittifera roscoffensis*; 37: *Danio rerio* + *Gallus gallus* + *Leucoraja erinacea* + *Xenopus tropicalis*; 47: *Gryllus bimaculatus* + *Rhodnius prolixus*.

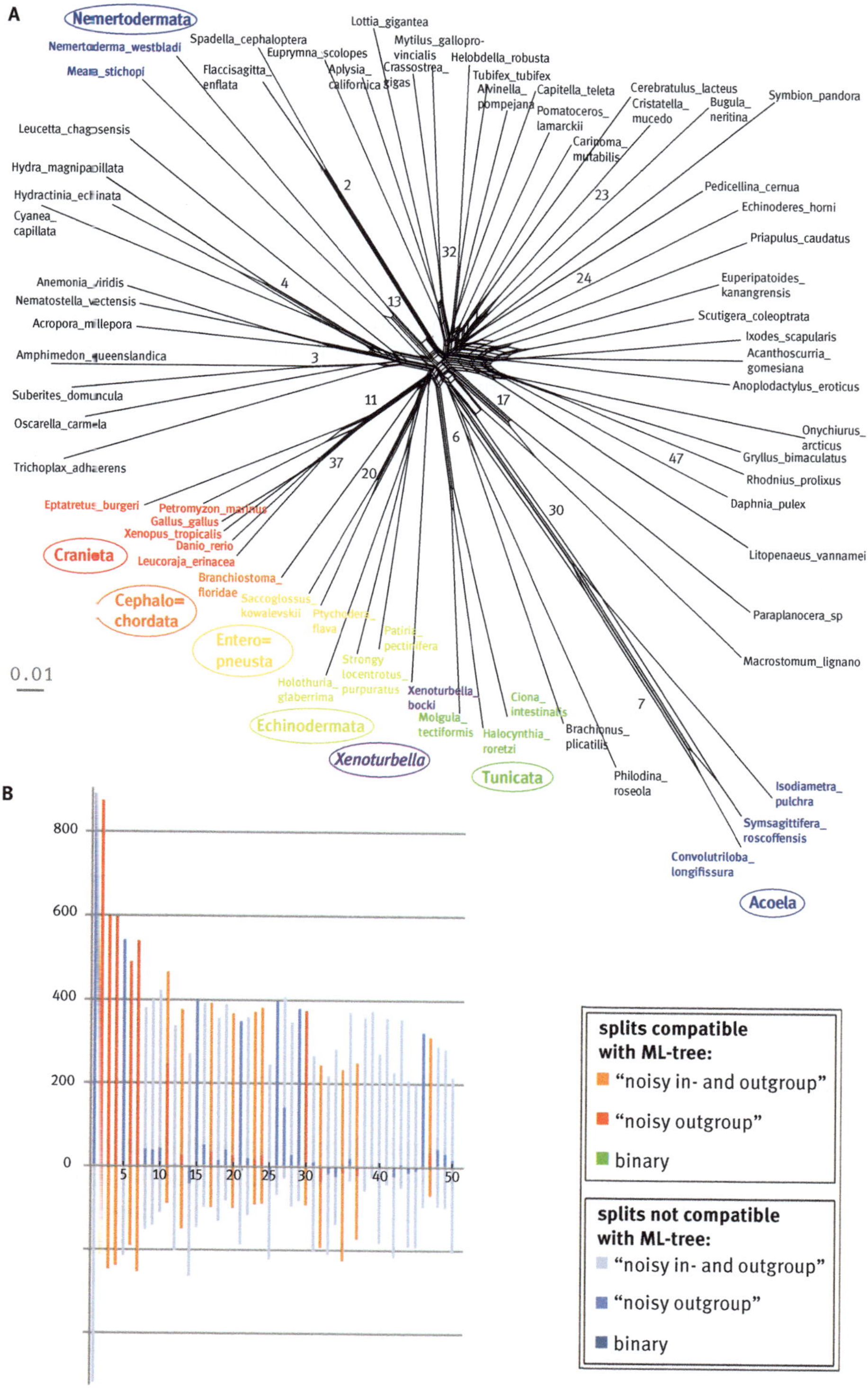
A
Nemertodermata
Nemertoderma_westbladi
Meara_stichopi
Spadella_cephaloptera
Flaccisagitta_enflata
Euprymna_scolopes
Aplysia_californica
Lottia_gigantea
Mytilus_galloprovincialis
Crassostrea_gigas
Helobdella_robusta
Tubifex_tubifex
Alvinella_pompejana
Capitella_teleta
Pomatoceros_lamarckii
Cerebratulus_lacteus
Cristatella_mucedo
Carinoma_mutabilis
Bugula_neritina
Symbion_pandora
Leucetta_chagosensis
Hydra_magnipapillata
Hydractinia_echinata
Cyanea_capillata
Pedicellina_cernua
Echinoderes_horni
Priapulus_caudatus
Anemonia_viridis
Nematostella_vectensis
Acropora_millepora
Amphimedon_queenslandica
Suberites_domuncula
Oscarella_carmela
Trichoplax_adhaerens
Euperipatoides_kanangrensis
Scutigera_coleoptrata
Ixodes_scapularis
Acanthoscurria_gomesiana
Anoplodactylus_eroticus
Onychiurus_arcticus
Gryllus_bimaculatus
Rhodnius_prolixus
Daphnia_pulex
Litopenaeus_vannamei
Paraplanocera_sp
Macrostomum_lignano
Eptatretus_burgeri
Petromyzon_marinus
Gallus_gallus
Xenopus_tropicalis
Danio_rerio
Leucoraja_erinacea
Craniota
Branchiostoma_floridae
Cephalo=chordata
Saccoglossus_kowalevskii
Ptychodera_flava
Entero=pneusta
Patiria_pectinifera
Strongylocentrotus_purpuratus
Holothuria_glaberrima
Echinodermata
Xenoturbella_bocki
Xenoturbella
Molgula_tectiformis
Halocynthia_roretzi
Ciona_intestinalis
Tunicata
Brachionus_plicatilis
Philodina_roseola
Isodiametra_pulchra
Symsagittifera_roscoffensis
Convolutriloba_longifissura
Acoela
2
32
13
4
3
23
24
11
17
6
37
20
47
30
7
0.01
B
800
600
400
200
0
5
10
15
20
25
30
40
50
splits compatible with ML-tree:
"noisy in- and outgroup"
"noisy outgroup"
binary
splits not compatible with ML-tree:
"noisy in- and outgroup"
"noisy outgroup"
binary

rampant problem in such enormous data sets and that the data set is not well structured; something one would expect if the phylogenetic signal were well preserved. Thus, at least currently, these large data sets are problematic in phylogeny reconstructions and it is encouraging that computer scientists and theoreticians are developing tools that on the one hand allow the evaluation of data quality before an analysis (Wägele and Mayer, 2007; Morrison, 2010; Trachana et al., 2011; Kück, 2012) and that will eventually enable researchers to include more levels of information that are present in their data, such as gene order (Bernt et al., 2007; Bernt and Middendorf, 2011), genome organization (Proost et al., 2009; Sims et al., 2009; Joseph, 2012), rare genomic events (Sims et al., 2009; Belinky, Cohen and Huchon, 2010; Telford and Copley, 2011), presence of micro-RNAs (Sperling and Peterson, 2009; Sperling et al., 2009), gene regulatory networks (Struckmann et al., 2008; Hughes and Friedman, 2009; Cross, Buchler and Skotheim, 2011; Fuellen, 2011; Krakauer et al., 2011), and that are currently not utilized to their full utility for phylogenetic inferences.

While analysis of morphological data suffers from other – sometimes severe – problems, such as complexity (Richter, 2005; Scholtz, 2005; 2010), mastering of specialized knowledge (de Carvalho and Ebach, 2009; Mooi and Gill, 2010), the "linguistic problem" (Vogt, Bartolomaeus and Giribet, 2010), and unclear data standards (Vogt, 2009; Deans, Yoder and Balhoff, 2012), the phylogenetic signal present in a data matrix based on morphological data is usually clearer as can be seen in spectral analyses of such data matrices (Stach, Braband and Podsiadlowski, 2010). Similarly, while in combined analyses morphological characters are usually severely outnumbered compared to nucleotide or amino acid positions, morphological characters disproportionally contribute to the outcome of the respective phylogenetic analysis (Giribet, 2010). If analyzed separately, morphological data and molecular data can either arrive at the same phylogenetic conclusion or contradict each other. While, in a Popperian sense agreement might convey knowledge about the phylogeny of the groups in question, contradiction should not be glossed over, but seen as an opportunity to investigate the reasons for discrepancy and a fresh stimulus for acquisition of additional data (Popper, 1984).

18.1.3 Fossils

The paleontological record is often considered a lynchpin in evolutionary considerations, and indeed, paleontology is indispensable in adding an absolute time frame to our understanding of evolution. Otherwise, the contribution of paleontology for the reconstruction of phylogeny as the backbone of evolutionary considerations remains highly controversial (Sudhaus, 2006). Certainly paleontologists do have a clear disadvantage compared to neontologists: they are interpreting (mostly) rocks. This of course means that many of the scientific tools that help neontologists to understand their research objects (such as ethology, histology, physiology, etc.) are to a large extent unavailable to a paleontologist. Still, empirical studies have shown that including fossils in phylogenetic data matrices can have an important influence on the outcome of a cladistic analysis (Donoghue et al., 1989; Cobbett, Wilkinson and Wills, 2007; Hermsen and Hendricks, 2008). However, while these studies are convincing under their respective premises, paleontology is also plagued with another disadvantage. The interpretation of the paleontological objects, the fossils, is necessarily always based on the morphology of living organisms. In cases where the fossil remains are quite similar to recent organisms the interpretation of the fossils seems straightforward, yet for other fossils the interpretation becomes difficult. This is the case for many of the fossils that have been claimed to be the earliest chordate, the earliest tunicate, or the earliest deuterostome (Donoghue and Purnell, 2009).

As an illustrative example, the case of *Yunnanozoon lividum* (Chen, 1995) might be instructive. With an age of 525 Million years *Y. lividum* was described in 1995 by Chen and co-authors as the oldest fossil remain of a chordate from the renowned Chengjiang-Lagerstätte in China (Chen, et al., 1995). Amongst its chordate features the authors suggested the presence of branchial arches, notochord, and endostyle. One year later Shu and co-authors (Shu, Zhang and Chen, 1996) interpreted the fossil remains of *Y. lividum* as a clearly tripartite member of Hemichordata, with "affinities" to enteropneusts. The key difference for the reinterpretation of *Y. lividum* in the two studies lay in the interpretation of a fossil element in Chen et al.'s study as a notochord and in Shu et al.'s study as a gut. Another interpretation of *Y. lividum*, published by Dewel (2000), emphasized similarities to Ediacaran fossils and to pennatulacean octocorals and concluded that *Y. lividus* should be interpreted as the sister-taxon to Bilateria. While this re-positioning of *Y. lividus* to different branches of the tree of life can be supported in the respective frames of interpretation of the different authors, the factual bases for their hypotheses would be easily testable in modern animals. Simple histology would suffice to distinguish between chordate notochord, enteropneust gut, or cnidarian-like rachis – alas this research avenue cannot be travelled and therefore uncertainty remains.

The unusual asymmetric fossils called calcichordates are a similar case, where the interpretation of substructures of the fossils as various anatomical features determines the phylogenetic hypothesis supported (Jefferies, 1986; Peterson, 1995; Dominguez Alonso and Jefferies, 2001; Clausen and Smith, 2005).

A different kind of problem associated with paleontological evidence can also be found among the many claims for the oldest fossil in deuterostomes, this time the oldest tunicate. In 2001, Shu and co-authors described a fossil, *Cheungkongella clava*, which they interpreted as a styelid ascidian (Shu et al., 2001). While the interpretation as a sessile ascidian seemed reasonable based on the examined material, other authors later claimed to have unearthed more complete fossils that in addition show a preserved tentacle apparatus. This finding prompted the authors to propose that *Cheungkongella clava* was a junior synonym of *Phlogites longus* and therefore probably a pterobranch (Chen et al., 2003) but see Xian-Guang et al. (2006).

One of the potential advantages of fossils is illustrated by *Herpetogaster collinsi*, a fossil that has been interpreted as a pterobranch (Caron, Conway Morris and Shu, 2010). If the interpretation of *H. collinsi* as a fossil pterobranch is correct, it does show two characteristics that are not found in present day pterobranchs. First, *H. collinsi* lacks a cephalic shield. Second, it shows segmentation. The lack of a cephalic shield seems to be an evident conclusion from the fossils (but remember the case of *Cheungkongella* now *Phlogites*: a more complete fossil might be waiting in Cambrian rocks), whereas the interpretation of repeated lines in the fossils as segments is not as straightforward (even when ignoring the difficulties of definition of "segment" in modern animals (e.g., Scholtz, 2002) for a moment). But, if we accept the interpretations of *H. collinsi* as a pterobranch-like animal and the lack of a cephalic shield, this fossil has the potential to alter our view of deuterostome evolution considerably, because it shows a character combination that is not found in modern animals. Accepting these premises would pose new questions regarding the evolution of a tripartite body organization and the evolution of a metameric segmentation. Both features present in combination in a pterobranch are not easily compatible with any of the currently competing phylogenetic hypotheses.

The examples above illustrate some of the problems and potentials associated with the fossil record. In order to avoid some of the pitfalls associated with fossils some researchers have argued that fossils should be placed on a phylogenetic hypothesis a posteriori based on explicit demonstration of synapomorphies for the respective stem lineage (Hennig, 1950; Sudhaus, 2006). This careful approach would then provide the minimum age of certain nodes in a cladogram ("terminus post quem non" of Hennig) as well as elucidating character evolution in cases where fossils show character combinations not found in present day representatives of the clades in question.

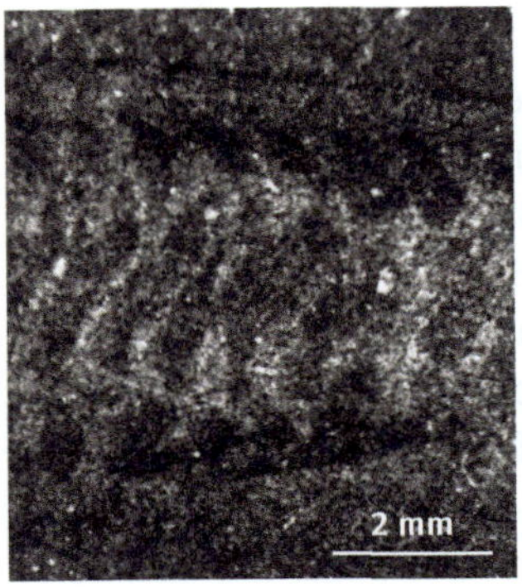

Figure 18.3: Sinusoid myomeres in the posterior part of the Cambrian fossil *Pikaia gracilens* (see also Conway Morris and Caron, 2012).

Using this approach it is possible to tentatively place one of what is recurrently called a key fossil of chordate evolution: *Pikaia gracilens*. While a detailed description of this fossil was promised over 30 years ago (Conway Morris and Whittington, 1979), it has only recently been published, while the current review was under review (Conway Morris and Caron, 2012). Personal examination of some of the fossils at the Smithsonian Institute (Washington, D. C.) revealed sinusoid shapes of the lines that are traditionally interpreted as remnants of myosepta[1] (Figure 18.3). The interpretation of these lines as myosepta is corroborated by the recent publication (Conway Morris and Caron, 2012). Detailed descriptions of the three-dimensional arrangement of the myosepta separating adjacent muscle segments are available for cephalochordates, basal agnathan craniates, and for osteichthyans. In the third recent chordate taxon, the tunicates, no myosepta exist (Burighel, Nunzi and Schiaffino, 1977; Burighel and Cloney, 1997; Stach and Kirbach, 2009). Instead, the tail muscles of appendiuclarians, of the tadpole larvae of ascidians and doliolids do not form muscle segments with a myocoel and separate cell types (dermatome and myotome) and the simple, mononucleate muscle cells border immediately to each other. While the myosepta in cephalochordates are somewhat cone-shaped (though slightly more complicated than an ice cream cone, Gemballa, Weitbrecht and Sanchéz-Villagra, 2003), all primarily aquatic craniates feature a more complicated myoseptum that is w-shaped in lateral aspect (Gemballa, 1998; Vogel and Gemballa, 2000). If the shallowly w-shaped lines in the fossil remains of *Pikaia gracilens* are indeed homologous remnants of myosepta, then these are clearly more similar to the w-shaped craniate structures than they are to the cone-shaped cephalochordate myosepta. Therefore, phylogenetic argumentation based on this structure as a synapomorphic character positions *P. gracilens* as a member of the stem lineage of craniates. This in turn means that the split between the stem lineage of cephalochordates and the stem lineage of craniates must be older than the age of the fossil *P. gracilens* that is dated to approximately 505 Million years (Collins, 2009). The marvelous and detailed description of *P. gracilens* (Conway Morris and Caron, 2012) contains a wider ranging discussion of the phylogenetic affinities of the fossils and prefers a position as a member of the chordate stem lineage while admitting uncertainties.

18.1.4 Molecular evo-devo results as cladistic characters

One of the most fascinating developments in recent years was the more widespread appearance of sophisticated molecular developmental studies in different (non-model) deuterostome organisms that informed us about the molecular mechanisms that regulate animal development (e.g., Sato, Bishop and Holland, 2008; Bertrand

1 It should be noted that species of the genus *Armandia* an opheliid polychaete are so similar to lancelets in appearance, that I once found them stored in the amphioxus section in one of the major Natural History Museums of the world.

and Escriva, 2011; Kugler, et al., 2011; Lemaire, 2011; Röttinger and Lowe, 2012). However, while gene expression patterns are usually mapped onto a molecular phylogeny to tell an evolutionary story and while gene expression patterns are still often regarded as proxies for morphological characters (e.g., “necessary and sufficient for vertebrate brain structures” (Pani et al., 2012)), such practices ignore the potential for phylogenetic information in these kinds of data per se and also confound levels of homology assessments.

How could results from evo-devo studies be utilized in a phylogenetic (cladistic) framework? Because I have not produced gene pattern analyses myself, I am limited to base my exemplary suggestion to published studies (which means I trust – among other things – the orthology hypotheses of the genes involved). Still, I would like to sketch possible ways of coding evo-devo results in a cladistic data matrix and would be pleased when better-qualified researchers versed in the field called evo-devo improve on the suggested approach. It is noteworthy that in this attempt several homology hypotheses are made explicit and therefore invite scrutiny. Two differing schools that have contributed to the growth of understanding of evolution could be brought together in such an attempt. Unfortunately, philosophies of these schools – “evo-devo” and “cladistics” – seem to be alien to one another, at least currently.

Nevertheless, there are also similarities. Like all evolutionary research both schools are and have to be comparative. The comparisons in evo-devo take place on numerous levels: orthology assessment of the genes, similarity in the local and temporal expression patterns, comparisons of the gene regulatory networks and their components, to name a few. Cladistic argumentation now suggests to explicitly propose hypotheses of homology and formally code these hypotheses in the form of a character matrix (Hennig, 1950; Kitching, et al., 1998; Wiley and Lieberman, 2011). This approach has several advantages. The hypotheses of homology can be evaluated. The different hypotheses are tested against each other to result in a phylogenetic hypothesis that is supported by the original data. Cladistic coding also has a valuable heuristic advantage because it reveals gaps in our knowledge.

Cladistic analysis is a quite sophisticated science and this is not the place to review the intricacies associated with considerations of weighing, binary versus multi-state coding, or different optimization approaches (Kitching et al., 1998; Wägele, 2001; Wiley and Lieberman, 2011). Instead, I will present an exemplary (and small) matrix based solely on different evo-devo characters and discuss some aspects of it (see Supplementary Tables 18.1 and 18.2). In the near future such a data matrix can be collaboratively coded and linked to images and literature in the matrix-module implemented in Morph-D-Base that is currently being tested as a beta-version (Vogt and Grobe, 2010; see also Chapter 20, this book).

Cladistic Hox-gene characters. The first part of the data matrix (see Supplementary Tables 18.1 and 18.2) consists of Hox-gene characters. These are commonly depicted in many primary and secondary publications and are a lynchpin for the success of evo-devo studies in an evolutionary context (Spagnuolo et al., 2003; Seo et al., 2004;

Aronowicz and Lowe, 2006; Cameron et al., 2006b; Ryan et al., 2007; Lowe, 2008; Zhong and Holland, 2011; Nielsen, 2012). Because of the numerous studies carried out on different aspects of Hox genes' role in ontogeny in different taxa, these genes are especially suitable for a comparative approach. Character coding is quite straightforward, because in most cases a simple absent/present coding can be applied. In cases where orthology of a gene cannot be decisively ascertained (e.g. Hox 10–13 in *Ciona* versus Hox10, Hox11, Hox12, Hox13 in *Xenopus*), the first of the potentially orthologous genes is picked. This strategy formulates a homology hypothesis that is counted once in the analysis. Binary (absent/present) coding does not suffice in the case of the numbers of paralogous groups present. Here, several problems arise. For one, the numbers probably reflect duplication events possibly with subsequent losses (Nolte, Ahn and Krumlauf, 2001; Pick and Heffer, 2012). Therefore a scheme that reflects this assumed evolution of paralogy groups could be implemented. In the present experimental matrix, I chose the approximate solution to simply code the number of paralogous genes and assign the character the status "ordered". Another problem concerns the linkage of the genes (Ferrier and Minguillión, 2003; Irimia et al., 2012). This is coded quite simple in the proposed matrix as either absent or present. With better knowledge of gene organization this could be improved e.g., the adjacency of certain genes could be coded as absent/present and with such a strategy the potential phylogenetic information present in the arrangement of genes in echinoids or the fragmentation of tunicate Hox genes could be accounted for. Coding expression patterns of Hox genes is beyond the scope of the presently suggested matrix, yet a coding strategy that codes spatial and temporal adjacency could be envisaged.

Dorsal/ventral axes determination genes. Another widely discussed example of "evo-devo" studies is the early ontogenetic dorsal/ventral-axis determination. Here, the respectively opposing expression domains in *Drosophila* and *Xenopus* of dorsalizing versus ventralizing transcription factors bmp/chordin, dpp/sog (De Robertis and Kuroda, 2004) have revived ancient ideas of the dorsal/ventral inversion of adult protostomes compared to chordates (Geoffroy Saint-Hilaire, 1822; Arendt and Nübler-Jung, 1995; Nübler-Jung and Arendt, 1999). Originally, it had been suggested that the expression domains, i.e., the "molecular axis" during ontogeny changed during evolution (similar to the numerous shifts of the position of the abdominal fins in teleosts or the position of the gills in mollusks), which is a subtly different concept from the image of an adult annelid turned upside down. This latter idea is in conflict with some homology hypotheses based on morphology and the cladistic approach to analyze such conflicts is to evaluate them compared to the competing hypotheses. For example, the hypothesis that the neurulated collar cord in enteropneusts could be homologous to the nerve cord in chordates (Kaul and Stach, 2010) can be combined in a data matrix that codes the expression domain of bmp dorsal in enteropneusts and ventral in chordates and of sog ventral in enteropneusts and dorsal in chordates. This is done in the suggested matrix based mainly on depictions of the expression domains of the genes presented by Lapraz and co-authors (Lapraz, Besnardeau and

Lepage, 2009). This study is chosen because in their schematic depiction the authors give homology hypotheses for several species. It immediately becomes obvious that the domains of expression can change during evolution, no matter how the respective embryos are oriented. Expression patterns for the two genes in tunicates have not (to my knowledge) been published. A simple absent/present coding resulting from the character depiction in Figure 12 in Lapraz, Besnardeau and Lepage (2009) is presented in characters 67–70 (see Supplementary Tables 18.1 and 18.2).

Results from more detailed studies of the gene regulatory networks that govern the dorsal/ventral axis determination in animals could also be coded. The depiction of the gene interaction by Kozmikova et al. (2011), e.g. could be used as the basis for an extensive absent/present coding: Vent1 induction by Bmp – absent, present; FoxD inhibition by Vent – absent, present; Goosecoid induction by Wnt/β-catenin – absent, present; etc. However, I have refrained from this coding presently, because the knowledge to propose homology hypotheses for other taxa seems too fragmentary.

Ancient brain signaling centers. To get it off my chest right away: molecular complexity should not be confused with structural complexity or behavioral complexity when hypothesizing homology. In an excellent comparative study of ontogenetic genes involved in nervous system determination Pani et al. (2012) demonstrated that several genes present in the ontogeny of the nervous systems of craniates are also present in enteropneusts but absent in tunicates and cephalochordates. Thus, these authors have conclusively shown that the molecular complexity revealed in the enteropneust and craniate is independent of the structural complexity of the central nervous system. This finding, interestingly, has been interpreted as evidence for a complex structural brain in the last common ancestor of craniates and enteropneusts (i.e., the last common ancestor of all deuterostomes) or as an indication that enteropneusts may have a repertoire of complex behaviors when we are not watching them in the dark mud (see Maxmen, 2012). While the latter hypothesis is testable, the former is simply not the most elegant explanation of our current knowledge. The most obvious explanation is that structural complexity is not explained by molecular complexity and that the respective levels of homology hypotheses need to be carefully considered. Or in other words: while the "brain signaling center" anterior neural ridge is "necessary and sufficient for the establishment of vertebrate-specific brain structures", this center is not equivalent to a structure called telencephalon in craniates. Indeed, having demonstrated that morphological structures are independent from so-called signaling centers is an argument in favor of the inclusion of such characters (i.e. homology of signaling centers) together with characters derived from comparative morphological analyses in a cladistic data matrix, because character independence is a logical prerequisite for cladistic analysis. Characters 72–73 code three signaling centers as depicted by Pani et al. (2012) as present or absent, thus uniting the categories absent and partially absent of the original authors. This can be justified because the more detailed information is coded in characters 74–78 in the form of the depicted expression patterns. While there is an interesting pattern there,

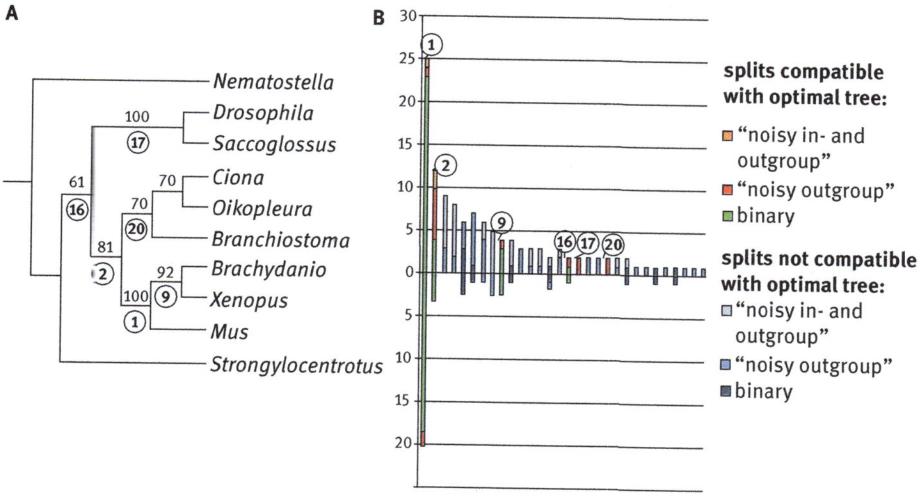

Figure 18.4: A. Single most parsimonious cladogram based on an experimental data matrix composed of "evo-devo" characters using PAUP 4.0 Beta 10 (Swofford, 2003). Branch and bound search, 10 taxa, 78 characters, 62 parsimony-informative characters, tree length: 128, consistency index: 0.80, rescaled consistency index: 0.68. Numbers above branches indicate jackknife percentages of 1000 replicates with 30 % character deletion. Circled numbers correspond to respective splits in B. **B.** Split composition. Numbers refer to the monophyletic groupings supported and present in the preferred phylogenetic hypothesis and marked in the network analysis depicted in A.

characters 74–76 are not informative in a cladistic sense, because they are uniform with the exception of one of the taxa. Thus, the formal aspect of a character matrix points to gaps in our knowledge and future studies should strive to fill the empty cells with homology hypotheses.

This short exemplary exercise in cladistic coding, however limited, already renders phylogenetic results directly derived from homology hypotheses implicit in published "evo-devo" studies. While the emerging phylogenetic hypothesis (Figure 18.4) is not overtly important due to the aforementioned limitations of the exercise, it is noteworthy that the outcome is (i) broadly in agreement with previous phylogenetic studies,[2] (ii) shows a relatively structured, and therefore informative, pattern in split analysis (Figure 18.4), (iii) explicitly formulates homology hypotheses that are evaluated against each other, and (iv) allows for the direct interpretation of which character state in an evolutionary transformation series is derived (apomorphic) and which is ancestral (plesiomorphic) at a certain node (this is facilitated considerably in a computer program like MacClade or Mesquite). These encouraging results indicate that a

2 with the exception of the position of *Drosophila* supported by one unambiguous character in the matrix: the similar expression pattern of Bmp (character 67).

cladistic approach to "evo-devo" studies will be fruitful and will increase the value of evolutionary interpretation of molecular ontogenetic mechanisms considerably.

18.1.5 Comparative morphology of chordate "key characters"

Hardly any of the numerous evolutionary studies on chordates published during the last decades fails to mention key chordate characters: post-anal tail, dorsal neural tube, and pharyngeal basket (Gerhart, Lowe and Kirschner, 2005; Brown, Prendergast and Swalla, 2008; Satoh, 2008; Horie et al., 2011). These characters attracted the interest of researchers early on and are now being re-investigated with molecular techniques. However, often these structures have defied detailed cladistic comparison, even on a morphological level. Characters discussed in the following paragraphs are coded for a cladistic analysis as can be seen in Supplementary Tables 18.3 and 18.4.

18.1.6 Tail tales

The post-anal tail of chordates is a locomotory structure that is a continuation of the trunk posterior to the anus. In cephalochordates and in basal[3] craniates it consists of a central notochord accompanied by lateral somites and a dorsal neural tube (Starck, 1975; Stach, 2000; Long et al., 2002; Stach and Kirbach, 2009; Satoh, Tagawa and Takahashi, 2012). While posterior to the anus in these two taxa there is no continuation of the endodermal intestine, the somites consist of lateral dermatomes and medial myotomes (Stach, 2000; Stickney, Barresi and Devoto, 2000; Bryson-Richardson and Currie, 2008) and the neural tube is recognizable due to its hollow nature with a neural canal that contains cilia and Reissner's fiber. The fin blade of the tail is made up of the epidermis and it is oriented vertically. Tunicates, in most cases their larvae, show a different yet similar post-anal tail (Burighel and Cloney, 1997; Stach, 2007). Situated posterior to the anus, the tail nevertheless contains a ventral row of endodermal cells that are in continuation with the intestine. The endodermal strand itself, however, does not contain an intestinal lumen. Like the post-anal tails of cephalochordates and chordates the tail of tunicates contains a central notochord, initially consisting of a row of cells, "stack-of-coins"-style (Ruppert, 1997; Stach, 1999). Laterally, the notochord in tunicates is accompanied by muscle cells (Figure 18.5) that do not form somites, i.e. they are not segmentally arranged units consisting of lateral dermatome, medial myotome, and ensheathed by tough collagenous myosepta. Dorsally, a hollow neural tube with cilia and Reissner's fiber stretches along

3 I am aware of the problematic nature of the term "basal". When using it, I see it as shorthand for: "being plesiomorphic in the trait/character discussed".

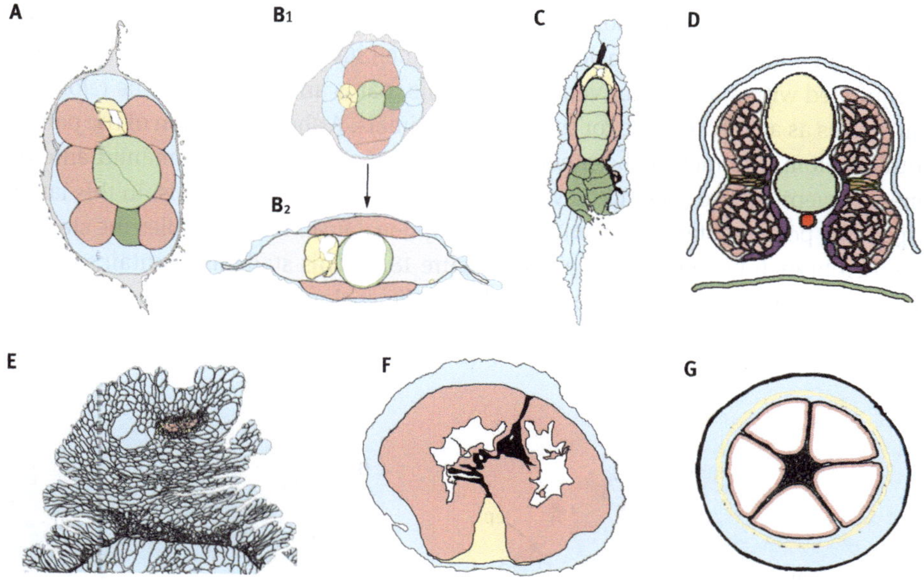

Figure 18.5: Schematic representations of cross sections through chordate tails and structures that have been proposed to be homologous to chordate tails. Blue – epidermis, yellow – nervous system, pink – muscle/myoepithelium, light green – notochord, dark green – endoderm, grey – tunic.
A. Larva of *Molgula occidentalis* (Stolidobranchiata, Tunicata). **B.** *Oikopleura dioica* (Appendicularia, Tunicata). B_1 – 4 hours after fertilization. B_2 – 24 hours after fertilization. **C.** Larva of *Branchiostoma lanceolatum* (Cephalochordata) 110 hours after fertilization. **D.** *Brachydanio rerio* (Teleostei, Craniota). **E.** Post-anal tail of *Saccoglossus kowalevskii* (Enteropneusta) at 122 hours after fertilization. **F.** Stalk of *Cephalodiscus gracilis* (Pterobranchia), see Figure 18.6. **G.** Stalk of *Comanthus japonica* (Crionoidea, Echinodermata) modified after (Grimmer, Holland and Kubota, 1984).

the notochord. The fin blade is oriented vertically in early developmental stages of tunicates, but in two higher tunicate taxa,[4] aplousobranch ascidians and appendicularians, the tail rotates through an arch of 90 degrees to the left as seen from behind during ontogeny, so the fin blade in these two taxa eventually becomes horizontal in orientation, similar to the fluke of a whale (Brien, 1948; Stach et al., 2008). The fin blade is made up of an extracellular cuticle that contains cellulose fibers in the tadpole larvae of tunicates (Stach, 2007; Sagane et al., 2010). The horizontally oriented fin blade of adult appendicularians consists of epidermal cells, different from the fin blade of their own larvae (Stach, 2007; Nakashima, Nishino and Hirose, 2011). Outside the chordates, the post-anal tail of harrimaniid enteropneusts had been suggested to be homologous to the ventral post-anal tail of chordates (Eaton, 1970; Gerhart, 2000). This post-anal extension is indeed situated posterior to the anus, but it is not present in other enteropneust families. Moreover, this short extension does

4 And also, probably convergently, in the phlebobranch Perophoridae.

not contain a central notochord, it does not possess a neural tube, and it does not contain an endodermal extension of the intestine (Stach and Kaul, 2011). Instead it is equipped with a multiciliar ventral epidermis that is an extension of the telotroch and serves as a creeping sole for the juvenile animals. A short extension of the paired trunk coelom, the metacoel, extends into the post-anal tail of harrimaniid enteropneust juveniles. The coelomic cavities are lined by myoepithelial cells. In this latter aspect the post-anal tail of harrimaniid enteropneust juveniles resembles the stalk of the pterobranch *Cephalodiscus gracilis* (Figure 18.6). This stalk also contains extensions of the trunk coelom, the metacoel and the coelomic lining again consists of myoepithelial cells. The myoepithelial cells are extremely extensible. The stalk does

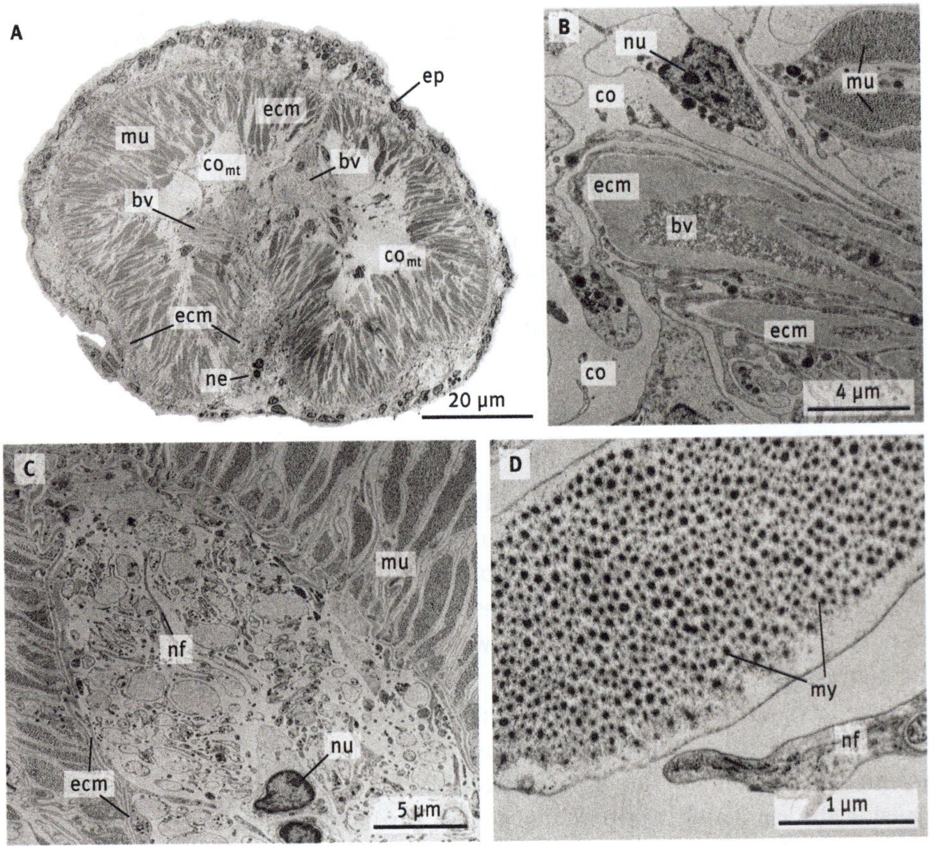

Figure 18.6: Transmission electron micrographs of cross sections through the stalk of *Cephalodiscus gracilis*. **A.** Overview, see also Figure 18.5F. **B.** Ventral blood vessel in the ecm of the central septum. **C.** Ventral intraepidermal nerve. **D.** Neurite ending in close proximity of a longitudinal muscle cell. *bv* blood vessel, co_{mt} coelomic cavity (posterior extensions of metacoel), *ecm* extracellular matrix, *ep* epidermis, *mu* muscle cell, *my* myosin, *ne* nerv, *nf* neurite fiber, *nu* nucleus.

not function in locomotion per se. During normal movements, the animals use their ciliated cephalic shields as a gliding sole. Contraction of the myoepithelial cells in the stalk enables the animals to quickly retract into the depths of their protective tubes. The stalk does not contain a separate skeletal element and it also lacks a neural tube. Nevertheless, the basiepidermal plexus is thickened along the ventral midline of the stalk. There are two longitudinal blood vessels running in the extracellular matrix between the adjacent coelomic compartments. Unfortunately, little is known about the ontogeny of *Cephalodiscus* species in general. Finally, another post-anal stalk can be found in deuterostomes in the form of the attachment stalk of crinoids, which is well documented in several species (Grimmer, Holland and Kubota, 1984, Grimmer, Holland and Hayami, 1985; Grimmer and Holland, 1990). These crinoid stalks also possess coelomic cavities, which in this case are five extensions of the single metacoel (called somatocoel in echinoderms) of the right body side. There is no notochord-like tissue in the stalk and endodermal tissue is also missing. Likewise, a neural tube is not present in the crinoid stalk.

Of course, this comparative overview concerning chordate tails and potentially homologous structures serves an illustrative example. Enough detailed data exist already to allow meaningful comparisons and therefore the tails can serve as examples for the strengths of a rigorous comparative method. Already the short review provided here should make it clear that a simple coding of post-anal tail absent versus post-anal tail present would not do the potential phylogenetic information justice that might be present in this more complex (but not too complex) character. Characters 21–40 in the morphological data matrix in Supplementary Tables 18.3 and 18.4 cover the comparative morphological aspects of the post-anal tails in deuterostomes.

18.1.7 Dorsal neural tube

As already mentioned, the chordates possess a dorsal neural tube with a fluid-filled canal into which cilia project and that contains a mucoid filamentous structure Reissner's fiber among other characteristics (Olsson, 1993). One of these other characteristics is the ontogenetic development of the neural tube. In all chordates the neural tube derives via a process called neurulation (e.g., Schoenwolf and Smith, 1990; Lowery and Sive, 2004; Saveliev, 2009). During neurulation a part of the epidermis along the dorsal midline flattens, becomes separated from the epidermis, and forms a hollow tube along the dorsal midline underneath the epidermis. This process is depicted in Figure 18.7 through cross sections in consecutive stages of an ascidian and a cephalochordate. Investigation into the morphogenesis of the collar cord in the enteropneust *Saccoglossus kowalevskii* revealed detailed similarities to chordate neurulation (Kaul and Stach, 2010). The collar cord derives from the dorsal midline epidermis in the collar region, from which it separates and forms a hollow tube. Similar to processes observed in craniates (Copp, Greene and Murdoch, 2003), dorsal cells equipped with filaments seem to be involved in the process of the formation of the tubular collar cord (Kaul and Stach, 2010). Detailed 3D-reconstruction of the collar cord of a juvenile *S. kowalevskii* revealed in addition the position and ultrastructure of some of the components of the collar cord. For example, an anterior group of giant neurons with contralateral neurite projections could be demonstrated and numerous peculiar synaptic connections between adjacent giant neurons (Bullock, 1944; Kaul and Stach, 2010). Especially the presence of synapses is indicative of an integration of nerve signals of some sort. Immunohistochemical analyses on the other hand do not show prominent concentrations of neurons in the collar cord (Nomaksteinsky et al., 2009) and the few available ethological observations demonstrate only a limited behavioral repertoire and a limited role of the collar cord (Bullock, 1940). Because all of these similarities are observed in a generally similar structural framework consisting of ventrally situated pharynx with lateral gill slits and laterally situated coeloms with myoepithelia,

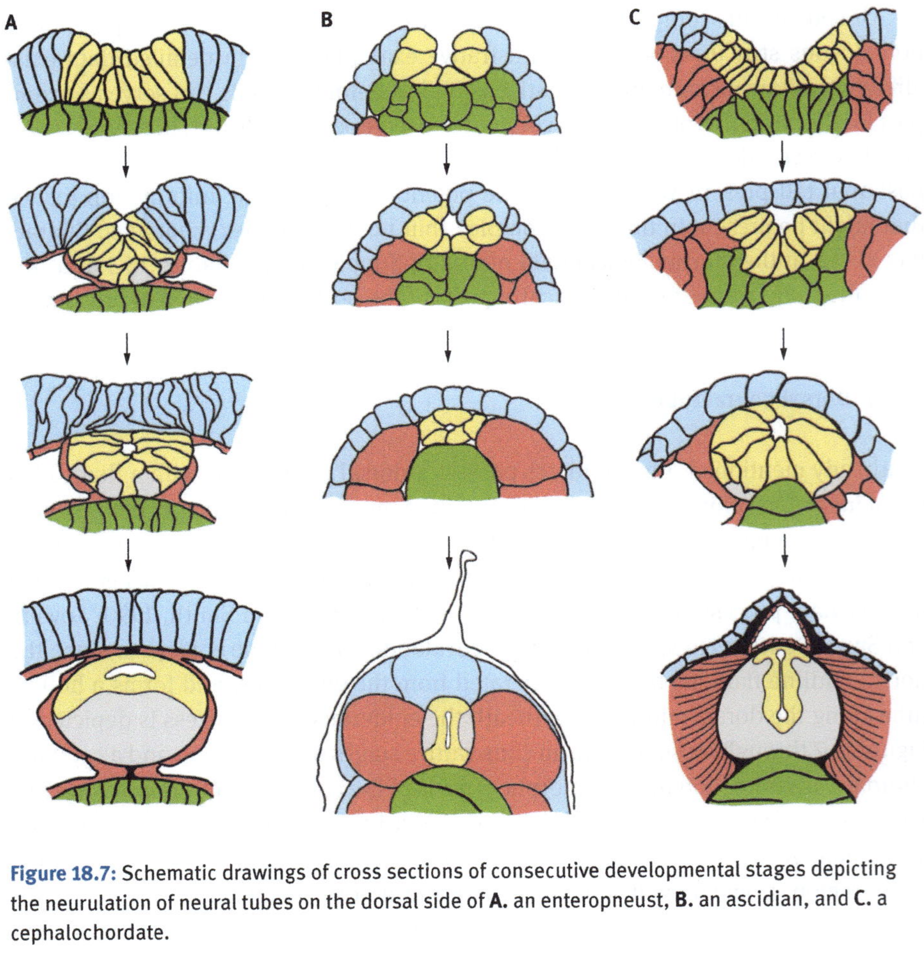

Figure 18.7: Schematic drawings of cross sections of consecutive developmental stages depicting the neurulation of neural tubes on the dorsal side of **A.** an enteropneust, **B.** an ascidian, and **C.** a cephalochordate.

in addition to an anterior excretory complex based on ultrafiltration through podocytic extensions, the process of neurulation in chordates is suggested to be homologous to neurulation in enteropneusts (Figure 18.7; character 43).

Neurulation of the nervous system can also be followed in the accelerated development of the miniaturized appendicularian *Oikopleura dioica* as the sinking in of certain cells (A6.3, A6.4, a7.9 and A6.3, A6.4, a7.9) using 4D-microscopy (Stach et al., 2008). This technique moreover would allow for numerous more detailed comparisons in a cladistic framework if more comparable data existed (see Figure S16 and Table S1 in the supplementary material to Stach et al., 2008).

Serotonergic cells etc.
Immunohistological investigations of deuterostomes have revealed a peculiar pattern of serotonergic cells in the main nerve center of planktonic larvae of echinoderms and enteropneust (Byrne et al., 2007), in adult pterobranchs (Stach, Gruhl and Kaul-Strehlow, 2012), in tunicates (Stach, 2005), in larval cephalochordates (Holland and Holland, 1993), and in larval lampreys (Hay-Schmidt, 2000). Serotonergic cells are clustered in the anterior part of the respective brains. In chordates and pterobranchs these serotonergic cells project neurites into a more posteriorly situated net, from which more posteriorly projecting nerves emerge. In enteropneusts serotonergic cells are more or less scattered throughout the basiepidermal plexus with a higher density in the proboscis (Nomaksteinsky et al., 2009). These character states are coded as homology hypotheses in character 53.

Other morphological structures have been coded for the present phylogenetic analysis as well, but these have either not been the focus of the studies within the frame of the Priority Program Deep Metazoan Phylogeny or they have been taken from the literature after careful consideration.

18.1.8 Resulting phylogenetic hypothesis

The matrix with the proposed hypotheses of homology of the morphological structures presently discussed, like the matrix of the “evo-devo-characters”, reflects a pattern of similarities found at the different observational levels for the living world. Cladistics explains this pattern by assuming an evolutionary process: descent with modification (Hennig, 1950; Wiley and Lieberman, 2011). Because with an increasing number of characters and taxa the pattern is not easily acknowledged, a tree-like representation of the information in the data matrix is often inferred according to an optimality criterion such as parsimony. The rationale for such an inference is, that based on an evolutionary process, homology hypotheses equate to a unique origin of the respective character. The resulting cladogram, with an appropriate root can then be interpreted as a phylogenetic tree and the character transformations implied by the tree (= pattern in the matrix = phylogenetic information) can be visualized. Both trees – the morphology-based tree and the evo-devo-based tree – derived from parsimony analyses of the respective matrices are broadly congruent in most of the overlapping monophyletic groups they support: e.g. monophyly of Craniota or of Chordata (Figure 18.8). A major difference exists within Chordata, where the experimental "evo-devo" data support a sister-group relationship between Cephalochordata and Tunicata, thus supporting the morphology-based "Atriozoa" hypothesis (Minot, 1897) that has not found favor during the last decades. Support in the "evo-devo" data matrix stems from the similarities found in the expression patterns of the brain center genes recently discovered by (Pani et al., 2012). The high jackknife supports in some of the nodes and the results of the split analysis indicate that the "evo-devo" data set is informative and therefore this data set clearly demonstrates the general suitability of these types of characters for cladistic analyses. However, the actual result per se may be judged cautiously bearing in mind that the data set was quite limited and experimental, intended as a proof of principle and the occurrence of *Drosophila* as a rogue taxon within an otherwise monophyletic Deuterostomia may vanish with the addition of more characters, more taxa, and more sophisticated coding strategies.

The analysis of the morphological data results in a more traditional phylogenetic hypothesis compared to currently favored sequence-based analyses (see Figures 18.1 and 18.2). Within Chordata, a sister-group relationship between Cephalochordata and Craniota is recovered with high statistical support reflecting the numerous and quite complex characters supporting this node as synapomorphies, such as the somites or the circulatory system. In contrast to many molecular phylogenies, the present morphology-based analysis recovers a monophyletic Chordata. The strict consensus tree (see Figure 18.8) indicates that the data preferentially support a monophyletic group consisting of Chordata and Enteropneusta, while not fully resolving the relationships between this clade, Pterobranchia, and Echinodermata. Overall, as in the case of the small experimental evo-devo matrix, the jackknife support and the split analysis indicate that the morphological matrix consists of a set of phylogenetically highly informative characters.

Because evolution consists of endless successions of ontogenies, analyses of the molecular interactions of development reveal aspects of the logic of evolutionary changes. At the same time, it is clear that selection can take place on every organismal level and therefore shapes characters on each of these levels – molecular, cellular,

Figure 18.8: A. Strict consensus of 354 equally parsimonious cladograms based on a data matrix composed of morphological characters. Branch and bound search, 36 taxa, 111 characters, 108 parsimony-informative characters, tree length: 179, consistency index: 0.63, rescaled consistency index: 0.58. Numbers above branches indicate jackknife percentages of 1000 replicates with 30 % character deletion. Circled numbers correspond to respective splits in B. **B.** Split composition. Numbers refer to the monophyletic groupings supported and present in the preferred phylogenetic hypothesis and marked in the strict consensus tree depicted in A. ▸

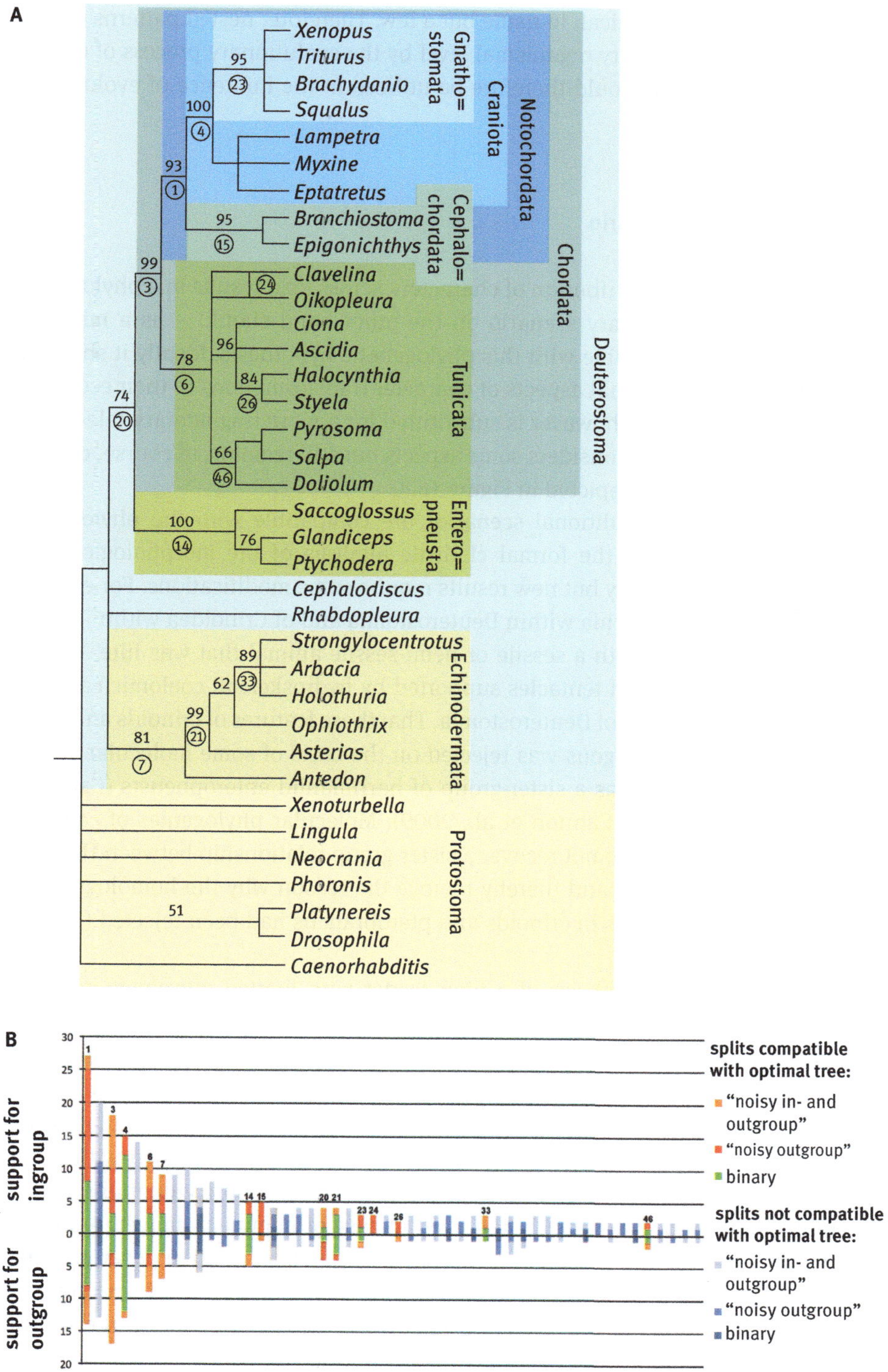
A
Xenopus
Triturus
Brachydanio
Squalus
Lampetra
Myxine
Eptatretus
Branchiostoma
Epigonichthys
Clavelina
Oikopleura
Ciona
Ascidia
Halocynthia
Styela
Pyrosoma
Salpa
Doliolum
Saccoglossus
Glandiceps
Ptychodera
Cephalodiscus
Rhabdopleura
Strongylocentrotus
Arbacia
Holothuria
Ophiothrix
Asterias
Antedon
Xenoturbella
Lingula
Neocrania
Phoronis
Platynereis
Drosophila
Caenorhabditis
Gnatho=stomata
Craniota
Notochordata
Cephalo=chordata
Chordata
Deuterostoma
Tunicata
Entero=pneusta
Echinodermata
Protostoma
B
support for ingroup
support for outgroup
splits compatible with optimal tree:
"noisy in- and outgroup"
"noisy outgroup"
binary
splits not compatible with optimal tree:
"noisy in- and outgroup"
"noisy outgroup"
binary

morphological, or ethological, to name but a few. Therefore, nested patterns of similarities are created on every organismal level by the evolutionary process of descent with modification and should therefore be utilized in the inference of evolutionary history.

18.1.9 Evolutionary scenario

I have argued that the distribution of characters across taxa results in a phylogenetic hypothesis. An evolutionary scenario on the other hand should – as a minimum requirement – be compatible with this phylogenetic hypothesis. Ideally it should in addition consider functional aspects of character transformations in their ecological contexts. Here, this ideal however is substituted for a more fragmentary tale of deuterostome evolution that considers some aspects but still remains, of course, compatible with the cladogram depicted in Figure 18.8.

In many respects traditional scenarios are compatible with the phylogenetic hypothesis derived from the formal cladistic analysis of the morphological data matrix in the present study but new results require some modifications. For example the position of Pterobranchia within Deuterostomia and of Crinoidea within Echinodermata is compatible with a sessile or hemi-sessile animal that was filter feeding with the aid of a crown of tentacles supported by hydroskeletal coelomic canals as the last common ancestor of Deuterostomia. That these features of crinoids and pterobranchs could be homologous was rejected on the basis of some molecular studies that found pterobranchs as a sister-group of harrimaniid enteropneusts (Cameron, Garey and Swalla, 2000; Cannon et al., 2009). Molecular phylogenies of complete mitochondrial genomes do not recover a sister-group relationship between Harrimaniidae and Pterobranchia and thereby remove the reason why the homology of the tentacle feeding apparatus in crinoids and pterobranchs had been rejected (Perseke et al., 2011).

In addition, the hypothesis of a filter feeder with limited movement capabilities at the base of Deuterostomia accounts for the position and organization of the central nervous system. The brain of pterobranch *Cephalodiscus gracilis* illustrates this point. It is situated dorsally in the mesosome at the base of the tentacles where it is ideally positioned to receive input from the sensory fibers of the tentacular nerves (Stach, Gruhl and Kaul-Strehlow, 2012). Tentacles need to be equipped with different sensory cells because they are the most exposed and most vulnerable part of the animal and need to be quickly retracted in case of imminent danger. A slightly asymmetric organization of the brain with an anterior concentration of serotonergic cells followed posteriorly by a fibrous network and posteriorly projecting nerves, corresponds to the ability of the animal to glide with the help of cilia on the ventral side of the oral shield, thereby defining a clear anterior-posterior orientation of the animals. While a similar – and potentially homologous – organization of serotonergic neurons in the central nervous system has been documented in tornariae (Hay-Schmidt, 2000; Nielsen and Hay-Schmidt, 2007), echinoderm larvae (Byrne et al., 2007; Burke, 2011), tunicates (Stach, 2005), cephalochordates (Holland and Holland, 1993), and lamprey (Hay-Schmidt, 2000), the pattern of immunoreactivity against serotonin in adult enteropneusts is different. There is no anterior concentration of serotonergic cell bodies in the collar cord, the centralized part of the nervous system in enteropneusts (Nomaksteinsky et al., 2009). However, numerous serotonergic cells are positioned within the epidermis of the proboscis of enteropneusts that project fibers into the collar cord. From their position these serotonergic cells correspond to similar cells in the epidermis of the pterobranch cephalic shield. Based on behavioral observations of experimentally mutilated animals the role of the collar cord does not seem to merit the term central nervous system (Bullock, 1940). Morphological and developmental studies on the other hand have shown that the collar cord is well structured, includes different cell types including giant neurons with contralateral neurite projections and glial cells, contains different types of synapses, and that the morphogenetic process closely resembles chordate neurulation (Bullock, 1944; Kaul and Stach, 2010). Molecular developmental studies (Pani et al., 2012) have in addition revealed an unexpected complexity on the molecular level, prompting one researcher to muse about complex enteropneust behavior when researchers are looking the other way (see Maxmen, 2012). In the context of the proposed phylogenetic hypothesis, the evolution of the complex ontogenetic process of neurulation can be seen as linked to an infaunal life habit: A part of the nervous system is protected from the potentially abrasive influence of the sediment. A distantly similar process in a different anatomical setup, clearly homoplasic in the proposed phylogenetic hypothesis, is seen in some echinoderms, the so-called Cryptosyringida (ophiurids, echinoids, and holothuroids). With the more vigorous burrowing capability of chordates, represented in the modern fauna by cephalochordates, a larger part of the nervous system is neurulated, opening up the possibilities of evolutionary change within the central part of the nervous system. Note that the homologization of neurulation in chordates and enteropneusts also effects the discussion concerning a dorsal ventral inversion in the stem lineage leading to the last common ancestor of Chordata (Gerhart, 2000; Lowe et al., 2006). As mentioned above, it is in my opinion more plausible to suggest a rearrangement (heterotopy) of molecular axes and infer evolutionary change of individual characters rather than assume an upside down twist of an adult animal during evolution. However, as also mentioned before, I strongly support the gathering of additional comparative evidence that might contribute to a denser evolutionary taxon sampling and character sampling.

The fish-like organization is a plesiomorphic condition in craniates and its hallmark feature is an undulatory swimming depending on a concerted functionality of an incompressible but flexible stiffening rod and laterally arranged muscle units. Primary physical constraints seem to govern the evolution of such a system. The Reynold's number (Re), which is a convenient coefficient allowing to roughly judge the behavior of a solid object (like an animal) in a fluid (such as sea water) depending on their respective parameters such as size, shape, or drag (Vogel, 1994). In a given system internal resistance of the fluid abruptly increases when the flow turns from laminar to turbulent at around Re = 1000. This is for example the case if the size of the object increases with everything else (e.g., speed, fluid, temperature etc.) remaining equal. In typical biological circumstances the threshold lies at linear dimensions of approximately 1 mm (Vogel, 1994). Thus it is interesting to note that in cephalochordate development a switch occurs from the larval ciliary locomotion to a fish-like muscle driven undulatory propulsion, when the larvae have attained a size of approximately 1.2 mm (Stokes and Holland, 1995a; 1995b; Stokes, 1997; Stach, 2000). That a functional physical constraint may favor such a muscle driven undulatory movement in water is supported by the observation of similar, clearly homoplastic cases of undulatory propulsion in cercariae of the parasitic fluke worms (Rees, 1971) or the large larvae of the echinoderm *Luidia sarsi* (Tattersall and Sheppard, 1934). These considerations make it unlikely that the simple epidermal extension called a post-anal tail in enteropneust is homologous to the chordate post-anal tail. It does not contain a central skeletal element and it also lacks muscle units capable of producing an undulatory movement. Other elements present in the chordate post-anal tail but lacking in enteropneusts are the dorsal neural tube and the ventral endoderm. Finally, so far no specific molecular network interactions typical for the ontogeny of the chordate post-anal tail has been reported to date (Satoh, Tagawa and Takahashi, 2012). However, it has to be noted that all this lack of evidence in favor of a homology hypothesis for the respective post-anal tails does not constitute proof for the lack of homology. One way to – at least temporarily – resolve such an ambiguity is to code the traits as homologous but also code the lack of homology for the respective (sub-)characters. This way a hypothesis of general homology is tested, while the more detailed evidence available is not neglected. The conclusion from this experiment is ambiguous, because the resulting phylogenetic hypothesis does not unambiguously reconstruct the presence of a post-anal tail in the last common ancestor of enteropneusts and chordates.

One of the most influential scenarios concerning the evolution of deuterostomes had been suggested by Walter Garstang (1928). In a nutshell, Garstang suggested that the fish-like organization of chordates originated in the larval stage of a deuterostome with a sessile adult. Using phylogenetic reasoning, several testable conclusions can be deduced from Garstang's hypothesis. For example, the sister-taxon of Chordata should possess a sessile adult with a dipleurula-like cilia propelled larval stage in its last common ancestor. This in turn would imply that Hemichordata were monophyletic and the sessility of pterobranchs homologous to the sessility of crinoids. This would be consistent with two phylogenetic hypotheses: either with a monophyletic Ambulacraria or a sister-group relationship between Hemichordata and Chordata. In both cases the similarities between enteropneusts and chordates (gill slits, iteratively arranged podocytes along the branchial basket, neurulation) would constitute homoplasies. The possibility that the sessility of ascidian tunicates comprises a homology to the sessility of pterobranchs and crinoids is alluded to by Garstang. However given the differences in feeding (internal filtration versus tentacle feeding) and organization (tentacle apparatus supported by coelomic cavity versus reduction of most coelomic cavities) this seems unlikely. While currently favored molecular phylogenies (Delsuc et al., 2006; Philippe et al., 2011b) support homoplasy of the different cases of sessility, the morphology-based phylogenetic hypothesis presented here does not rule out the possibility of homology. This should be seen as an incentive to investigate morphological processes that lead to sessility in crinoids, pterobranchs, and ascidians comparatively in more detail.

Another corollary of Garstang's hypothesis concerns the role of heterochrony during the evolution of a fish-like chordate ancestor. According to Garstang the last common ancestor arose via heterochrony and thus the speed of development should be faster compared to that observed in outgroups. Phylogenetic consideration does not support this hypothesis, because the developmental speed of cephalochordates is similar to that observed in echinoderms and enteropneusts and in craniates development is rather slowed instead of accelerated (Stach et al., 2008). The case for heterochrony can be made in the evolution of tunicates. Unfortunately, the internal phylogeny of tunicates remains highly uncertain and the position of Tunicata itself within Chordata is, in my view, also unsettled (Zeng and Swalla, 2005; Goloboff et al., 2009; Tsagkogeorga et al., 2009; 2010). These difficulties notwithstanding, the role of heterochrony in the form of neoteny can tentatively be outlined based on current knowledge. A clear case for neoteny can be made for Appendicularia. These entirely planktonic animals show an incredible developmental speed with the hatching of an actively swimming little chordate from the fertilization envelope after merely 4 hours following fertilization in the well-studied appendicularian species *Oikopleura dioica* (Delsman, 1910; Stach et al., 2008). This drastic developmental speed goes along with evolutionary changes on many organismic levels. For example, the Hox gene linkage is broken down (Seo et al., 2004), histone metabolism is unusual (Chioda et al., 2004), an inner cellular strengthening is present in the epidermis (Nakashima, Nishino and Hirose, 2011), the number of gill pores is reduced. Although the sister-taxon of Appendicularia is currently under debate, these traits constitute apomorphic characters for Appendicularia, because all potential sister-taxa do not share them. The small size of appendicularians, the reduction of the blood-vascular system, or the perseverance of endodermal cells in the tail are all indications that neoteny can reasonably be assumed for the stem lineage of Appendicularia. Ascidians, on the other hand, also seem to be characterized by a fast, stereotypic development (Nishida, 1987; Lemaire, 2011). The main difference to appendicularians and also to the non-tunicate chordates is the occurrence of a dramatic metamorphosis and a subsequent adult sessile stage. Thus, acceleration of development can be observed in ascidians up to the tadpole larval stage. This development of the free-swimming fish-like larva is also quite rapid, typically taking around 10–12 hours in solitary ascidians. These observations are consistent with a scenario where developmental acceleration occurred in the stem lineage of Tunicata and was even further pronounced in the stem lineage of Appendicularia. While the change of habitat from the benthos to the plankton is known to exert evolutionary pressure towards developmental acceleration (Strathmann, Staver and Hoffman, 2002), another tunicate autapomorphy surely plays a role in the general tendency towards developmental acceleration. This is the name-giving character for the group: the acquisition of an outer protective cover – the tunic. The unique ability in the animal kingdom to secrete cellulose as an integral part of the tunic was most probably acquired through lateral gene transfer from a bacterium (Mathysse et al., 2004; Nakashima et al., 2004). While the molecular consequences of such a lateral gene transfer for tunicates are not clear,[5] a direct consequence of this newly acquired tunicate ability seems obvious. It should be remembered here that the tunic completely covers also the larval stage, thus precluding larval feeding. This exclusion from external nutritional supplies will, in general, accelerate development (though of course many evolutionary factors simultaneously shape development: higher mortality in the plankton, production of predator deterrents, geographic propagation, etc.). It should be noted that in this scenario, it would be more parsimonious to locate appendicularians as derived within Tunicata emphasizing an already established trend towards accelerated development. This hypothesis is in accordance with the recently discovered larval tunic and traces of sessility in appendicularians (Stach, 2007; Sagane et al., 2010). Interestingly, the cellulose synthase gene acquired by lateral gene transfer in the tunicate stem lineage, duplicated in the appendicularian stem lineage. The two secretion products of the duplicated genes seem to correspond to the larval tunic and the adult filtering house respectively.

The more than 2,500 tunicate species show a very interesting variety of life cycles (Paffenhöfer, 1973; Svane, 1984; Svane and Young, 1989; Paffenhöfer and Gibson, 1999) and reproductive strategies that could serve as an example for the flexibility of ontogenies and life history consequences. A thorough understanding of tunicate phylogeny based on many biological levels is needed in order to take advantage of the opportunities offered by this intriguing group of animals. Tunicate phylogeny as a prerequisite to reconstruct the last common ancestor of Tunicata, including its life cycle, is the central scientific problem that needs to be solved before substantial progress in our understanding of chordate evolution can be achieved.

5 Of course one is immediately reminded of the many unusual features of tunicate DNA sequences (see e.g. Stach et al. 2010; Tsagkegeorga et al. 2010; Seo et al. 2004).

Supplementary material

Supplementary Table 18.1: Character list of data matrix based on "evo devo" characters exported form MacClade 4.08 OSX (Maddison and Maddison, 2000). See text for details. Original MacClade file available from the author upon request: tstach@zoosyst-berlin.de

No	Character	Type	Weight	States	Steps
1	Number of Hox1 clusters	ordered	1	3	3
2	Hoxa1	unordered	1	1	0
3	Hoxb1	unordered	1	2	1
4	Hoxc1	unordered	1	1	0
5	Hoxd1	unordered	1	2	1
6	Number of Hox2 genes	ordered	1	2	1
7	Hoxa2	unordered	1	1	0
8	Hoxb2	unordered	1	2	1
9	Hoxc2	unordered	1	1	0
10	Hoxd2	unordered	1	1	0
11	Number of Hox3 genes	ordered	1	5	8
12	Hoxa3	unordered	1	2	2
13	Hoxb3	unordered	1	2	1
14	Hoxc3	unordered	1	2	1
15	Hoxd3	unordered	1	2	1
16	Number of Hox4 genes	ordered	1	3	4
17	Hoxa4	unordered	1	2	1
18	Hoxb4	unordered	1	2	1
19	Hoxc4	unordered	1	2	1
20	Hoxd4	unordered	1	2	1
21	Number of Hox5 genes	ordered	1	4	5
22	Hoxa5	unordered	1	2	2
23	Hoxb5	unordered	1	2	1
24	Hoxc5	unordered	1	2	1
25	Hoxd5	unordered	1	2	1
26	Number of Hox6 genes	ordered	1	4	5
27	Hoxa6	unordered	1	2	2
28	Hoxb6	unordered	1	2	1
29	Hoxc6	unordered	1	2	1
30	Hoxd6	unordered	1	1	0
31	Number of Hox7 genes	ordered	1	3	4
32	Hoxa7	unordered	1	2	3
33	Hoxb7	unordered	1	2	1
34	Hoxc7	unordered	1	1	0
35	Hoxc7	unordered	1	1	0
36	Number of Hox8 genes	ordered	1	4	4
37	Hoxa8	unordered	1	2	2

No	Character	Type	Weight	States	Steps
38	Hoxb8	unordered	1	2	1
39	Hoxc8	unordered	1	2	1
40	Hoxd8	unordered	1	2	2
41	Number of Hox9 genes	ordered	1	5	6
42	Hoxa9	unordered	1	2	2
43	Hoxb9	unordered	1	2	2
44	Hoxc9	unordered	1	2	1
45	Hoxd9	unordered	1	2	1
46	Number of Hox10 genes	ordered	1	5	5
47	Hoxa10	unordered	1	2	1
48	Hoxb10	unordered	1	2	1
49	Hoxc10	unordered	1	2	1
50	Hoxd10	unordered	1	2	1
51	Number of Hox11 genes	ordered	1	4	6
52	Hoxa11	unordered	1	2	1
53	Hoxb11	unordered	1	1	0
54	Hoxc11	unordered	1	2	1
55	Hoxd11	unordered	1	2	1
56	Number of Hox12 genes	ordered	1	3	3
57	Hoxa12	unordered	1	1	0
58	Hoxb12	unordered	1	1	0
59	Hoxc12	unordered	1	2	1
60	Hoxd12	unordered	1	2	1
61	Number of Hox13 genes	ordered	1	5	7
62	Hoxa13	unordered	1	2	1
63	Hoxb13	unordered	1	2	2
64	Hoxc13	unordered	1	2	1
65	Hoxd13	unordered	1	2	1
66	Hox genes linked(0) or "broken"(1)	unordered	1	2	2
67	Bmp-expression on prospective oral side	unordered	1	2	1
68	Bmp-expression on prospective aboral side	unordered	1	2	2
69	Sog-expression on prospective oral side	unordered	1	2	1
70	Sog-expression on prospective aboral side	unordered	1	2	1
71	Anterior neural ridge absent/present	unordered	1	2	1
72	Zona limitans intrathalamica absent/present	unordered	1	2	1
73	Isthmic organizer absent/present	unordered	1	2	1
74	Anterior expression of sfrp1	unordered	1	2	1
75	Anterior expression of fgf8/17/18	unordered	1	2	1
76	Posterior expression of fgf8/17/18	unordered	1	2	1
77	mid-body expression of hh,Shh	unordered	1	2	1
78	More posterior expression ofWnt1	unordered	1	2	2

Supplementary Table 18.2: Data matrix of "evo devo" characters. See text for details. Original MacClade file available from the author upon request: tstach@zoosyst-berlin.de

\Character# **Taxon**	**1**	**10**	**20**	**30**	**40**	**50**	**60**	**70**
Nematostella	110001100	0000000000	0000000000	0000000000	0110000000	0000000000	0000000111	0????1110
Drosophila	110001100	0310001100	0110001100	0110001100	0110000000	0000000000	0000001011	0???11111
Strongylocentrotus	110001100	0110000000	0110001100	0110001000	0110000000	0100000000	0000000101	0???1??1?
Saccoglossus	110001100	0110001100	0110001100	0110001100	0100000000	0100000000	000000?011	011111111
Ciona	110001100	0110001100	0110001100	0000000000	0000001100	0110001001	0110001100	100010100
Oikopleura	110001100	0000001100	0000000000	0000000000	0211001100	0110001001	0110001???	?????????
Branchiostoma	110001100	0110001100	0110001100	0110001100	0110001100	0110001001	0110000100	100001000
Brachydanio	411012110	0511114111	1411104111	0101003011	0511115111	1510112001	0611110100	111111111
Xenopus	311012110	0411114111	1311103111	0211003011	1411114111	1310111001	0310110100	111111111
Mus	311012110	0311014111	1311113111	0211002001	1411113101	1310112001	1411110100	111111111

Supplementary Table 18.3: Character list of data matrix based on morphological characters. Original MacClade file available from the author upon request: tstach@zoosyst-berlin.de

1. Adults sessile (Ax, 2001): (0) absent; (1) present.
2. Pentamerous radial symmetry in adults (Ax, 2001): (0) absent; (1) present.
3. Body division trimeric in adults (Bartolomaeus, 2001; Ruppert, 2005): (0) absent; (1) present.
4. Atrial openings at opposing ends of body (Ax, 2001): (0) absent; (1) present.
5. Mesodermal calcareous stereome skeleton (Ax, 2001): (0) absent; (1) present.
6. Mesoderm developing directly from archenteron (enterocoely sensu Lüter, 2000; Ruppert, 2005): (0) absent; (1) present.
7. Circular skeletal stereome element surrounds anterior pharynx (Cameron, 2005): (0) absent; (1) present.
8. Colonies formed through asexual budding (Nielsen, 2001; Cameron, 2005): (0) absent; (1) present.
9. Radial nerves interior, forming an epineural canal (Cameron, 2005): (0) absent; (1) present.
10. Polian vesicles and Tiedemann's bodies present on radial canal (Cameron, 2005): (0) absent; (1) present.
11. Ambulacral tube feet used in locomotion (Cameron, 2005): (0) absent; (1) present.
12. At least a single pair of gill slits (Ax, 2001): (0) absent; (1) present.
13. More than a single pair of gill slits (Ax, 2001): (0) absent; (1) present.
14. Dorsal organ in branchial basket: (0) trough-like; (1) languets/tentacles; (2) continuous band; (3) flat ciliary epithelium.
15. Atrium: (0) absent; (1) present.
16. Atrium developing dorsal during ontogeny (Stach and Turbeville, 2002; Stach unpubl.): (0) absent; (1) present.
17. Paired peripharyngeal ciliary bands (Ruppert, 2005): (0) absent; (1) present.
18. Branchial tentacles: (0) absent; (1) present.
19. Branchial tentacles: (1) simple; (2) branched
20. Post-anal extension (Ax, 2001): (0) absent; (1) present.
21. Urochord in larval stages (sensu Nielsen, 2001): (0) absent; (1) present.
22. Urochord (sensu Nielsen, 2001) or homologous structure in adult stage: (0) absent; (1) present.
23. Notochord (sensu Nielsen, 2001): (0) absent; (1) present.
24. Notochord in larval stage (Stach, 2007): (0) absent; (1) present.
25. Notochord or homologous structure in adult stage (Stach, 2007): (0) absent; (1) present.
26. Notochord in adults almost completely replaced by chondrous or bony segmental skeletal elements (Starck, 1975): (0) absent; (1) present.
27. Locomotory tail musculature in ontogenetic stages (Starck, 1975;Stach, 2007; Stach and Kirbach, 2009): (0) absent; (1) present.
28. Orientation of early larval tail (Stach, 2007; Stach et al., 2008): (0) absent; (1) present.
29. Orientation of late larval tail: 0 = vertical; 1 = horizontal (Stach & Turbeville 2002): (0) absent; (1) present.
30. Rotation of the tail to the left during ontogenesis from vertical to horizontal: (0) absent; (1) present.
31. Notochord with extracellular fluid filled spaces (Stach, 1999) : (0) absent; (1) present.
32. Notochordal cells with intracellular vacuoles (Stach, 1999): (0) absent; (1) present.
33. Post-anal endodermal tissue next to notochord: (0) absent; (1) present.
34. Epidermis constitutes adult tail fin blade (Stach, 2007; Starck, 1975): (0) absent; (1) present.

35. Tunic constitutes larval fin blade (Stach, 2007): (0) absent; (1) present.
36. Epidermis constitutes larval tail fin blade (Stach, in press): (0) absent; (1) present.
37. Adults move by undulatory tail beat (Stach and Turbeville, 2002): (0) absent; (1) present.
38. Somites, segmented musculature separated by ecm (Starck, 1975, Stach, 2000): (0) absent; (1) present.
39. Myocoel (Starck, 1975, Stach, 2000): (0) absent; (1) present.
40. Dermatom (Starck, 1975, Stach, 2000): (0) absent; (1) present.
41. Sarcotubular invaginations in tail muscle cells (Stach and Kirbach, 2009): (0) absent, (1) tubular; (2) laminar
42. Anterior dorsal brain: (0) absent; (1) present.
43. Neurulation: (0) absent; (1) present.
44. Dorsal neural tube in early ontogenetic stages: (0) absent; (1) present.
45. Dorsal neural tube as adults (Burighel and Cloney, 1997; Ruppert, 1997; Cameron, 2005): (0) absent; (1) present.
46. Reissner's fiber (Olsson, 1993): (0) absent; (1) present.
47. Coronet cells in the central nervous system (Ruppert, 2005): (0) absent; (1) present.
48. Basiepithelial nerve net (plexus) constitutes overwhelming part of the nervous system (Nielsen, 2001; Holland, 2003): (0) absent; (1) present.
49. Central nervous system with telencephalon (Starck, 1975): (0) absent; (1) present.
50. Solid ganglion-like brain in adults (without neural canal or similar): (0) absent; (1) present.
51. Immunoreactivity against Salm-F-Amid-1 (Stach et al., 2005): (0) absent; (1) present.
52. Immunoreactivity against Salm-F-Amid-2 (Stach et al., 2005): (0) absent; (1) present.
53. Serotonergic nervous system with anterior cell concentration, nerve net, and posterior fibers (Stach, 2005): (0) absent; (1) present.
54. Central nervous system with diencephalon (Starck, 1975): (0) absent; (1) present.
55. Central nervous system with messencephalon (Starck, 1975): (0) absent; (1) present.
56. Central nervous system with metencephalon (Starck, 1975): (0) absent; (1) present.
57. Central nervous system with myencephalon (Starck, 1975): (0) absent; (1) present.
58. Central nervous system with spinal ganglia (Starck, 1975): (0) absent; (1) present.
59. Ventral root of spinal ganglia contains visceromotoric fibers (Starck, 1975): (0) absent; (1) present.
60. Hepatic portal vein and Ductus cuvieri present (Rähr, 1979): (0) absent; (1) present.
61. Caudal vein: (0) absent; (1) present.
62. Caudal aorta: (0) absent; (1) present.
63. Blood vascular system equipped with endothelial lining (Starck, 1975): (0) absent; (1) present.
64. Heart with atrium, ventricle and Conus arteriosus (Starck, 1975): (0) absent; (1) present.
65. Heart beat reversal (Stach and Turbeville, 2002): (0) absent; (1) present.
66. Internal longitudinal blood vessels in branchial basket (Stach and Turbeville, 2002): (0) absent; (1) present.
67. Preoral ciliary organ (Cameron, 2005): (0) absent; (1) present.
68. Tongue bars separate original gill slits (Cameron, 2005): (0) absent; (1) present.
69. Madreporite present (Cameron, 2005): (0) absent; (1) present.
70. Secernation of iodotyrosin (Cameron, 2005): (0) absent; (1) present.
71. Synapticles (Cameron, 2005; Rychel et al., 2006): (0) absent; (1) present.
72. Multiciliary epithelium (Nielsen, 2001): (0) absent; (1) present.
73. U-shaped intestinal tract (Nielsen, 2001): (0) absent; (1) present.
74. Suspension feeders using internal mucus filter (Nielsen, 2001): (0) absent; (1) present.

75. Tentacles supported by mesodermal coelomic cavities (Nielsen, 2001): (0) absent; (1) present.
76. Mutable connective tissue (Nielsen, 2001, Cameron 2005): (0) absent; (1) present.
77. Telotroch (Peterson and Eernisse, 2001): (0) absent; (1) present.
78. Neotroch (Peterson and Eernisse, 2001): (0) absent; (1) present.
79. Paired openings of the mesocoel to the exterior (Peterson and Eernisse, 2001): (0) absent; (1) present.
80. Anus formed in region of blastopore (Ruppert, 2005): (0) absent; (1) present.
81. Tight junctions (Ruppert, 2005): (0) absent; (1) present.
82. Primary gill slits in adults (sensu Ruppert, 2005): (0) absent; (1) present.
83. Endostyle/thyroid gland (Starck, 1975): (0) absent; (1) present.
84. Jaws supported by skeletal elements (Starck, 1975): (0) absent; (1) present.
85. Epidermis multilayered (Starck, 1975): (0) absent; (1) present.
86. Paired anterior labyrinth organ (Starck, 1975): (0) absent; (1) present.
87. Proboscis skeleton (van der Horst, 1939; Cameron, 2005): (0) absent; (1) present.
88. Adenohypophysis consisting of ectodermally and endodermally derived tissue (Stach, 1996; Ruppert, 2005): (0) absent; (1) present.
89. Unpaired, preoral excretory complex (Stach, 2002): (0) absent; (1) present.
90. Position of gonads (Stach and Turbeville, 2002): (0) at the atrial epithelium (pleurogon; (1) close to intestinal tract (enterogon)
91. Largest part of intestinal tract (Stach and Turbeville, 2002): (0) posterior to the branchial basket; (1) lateral to the branchial basket
92. Opening of neural gland (Stach and Turbeville, 2002): (0) simple; (1) more complex
93. Pyloric gland (Stach and Turbeville, 2002): (0) absent; (1) present.
94. Muscle bands functioning in jet propulsion (Stach and Turbeville, 2002): (0) absent; (1) present.
95. Tunic in adult stages present (Stach and Turbeville, 2002, Stach 2007): (0) absent; (1) present.
96. Obligatory metagenesis (Stach and Turbeville, 2002): (0) absent; (1) present.
97. Hermaphroditism (Stach and Turbeville, 2002): (0) absent; (1) present.
98. Branchiomeric podocytes (Pardos and Benito, 1987; Brandenburg and Kümmel, 1961; Stach and Eisler, 1998): (0) absent; (1) present.
99. Pericard cells form myofilaments during early embryogenesis (Hirakow, 1989; Stach, 1998): (0) absent; (1) present.
100. Paired anterior directed nerves leaving anterior central nervous system (Lacalli and Holland, 1998; Stach and Hessling, 2001): (0) absent; (1) present.
101. Liver caecum developing from intestine (Ruppert, 1997; Stach, 2000): (0) absent; (1) present.
102. Protocoel with unpaired opening to the exterior (Ruppert and Balser, 1986, Stach, 2002): (0) absent; (1) present.
103. Body division trimeric in early ontogenetic stages: (0) absent; (1) present.
104. Body division metameric in adults: (0) absent; (1) present.
105. Suspension feeder using tentacles as adults: (0) absent; (1) present.
106. Suspension feeder using cilia as larvae: (0) absent; (1) present.
107. Anterior right coelome forms pulsatile element to propagate blood: (0) absent; (1) present.
108. Paired anterior chemical sensory organ (nose) (Starck, 1975): (0) absent; (1) present.
109. Paired anterior light sensitive organ with lens (eyes) (Starck, 1975): (0) absent; (1) present.
110. Podocytes: (0) absent; (1) present.
111. Calcitic elements in suckers of ambulacral feet: (0) absent; (1) present.

Supplementary Table 18.4: Data matrix of morphological characters. See text for details. Original MacClade file available from the author upon request: tstach@zoosyst-berlin.de

\Character# Taxon\	1	10	20	30	40	50	60	70	80	90	100	110
Xenopus	000-01000	0011-0000-	1111111100	0010101111	1111111101	0??1111111	1111100000	1000000000	11-1111010	-000000011	1100100?11	10
Triturus	000-01000	0011-0000-	1111111100	0010101111	1111111101	0??1111111	1111100000	1000000000	11-1111010	-000000011	1100100?11	10
Brachydanio	000-01000	0011-0000-	1111111100	0010101111	1111111101	0??1111111	1111100000	1000000000	1111111010	-000000011	1100100?11	10
Squalus	000-01000	0011-0000-	1111111100	0010101111	1111111101	0??1111111	1111100000	1000000000	1111111010	-000000011	1100100?11	10
Lampetra	000-01000	0011-0001-	1111110100	0010101111	1111111101	0??1111110	1111100000	1000000000	1111011010	-000000011	1100100?11	10
Myxine	000-01000	0011-00000	1111110100	0010101111	1111111101	0??1111110	1111100000	1000000000	1111011010	-000000011	1100100?11	10
Eptatretus	000-01000	0011-0000-	1111110100	0010101111	1111111101	0??1111110	1111100000	1000000000	1111011010	-000000011	1100100?11	10
Branchiostoma	000001000	0011010110	1111110100	0110101111	1011111000	0??1000000	1110000010	1100100000	1001000011	0000000011	1111101100	10
Epigonichthys	000001000	0011010110	1111110100	0110101111	1011111000	0??1000000	1110000010	1100100000	1001000011	0000000011	1111101100	10
Clavelina	100001010	0011111110	1100100101	1111-10000	0111101100	1??000000-	0000010000	1011100000	1101000010	10010101-1	1000000100	00
Oikopleura	000001000	0010?10110	1110100101	1111110100	0111111100	1??000000-	0000010000	1011100000	1111000010	10010101-1	1000000100	00
Ciona	100001000	0011111110	1100100100	0111-10000	0211101100	100100000-	0000011000	1011100000	1101000010	11110101-1	1000000100	00
Ascidia	100001000	0011111110	1100100100	0111-10000	0211101100	100100000-	0000011000	1011100000	1101000010	11110101-1	1000000100	00
Halocynthia	100001000	0011211111	1100100100	0111-10000	0211101100	1??100000-	0000011000	1011100000	1101000010	01110101-1	1000000100	00
Styela	100001000	0011211110	1100100100	0111-10000	0211101100	1??100000-	0000011000	1011100000	1101000010	01110101-1	1000000100	00
Pyrosoma	00010?010	0011111110	00-000----	--0----000	0-1??0??00	1???00000-	0000011000	1011100000	1101000?0	10010111-1	1000000100	00
Doliolum	00010?010	001131110-	1100100100	0??0-10000	001?10??00	1??100000-	0000010000	1011100000	11?10000?0	10011111-1	1000000100	00
Salpa	00010?010	001131110-	00-000----	--0----000	0-1?10??00	1??100000-	0000010000	1011100000	11?10000?0	10011111-1	1000000100	00
Saccoglossus	001-01000	001130000-	10-000-0--	--00---000	0-11110010	011000000-	0000000110	1110000101	1000000101	-0-0000010	0011001100	10
Glandiceps	001-01000	001130000-	00-000-0--	--0----000	0-11110010	0??000000-	0000000110	1110000111	1000000101	-0-0000010	0011001100	10
Ptychodera	001-01000	001130000-	00-000-0--	--0----000	0-11110010	0??000000-	0000000110	1110000111	1000000101	-0-0000010	0011001100	10
Cephalodiscus	101-01010	001030000-	10-000-0--	--0----000	0-10000010	1??100000-	0000000000	0011010001	1010000001	-0-0000000	001101?100	10
Rhabdopleura	101-01010	000030000-	10-000-0--	--0----000	0-10000010	1??100000-	00000000-0	0011010001	10-0000001	-0-0000000	001101?100	10
Strongylocentrotus	010-11101	1100-0000-	00-000-0--	--0----000	0-00000010	-11000000-	000000-0-1	0000001010	10-0000001	---0000000	0011001100	11
Arbacia	010-11101	1100-0000-	00-000-0--	--0----000	0-00000010	-11000000-	000000-0-1	0000001010	10-0000001	---0000000	0011001100	11
Holothuria	010-11101	1100-0000-	00-000-0--	--0----000	0-00000010	-??000000-	000000-0-1	0000001010	10-0000001	---0000000	0011001100	11
Asterias	010-11000	1100-0000-	00-000-0--	--0----000	0-00000010	-11000000-	000000-0-1	0000001010	10-0000001	---0000000	0011001100	10
Ophiothrix	010-11001	1100-0000-	00-000-0--	--0----000	0-00000010	-11000000-	000000-0-1	0000001010	10-0000001	---0000000	0011001100	10
Antedon	110-11000	0000-0000-	00-000-0--	--0----000	0-00000010	-??000000-	000000-0-0	0001011010	10-0000001	---0000000	0011011100	10
Xenoturbella	000-0?000	0000-0000-	00-000-0--	--0----000	0-00000010	-11000000-	00000--0-0	0010000000	?0-0000000	---0000000	0010000000	00
Lingula	100-0?000	0000-0000-	00-000-0--	--0----000	0-10000010	1??000000-	000000-0-0	0001010000	00-0000000	---0000000	0000011000	10
Neocrania	100-01000	0000-0000-	00-000-0--	--0----000	0-10000010	110000000-	000000-0-0	0001010000	00-0000000	---0000000	0000011000	10
Phoronis	100-00000	0000-0000-	00-000-0--	--0----000	0-10000010	1??000000-	000000-0-0	0001010000	00-0000000	---0000000	0000011000	10
Platynereis	000-00000	0000-0000-	00-000-0--	--0----001	0-10000000	11?000000-	000000-0-0	00100000?0	00-0000000	---0000000	000?101000	10
Drosophila	000-00000	0000-0000-	00-000-0--	--0----000	0-10000000	1??000000-	000000-0-0	0000000000	00-0000000	---0000000	0000100000	-0
Caenorhabditis	000-00000	0000-0000-	00-000-0--	--0----000	0-10000010	1??000000-	000000-0-0	0000000000	00-0000000	---0000000	0000000000	-0

Lars Podsiadlowski, Adina Mwinyi, Peter Lesný, and Thomas Bartolomaeus

19 Mitochondrial gene order in Metazoa – theme and variations

Abstract: Mitochondrial (mt) genomes are often used as a source of data sets for phylogenetic analyses on various taxonomic levels of animals. Mt gene arrangement is variable among Metazoa, but to some degree ground pattern hypotheses can be deduced by comparative analysis. We aim here to give an overview of mt gene order variation in metazoan evolution. Focusing mainly on Bilateria, we suggest ground patterns for Ecdysozoa, Lophotrochozoa and Deuterostomia and list the major derivations from these patterns. Although a number of phyla exhibit variations in gene order and cases of homoplasious gene rearrangements are known, there are some gene blocks common between major branches of Bilateria, enabling statements on the bilaterian ground pattern. In a few cases mt gene arrangements support one out of a set of competing phylogenetic hypotheses. Prominent examples are the placement of *Xenoturbella* with Deuterostomia and the inclusion of Myzostomida in Annelida. One notable example of homoplasious gene arrangement is provided by the large inversion found in some mollusks, phoronids and entoprocts.

19.1 Introduction

19.1.1 Mitochondrial genome structure

Mitochondrial (mt) genomes are interpreted as relicts of bacterial genomes which underwent strong reduction during long-standing close intracellular symbiotic relationship to eukaryotes. In bilaterian animals there are typically less than 40 genes present in the mt genome, 13 coding for protein subunits (*atp6*, *atp8*, *cox1–3*, *cob*, *nad1–6*, *nad4L*), two ribosomal RNAs (*rrnS*, *rrnL*) and a set of 22–24 tRNAs (Wolstenholme, 1992; Boore, 1999). In the mt genomes of non-bilaterian animals there are some additional genes present (see Section 19.2.1). A few taxa show loss of genes. From protein-coding subunits, *atp8*, the smallest and most variable sequence, is lost independently in several phyla. However, due to its short, variable sequence this gene is not easily identifiable by the BLAST search and a lack of this gene in a GenBank entry may in some cases be due to an incomplete annotation. In some taxa, like Cnidaria and Chaetognatha, almost all tRNAs are missing. The exceptions are *trnM* and *trnW* (being one of the exceptions where animal mitochondrial genetic codes are different from the "universal code"). Missing tRNAs are reported as well from several other taxa, in some cases this again may be due to problems in gene annotation. In some cases extensive tRNA editing occurs (Lavrov, Brown, and Boore, 2000; Segovia et al., 2011), in other cases tRNA remolding was reported (Rawlings, Collins, and Bieler, 2003; Wang and Lavrov, 2011). Both mechanisms may lead to problems in tRNA identification. Duplications of partial or complete genes and even of bigger blocks of adjacent genes are other modifications frequently found in mt genomes (e.g. Endo et al., 2005; Podsiadlowski, Kohlhagen, and Koch, 2007).

Predominantly the amino acid sequences of mt protein-coding genes were used for phylogenetic analysis of metazoan interrelationships. However, it became apparent early on that this data set is problematic for resolving phylum-level metazoan relationships, mainly due to strong differences in substitution rates and long-branch attraction, affecting e.g. Platyhelminthes, Syndermata and Nematoda (e.g. Steinauer et al., 2005; Podsiadlowski, Braband, and Mayer, 2008). Besides sequence information there are a few examples where gene arrangements were successfully used to support phylogenetic hypotheses (e.g. Boore, Lavrov, and Brown, 1998; Lavrov, Brown, and Boore, 2004; Bleidorn et al., 2007). In this review article we aim to give an overview of mt gene order variation in Metazoa, present some hypotheses for ancestral gene orders and some scenarios for mt genome rearrangements during the evolution of the bilaterian phyla.

19.1.2 Mechanisms of genome rearrangements

Changes in the gene order of mt genomes can be assigned to a few types of rearrangement events: inversions, transpositions and TDRL (tandem duplication followed by random loss of genes) (Mueller and Boore, 2005). Inversions are probably the result of intrachromosomal recombination, while tandem duplication of a partial mt genome is best explained by an erroneous replication process, e.g. slipped strand mispairing (Broughton and Dowling, 1997). Transpositions and inverted transpositions are probably due to interchromosomal recombination (as there is more than one genome in a mitochondrion) or a combination of TDRL and inversion events. Hypotheses of the most parsimonious sequence of rearrangement events must consider these elementary steps. Especially TDRL events are not at first sight distinguishable from more complicated scenarios. In our approach to compare gene arrangements we tried to find the shortest chain of events leading from one arrangement to another. With slightly different gene orders we simply counted the number of apparent transpositions. In addition we used the CREx tool (Bernt et al., 2007) which proposes the most parsimonious sequence of events leading from one arrangement to another, considering the elementary events mentioned above. Ground pattern hypotheses were reconstructed by selecting conservative gene orders from several taxa and generating a pattern that connects these with the shortest number of events.

19.1.3 Mitochondrial genomes in other eukaryotes

Mt genomes of higher plants (Embryophyta) are strikingly different from those of animals (Palmer, Soltis, and Soltis, 1992; Kubo and Newton, 2008). Plant mt genomes are much larger (100 kb up to several Mb) and are much more variable in gene content (43–196 genes). Extreme examples are *Silene conica*, with an mt genome of 11.3 Mbp arranged in 128 circular minichromosomes (Sloan et al., 2012) and *Zea mays* with a 570 kbp mt genome coding for 163 putative proteins and 33 RNA genes (Clifton et al., 2004). There is a high degree of gene order variation even between closely related plant species. Sequence substitution rate is lower than in chloroplast and nuclear genomes, a striking contrast to mt genomes of bilaterian animals where the substitution rate is much higher than in the corresponding nuclear genomes. Thus in plant phylogenetics chloroplast genomes are much more often used as molecular markers than mt genomes. Fungi, protists and algae usually have smaller mt genomes than Embryophyta (20–120 kb, 30–80 genes). Apicomplexa form an exceptional case, with the smallest known mt genomes (6 kb) with only three protein-coding subunits and fragmented ribosomal RNA genes (Omori et al., 2007).

Choanoflagellida are putatively the closest relatives of Metazoa. The choanoflagellid species *Monosiga brevicollis* has an mt genome of 76.5 Kb, containing information for 32 protein subunits and 27 non-coding RNAs, including all of the genes known from metazoan mt genome (Burger et al., 2003). Compared to the mt genomes of protists and fungi, metazoan mt genomes appear to be reduced in size and gene content while higher plant mt genomes seem to be expanded in size.

19.2 Metazoan mitochondrial genomes

19.2.1 The non-bilaterian taxa

Genes or ORFs not present in bilaterian mt genomes are found in the non-bilaterian taxa Placozoa, Porifera and Cnidaria, pointing towards the reduced nature of bilaterian mt genomes (Dellaporta et al., 2006). The presence of introns was reported in a few cases, predominantly in non-bilaterian taxa (Signorovitch, Buss, and Dellaporta, 2007; Valles, Halanych, and Boore, 2008).

Mt genomes of Placozoa gave evidence that this taxon comprises more than just the sole described species *Trichoplax adhaerens*. Five mt genomes have been published by now, which have sizes between 32–43 kb and have variation in gene order (Dellaporta et al., 2006; Signorovitch, Buss, and Dellaporta, 2007; Miyazawa et al., 2012). All five mt genomes show different gene arrangements with at least a few single gene translocations and in some taxa an inversion of a large part of the mt genome.

Poriferan mt genomes from the subtaxa Demospongiae and Homoscleromorpha share some ancient features present as well in Cnidaria and Placozoa, while mt genomes from Hexactinellida have a few features in common with Bilateria rather than with the other Porifera. They share a derived mt genetic code (codons AGG/AGA code for serine instead of arginine) and an unusual tRNA structure for one of the two *trnS* (Haen et al., 2007). Recently the first mt genome from Calcarea was sequenced: *Clathrina clathrus* shows a highly derived genome structure, partitioned into six linear chromosomes and with a fragmented organization of rRNA genes (Lavrov et al., 2013).

In Cnidaria there are some exceptions to the typical organization of mt genomes. Linear mt genome molecules are reported from medusozoan Cnidaria, while Anthozoa have retained the circular organization (Bridge et al., 1992). Among Medusozoa, Scyphozoa and many Hydrozoa the complete mt genome is formed by one linear chromosome, while *Hydra* species have two linear fragments, and Cubozoa have fragmented mt genomes in the form of eight linear DNA molecules (Kayal and Lavrov, 2008; Kayal et al., 2012). Many Cnidaria have open reading frames additional to the usual set of mt proteins. In Medusozoa there are DNA polymerase *polB* (Shao et al., 2006) and *ORF314* (Kayal et al., 2012), in Octocorallia *MutS* (Pont-Kingdon et al., 1995), and in Hexacorallia some unidentified ORFs (Medina et al., 2006). Only two tRNA genes are present in mt genomes of most Cnidaria, *trnW* and *trnM*, in Octocorallia only the latter is retained.

The first sequenced mt genome from Ctenophora (Pett et al., 2011) represents the smallest animal mt genome ever recorded (*Mnemiopsis leidyi*, 10326 bp). No tRNA genes are detectable and *atp8* and *atp6* are lost, thus leaving a set of only 13 genes, 11 coding for protein subunits and the two ribosomal RNAs. Except for the pair *rrnS-rrnL* mt gene order has no obvious similarity to other taxa.

19.2.2 Basal splits of the Bilateria and some taxa with uncertain position

The extensive use of molecular sequence data led to a "new view" of animal phylogeny, with Bilateria split into three large subtaxa (Lophotrochozoa, Ecdysozoa and Deuterostomia), but still leaving a few taxa with uncertain affinities (Adoutte et al., 2000; Halanych, 2004). Mt gene order characters added some support to this new

Figure 19.1: Mitochondrial gene arrangements of Bilateria. Ground pattern hypotheses of Arthropoda (Staton et al., 1997), Deuterostomia (Bourlat et al., 2009) and Lophotrochozoa (this publication), alongside of a couple of further examples of gene orders. The arrangements of Deuterostomia and Lophotrochozoa are not realized in extant species, but rather represent interpolations between realized ingroup and outgroup arrangements. For missing tRNA genes in the lophotrochozoan and annelid ground pattern there is not only one most parsimonious hypothesis because the positions of these tRNA genes are variable among ingroup taxa and differ from positions in outgroup taxa. Colors depict blocks of genes showing identical gene order (or only minor derivations) in a couple of taxa. Note that the light blue and yellow blocks are adjacent due to the circular character of the mt genomes and form actually one conserved block. Transfer RNA genes are abbreviated by the one letter amino acid code, ncr depicts the major non-coding region. ►

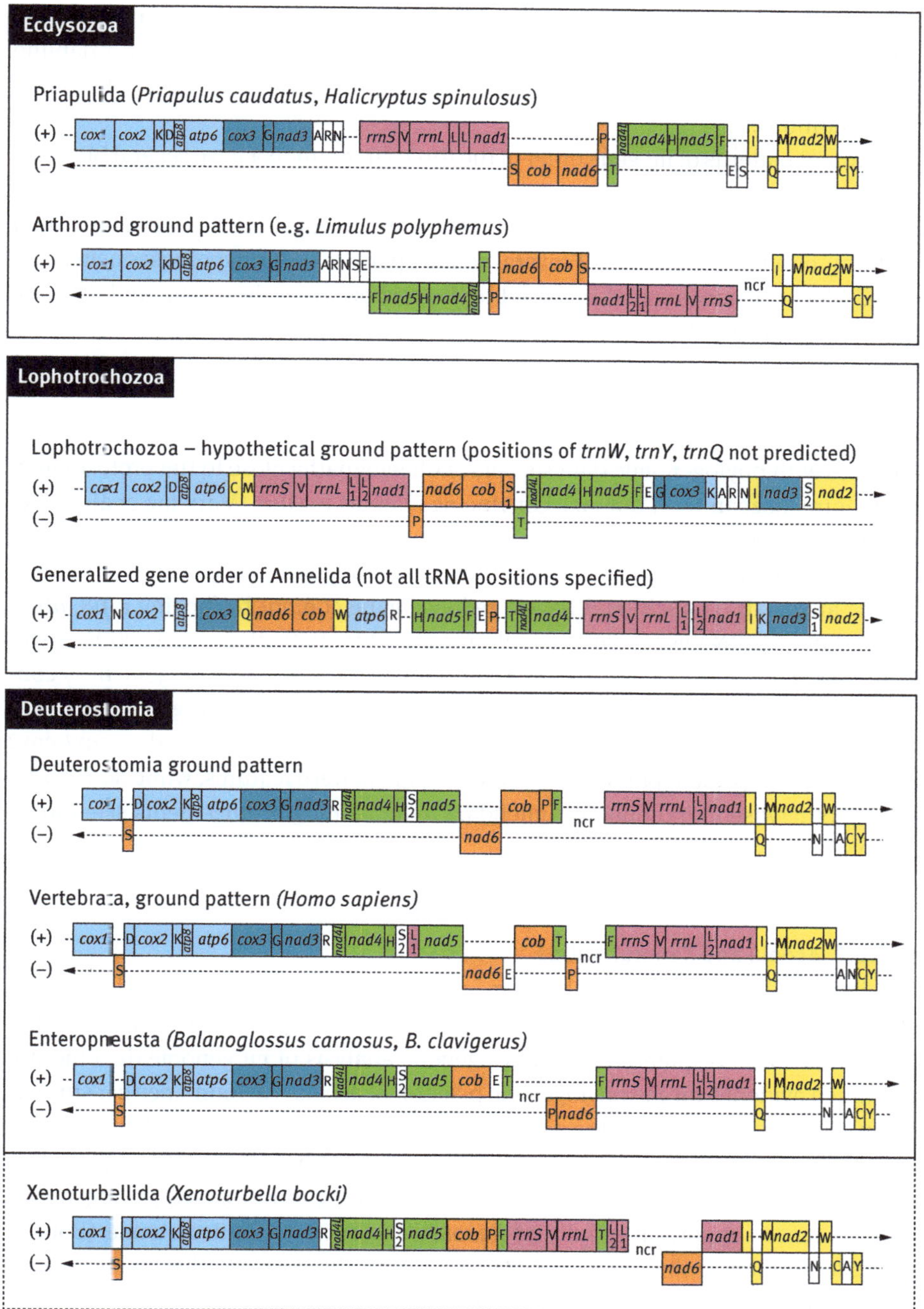
Ecdysozoa
Priapulida (Priapulus caudatus, Halicryptus spinulosus)
(+)
(–)
Arthropod ground pattern (e.g. Limulus polyphemus)
(+)
(–)
ncr
Lophotrochozoa
Lophotrochozoa – hypothetical ground pattern (positions of trnW, trnY, trnQ not predicted)
(+)
(–)
Generalized gene order of Annelida (not all tRNA positions specified)
(+)
(–)
Deuterostomia
Deuterostomia ground pattern
(+)
(–)
ncr
Vertebrata, ground pattern (Homo sapiens)
(+)
(–)
ncr
Enteropneusta (Balanoglossus carnosus, B. clavigerus)
(+)
(–)
ncr
Xenoturbellida (Xenoturbella bocki)
(+)
(–)
ncr

view, e.g. the high similarity of the molluskan and brachiopod mt gene arrangement (Stechmann and Schlegel, 1999) and the similarity of the arthropod and priapulid gene order (Webster et al., 2006), although these similarities may reflect plesiomorphic conditions.

Ground patterns of mt gene order of the three large subtaxa have several blocks of adjacent genes in common (Figure 19.1). A hypothesis of a ground pattern of Bilateria was already presented (Lavrov and Lang, 2005). We do not share this view, but rather see this ground pattern as one of several equally parsimonious hypotheses using conserved blocks of genes found in extant lophotrochozoan, ecdysozoan, and deuterostomian taxa (Figure 19.1, conserved blocks share the same color).

19.2.2.1 Chaetognatha

Analyses with phylogenomic data sets place Chaetognatha close to the protostomian or lophotrochozoan root, but lack convincing support for this (Marletaz et al., 2006; Matus et al., 2006; Hejnol et al., 2009). Mt sequences do not add much evidence for resolving their phylogenetic position. Similar to Ctenophora, mt genomes of Chaetognatha are small (11–13kb) and have a reduced gene content by lacking *atp8*, *atp6*, and all but one tRNA gene, *trnM* (Papillon et al., 2004; Helfenbein et al., 2004; Miyamoto, Machida, and Nishida, 2010). About half of the genome arrangement is conserved among the five species so far characterized for complete mt genomes. Genes of the more variable part are encoded on the opposite strand. Gene orders *cox1-cox2* and *nad6-cob* represent putative bilaterian ground pattern features, while *nad1-nad3-nad2* is otherwise found only in Annelida. The remaining arrangements seem to be unique for Chaetognatha.

19.2.2.2 Acoela and Xenoturbellida

The first complete mitochondrial genome sequenced from Acoela is that of *Symsagittifera roscoffensis* (Mwinyi et al., 2010). Gene order is completely different from that of other Metazoa. Sequence-based phylogenetic analysis of mt genome data did not give much evidence for the phylogenetic position of acoels. We found neither support for an affinity to platyhelminths nor to deuterostomes or *Xenoturbella* (Mwinyi et al., 2010). In contrast, the phylogenetic position of Xenoturbellida is well supported by mt genome sequence data and gene order. Conforming to EST-based phylogenomic analysis those support an affinity of *Xenoturbella bocki* to Deuterostomia (Bourlat et al., 2006). Concerning mt gene order the main difference to other deuterostome patterns is the position of *nad6* (Figure 19.1), which is likewise variable in the remaining deuterostome lineages (Perseke et al., 2007; Bourlat et al., 2009).

19.2.3 Lophotrochozoa

19.2.3.1 Platyzoa

Platyhelminthes, Rotifera, Acanthocephala and some smaller taxa are combined to the taxon Platyzoa (Cavalier-Smith, 1998), forming one clade of Lophotrochozoa. Mt genomes of Platyzoa have in common that all genes are transcribed from the same strand and that *atp8* is missing. Rotifers and acanthocephalans form the monophyletic group Syndermata (Ahlrichs, 1997), supported by morphological and molecular data sets (Witek et al., 2008; Witek et al., 2009). Mitochondrial gene order in acanthocephalan species are similar to each other with respect to protein-coding and rRNA genes, while tRNA genes have variable positions (Steinauer et al., 2005; Gazi et al., 2012; Weber et al., 2013). Acanthocephalan mt gene order shares the segment *nad4L-nad4-trnH-nad5* with the bilaterian ground pattern, but has otherwise nothing in common with other metazoan gene orders, except for bdelloid rotifers (Min and Park, 2009; Weber et al., 2013). Concerning protein-coding and rRNA genes, only three out of 14 genes have different relative positions in the two groups. The mt genome of the monogonont rotifer *Brachionus plicatilis* has very special features. It is divided into two circular subgenomes (11 kb, 12.5 kb) which both contain large regions of non-coding sequence (Suga et al., 2008).

All but one of the 39 platyhelminth mt genome records in GenBank are from parasitic species, altogether forming the clade Neodermata. Mt gene orders of Neodermata are not much different from each other. There are translocations of protein-coding genes in some *Schistosoma* species (Le et al., 2000) and the monogenean species *Microcotyle sebastis* (Park et al., 2007) and *Polylabris halychoeres* (Zhang et al., 2011). The neodermatan gene orders have almost nothing in common with the proposed ground patterns of Lophotrochozoa, Ecdysozoa and Deuterostomia. Recently a mitochondrial genome sequence of a non-parasitic flatworm was published via GenBank (*Dugesia japonica*, NC_016439), having a gene order which is completely different from both the parasitic flatworms and from other Metazoa. Thus, Platyzoa seem to have a long history of gene order rearrangements resulting in highly derived gene orders.

19.2.3.2 The lophophorate taxa and Entoprocta

Four mitochondrial genomes are known from Brachiopoda, each with a completely different gene order. Gene order of *Terebratulina retusa* (Stechmann and Schlegel, 1999) is very similar to that of nemerteans and some mollusks (Figure 19.2). The other brachiopod mt genome arrangements have no significant similarity to any other metazoan mt genomes (Noguchi et al., 2000; Helfenbein, Brown, and Boore, 2001). *Lingula anatina* (Endo et al., 2005) has an unusually large mt genome (28 kb) with large non-coding sections and several repeat regions including complete and partial genes.

Recent studies suggest a close affinity of phoronids to brachiopods. The mt genome arrangements of *Phoronis psammophila* (Helfenbein and Boore, 2004) and *Phoronopsis harmeri* (GenBank JN832704) differ only by the relative position of *atp6* and some tRNAs from the mollusk *Katharina tunicata* (Figure 19.2). However, the large inversion common to phoronids and *Katharina* probably occurred independently in both taxa (discussed in Section 19.2.3.3).

Ectoprocts are among the taxa with the highest variation in gene order among Metazoa (along with Copepoda and Tunicata). Each species has a different gene order and none of these shows significant similarities outside of Ectoprocta (Waeschenbach et al., 2006; Sun et al., 2009; Jang and Hwang, 2009; Sun et al., 2011; Nesnidal et al., 2011; Shen et al., 2012). Phylogenomic analyses resulted in weakly supported affinities to either Platyzoa or Entoprocta (Hausdorf et al., 2007; Dunn et al., 2008; Hejnol et al., 2009; Hausdorf et al., 2010) and excluded a close relation to brachiopods and phoronids.

Two entoproct mt genomes were described, which are very similar to each other and have similarities in gene order with other lophotrochozoan taxa (Yokobori et al., 2008). Entoproct gene order is derivable from the *Katharina* pattern assuming a large inversion of the part with *rrnL*, *cox3*, *nad3*, and *nad2* (Figure 19.2, orange segment).

19.2.3.3 Mollusca

Mt genomes of mollusks exhibit a great variation in gene orders and several taxa have accelerated substitution rates, resulting in long branches and polyphyletic Mollusca in analyses based on mt genome sequence data (e.g. Boore, Medina, and Rosenberg, 2004; Yokobori et al., 2007; Grande, Templado, and Zardoya, 2008). Some of the molluskan mt gene arrangements are similar to those of other lophotrochozoan taxa. The most probable ground pattern of Mollusca (the only pattern present in more than one class) is the pattern found in the polyplacophoran *Katharina tunicata*, some gastropods (e.g. *Haliotis tuberculata*) and cephalopods (e.g. *Octopus vulgaris*). This pattern is characterized by an inversion covering about half of the genome. The same part of the mt genome was inverted in Phoronida and probably as well in Entoprocta, here complicated by a second inversion (Figure 19.2). A gene order different from *Katharina* only by the lack of that inversion is found in nemerteans, in the brachiopod *Terebratulina retusa*, and in some other gastropods (e.g. *Ilyanassa obsoleta*, Figure 19.2). There is no way to explain the distribution of the patterns by a single inversion event under the current views of lophotrochozoan and molluskan phylogeny. The molluskan ground pattern very probably includes the inverted part. If this is assumed, the gene order of *Ilyanassa* is probably the result of a secondary reversal of that inversion. This hypothesis is supported by a block of tRNA genes kept on the other strand in *Ilyanassa* (Figure 19.2, blue genes), while this block is not found in patterns from outgroups without inversion (Nemertea, *Terebratulina*). Outgroups sharing the inversion seen in the molluskan ground pattern may be candidates for the sister-group of

Mollusca. Currently mt gene orders of Phoronida and Entoprocta show evidence for this large inversion. While entoprocts were discussed as putative sister-group to Mollusca, there is no current hypothesis promoting a close relationship of Mollusca and Phoronida. Differences in the tRNA cluster between *rrnS* and *cox3* (position of *trnG* retained in *Phoronopsis*) are hints to the idea that the big inversion occurred independently in phoronids and mollusks. Entoprocta cannot easily be included into this discussion as their mt gene order probably was subject to a second inversion involving some of the genes inverted by the first one. The tRNA cluster *trnC-trnY-trnW-trnQ* is shared by entoprocts and *Katharina*, providing weak support for a closer relation of entoprocts and mollusks.

19.2.3.4 Nemertea

Currently seven nemertean species have mt genome information in GenBank, covering three of the four major clades of this phylum. Gene order is similar to that of the mollusk *Katharina*, the brachiopod *Terebratulina* and the phoronids. Except for two tRNAs all genes are on the same strand. Inside Nemertea there is gene order variation concerning the positions of *nad6* and *atp8*. Here *Cephalothrix* species show a derived condition (Chen et al., 2009; Chen et al., 2011), while the other nemerteans show a pattern which is more similar to outgroup taxa like the brachiopod *Terebratulina retusa* (Podsiadlowski et al., 2009).

19.2.3.5 Annelida (sensu lato)

Sipuncula, Echiura, Pogonophora and Myzostomida are treated as separate phyla in older textbooks. Currently, these are found to be closely related or included into Annelida (Struck et al., 2011). Analyses of mt genome data give additional support to this result. Traditional (segmented) annelids have a clearly modified gene order among Lophotrochozoa (Figure 19.2), with protein coding and rRNA genes retaining their relative positions, while tRNA genes are variable (e.g. Zhong, Struck, and Halanych, 2008). Pogonophora (Boore and Brown, 2000) and Myzostomida (Bleidorn et al., 2007; Bleidorn et al., 2009a) share many derived features with this annelid pattern. Sipuncula (Boore and Staton, 2002; Mwinyi et al., 2009; Shen et al., 2009) and Echiura (Boore, 2004; Wu et al., 2009) have derived gene orders, but can be converted with a few steps, and share derived features with segmented annelids (for more details see Mwinyi et al., 2009).

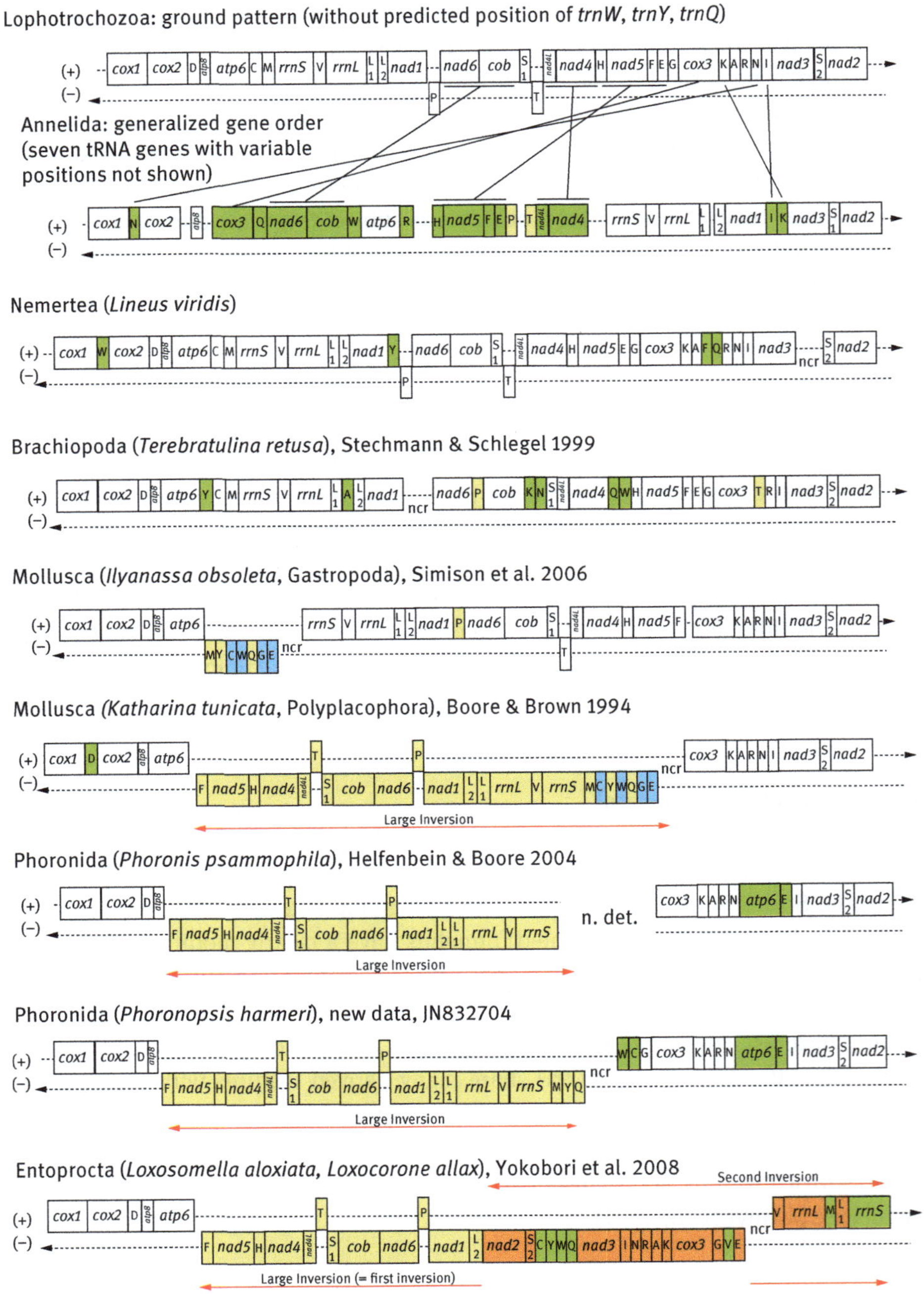
Lophotrochozoa: ground pattern (without predicted position of trnW, trnY, trnQ)
(+) cox1 cox2 D atp8 atp6 C M rrnS V rrnL L 1 L 2 nad1 nad6 cob S 1 nad4L nad4 H nad5 F E G cox3 K A R N I nad3 S 2 nad2
(−) P T
Annelida: generalized gene order
(seven tRNA genes with variable
positions not shown)
(+) cox1 N cox2 atp8 cox3 Q nad6 cob W atp6 R H nad5 F E P T nad4L nad4 rrnS V rrnL L 1 L 2 nad1 I K nad3 S 1 nad2
(−)
Nemertea (Lineus viridis)
(+) cox1 W cox2 D atp8 atp6 C M rrnS V rrnL L 1 L 2 nad1 Y nad6 cob S 1 nad4L nad4 H nad5 E G cox3 K A F Q R N I nad3 ncr S 2 nad2
(−) P T
Brachiopoda (Terebratulina retusa), Stechmann & Schlegel 1999
(+) cox1 cox2 D atp8 atp6 Y C M rrnS V rrnL L 1 A L 2 nad1 ncr nad6 P cob K N S 1 nad4L nad4 Q W H nad5 F E G cox3 T R I nad3 S 2 nad2
(−)
Mollusca (Ilyanassa obsoleta, Gastropoda), Simison et al. 2006
(+) cox1 cox2 D atp8 atp6 rrnS V rrnL L 1 L 2 nad1 P nad6 cob S 1 nad4L nad4 H nad5 F cox3 K A R N I nad3 S 2 nad2
(−) M Y C W Q G E ncr T
Mollusca (Katharina tunicata, Polyplacophora), Boore & Brown 1994
(+) cox1 D cox2 atp8 atp6 T P ncr cox3 K A R N I nad3 S 2 nad2
(−) F nad5 H nad4 nad4L S 1 cob nad6 nad1 L 2 L 1 rrnL V rrnS M C Y W Q G E
Large Inversion
Phoronida (Phoronis psammophila), Helfenbein & Boore 2004
(+) cox1 cox2 D atp8 T P n. det. cox3 K A R N atp6 E I nad3 S 2 nad2
(−) F nad5 H nad4 nad4L S 1 cob nad6 nad1 L 2 L 1 rrnL V rrnS
Large Inversion
Phoronida (Phoronopsis harmeri), new data, JN832704
(+) cox1 cox2 D atp8 T P W C G cox3 K A R N atp6 E I nad3 S 2 nad2
(−) F nad5 H nad4 nad4L S 1 cob nad6 nad1 L 2 L 1 rrnL V rrnS M Y Q ncr
Large Inversion
Entoprocta (Loxosomella aloxiata, Loxocorone allax), Yokobori et al. 2008
Second Inversion
(+) cox1 cox2 D atp8 atp6 T P V rrnL M L 1 rrnS
(−) F nad5 H nad4 nad4L S 1 cob nad6 nad1 L 2 nad2 S 2 C Y W Q nad3 I N R A K cox3 G V E ncr
Large Inversion (= first inversion)

19.2.4 Ecdysozoa

19.2.4.1 Arthropoda

Gene order of Arthropoda can be derived from a ground pattern which was first described in *Limulus polyphemus* (Staton, Daehler, and Brown, 1997) and is retained in some other chelicerates. Many hexapods and crustaceans differ from that ground pattern only in the position of *trnL2* (Boore, Lavrov, and Brown, 1998), an apomorphy for Pancrustacea. More or less variation in gene order is found in members from all large subtaxa. In Myriapoda almost all sequenced species show unique derivations from the arthropod ground pattern (Lavrov, Brown, and Boore, 2000; Lavrov, Boore, and Brown, 2002; Negrisolo, Minelli, and Valle, 2004b; Woo et al., 2007; Podsiadlowski, Kohlhagen, and Koch, 2007; Gai et al., 2008). Among "Crustacea" mainly decapod species have retained the pancrustacean ground pattern (the arthropod ground pattern, except for the different position of *trnL2*), while most other taxa have derived gene arrangements (Lavrov, Brown, and Boore, 2004; Kilpert and Podsiadlowski, 2006; Kilpert, Held, and Podsiadlowski, 2012). Some pycnogonids follow the arthropod ground pattern (Park, Lee, and Hwang, 2007; Masta, McCall, and Longhorn, 2010), while e.g. *Nymphon gracile* has a strongly rearranged gene order (Podsiadlowski and Braband, 2006).

In arachnids, recently first data were presented from Pseudoscorpiones (Ovchinnikov and Masta, 2012), Opiliones (Masta, 2010; Podsiadlowski and Fahrein, 2010), Amblypygi (Fahrein, Masta, and Podsiadlowski, 2009), Uropygi (Masta and Boore, 2008), Ricinulei (Fahrein et al., 2007), and Solifugae (Masta, Klann, and Podsiadlowski, 2008). Thus the arachnid taxon sampling reaches coverage of all major branches, except Palpigradi. Changes in gene order, although common, do not help to resolve interrelationships of the arachnid "orders" (Fahrein et al., 2007; Masta, 2010), as there are no derived gene orders known which are shared by species from different orders. On lower level they might become helpful, e.g. in spider phylogeny tRNA translocations are good apomorphic characters for some branches (Qiu et al., 2005; Fahrein et al., unpubl.).

In arthropods, there are some examples of unusual mt genome organizations. Linearized mt genomes are reported from the isopod *Armadillidium vulgare*, forming linear monomers and circular dimers in the same individual (Rigaud et al., 1999; Marcade et al., 2007). Several species of lice (Hexapoda: Phthiraptera) have fragmented mt genomes (Cameron et al., 2011), e.g. in the human body louse *Pediculus humanus* it is formed by 18 minicircle molecules (Shao, Kirkness, and Barker, 2009).

◂ **Figure 19.2:** Variation in gene order of lophotrochozoan mitochondrial genomes. Genes shaded in color indicate relative positions different from the ground pattern hypothesis. Green genes depict individual gene translocations. For the annelid gene order lines indicate the possible translocation events. Missing tRNAs in annelid species are found in the "gaps" indicated by the dotted line. Blue, yellow and orange genes were subject to inversions (see text, Section 19.2.3).

19.2.4.2 Onychophora and Tardigrada

Gene order in Onychophora is highly variable (Podsiadlowski, Braband, and Mayer, 2008; Rota-Stabelli et al., 2010; Braband et al., 2010a; Braband et al., 2010b). Two of the sequenced species, *Opisthopatus cinctipes* and *Peripatoides* spec. have retained the arthropod gene order with respect to the protein-coding and ribosomal RNA genes. In the latter species tRNA genes also have the same relative position as in arthropods, except for an inversion concerning *trnQ*. The mt genomes of the tardigrade species *Macrobiotus hufelandi* and *Thulinia* spec. differ from the arthropod ground pattern only with respect to the position of *trnI* and the major non-coding region (Rota-Stabelli et al., 2010).

19.2.4.3 Cycloneuralia

The mt gene orders of the priapulids *Priapulus caudatus* and *Halicryptus spinulosus* are very similar to that proposed for the arthropod ground pattern (Webster et al., 2006; Rota-Stabelli et al., 2010). The only difference is an inversion affecting a block of 18 adjacent genes. The strong GC bias in arthropods and the more moderate one in priapulids suggest a more recent inversion event in the priapulid lineage. Nematode gene orders are highly diverse and strikingly different from those of Priapulida and Panarthropoda. Only the gene order of *Trichinella spiralis* has some common gene adjacencies (*cox1-cox2, nd6-cob, rns-trnV-rnL, atp6-cox3*) to the proposed ecdysozoan ground pattern (Lavrov and Brown, 2001). Nematodes of the genus *Globodera* have six circles bearing different genes of the mt genome (Gibson et al., 2007). Members of the Loricifera seem to be the only animals without an mt genome (Danovaro et al., 2010). These tiny marine animals live in anoxic, sulfidic environments and their cells have hydrogenosome-like organelles, probably degenerated mitochondria. For Nematomorpha and Kinorhyncha there is currently no sequence information from mt genomes.

19.2.5 Deuterostomia

Among Enteropneusta the two *Balanoglossus* species differ only in a few gene positions (interchange of *nad6/cob* and four tRNA gene translocations, Figure 19.1) from the vertebrate pattern (Castresana et al., 1998; Perseke et al., 2010). *Saccoglossus kowalewskii* (NC_007438) has a different gene order, with interchange of *nad1/nad2* and much more shuffled tRNA genes. The first published mt genome from the pterobranch *Rhabdopleura compacta* has four protein-coding genes (*cox3, nad6, nad1, nad2*) in different relative positions compared to the deuterostome ground pattern as well as to the Enteropneusta (Perseke et al., 2011). This genome has one of the most extreme strand-specific nucleotide biases.

Gene order in Echinodermata is also variable and must have been modified by a number of different rearrangement types (Scouras and Smith, 2001; Scouras et al., 2004; Scouras and Smith, 2006; Perseke et al., 2008; Perseke et al., 2010).

Tunicata have a very high interspecific variation of gene order (Gissi et al., 2010), making it impossible to derive a scenario of gene order changes with the current taxon sampling. This is correlated with high substitution rates on nucleotide and amino acid level and therefore unreliable results in analyses concerning the phylogenetic position of Tunicata (Stach, Braband, and Podsiadlowski, 2010).

Most species of Cephalochordata differ in only four tRNA positions from the vertebrate ground pattern (Boore, Daehler, and Brown, 1999), but some species show a strongly derived mt gene order (Nohara et al., 2005; Kon et al., 2007).

The mt gene arrangement in vertebrates shows little variation, e.g. humans and hagfishes share exactly the same gene order (Rasmussen, Janke, and Arnason, 1998; Delarbre et al., 2002). Nevertheless a number of derivations from the vertebrate ground pattern are found. A few teleost fishes show novel gene orders compared to other vertebrates, notably members of the Anguilliformes (Inoue et al., 2001; Inoue et al., 2003), Osmerifomes (*Galaxiella nigrorostrata*, NC_008448), and Batrachoidiformes (Miya, Satoh, and Nishida, 2005). Among Amphibia there are derived gene orders found in Gymnophiona (San Mauro et al., 2006), plethodontid salamanders (Mueller et al., 2004), and frogs from the families Mantellidae (Kurabayashi et al., 2008), Dicroglossidae (Alam et al., 2010), and Ranidae (Kurabayashi et al., 2010). Sauropsid reptiles show mt genome rearrangements in Amphisbaenida (Macey et al., 2004) and in some monitor lizards (Kumazawa, 2007). All birds have a gene order different from the vertebrate ground pattern with an exchange of positions of *nd6* and *cob* (Desjardins and Morais, 1990). In addition there is some variation concerning multiple independent duplications of the control region and adjacent tRNA genes (Mindell, Sorenson, and Dimcheff, 1998; Gibb et al., 2007; Sammler, Bleidorn, and Tiedemann, 2011).

19.3 Conclusions

The frequently observed mix of ancient, moderately and highly derived patterns in mt gene order gives the impression that gene order evolves in a non-clock-like fashion. In some taxa the gene arrangement underwent no or only a few changes since the last common ancestor of Ecdysozoa, Lophotrochozoa or Deuterostomia (many deuterostomes, hexapods, *Limulus*, *Priapulus*, nemerteans, *Terebratulina*), while other species have a completely different gene order. Some taxa seem to undergo a high degree of rearrangements (Tunicata, Ectoprocta, Copepoda), as there are merely two species with the same arrangement. A few records of examples with linear and fragmented mt genomes hint to the fact that there is still not much known about the molecular mechanisms of mt genome rearrangements. Several more examples of aberrant mt genomes might emerge with the better taxon sampling of genomic sequencing approaches. Gene order is currently only a small help in phylogenetic analysis, restricted to a few cases where unequivocally derived gene orders are shared by different taxa, but homoplasious events may as well occur in gene rearrangements.

Acknowledgments

This study was supported by the German Research Foundation through grants DFG BA 1520/10-1,2 and DFG PO 765/4-3 in the frame of priority program 1174 "Deep Metazoan Phylogeny".

Part II: **New Tools and Methods**

Peter Grobe and Lars Vogt

20 Documenting Morphology: Morph·D·Base

Abstract: As the source for phenotypic data and by providing a terminology that functions as a basic reference system for all supramolecular biological entities, morphology plays a central and integrative role within life sciences. Unfortunately, the Linguistic Problem of Morphology and the problem of data loss due to the lack of adequate data repositories in morphology confine the impact of morphological research. After (i) providing a clear definition of what morphological data and metadata are and why media are not data but 'raw data' and (ii) claiming that in order to meet the requirements of eScience modern data and metadata standards must comprise four aspects (i.e. content, nomenclature, concept, format), we introduce *Morph·D·Base*, an online web portal and data repository for morphology. *Morph·D·Base* is intended to serve as a central repository for all kinds of morphological research. It allows the archiving, sharing and dissemination of morphological metadata and media, including 3D image stacks and their annotation using ontologies. It manages entries about taxa, biological specimens, media, literature, and morphological character matrices. Where appropriate, information is uploaded using controlled vocabularies. Entries can be released and receive a unique accession id under which they are openly available and freely accessible through the internet. *Morph·D·Base* allows the interlinking of entries and cross-linking to external resources. It enables collaborative editing of yet unreleased entries by assigning personalized *read only* or *editing* rights. We provide a brief introduction to the web interface, its basic functionality, the underlying design principles, the different entry types and how they can be interlinked.

20.1 Introduction

With the development of the so-called *high-throughput techniques*, data are being produced in quantities like never before. As a consequence, scientists often can no longer themselves look at all the data they produce. Instead, such *data-intensive science* uses computers and software applications to sift through data and evaluate what is relevant and qualitatively and quantitatively adequate for a given research program. Thus, research practice has changed dramatically within the last two decades. Whereas experimentation and theoretical modeling represent the two traditional driving forces for progress in science, and the development and availability of sufficient computer power allowed for computer simulations to significantly contribute as well (Gray, 2009), with the emergence of data-intensive science, *data exploration* emerged as an additional driving force (i.e. the *fourth paradigm*, see Gray, 2009). It represents a new and independent approach to research, called **eScience**, which is becoming increasingly important in all fields of empirical research and especially in the life sciences.

eScience depends on the development of adequate algorithms for the analysis and processing of information and requires data to be openly available, easily accessible via the Internet and compatible with each other (Hanson, Sugden, and Alberts, 2011; Lynch, 2009). Applications and services required for eScience usually focus on data capture, data curation, data mining, data analysis, and data visualization (Gray, 2009) and are used for data and text mining, inferencing, information retrieval and filtering, and the merging and integration of different data sets (Lynch, 2009). In order to be useful for the scientific needs of everyday research practice, the applications require data to be codified in a standardized way that makes data computer parsable, but also sufficiently semantically transparent. Thereby they typically utilize the Semantic Web (i.e. "Web 3.0"), *data repositories*, web services and data harvesters.

Along with the increasing importance of eScience, data repositories are becoming increasingly important in everyday research practice. Therefore, if a scientific community wants to participate in eScience and benefit from the increasing possibilities for (interdisciplinary) collaboration provided by the respective data repositories and software applications, they must make their data openly available and easily accessible to a broader scientific community via the Internet by means of uploading them to adequate data repositories. Many such data repositories are already available and extensively used by biologists, for instance NCBI GenBank (http://www.ncbi.nlm.nih.gov/genbank/) for sequence data or various data repositories for data about model organisms, for instance FlyBase (http://flybase.org/), Zebrafish Model Organism Database (http://zfin.org/), Worm-Base (http://www.wormbase.org/) and Saccharomyces Genome Database (http://www.yeastgenome.org/).

Within the last decades, a lot of new morphological methods and techniques for data production have been developed that bring a new quality to morphological data (see also Ramírez et al., 2007). These methods generally concern new sources of data and their representation and documentation using the possibilities that modern computers provide. They include, for instance, transmission electron microscopy, laser scanning microscopy, 4D microscopy, 3D reconstructions, molecular developmental biology and cell lineage tracing (e.g., Giribet, 2002; Jenner, 2004b). Consequently, the need for the development of a database for the collection, storage, management, dissemination and sharing, the organization, visualization, annotation and analysis of morphological data and their accompanying metadata and media is becoming increasingly urgent. Many projects that include morphological research are collaborative, with many researchers from different laboratories being involved. This requires secure management of personal information as well as differentiated and controlled data access rights. Moreover, if morphologists want to participate in eScience as well, they too must make their data and metadata openly available. All this requires the morphological community to establish adequate data repositories and to develop modern data and metadata standards for their domain that not only guarantee comparability and compatibility of data, but also computer parsability.

To address the needs for a data repository for morphology, we have developed the online relational database ***Morph·D·Base*** (https://www.morphdbase.de/) for morphological metadata and media contents. *Morph·D·Base* provides an interface for data entry via web forms, advanced data access and user rights management. It was originally developed for the needs of a particular multi-site project (Deep Metazoan Phylogeny, http://www.deep-phylogeny.org/) and subsequently extended and generalized as a freely accessible online morphological data repository. Before we provide an introduction to the main features of *Morph·D·Base*, however, we will briefly discuss the role of morphology in the life sciences in general, will comment on the distinction of data in metadata in morphology and the role of media and phylogenetic character matrices regarding the documentation of morphological data. We also discuss how a morphological data repository contributes to solving old problems of morphology and facing its new challenges and which components and criteria modern standards of data and metadata require for eScience.

20.2 The role of morphology in the life sciences

Morphology is a branch of biology that deals with the form of living organisms and the spatio-structural arrangement of their supramolecular parts. This includes the study of the outward appearance of organisms, for instance their shape, size, surface texture and coloring, but also of the internal parts of organisms, their intrinsic qualities and topological interrelationships.

Morphology is traditionally contrasted with physiology, which is a branch of biology that deals with the functions of organisms and their parts. In purely morphological studies, the active participation of organisms and their parts in causal processes and their role in biological systems is ignored. In other words, morphology studies biological material entities in the framework of the three dimensions of space while blending out the dimension of time. One could even argue that morphology merely represents a set of methods and techniques for producing descriptions of mind-independent and thus objective facts about the anatomy and spatio-structural organization of organisms. As a consequence, morphology is not a discipline that develops mechanismic explanations in the form of causal theories, but, instead, is *primarily* a **descriptive science** that is not explanatory per se. In short: morphology produces morphological data but no causal theories.

Although morphology is primarily a descriptive science, it nonetheless receives explanatory relevance through relating morphological data to causal theories, which in turn are typically provided by other biological disciplines. As a consequence, morphology provides data for research fields as diverse as ecology, physiology, neurosciences, and developmental biology (all of which primarily work *experimentally*), or medicine, phylogenetics, and historical evolutionary biology (which primarily work *diagnostically*). Morphological knowledge is also used in many practical questions of everyday research, for instance whenever structures (e.g. cell types) have to be identified or species membership has to be determined via diagnostic characters (Masci et al., 2009). In fact, due to the interdependence of form and function and the therewith involved fundamental role of anatomy in evolution, morphology provides in some way or other crucial structural background knowledge for every biological discipline. Thus, morphology is *"[...] one of the covering disciplines that spans every single entity in any biological organism"* (Gupta et al., 2007, p. 65; emphasis in the original). Morphological terminology takes over a central function in this context, as it not only has to provide adequate terms and concepts for the representation, communication and documentation of morphological data, but a well-organized inventory of concepts of different types of anatomical forms and their organization, and with this the terminology and the descriptive framework for the supramolecular domain of biology in general. Similar to biological taxonomy and nomenclature, which provide the basic reference systems for generalizing about individual organisms, a taxonomy of anatomical terms and concepts provides a **basic reference system for all supramolecular biological material entities**. Morphology therefore takes a central and integrative position among the otherwise rather diverse and heterogeneous field of the life sciences.

Therefore, it is not surprising that due to its central role in structuring and organizing bio-medical information, morphology also plays an important role in current efforts to develop knowledge representations of various biological disciplines, providing a common backbone for biological information integration (Bard, 2003; Bard and Rhee, 2004; Gupta et al., 2007; Rosse and Mejino, 2003; Smith et al., 2007; Stevens, Goble, and Bechhofer, 2000; Vogt, 2010). Thus, morphology is not only one of the oldest biological disciplines (Nyhart, 1995), but also plays a central role in the life sciences in general and substantially contributes to the conceptual foundation of many biomedical disciplines.

20.3 Data and metadata in morphology

Making observations and measurements about particular material entities and the processes they are involved in, either in the field or under a controlled environment within a laboratory, and recording and documenting the results represents an essential procedure in every empirical science. Usually, the results are reported in the form of **descriptive propositional statements**, and thus as texts, symbols and graphs (i.e. textual or graphic representational artifacts; Smith et al., 2006; Vogt, 2011). Reports that refer to a real correlate in nature and can be legitimately considered to represent mind-independent objective facts about a particular real material entity or event and can be referred to as **data** about that material entity or event. Thus, data are always existence or event statements that are based on and substantiated by observational judgments and empirical measurements. In morphology, data are *descriptive propo-*

sitional statements about the existence of particular organisms and/or their supramolecular part and their specific intrinsic material and topological properties or specific behavior they exhibit under certain conditions (Vogt, Bartolomaeus, and Giribet, 2010; Vogt, 2011).

Metadata, in contrast, represent all information that is required for a third person to reproduce the same data. This covers all relevant information about the *what, where, when, who* and *how* of the experiments and observations conducted and usually includes information about the experimental design, the source, preparation and treatment of the biological materials being studied, the methods, procedures and concepts applied, and the parameters, values, techniques and instruments used during data production. This information must also be documented in order to guarantee **reproducibility** and **procedural transparency** for a given research result.

Obviously, the distinction between data and metadata is not universal, but instead research-dependent – sampling location coordinates of specimens, for instance, represent data for biogeographical studies, but usually only metadata in phylogenetics.

20.3.1 Media are not data, but important nonetheless

An important property of data is that data always go beyond the private phenomenal world of an individual scientist's sensory experience – they go beyond the necessarily subjective descriptions of particular sensory phenomena. In fact, data represent first order **hypotheses about the existence of things and events in reality**, which in turn must be based on and **substantiated by observational judgments** (Vogt, Bartolomaeus, and Giribet, 2010) and thus **grounded in experience** that can be shared by all scientists in one way or another. As descriptive propositional statements and hypotheses about reality, data cannot be recorded and documented in the form of media files and images (i.e. *perceptual* or *aesthetic nonconceptual content*; Vogt, 2011), but as textual or symbolic propositions (i.e. *textual* or *semantic conceptual content*; Vogt, 2011). As a consequence, properly recording, documenting and organizing data requires a particular scientific terminology that enables reliable sharing and communication of the underlying contents in a form that (i) can be subjected to criticism and correction in principle and (ii) implemented for analyses using relevant algorithms and software. In other words, media recordings, images and drawings, all of which carry information in the form of perceptual contents, do not represent data in this sense, but can be referred to as **'raw data'**. Idiosyncratically annotated measurement curves and diagrams that are not publicly accessible in some enduring form also belong to this category. In contrast to data, media contents as such cannot be analyzed and first have to be translated into propositions and thus into data. Images, audio and video recordings, as well as type exemplars and voucher specimens, nevertheless play an important role in science, as they carry perceptual contents that can substantiate the existence hypotheses entailed in data by providing a link between the phenomenal realm of perception and the linguistic theoretical realm of the descriptive propositional statements of data through the reference to experience. By making relevant image plates, audio or video recordings openly available, scientists provide the empirical grounds for their data, thereby making the relevant perceptual contents openly available, so that other scientists can reconstruct and validate the perceptual judgments implicitly entailed in the data by themselves (Vogt, 2011). Data repositories play an important role with respect to the storing, archiving, communicating and disseminating of such 'raw data'. However, in order to be easily accessible, they must be properly annotated.

20.3.2 Phylogenetic character matrices are not morphological data either

Recently, it has been suggested to record and represent morphological data as discrete characters and character states (e.g. Edgecombe, 2008; Ramírez et al., 2007; see also Sereno, 2007). This has also become the current practice in some morphological databases, for instance MorphoBank (http://www.morphobank.org/) and Morphbank (http://www.morphbank.net/). This is insofar problematic, as phylogenetic character matrices represent sets of taxic and transformational homology hypotheses (Brower and Schawaroch, 1996; *character* and *character state homology hypotheses* sensu Freudenstein, 2005), which must be based and substantiated on morphological data themselves. Whereas morphological data is descriptive, morphological phylogenetic characters are explanatory, because they go beyond the perceptually given and provide possible evolutionary explanations for the perceivable sameness of anatomical entities (Vogt, Bartolomaeus, and Giribet, 2010). Sameness of anatomical entities is not a sufficient criterion for homology since sameness is not necessarily caused by common evolutionary history – otherwise we would not require the concept of homoplasy. Morphological data retain their validity independent of changes of homology hypotheses. In other words, morphological data does not depend on homology hypotheses, but homology hypotheses and thus character matrices depend on morphological data. Therefore, morphologists should be able to express and communicate a perceived sameness of anatomical entities independent of considerations of their underlying evolutionary history. Moreover, phylogeneticists should document in detail all morphological data relevant for a given character matrix. This would significantly increase the transparency and reproducibility of morphological phylogenetic character matrices and thereby meet one of the main criticisms of the use of morphological character matrices in phylogenetic publications – that they lack transparency and sufficient documentation (Hillis and Wiens, 2000; Jenner, 2002; 2003; 2004b; 2004c; 2004d; Poe and Wiens, 2000; Wiens, 2001; 2004).

20.4 Old problems and new challenges

20.4.1 The Linguistic Problem of Morphology

Unfortunately, morphology suffers from various theoretical, conceptual, and organizational problems and shortcomings. The overwhelming variety and diversity of anatomical forms and organizations, the long history of anatomical research and the influence of essentialist thinking, the diversely organized morphological communities with their various disparate traditions that are usually associated with the study of a specific taxonomic group, and the poor theoretical analysis and conceptual comprehension of key concepts of morphology resulted in **the Linguistic Problem of Morphology** (Vogt, 2008; 2009; Vogt, Bartolomaeus, and Giribet, 2010):

1. the lack of a commonly accepted, **standardized, taxon-independent** morphological terminology with concepts defined **without reference to underlying homology** assumptions;
2. the lack of a commonly accepted **standardized and formalized method for describing anatomical entities**;
3. the lack of a **rationale for the delimitation of morphological units.**

The Linguistic Problem of Morphology can lead to identical coding of traits in a phylogenetic character matrix that anatomically are actually significantly different. Moreover, it hampers the cooperation of morphologists from different taxonomic backgrounds and sometimes forces morphologists to trace back the different uses of a specific term by different authors in different time periods, thereby significantly slowing down scientific progress in this field. It is also responsible for a low degree of comparability of morphological data due to their consequently low degrees of semantic transparency and communicability, which affects studies of broader taxonomic range in particular, and often results in the necessity to re-examine what already has been studied before. The Linguistic Problem of Morphology thus prevents the integration of the morphological record (i.e. publications, morphological data and their underlying 'raw data'), and thus impedes the accessibility of morphological data to non-morphologists. Thereby it often leaves the impression that morphological research is more prone to subjectivity than many other biological disciplines (e.g. genetics, physiology – so called 'hard sciences'), thus weakening the position and reputation of morphology within the biological community (Vogt, Bartolomaeus, and Giribet, 2010). Ultimately, it prevents morphology from taking part in eScience.

Fortunately, the need for developing a taxon-independent morphological terminology that is free of implicit homology assumptions has now been recognized and involves the use of **ontologies** (Dahdul et al., 2010a; b; Deans, Yoder, and Balhoff, 2011; Haendel et al., 2009; O'Leary and Kaufman, 2011; Serna, Bolton, and Mackay, 2011; Vogt, 2008; 2009; 2011; Vogt, Bartolomaeus, and Giribet, 2010; Yoder et al., 2010).

20.4.2 Data loss and data repositories

Probably due to the terminology problems resulting from the Linguistic Problem of Morphology, no morphological data repository exists today that allows the archiving and dissemination of anatomical descriptions (i.e. data) and their underlying 'raw data' (e.g. image stacks, 3D reconstructions) for a larger taxonomic group, for instance the Metazoa. As a consequence, the scientific record for comparative anatomy, i.e., the **morphological record**, is currently still restricted to scientific publications. Michener (2006) points to the fact that data contributed by a particular publication degrades with time and becomes ultimately lost, strongly peaking after retirement and death, or after an early accident of its author. Thus, zoomorphology suffers considerably from data loss. Loss of data represents a major problem of science and does not only concern published data, since a lot of data generated in funded research are not analyzed, not published or, if at all, only presented in talks (Easterbrook and Matthews, 1992). Therefore, **data repositories play an important role for the prevention of data loss.** The development of adequate repositories in turn depends on the development of respective data and metadata standards. Thus, not only scientific communities but also funding agencies should have an interest in developing working solutions to the problem of data loss – and data repositories and standards are part of this solution.

20.5 Modern standards of documentation and communication of data and metadata

When standardizing data and metadata, one obviously must consider the role and function of scientific terminology in everyday research practice, and the specific problems connected to the use of language for communicating information in general (Vogt, 2011). In order to meet the requirements of eScience, a modern standard must not only establish (i) operational transparency for increasing reproducibility of data and (ii) compatibility and computer parsability of data and metadata for their use in data repositories and the Internet, but also (iii) semantic transparency for improving their communicability and comparability. Therefore, specifying relevant contents (see *minimum information convention*, Brazma, 2001; Brazma et al., 2001; Sansone et al., 2006) and preferable formats (Wang, Gorlitsky, and Almeida, 2005) for the development of a modern data and metadata standard will not be sufficient. Instead, when developing standards for the demands of eScience, one has to refer to the following four components (Vogt, 2009):

1. **Content Standard.** Content standards specify which metadata must be provided to enable a third person to reproduce the same results. It is, thus, the role of content standards to increase the overall degree of **reproducibility** and **operational transparency** of data production. Minimum information checklists

provide metadata content standards for a specific data type. However, common practice indicates that they may not necessarily cover the amount of metadata required for a full resolution of the analysis of the data.

2. **Nomenclatural Standard**. A nomenclatural standard establishes a **stable link** between a term, all its relevant synonyms, and its corresponding concept or the real entity it refers to. It turns out that with respect to nomenclatural standards one must distinguish between proper names, i.e. terms that refer to particular entities, and general terms, i.e. terms that refer to a certain type of entity. A nomenclatural standard for a proper name provides a stable and unambiguous *direct* referential link between the term and one particular material entity, a specific set of such entities, or a particular digital resource. In the context of the Internet, **persistent identifiers**, for instance globally unique identifiers (GUIDs), life science identifiers (LSIDs) and digital object identifiers (DOIs) establish direct reference and thus provide modern nomenclatural standards for proper names (Vogt, 2011). A nomenclatural standard for a general term, on the other hand, can only establish an unambiguous link between the term and a particular concept through an intensional definition – it cannot establish a direct reference to a real correlate, because general terms do not directly represent universal classes, natural kinds or types (Vogt, 2011).
3. **Concept Standard**. A concept standard specifies the **semantic content** of a concept. Besides an intensional definition specifying a set of ontologically essential properties that are often provided in the form of *Aristotelian definitions* (Vogt, 2008; 2009), concept standards should include a set of epistemological recognition criteria that enable a scientist to properly apply the term (Vogt, 2011), ideally without requiring further assistance. Whereas the former provides the meaning to a general term, only the combination of both provides unambiguous reference of a general term to its real correlate. Many scientific **ontologies** and **controlled vocabularies** provide modern nomenclatural and concept standards for general terms.
4. **Format Standard**. Format standards specify the **scheme** in which information should be recorded and presented. Format standards facilitate communication between humans, between humans and machines, and between machines by specifying a particular **formal syntax** that must be used for the transmission and communication of scientific content. They usually include the specification of **file** and **data exchange formats** for the recording, storage, and communication of data and metadata. Especially for the communication between data repositories and data harvesters such as the Global Biodiversity Information Facility (GBIF; http://www.gbif.org), or between specific instruments or software applications, data and metadata must be recorded in a **computer parsable** way and thus in a highly formalized format. By combining some kind of logic that provides operators (e.g. quantifiers and modal operators) with formal semantics for applying functions to symbols that, in turn, facilitate the drawing of new conclu-

sions from known knowledge, **knowledge representation languages** provide such format standards in the context of information management. The **Resource Description Framework** (RDF) and the RDF-based semantically extended **Web Ontology Language** (OWL) are frequently used knowledge representation languages for representing relationships between (web) resources.

Only the consideration of all four aspects provides data and metadata standards that meet the requirements of eScience. Fortunately, some standards that are relevant to morphology already exist and can be utilized. For instance the *DarwinCore* (http://www.tdwg.org/activities/darwincore) controlled vocabulary for specimen data and the Access to Biological Collections Data (ABCD; http://www.tdwg.org/activities/abcd/) controlled vocabulary for biological collection units. Unfortunately, these controlled vocabularies do not completely meet the above mentioned criteria. More serious, however, is the fact that morphology completely lacks modern data standards, because a morphological data standard necessarily requires a proper morphological terminology. This terminology, as has been discussed above, is still lacking due to the yet unsolved Linguistic Problem of Morphology (Vogt, 2008; 2009; Vogt, Bartolomaeus, and Giribet, 2010). In order to develop modern data and metadata standards for zoomorphology, the ***International Consortium of Zoomorphology Standards*** (ICZS; http://iczs.org/) was founded in 2011. The ICZS is a non-profit association that has been formed to establish international collaboration among the creators (i.e. morphologists), managers and users of morphological data about multicellular animals. The **designated goals of ICZS** are (i) to develop, adopt and promote standards and guidelines for the recording and exchange of morphological data and metadata concerning Metazoa, (ii) to promote the use of ICZS standards, act as a forum for the discussion of all topics relating to zoomorphology standards, and (iii) to provide a collaborative infrastructure for efficiently supporting standards' development. However, until adequate modern data and metadata standards have been developed, the morphological community does not profit from the benefits that modern standards bring, for instance with respect to data comparability, exchangeability, and overall dissemination. As a consequence, morphology is presently not prepared for eScience.

It is important to note, however, that data and metadata standards are not necessarily unique or exclusive. Like science itself, they evolve and frequently change. Moreover, a community usually does not agree on a single common standard and thus deals with several concurring standards. However, standardization of data and metadata is nonetheless important, as it provides the necessary reference framework for consistently documenting and communicating data. Thus, whenever data and metadata are recorded, documented or communicated, it is essential for their reliable communication to actually use and also specify the respective standards.

20.6 Morph·D·Base: A modern data repository for morphology

Morph·D·Base constitutes an online relational database for recording, archiving, managing, and sharing all kinds of 'raw data' (i.e. media files), data and metadata relevant to morphology according to modern standards and by using controlled vocabularies like *DarwinCore*. It provides the computer hardware and software for storage and easy access of large amounts of all kinds of media (e.g., traditional scientific graphics and drawings, TEM and SEM photos, photos of whole organisms and histological dissections, movies, virtual 3D and 4D models) and the recording of relevant metadata (e.g., specimen information, methods and techniques applied during specimen preparation). *Morph·D·Base* is freely accessible through a web interface in the Internet and allows the dissemination of morphological contents through the Internet all over the world. Access to *Morph·D·Base* is thus independent of a researcher's current location. Moreover, it is free of charge and no additional program (besides a web browser) is required for working with *Morph·D·Base*. Due to its sophisticated accession rights systems and the specification of user-defined groups of users, *Morph·D·Base* can also be utilized as a convenient tool for sharing morphological contents that are yet not openly accessible.

20.6.1 Historical background

From discussions with various morphologists we realized that working on a single and continuously growing character matrix usually involves fundamental conceptual problems, because character matrix entries tend to frequently change with the addition of new morphological information. This is due to the fact that phylogenetic character matrices represent sets of homology hypotheses and not morphological data sets (see above). Moreover, contrary to morphological data, phylogenetic character hypotheses are not epistemologically independent from each other – the addition of character states for a new operational taxonomic unit (OTU) to an existing character matrix can result in substantial re-coding of the already coded part of the matrix, because it frequently happens that the underlying homology hypotheses must be reevaluated with the addition of new morphological information. We realized that character matrices suffer from this epistemological interdependence of phylogenetic characters, whereas the underlying morphological descriptions are less prone to reevaluation with the addition of descriptions of hitherto not considered OTU. Therefore, we concluded that it is necessary to record and document morphological descriptions (i.e. morphological data) within a data repository, just like DNA sequences in NCBI GenBank. This should enable a clear distinction between data (i.e. descriptions) and subsequent conclusions (i.e. character and character state hypotheses) and should thus increase transparency and reproducibility of a morphological phylogenetic argumentation. In case of disagreement regarding specific phylogenetic hypotheses, the data repository should aid in analyzing the source of incongruence (i.e. is it on the level of the character hypotheses or on the level of the considered morphological data) and provide a platform for communicating and discussing it. In order to enable proper documentation and empirical substantiation for descriptions, the data repository should also allow the uploading of media files and relevant morphological metadata. This represents the main reason why we started developing *Morph·D·Base*: we were convinced that morphology requires a tool that enables a group of morphologists to continuously work on a data matrix that is directly linked to its underlying descriptions, while the descriptions themselves are understood as the actual morphological data.

Since we wanted morphological descriptions to be documented in a somewhat organized way, we conceived a basic scheme for recording morphological data in the form of free-text descriptions that are differentiated into several fundamental categories. These biological and cognitive categories correlate with different fundamental types of properties of an anatomical entity and represent the basic descriptive dimensions of every anatomical structure (see Vogt, 2008; Vogt, Bartolomaeus, and Giribet, 2010). We programmed a first version of *Morph·D·Base* in 2003 and tested it during a practical university course that was aimed at communicating the problems of morphological characters in phylogenetics and the difficulty of producing good (i.e. transparent and preferably reproducible) morphological descriptions. This first beta version of *Morph·D·Base* was a web platform for storing and documenting morphological data and for collaborating, sharing information and discussing structures and their possible interpretations in terms of character and character state hypotheses.

In 2004 we proposed *Morph·D·Base* as an integrative tool for morphologists for the priority program *Deep Metazoan Phylogeny* of the German Research Foundation (DFG; SPP 1174), which was in the final stage of planning. Unfortunately, the evaluation of feedback from using the first beta version of *Morph·D·Base* in further university courses showed that the scheme in which we represented morphological descriptions did not significantly improve on the Linguistic Problem of Morphology. The use of free text still brings about all the problems of the lack of a standardized taxon-independent morphological terminology, and many of the terms used referred to concepts defined with reference to underlying homology. This resulted in the problems associated with the Linguistic Problem of Morphology (Vogt, Bartolomaeus, and Giribet, 2010). Moreover, the proposed scheme of differentiating information into fundamental biological and cognitive categories did not provide sufficient standardization and formalization for the method of descriptions. We realized that the solution to the linguistic problem involves the use of controlled vocabularies and knowledge management systems, as they are commonly used in ontologies (Vogt, 2008; 2009; Vogt, Bartolomaeus, and Giribet, 2010). Unfortunately, using ontologies for comparative anatomy across a broader taxonomic range is a rather complex task that involves many conceptual and theoretical problems that must be solved before a proper software application can be developed (see e.g. Vogt, 2010; Vogt, Bartolomaeus, and

Giribet, 2011; 2012). Therefore we removed the description module from the beta version of *Morph·D·Base* – a new solution to the linguistic problem had to be developed involving ontologies, which, unfortunately, did not represent a short-term goal but required further theoretical and conceptual research. Thus, we had to change our short-term focus on the storing, representation and documentation of morphological 'raw data' (i.e. media files) and their accompanying metadata. October 25, 2006, the first working version of *Morph·D·Base* went online. It contained the working modules for entering and browsing information about specimen, taxa, and media as well as a taxonomy browser. Since then, *Morph·D·Base* has been constantly improved and several changes to the interface and additions of new modules, for instance for literature and for phylogenetic matrices, have been implemented.

20.6.2 Types of entries in Morph·D·Base

Five different categories of entries are distinguished in *Morph·D·Base*: taxa, specimens, literature, media, and phylogenetic character matrices. Content is uploaded to *Morph·D·Base* in the form of single entries consisting of several input fields, drop-down lists and check boxes that are organized into input forms. Whereas for each type of entry in *Morph·D·Base* a variety of different aspects of information can be provided by a user through the respective input form, the amount of information necessary for creating an entry is limited and depends on the type of entry. Where possible and appropriate, controlled vocabularies are used, either in the form of keywords, with which the uploaded information are associated, or in the form of drop-down lists or check-boxes. The use of controlled vocabularies provides a basic standardization of the uploaded contents and increases their accessibility and relocation (for a detailed scheme of the different entry types see: entry scheme under help in http://www.morphdbase.de).

20.6.2.1 Taxa

For the purposes of easy data exploration and the general accessibility of uploaded contents, *Morph·D·Base* uses a particular taxonomy as a *reference backbone*. All information in *Morph·D·Base* that is associated to a specific taxon, for instance a particular specimen entry or an image depicting a representative of a specific taxon, is accessible through this reference backbone. The taxonomy is mirrored from the **ITIS project** (http://www.itis.gov) and is not intended to represent a valid phylogenetic system, but merely an information reference and storage system. Like any other available taxonomy, however, the ITIS taxonomy is not complete. Therefore, users of *Morph·D·Base* can add additional taxa within *Morph·D·Base*, following the rules for adding a taxon provided by ITIS (ITIS Data Requirements; http://www.itis.gov/submit_guidlines.html). In order to create a taxon entry in *Morph·D·Base*, a user must provide the following information: to which kingdom the taxon belongs, its name, author(s) and year of first description, its taxonomic rank and its parent taxon within ITIS. All additional information that can be provided through the taxon input form of *Morph·D·Base* is not mandatory for creating a taxon entry. After its creation, the user is able to upload this information to the original ITIS database. ITIS will review and verify the information and the respective taxon will receive a unique *Taxonomic Serial Number* (TSN). Independent of whether the new taxon has been uploaded to ITIS, however, the user can immediately use the new taxon entry in *Morph·D·Base*, for instance, by linking it to already uploaded media files or by creating specimen entries for it. After ITIS has provided a TSN to the new taxon, thereby making its name publicly available through their taxonomy, *Morph·D·Base* will download and merge the new taxonomy with its older copy in *Morph·D·Base*.

20.6.2.2 Specimens

Information about particular specimens a user has investigated can be entered into *Morph·D·Base* as well. Such information includes, i.a., a specimen's taxonomic determination, sex specification, developmental stage, sampling data, or museum's collection information. Such additional information is, where possible, restricted to a closed vocabulary. The following information must be necessarily provided in order to create a new specimen entry in *Morph·D·Base*: a specimen label as an identifier for the real specimen, the number of individuals to which the entry refers, and their taxonomic determination in the form of a taxon name of the lowest taxonomic rank from ITIS that can be applied to the specimens. This guarantees that every specimen entry is linked to a taxon within *Morph·D·Base*'s ITIS taxonomy. The specimen input form is adapted to meet the common standard of specimen metadata annotation **DarwinCoreV2** (DwC 1.4 schema; http://rs.tdwg.org/dwc/tdwg_dw_core.xsd).

20.6.2.3 Media

Various types of media files can be uploaded to *Morph·D·Base*, including bitmap and vector graphics, series of images, audio and movie files. All uploaded images receive a watermark indicating the authorship of the respective media entry. The author of the entry can allow access to different resolutions of the original file, up to its original resolution, which will be accessible by clicking on the respective resolution index. After upload, the files can be viewed online in these resolutions, videos are streamed online, and series of images can be scanned by sliding a ruler back and forth (Figure 20.1). The following information must be necessarily provided in order to create a new media entry in *Morph·D·Base*: besides uploading the media file, the author must specify a title for the entry and a caption for the media file and the author of the media file must be specified. Besides uploading a single media file, it is also possible to upload several media files at once, e.g. images from a series of histological sections. A user only has to check the checkbox 'Upload series' to create a single media entry that will contain all files the user has uploaded in this step. If the checkbox is not checked, a separate media entry will be generated for each uploaded file. In order to prevent abuse, images automatically receive a watermark or a copyright statement showing the copyright information for download and presentation in *Morph·D*·Base. The author of the media entry can chose between watermark and copyright statement.

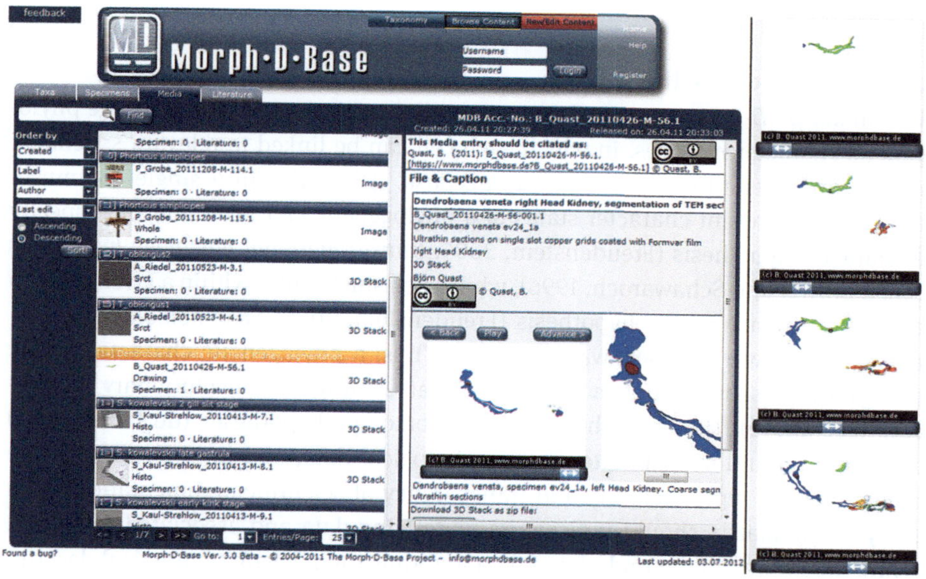

Figure 20.1: Media entry in *Morph·D·Base*. **Left:** Screenshot showing a released 3D stack (https://www.morphdbase.de?B_Quast_20110426-M-56.1). The interface is organized in a header containing the main navigation and a main display window. **Right:** Detail from the same stack, showing different images which are obtained by sliding the ruler located below the stack.

Additionally, but not mandatory, the files can be classified into general categories, such as CLSM, µCT, light microscopy, whole organism, diagram, or table. The range of objects that can be distinguished and identified according to their respective level of organization can be specified (i.e., from molecules to organelles, cells, epithelia, body regions and whole multicellular organisms). All annotated information is, where possible, restricted to a closed vocabulary. This enables subsequent searching for specific types of media entries, for instance CLSM or µCT media. Furthermore, in case a media file corresponds to another media entry, a specimen or a taxon entry in *Morph·D·Base*, the media entry can be cross-linked to this entry.

20.6.2.4 Literature

The literature module comprises all the information typical for literature databases. It is possible to upload existing literature data sets. *Morph·D·Base* reads the common formats, for instance RIS (e.g., used in Reference Manager, Endnote and Connotea [http://www.connotea.org]) and BibTeX (e.g., used in citeulike [http://www.citeulike.org]). Various entries in the different modules of *Morph·D·Base* can be linked to literature entries, in order to meet scientific conventions of citation and referencing. Each user maintains his/her personal literature database, only cross-linked literature entries are visible to other users and to the public.

20.6.2.5 Character matrix

Within *Morph·D·Base*, groups of users can collaboratively create and update phylogenetic character matrices, in which each cell can be linked to relevant specimen, media, and literature entries in *Morph·D·Base*. All entries in a single column of the matrix represent character states of a phylogenetic character and thus specify a character hypothesis (Freudenstein, 2005; *transformational homology hypothesis* sensu Brower and Schawaroch, 1996), while each single entry within a given column assigns a character state hypothesis (Freudenstein, 2005; *taxic homology hypothesis* sensu Brower and Schawaroch, 1996). Thus, a phylogenetic character matrix does not represent a set of morphological data, but a set of evolutionary hypotheses (see above). However, these primary homology hypotheses (de Pinna, 1991) are based on and substantiated by morphological data, and in order to meet the requirement of reproducibility and transparency of scientific argumentation, each entry in the matrix should indicate its underlying data and associated metadata. While the linking to the underlying data and thus the relevant anatomical descriptions must wait until an ontology-based description module has been developed for *Morph·D·Base* (see Section 20.6.5 *Further Development of Morph·D·Base*), all relevant metadata can already be linked (Figure 20.2). This is accomplished by giving the character and character state definitions as 'free text' statements, while each cell in the matrix entry can be cross-linked to media, specimens, and literature entries on which it is based upon.

The matrix module possesses many additional features for managing and co-operatively editing large character matrices online. This includes an efficient way of loading only those parts of the matrix that are currently displayed in the interface of the user; **co-operative live editing**, with *on-the-fly* saving and dynamic (i.e. immediate) update for all collaborators currently online; import and export of NEXUS files; color labeling of single cells and entire rows or columns for various purposes, for instance for tracking progress markers (i.e. 'new', 'in process', 'completed'), for a schedule-notification system (e.g. contributor *X* must complete cells/columns/rows *XYZ* of the matrix until date *Z*), or for tracking changes, etc.

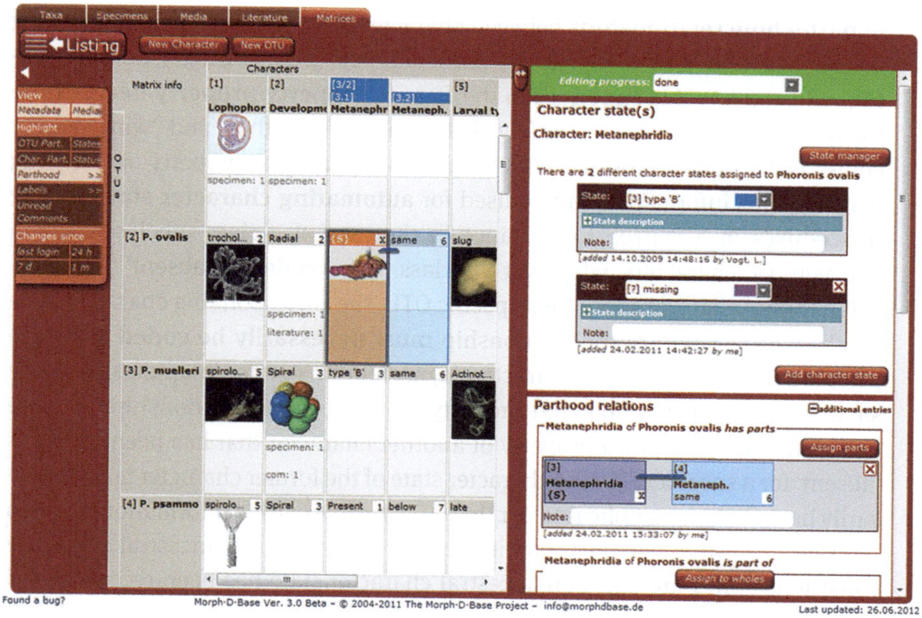

Figure 20.2: Matrix entry in *Morph·D·Base*. The editing window is divided into three sections which serve different purposes. The leftmost section holds buttons to change the view on the matrix, e.g., displaying thumbs of cross-linked media entries and switching on/off information regarding number and kind of cross-linked metadata for each cell. In the center section the matrix is displayed where a cell can be selected for editing content. The rightmost section displays details from the selected cell and allows the definition of character states, parthood relations, cross-links to other entries within *Morph·D·Base* and the creation of color labels for workflow direction. In the top the editing status of this cell can be changed.

Two features are new and innovative for phylogenetic character matrices:

1. The matrix module of *Morph·D·Base* will allow for **coding of alternative characters**, which can be used in case the (topographical) correspondence between anatomical entities of two or more specimens belonging to different operational taxonomic units (OTUs) cannot be unambiguously assessed and, as a consequence, a single character homology hypothesis (i.e. *transformational homology hypothesis* sensu Brower and Schawaroch, 1996) cannot be assigned to the respective anatomical entity, because more than one character homology hypothesis must be considered. This feature is currently under development and will be implemented with an upcoming new version of *Morph·D·Base*.
2. One can specify **dependences between character states belonging to different characters**. These dependences cover (i) **ontological subsumption dependences** in the form of **class-subclass relationships** between characters (e.g. between the characters 'brain' and 'syncerebrum' because a syncerebrum is a specific type of brain), (ii) **biological parthood dependences** in the form

of **part-whole relationships** between characters (e.g. between 'wing' and 'wing feather', because a wing feather is a particular part of some wings), and (iii) **biological property dependences** in the form of **property-property bearer relationships** between characters (e.g. between 'wing feather' and 'wing feather color', because the color of a wing feather is a particular property of this wing feather). This information can be used for **automating character state coding procedures** in a matrix: for instance in the case of a class-subclass relation between two characters, whenever the class part is coded as 'absent' (e.g. character: 'brain'; state 'absent') for a specific OTU, the corresponding character state for the subclass part of the relationship **must necessarily be coded as 'inapplicable'** for this OTU (e.g. character: 'syncerebrum'; state: 'inapplicable'). The same applies to part-whole and property-property bearer relationships: in case one character is part of/a property of another character that has been coded as 'absent' for a specific OTU, the character state of the former character must necessarily be coded as 'inapplicable' for this OTU. Moreover, this information can also be used for testing whether combinations of reconstructed ancestral states are consistent – they are not if an ancestral character state has been reconstructed as having a specific value although, according to the above mentioned rules, it necessarily had to be 'inapplicable'.

While the matrix module of *Morph·D·Base* is still under development, a beta version is already functional and online. Its functionality allows the editing of matrices and within the priority program "Deep Metazoan Phylogeny" a large matrix, containing 122 × 131 cells is currently being edited by 23 different users. This matrix contains character hypotheses from the neuroanatomy of different Metazoa (see Chapter 21 by R. Lösel and S. Richter: Neurophylogeny – from description to character analysis)

20.6.2.6 Linking contents: Internal and external cross-links

As already mentioned above, in many types of entries in *Morph·D·Base* cross-links can be assigned that create two-way referential links between different entries within *Morph·D·Base* (Figure 20.3). For instance, in the case where a specific image depicts a particular specimen, the author of the respective media entry can create a cross-link to the respective specimen entry, thereby cross-linking these two entries. As a consequence, formerly unrelated entries become interconnected. The cross-links thereby can document the underlying biological relations of different entries within *Morph·D·Base*. All relevant information for a cross-link is displayed in a standardized way, while still allowing comments or the posting of specific information that only refers to the particular cross-link itself (e.g. a new figure caption for a cross-linked media entry, Figure 20.3b).

Besides these internal two-way cross-links, *Morph·D·Base* also allows the creation of various one-way cross-links to particular external web resources. These can

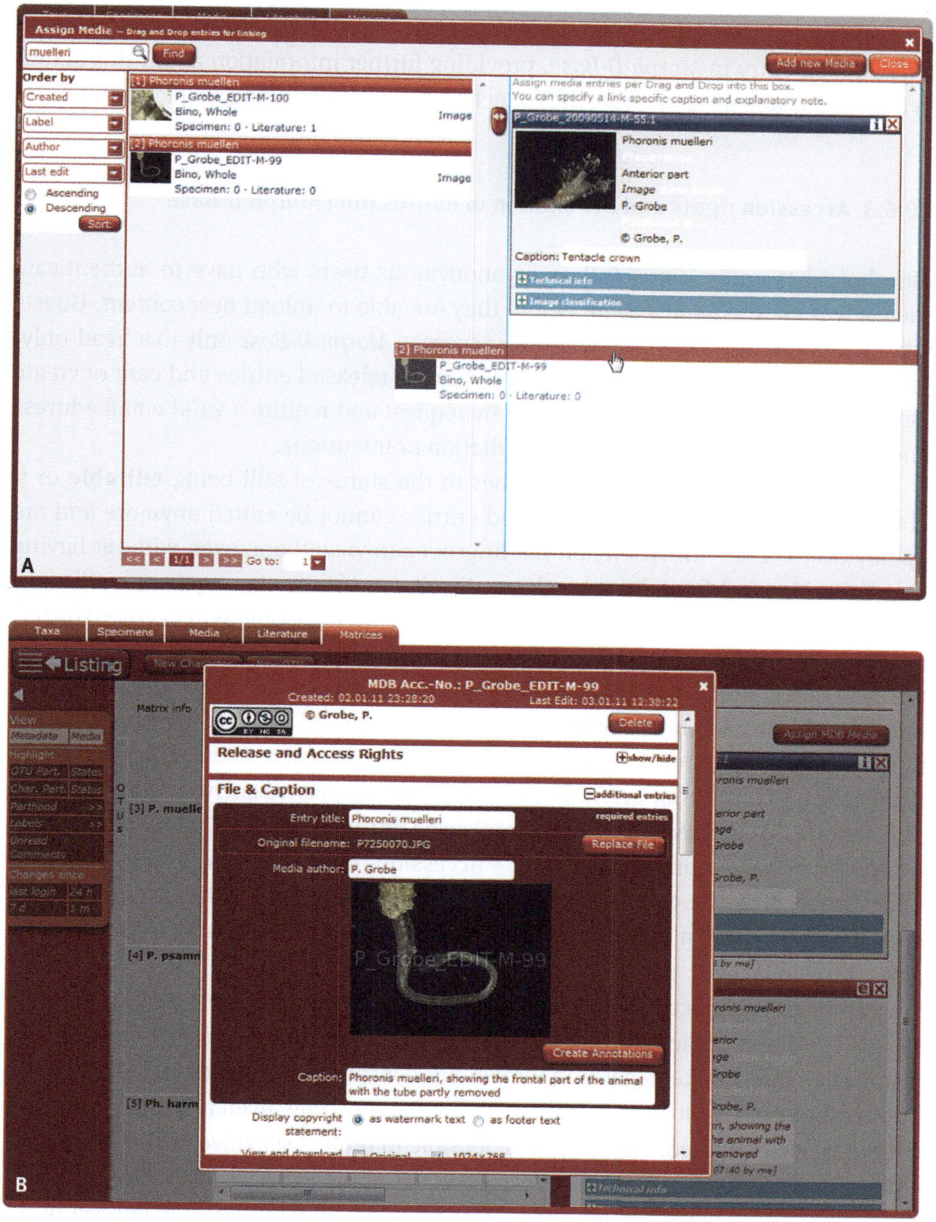

Figure 20.3: Cross-link media in a matrix entry: **A.** Entries can be cross-linked in a popup window by simply dragging and dropping the respective entry from the left browsing list into the field on the right. Cross-linked entries are immediately saved and a cross-link specific caption can be given. In the assignment view all relevant information of the entry are shown. **B.** Cross-linked entries can be edited directly by clicking on the [e] button – providing one has the appropriate editing rights.

be used for additional documentation, for instance for linking an external website to a new taxon entry in *Morph·D·Base*, providing further information about this taxon, or a museum web page of a specific collection to a corresponding specimen entry.

20.6.3 Accession rights and the citation of entries from Morph·D·Base

Morphologists enter *Morph·D·Base* as anonymous users who have to authenticate themselves via the login system before they are able to upload new content. Guests who do not possess a user account have access to *Morph·D·Base* only in a 'read-only' mode with restricted rights: they can only access released entries and cannot create new entries. User accounts are granted by request and require a valid email address and evidence of respective scientific affiliation or intentions.

Each entry in *Morph·D·Base* is either in the status of still being **editable** or it already has been **released**. All released entries cannot be edited anymore and are openly accessible – everybody on the Internet can view them, even without having to login to *Morph·D·Base*. Prior to release, the owner of the entry must apply a license to the entry. The user can choose from six different Creative Common licenses (http://creativecommons.org) and assign one of them to the entry (Figure 20.4b). Every released entry receives a unique accession id, which is composed of the username, the date of release, an acronym for the content (e.g. 'S' for Specimen), a running number and a version number (see Figure 20.4a). This accession number enables the creation of links from external resources to the respective entry in *Morph·D·Base*. This feature, *MDB Direct*, makes every released entry available by using a short URL composed of www.morphdbase.de + ? + Accession Number (e.g., www.morphdbase.de?P_Grobe_20061115-S-79.1). Through the accession id and *MDB Direct*, released entries can be cited in publications and on the web and can be accessed through a direct link.

Although entries cannot be edited anymore after they have been released, in the case of specimen entries *Morph·D·Base* allows its owners to still change the visibility status of all information regarding the sampling locality of a specimen. This sensitive information can be blocked for all other users at any time, even after the respective entry has been released. This way, it is possible to protect sensitive information from abuse.

As long as an entry has not been released, it is still editable. To share yet unreleased information among users within *Morph·D·Base*, users can create groups of users and invite other users to join these groups. Access rights to an editable entry can then be granted to single groups and the respective right applies to all its members (Figure 20.4b). Access rights for editable entries are either **read only**, allowing all group members to read but not to change the entry (though the owner of the entry can still edit it), or they are **editable**, allowing both the owner and all group members to edit the entry. Only registered users can create or join groups – guests cannot. Only the owner of the entry is allowed to change the access rights of an entry, release it to the public or completely delete the entry from *Morph·D·Base*.

In other words, only logged-in users have access to the section of the interface in which new entries can be created or existing but yet unreleased entries can be edited. Users have access to all the entries they own as well as all entries others have granted them *read only* or *editable* rights. Thus, *Morph·D·Base* uses a flexible group management and access rights system that allows users to archive all information uploaded to *Morph·D·Base*, keep entries that are not ready for release hidden from all other users, share entries that a user wants to share with specific colleagues and only them through the specification of user-defined groups, and release entries that the user wants to be accessible to the general public.

20.6.4 Interface design and usability of Morph·D·Base

During the development of *Morph·D·Base*, we put a lot of effort into designing a proper web application with a user-friendly interface. Thereby, we followed basic design principles, for instance consistently applying a specific layout over the entire interface, never letting the interface be overloaded with content, always keeping the navigation tree flat, directly mapping the screen flow to the task of the users, communicating basic database meanings with a consistent color coding scheme of the interface, using short but meaningful labels, making the content of the interface easily discriminable and comprehensible to the users, keeping the main tasks easy to accomplish, and providing instant gratification whenever possible.

Where appropriate, we also provide 'best practice' advice for data input to avoid inconsistencies or incompleteness, for instance when creating a new taxon entry without providing a literature cross-link for the first description of the taxon – while this information is not necessarily required for creating a new taxon entry in *Morph·D·Base*, it nevertheless represents valuable information that should be provided by the user if available.

20.6.4.1 The web interface: General Organization

The interface of *Morph·D·Base* is organized into two sections: the header at the top that contains the main navigation tool and the content window that displays all the information content of the database (Figure 20.1). Short information messages (i.e., tooltips) about the functionality of a button, navigation links or the meaning of an input field appear dynamically when hovering with the mouse pointer over the respective area of the interface.

The header itself is divided into two areas. The right area refers to user specific information, profile settings and general information about *Morph·D·Base*. The top area of the header contains the main navigation links to the actual entry modules of

Morph·D·Base. A login form is also located in the header, allowing login to registered users anytime during browsing.

The content window of *Morph·D·Base* takes up most space of the interface and is situated below the header (Figure 20.1). Depending on the color, users are in the 'read mode', in which they either look for general information about *Morph·D·Base*, including help functions and user-specific information (grey-blue) or browse and search particular entries (dark blue), or they are in the 'write mode', in which they can upload new content to *Morph·D·Base* (i.e. create new entries) or edit existing but yet not released entries (red).

Entries in *Morph·D·Base* are accessible through the dark blue **'Browse Content'** button in the case of released entries and through the red **'New/Edit Content'** button in the case of still editable entries. Users only have access to those editable entries in *Morph·D·Base* to which they possess the respective accession rights (see above). All accessible entries are depicted in an ordered list in the left part of the content window, from which single entries can be selected in order to access their information, which in turn is depicted in the right part (Figure 20.4).

20.6.4.2 The web interface: Creating and editing content

In addition to the above mentioned general design principles for developing the interface of *Morph·D·Base*, we also followed specific design principles for establishing procedures and workflows for creating and editing entries in *Morph·D·Base*, and we adapted our interface design to optimally support them. For instance, we wanted the interface to support incremental construction, i.e. being able to make a series of small changes instead of having to completely create something in a single session. In *Morph·D·Base* we achieved this by providing input forms with a small number of fields that support instant saving instead of a single and rather large form of multiple input fields and a single save button at the bottom of a web page.

We also wanted the interface and the workflows in *Morph·D·Base* to allow for changes in midstream. It is a problem of many currently available biological online databases that information that is actually interlinked (e.g., specimen information and images of the specimen) has to be uploaded through separate web pages (e.g., specimen information through a 'Specimens' page and image files and their annotations through a 'Media' page). Usually, users have to finish creating one entry before creating another one. This is inconvenient, especially in those cases in which, for instance, one realizes during the procedure of creating a new specimen entry that the relevant media files have not yet been uploaded. It would thus be very good if one

Figure 20.4: Color coding the workflow in *Morph·D·Base*. **A.** The 'Browse Content' and 'Taxonomy' buttons in the header gives access to all released entries and to the reference taxonomy. **B.** The 'New/Edit Content' button gives access to the section in which new entries can be uploaded, existing ones edited and released. ►

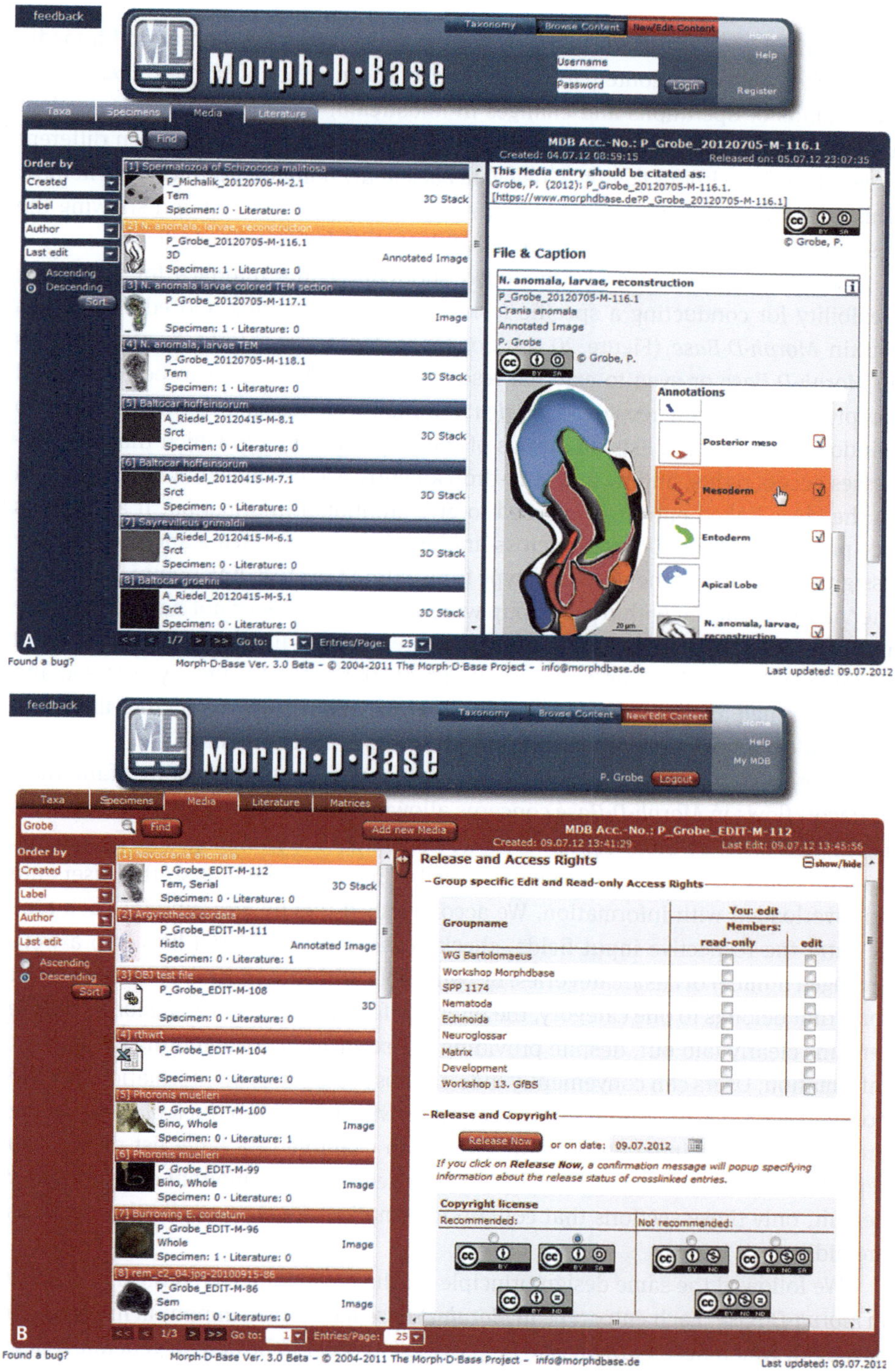
feedback
Morph·D·Base
Taxonomy
Browse Content
New/Edit Content
Home
Help
Register
Username
Password
Login
Taxa
Specimens
Media
Literature
Find
Order by
Created
Label
Author
Last edit
Ascending
Descending
Sort
MDB Acc.-No.: P_Grobe_20120705-M-116.1
Created: 04.07.12 08:59:15
Released on: 05.07.12 23:07:35
This Media entry should be cited as:
Grobe, P. (2012): P_Grobe_20120705-M-116.1.
[https://www.morphdbase.de?P_Grobe_20120705-M-116.1]
© Grobe, P.
File & Caption
N. anomala, larvae, reconstruction
P_Grobe_20120705-M-116.1
Crania anomala
Annotated Image
P. Grobe
Annotations
Posterior meso
Mesoderm
Entoderm
Apical Lobe
N. anomala, larvae, reconstruction
[1] Spermatozoa of Schizocosa malitiosa
P_Michalik_20120706-M-2.1
Tem
Specimen: 0 · Literature: 0
3D Stack
[2] N. anomala, larvae, reconstruction
P_Grobe_20120705-M-116.1
3D
Specimen: 1 · Literature: 0
Annotated Image
[3] N. anomala larvae colored TEM section
P_Grobe_20120705-M-117.1
Image
[4] N. anomala, larvae TEM
P_Grobe_20120705-M-118.1
[5] Baltocar hoffeinsorum
A_Riedel_20120415-M-8.1
Srct
[6] Baltocar hoffeinsorum
A_Riedel_20120415-M-7.1
[7] Sayrevilleus grimaldii
A_Riedel_20120415-M-6.1
[8] Baltocar groehni
A_Riedel_20120415-M-5.1
A
Go to:
Entries/Page:
Found a bug?
Morph·D·Base Ver. 3.0 Beta - © 2004-2011 The Morph·D·Base Project - info@morphdbase.de
Last updated: 09.07.2012
My MDB
P. Grobe
Logout
Matrices
Add new Media
MDB Acc.-No.: P_Grobe_EDIT-M-112
Created: 09.07.12 13:41:29
Last Edit: 09.07.12 13:45:56
Release and Access Rights
show/hide
Group specific Edit and Read-only Access Rights
Groupname
You: edit
Members:
read-only
edit
WG Bartolomaeus
Workshop Morphdbase
SPP 1174
Nematoda
Echinoida
Neuroglossar
Matrix
Development
Workshop 13. GfBS
Release and Copyright
Release Now
or on date: 09.07.2012
If you click on Release Now, a confirmation message will popup specifying information about the release status of crosslinked entries.
Copyright license
Recommended:
Not recommended:
[1] Novocrania anomala
P_Grobe_EDIT-M-112
Tem, Serial
[2] Argyrotheca cordata
P_Grobe_EDIT-M-111
Histo
[3] OBJ test file
P_Grobe_EDIT-M-108
[4] rthwrt
P_Grobe_EDIT-M-104
[5] Phoronis muelleri
P_Grobe_EDIT-M-100
Bino, Whole
[6] Phoronis muelleri
P_Grobe_EDIT-M-99
[7] Burrowing E. cordatum
P_Grobe_EDIT-M-96
Whole
[8] rem_c2_04.jpg-20100915-86
P_Grobe_EDIT-M-86
Sem
B

could switch back and forth between, for example, a specimen entry and a media entry, when uploading content to *Morph·D·Base*. The interface of *Morph·D·Base* allows these kinds of operations and changes in midstream, thereby enabling convenient upload and editing procedures: users can switch back and forth between different types of entries without having to navigate back to the point they left when switching. This is accomplished by using tabs for navigation, i.e., by hiding and showing the respective pages. These tabs are situated at the top of the content window.

Wherever possible we also included shortcuts in the interface for increasing flexibility for conducting a specific task, for instance creating and editing entries within *Morph·D·Base* (Figure 20.3c). So, for instance, cross-links to other entries in *Morph·D·Base* or even to external resources (e.g. web pages) can be established simply by clicking the respective '*Assign X*' button (e.g. '*Assign Literature*'). A popup window opens, which lists all entries of the respective type to which the user possesses access rights, and by drag-and-drop an entry is cross-linked. A comment field in the cross-linked entry can be used to give additional information. If an entry is not present in *Morph·D·Base* for cross-linking, the user can create and immediately assign a new entry of the requested type from within the cross-link popup by clicking the '*Add new*' button within the popup window – the user does not have to go to the page of the respective entry type. Moreover, already cross-linked entries that are still editable and for which a user possesses *editable* rights can be directly accessed and edited through an '*e*' button from the page of the entry it has been cross-linked to, which allows convenient direct editing of the cross-linked entry itself.

Another design principle that we have followed for the interlinking of interface and workflows in *Morph·D·Base* concerns allowing for the deferring of choices. This is accomplished by breaking down input forms into better manageable fragments, thereby addressing as few aspects as possible in each fragment so that users are not overloaded with information. We accomplished this by grouping for each type of entry the respective input fields, check boxes and drop-down lists into a well-arranged number of basic categories. By being able to hide or show the part of an input form that belongs to one category, the interface for editing an entry in *Morph·D·Base* remains clearly laid out, despite providing for extensively documenting all relevant information. Users can conveniently hide all those categories of information that are not relevant for their current purpose and this way better keep track of their editing progress. The specimen entry input form is an example of a large list of over 110 separate information entries, which is categorized into nine different categories. Per default, only those sections that contain information are shown, all other sections are hidden.

We followed the same design principle for the procedure of creating a new entry in *Morph·D·Base*. Each entry type in *Morph·D·Base* has its own minimum information specified, which necessarily has to be provided in order to be able to generate a new entry of this type in *Morph·D·Base* (see above). New entries can be created by pressing the '*Add new X*' (e.g. '*Add new Specimen*') button on the respective entry type page in the red '*New/Edit Content*' section of the interface, and a popup window opens that only contains the input fields for the minimally required information for the respective entry type.

20.6.5 Further Development of Morph·D·Base

As the capacity of the storage space in *Morph·D·Base* is limited, we will implement a financing model for managing the storage space that uploaded media files take up. According to this model, *Morph·D·Base* will grant each user a specific amount of free storage space for their uploaded media files. Individual users, but also groups of users and projects, can extend their storage space by buying additional space from *Morph·D·Base*. This will be especially necessary when uploading multiple large files of image stacks. With this financing model, the currently available storage space would be evenly distributed among the users and additional space would be bought according to demand.

This is, of course, not the only idea we have of how we could further improve *Morph·D·Base*. For instance, currently we are developing an annotation tool for 3D image stacks. Annotations are shapes drawn on single images with labels explaining the shape. The tool will allow linking these labels to an ontology, which can be loaded from one of the many existing OBO ontologies (http://obofoundry.org). This annotation module will also enable importing different annotation formats made with other software packages (e.g., ImageJ, http://rsb.info.nih.gov/ij/) together with the corresponding image stacks to *Morph·D·Base*.

Dependent on further funding is the development of a tool for conducting highly formalized ontology-based morphological descriptions based on RDF triples (e.g. Vogt, Bartolomaeus, and Giribet, 2010), which would accompany traditional natural language descriptions in a newly developed description module. This tool would establish an ontology-dependent local standard for morphological data within *Morph·D·Base*. The addition of an ontology-based description module to *Morph·D·Base* would also increase *Morph·D·Base*'s overall functionality. By providing the possibility to enter and disseminate morphological data independent of any phylogenetic hypothesis (i.e. no character or character state hypotheses), *Morph·D·Base* would close the gap between the morphological work in the field and laboratories and the subsequent use of morphological data in various applications and biological studies, and would **sustainably store morphological data** that could be used and **re-used** for all kinds of investigations, not only for phylogeny reconstruction. Especially the utilization of ontologies will increase the general value of morphological data, since the respective descriptions would be highly standardized and formalized and therefore also **better accessible to non-morphologists**. With this module, *Morph·D·Base* could efficiently mediate between morphological investigations, data production, and the various disciplines that use morphological data.

The utilization of anatomy ontologies in *Morph·D·Base* for annotating images and conducting morphological descriptions will also substantially improve possibilities for conducting well-directed searches for specific morphological data entailed within media and description entries – for instance by utilizing the *part_of* property relation (see Vogt, 2008; 2009).

20.6.6 Technique

Morph·D·Base is freely accessible via the Internet through a web browser based graphical user interface. The main components of the architecture are separated into (i) a software middleware controlling and coordinating data access and data flow between user interface and storage, (ii) the storage system for database entries and media files on the files system, and (iii) the web browser based user interface for data input and editing.

Morph·D·Base is built within the web application platform Zope (http://www.zope.org) and uses the object database ZODB and the relational database MySQL (MySQL 5.5; http://www.mysql.com) for storing the metadata and user profiles (Figure 20.5). The media files (i.e. the 'raw data') are stored on the file system. The file system storage consists of a hardware RAID system with a current capacity of 50 TB and an additional cache space with a capacity of 10 TB. The cache space is used for fast access to recently used media files and as a buffer to provide smaller views of the media files uploaded to *Morph·D·Base*. To guarantee maximum performance and high availability together with high security, the system is mirrored between two servers.

The hardware components consist of several HP Proliant Servers. One server acts as web proxy, directing the traffic to the software and database server framework behind. A separate image server processes the uploaded media and renders smaller views, which are generated on first access and saved in the file cache for quicker access. A file server controls the access to the 50 TB disk space, which constitutes the long-term storage for all files uploaded to *Morph·D·Base*. All hardware components are housed by the Computing Center of the University of Bonn and the Zoological Research Museum Alexander Koenig in Bonn.

In order to fulfill the design principles mentioned above, the whole interface is implemented using Web 2.0 techniques: changes conducted on data in the interface are transferred to the server in the background using AJAX (Asynchronous JavaScript and XML) web technology, without causing noteworthy interference with the interface – except for giving feedback to the user about the success or failure of the data transfer and storage.

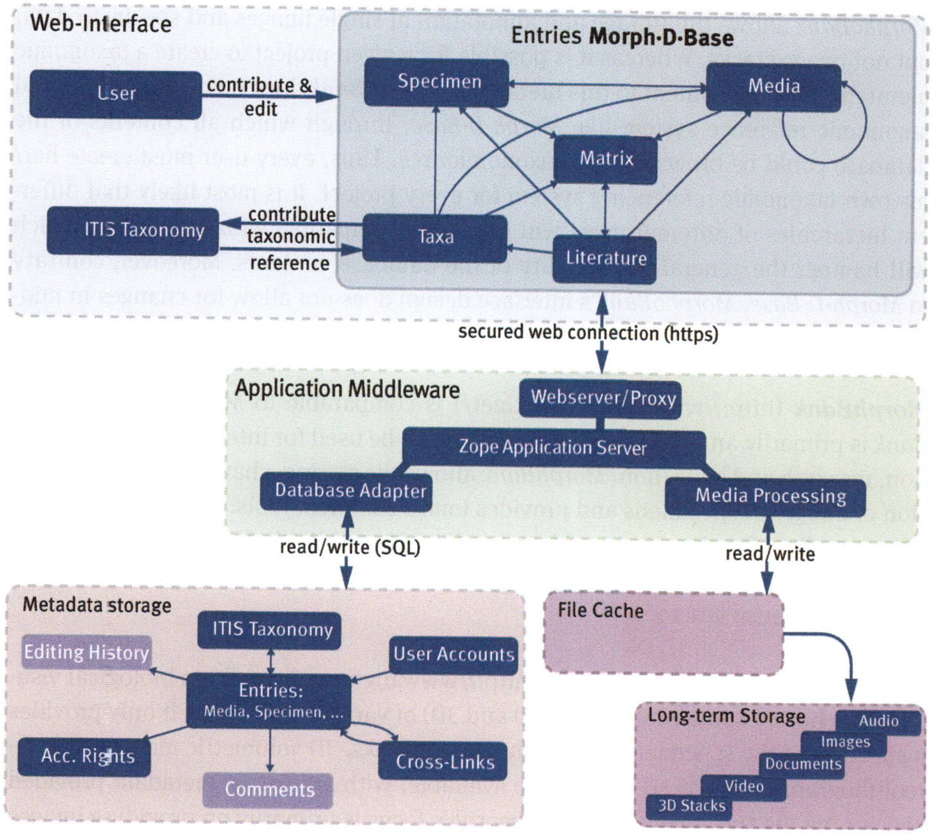

Figure 20.5: Architecture of *Morph·D·Base* and data flow scheme: The interface allows editing and uploading of new content to the different entry types, which can be cross-linked with each other. The application middleware receives data from the interface and gives controlled access to the metadata storage as well as to the file cache and long-term storage system.

20.6.7 Similar Databases

Most biological databases that contain morphological data are restricted to a specific model organism or model taxon (e.g. *e-Mouse Atlas Project*, http://www.emouseatlas.org/emap/home.html; *Flybase*; http://www.flybase.com/; Zebrafish Model Organism Database, http://zfin.org/). Not many databases focus on a broader taxonomic range.

MorphoBank is *"a web application for conducting phylogenetics or cladistics research on morphology"* (http://www.morphobank.org/). *MorphoBank* provides a platform for the cooperative generation and annotation of morphological character matrices (i.e. according to *MorphoBank*, "morphological data"). It is not possible to enter morphological descriptions and thus morphological data sensu stricto in *MorphoBank*.

MorphoBank allows the storage and annotation of single images and specimen data, but not image stacks. Whereas it is possible for a given project to create a taxonomic hierarchy and link content to this hierarchy, *MorphoBank* does not possess a general taxonomic reference system like *Morph·D·Base*, through which all contents of the database could be browsed via a taxonomic tree. Thus, every user must create her/his own taxonomic referencing system for every project. It is most likely that different hierarchies of different users will not all be compatible with each other, which will hamper the general accessibility of the database contents. Moreover, contrary to *Morph·D·Base*, *MorphoBank*'s interface design does not allow for changes in midstream. At present, *Morph·D·Base* provides a similar basic functionality.

MorphBank (http://www.morphbank.net/) is comparable to *MorphoBank*. *MorphBank* is primarily an image repository, which can be used for international collaboration, research and education. *MorphBank* allows the storing, sharing, and the discussion of images of specimens and provides image comment tools. Again, the interface does not allow for changes in midstream. Like *MorphoBank*, *MorphBank* allows the storing of phylogenetic character information, but it is not possible to enter morphological data sensu stricto.

Digimorph (i.e. Digital Morphology; http://www.digimorph.org/) is a biological visualization library containing media (2D and 3D) of various organisms. It only provides a short list of the specimens for which images stacks, 3D volumetric movies and stereolithography files of 3D objects are available, with almost no metadata provided concerning the specimens. Thus, *Digimorph*'s focus is primarily on providing images and other media. Unfortunately, it is neither possible to download the original media files nor to upload new files through *Digimorph*'s interface.

20.7 Conclusions

Morph·D·Base allows archiving of one's own data and provides ubiquitous data access, which is especially interesting for nomadic users. It is the only general morphological data repository that allows the storing and dissemination of image stacks. Moreover, *Morph·D·Base* can be used for self-promotion and enables exchange of yet unpublished contents, which facilitates co-operation and collaboration. Thus, individual morphologists benefit in various ways from using *Morph·D·Base*. *Morph·D·Base* moreover provides a platform for detailed documentation of metadata and images (i.e. 'raw data'), which can be used as external supplements in morphological publications. This way, *Morph·D·Base* could have an effect on the overall quality of data and metadata in morphology and provide examples for best practice of data and metadata documentation. Last, but not least, *Morph·D·Base* can be used in education and teaching, since a supervisor could use it to efficiently monitor the work in her/his lab.

The next important step for further improving *Morph·D·Base* would be the development of an ontology-based description module. The use of ontologies in descriptions would significantly increase the communicability and comparability of morphological data. Together with the possibility to link descriptions to relevant metadata concerning the studied specimens, relevant literature, and the methods applied during data production, as well as to relevant media content that serves as 'raw data' for empirically substantiating the description, *Morph·D·Base* would close the methodological gap between data production and data analysis and contribute to a solution to the Linguistic Problem of Morphology. Most importantly, however, **respective morphological data would be sustainably stored, archived, shared and disseminated and made accessible also to non-experts**. This would facilitate interdisciplinary studies (e.g. questions concerning gene-phene interactions) and would significantly improve the reputation of morphology within the biological community.

However, while there are still many things that could be done to further improve *Morph·D·Base*, at its present stage it already facilitates not only the collaboration of morphologists all over the world, independent of the country they are actually living in, but also the co-operation of specialists of different groups of organisms or the co-operation of morphologists with biologists from other biological disciplines.

Acknowledgements

We thank numerous users who contributed to the development of *Morph·D·Base* for their comments, suggestions and criticism. We are solely responsible for all the arguments and statements in this paper. This study was supported by the German Research Foundation DFG (VO 1244/3-2 and VO 1244/3-3). We are also grateful to the taxpayers of Germany.

Rudi Loesel and Stefan Richter

21 Neurophylogeny – from description to character analysis

Abstract: 'Neurophylogeny' refers to a sub-discipline of phylogenetics dealing with the description of neuroanatomical structures, the conceptualization of neuronal characters, and the analysis of these characters separately or in combination with other characters, and finally, the interpretation of the evolutionary transformations of neuronal characters. We describe the joint efforts of our consortium of 'neurophylogeneticists' in the DFG priority program "Deep Metazoan Phylogeny" of describing nervous systems of various metazoans, defining neuroanatomical terms in a glossary, compiling a data matrix for neuronal characters and finally analyzing the data separately and in combination. For character conceptualization as well as compilation of the character matrix the openly available and freely accessible web portal and data repository Morph-D-Base has been used.

21.1 History and concepts

The higher-level phylogenetic relationships between multicellular animals (Metazoa) pose one of the most challenging riddles in evolution and have done since Haeckel's first attempts to unravel them in 1866. For many decades, morphological descriptions were the only source available from which phylogenetic relationships could be inferred, and they continue to be the logical prerequisite for understanding the evolutionary transformations undergone by body plans and organ systems. Few anatomical traits are so ubiquitous that they provide characters for comparison at the highest taxonomic levels and hence for phylogenetic studies encompassing all multicellular animals, but those that are include developmental patterns, body cavity organization, the digestive system, the nephridial system and the nervous system (Peterson and Eernisse, 2001).

Neurons (nerve cells) are found in all but the simplest metazoans such as sponges and placozoans (but see Nickel, 2011). Via synaptic connections they form a functional entity known as the nervous system. The morphological complexity of nervous systems can vary dramatically between groups of animals from relatively simple, sparsely meshed arrangements to the highly complicated neuroarchitectures found in arthropods, cephalopods and vertebrates (Bullock and Horridge, 1965).

This complexity has fascinated anatomists for over a century. Early researchers such as Félix Dujardin (1801–1860) and Johann Heinrich Ludwig Flögel (1834–1918) produced graphical reconstructions of the insect brain which are remarkable not only for their accuracy but also for their artistic value. One of the most important neuroanatomical findings in these early days was Santiago Felipe Ramón y Cajal's (1852–1934) discovery that the nervous system consists of individual cells rather than being a contiguous structure like the vascular system as was the predominant view at the time. Ramón y Cajal's findings paved the way for subsequent comparative studies based on descriptions of the anatomy of single cells. An excellent overview of the early history of invertebrate neuroanatomy research has recently been published by Strausfeld (2012).

One of the first researchers to utilize the morphology of neurons to address evolutionary questions was the Russian anatomist Alexei Alexseevich Zavarzin, who in the early 1900s formulated ideas about parallel evolution based on similarities between invertebrate and vertebrate nervous systems. Later, two Swedish neuroanatomists, Nils Holmgren and his student Bertil Hanström, issued elaborate descriptions of the brain anatomy of all major invertebrate taxa. These studies, published in the first half of the 20th century, were the first to derive phylogenetic relationships exclusively on the basis of neuroanatomical data. Founded on his observations, Hanström (1928) constructed a tree in which he placed insects within Crustacea, an idea which has recently gained wide support from a number of molecular studies (e.g., Regier et al., 2010; Meusemann et al., 2010; von Reumont et al., 2012 for the latest approaches).

The classic histological techniques employed by early anatomists are still in use today, but the introduction of modern methods over the last two decades has brought new momentum to the field of comparative neuroanatomy. Cutting-edge techniques such as immunocytochemistry and molecular fingerprinting make it possible to analyze transmitter content, surface markers and gene expression patterns in neurons of interest. Along with recent developments in imaging methods such as confocal laser scanning microscopy and computer-aided 3D reconstruction, these give us the tools to analyze the architecture of the nervous system at various levels of complexity and in a much higher number of species than previously, opening doors to a deeper understanding of its morphological disparity.

Once the data have been acquired, the findings for different taxa have to be compared. The question is how to go about this objectively. Each nervous system consists of a large number of individual morphological units, the neurons, which together form complicated arrangements. In order to make objective and comprehensible statements, strict standards and well-defined criteria for identifying correspondences are required; otherwise neuroanatomical patterns would only exist in the eye of the beholder and would not be reproducible for other researchers. Criteria covering the whole spectrum of structural resolution – from the gross morphology of the nervous system down to subcellular details of individual neurons – were put forward by Kutsch and Breidbach in 1994. These criteria should form the basis for any comparative study and have themselves been refined with the advance of histological techniques. The following recalls some of the criteria that can be used to describe and compare features of the nervous system:

- General layout of the nervous system (diffuse, centralized, rope-ladder-like etc.)
- If there is a brain, are neuronal processes homogenously distributed, or is the brain divided into discrete neuropils?

- Number and arrangement of neuropils in the brain
- Connections of a given neuropil to other areas in the nervous system
- Number and three-dimensional arrangement within the neuropil of subunits such as layers and tracts
- Presence and distribution of molecular markers (neurotransmitters, surface molecules, gene products)
- Developmental patterns

Other criteria extend to the physiological properties and behavioral relevance of nervous system components.

Correspondences are usually interpreted as the result of a common evolutionary origin and are thus considered as homologies, but this need not necessarily be the case. Correspondences can equally result from functional identity or similarity, analogies or convergences. Only where the correspondences are detailed enough (or outweigh the differences) does homology become a more plausible explanation than convergence (e.g., Richter, 2005; Scholtz, 2005).

Does this approach really work? Is there sufficient phylogenetic signal in neuroarchitectural characters to make them useful for reconstructing phylogenetic trees or testing existing hypotheses? The answer is that yes, there is. The field of neurophylogeny – a term coined by the Canadian neurobiologist Dorothy Paul in 1989 and popularized by Steffen Harzsch (2006; 2007) – has made significant advances in recent years. The next chapter will summarize some of the progress that has been made in this field. Neurophylogeny is understood herein in a fairly comprehensive sense, extending to the process of conceptualizing characters on the basis of neuroanatomical descriptions, finding evidence to justify homologizing these characters, using them to reconstruct phylogenies and, finally, analyzing the evolutionary transformations of characters whose polarity is revealed by the phylogenetic analysis (following the general approach delineated by Evolutionary Morphology sensu Wirkner and Richter, 2010).

21.2 Neuroanatomical characters and phylogenetic trees

The majority of modern neurophylogenetic studies deal with arthropod relationships, primarily those between the four subordinate taxa Hexapoda, Crustacea, Myriapoda and Chelicerata. The first, pragmatic reason for this is that a wealth of anatomical studies exists on the brains of arthropods from which a large amount of data can be drawn. The second reason is that the relationships between the four major arthropod groups have been under debate for decades, with trees featuring almost all thinkable permutations in topology having been suggested. Tetraconata and Tracheata (Atelocerata) are two such competing hypotheses, the first proposing a monophyletic clade including Hexapoda and Crustacea, the second including Hexapoda and Myriapoda.

The position of Chelicerata has long been unresolved, too. The Mandibulata hypothesis suggests that the chelicerates are the sister-group to the remaining arthropods, while the Myriochelata (Parodoxopoda) concept proposes a sister-group relationship between chelicerates and myriapods (recently summarized by Giribet and Edgecombe, 2012).

The first attempts to resolve this problem using neuroanatomical methods were performed with relatively few characters and without the aid of computers. Investigators were initially interested in character transformations, i.e., the question of how certain complex cerebral features evolved gradually during arthropod radiation. Neuronal characters were mapped onto the tree which best fitted the data. The set of characters used was derived either from specific parts of the brain such as the central complex (Loesel, Nässel, and Strausfeld, 2002) or the optic lobes (Sinakevitch et al., 2003), or from the brain as a whole (Fanenbruck and Harzsch, 2005). The neuroanatomical data clearly supported Tetraconata and Mandibulata in each case.

Neuronal characters obviously provide phylogenetic information for higher taxonomic levels (Loesel, 2005; 2011) but not for lower ones. Attempts to resolve interordinal relationships among insects, for example, failed because the nervous system within Hexapoda is too conserved to provide satisfactory resolutions. The small differences which are found between taxa are correlated with the behavioral ecology of a given species and not with its position in the phylogenetic tree (Loesel, 2006).

Over the last few years, increasingly large sets of data have been computed and phylogenetic trees exclusively inferred from neuroanatomical characters have been calculated anew. The first tree to be computer generated and not assembled 'by hand' was published by Strausfeld in 1998. It was based on a matrix of 100 independent characters identified in adult arthropod brains that were scored as present or absent in 28 taxa. Since then, as the number of characters has increased, and additional taxa have been added, modifications of this tree have been put forward by Strausfeld and his team (Strausfeld et al., 2006a; Strausfeld and Andrew, 2011). Strausfeld and Andrew call their approach 'neural cladistics' – a process that could be considered the analytical part of neurophylogeny in the broader sense suggested herein.

Tetraconata and Mandibulata have high support in all of these neuroanatomical phylogenetic hypotheses. With regard to the position of hexapods within crustaceans, trees based on neuroanatomical characters naturally show a close relationship between Hexapoda and those crustacean taxa with a more complex brain, malacostracans and remipedes (Fanenbruck and Harzsch, 2005; Strausfeld and Andrew, 2011). Andrew (2011) and Strausfeld and Andrew (2011), however, have pointed out that this might be incorrect. An EST study they performed in order to be able to compare their latest neuroanatomical tree with a molecular phylogeny of the same set of species suggests that hexapods are more closely affiliated with branchiopods, which have a relatively simple brain (a result supported by all recent molecular studies, e.g., Regier et al., 2010; Meusemann et al., 2010; von Reumont et al. 2012). This discrepancy highlights a major pitfall of neurophylogenetic/neural cladistic studies: the inability to resolve correctly the position of taxa that have undergone an overall reduction in neuronal complexity during evolution. While this is generally true for all phylogenies based on morphological data, some neuronal characters seem to be especially vulnerable to secondary reductions (see the discussion in Heuer et al., 2010, Loesel, 2011). This is the reason why myriapods emerged as polyphyletic and diplopods were relegated to a basal position within Arthropoda in a study by Loesel et al. (2002). The obviously incorrect topology is most likely due to a simplification of diplopod brains, which have lost a number of complex characters. This problem, which generally concerns phylogenetic studies based on restricted data sets, shows on the one hand the potential strength of total evidence approaches (though these may raise other problems), and on the other hand, even more crucially, the importance of a detailed discussion when it comes to character transformations based on phylogenetic analyses.

Phylogenetic trees based exclusively on neuronal data have been restricted to date to (pan)arthropods, possibly because it is not easy to compare the extensive knowledge available on arthropod brain anatomy with findings in other phyla. Data was either missing, or taxa were undersampled. Efforts, particularly in the frame of the Deep Metazoan Phylogeny program, have only lately been made to analyze the neuroarchitecture of hitherto insufficiently investigated groups such as Chaetognatha (Rieger et al., 2010), basal Mollusca (Faller et al., 2012), Annelida (Heuer et al., 2010), Gastrotricha (Rothe and Schmidt-Rhaesa, 2009) and Priapulida (Rothe and Schmidt-Rhaesa, 2010a).

Another problem that arises when comparing new findings with data published earlier is that researchers working on different groups of animals have created their own terminologies for describing the nervous system. Since it is usually not possible to present the entire neuronal complexity of a species in image plates, comparative anatomists often have to rely on written descriptions. Inconsistencies in nomenclature between different schools of anatomy complicate the task of incorporating the available data into a broad taxonomic comparison.

21.3 The problem of terminology or 'What is a brain?'[1]

Due partly to the long history of anatomical research and partly to the diversity of nervous systems, the terminology used today is very rich and often confusing. Even common terms such as 'brain', 'nerve' and 'eye' have been used in various ways in different animal groups, and no consensus on their exact meaning exists. This hinders our understanding of the architecture of the invertebrate nervous system in general

1 This section is largely based on the Introduction in Richter et al. (2010).

and of evolutionary transformations of nervous system characters between different taxa in particular. Without a common terminology, this problem will not be overcome. The aim here must be to increase the transparency, reproducibility and communicability of morphological data (Vogt, Bartolomaeus, and Giribet, 2010).

Various proposals have been made on how best to approach such a task. First and foremost, the terminology used to describe structures needs to be free both of any kind of assumption about homology and of restrictions to specific taxa (Vogt, 2008; Vogt Bartolomaeus, and Giribet, 2010). Only conceptualizing morphological features as characters in order to analyze them phylogenetically inevitably involves homologizing them. To minimize the risk of error at such an early stage of a phylogenetic analysis, characters need to pass a test of primary homology (based on structural correspondence) (Richter, 2005). Only then can the character in question, and its name, be applied to two different taxa (in cladistic terminology these characters are usually called character states). When dealing with characters that are assembled from different literature sources in particular, it is important to be aware of the 150 years history of anatomical research. Morphological terminology is not a pristine field, and every morphological description and term is involuntarily affected by evolutionary interpretations.

Several glossaries containing neuroanatomical definitions are already in existence, but these are restricted to limited groups of animals such as Hymenoptera or Amphibia, or to individual species, e.g., *Drosphila melanogaster* (the Virtual Fly Brain) (see NCBO BioPortal: http://bioportal.bioontology.org/). One major piece of progress to have been achieved in the last few years is that modern anatomical terminologies (and more importantly the concepts behind the terms) are now being organized as ontologies, meaning that their language is highly standardized and therefore computer readable (Vogt, 2009). What was lacking until recently, however, was a uniform conceptual framework and vocabulary fulfilling the criteria of a modern ontology and applicable to all the Metazoa, with a focus on invertebrates. In 2010 a consortium of scientists working together in the German Research Foundation (DFG) priority program "Deep Metazoan Phylogeny" made a first attempt to define neuroanatomical structures and come up with a precise and consistent terminology that is taxon-independent and free of homology assumptions. This terminology can be applied to all invertebrate taxa, even those that are poorly investigated and thus without their own tradition of nomenclature (Richter et al., 2010). The intention was to form a basis for new morphological descriptions and a starting point for the conceptualization of characters in phylogenetic character matrices.

In this glossary, each entry consists of a definition, discouraged terms and a background/comment section. The defined terms are organized hierarchically, i.e., information is provided at the level of complexity at which a given term is situated within the framework of the nervous system. The term 'brain', for example, is subordinated to the term 'cluster of neurons'. It is thus a prerequisite that a brain be composed of more than just one cell, but not every cluster of neurons is necessarily a brain since there are additional terms subordinated to 'cluster of neurons', such as 'ganglion' or 'nerve cord'. What defines a brain as such and distinguishes it from the other clusters of neurons is then specified in the definition, which looks like this:

A brain is a cluster of → **neurons**. It is part of a → **nervous system**. It is the most prominent anterior condensation of neurons and may also include other types of cells, including → **glial cells** and **pigment cells**.

The first sentence establishes what type of structure the defined term is, while the second sentence identifies the superordinate structure of which the structure is a part. Since there might be several terms that fulfill the same 'type of' and 'part of' criteria, additional specifications are provided in the part of the definition which follows ('most prominent anterior condensation'). Other defining properties may then be specified. Those neuroanatomical terms which are printed in bold with an arrow are main entries and have a definition devoted specifically to them. Those neuroanatomical terms which are printed in bold without an arrow are side entries. No specific definition is given for them but they are equally important for neuroanatomical descriptions and are listed in the register.

As mentioned above, the definition of each main entry is followed by a list of synonymous terms that have accumulated in the literature over time. The use of these terms is discouraged. Finally, a background/comment section is provided for every main entry in which additional information outlining the history of the term and its role to date in the discussion of nervous system evolution is given. Each definition is accompanied by one or several images, either original photographs of the structure or schematic drawings. This helps the reader, especially students or newcomers to the field, to form an easy understanding of neuroanatomical terms.

21.4 Conceptualizing characters and constructing a matrix

Once neuronal structures have been described in an inter-subjective way using a clearly defined vocabulary and published findings have been critically evaluated, neuroanatomical structures can be conceptualized as characters (i.e., evolutionary units) for use in phylogenetic studies. That structures (as a purely descriptive term) and characters (as an evolutionary term) are not necessarily congruent becomes obvious if we consider bilaterally paired structures, which in most cases are treated as a single character only. After the process of character conceptualization is complete, putatively homologous character states are tested for (primary) homology and then subordinated to characters (in the cladistic terminology) (see, for example, Richter, 2005; Wirkner and Richter, 2010; see also Wagner, 2001). The results of this step form the primary data for use in phylogenetic analyses and can, essentially, be computed in one of three ways. The first method is to map the available character states onto an existing phylogenetic tree, an approach which can provide fundamental insights into the polarity of character (state) transformation, but which very much depends on the availability of a well-resolved, generally accepted tree. The second method is to combine neuroanatomical data with other data sets in a so-called 'total evidence' approach. In a comprehensive study into arthropod phylogeny by Giribet et al. (2005), molecular data was combined with morphological characters, with the authors concluding that incorporating anatomical data into a molecular tree significantly enhances its resolution. Finally, as mentioned above, another method is to compute phylogenies exclusively on the basis of neuronal characters.

In all of these approaches a data matrix needs to be constructed in which characters and character states are scored for terminal taxa. A few neuronal character matrices have recently been published, e.g., for tunicates (Stach and Turbeville, 2002), and panarthropods (Strausfeld et al., 2006a; Strausfeld and Andrew, 2011). All of these studies are restricted to specific taxa, however, for reasons which are predominantly practical. First of all, many researchers are primarily interested in the phylogeny of the group of animals they are working with. Secondly, due to the Babylonian confusion in neuroanatomical terminology mentioned earlier, assembling and conceptualizing characters from different taxa has simply not been a feasible task. And finally, and maybe most importantly, the main factor hampering an analysis that encompasses all (or the majority of) the Metazoa is the sheer amount of work involved in constructing a large enough matrix. A matrix covering 50 terminal taxa and featuring 100 characters, for example, contains 5000 cells that have to be filled by hand. Ideally, the character state in each cell also has to be documented and supported by literature sources – time-consuming work that increases exponentially as taxa and characters are added.

If we are to be able to construct a neuronal character matrix that covers the majority of Metazoa, certain requirements clearly have to be met. A number of researchers working on different taxa have to get together, a uniform and standardized terminology and system of character conceptualization has to be agreed on, and the matrix preferably has to be set up in a way that permits continuous editing and the addition of more characters and terminal taxa as our knowledge of neuroanatomy grows. Internet-based platforms are the ideal venue for this endeavor.

The tool for creating a matrix of this nature has lately been installed on the Morphological Description Database 'MorphDBase' website (www.morphdbase.de, see Grobe and Vogt, 2008; also Vogt and Grobe, Chapter 20, this book), which is accessible to registered users. Currently, a group of some twenty neuroanatomists are working on a matrix covering more than 120 terminal taxa from all the major metazoan groups and featuring over 130 characters. This yields a total of nearly 16,000 cells that are being filled and documented cooperatively. An exemplar approach (Prendini, 2001) is being taken, although some concatenation of taxa on a low hierarchical level might not be avoidable. The novelty of this database is that investigators can work on the matrix at any time wherever internet access is available without actually meeting in person. Each character and character state is accompanied by a definition based on the glossary by Richter et al. (2010), and the character state in each cell is documented

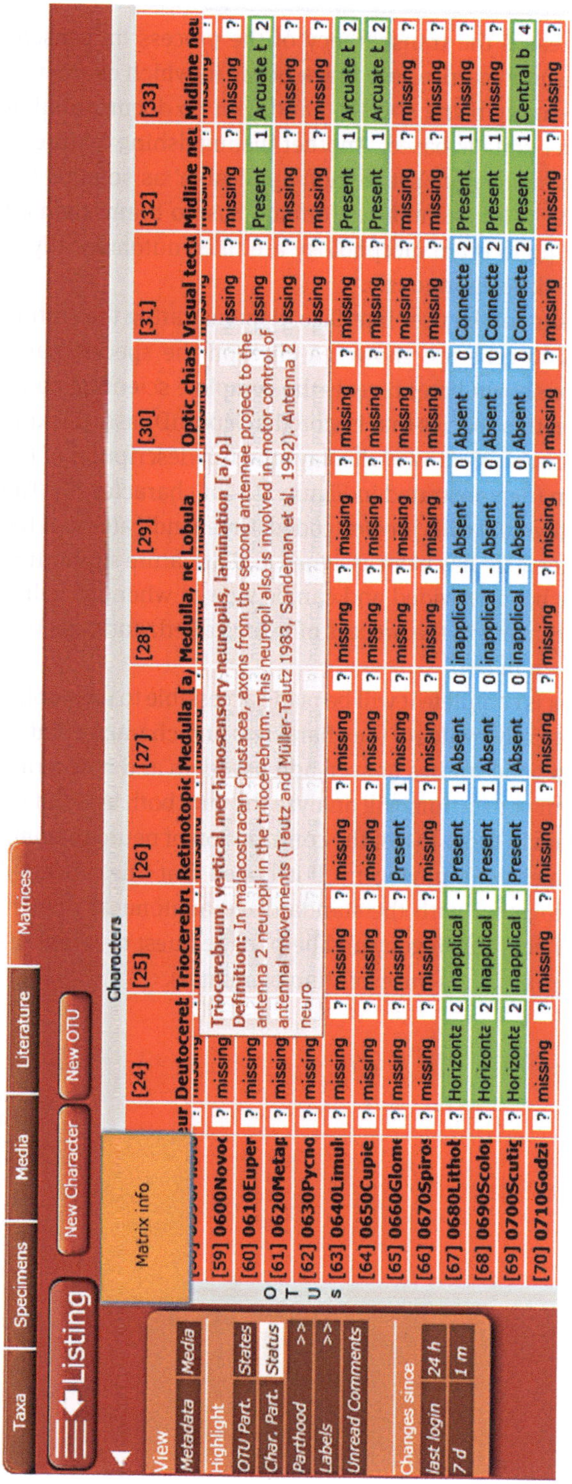

Figure 21.1: Screenshot of the user interface of the matrix module of MorphDBase. 120 of a total of 16,000 cells are shown.

by literature sources, which increases the transparency of the process immensely. This is an important point since some matrices are black boxes in which characters are scored without their exact meaning being specified and with no documentation of the data on which the character state is based. Another distinguishing feature of the MorpDBase matrix is that characters are set to 'non-applicable' as soon as the superordinate structure is absent. If, for example, an animal has no brain, then all cells referring to characters relating to brain morphology are filled automatically, a feature which speeds up the procedure enormously.

Figure 21.1 provides a screenshot of the user interface, demonstrating the editing process. 120 of the matrix's 16,000 cells are shown. A click on the species name (y-axis, species number 59–70 are shown here) reveals the complete scientific name and any additional information on this species that is stored in MorphDBase. Clicking on the character (x-axis, characters 24–33 shown) displays the full description of the character along with its definition (demonstrated in Figure 21.1 for character 25). The status of the editing process of individual cells is color-coded. Green indicates that the cell is complete, that is, a character state has been assigned and literature supporting the applicability of that state has been provided and can be viewed when the cell is clicked on. Blue designates cells that are in the process of being edited, and cells that have not been worked on are shown in red.

When the MorphDBase matrix is complete it will be made accessible to the public, thus enabling any scientist to critically evaluate the characters and character states, use the data for their own phylogenetic studies or even add new data and new terminal taxa to enhance the matrix. Our consortium will analyze the data both in isolation and in combination with other data in order to test the contribution of neuroanatomical characters to resolving deep metazoan phylogeny. It is hoped that this concerted and cooperative approach will, together with the standardized anatomical nomenclature used, advance the field of neurophylogeny and help in the quest to reveal the early evolution of nervous systems and thus the evolution of animals.

Matthias Bernt, Daniel Merkle, Martin Middendorf,
Bernd Schierwater, Martin Schlegel, and Peter F. Stadler

22 Computational methods for the analysis of mitochondrial genome rearrangements

Abstract: The evolution of gene orders by genome rearrangements is an important source of phylogenetic information. A large number of mitochondrial gene orders is nowadays available for phylogenetic analyses. On the one hand their small size and nearly perfectly conserved gene content makes them suitable for many approaches. But the different types of relevant rearrangement operations, on the other hand, complicate the analysis. In this chapter we review approaches for genome rearrangement analysis suitable for mitochondrial gene orders. This includes the analysis of pairs of gene orders as well as of larger sets of gene orders with and without a given phylogenetic tree. Particular attention is payed to the analysis of the tandem duplication random loss rearrangement.

22.1 Introduction

The analysis of gene order rearrangements is a promising approach for the reconstruction of the phylogeny, provided the corresponding species share approximately the same gene complement. This is the case in particular for the more than 3000 mitochondrial genomes that have been published for a wide variety of animal taxa. Metazoa mitochondria have small (≈ 16.5kbp) circular chromosomes usually harboring the same 37 genes (Boore, 1999). These are 13 protein-coding genes, two ribosomal genes, and 22 tRNAs genes. The latter consist of one tRNA gene for each amino acid with the exception of Leucine and Serine, both of which are represented by two tRNAs that can be differentiated by their anticodon.

In this chapter we give an overview on computational methods for the analysis of gene order rearrangements with an emphasis on methods that fit for the metazoan mitochondrial genomes. We focus in particular on methods that have been developed within the project "Refined Methods for Inference of Metazoan Phylogeny Using Mitochondrial Genome Arrangement Data" that was funded within the German Research Foundation (DFG) Priority Program SPP 1174 "Deep Metazoan Phylogeny".

For the analysis of gene order differences one usually considers lists of the genes in their order of appearance in the genomes. The conserved gene content of the metazoan mitochondrial genomes simplifies their formal and algorithmic treatment since their gene orders can be represented by *signed permutations*, where each gene is represented by a number and its sign (minus or plus) is used to indicate the orientation of the corresponding gene. Here we give a less formal "field manual" for gene order analysis methods and refer to Fertin et al. (2009) for a formal treatment.

Different types of mutations, so-called *rearrangement operations*, can modify the order of genes on a genome during evolution. For metazoan mitochondrial genomes at least four types of rearrangement operations appear to be relevant: inversions, transpositions, inverse transpositions, and tandem duplication random loss mutations (see Figure 22.1). An *inversion* reverses the order of the genes in a segment of the genome and changes their strandedness, e.g. Smith et al. (1990). A *transposition* moves a segment of the genome to a different position, e.g. Macey et al. (1997). An *inverse transposition* moves a segment of the genome to a different position and in addition reverses the segment, e.g. Boore, Lavrov, and Brown (1998). A *tandem duplication random loss* (TDRL) consists of the tandem duplication of a segment of the genome (i.e. the segment and its copy are adjacent) and the subsequent random loss of one of the two copies of each duplicated gene (Boore, 2000). TDRL operations are often considered to be atomic, i.e. the duplication and deletion processes are not separable (see Bernt et al., 2013b, for a review of potential molecular mechanisms generating the different rearrangement operations).

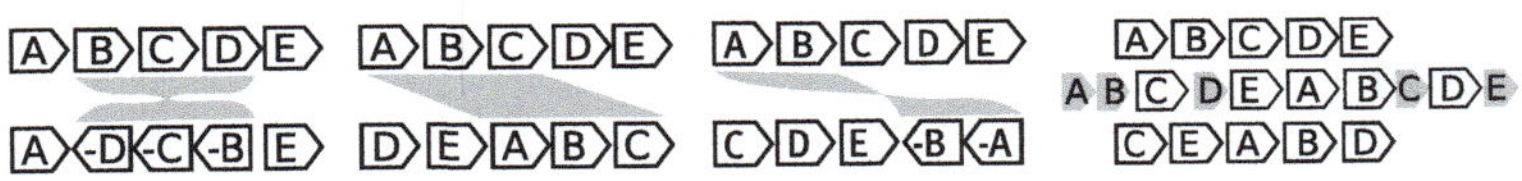

Figure 22.1: The four types of rearrangement operations exemplified on a gene order of length five; from left to right: inversion, transposition, inverse transposition, and TDRL.

In order to extract phylogenetic information from gene orders two different approaches have been developed, which may be classified as model-based and model-free. In the *model-based* framework, pioneered by Watterson et al. (1982) and Sankoff (1992), a fixed set of admissible types of rearrangement operations is given, e.g. a subset of the rearrangement operations introduced above. This set is called the *rearrangement model*. The goal is to find a parsimonious common rearrangement scenario for a set of given gene orders that is consistent with the rearrangement model, i.e. only admissible rearrangement operations are used to explain the differences between the observed gene orders. In a simple case only two gene orders are given and one searches for a parsimonious *scenario*, i.e. a shortest sequence of admissible rearrangements transforming one given gene order into the other. Such a scenario is called a *sorting* scenario and the problem of finding a sorting scenario is the *sorting problem*. The length of a sorting scenario may be used as a *distance* measure for gene orders. The possibility that different sorting scenarios might exist raises the problem of enumerating and generating them. The knowledge of their number could help in estimating the confidence in a specific reconstruction. For the "inversions-only" rearrangement model, efficient algorithms to compute a sorting scenario or just the distance between the two given gene orders are known (Hannenhalli and Pevzner, 1995). The set of alternative solutions for this rearrangement model was considered in Braga et al. (2008). When the rearrangement model allows only transpositions the sorting problem is NP-hard[1] (Bulteau, Fertin, and Rusu, 2011). Hence, one practical approach to solve the sorting problem when transpositions (and other rearrangement operations) are included in the rearrangement model is to design efficient approximation algorithms, e.g. Bader and Ohlebusch (2007). Another promising approach is to use generalized rearrangement operations (Yancopoulos, Attie, and Friedberg, 2005).

For more than two given gene orders the *maximum parsimony* problem is to reconstruct a phylogenetic tree with the given gene orders at the leaves and ancestral gene orders at the inner nodes so that the sum of the rearrangement distances along all edges of the tree is minimal. In a related problem, the phylogenetic tree is already given and only the ancestral gene orders are sought.

The special case of the maximum parsimony problem with exactly three gene orders is known as the *median problem*. In this case only one unrooted phylogenetic tree exists and only a single ancestral gene order – called median – is sought. For inversion operations the median problem was shown to be NP-hard (Caprara, 2003). The most widespread approach is to find a solution by the iterative application of inversions. The inversions may be chosen heuristically, e.g. greedily by minimizing the inversion distances (Bernt, Merkle, and Middendorf, 2006a; Bourque and Pevzner, 2002), or they may be found by an exhaustive enumeration, e.g. by using a branch and bound algorithm (Siepel and Moret, 2001). Alternatively, a median is constructed by appending the genes one after the other in a branch and bound scheme (Caprara, 2003).

All available algorithms for solving the parsimony problem are based on algorithms that solve the median problem. For a given tree topology GRAPPA (Moret, Tang, and Warnow, 2005) replaces the gene orders at the inner nodes of the phylogeny by a median of the gene orders at the three adjacent nodes if this reduces the distance. This is repeated as long as improvements are made. This method is applied to all rooted tree topologies by branch and bound enumeration. Algorithms MGR (Bourque and Pevzner, 2002) and amGRP (Bernt, Merkle, and Middendorf, 2007) build up a tree by sequential addition. In each step an existing branch of the partially constructed tree is removed and the two incident gene orders and a third so far unconnected gene order are connected by adding their median. Ahead of this MGR uses a method for generalizing its median finding procedure, i.e. inversions are applied to those gene orders which then reduce their distance to the other gene orders until no more such inversions exist. In Bernt, Merkle, and Middendorf (2007) it has been shown

---

1 NP hardness describes a very large class of problems in computational complexity theory for which it is unknown if they can be solved efficiently, i.e. in polynomial time. If an efficient algorithm were to exist for one NP hard problem, then all NP hard problems could be solved by efficient algorithms. It is generally believed, but still unproved, that no such efficient algorithm can exist.

that improved solutions for the parsimony problem can be found when in each step all solutions of the median problem are considered. These alternative solutions are incorporated by a tree search algorithm that branches over all equally good medians as well as all equally good choices of nodes that are added next. An efficient approach for incorporating inversions and transpositions in a weighted fashion has been proposed in Bader, Abouelhoda, and Ohlebusch (2008). The efficiency of this approach is due to a two-phase algorithm that constructs an initial tree without using time consuming median solvers and subsequently improves this tree by local rearrangements of the tree topology and modification of the inner nodes by solving median problems.

Approaches for solving the parsimony problem that are solely based on inversions may give useful results for data sets where the used distance measure approximates phylogenetic relatedness, e.g. for a few distantly related metazoans (Bourque and Pevzner, 2002). For data sets with gene orders that evolved by various types of rearrangements, however, the results have been unsatisfactory, e.g. Grande, Templado, and Zardoya (2008).

The *model-free* approach, e.g. Adam et al. (2007), tries to extract phylogenetic information directly from the gene orders and does not make assumptions on the types of rearrangement operations. Instead, properties of gene orders are used as phylogenetic characters. Implicitly, assumptions on the phylogenetic significance of these properties are made. A simple example is the absence or presence of a gene adjacency or, more generally, whether or not a set of genes occurs consecutively. In this way a collection of binary characters is extracted from the gene orders that can then be analyzed with standard phylogenetic tools. Using maximum parsimony, for instance, ancestral binary states can be reconstructed. These can in turn be used for inferring ancestral gene orders (Chauve and Tannier, 2008; Stoye and Wittler, 2009). An alternative method is the alignment of gene orders (see Section 22.5).

The following section introduces the concept of gene clusters and the strong interval tree data structure (SIT) which is the basis of algorithms introduced in the following. Section 22.3 reviews several algorithms for analyzing mitochondrial gene orders. That is, the algorithm CREx (Section 22.3.1) for reconstructing rearrangement scenarios for pairs of gene orders, TreeREx that reconstructs the rearrangements on the edges of a given phylogenetic tree (Section 22.3.2), and finally a method that builds on these approaches to compute an inventory of rearrangement operations for a set of gene orders (Section 22.3.3). An overview of the approaches for treating TDRLs is given in Section 22.4. In Section 22.5 a model-free reconstruction approach based on circular alignments is presented.

22.2 Background material: Gene clusters and strong interval trees

A *gene cluster* is a set of genes that satisfy certain proximity constraints. Gene clusters that are relevant for phylogenetic analysis are mainly those that have been conserved during evolution and thus can be found in several gene orders. Different formal definitions of gene clusters have been proposed in the literature, see Bergeron, Chauve, and Gingras (2007). One of the simplest notions is the *common interval.* A common interval of a set of gene orders is a subset of genes that occur consecutively in every one of the gene orders (Heber and Stoye, 2001). Here, the orientation of the genes is disregarded. A common interval may appear for several reasons, e.g. evolutionary inertia and evolutionary constraints, or it could have occurred in the different gene orders independently by chance. Approaches to compute the probability of an independent appearance have been explored e.g. in Sankoff (2005).

If gene clusters have formed in response to specific functional or gene regulatory mechanisms, it is important to consider this fact in the reconstruction of the phylogeny and the ancestral states. In particular, the corresponding common intervals of the contemporary gene orders should also occur consecutively in the intermediate states. Rearrangements that do not destroy the common intervals are called *preserving.* The additional constraint that only preserving rearrangements are allowed leads to modified versions of the optimization problems introduced above, namely the computation of preserving rearrangement distances, preserving sorting scenarios, and preserving medians. So far, mostly inversions have been considered under this constraint. It was shown that the preserving sorting problem is an NP-hard problem (Figeac and Varré, 2004). But for some types of problem instances solutions can be found efficiently (Bérard et al., 2007). The SIT, which is defined below, data structure plays a crucial role here.

Several algorithms for the preserving inversion median problem have been proposed. rEvoluzer (Bernt, Merkle, and Middendorf, 2006a) is a heuristic algorithm. Similar to MGR, rEvoluzer tries to preserve the *conserved intervals* (Bergeron and Stoye (2006), i.e. the common intervals that are framed by the same two elements in the same or opposite orientation. A parallel version of rEvoluzer has been presented in Bernt, Merkle, and Middendorf (2005). Also, several algorithms for computing exact solutions for the preserving inversion median problem for common intervals have been developed. The first approaches, CIP and ECIP (Bernt, Merkle, and Middendorf, 2006b), are based on Caprara's inversion median solver (Caprara, 2003) which is modified such that the addition of an element to the partial solution is only allowed if it is consistent with the common intervals to be preserved. The latest development, called TCIP (Bernt, Merkle, and Middendorf, 2008a), is an efficient exact algorithm that is based on an extended version of the SIT data structure, which is described below. The success of TCIP stems from the fact that subproblems which are defined by the SIT can be treated independently. This leads to an algorithm that has for many biologically relevant problem instances a linear run time.

A *strong common interval* (Bérard et al., 2007) of a set of gene orders is a common interval that does not overlap with any other common interval. Here, two intervals are considered as overlapping if they share genes but none of the two intervals is completely contained in the other. Strong common intervals therefore form a hierarchy, which is captured by the *strong interval tree* (SIT) (Bérard et al., 2007) data structure where each node represents one of the strong intervals and is connected to the node corresponding to the smallest strong interval that includes it. The leaves of the SIT correspond to the strong common intervals that contain only a single gene (singletons) and the root to the strong common interval that contains all genes. The SIT is considered to be an ordered tree, i.e. the children of a node are ordered by their order in one of the given gene orders (without loss of generality the first gene order in the data set can be chosen).

For each node of the SIT the relative order of the strong common intervals that correspond to its child nodes can be computed for a pair of the given gene orders. There are three possible cases: *linear increasing*: the child nodes are in the same order, *linear decreasing*: the child nodes are in opposite order, and otherwise *prime*. This is represented in the SIT by the type of a node. Furthermore each node gets assigned a sign: + or – for linear increasing and decreasing nodes, respectively. A leaf gets a + sign if the corresponding gene has the same orientation and a – sign otherwise. Finally, a prime node with linear parent inherits the sign from its parent.

Consider the gene orders (6 9 7 10 8 1 –4 5 –3 –2) and (1 2...10) where each number represents one gene. The gene orders have the following common intervals: {7,8,9,10}, {6,7,8,9,10}, {2,3}, {2,3,4,5}, {3,4,5}, {4,5}, {1,2,3,4,5}, the singletons, and the set of all elements. The common intervals {2,3} and {3,4,5} overlap (they share element 3) and are consequently not strong. All other common intervals are strong because they are either included in one of the other common intervals or vice versa. These strong common intervals define the nodes of the strong interval tree of the two gene orders (see Figure 22.3 left). The node corresponding to the strong common interval {1,...,5} has two child nodes corresponding to the strong common intervals {1} and {2,3,4,5}. Since these intervals appear in the same order in both gene orders the node is called a linear increasing node. In contrast, the node {2,3,4,5} is called linear decreasing since its three children are in opposite order in the two gene orders. For node {7,8,9,10} the children are neither in the same nor in opposite order and therefore this node is called a prime node. The leaf node corresponding to element 1 (2) has a positive sign since the element has the same (opposite) orientation in the gene orders.

The keys to the success of the strong interval tree data structure are the following. First, they can be computed efficiently (Bergeron, Chauve, and Gingras, 2008). Second, they represent all common intervals in a memory efficient manner. This is due to the fact that a set of genes is a common interval if and only if it is strong (i.e. a node of the strong interval tree) or the union of consecutive children of a linear node (Bérard et al., 2007).

22.3 Exploring mitochondrial rearrangements

A comprehensive analysis of mitochondrial gene orders should consider all four types of rearrangement operations introduced above. In this section we discuss such methods.

22.3.1 Pairs of gene orders

The Common interval Rearrangement EXplorer (CREx) is a heuristic algorithm for computing rearrangement scenarios for pairs of gene orders (Bernt, Merkle, and Middendorf, 2007). CREx was developed with a focus on metazoan mitochondrial gene orders. It considers all four relevant rearrangement operations, but it might also be applied to gene orders of other species that evolve under a similar rearrangement regime.

The CREx algorithm uses the SIT data structure restricted to pairs of gene orders, which has also been employed for the problems of sorting by preserving inversions (Bérard et al., 2007) and by inversions and transpositions (Parida, 2006). As a consequence of the data structure used, CREx requires gene orders that (i) have the same set of genes and (ii) for which each gene occurs exactly once per gene order. In order to also be applicable to other gene orders, CREx removes all genes that are duplicated or deleted in any of the input gene orders.

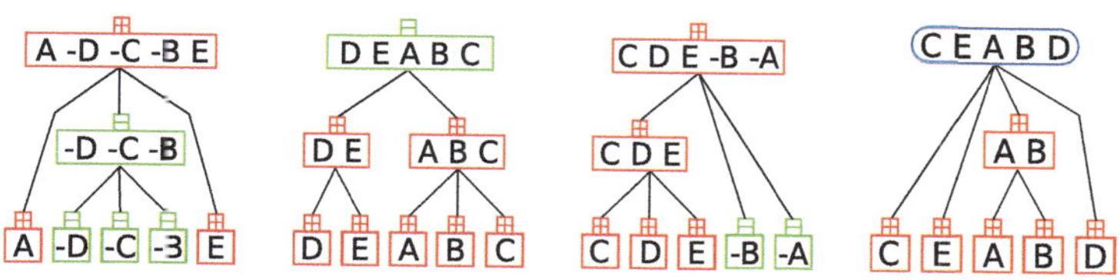

Figure 22.2: The strong interval trees (SITs) for the gene orders given in Figure 22.1 resulting from a single rearrangement; from left to right: inversion, transposition, inverse transposition, and TDRL; the strong interval trees (SIT) are shown with linear increasing (decreasing) nodes as *green (red)* rectangles with added + (–) and prime nodes as *blue* ellipses.

CREx exploits the fact that each of the four types of rearrangement operations generates a specific pattern in the strong interval tree. Here, pattern refers to specific configurations of the types of a node and its neighbors (see also Figure 22.2). The four rearrangement patterns are:

inversion: a linear node whose orientation is opposite to its parent's orientation

transposition: a linear node with two children that both have the opposite of the node's orientation

inverse transposition: a linear node where either the first or the last child node has the opposite sign

TDRL: a prime node.

For a given pair of gene orders CREx computes the SIT and searches for the different rearrangement patterns. It outputs the scenario that consists of the rearrangement operations corresponding to all the identified patterns. The rearrangement scenarios that are reconstructed by CREx are preserving. An example for an artificial gene order is shown in Figure 22.3.

Since prime nodes might also occur due to an accumulation of several overlapping rearrangements, e.g. at rearrangement hot spots, and TDRLs cannot explain differences in the orientation of the child nodes, special care has to be taken in this case. The approach that is implemented in CREx is to search for a short rearrangement scenario with a relative order and orientation of the child nodes incorporating inversions and TDRLs in which all inversions either precede or succeed the TDRLs.

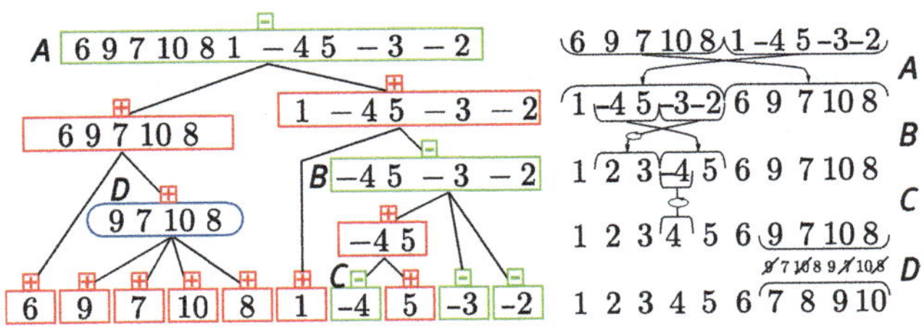

Figure 22.3: Example of the function of CREx for the artificial example of the gene orders (6 9 7 10 8 1 −4 5 −3 −2) and (1 2...10); **left:** the strong interval tree (SIT) of the two gene orders (node types are depicted as in Figure 22.2); **right:** the inferred rearrangement scenario consisting of transposition *A*, inverse transposition *B*, inversion *C*, and TDRL *D*; the nodes of the SIT for which the corresponding rearrangement patterns matched are indicated.

CREx is available online at http://pacosy.informatik.uni-leipzig.de/crex including a tutorial, documentation, and examples. Some phylogenetic studies which have used CREx are listed in Table 22.1. Run time and simulation results for CREx can be found in Bernt and Middendorf (2011).

22.3.2 Gene orders with a given phylogeny

The aim of the TreeREx (Tree Rearrangement Explorer) software (Bernt, Merkle, and Middendorf, 2008b) is to reconstruct the rearrangements on the branches in a given phylogenetic tree with given gene orders at the leaves. TreeREx also reconstructs the

Table 22.1: Usage of CREx for metazoan taxa.

Taxon	Reference
Echinodermata	Perseke et al. (2008); Perseke et al. (2010)
Hemichordata	Perseke et al. (2011)
Tunicata	Stach, Braband, and Podsiadlowski (2010)
Copepoda	Minxiao et al. (2011)
Amphipoda	Bauzà-Ribot et al. (2009); Krebes and Bastrop (2012)
Acari	Dermauw et al. (2009)
Bryozoa	Nesnidal et al. (2011)
Annelida	Bleidorn et al. (2009b)
Platyhelminthes	Webster and Littlewood (2012)
Acoelomorpha	Mwinyi et al. (2010)

putative ancestral gene orders at the inner nodes of the input tree. The basic idea of TreeREx can be summarized as follows: Assume that the given phylogenetic tree is correct and that a perfect oracle method for the reconstruction of the true pairwise rearrangement scenarios exists, i.e. exactly those rearrangements that led from the common ancestor to the pair of considered gene orders. Then the rearrangements on a particular edge of the tree are exactly those that occur in each pairwise scenario from any gene order that is not in the subtree below the edge to any gene order within this subtree. This approach would result in the correct reconstruction, apart from the assignment to the branches at the root node which needs additional outgroup information.

Of course, a perfect oracle method does not exist. Instead, TreeREx relies on a method for parsimonious reconstructions and employs CREx for this purpose. TreeREx tries to reconstruct the ancestral gene orders and the rearrangements on the edges of the phylogenetic tree by considering small subtrees that have only three or four leaves. This is done in a bottom-up fashion in order to reconstruct the rearrangements for the complete phylogenetic trees (see Figure 22.4). The bottom-up approach is chosen since extensive simulations showed that pairwise CREx reconstructions tend to be more often correct for more closely related species (Bernt and Middendorf, 2011). In addition, TreeREx tests the validity of the rearrangements that are to be assigned to the edges by checking whether they indeed lead to a common ancestral gene order. If so, the reconstruction, which is called *consistent* in this case, is accepted. Otherwise, the reconstruction is called *non-consistent* and TreeREx provides two alternative strategies to reconstruct an ancestral gene order as described in the following.

One possible cause for a non-consistent reconstruction is that some of the pairwise rearrangement scenarios are wrong. Therefore, TreeREx tries to discard a small number of pairwise comparisons such that a consistent reconstruction is possible with the remaining pairwise comparisons. If one or more consistent reconstructions are found after discarding *k* pairwise comparisons a parsimonious solution is chosen. Such a reconstruction is termed *k-consistent*. If no such reconstruction can be found for a small value of *k*, a fallback method is implemented in TreeREx. This method considers the set of possible intermediate gene orders on the rearrangement scenario for the pairs of sister leaf taxa as possible ancestral gene orders. For the case of a subtree with three leaves, a gene order is chosen that minimizes the rearrangement distance to the outgroup of the considered pair. For the case of subtrees with four leaves, for each of the two inner nodes a gene order is chosen such that the rearrangement distance between the two gene orders is minimal. Here the rearrangement distance is measured as the number of events predicted by CREx.

Since TDRLs are asymmetric operations, special care has to be taken. First, the rearrangement scenarios need to be constructed in the correct direction, i.e. toward the leaf node. For the consistency check, the inverse operation of the TDRL must be applied. Recall that, for the other types of rearrangement operations, there is no such difference between the operation and its inverse.

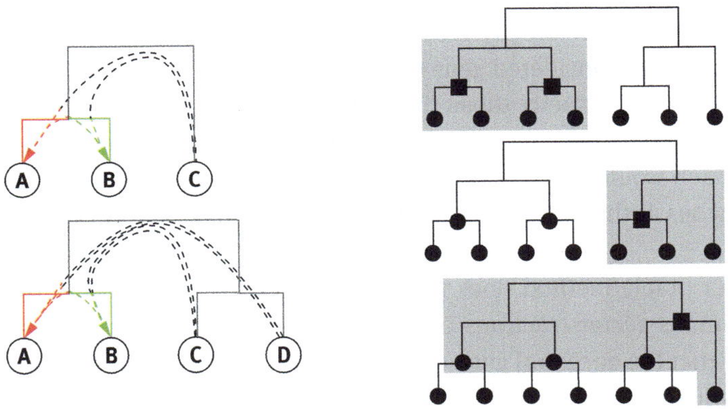

Figure 22.4: left: Approach of TreeREx for determining the rearrangements on the branches leading to two sister leaf taxa (*A*,*B*) for subtrees with three and four leaves; the pairwise rearrangement scenarios used are indicated as dashed lines; the common parts of these scenarios are colored as the corresponding branch which they are mapped on; **right:** Illustration of the recursive bottom-up procedure of TreeREx (from top to bottom); *shaded area*: considered subtree; *circles*: nodes with known or reconstructed gene order; *squares*: nodes that are reconstructed in the current step.

TreeREx has been successfully applied for rearrangement analysis of the mitochondrial gene orders of teleost fishes and echinoderms (Bernt, Merkle, and Middendorf, 2008b), and Fungi (Duò et al., 2012). TreeREx is available for download at http://pacosy.informatik.uni-leipzig.de/TreeREx.

22.3.3 The rearrangement inventory graph

As has been shown in several studies, CREx and TreeREx (see Subsections 22.3.1 and 22.3.2) are well suited for the analysis of rearrangement scenarios. The manual analysis of all pairwise CREx rearrangement scenarios becomes impractical, however, when the number of input gene orders becomes large. For the application of TreeREx a phylogenetic tree is required. Typically, such a tree is constructed by using sequence data, morphological data, or other biological information on the corresponding species. Hence, a pure rearrangement-based phylogeny reconstruction approach for a larger number of species is difficult with CREx and TreeREx.

As a basis for a purely rearrangement-based phylogenetic approach, the *rearrangement inventory graph* (RIG) has been introduced in Bernt and Middendorf (2011). The goal is to find for each of the given gene orders at least one potential evolutionary explanation for its current state of evolution, i.e. a rearrangement scenario leading to another gene order. To this end, a subset of all pairwise rearrangements between the considered gene orders is determined that are likely to be correct. These rearrangements form the basis for further analysis. The identified rearrangements are organized in a graph for a convenient overview. The selection of the rearrangement scenarios that are included in RIG is based on the following findings that have been gained from a simulation study (Bernt and Middendorf, 2011) that involved different rearrangement models: (i) most importantly, the absence of prime nodes in the strong interval tree of the corresponding gene order pair is a good indicator for a correct reconstruction, and (ii) reconstructions of short rearrangement scenarios are more often correct than longer ones.

To construct the RIG, in a first step only pairs of gene orders are considered that have no prime node in the corresponding strong interval tree. The rearrangement scenario for two gene orders is included in the RIG if for none of them a shorter rearrangement scenario to another gene order is known. Since the comparisons are limited to gene order pairs with a prime node free strong interval tree, TDRLs are excluded from the RIG constructed so far. In a second step, prime node comparisons are added restrictively. Comparisons with reconstructions incorporating exactly one TDRL are added if they connect different components of the preliminary RIG that was constructed in the first step. Again, the edge is only considered if there is no shorter connection of the adjacent nodes to a third one, but now considering the prime node involving comparisons as well.

A RIG has been constructed for all the 185 unique mitochondrial gene orders included in RefSeq release 36 having the standard set of 37 genes (see Figure 22.5). The whole computation needs only 30 seconds on a standard laptop computer. The resulting RIG, called metazoan RIG in the following, has several connected components: several small components (nine with two nodes and six with three nodes each), two medium sized components including five and eight nodes, respectively, and two large components connecting 45 and 62 gene orders, respectively. The meta-

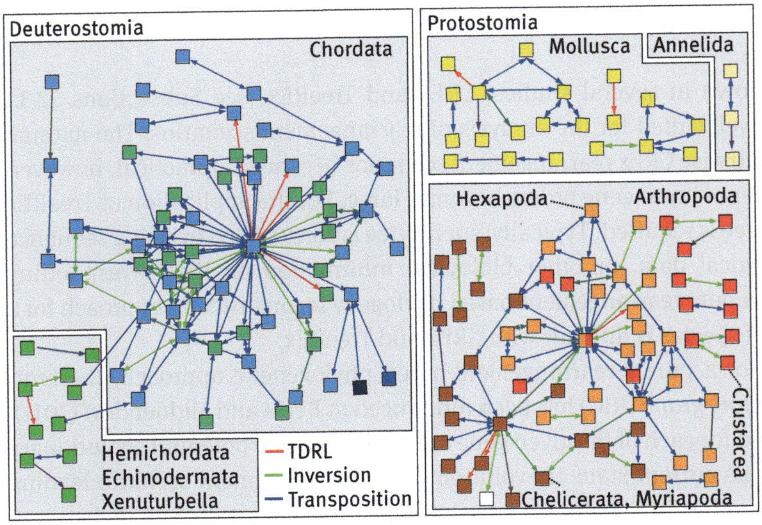

Figure 22.5: metazoan RIG; each node represents one unique gene order; node color represents taxonomic grouping; edges represent rearrangement scenarios; edge color indicates the composition of the corresponding scenario; directionality is indicated by arrowheads; RIG components of size one are omitted.

zoan RIG has several nodes with very high degree which have a direct connection to many of the other nodes in their component. Furthermore, each of these nodes represents many species with the same gene order. The central node in the chordate component, for instance, represents the typical vertebrate gene order that is shared by most vertebrate species.

The fact that the components of the metazoan RIG connects gene orders in a taxonomic/phylogenetic meaningful way, i.e. within phyla, shows the principal usefulness of the approach. Another peculiar feature is that the metazoan RIG contains cycles, i.e. not all components are trees, which contain the same rearrangements multiple times. This might represent alternative explanations of the rearrangement evolution, cases of convergent evolution, or just show that the effects of an ancient rearrangement that can still be observed in the distantly related groups.

A detailed analysis of the rearrangements occurring in the components that include gene orders from Protostomia can be found in Bernt (2010) and Bernt and Middendorf (2011). Most of the rearrangements that are included in these components have been shown to be correct. More than 50 % have been confirmed in the literature, nearly 30 % of the rearrangements have not been analyzed in the literature and may be considered as new, and the remaining almost 20 % have been reported differently in the literature or are caused by annotation errors contained in the RefSeq. The majority of the reconstructed rearrangement events are transpositions (63 %) whereas inversions, inverse transpositions, and TDRLs each account for a much smaller share (14 %, 13 %, and 10 %, respectively). Note that these fractions are similar to those presented in Dowton et al. (2009) which include the manual comparisons of hymenopteran gene orders with their putative ground pattern.

22.4 Tandem duplication random loss

The tandem duplication random loss operation (TDRL) is one of the major players in metazoan mitochondrial genome rearrangement (Boore, 2000, see also Jühling et al., 2012a). Nevertheless, the formal properties of this rearrangement operation have been studied only recently. A key element for the theory of TDRL sorting problems is the notion of a *chain*. Given a gene order S and a target gene order T a chain is a maximal sequence of genes that occur in the same order in both sequences and where it holds that the genes are consecutive in T. The TDRL distance of a pair of gene orders with ρ chains is $\lceil \log_2 \rho \rceil$ and an efficient method to construct a sorting TDRL scenario exists (Chaudhuri et al., 2006). The distance result is based on the insight that it is always possible to halve the number of chains by a single TDRL but a larger reduction is not possible. Such a TDRL can be found by numbering the chains in the order of their appearance in the target gene order and keeping the elements of the odd (even) chains in the first (second) copy. Consider for instance the artificial gene order $S = (DBFGCAHE)$ and $T = (AB...H)$. S has four chains with respect to T: $c_1 = (A)$, $c_2 = (B,C)$, $c_3 = (D,E)$, and $c_4 = (F,G,H)$. Applying the TDRL which keeps D, A, E in the first copy and the remaining elements in the second, halves the number of chains, i.e. it is sorting.

In order to evaluate the significance of a reconstructed TDRL scenario it is important to know how many different sorting TDRL scenarios exist. In Bernt et al. (2011) a closed formula has been derived to compute the number of sorting TDRLs, i.e. all TDRLs that reduce the TDRL distance between two gene orders by one:

$$\sum_{i=0}^{2^{\lceil \log_2(\rho) \rceil} - \rho} \binom{n}{i}.$$

Here n is the number of genes in the gene order and ρ is the number of chains. If ρ is a power of two, only a single sorting TDRL exists (because the upper bound of the summation index is zero, the sum is equal to $\binom{n}{0} = 1$. Since the only sorting TDRL divides the number of chains by two, the resulting permutation has again a number of chains that is a power of two. Consequently, the sorting scenario is unique. This is also the case in the example given above.

A peculiar feature of the TDRL operation is its asymmetry, which was also analyzed in Chaudhuri et al. (2006). That is, when applying one TDRL to a gene order it is in general not possible to recover the resulting gene order back with a single TDRL operation. This is in contrast to the other types of rearrangement operations – inversion, transposition, and inverse transposition – which are all symmetric. In fact, the only symmetric TDRL operations are those mimicking a transposition. Therefore, care has to be taken regarding the directionality of the comparisons when TDRL operations are considered. The asymmetry of the TDRL operation also offers interesting possibilities for phylogenetic studies. When comparing sister-taxa in a phylogeny one can usually not assign the rearrangement operations between the corresponding gene orders to the two branches without further outgroup information. For an asymmetric operation such as TDRLs, however, parsimony alone is sufficient because the assignment of the TDRL(s) to one branch inverts the orientation of the comparison and thereby implies different numbers of rearrangements in most cases. This argument was used for instance to reconstruct the rearrangement history of the echinoderms (Perseke et al., 2008).

Compared to the other types of rearrangement operations a TDRL can rearrange the gene orders more severely. A single TDRL can destroy all of the adjacencies of a gene order whereas a single rearrangement of one of the other types can destroy at most three. This should be regraded for the reconstruction of a phylogeny and the ancestral gene orders. The model-free approaches, for instance, will fail if only adjacencies are used. Furthermore, after $\frac{3}{2}\log(n)$ TDRL operations the resulting gene order is randomized and no phylogenetic information is left – as implied by results on riffle shuffles which are the inverse operation of TDRLs (Bayer and Diaconis, 1992).

It is important to note that there also exist cases of so-called tandem duplication non-random loss (Beckenbach, 2012; Lavrov, Boore, and Brown, 2002) mutations. In these cases the loss of genes that are located within the same transcript becomes much more likely if the promotor of the transcript is lost.

22.5 Character-based approaches

Although alignments are most commonly used to compare sequences over small alphabets, e.g. the different types of nucleotides or amino acids, one can apply alignment algorithms and the closely related chaining algorithms (Morgenstern, 2002) as a means of comparing any type of ordered lists of objects. In a bioinformatics context, such an approach was proposed, e.g. by Fried et al. (2004) to extract a co-linear set of "phylogenetic footprints" from multiple sequences. In the case of mitochondrial gene orders, the situation is further complicated by the circularity of the mitogenomes. Pairwise alignments of circular sequences require more computing time than linear sequence alignments, in particular when non-additive gap cost functions are used (Maes, 1990; Mosig, Hofacker, and Stadler, 2006). The *circular alignment* algorithms still have polynomial run time and due to the limited number of genes, mitogenomic data are easily tractable in practice.

The scoring system is quite simple: matches are not penalized while mismatches are forbidden entirely. Since insertions and deletions correspond to intervals of genes that have changed their genomic position, the differences in the "mobility" between tRNAs, rRNAs, and proteins is modeled by an in/del scoring function that depends on the lengths and the gene content of the inserted or deleted regions. Genes with opposite reading directions are encoded as distinct objects. An advantage over permutation-based methods is that gene duplications are handled without extra effort. A disadvantage is that the cost model does not represent a molecular mechanism. A region that is deleted from a mitogenome may be broken up and inserted in several distinct locations in the other genome. The alignment cost thus is a lower bound on the number of "cut & paste" events.

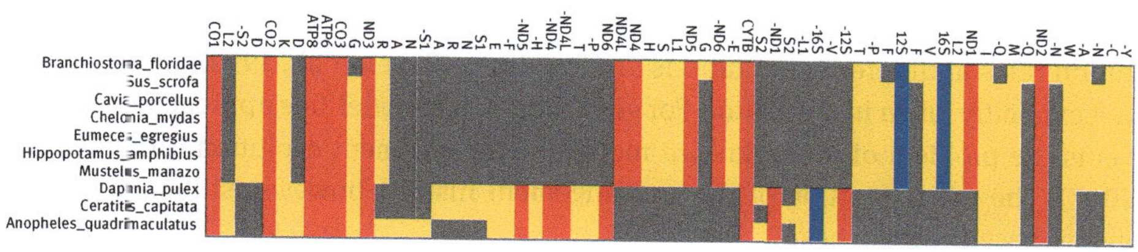

Figure 22.6: Character table computed with circal from the gene order of several animal mitogenomes. Presence of protein-coding genes, tRNAs, and rRNA are shown in *red*, *yellow*, and *blue*, respectively. Only presence and absence is encoded.

Multiple alignments can be constructed from gene order data in the same way as for sequences. The software tool circal (Fritzsch, Schlegel, and Stadler, 2006) implements a progressive alignment scheme. A multiple alignment can be interpreted as a binary character table that encodes the presence or absence of a gene in a particular context, see Figure 22.6. As a consequence, maximum parsimony methods can be applied to these data to reconstruct phylogenetic trees, to estimate consensus arrangements, and to infer ancestral gene organizations. We note that it is left to the user whether the two serine and leucine tRNAs should be distinguished or treated as duplicate genes. A practical application of the circal approach can be found, e.g. in an analysis of the phylogenetic position of the Myzostomida (Bleidorn et al., 2007).

22.6 Concluding remarks

In many aspects animal mitogenomes differ substantially from other genomic data, e.g. nuclear genomes (Boore, 1999 and review in Bernt et al., 2013b). Like small RNA and DNA viruses they are small enough to be accessible to accurate but expensive algorithms. In contrast to the latter, their gene content is virtually invariable so that mitochondrial gene orders are (almost) permutations of each other even when deep phylogenetic branches are considered. The rapidly growing collection of mitochondrial sequence data contains highly valuable phylogenetic information, which is becoming easily accessible with the advent of large-scale automatic annotation pipelines such as MITOS (Bernt et al., 2013c).

Adequate computational tools that are geared to the specifics of mitogenomic sequences, however, are required to unlock this treasure trove. While much progress has been made in this direction, see Bernt et al., (2013d), for a recent review on mitogenome-specific software, much remains to be done. Most importantly, at present we still lack a comprehensive mechanistic understanding of mitochondrial rearrangements and the interplay of mitochondrial gene orders and the evolution of the underlying features. At the sequence level, evolutionary processes are treated by the methods to derive models for sequence evolution. These models are encapsulated by mitochondria-specific substitution models. The asymmetries of nucleotide substitution rates at different orientations and locations (Reyes et al., 1998), however, are not explicitly taken into account. For gene orders, the model-free approaches circumvents the problem of the unknown mode of rearrangement evolution (Adam et al., 2007). The reconstruction of the rearrangement history, however, still calls for the model-based approaches.

The connection between the formal studies of genome rearrangement and the molecular mechanisms needs to be developed. First, in order to choose between alternative solutions of the rearrangement problems and second for drawing conclusions on the rearrangement processes. Weighting schemes and algorithms allowing for their incorporation seem to be another promising path to improved reconstructions.

Acknowledgments

This work was supported by the German Research Foundation (DFG) through the project “Deep Metazoan Phylogeny” within SPP 1174 (projects STA 850/3-2, STA 850/2, Schi-227/20, Schi-227/26-1, SCHL 229/14-1).

Roman R. Stocsits, Harald Letsch, Karen Meusemann, Björn M. von Reumont, Bernhard Misof, Jana Hertel, Hakim Tafer, and Peter F. Stadler

23 RNA in phylogenetic reconstruction

Abstract: Ribosomal RNA (rRNA) genes are the most widely used data source in phylogenetic reconstruction. They are highly structured, with large parts of the molecules exhibiting very strong conservation of their base pairing patterns. Since also sequence conservation varies dramatically between different regions of rRNA genes, these data are informative on a wide range of phylogenetic timescales, ranging from recent to ancient splits. But individual columns of rRNA alignments are not independent. This is a consequence of biologically functional secondary structures with highly varying degrees of conservation. Therefore, any correlations within sequence alignments of structurally functional RNA will distort the phylogenetic signal and/or lead to gross overestimates of tree stability. On the other hand, alignment accuracy can be improved substantially by incorporating secondary structure conservation. Maximum Likelihood and Bayesian approaches are amenable to using RNA-specific substitution models that treat conserved base pairs appropriately, but they require accurate secondary structure models as input. Structure prediction algorithms for single RNA sequences are well known and widely used; but it is not straightforward to apply these predictions for phylogenetic purposes. The main limitation is that accuracy of thermodynamic folding algorithms declines sharply as the length of the RNA increases, this is in part due to inaccuracies within the thermodynamic folding parameters, in part caused by the kinetics of folding process and tertiary interactions, and even more in the fact that RNA and protein components of the ribosome are tightly packed and thus mutually influence their folds. The functional rRNA structures, therefore, cannot reasonably be expected to be identical with the minimum free energy structures of isolated rRNAs computed by thermodynamic folding algorithms. Our RNAsalsa approach has been the first accurate and easy-to-use tool to address those implications mentioned. It is available for computing structure-aware alignments and consensus structures and can deal with large sets of large rRNAs. Multiple pairwise sequence alignments, intersections with individualized structural constraint modeling, and computationally derived collections of optimized subgroup consensus structures allow to extract a refined individualized structure constraint for each single sequence by means of a majority voting procedure. That way an extensively optimized set of individual constraints serves in minimum energy folding to obtain the final structural annotation for each input sequence. Further, the structure-annotated input sequences are aligned by a structure-aware alignment method, resulting in a final alignment and a corresponding consensus structure which are ready for phylogenetic analysis based on RNA-specific substitution models. The difference to DNA substitution models is that structure-aware models consider pairs of nucleotides as their elementary states rather than single nucleotides. Further, studies suggest that RNA secondary structures are by themselves also phylogenetically informative since structural variations are specific to individual clades. This has been reported for a wide variety of non-coding RNAs, including rRNAs, 7SK RNAs, and mitochondrial tRNAs. But in all cases the massive epistasis needs to be taken into account explicitly which is introduced by strong selection on secondary structure features.

23.1 Introduction

Ribosomal RNAs are the most widely used source of phylogenetic information. At least parts of the small and large subunit (SSU and LSU) rRNA genes have been sequenced for thousands of taxa throughout the metazoan kingdom, providing by far the densest taxon coverage, rivaled only by mitochondrial sequence data. Since sequence conservation varies dramatically between different regions of rRNA genes, these data are informative on a wide range of phylogenetic timescales, ranging from recent to ancient splits (Hillis and Dixon, 1991; Woese, 1987). But this variation in substitution rates is at the same time also a major technical obstacle for using rRNA gene sequences in molecular phylogenetics: it is impossible in practice to construct unambiguous alignments of the more variable regions by means of standard sequence alignment algorithms. Ribosomal RNAs are further highly structured molecules, with large parts of the molecule exhibiting very strong conservation of base pairing patterns. Alignment accuracy can thus be improved substantially by incorporating knowledge about secondary structure conservation (Buckley et al., 2000; Kjer, 1995; Letsch et al., 2010; Mallatt and Sullivan, 1998; Mallatt and Winchell, 2007; Misof et al., 2006).

In practice, however, the application of this idea has until recently remained a hard and very tedious task. The main issue is that it is very difficult to obtain good structural annotation. Given reliable secondary structure predictions, any one of the many approaches for combined sequence-structure alignments (Dalli et al., 2006; Hochsmann, Voss, and Giegerich, 2004; Katoh and Toh, 2008b; Lindgreen, Gardner, and Krogh, 2007; Reeder, Steffen, and Giegerich, 2005; Tabei et al., 2008) would readily provide a high-quality alignment and an associated consensus secondary structure. The alternative, a Sankoff-style simultaneous computation of alignment and consensus structure (Mathews, 2005; Torarinsson, Havgaard, and Gorodkin, 2007) unfortunately is computationally too demanding to be of practical use for applications to rRNAs.

An important consequence of the strong stabilizing selection acting on the secondary structure is a high level of epistasis between paired bases (Misof et al., 2002; Stephan, 1996). These epistatic interactions dramatically violate the assumption that alignment columns provide independent phylogenetic information. This observation has led to the construction of a variety of RNA-specific substitution models over the last two decades (Parsch, Braverman, and Stephan, 2000; Rzhetsky, 1995; Schoeniger and Von Haeseler, 1994; Tillier and Collins, 1995; Tillier and Collins, 2002). Nevertheless, most phylogenetic analyses using rRNA sequences employed simpler substitution models, presumably because the structural information was not readily available and few software tools, with the notable exception of PHASE (Gowri-Shankar and Jow, 2006; Hudelot et al., 2003; Jow et al., 2002) and MrBayes (Huelsenbeck and Ronquist, 2001; 2003) were capable of processing secondary structure annotations. A handful of papers already demonstrated the effect on tree inference and robustness using adequate substitution models in rRNA-based studies (Fleck et al., 2008; Kjer, Baldridge, and Fallon, 1994; Letsch et al., 2009; Misof et al., 2007; Reumont et al., 2009; Telford, Wise, and Gowri-Shankar, 2005). They clearly show that the explicit modeling of the strong epistatic effects between paired bases provides a much more realistic picture of the phylogenetic signal. However, this increase in realism can also mean a decrease in tree robustness using resampling techniques. It is exactly this information, realistic estimates of phylogenetic signal, on which phylogenetic tree inference is most dependent. The solutions of many phylogenetic problems still rely on extensive species sampling, which in practice is still only possible for rRNA gene sequences. Therefore, all of these studies have also to rely on a proper implementation of sequence-structure alignment algorithms as well as RNA-specific substitution models. Both tools mentioned are available nowadays but still not in routine use.

23.2 RNAsalsa: Improved alignments of ribosomal RNA

Although structure prediction algorithms for single RNA sequences are readily available in well-known software packages such as mfold and the Vienna RNA Package (Hofacker et al., 1994; Zuker and Sankoff, 1984; Zuker, 2003), it is not straightforward to apply their predictions for phylogenetic purposes. The main problem is that the accuracy of thermodynamic folding algorithms declines sharply as the length of the RNA increases beyond about 100 nucleotides. This limitation is in part due to inaccuracies in the "nearest neighbor model" and its parameters, in part caused by the kinetics of folding process and tertiary interactions (Ding et al., 2008; Doshi et al., 2004; Layton and Bundschuh, 2005). Even more importantly, the RNA and protein components of the ribosome are tightly packed and thus mutually influence their folds (Yusupov et al., 2001). The functional rRNA structures, therefore, cannot reasonably be expected to be identical with the minimum free energy structure models of isolated rRNAs computed by thermodynamic folding algorithms. Patterns of sequence covariation provide information on evolutionarily conserved base pairs and thus capture at least in part the functionally relevant differences between biologically relevant and thermodynamically optimal secondary structure. Computational tools such as RNAalifold (Bernhart et al., 2008; Hofacker, Fekete, and Stadler, 2002) have demonstrated that the accuracy of biological structure predictions can be increased to acceptable levels by using the consensus structures of a set of closely related sequences and by explicitly taking information on base-pair covariation into account. Designed to compute a consensus secondary structure for relatively closely related sequences, this group of tools require a sequence alignment as input. The software RNAsalsa (Stocsits et al., 2009) has been designed to overcome those limitations by combining the construction of consensus structures of closely related sequences with prior knowledge that allows to constrain the set of acceptable structures. The basic idea is to use consensus structures for groups of related sequences to generate high quality alignments by funneling the structure information into the alignment scoring function. Thus RNAsalsa uses both, structure information for adjusting and refining the sequence alignment, and sequence information contained in the alignment to refine the secondary structure prediction, see Figure 23.1 for details. The quality of both, the predicted consensus secondary structure and the sequence alignment, compare favorably with other approaches (Letsch et al., 2010; Stocsits et al., 2009). RNAsalsa is freely available (http://www.rnasalsa.zfmk.de/) and it was primarily designed with phylogenetic applications in mind.

In that context, RNA secondary structure is of importance at two levels: First, changes in secondary structures can be useful phylogenetic markers in their own right (Caetano-Anollés, 2002a; 2002b; Collins, Moulton, and Penny, 2000). This type of approach is still in its infancy, see the chapter on Molecular Morphology. The second, much more direct application of consensus secondary structures is to inform a structure-aware substitution model for Maximum Likelihood or Bayesian methods. In the remainder of this chapter we focus on the latter topic.

Figure 23.1: Overview of the RNAsalsa workflow. Starting from an initial sequence alignment A^0 and a structure constraint σ^0, a relaxed consensus constraint and then constraints for each input sequence are produced. Pairwise sequence alignments and the intersections of the two individualized constraints are used to compute a collection of constrained pairwise consensus structures τ_{ij}, from which a refined individualized structure constraint τ^i is extracted by means of a majority voting procedure. Then τ^i is used as constraint in minimum energy folding to obtain the final structural annotation ψ^i for input sequence χ^i. The structure-annotated input sequences are now aligned with a structure-aware alignment method, resulting in the final alignment B and a corresponding consensus structure ω (Adapted from Stocsits et al., 2009).

23.3 Substitution models for structured RNAs

The rationale of RNA substitution models is rooted in the effect of covariation in paired sites of rRNA sequences. Slightly deleterious substitutions at one site of a helix which disrupt the structure are compensated by a second substitution at the pairing site and restore the pairing ability (Savill, Hoyle, and Higgs, 2001). This leads to a strong correlation of paired positions within rRNA sequences, which therefore do not display independent phylogenetic information anymore, and are violating one central assumption in molecular systematics: the independent evolution of each single sequence site. Consequently, if paired sites are strongly correlated but treated as independent, phylogenetic information is scored twice, and therefore leads to unjustified high support for in itself strongly supported trees and a false low support for alternative trees. The necessary structure-aware substitution models are classical Markovian models of site evolution. They are principally designed in the spirit of conventional DNA substitution models routinely used in phylogenetic inferences (Felsenstein, 2004). The difference to DNA substitution models is that structure-aware models consider pairs of nucleotides as their elementary states rather than single nucleotides.

These specific RNA substitution models can further be divided into two distinct classes, according to the way they account for compensatory mutations in paired sites. Models which are part of the first class assume a two-step process of compensatory substitution in paired positions, considering two single substitutions within nucleotide pairs, with a probability of zero for all double substitutions, whereas models that are part of the other class assume a one-step process of compensatory substitution and thus allow for double substitutions (e.g. AU → GC). Both approaches have been explained on the basis of population genetics (Higgs, 2000). In this context, a mutation in a base pair (e.g. AU → GU) may have led to slightly deleterious UG or GU pairs. Fixation of this intermediate state then leads to a high frequency in a population, before a second mutation occurs at the corresponding site. This is resembled by two-step RNA models. In contrast, one-step models assume a strong selection against these intermediates, leading to a rather low frequency in the population. If a second mutation occurs at the corresponding site (e.g. GU → GC) a drift in gene frequency may lead to a domination of this new base pairing in the population.

One-step models can be further discriminated by their treatment of mismatches: 6-state models completely ignore these pairings, 7-state models lump all mismatches in one category, whereas 16-state models apply an individual substitution probability to each possible mismatch according to the Watson–Crick pairs. Several variations of these structure-aware models, with different levels of character state complexity, have been implemented in the PHASE software (Gowri-Shankar and Rattray, 2007) with a Bayesian background and in RAxML within a likelihood frame (Stamatakis, 2006; Stamatakis, Hoover, and Rougemont, 2008), whereas two-step models are used in MrBayes. Savill, Hoyle, and Higgs (2001) could identify a slightly superiority of substitution models considering double substitutions, but both implementations have demonstrated their effectivity in several studies dealing with rRNA gene sequences (Dohrmann et al., 2008; Meyer et al., 2010; Niehuis, Naumann, and Misof, 2006). However, a recent study on rRNA genes as phylogenetic markers (Letsch and Kjer, 2011) could identify a generally inferior performance of a mixed RNA/DNA model setup, compared to a simple DNA model setup, if the unpaired regions of the used rRNA sequences have already experienced substitutional saturation (Figure 23.2). Application of these RNA models in tree reconstruction virtually down-weigh the impact of paired positions and augment unpaired positions, which is clearly not wanted if the latter are homoplastic. However, as the application of RNA models might be in general a trade-off between an optimized estimation of covariation in paired sites and the risk of magnifying potentially homoplastic unpaired positions, a separate test for substitutional saturation in paired and unpaired partitions prior to the tree reconstruction is highly recommended.

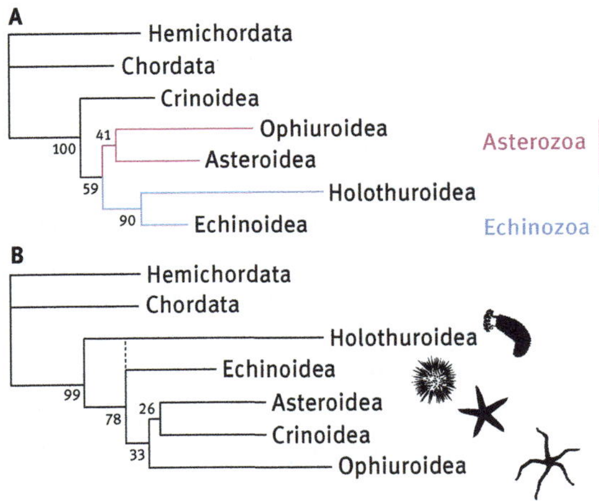

Figure 23.2: Phylogeny of echinoderm classes. 18S rRNA: Maximum Likelihood + bootstrap support values. Substitutional saturation was detected in unpaired regions of the rRNA alignment. **A.** Tree-based on a standard GTR model setup, showing basal Crinoidea and Eleutherozoa divided into Asterozoa and Echinozoa. **B.** Tree-based on mixed RNA/DNA model setups, showing Holothuroidea as sister-taxon to the rest of Echinodermata. The latter clearly contradicts current expectations of echinoderm phylogeny (Adapted from Letsch and Kjer, 2011).

23.4 Practical applications of structured RNAs in molecular phylogenetics

The PHASE package was the first practically workable tool of secondary structure-aware phylogenetics. It implements a Bayesian MCMC algorithm. Among its early applications is a phylogeny reconstruction of Eutheria using mitochondrial tRNAs and rRNAs only, which is surprisingly close to recent genome-based trees. A few studies successfully used short structured RNAs, for which reliable alignments and consensus structure can be constructed rather easily with the help of infernal (Nawrocki, Kolbe, and Eddy, 2009) and the Rfam covariance models (Gardner, 2009). In Niehuis et al. (2012) for instance, they used a RAxML-based analysis of a set of 13 structured ncRNAs (among them some microRNAs and snRNAs) to obtain an additional independent confirmation for the position of Strepsiptera as sister-group to the Coleoptera.

By far the most common application of structured RNAs in molecular phylogenetics, however, concerns ribosomal RNAs (rRNAs). The list is endless due to the ease of amplifying rRNAs of nuclear and mitochondrial resources. The taxonomic scale varies from studies addressing species relationships within genera to the analyses of deep metazoan relationships. The hallmark of these studies is that applying adequate structure-based alignment algorithms and structure-aware substitution models led to phylogenetic hypotheses congruent to hypotheses derived from transcriptome or other multi-gene data. For example, Kjer (2004) and Letsch et al. (2010) could demonstrate that phylogenetic analyses of SSU and LSU non-coding rRNA sequences of hexapods could extremely benefit from the secondary structure consideration – especially in the alignment process. A handful of studies suggest that RNA secondary structures are by themselves also phylogenetically informative (Collins, Moulton, and Penny, 2000; Letsch et al., 2009; Misof et al., 2006). Structural variations that are specific to individual clades have been reported for a wide variety of non-coding RNAs, including rRNAs (Caetano-Anollés, 2002a; 2002b), 7SK RNAs (Gruber et al., 2008), and mitochondrial tRNAs (Jühling et al., 2012b); see also Donath et al. (2010) for a review.

So far, systematic studies to assess the value of secondary structure derived characters have not been conducted, at least in part due to the lack of convenient computational tools. The combination of primary sequence-based analyses with structure-based analyses has also not been explored sufficiently despite the obvious merits (Fleck et al., 2008; Letsch et al., 2009; 2010; Misof and Fleck, 2003; Misof et al., 2002; 2006; Niehuis, Naumann, and Misof, 2006).

23.5 Concluding Remarks

Difficult phylogenetic questions can be tackled with the help of RNA data. In particular ribosomal RNA sequences are readily and easily accessible for a wide variety of species. However, the massive epistasis introduced by strong selection on secondary structure features needs to be taken into account explicitly. To this end, high quality consensus structures are required. At present, the associated computational problems are solved only in part, leaving ample potential for future developments to produce further improvements of rRNA-based phylogenies despite the limited number of characters. An adequate modeling of the substitution processes is of utmost importance in large-scale phylogenetic analyses relying on rRNA gene sequences. This observation can immediately be ported to guide tree inference based on large sets of protein-coding genes. Currently, due to a lack of knowledge, site interactions in protein-coding genes are largely ignored in primary sequence-based tree inferences, with entirely unknown effects. Work with rRNA gene sequences has paved the road to a much more subtle treatment of site interactions or epistatic effects in any kind of primary sequence-based tree inference.

Acknowledgments

This work was supported by the DFG under the auspices of the SPP-1174 Deep Metazoan Phylogeny, projects MI-649/7 (to BM), STA850/2 and STA 850/3 (to PFS), and WA-530/34 (BMvR).

Jörg Lehmann, Carina Eisenhardt, Veiko Krauss, and Peter F. Stadler

24 Intron positions and near intron pairs

Abstract: The relatively rare changes of spliceosomal intron positions have been used both for the estimation of intron evolution and phylogenetic relationships, however with variable success. A specific subset of changes at intron positions is here introduced as a novel phylogenetic marker, termed near intron pair (NIP). It is defined here as a pair of mutually-exclusive intron presences at two (different) orthologous sequence positions in close proximity. The idea that NIPs are powerful characters is based on the assumption that both very small exons and multiple gains at the same position are rare. The results of several studies briefly summarized in this chapter demonstrate the successful usage of the NIP character to reconstruct the tree of life and to deliver new insights into mechanisms of intron gain. In a first NIP-based application the phylogenetic position of major orders of holometabolic insects was evaluated supporting Hymenoptera as resulting from their basal split, in agreement with other studies, but contradicting the previously established view. As part of the genome paper describing a new species of twisted-wing parasites (Strepsiptera), the NIP method was applied to help to resolve the phylogenetic position of them within holometabolic insects. However, a NIP-based reconstruction of the metazoan tree covering a broad selection of representative species also identified some weaknesses of the NIP approach that may suffer e.g. from alignment/ortholog prediction artifacts (also depending on the depth of range of taxa) and long-branch attraction (unequal evolutionary rates of intron gain/loss; maximum parsimony based analysis). In a further study, the identification of NIPs within the genus *Drosophila* could be used to characterize recent intron gain events that apparently involved several cases of intron sliding and tandem exon duplication. Summarizing, the NIP marker could be established as a novel phylogenetic character, in particular dedicated to complementarily explore the wealth of genome data for phylogenetic purposes and to help to resolve open questions of intron evolution.

24.1 Introduction

Spliceosomal introns are characteristic parts of eukaryotic gene structures. The position of these introns within the coding sequence (CDS) is often well conserved, even between genes of distantly related species such as plants and animals (Rogozin et al., 2005). Compared to substitution patterns of primary sequences and, similar to other genomic characters such as transposable element insertions/deletions, intron *positions* have been shown to evolve more slowly and largely independent of the primary gene sequences (Yandell et al., 2006). Thus patterns of conserved intron positions should contain a valuable phylogenetic signal suitable for tracing deeper evolutionary splits.

The evolution of gene structures by intron loss and gain has been investigated in a limited number of studies employing intron presence/absence as phylogenetic character (e.g. Venkatesh, Ning, and Brenner, 1999; Roy and Gilbert, 2005; Nguyen, Yoshihama, and Kenmochi, 2005; Roy and Irimia, 2008a). These studies have been plagued mainly by multiple independent losses of intron positions during evolution and the high variability of intron loss and gain rates, which render presence/absence data for introns rather unreliable phylogenetic markers (Cho et al., 2004; Krauss et al., 2005; Carmel et al., 2007; Irimia and Roy, 2008).

This chapter introduces to a novel kind of rare genomic change marker that traces changes of spliceosomal intron positions across short distances of CDS, so-called near intron pairs (NIPs). We briefly summarize here several studies demonstrating that the usage of NIP characters (1) could improve reconstructions of the tree of life and (2) could deliver new insights into the mechanisms of intron gain. Starting from the evaluation of a phylogenetic hypothesis, the NIP character approach was then applied to help resolve the phylogenetic position of twisted-wing parasites within holometabolic insects, to infer the relationships of a representative collection of metazoan genomes, and to characterize recent intron gain events during the radiation of the genus *Drosophila*.

24.2 Near intron pairs

In a case study by Krauss et al. (2005), the phylogenetic distribution of spliceosomal intron positions in the γ subunit of the eukaryotic translation initiation factor 2 gene (eIF2) was evaluated. Notably, "several cases of successive losses and gains of only slightly different intron positions" were observed, leading to the proposal of such "near intron pairs" as a new type of marker for phylogenetic analyses. NIPs derive their power from the connection between the distribution of exon lengths and intron positions that makes very short exons, and hence the co-existence of very closely spaced introns, extremely unlikely. As a first application, a sister-clade relationship between Hymenoptera and the remaining holometabolic insects was proposed based on two such cases of successive intron loss and gain within the eIF2γ gene (Krauss et al., 2005).

In general, a short distance (e.g. < 50 nt) between any two intron positions present within orthologous coding genes usually should exclude the simultaneous existence of both introns for a single gene. This is based on the observation that short internal exons of corresponding lengths are relatively rare within known species (Saeys, Rouze, and van de Peer, 2007) and functionally detrimental for the splicing process (Weir, Eaton, and Rice, 2006). For instance, for the ortholog data set underlying the NIP search in Metazoa (Figures 24.2 and 24.3, and Lehmann, Stadler, and Krauss (2013)), about 4.5 % and 1.5 % of the internal exons are shorter than 50 and 32 nucleotides, respectively. The fraction of short exons is strongly species dependent, ranging from 0.5 % (*Caenorhabditis brenneri*) to 17.4 % (*Meloidogyne hapla*) for introns < 50 nt, and from 0.09 % to 8.3 % for introns < 32 nt. Under the assumption that multiple gain of an intron at the same CDS position is rare, the shared presence of introns at the same position can be traced back to a common ancestral position. When observing pairs of such conserved intron positions that are located very close to each other in an alignment of the CDS, one can then confidently infer an edge in the phylogenetic tree separating the species sharing one of the positions of the pair from those that share the other, along which intron loss followed by intron gain must have occurred. Alternatively, the two changes of intron positions could have happened simultaneously by the process of intron sliding. In contrast to evaluating shared intron presence/absence of individual positions, the subset of intron positions found at small distances from each other and the usage of a pair as a single character (with two equally complex states, cf. Telford and Copley (2011)) avoids the problem of erroneously inferring an intron gain instead of independent intron losses in different lineages.

Given a multiple alignment of orthologous CDS, a near intron pair (NIP) is defined as a pair of mutually-exclusive intron presences at nearby positions of the CDS. Pragmatically, the pair of positions is required to be located less than 50 nucleotides from each other. The most appropriate threshold to avoid the inclusion of NIPs that likely represent cases of short internal exons can be inferred from the distributions of exon lengths in all the species included in a particular phylogenetic study. To avoid the inclusion of NIPs that are actually caused by alignment artifacts, a number of constraints upon the quality of the CDS alignment around intron position loci may be applied, such as a gap-free amino acid alignment window with similarity scores exceeding a selected threshold. In addition, the amino acid alignment may be optimized for the recognition of conserved intron positions prior to NIP extraction, e.g. by using the realignment tool of Csűros, Holey, and Rogozin (2007).

NIPs are encoded as binary characters with state ‘1’ denoting species having the upstream intron and ‘2’ for species having the downstream intron of the pair. The fully intron-less state is coded as missing data, ‘?’, as it cannot be decided a priori whether the absence of an intron is ancestral or the result of a secondary loss of either one of the two intron positions. Figure 24.1 displays an exemplary NIP both as binary state column and within the context of its harboring CDS alignment.

An issue that is not fully resolved at present is that NIP characters are not strictly independent from each other. The simplest case of such “overlapping” NIPs consists of three intron positions x,y, and z so that x,y and y,z form NIPs. Assuming that x,z do not form a NIP because their distance is too large, only the combinations 1?, 12, 21, ?2, and ?? are possible for the two NIPs x,y and y,z, while 11, 22, ?1, and 2? cannot be realized. If x,z is also a NIP, then 12 is also impossible. Under the assumption of character independence, however, the impossible states may be reconstructed as ancestral states. One way of avoiding this problem would be to re-code overlapping NIPs as complex multi-state characters with restricted transition rules. So far, this strategy has not been explored in practice, however. Overlapping NIPs are not rare,

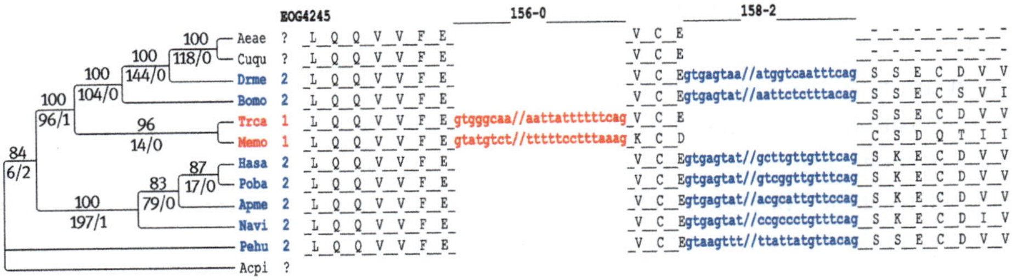

Figure 24.1: NIPs support a monophyletic group of Strepsiptera and beetles (Niehuis et al., 2012). **L.h.s.:** Cladogram inferred from 1,173 parsimony-informative NIP characters from a data set consisting of 12 taxa of holometabolic insects including the newly sequenced Strepsiptera species. Bootstrap support values are shown above branches. Below branches, the numbers of supporting and of incompatible NIP characters are shown. **R.h.s.:** NIP 'EOG4245.156-0_158-2' as one of the 1,173 parsimony-informative NIP characters. It is encoded as a single binary state character (first column). Codons have been replaced by the corresponding amino acid one letter symbols to simplify the presentation. The two alternative intron positions are separated by only 8 nt. Species abbreviations: Aeae = *Aedes aegypti*, Cuqu = *Culex quinquefasciatus*, Drme = *Drosophila melanogaster*, Bomo = *Bombyx mori*, Trca = *Tribolium castaneum*, Memo = *Mengenilla moldrzyki*, Hasa = *Harpegnathos saltator*, Poba = *Pogonomyrmex barbatus*, Apme = *Apis mellifera*, Navi = *Nasonia vitripennis*, Pehu = *Pediculus humanus*, Acpi = *Acyrthosiphon pisum*.

hence they may affect phylogenetic reconstruction in practice. In a large data set covering 45 metazoans about 16 % of the informative introns are contained in more than one NIP.

24.3 Phylogenetic applications of NIPs

24.3.1 Holometabolic insects

Since NIPs are relatively rare their use as phylogenetic markers requires the collection of genome-wide data sets. The first purely NIP-based phylogenetic study was concerned with relationships among the major clades of the holometabolic insects (Krauss et al., 2008). It was based on 758 sets of intron-containing one-to-one orthologous genes between the bee *Apis mellifera* and the beetle *Tribolium castaneum* containing at least one additional intron position within a human and the *Tribolium* ortholog that is not present within the *Apis* copy. To this end, the NIP characters were manually extracted from both automatically generated and manually improved alignments. A final set of 189 *Apis*-*Tribolium* pairs of introns separated by less than 50 nt was extended to include nine metazoan outgroup taxa and five additional holometabolic insect genomes. 54 out of the initial 189 NIP characters were excluded from the data set due to insufficient sequence conservation at the NIP locus or due to apparent gene duplications during metazoan evolution. The final NIP data set contained 135 characters between *Apis* and *Tribolium* originating from 118 different genes.

Since the manual construction of NIP character sets is an extremely tedious and time-consuming endeavor, all crucial steps were implemented in Perl scripts. The first computational step consisted of an automated adaption of CDS annotations correcting the splice site position according to U2 consensus dinucleotides (GT/AG and GC/AG). Second, multiple alignments of the annotated CDS were created with the help of DIALIGN2 (Morgenstern et al., 2006), transAlign (Bininda-Emonds, 2005), and Clustal W (Larkin et al., 2007). Intron positions were then mapped onto the multiple alignment and labeled according to the reference protein sequence (here *Apis*). All alignment regions containing NIP characters as determined by the distance of intron positions within the alignment were extracted together with mapped intron positions indicated by the start (8 nt) and end (14 nt) of the corresponding intronic sequences for each species, respectively. This automated approach completely agreed with the manual prediction of all 135 selected *Apis-Tribolium* NIPs, showing that the extraction of NIP character data from genomic information can at least in principle be automatized.

Of the 135 NIPs, 22 supported a sister-group relationship of the Hymenoptera (including *Apis*) and the remaining Holometabola (including *Tribolium*, Lepidoptera, and Diptera) without any contradictory evidence. This is in agreement e.g. with analyses based on morphological data (Kukalova-Peck J., 2004), paleontological data (Rohdendorf and Rasnitsyn, 1980), EST data (Savard et al., 2006), genomic sequences (Zdobnov and Bork, 2007; Wiegmann et al., 2009), and 18S ribosomal sequences (Misof et al., 2007), and demonstrates that NIPs could offer a phylogenetic signal complementary to substitution patterns of primary sequences.

Additional to Hymenoptera and Coleoptera, other groups are potentially involved in the basal evolutionary splitting event of holometabolic insects (Whiting, 2002b; Beutel and Pohl, 2006). These groups are Neuropterida, Strepsiptera, and Mecoptera. In order to resolve their phylogenetic position, the distribution of 26 phylogenetically informative NIPs (Krauss et al., 2005, 2008; Lehmann, Stadler, and Krauss, 2013) within representative species of these groups was investigated. For this purpose, specimens of Megaloptera, Neuroptera, Raphidioptera, Mecoptera, and Strepsiptera were collected. Additionally, genome assemblies of *Priacma serrata* and *Mengenilla moldrzyki* (Niehuis et al., 2012) were included in the analysis.

In order to obtain orthologous sequences surrounding the selected intron pairs genomic DNA was amplified by PCR with degenerate primer pairs. Degenerate primers were derived from nearby conserved CDS using amino acid alignments of insect proteins. Touch-Down-PCR regimes implementing nested degenerated primer sets were used. The resulting PCR fragments were cloned and sequenced. Among the analyzed 26 NIPs, we found several synapomorphies for specific taxa. For example, 4 NIPs support the monophyly of polyphagan beetles. Importantly, 2 other NIPs support the proposed group Neuropteriformia (Coleoptera + Neuropterida) (Ax, 1999), consistent with most newer studies (Büning, 2005; Krenn, 2006; Wiegmann et al., 2009). Alternatively, Whiting (2002) proposed a group consisting of Hymenoptera, Neuropterida, Mecoptera, and Siphonaptera as sister-group to all other holometabolic insects. However, 5 additional NIPs were found to contain a shared derived intron position of Coleopterida (Coleoptera + Strepsiptera) (Niehuis et al., 2012), Mecopterida, and Neuropterida, and thus also argue against the latter hypothesis.

Because of their successful application in studies of holometabolic insects, NIPs were also employed as part of a recent study on the phylogenetic position of twisted-wing parasites (Strepsiptera) (Niehuis et al., 2012). Based on a data set extracted from 4,485 orthologous protein-coding genes for 13 arthropod species including the twisted-wing parasite *Mengenilla moldrzyki*, whose genome was specifically sequenced for this study, about 1,200 parsimony informative NIP characters resulted in exactly the same topology as inferred from the primary sequence data: Strepsiptera could be placed as the closest-living relative of beetles, see also Figure 24.1. Thus, NIPs could contribute to solve the long-standing enigma of the evolutionary roots of Strepsiptera.

24.3.2 NIPs and the metazoan tree

Phylogenetically informative NIPs observed for other branches of the metazoan tree within the first NIP studies pointed to a general usability of NIP markers, not only for the evaluation of particular phylogenetic hypotheses, but also for character-based phylogenetic reconstructions of the eukaryotic tree.

Whereas the major splits of bilaterian animals are now relatively well settled, major open questions in animal phylogeny concern the order of the earliest branches separating the lineages of Bilateria, Cnidaria, Placozoa, Ctenophora, and Porifera (Edgecombe et al., 2011), but also e.g. the relationships within Lophotrochozoa (Hejnol, 2010). Inconsistent results from numerous studies based on different character data types, either based on morphological or molecular data, show the need for alternative methods and complementary markers.

In contrast to applying NIPs to specific metazoan subtrees, the reconstruction of the whole metazoan phylogeny is a computationally more challenging enterprise. A first NIP-based study of 48 genomes comprising 12 metazoan phyla was completed recently (Lehmann, Stadler, and Krauss, 2013). It used an improved pipeline to extract and process NIP characters, see Fig. 24.2. This pipeline includes filtering of NIPs with intron position CDS distances below 32 nt and sufficient alignment quality around intron loci (gap-free alignment window exceeding a conservation threshold) to reduce alignment artifacts.

The broad range of taxa in this metazoan study increases the importance of a reliable ortholog data set and of the corresponding alignments underlying the NIP characters. Lehmann, Stadler, and Krauss (2013) compiled the data set based on a combination of already available genomic and ortholog annotation data from Ensembl and

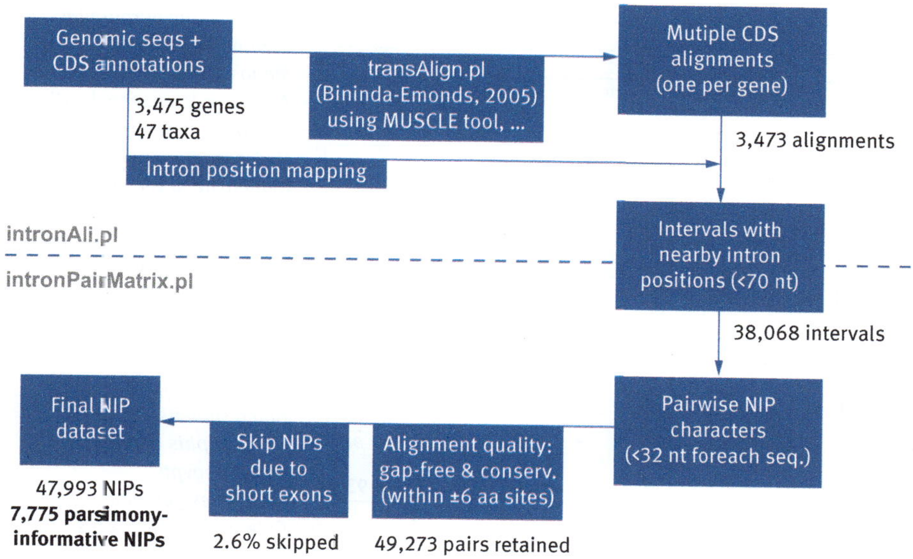

Figure 24.2: NIP extraction workflow. Numbers and settings for a preliminary metazoan data set are shown next to the workflow.

Ensembl Compara databases and a prediction of additional orthologs either by targeted ortholog search using the HaMStR tool (Ebersberger, Strauss, and von Haeseler, 2009) on annotated genomes or using a BLAST-based (Camacho et al., 2009) search together with a subsequent annotation of gene structures with the help of exonerate (Slater and Birney, 2005).

Being restricted to completely sequenced genomes, the taxon sampling of this study is still very limited in some lineages (e.g. Lophotrochozoa) and sometimes includes rather uncommon representatives of taxonomic groups (e.g. Schistosoma). Furthermore, the coverage of ortholog genes in the derived data set greatly varied between species (*A. californica* was present in a fifth of the genes only) and differences in intron densities between more distantly related species further reduce the number of usable NIPs.

Despite these complications, the maximum parsimony analysis of NIP data (7,775 parsimony-informative characters from a preliminary data set of 47 taxa, excluding the ctenophore *Mnemiopsis leidyi*) supported well-established clades such as Bilateria, Protostomia, Deuterostomia, Ecdysozoa, Platyhelminthes, Pancrustacea, Chordata, and Vertebrata, Figure 24.3. Arthropoda, however, were not recovered, as the mite *Ixodes* was placed as sister to remaining ecdysozoans. Some additional problematic branchings could be explained by paucities of synapomorphic changes of intron positions shared between some taxa (e.g. no clade of mollusks). In addition, the NIP-based MP topology predicted an assemblage of Cnidaria, Porifera, and Placozoa as sister-group of bilaterian animals, however, not with conclusive evidence. Using an

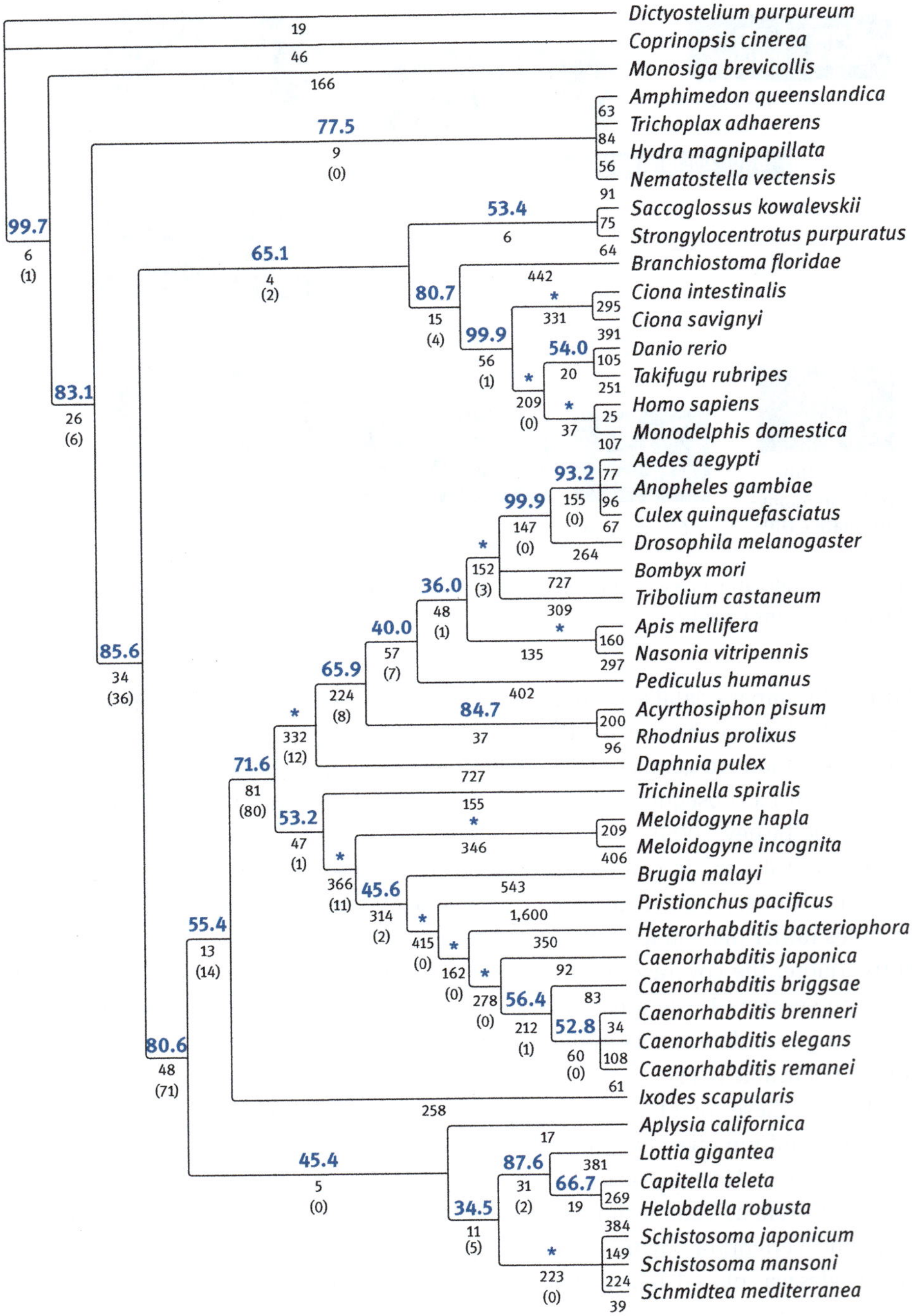
Dictyostelium purpureum
Coprinopsis cinerea
Monosiga brevicollis
Amphimedon queenslandica
Trichoplax adhaerens
Hydra magnipapillata
Nematostella vectensis
Saccoglossus kowalevskii
Strongylocentrotus purpuratus
Branchiostoma floridae
Ciona intestinalis
Ciona savignyi
Danio rerio
Takifugu rubripes
Homo sapiens
Monodelphis domestica
Aedes aegypti
Anopheles gambiae
Culex quinquefasciatus
Drosophila melanogaster
Bombyx mori
Tribolium castaneum
Apis mellifera
Nasonia vitripennis
Pediculus humanus
Acyrthosiphon pisum
Rhodnius prolixus
Daphnia pulex
Trichinella spiralis
Meloidogyne hapla
Meloidogyne incognita
Brugia malayi
Pristionchus pacificus
Heterorhabditis bacteriophora
Caenorhabditis japonica
Caenorhabditis briggsae
Caenorhabditis brenneri
Caenorhabditis elegans
Caenorhabditis remanei
Ixodes scapularis
Aplysia californica
Lottia gigantea
Capitella teleta
Helobdella robusta
Schistosoma japonicum
Schistosoma mansoni
Schmidtea mediterranea

◂ **Figure 24.3:** Maximum parsimony strict consensus tree constructed from 7,775 parsimony informative NIPs, derived from a preliminary metazoan data set (47 taxa). Bootstrap support (in percent) is given above the branches (*blue*), the number of NIP characters that support the branch are shown below each branch. These numbers represent synapomorphies and autapomorphies at internal and terminal branches, respectively. In parentheses, the number of characters inconsistent with the branch is given. A star denotes 100 % bootstrap support. The amoebozoan *Dictyostelium purpureum* was used to root the tree. CI = 0.886, RI = 0.930.

extended data set that also includes the ctenophore *Mnemiopsis leidyi* (Lehmann, Stadler, and Krauss, 2013) resulted in a slightly different tree topology, and did not recover some more major clades (Bilateria, Ecdysozoa, Deuterostomia) as a consequence of unusual positions of *Mnemiopsis*, *Ixodes*, and Ambulacraria.

24.4 NIPs and the mechanisms of intron gain

NIPs can be used to differentiate between intron gain and recurrent intron losses within independent lineages. They furthermore provide a means of unambiguously identifying recently inserted introns. A comparative analysis of NIP-associated exon-intron structures within the fruit fly genus *Drosophila* was used to evaluate possible mechanisms of intron origin (Lehmann et al., 2010). Starting point for the analysis were the 12 *Drosophila* genomes described in *Drosophila* 12 Genomes Consortium (2007) whose gene and orthology annotation data are readily available to the community (Tweedie et al., 2009). A data set of 31 NIPs with *Drosophila melanogaster* as reference was extracted after stringent computational filtering, extensive manual curation, and PCR-based validation, see Lehmann et al. (2010) for details. Orthologous sequences from arthropod outgroup species allowed to unambiguously polarize 17 NIPs into one ancestral and one novel position.

Based on the conservation of intronic sequences, but also on evaluating properties of the NIP pattern such as the absence of an intermediate taxon with intron-less state, the NIP distance (multiple of 3 nt), the nature of splice sites (tandem), or the existence of additional cryptic splice sites, 9 out of 31 NIPs were identified as the result of intron sliding. This is surprising since earlier studies predicted intron sliding to be a very rare event (Rogozin, Lyons-Weiler, and Koonin, 2000). However, frequent intron sliding can be expected as a consequence of the high abundance of tandem splice sites (e.g. GYRGYR/NAGNAG) in eukaryotic genes (Hiller et al., 2007; Sinha et al., 2010). In 7 cases there was evidence for intron gain by tandem duplication of exonic DNA containing a "proto-splice site", i.e., the sequence pattern 5'-AGGY-3', which was proposed as preferential exonic site for intron gain or loss by Dibb and Newman (1989). This mechanism has been evaluated previously without reaching a valid conclusion about the abundance of such intron insertions (Coulombe-Huntington and Majewski, 2007; Venkatesh, Ning, and Brenner, 1999; Zhuo et al., 2007; Roy and Irimia, 2008b).

The mechanisms causing more than half of the NIPs identified in the *Drosophila* study remained unknown. In particular, and contrary to expectations, no evidence was found for intron gain by reverse-splicing and subsequent reverse-transcription and recombination as proposed e.g. by Sverdlov et al. (2004). It appears that intron gain is predominantly caused by local mutations rather than insertions of reverse-transcribed sequences.

24.5 Conclusion

Near intron pairs clearly form a valuable type of phylogenetic marker. Computational pipelines such as the one outlined in Figure 24.2 make the amount of work that is necessary to collect the character matrix from dozens of genomes manageable and hence guarantee practical applicability. NIPs already have been accepted as useful means of re-evaluating sequence-based phylogenies (Kumar et al., 2012; Niehuis et al., 2012).

Like any other set of phylogenetic markers investigated so far NIPs are not perfect markers and come with their own set of problems, however. Intron sliding, for instance, which accounts in particular for many of the very closely spaced NIPs, can generate homoplasies. The lack of independence of overlapping NIPs, furthermore, can lead to an over-estimate of confidence and reliability measures. The lack of a well-parametrized model of NIP formation, furthermore, precludes the use of modern stochastic methods of phylogeny reconstruction and dictates the use of parsimony methods, which are vulnerable to artifacts such as long-branch attraction.

Nevertheless, NIPs have turned out to be useful and informative at class and phylum level in animal phylogeny. Clearly, they lose power both at small time scales (because intron insertion becomes very rare e.g. within the Drosophila genus) and at very large time scales (because sequences become so divergent that homologous CDS positions cannot be determined with sufficient reliability). Methodological improvements are conceivable in several directions. Minimal exon sizes, for instance, may vary by clade so that the maximal distance of NIPs could be adjusted to minimize the chance of incorporating co-existing intron positions. Alternatively, NIPs could be weighted according to the distances. Further improvements in the computational pipelines could address both orthology assignments and the re-alignment algorithms.

Acknowledgments

This work was supported by the Deutsche Forschungsgemeinschaft, Proj. Nos. KR2065/2 and STA850/6.

Alexander Donath and Peter F. Stadler

25 Molecular morphology: Higher order characters derivable from sequence information

Abstract: With the rapid technological developments in high-throughput sequencing, large amounts of genomic sequence data have become available also for non-model organisms. Phylogenomics has been highly successful in utilizing this wealth of information but by no means all open phylogenetic problems have been solved with sequence-based methods alone. Genome sequences implicitly contain a plethora of higher-level features that convey phylogenetic information in their own right. The long list includes genes, protein domains, microRNAs, introns and their relative positions within genes, regulatory DNA elements, insertion points of transposable elements, patterns of insertions and deletions of sequence elements, and the relative order of such annotation items along the genomic DNA. Molecular morphology attempts to systematically extract them from raw data and harnesses them for phylogenetic purposes. This poses a series of practical and theoretical problems, ranging from the issues of obtaining unbiased and comparable genome annotations, to questions on how features are to be treated that can be defined only in a comparative setting, to the mathematics of data types that cannot be captured by character matrices.

25.1 Introduction

"More data, more data" has been the mantra of Molecular Phylogenetics, and it has certainly been a highly successful program, in particular with the advent of EST-based phylogenomics. The rise of genome sequencing, however, has forced the realization that there are inherent fundamental limits to the amount of sequence information set by the size of the genome that cannot be overcome. This has led to the suspicion that genome-wide data, dense taxon sampling, and more sophisticated (and thus computationally more expensive) models of sequence evolution alone will not be sufficient to eventually resolve all open questions in phylogenetics (Rokas and Carroll, 2006). Of course, more careful sampling and improved modeling of the data (Philippe et al., 2011a), and a more systematic analysis of possible artifacts (Kück et al., 2012) have helped to resolve many of the long-standing questions.

The flood of genomic data, on the other hand, also brings with it the chance to make systematic use of information beyond mere sequenced-based characters. Rokas and Holland (2000) proposed rare genomic changes (RGCs) as (nearly) perfect phylogenetic characters, see also (Boore, 2006; Boore and Fuerstenberg, 2008). Events such as insertions of transposable elements, intron gain and loss, changes in gene order, duplications, fusions or fissions of domains, and the emergence of novel genes or domains appear to have a sufficiently large state space to allow a nearly unambiguous determination of homology. Nevertheless, homoplasies have been observed also for RGC characters (Han et al., 2011; Ray et al., 2006). Some of the RGCs, in particular the insertion of transposable elements (Stewart et al., 2011) or changes in intron structure (Cho and Doolittle, 1997), are thought to have little functional impact and thus to follow simple statistical principles. Other changes, such as the emergence of novel combinations of proteins, gene duplications, and innovation of novel protein domains or RNA genes, on the other hand, are likely to have phenotypic consequences and thus to be subject to strong selection. At least at present we do not have good statistical models describing the evolutionary dynamics of many of these structural features, effectively limiting their analysis to parsimonious arguments.

We use the term Molecular Morphology to refer to the diversity of structural features that shape the genomic DNA sequence. Molecular morphology is also an emerging field concerned with the systematic annotation of genomic structure and their application in a phylogenetic context. In this chapter, we give an overview of the state of the art and discuss some of the theoretical and practical problems associated with this approach.

One major difficulty in resolving animal phylogeny is the resolution of the comparably short branches that separate the deep nodes associated with and preceding the Cambrian radiation. Selectively neutral or nearly neutral characters thus did not have much time to deposit information on these short branches and the little signal there was had plenty of time to degrade and fade into oblivion. The stunning divergence of body plans and morphological features that distinguish the major animal lineages, on the other hand, suggests that at least some of these short branches have seen episodes of rapid adaptive evolution that later turned into strong stabilizing selection that preserves at least parts of these adaptive innovations. Short, deep branches, furthermore, may in some cases also be artifacts of the data analysis that fails e.g. to disentangle divergent adaptive modes of sequence evolution with common reversible models.

It thus becomes crucial also for phylogenetic questions to gain a better understanding how the phenotypic innovations are represented in the comparably easily accessible sequence data. To a first approximation, the overall gene content has changed surprisingly little between fungi and animals and even most of the animal-specific gene families already emerged in their single-celled ancestors. The morphological simplicity of basal animals, and the great differences in morphology of the metazoan phyla therefore, cannot be explained by the presence or absence of major protein-coding gene families but instead must be attributed to differences in the temporal and spatial deployment of essentially the same complement of protein-coding genes (Erwin et al., 2011). The major processes that play a role here are, thus, (1) functional innovations in existing (regulatory) genes and (2) the emergence of additional regulators.

The molecular consequences of radiations have been studied in some detail e.g. for Hox gene clusters (Wagner, Amemiya, and Ruddle, 2003). Following the fish-specific genome duplication, for instance, the radiation of the teleosts (Crow et al., 2006) is coincident with an episode of rapid adaptive evolution in developmentally important genes, see Figure 25.1 for an example. Here the time scales are so short that synonymous substitutions are virtually absent.

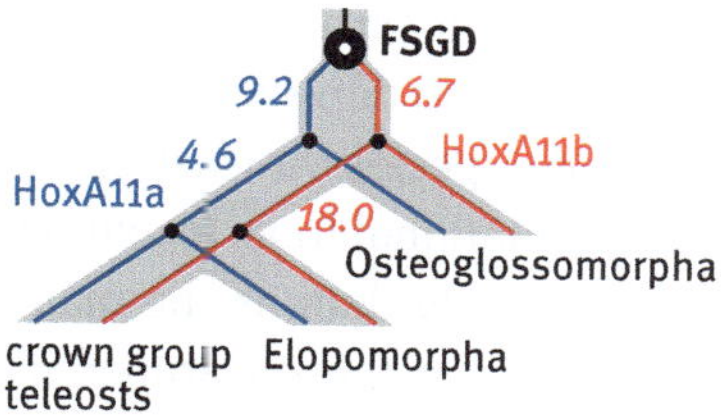

Figure 25.1: Example of strongly divergent trends in adaptive evolution. A maximum-likelihood reconstruction (Crow et al., 2006) of the fate of the HoxA11 gene in the wake of the Fish Specific Genome Duplication (FSGD) estimates substantial numbers of non-synonymous substitutions (indicated at the branches) but not a single synonymous substitution for the two emerging paralogous genes HoxA11a and HoxA11b. Note that the rates of adaptive evolution also differ strongly between the paralogs.

The most obvious genomic features that are likely to have a functional impact are innovations of novel proteins, albeit this is very rarely a de novo process. Much more frequently, new functionalities are obtained through the emergence of new combinations of functional protein domains (Buljan and Bateman, 2009; Orengo and Thornton, 2005) and other forms of tinkering such as the recruiting and adaptation of smaller fragments of adjacent DNA (Bornberg-Bauer, Huylmans, and Sikosek, 2010).

Functional changes are not only reflected in the sequences of protein-coding genes. A substantial part of the evolutionary adaptations appears to be brought about by the innovation of regulatory DNA elements (transcription factor binding sites). Although hard to determine in practice, they can also be phylogenetically informative. Somewhat surprisingly, large-scale ChiP-seq data do not make the problem easier at least for phylogenetic applications: most of the detected transcription factor binding sites are subject to extremely rapid turn-over (Schmidt et al., 2010), suggesting that only a small fraction conveys biological function. Presence/absence information on entire clusters of conserved binding sites, nevertheless, was used in (Prohaska et al., 2004) to help disentangle the relationships among vertebrate Hox clusters. In the same spirit, ultra-conserved DNA regions, which are regarded as (typically not protein-coding) DNA regions under extreme stabilizing selection (Bejerano et al., 2004; Stephen et al., 2008) can be utilized for phylogenetic purposes (Crawford et al., 2012; Faircloth et al., 2012). In contrast to protein-coding genes, furthermore, there is good evidence that at least certain types of small regulatory RNAs, in particular microRNAs, are introduced into plant and animal genomes by a process of de novo innovation (Hertel et al.,

2006). We will return to their use in phylogenetics in the work of Kevin Peterson and collaborators (Peterson, Dietrich, and McPeek, 2009) in the following section.

There are three major practical obstacles to using RGC characters in phylogenetic studies. First, there are still only a small number of completely sequenced genomes. This limitation appears to lose its importance as sequencing costs rapidly decrease. The genome of the strepsipteran *Mengenilla moldrzyki* and the basal beetle *Priacma serrata* were sequenced within the DMP primarily for phylogenetic purposes (Niehuis et al., 2012). In just a few years' time, it will probably be natural to deposit a genome sequence together with a morphological description of a new species. At least in the long run, this leaves the lack of efficient and accurate computational tools to extract features of the "genomic morphology" from genome sequencing data as the main practical problem. A third, more theoretical issue is that some of these features, in particular gene order data, do not behave like phylogenetic characters in the classical sense and hence cannot be analyzed by well-established methods of computational phylogenetics.

25.2 Characters and "pseudo-characters"

The Character Concept, maybe surprisingly, is among the most complex and most disputed issues in evolutionary biology (Wagner, 2001). It has received most attention in its relation with the notion of homology see e.g. (Patterson, 1988). At least for phylogenetic purposes there seems to be a common consensus that homology means "derived from a common ancestor". Most authors agree that characters must be directly observable and the task of homologizing characters is separate from the task of identifying characters. This is most clearly evident in projects such as MorphDBase (Vogt, Bartolomaeus, and Giribet, 2010; see also Vogt and Grobe, Chapter 20, this book), which strives to establish a morphological structure concept that is independent of evolutionary considerations.

In the context of the sequence-based methods of molecular systematics, individual nucleotides or amino acids are viewed as characters and their homology is pragmatically established by means of sequence alignments. In practice, thus, alignment columns identify characters and their states. Large-scale multiple sequence alignments of genomes are becoming readily available as an intermediate result of studies into computational genomics. At least at a pragmatic level, this source of data provides homology information: the fine-grained level of nucleotides that can immediately be converted into a homology statement for genomic regions and genomic features without the need to first identify particular features in the individual genomes as characters. This route is often taken in practice. Sequence-based similarity search with tools such as BLAST (Altschul et al., 1990) is used to identify a previously unannotated sequence as a homolog of the query sequence. In EST-based phylogenomics, of course, one has at least the additional evidence that one identifies only expressed mRNAs as possibly homologous genes for the subsequent phylogenetic analysis. Higher-level features such as microRNAs, protein-coding genes, or introns thus relax the classical requirement to separate character identification and homologizing.

The most extreme case of alignment-derived "characters" are gap patterns. Corresponding to insertions or deletions they are by definition not observable in data from an individual taxon. Nevertheless, their pattern of occurrence confirms all the other salient properties of characters, including an at least approximate independence of non-overlapping gaps and a mathematical product structure inherited from the alignment that allows us to encode them in the form of character matrices. Several studies have shown that gap patterns are indeed phylogenetically informative (Lloyd and Calder, 1991; Murphy et al., 2007).

Not every phylogenetically informative structure is necessarily a character in the stricter sense of Lewontin and others, who view characters as phenotypic building blocks of an organism (Lewontin, 1974): in this approach, evolutionary quasi-independence is required as an additional condition, stipulating that the states of one character can vary independently of those of other characters. This structural independence condition does not imply that characters vary statistically independently in any given data set. The much stronger assumption of statistical independence, however, is usually made in the practice of molecular phylogenetics: alignment columns are treated nearly ubiquitously as if they evolve statistically independently from each other. Only a few approaches to RNA-based phylogenetics treat base pairs in conserved regions as units, i.e., as single characters, see e.g. (Hudelot et al., 2003; Jow et al., 2002) and Bernt et al. (Chapter 22). The price of assuming statistical independence, which nearly always is an approximation at best, is a systematic overestimation of the amount phylogenetic information.

Independence assumptions bring with them the mathematically necessity for an underlying product structure on the space of structural descriptions of the taxa (Wagner and Stadler, 2003). This mathematical requirement is also the basis for the representation of both morphological and sequence data in the form of character matrices and alignments and implicitly underlies computational methods that infer trees from the character matrices. For example, the column-wise solution of the "small parsimony problem" is mathematically justified only if the columns are structurally independent.

Not all types of phylogenetically useful information satisfy this precondition and hence cannot be represented in the form of character matrices. The best-known example is gene order data. Gene orders are naturally encoded as permutations, which utterly lack a product structure. Their analysis, e.g. in the context of mitochondrial genome rearrangements, thus requires entirely different mathematical approaches (see Podsiadlowski et al., Chapter 19, this book). It is possible to derive character-like descriptions from such data. For example gene orders can be encoded as the presence or absence of particular adjacencies or clusters of genes. The resulting "characters", however, retain strong combinatorial dependencies and it remains an open question at this point to what extent the phylogenetic information can be distorted by this type of embedding. Such re-codings, furthermore, raise the question in what sense, if any, the derived "characters" are homologous.

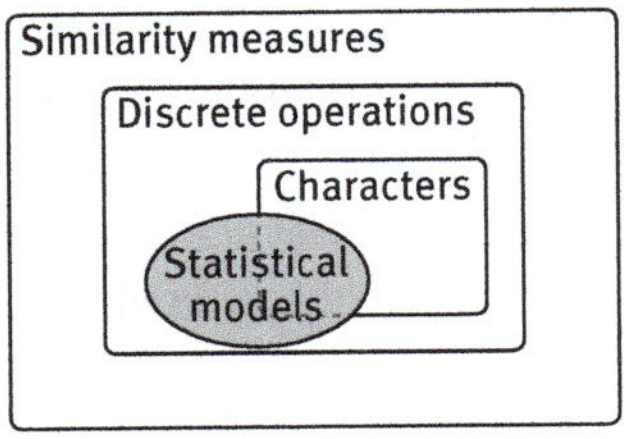

Figure 25.2: Types of phylogenetic information. Only a subset of high-level features can be meaningfully encoded in a string-like manner to yield alignments or character tables. More generally, at least possible evolutionary operations are known as in the case of gene order data. The most general setting allows for similarity measures only. Statistical approaches (Maximum Likelihood or Bayesian methods) are available for a subset of feature types.

Molecular morphology therefore not only expands the sources of phylogenetic information but also poses methodological challenges arising from the more general data types. Figure 25.2 summarizes the situation: Statistical models, and hence Maximum Likelihood and Bayesian methods can be constructed only on a mechanistic understanding of the underlying evolutionary process, including, of course, sequence characters and, to a certain extent, gene order data. Presence/absence data of microRNAs, introns, or mobile elements, as well as most views on mitochondrial rearrangements, are at least amenable to methods that strive to explain the data in terms of minimal numbers of changes. In the most general setting, only similarity or distance information is available. Despite some progress, also resulting from DMP projects, it remains an important challenge, both methodologically and computationally, to analyze non-character-like data and determine if and when they can be recast as "pseudo-characters" that can be treated with the well-established machinery of computational phylogenetics.

25.3 Practical approaches to molecular morphology

25.3.1 Gene content

Probably the simplest phylogenomic approach is the comparison of gene content. Such data indeed yield an overall good separation between the major kingdoms and consistently group closely related species together (Snel, Huynen, and Dutilh, 2005). It seems to be less promising for the intermediate regime, however. Problematic issues include the difficulty to determine orthology due to structural changes in protein organization such as loss or gain of individual domain, fusion of ancestral proteins, and fission events. Horizontal gene transfer and secondary loss are further sources of false hypotheses. A possibly remedy is the use of “signature genes”, i.e., a core set of genes that is present in every daughter lineage of a given clade but not outside of it (Dutilh et al., 2008). The justification is parsimonious: if a gene can be found at the root of a clade and all its descendant lineages, then it has been retained because it possesses an important function for the species in that clade.

25.3.2 Metabolic networks

An extension of gene content-based approaches is to consider metabolic pathways or networks. Here, sets of enzymes are grouped into pathways or network modules. Instead of orthology of the underlying changes one strives to compare the functional aspects of the pathways, i.e., functional analogy is explicitly accepted. Often distance-based methods are employed using measures of dissimilarity of both substrates and enzymes (Forst and Schulten, 2001; Heymans and Singh, 2003). The Network of Interacting Pathways introduced by Mazurie et al. aggregates the available information even further to a graph whose nodes are entire pathways connected by edges when they share metabolites (Mazurie et al., 2008). Phylogenetic trees are then derived from distances measures between these graphs.

Pathway and network methods suffer from the same problems as approaches based on gene content. In addition, pathways are only known from “simple” organisms, such as Bacteria and Archaea, and few model organisms, e.g., human, rat, and *Drosophila*. Retrieval of the interaction data is not trivial and depends on the enzyme classification and on reference pathway sources such as KEGG (Kanehisa et al., 2006).

25.3.3 Repetitive elements, introns, and NUMTs

The most commonly used RGC characters are insertions of repetitive elements, in particular SINEs (short interspersed nuclear elements) (Shedlock and Okada, 2000). These appear to be applicable mostly to relatively shallow phylogenetic relationships such as those among turtles (Sasaki et al., 2004) or in issues of mammalian phylogeny, see e.g. (Schröder et al., 2009). They are less useful for deep phylogeny since repetitive DNA is typically not subject to strong stabilizing selection. The sequence information required for the identification of a particular mobile element therefore is degraded at a rate close to the background mutation rate.

A special type of characters restricted to Eukarya are nuclear mitochondrial sequences (NUMTs, pronounced “new-mights”), that is, insertions of mitochondrial DNA in the nuclear genome (Hazkani-Covo, Zeller, and Martin, 2010; Lopez et al., 1994). In contrast to mobile elements they do not proliferate but are replenished from the mitogenome. From a phylogenetic point of view they can be treated like repeat characters. Presence/absence patterns of NUMTs have been valuable to disentangle a phylogenetic branching order of primates (Hazkani-Covo, 2009). They are not applicable for deep phylogeny problems, however, since they are not under stabilizing selection.

The vast majority of genes in most eukaryotes contain spliceosomal introns. Their positions usually evolve slowly, in particular within coding regions (Rogozin et al., 2005), but to a lesser extent also in long non-coding RNAs (Hiller et al., 2009; Nitsche et al., 2012). Several studies thus have used presence/absence patterns of introns for phylogenetic purposes. Although splicing is linked to the efficiency of gene expression (Kornblihtt et al., 2004), it appears that the fitness effects associated with the loss of a particular intron are usually small, so that intron data sets are plagued mainly by multiple independent losses and large variations in gain/loss rates, see e.g. (Irimia and Roy, 2008; Krauss et al., 2005). To overcome this issue, near intron positions (NIPs), i.e., the presence of alternative, mutually exclusive intron positions, have been proposed as homoplasy-poor characters (Krauss et al., 2008). NIPs and their application to animal phylogeny are covered in detail in a separate chapter of this book (see Lehmann et al., Chapter 24, this book).

25.3.4 Genome rearrangements

Nearly all the literature on gene orders as phylogenetic marker deals with mitochondrial genomes, see e.g. (Boore and Brown, 1998; Lavrov, 2007). Due to their small size, their nearly invariant gene content (37 genes, of which 22 are tRNAs, 2 are ribosomal RNAs, and 13 code for proteins), and the large number of taxa (more than 3,000 mitogenomes have been published in GenBank so far) they have become a very convenient source of data. Mitogenomes can be annotated exhaustively with high accuracy (Bernt et al., 2013c). Because gene duplications and gene losses are absent in most cases, it is possible to model mitochondrial gene orders as signed permutations. Gene order data are the prime example of a data type, permutations, that cannot be treated in the form of character matrices. Specialized computational methods therefore have been developed to analyze mitogenomic data, reviewed e.g. in (Bernt et al., 2013d). A separate chapter in this book (see Bernt et al., Chapter 22, this book) highlights in particular the contributions of the DMP program to the technical advance. Within the Deep Metazoan Phylogeny program a comprehensive analysis of metazoan mitogenomes explored the use and limitations of mitogenomic data (Podsiadlowski et al., Chapter 19).

Gene order data from chloroplasts have been used occasionally in plant phylogenetics (Karol et al., 2010; Mishler and Kelch, 2009). A major difficulty compared to mitochondria are the much larger numbers of insertions and deletions. The feasibility of gene-order based phylogeny inference using nuclear genomic data is shown e.g. in

(Huang et al., 2010; Lin, Rajan, and Moret, 2012). An interesting variation is the use of overlapping genes to process partial gene order information (Cheng et al., 2010). So far, practical applications of nuclear gene orders are virtually lacking, in part due to the size of the data sets, although this is not primarily an algorithmic problem. A practical limiting factor appears to be the notorious incompleteness of genome annotation and the difficulty to obtain reliable orthology assignments.

25.3.5 MicroRNAs as phylogenetic marker

MicroRNAs are short, about 22 nt long, non-coding RNAs that act primarily as post-transcriptional regulators of gene expressions. Animal microRNAs are processed, with a few exceptions, from long pol-II transcripts by extraction of a hairpin-shaped precursor by Drosha and subsequent Dicer cleavage of the double-stranded region of the precursor hairpin (see Okamura, 2012). The short functional sequence and the abundance of approximately hairpin-shaped RNA secondary structure explains why novel miRNA families keep being invented throughout evolution: The abundance of non-coding transcription in all crown group Eukaryotes and the ease with which the basic structural requirements are satisfied ensures that candidates for innovations are frequently funneled into the miRNA processing pathway (Campo-Paysaa et al., 2011; Tanzer and Stadler, 2004). If it turns out to be useful, it will readily be picked up by first directional and then stabilizing selective pressures. Once miRNAs are established, they appear to acquire multiple targets, thus increasing the amount of stabilizing selection and reducing the chance of compensatory substitutions. Mature miRNA sequences thus show very little sequence variation and often none at all (Wheeler et al., 2009). As a consequence, they are fairly easily detected and identified both in genomic sequencing data and in short RNA-seq experiments.

Novel microRNA families have appeared throughout the evolution of both animals (Heimberg et al., 2007; Hertel et al., 2006; Prochnik, Rokhsar, and Aboobaker, 2007; Sempere et al., 2006) and plants (Axtell, Snyder, and Bartel, 2007; Sunkar and Jagadeeswaran, 2008; Zhang et al., 2006), albeit in the latter there are fewer conserved families. The expanding repertoire of both miRNA genes and their putative targets (Lee et al., 2007) appears to be correlated with major body-plan innovations (Berezikov, 2011; Peterson, Dietrich, and McPeek, 2009). Beyond the continuing innovation of novel families, evolution spawns new members of miRNA families by duplication. As other genes, miRNAs are of course also subject to gene loss (Hertel et al., 2006; Tanzer and Stadler, 2004). MicroRNA paralogs are often impossible to distinguish based on the mature sequence alone. Phylogenetic applications thus typically use presence/absence of families rather than individual miRNA genes since the complete loss of entire miRNA families appears to be rare in most lineages. This is not always the case, however. The morphological simplification in urochordates, for instance, is accompanied by the loss of numerous highly conserved bilaterian miRNAs and a reorganization of their miRNAs that clearly sets them apart from the other chordate lineages (Fu, Adamski, and Thompson, 2008). A similar massive reorganization of the miRNA repertoire seems to have affected the flatworm lineage. Most of the 71 miRNAs reported in *Schmidtea mediterranea* (Palakodeti et al., 2006), for instance, cannot be homologized unambiguously with the miRNA families of other protostomes. Nevertheless, miRNAs still correctly place ascidians as sister-group of vertebrates and identify Schmidtea as a lophotrochozoan protostome.

In addition, miRNA characters have been strikingly informative for many of the difficult problems in metazoan phylogeny. Examples include support for monophyletic Lophotrochozoa (Wheeler et al., 2009) and Panarthropoda (Campbell et al., 2011), the inclusion of Acoelomorpha and other "enigmatic worms" as a basal branch within the Deuterostomes (Philippe et al., 2011b), and a resolution of annelids (Sperling et al., 2009), see also Figure 25.3.

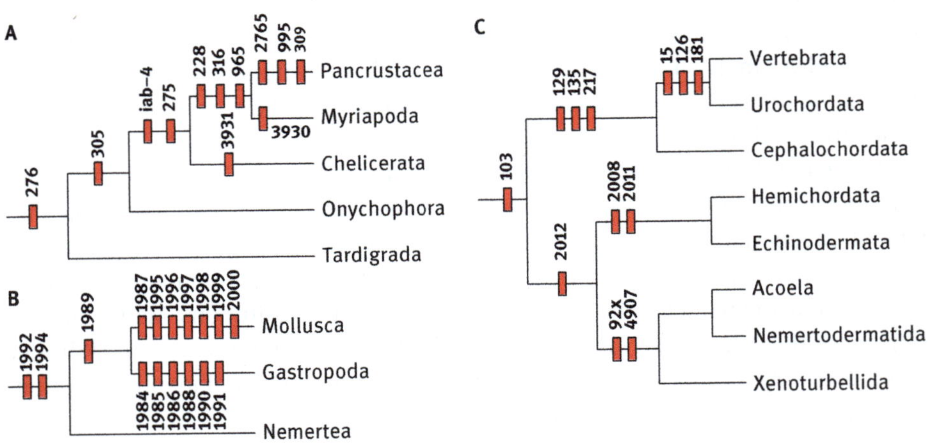

Figure 25.3: Examples of miRNA-based deep phylogenies. **A.** Ecdysozoa (Campbell et al., 2011). The nematodes form the sister-group to the subtree shown here. **B.** Lophotrochozoa (Wheeler et al., 2009). **C.** Deuterostomia (Philippe et al., 2011b) supporting Olfactores as sister-group of the Cephalochordata and a sister-group relation of Ambulacraria with a clade of "enigmatic worms". 103 refers to the deuterostome-specific miR-103/107/2013 family, 92x denotes a clade-specific paralog of miR-92. The panels have been updated to reflect current miRNA nomenclature.

MicroRNAs are not infallible, however. In many cases they are not informative, as in the example of class-level relationships within Echinodermata (Pisani et al., 2012). In some cases they even appear to be misleading, as exemplified by the recent proposal of a turtle+lizard clade (Lyson et al., 2012) that is rejected so far by all other phylogenetic markers, see (Crawford et al., 2012) for a detailed discussion. Apart from unexpectedly high loss rates in some clades, miRNA gain is also not entirely free of homoplasies. While the true de novo innovations most likely do not exhibit homoplasy at a practically relevant level, there is also a subgroup of small RNAs that arose by exaptation and are annotated as miRNAs. In most cases the evolutionary sources are abundant repeat families (Smalheiser and Torvik, 2005; Yuan et al., 2011). The same repeat family may have given rise to multiple miRNAs, rendering these families effectively uninformative. In a few cases, putative "microRNAs" also arose from pseudogenes. A well-known example of the latter class is the mir-220 family (Devor, 2006), which is associated with tubulin pseudogenes and has meanwhile been removed altogether from the miRNA annotations (Chiang et al., 2010). It has been argued that most if not all repeat-derived miRNAs are misannotations (Sperling et al., 2009) and the independent innovation of the same miRNA in two different taxa has not been demonstrated to-date. The question which microRNA-sized RNA products should be classified as bona fide miRNAs is a question of open debate. The situation has become more complicated in the last few years as multiple alternative processing pathways for small RNAs have been discovered (Yang and Lai, 2011). For phylogenetic purposes a restrictive definition would certainly be desirable. Pragmatically, additional curation of data taken from repositories such as miRBase (Kozomara and Griffiths-Jones, 2011) is advisable.

25.3.6 Protein domains

Much can be said about a protein's function when its functional domains are known, see e.g. (Forslund and Sonnhammer, 2008). The genomic inventory of protein domains is thus an important indicator of an organism's regulatory and metabolic capabilities. Presence/absence of protein domains therefore is a valuable source of phylogenetic information in particular in deep phylogeny (Yang, Doolittle, and Bourne, 2005), where precise statements on the orthology of entire proteins become problematic because of frequent changes in domain content and organization. It is still a difficult problem to obtain a comprehensive annotation of protein domains and in particular of protein domain co-occurrences (Parikesit, Stadler, and Prohaska, 2011). Problems can arise from incomplete protein annotations for newly sequenced genomes. Protein domains evolve, albeit slowly, so that at large distances from the reference sequences they may become undetectable. A possible remedy is to use domain families only. An interesting variation on the theme is the use of single-copy protein domains in a recent study of opisthokont phylogeny (Torruella et al., 2012). The order of domains in complete proteomes is used in (Fukami-Kobayashi et al., 2007) to estimate a full tree of life. Protein domains have also been used on much smaller time-scales, e.g. in a recent study within DMP (Niehuis et al., 2012) that placed Strepsiptera a sister-group to Beetles within the Holometabola.

25.3.7 RNA secondary structure elements

RNA secondary structure can also serve as a source of phylogenetic information. One of the first applications inferred a gene phylogeny based on the RNase P and RNase MRP RNA components (Collins, Moulton, and Penny, 2000). Systematic investigation into the structural evolution of non-coding RNA families highlight apparently clade-specific insertion domains and losses of entire structure components, see e.g. (Donath et al., 2010) and the references therein for a detailed review. This is also the case for ribosomal RNAs, which are readily available for a very wide variety of species. A catalogue of taxon-specific secondary structure variations is presented in Misof et al. (2006) for nuclear SSU rRNA of Hexapoda, discussing in detail subgroup-specific helices. Earlier work (Misof and Fleck, 2003) on LSU rRNAs in Odonata suggested that secondary structure variations can be homoplastic, presumably because only a quite limited number of structural alternatives can be accommodated in the complex ribosomal RNA structures. Secondary structure models also play an important role as a priori information for RNA-specific substitution models. In this setting, structural information can provide a major improvement in sequence-based phylogenies (Letsch et al., 2010; Stocsits et al., 2009).

Two major technical problems seem to hamper the use of secondary structure itself: the problems involved in computing sufficiently accurate secondary structure models, a topic that is discussed in detail in the (see Stocsits et al., Chapter 23, this book), and the problems involved in defining characters from the predicted secondary structures. The main issue here is that it is still unclear what exactly should be used as the units of structural variation. Conceptually the simplest approach is to identify all possible helices in a sequence/structure alignment and use the presence or absence information of each of these helices as a binary character (Caetano-Anollés, 2002b). The NOBAI package (Knudsen and Caetano-Anollés, 2008) implements this idea in practice, although the resulting "characters" can be partially overlapping and are typically far from structurally independent of each other. Structurally diverse regions, furthermore, are often overrepresented and hence may introduce substantial noise. Algorithms to compare RNA structures as tree editing or tree alignment are available that appear to better fit the problem, see e.g. (Gardner and Giegerich, 2004). Nevertheless, applications to several phylogenetic problems show that RNA secondary structure can be phylogenetically informative on its own.

25.3.8 Alignment gaps as pseudo-characters

While presence/absence characters for complex genomics features are readily used, the analogous information is rarely utilized at the sequence level itself. Insertions and deletions (in/dels) at sequence level can be detected only in pairwise or multiple sequence alignments. This suggests that the number and position of gaps is dependent on the choice of the alignment algorithm. In particular, the detection of gaps necessarily requires that homology statements are made before the definition of the characters. Nevertheless, alignment gaps provide meaningful phylogenetic information. One of the first studies employing in/dels was conducted in birds, where several issues in their phylogeny could be resolved (Groth and Barrowclough, 1999; Johansson et al., 2001; Paśko, Ericson, and Elzanowski, 2011). Subsequently, shared multi-residue deletions have been used to support hypotheses derived from molecular data in single gene analyses, e.g. for arthropods (Vogt, 2002), bats (Teeling et al., 2005), and mammals in general, providing support in particular for the Atlantogenata hypothesis (sister-group relation of Xenarthra and Afrotheria) (Murphy et al., 2007).

Rigorous approaches for phylogenetic reconstruction incorporate gaps as 5th character state for nucleotides (or 21st character state for amino acids). A more sophisticated technique is the application of a coding scheme. While different possibilities have been suggested, one of the most common is the replacement of the gapped regions by a binary matrix that codes presence/absence of the respective in/del (Simmons and Ochoterena, 2000). The simple in/del coding (SIC) approach adds the binary matrix to the ungapped sequence data which can be subsequently employed in tree inference by maximum parsimony. SIC not only dismisses regions where one in/del completely overlaps an in/del contained within another sequence, it also underestimates in certain cases the steps that are biologically possible for the transformation of one in/del into another. Furthermore, SIC can unnecessarily incorporate missing data, leading to ambiguous character optimizations. Modified complex in/del coding (MCIC) maximizes the amount of phylogenetic information coded in such a matrix by incorporating all possible gap-nucleotide patterns (Müller, 2006). A technical problem that arises from coding in/dels as multi-state characters and describing a distance between two states is the violation of the triangle inequality. In certain cases, this results in an overestimation of necessary transformation steps. Indeed, newer studies indicate the existence of regions where the frequency of in/del events is about 50 % higher than expected (Snir and Pachter, 2011; Taylor, Ponting, and Copley, 2004).

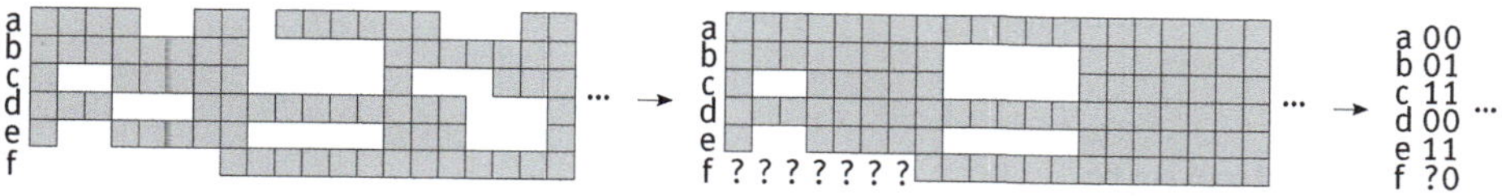

Figure 25.4: Coding of split-inducing in/dels from an arbitrary alignment. Gaps which occur in at least two sequences and define a reasonable split of the data set are coded in a presence/absence matrix. Ambiguous in/dels are excluded from the re-coding. While gaps introduced at the beginning or end of an alignment have to be treated as missing data.

A more reliable coding scheme would be based on a subclass of in/dels that define a binary split among the set of taxa (Donath and Stadler, 2011), see Figure 25.4. The definition and inference of such split-inducing in/dels is based on two basic principles that are largely accepted in the literature. First, in/dels at the same position, i.e., sharing the same end points in two sequences, are likely homologous. Second, independent single-residue in/dels tend to occur more frequently than multi-residue in/dels. Hence they are expected to contribute a noisier signal and are therefore disregarded in the coding. Thus, instead of taking into account all available in/del information, the approach focuses on patterns which can be thought of as being almost homoplasy-free.

25.4 Concluding Remarks

We argue here that rare genomic changes that are associated with strong selection are of particular interest for the resolution of difficult phylogenetic problems. Such characters, of which microRNAs are a prime example, provide a link between molecular and phenotypic evolution.

Despite the high hopes of finding perfect characters among the rare genomic changes, so far none of the RGCs studied in some detail is completely homoplasy-free. Nevertheless, they provide highly valuable phylogenetic information since modern techniques of computational phylogenetics can tolerate a small fraction of homoplastic characters provided that the correct information dominates the data set. Thus large, and hence automatically generated, data sets are required also for the successful use of molecular morphology approaches.

Improved computational methods for some types of markers, including introns and mitogenomic gene orders, as well as novel approaches e.g. utilizing the gap patterns in genome wide alignments, have been developed as part of the Deep Metazoan Phylogeny priority program. In general, the extraction of phylogenetically informative features from genomes, either directly from sequences or from genome-wide alignments, remains the bottleneck for the application of molecular morphology methods and still presents a challenge to bioinformatics research.

Acknowledgments

This work was supported by the DFG under the auspices of the SPP Deep Metazoan Phylogeny, projects STA 850/2 and 850/3-2 (to PFS).

Patrick Kück, Bernhard Misof, and J. Wolfgang Wägele

26 Systematic errors in maximum-likelihood tree inference

Abstract: Recent improvements of tools for inference of phylogenies from molecular data could not impede that tree topologies estimated from different data sets are often mutually incompatible (see e.g. recent discussions about placement of Porifera in the metazoan tree). There are obviously causes of error that remain undetected. Fundamental discrepancies between molecular phylogenies and morphological data (see e.g. Wägele and Kück, this book) nourish major doubts concerning the reliability of established methods.

A series of simulations with an 11-taxon setup designed to study conditions (branch length ratios, among-site variation, alignment lengths) that are prone to long-branch effects prove that three different classes of artifacts (Wägele and Mayer, 2007) can be discerned. Maximum likelihood inference is more successful if among-site variation is considered with a mixed-distribution model, however reconstruction fails when differences in branch lengths are very large, especially in topologies with short stem lineages. Long inner branches can lead to a systematic error with high support for the wrong tree caused by plesiomorphies even when long alignments and the correct model are used. Other errors are caused by signal erosion and by chance similarities. We present for a set of tree topologies the parameter ranges for which correct topologies can be obtained.

26.1 Introduction

We present here new observations on errors that distort tree topologies estimated from sequence data using maximum likelihood and model assumptions, and we review previous publications, including especially results recently published by Kück, Wägele and Misof (2012).

26.1.1 Choice of an appropriate tree reconstruction method

One of the most important steps during phylogeny inference using sequence data is the choice of a tree reconstruction method which is robust against different model violations and efficient in finding the correct topology of underlying data. In general, there are four main groups of reconstruction methods which are commonly used in phylogenetic analyses: Distance methods (e.g. neighbor joining tree inference (NJ)), Maximum Parsimony (MP), Maximum Likelihood (ML), and the Bayesian approaches.

Considering the different heuristics of these methods, ML is certainly more robust and more efficient than other methods (Anderson and Swofford, 2004; Bruno and Halpern, 1999; Felsenstein, 1978; Fischer and Steel, 2009; Gaucher and Miyamoto, 2005; Gaut and Lewis, 1995; Huelsenbeck and Hillis, 1993; Sullivan and Swofford, 1997; 2001; Swofford et al., 2001; Yang, 1993; 1995; 1997). The advantages of ML towards other tree reconstruction methods have led to a widespread application and acceptance of ML in molecular phylogenetics.

Bayesian analyses tend to overestimate data signal, which leads to high support values even if the data is completely uninformative (Suzuki, Glatzko and Nei, 2002; Erixon et al., 2003). They have a much higher type I error rate than ML, especially in cases of model misspecification (Erixon et al., 2003). Furthermore, the way in which subjective prior assumptions can negatively influence Bayesian analyses is still unclear.

While Bayesian approaches often overestimate data signal, parsimony analyses tend to do the opposite. MP does not account for multiple substitutions and among-site rate variation (ASRV) and therefore becomes inconsistent if evolutionary rates are heterogeneous. This is true especially for distantly related sequences (Rannala et al., 1998; Yang, 1996). Although the statistical properties of MP are not completely understood, it is commonly assumed that parsimony will find the correct topology under a finite number of characters when the evolutionary rate is constant (Felsenstein and Sober, 1986). Nevertheless, parsimony can also be inconsistent under that condition, because the probability of a single substitution on a short interior branch is often lower than the occurrence of multiple substitutions along longer branches (Takezaki and Nei, 1994). This inconsistency will also appear when sequence divergence is small.

In contrast to Bayesian approaches and parsimony tree inference, ML is statistically consistent for binary trees with finite branch lengths under correct model and model parameter assumptions as sequence length increases to infinity (Chang 1996; Felsenstein, 1973; 1978; 1981; Rogers, 1997). Therefore, the probability of finding the correct tree increases as more and more data are accumulated (Rogers, 1997; Swofford et al., 2001). It has been further shown that ML has a high tolerance towards different kinds of model violations (Anderson and Swofford, 2004; Bruno and Halpern, 1999; Felsenstein, 1978; Gaut and Lewis, 1995; Kelchner and Thomas, 2006; Kolaczkowski and Thornton, 2004; Swofford et al., 2001). If branch lengths are well balanced, even oversimplified maximum likelihood models will find the correct tree in most instances (Yang, Goldman and Friday, 1994).

While most of the knowledge about ML robustness and efficiency is based on four-taxon tree simulations, studies with more than four taxa are comparatively rare. However, such studies are required for a better understanding of taxon misplacements in multiple taxon trees, like the attraction of distantly related taxa across several internal nodes. Possible interactions between branches of varying lengths across several nodes and the phenomenon of taxon-slippage cannot be tested on four-taxon trees. Therefore, we analyzed the robustness and efficiency of ML to different long-branch effects by using an 11-taxon setup and different internal and terminal branch lengths. We show that ML is indeed very robust against different model parameter violations and very efficient in finding the correct topology over a wide range of different branch length ratios. However, we discovered instances in which ML methods reconstruct the wrong topology for relatively long alignments even under correct model assumptions if branch length ratios differ strongly across several nodes. These effects have not been extensively investigated in previous studies, but are potentially common in analyses of empirical data. Furthermore, we analyzed how the ML reconstruction success is affected by deviations from the simulated Γ parameter. Another point which we have further addressed is the important question whether ML tree inferences with mixed-distribution models (Γ + I) are superior to pure Γ models and whether these parameters could be estimated correctly for a small taxon set of only 11 taxa.

26.1.2 Long-branch artifacts (LBA)

Long-branch attraction was first demonstrated by Felsenstein (1978) on a four-taxon setup and is generally defined as an incorrect sister-group relationship between two long branches due to a high accumulation of random character similarities which are erroneously identified as shared synapomorphies. Long-branch attraction is particularly a problem of molecular data because of the small number of possible character states (Bergsten, 2005; Jenner, 2004c) and occurs especially if species radiated quickly or if stem lineages are short (Wägele and Mayer, 2007). However, the Felsenstein effect is not the only one caused by long branches. Three types of LBAs can be distinguished (Figure 26.1) according to different causes for the misplacement (Wägele and Mayer, 2007).

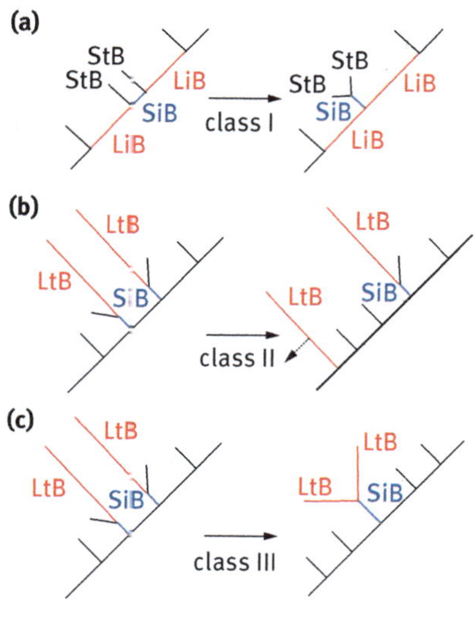

Figure 26.1: Long-branch effects. **(a)** Class I effect (attraction due to symplesiomorphies): two short terminal branches (StB) appear as wrong monophylum due to shared homologies. This effect is caused by substitution of synapomorphies and "back mutations" to plesiomorphic states along the second long internal branches (LiB). **(b)** Class II effect: at least one of the two long terminal branches (LtB) slides down the tree or appears elsewhere in the inferred tree topology, mainly due to signal erosion in which synapomorphies supporting the correct clade are substituted along a terminal long branch. **(c)** Class III effect: two long terminal branches (LtB) separated by at least one internal branch are attracted due to the dominance of chance similarities over homologies (modified from Kück, Wägele and Misof 2012).

26.1.2.1 Three different classes of LBA

Class I effects consist of the attraction of terminal taxa adjacent to short inner branches. It is caused by substitution of synapomorphies and "back mutations" to plesiomorphic states along long internal branches next to short inner branches (Figure 26.1a). This increases the ratio of plesiomorphies to conserved apomorphies shared between paraphyletic taxa ("StB" in Figure 26.1a). As a consequence, this ratio determines if these taxa appear as paraphylum or as monophylum. The artifact is caused by true homologies (plesiomorphic states, which can only be identified as such in rooted topologies) (Wägele and Mayer, 2007), in contrast to the chance similarities typical for the Felsenstein effect.

Class II effects are mainly caused by strong "signal erosion". Synapomorphies can be substituted along a single long branch (Figure 26.1b). Thus it can happen that a clade shares only character states with distantly related taxa and the long branch slides down the tree or appears elsewhere in the resulting tree topology (slipping "LtB" in Figure 26.1b).

Class III effects consist of the attraction of two long terminal branches due to the dominance of chance similarities over homologies (Figure 26.1c). This well-known and widely studied effect of LBA was first noticed by Felsenstein (1978) and affects mainly the maximum parsimony method.

26.1.2.2 LBA and empirical data

Less than a decade ago it was doubted that LBAs could occur in analyses of empirical data sets. Because no model can be assumed to be entirely correct for empirical data (Sullivan and Swofford, 2001), the occurrence of long-branch artifacts is not only a theoretical problem, but also a real phenomenon (Huelsenbeck, 1997; Huelsenbeck and Crandall, 1997; Wägele and Mayer, 2007) for which many examples have been studied (Bergsten, 2005; Inagaki et al., 2004; Omilian and Taylor, 2001; Wägele and Mayer, 2007). Therefore it is necessary to develop reliable methods that can deal with this problem (Bergsten, 2005; Wägele and Mayer, 2007). Although LBAs are commonly invoked as a source of inconsistency, it is still unclear if conditions producing artifacts in empirical analyses can be clearly defined (Huelsenbeck, 1998; Omilian and Taylor, 2001; Sanderson et al., 2000; Wiens and Hollingworth, 2000). Even when potential biases can be identified, it is difficult to determine if the inferred phylogeny results from a bias or from evolutionary history (Huelsenbeck, 1998; Steel, Hudson and Lockhart, 2000; Whiting, 1998). Usually, the historical effects of evolution on sequences remain forever unknown. It cannot be tested if model assumptions reflect the historical reality. Therefore, substitution models are only averaged approximations and data has to be evaluated to detect sampling errors and differences in data quality before any conclusions are derived (Wägele and Mayer, 2007).

26.1.2.3 Methods to avoid LBAs

In many studies LBAs are not addressed at all. Other studies suggest to avoid LBAs by using ML tree inference which is more resistant to LBAs than MP, but not immune (Bergsten, 2005; Cunningham, Zhu and Hillis, 1998; Fuellen, Wägele and Giegerich, 2001; Gaut and Lewis, 1995; Hillis et al., 1992; Huelsenbeck, 1995; Lockhart et al, 1996; Pol and Siddal, 2001; Sanderson et al., 2000; Wägele and Mayer, 2007). To reduce the risk, among-site rate variation should be considered. Models that incorporate among-site rate variation (ASRV) fit the data much better (e.g. Lartillot et al, 2007). ASRV exists in almost all genes, with the possible exception of some pseudo-genes or "junk" DNA, based on different functional and/or structural constraints at different sites (Yang, 1996). MP does not account for ASRV, and therefore strongly underestimates substitution rate heterogeneity at sites, especially for distantly related taxa. Consequently, MP has a strong tendency to group long branches of strongly derived taxa in sister-group relationships, sometimes in the right direction (Farris Zone), sometimes in the wrong direction (Felsenstein Zone) (Bruno and Halpern, 1999). ML models take ASRV into account by using a discrete gamma distribution, but can also induce biased estimates of phylogenies when the true evolutionary model is strongly violated (Gaut and Lewis, 1995; Yang, 1996). In any case, ML tree inference is much more robust towards LBAs than MP, even if assumed model parameters are inadequate (Philippe and Germot, 2000).

Another possibility to avoid LBA could be the exclusion of strongly derived taxa or sequence positions (like the third codon position) (Dacks et al., 2002; Farias et al., 2001; Hanelt et al., 1996, Lyons-Weiler and Hoelzer, 1997). A disadvantage is the potential loss of important taxa and/or the loss of too much phylogenetic signal, which can lead to lower tree resolution (Kim, 1996; Sullivan and Swofford, 1997). Removing strongly derived taxa is also contrary to many studies which suppose that accuracy of topologies generally increases with higher taxon samplings (Graybeal, 1998; Huelsenbeck and Lander, 2003; Kim, 2003; Poe, 1998; Poe and Swofford, 1999; Pollock et al., 2002, Rannala et al., 1998; Zwickl and Hillis, 2002).

The exclusion of fast evolving genes can help to overcome signal erosion along long branches, but could also lead to an increasing number of shared conserved (plesiomorphic) character states which probably increase the possibility of class I effects. Moreover, the use of exclusively conservative genes can be associated with a lack of signal for less derived (younger) taxon groups. Finally, more conservative genes will not necessarily correct biases and effects of wrong substitution models (Stefanoviae, Rice and Palmer, 2004; Lartillot et al, 2007; Wägele and Mayer, 2007), namely when systematic errors (not caused by noise in the data) occur.

One of the most commonly suggested approaches to avoid LBAs is the inclusion of additional taxa to break up long branches (Graybeal, 1998; Hendy and Penny, 1993). Additional taxa can actually help to shorten branches, but not to recover phylogenetic signal which has been substituted along long branches (class II effect). Unfortunately, often the required taxa are either hard to determine, not easy to find or simply extinct, which can make the addition of taxa impossible (Bergsten and Miller, 2004).

26.1.2.4 Methods to detect LBAs

The unknown evolutionary history of underlying data sets makes it nearly impossible to distinguish between true evolutionary relationships and incorrect node resolutions. The placement of two long branches in a single node can be either due to topological bias or to common evolutionary history (Siddal, 1998; Wägele and Mayer, 2007; Whiting et al., 1997). A widely used approach to detect wrong taxon relationships in phylogenetic trees is the comparison of molecular and morphological results. Conflicting results for the placement of strongly derived taxa point to potential LBAs (Bergsten and Miller, 2004; Clements, Gray and Choat, 2003; Wiens and Hollingworth, 2000).

Another attempt commonly used to identify the classical case of long-branch attraction (class III effects) is the inclusion or removal of one attracted long branch while keeping the other one in the data set. If the topology is correct, the position of the remaining branch will not change, otherwise a class III LBA can be suspected. But even if the positions of long branches do not change, there is no reason why this branch placement should be indeed correct (Bergsten, 2005). This is also a main criticism to the parametric simulation approach (Huelsenbeck, 1997) which has been applied in several other studies (Maddison, 1999; Omilian and Taylor, 2001; Sanderson et al., 2000; Tang et al., 1999; Tourasse and Gouy, 1999; Wiens and Hollingsworth, 2000; Wilcox et al., 2004). Furthermore, inclusion or removal of long branches is not helpful to identify potential cases of signal erosion along a single long branch (class II effects) or signal substitutions of synapomorphies due to the class I effect.

The random outgroup approach (Gaut and Lewis, 1995; Philippe and Forterre, 1999; Qiu et al., 1999; Stiller and Hall, 1999; Stiller, Riley and Hall, 2001; Sullivan and Swofford, 1997) which is often used to identify LBAs between outgroup and ingroup taxa, is not a recommendable option, since there is no reason why the true root cannot be on the longest branch in the ingroup (Bergsten, 2005).

RASA (Lyons-Weiler, Hoelzer and Tausch, 1996), a relative apparent synapomorphy analysis which was originally developed to measure phylogenetic signal (Lyons-Weiler and Hoelzer, 1999) is not an appropriate method to detect LBA (Simmons et al., 2002; Wägele and Mayer, 2007; Xiang et al., 2002). It was shown that RASA can erroneously identify non-problematic long branches as hazard sources (Faivovich, 2002).

Finally, split decomposition and related spectral analysis are tree independent methods for detecting conflicting signals in phylogenetic data. Both methods can be used to identify potential LBAs if LBAs are due to overwhelming convergent signal (Clements, Gray and Choat, 2003; Flook and Rowell, 1997, Kennedy et al., 1999; Lockhart and Cameron, 2001; Waddell et al., 1999; Wägele and Mayer, 2007).

26.1.3 The influence of model and parameter choice on ML tree inference

26.1.3.1 Among-site rate variation (ASRV)

It is well known that ignoring ASRV increases the probability of LBA occurrence in ML tree inferences (Chang 1996; Gaut and Lewis, 1995; Huelsenbeck and Hillis, 1993; Huelsenbeck, 1995; Kelchner and Thomas, 2006; Lartillot et al, 2007; Lockhart et al., 1996; Sullivan and Swofford, 1997; Sullivan and Swofford, 2001; Swofford et al., 2001; Yang, Goldman and Friday, 1995; Yang, 1996). If ASRV is not recognized, the ML approach underestimates substitution rates, which becomes progressively worse with increasing evolutionary distances between taxa. Therefore, the consideration of ASRV is one of the most powerful improvements brought to model-based ML reconstruction methods. In general, there are three possibilities to account for rate variation among sites: The "invariant sites model" (I), the "Γ-distributed rates model", and a combination of both models (Γ + I).

With the invariant sites parameter (I) alone, ML estimates the fraction of invariable sites while remaining sites follow an equal substitution rate (two categories of substitution rates). A main disadvantage of an equal substitution rate for all variable sites is the rate underestimation for highly variable sites and the rate overestimation for slowly evolving sites.

Under the Γ-distributed rates model, substitution rate heterogeneity among sites is modeled by using well-chosen rate classes to approximate the continuous gamma distribution. A bell shaped Γ-distribution caused by an α value greater than 1 implies that most sites have intermediate rates while few sites have very low or very high rates. The larger the α value, the higher the number of sites with a constant substitution rate whereas a reverse-J shaped Γ-distribution caused by α values lower than 1 means that most sites have very low substitution rates while a few sites exist with very high rates (Yang, 1996).

Recently published studies relied on the application of the restricted Γ-distributed rates model without additional consideration of completely invariant sites (I) (e.g. (Murienne, Edgecombe and Giribet, 2010; Regier et al., 2010; Rota-Stabelli et al., 2010; Sanderson et al., 2000; Savard et al., 2006). A main argument is the dependency of both parameter optimizations which can lead to multiple optima in the likelihood function (Mayrose, Friedman and Pupko, 2005; Ren, Tanaka and Yang, 2005). The estimations of Γ + I are indeed strongly correlated and subject to large sample variance, which makes it difficult to distinguish between truly invariable sites and slowly evolving sites, especially if alignments consist of a very small number of taxa and/or character positions.

Earlier studies argued that a combination of both models to a mixed-distribution model (Γ + I) significantly improves the estimation of substitution rate heterogeneity compared to a single estimate of the Γ-distribution (Anderson and Swofford, 2004; Gu, Fu and Li, 1995; Philippe and Germot, 2000; Sullivan and Swofford, 1997; Sullivan, Swofford and Naylor, 1999). It was shown by earlier studies that already a moderate number of taxa (N > 20) can lead to reliable estimates of both parameters under the mixed-distribution model.

One advantage of the mixed-distribution model is that erroneous estimates of one parameter can be compensated by the other. It has been further shown that erroneous estimates of both parameters together can still fit the data in such a way that the likelihood score changes only marginally (Sullivan, Swofford and Naylor, 1999).

26.2 Materials and methods

26.2.1 Simulations

We designed six different sets of simulations under different topologies to detect the three long-branch artifacts caused by long terminal branches (Topology A) or by long internal branches (Topology B and its variants C–F), where branch lengths reflect the amount of expected substitution rates per site for corresponding lineages. The first set used topology A (Figure 26.2), which was characterized by a stepwise elongation of two terminal non-neighboring branches (BL2) which are separated by more than one internal branch. The ancestral branch lengths of the two long branches were kept

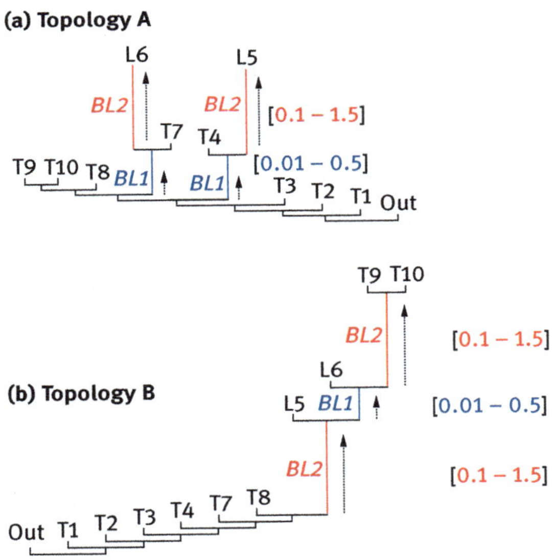

Figure 26.2: Two sets of simulations. **(a)** Topology A: two terminal branches (BL2) and shorter ancestral branches (BL1) are stepwise elongated. **(b)** Topology B: two long internal branches (BL2) are stepwise elongated under different lengths of an intermediate short branch (BL1). Topology A was used to identify class II and III effects and Topology B was used to identify class I effects (modified from Kück, Wägele and Misof 2012).

short, but also varied in length. This setup can potentially produce cases of class II and class III long-branch artifacts. Topologies B–F were characterized by a stepwise elongation of either one or two internal branches (BL2) for different lengths of intermediate and adjacent internal branches (BL1). These topologies were used to investigate mainly class I effects and rare class II effects. All topologies consist of 11 taxa.

In a first round of analyses, we increased stepwise the length of BL1 for Topology A and B (0.01, 0.05, 0.1, 0.3, 0.5) (Figure 26.2). For each length of BL1, we further increased the length of BL2 from 0.1 to 1.5 in steps of 0.2. Thus, branch length ratios of BL2/BL1 ranged from one-fifth to 150. Branches which are neither assigned to BL1 nor to BL2 are kept constant (L = 0.1).

In a second round, we tested if remaining branches of the "background" taxa, which are neither assigned to BL1 nor to BL2, can have a potential effect on the tree reconstruction success of topology B. For this purpose, we excluded five stem-lineage taxa at one end of the tree ("T1" to "T7"). Branch length elongation steps for BL1 and BL2 are performed for the reduced Topology B to the same extent as before for Topologies A and B (Figure 26.3).

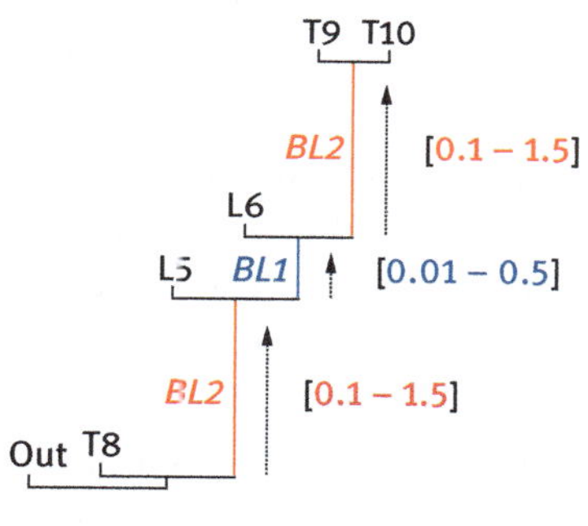

Figure 26.3: Variation of Topology B with shortened stem lineage.

After evaluation of the tree reconstruction success for Topologies A and B, we analyzed in a third round further variations of Topology B (Topologies C–F) to gain additional insight about the impact of different branch length relationships in relation to class I effects. For this purpose, we increased for Topologies C–F the length of BL2 from 0.9 to 1.5 in steps of 0.2 while the length of BL1 was kept constantly short (BL1 = 0.01). We concentrated our analyses in the third round on a smaller branch length range of BL2 (0.9 to 1.5) as this appears to be the tipping range in which class I effects mainly start to predominate in ML tree reconstructions. In these analyses, the length ratios of BL2/BL1 ranged from 90 to 150.

All sequence alignments were generated with INDELible v.1.01 (Fletcher and Yang, 2009) using the Jukes–Cantor model (JC) of sequence evolution and a mixed-distribution model of Γ + I for ASRV. All data were simulated with a shape parameter $\alpha = 1.0$ and a proportion of invariant sites $\theta = 0.3$. ASRV was modeled using a continuous Γ-rate distribution while indel events were not simulated. For each branch length combination of BL1 and BL2, we simulated the evolution of 100 data replicates for each sequence length (2,000, 3,000, 4,000, 10,000 and 100,000 bp) of Topology A and B. For Topology "B reduced" (with only six taxa) we only simulated 100 data replicates of 100,000 bp, and for the Topologies C–F of 10,000 bp for each branch length combination. The JC model was chosen for the simulations because it is well defined and we wanted to keep the model parameter space as small as possible. Due to the simple model assumptions, each base in the sequence has an equal probability of changing, which results in equal frequency of the four bases. Therefore, the reconstruction success of ML is directly linked to simulated branch length conditions, sequence lengths, and the ASRV parameters used in each ML analysis.

Note that we can talk of homologies, plesiomorphies and apomorphies because the nucleotide patterns produced in simulations are the result of sequence evolution along a rooted tree, the root being the starting point of the simulation.

26.2.2 Maximum Likelihood analyses

Trees were reconstructed with ML and the Jukes–Cantor (JC) model under different parameter settings using PhyML-3.0-linux64 (Guindon and Gascuel, 2003; Guindon et al., 2010, Table 1).

For Topologies A and B, we analyzed the data either with a mixed-distribution model ($JC + \Gamma + I$), or with Γ-distributed rates without additional consideration of complete invariable sites ($JC + \Gamma$), or without consideration of ASRV (JC). Using the mixed-distribution model ($JC + \Gamma + I$), the Γ-shape parameter α and the fraction of invariant sites were either estimated or set equal to the simulated values ($\alpha = 1.0$ and $\theta = 0.3$). Using the Γ-distribution model ($JC + \Gamma$), the α shape parameter of the Γ-distribution was always estimated from the data. As approximation to a constant substitution rate without consideration of ASRV (JC), the α shape parameter was set to 100 ($\alpha = 100$).

For the alignment length of 100,000 bp simulated for Topologies A, B, "B reduced" as well as for the alignment length of 10,000 for the Topologies C–F, the tree reconstruction process was only performed with the correct model parameter assumptions ($\alpha = 1.0$ and $\theta = 0.3$).

With the discrete gamma model, the number of relative substitution rate categories was set to four ($c = 4$) and tree topologies and branch lengths were optimized by the heuristic search algorithm of the PhyML software package. Maximum Likelihood analyses were performed and evaluated with a Perl pipeline, and ran for three months on a Linux Cluster with HP ProLiant DL380 G5 blades (Dual quad core Intel Xeon E5345, 2.33 GHz, 2 × 4MB L2-cache, 1333 MHz Bus, 32 GB RAM).

26.2.3 Scoring

Incorrect tree topologies reconstructed from simulated alignments were classified into LBA classes I, II or III (Figure 26.1). Wrong topologies for which we found a paraphyletic grouping of the two terminal "non-long branches" in Topology B (and its variants) were summarized as class I effects. Wrong topologies which showed an attraction of the two long terminal branches in Topology A were sampled as class III effects. Wrong topologies for which only one long branch had been misplaced in Topology A and E (probably due to signal erosion) were collectively classified as class II effects.

Topologies that did not fit any of these categories like incorrect placements of "background" taxa have not been found in any of our analyses.

26.3 Results

26.3.1 Reconstruction success for Topology A

Topology A (Figure 26.2a) was chosen as model topology to analyze the robustness and efficiency of ML tree inference regarding LBAs of class II (signal erosion) and class III (attraction based on chance similarities).

If for tree inference the model parameter values of the mixed-distribution model are given correctly or if they are estimated, ML is able to reconstruct predominantly correct topologies if ancestral branch lengths (BL1) are longer than 0.01 (BL1 ≥ 0.05) even if terminal branches are extremely long (BL2 ≤ 1.5) (Figure 26.4a). This range of harmless branch lengths corresponds to a branch length ratio of BL2/BL1 ranging from 0.5 to 30. Nearly identical reconstruction success is also found if the proportion of completely invariable sites is not considered as an extra parameter value (JC + Γ).

If ancestral branch lengths are very short (BL1 = 0.01), we found predominance of class II LBAs (signal erosion) in which at least one single long terminal branch appears elsewhere in the topology. For these branch length conditions, predominance of class II effects is found even under moderate lengths of terminal branches (BL2 ≥ 0.7) (Figures 26.4a and 26.5a) equivalent to a branch length ratio ranging from 70 to 150. This implies weak signal supporting internal nodes.

Long-branch artifacts due to chance similarities (class III effects) are only rarely seen and appear whenever terminal branches are distinctly long (BL2 ≥ 1.1) if ASRV is considered for the analyses, but appear more often if JC is used with the Γ-distribution model alone (JC + Γ) (Figure 26.5a).

As expected, in contrast to the tree reconstruction success of ML with consideration of ASRV, ML performs much worse without any consideration of ASRV (Figure 26.4a). Under this ML setting especially class III effects, which are based on the attraction of both long terminal branches due to chance similarities, predominate under a wide range of simulated ancestral branch lengths (BL1 ≤ 0.3). This corresponds to a branch

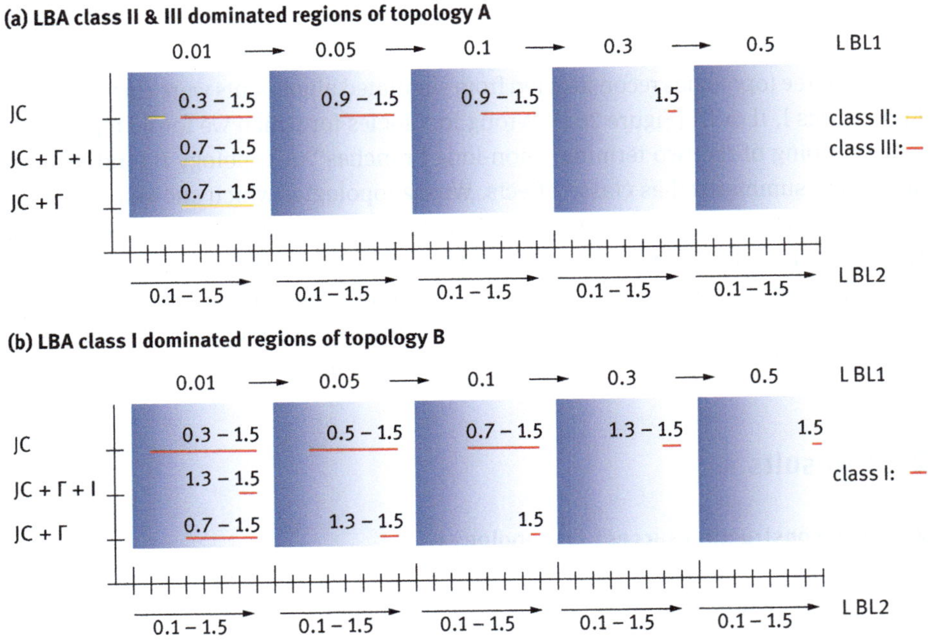

Figure 26.4: Occurrence of long-branch artifacts (LBAs) when model assumptions (JC, Γ, I), long branch lengths (lower scales: from 0.1 to 1.5) and short branch lengths (blue boxes: from 0.01 to 0.5) vary. Horizontal bars (yellow: class II artifact; red: class III or class I artifact) indicate the range of parameters in which LBAs predominate. **(a)** Predominant occurrence of class II and class III effects are found in Topology A (Figure 26.2a). **(b)** The Class I effect is more frequent in Topology B (Figure 26.3b). Corresponding branch lengths of BL1 are shown above each bar plot. The tree reconstruction success was found to be identical when estimated or correct model assumptions of JC + Γ + I were used.

length ratio of BL2/BL1 that ranges from five to 150. The range of class III predominance decreases continuously with increased branch lengths of the short internal branches BL1 (Figure 26.4a) until correct topologies are resolved in the majority of tree reconstructions (BL1 = 0.5), which corresponds to a branch length ratio of BL2/BL1 that ranges from one-fifth to three. In this range of branch length ratios, correct topologies also predominate tree reconstruction if internal branches are smaller (BL1 ≥ 0.01) which implies better signal support for inner nodes within this range.

While class II effects (signal erosion) predominate ML tree reconstructions when internal branches are very small (BL1 = 0.01), even if model assumptions are correct ($\alpha = 1.0$ and $\theta = 0.3$) and alignment lengths moderate (10,000 bp), ML is able to resolve nearly all topologies correctly under this setting when sequence lengths are increased to 100,000 bp (Figure 26.6a).

In general it can be concluded that the performance of ML inference in our simulations is mostly afflicted by large branch length differences, less so by wrong model assumptions.

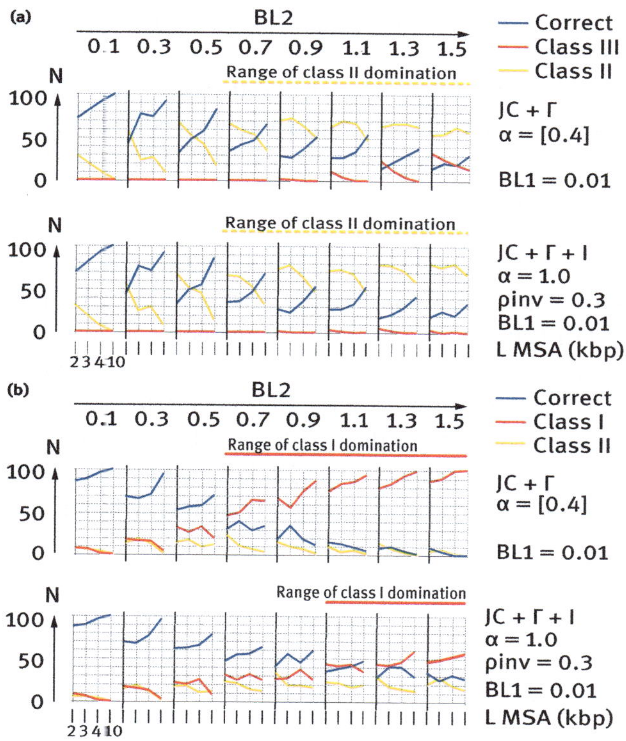

Figure 26.5: Selected reconstruction results of ML tree inference using either a mixed-distribution model (JC + Γ + I) or a Γ-distributed model (JC + Γ). Class I, class II, and class III effects inferred from 100 simulation repeats for each branch length combination and alignment length. For this example, short branch length BL1 = 0.01 is constant. Reconstruction scheme for the models JC + Γ (α estimated to 0.4) or JC + Γ + I (α = 1.0; θ = 0.3). Long branch length differences increase from left to right by increasing branch BL2 in discrete elongation steps (0.1–1.5, upper scale). Four successive data points (belonging to one cell in each plot) correspond to four alignment lengths (2,000, 3,000, 4,000, 10,000 base pairs, lower scale). The y-axis depicts the reconstruction success of the 100 simulation repeats (N) for (a) Topology A (Figure 26.2a) and (b) Topology B (Figure 26.2b).

26.3.2 Reconstruction success for Topology B

Topology B (Figure 26.2b) was chosen as model topology to analyze the robustness and efficiency of ML tree inference in regard to LBA class I effects. While in Topology A the two long terminal branches are parallel, the evolving sequence passes through two long internal branches in Topology B.

Even under correct model assumptions (α = 1.0 and θ = 0.3), ML is not able to reconstruct the correct topologies in the majority of simulations if the length of the two long internal branches is large (BL2 ≥ 1.3) and the length of the intermediate internal branch is short (BL1 = 0.01) independent of the model parameter assumptions

(Figures 26.4b and 26.5b). The branch length differences in this range have a ratio of BL2/BL1 that reaches from 130 to 150.

Except for the larger alignment length of 10,000 bp, class I effects start to predominate tree reconstruction of topology B under the mixed-distribution model (JC + Γ + I) if the length of the long internal branches is large (BL2 = 1.1). If the length of the two long internal branches increases (BL2 ≥ 1.3), class I effects are also found for alignment lengths of 10,000 bp (Figures 26.4b and 26.5b).

The frequency in which class I effects occur is even higher if the proportion of invariant sites is not considered in the tree reconstruction process (JC + Γ) (Figures 26.4b and 26.5b). In this case, class I effects already predominate tree reconstructions under a moderate length of long internal branches (BL2 ≥ 0.7) if the intermediate, internal branch is small (BL1 = 0.01), implying a BL2/BL1 range from 70 to 150.

With an internal branch length (BL1) larger than 0.01 we still found predominance of class I effects, but only for very long lengths of BL2 (BL2 ≥ 1.3 for BL1 = 0.05 and BL2 = 1.5 for BL1 = 0.1) (Figure 26.4b), corresponding to branch length ratios of BL2/BL1 of 26 to 30 and 26 to 15.

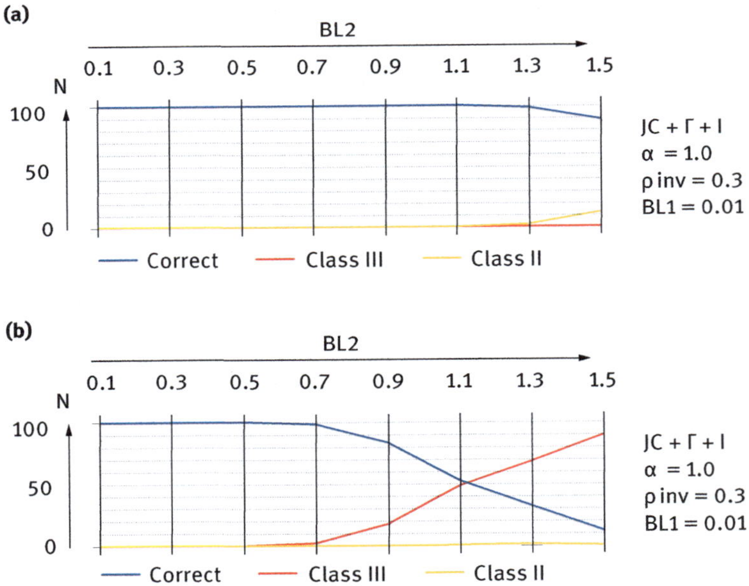

Figure 26.6: Reconstruction success of ML as in Figure 26.5, but for alignment lengths of 100,000 base positions. **(a)** Topology A (Figure 26.2a) and **(b)** Topology B (Figure 26.3b) under alignment lengths of 100,000 base positions if model assumptions are identical to the simulated parameters (α = 1.0; θ = 0.3). Branch length differences increase from left to right by increasing BL2 in discrete steps (0.1–1.5, upper scale) while the length of BL1 is kept constant (0.01). The y-axis depicts the reconstruction success of the 100 simulation repeats (N). Note that topology B is much more susceptible to class I LBA than topology A.

Without consideration of ASRV, the tree reconstruction success for Topology B is even worse than for Topology A (Figure 26.4). In this case, we found predominance of incorrect class I topologies at a branch length ratio which ranges from three to 150 instead of five to 150 which we found for Topology A.

Furthermore, in contrast to Topology A, the high frequency of class I effects does not disappear in Topology B under correct model assumptions when sequence length rises to 100,000 bp (Figure 26.6b). This is in agreement with the fact that the class I effect (signal substitution of synapomorphies) is a systematic error, not only caused by random variations but enforced by shared homologies. A better taxon sampling shortening the long branches can help to overcome this effect (Wägele and Mayer, 2007).

26.3.3 Reconstruction success of the reduced taxon set of Topology B

The reduced 6-taxon set of Topology B (Figure 26.7), in which five "background" taxa are excluded on the stem lineage, was chosen to investigate whether a decreased overall length of the stem lineage of the small intermediate branch (BL1) will reduce class I long-branch artifacts. For this purpose, we compared the ML reconstruction success of both topologies (with long and with short stem lineage) for an alignment length of 100,000 bp using the correct model assumptions (α = 1.0 and θ = 0.3).

While correct topologies weakly predominate in the 11-taxon tree reconstructions under a relatively long internal branch length of BL2 (BL2 = 1.1) (Figure 26.6b), incorrect topologies outweigh correct topologies in the 6-taxon case under this length condition (Figure 26.7).

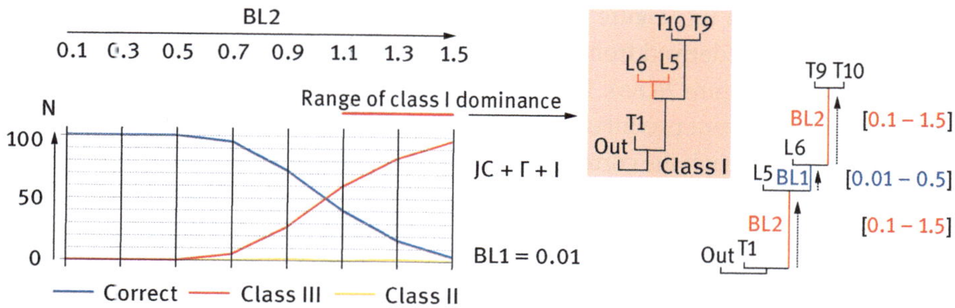

Figure 26.7: The reconstruction success of a reduced 6-taxon set of Topology B from which five stem-lineage taxa are excluded (Figure 26.3). Alignment length: 100,000 base positions. Model assumptions are identical to the simulated parameters (α = 1.0; θ = 0.3). Compared to Figure 26.6b (Topology B with more basal taxa), the number of incorrect topologies is more frequent and outweigh correct topologies in the 6-taxon case under large branch length differences.

26.3.4 Reconstruction success of topologies C–F

Topologies C–F (Figure 26.8) have tree shapes identical to Topology B (Figure 26.2b), but different lengths for internal branches. All topologies are analyzed with ML under correct model assumptions ($\alpha = 1.0$ and $\theta = 0.3$) for the alignment length of 10,000 bp.

26.3.4.1 Topologies C and D

Topologies C and D are designed to gain additional insight about the influence of strong branch length differences on the occurrence of class I effects in ML tree reconstructions when very short internal branches stretch across several nodes between long internal branches (Figures 26.8a and 26.8b). For both topologies we find a predominance of incorrect topologies if the ratio of the two long internal branches (BL2) to the two or three short intermediate branches (BL1) exceeds a value of 90 (BL2/BL1 > 90). The number of incorrect trees strongly increases for both topologies with increasing branch length differences. For Topology C, in which two long internal branches are separated by two intermediate branches, incorrect topologies dominate in ML tree reconstructions even when BL2/BL1 = 90 (Figure 26.8a). The lower reconstruction success of Topology C indicates that substitutions along both long internal branches have a stronger impact on the intermediate branching zone in Topology C than in Topology D, probably because there are fewer supporting characters for (L6 + T9 + T10).

26.3.4.2 Topology E

Topology E has been designed to analyze the different impact of class I and class II effects on ML tree reconstructions. For this purpose, we did not elongate the internal branch between taxon T7 and taxon T8 of Topology C. Instead, we kept this branch length constant with 0.1 (Figure 26.8c). Through the exclusion of one long internal branch, Topology E (derived from C) achieves a much higher reconstruction success than Topology D (Figure 26.8a). Even if the single long internal branch is 130 times higher than the adjacent two internal branches on the left side of the long branch, ML is still able to reconstruct the correct phylogeny of Topology E in the majority of repeat steps (Figure 26.8a), whereas the opposite occurs if we analyze the reconstruction success of Topology D (Figure 26.8b). Therefore, it can be concluded that the impact of signal loss along a single internal long branch is much lower, because the misleading homologies that evolve on the first long branch and act as symplesiomorphy for taxa T8, L5 and L6 in C are missing.

26.3.4.3 Topology F

Topology F (Figure 26.8d) has been designed to analyze the impact of two long internal branches on the ML reconstruction of the intermediate branching events if one of the two short intermediate branches of topology C is 10 times longer than the other short one. In this case, we find a much lower number of incorrect topologies. Even if both long branches are 150 times larger (BL2 = 1.5) than the shortest intermediate edge, ML was able to reconstruct correct topologies in the majority of repeat steps (Figure 26.8d). This indicates that a longer branch between the nodes of the intermediate zone basal to a long branch reduces the negative impact of this long branch, because there is more signal supporting the terminal monophylum L6 + T9 + T10.

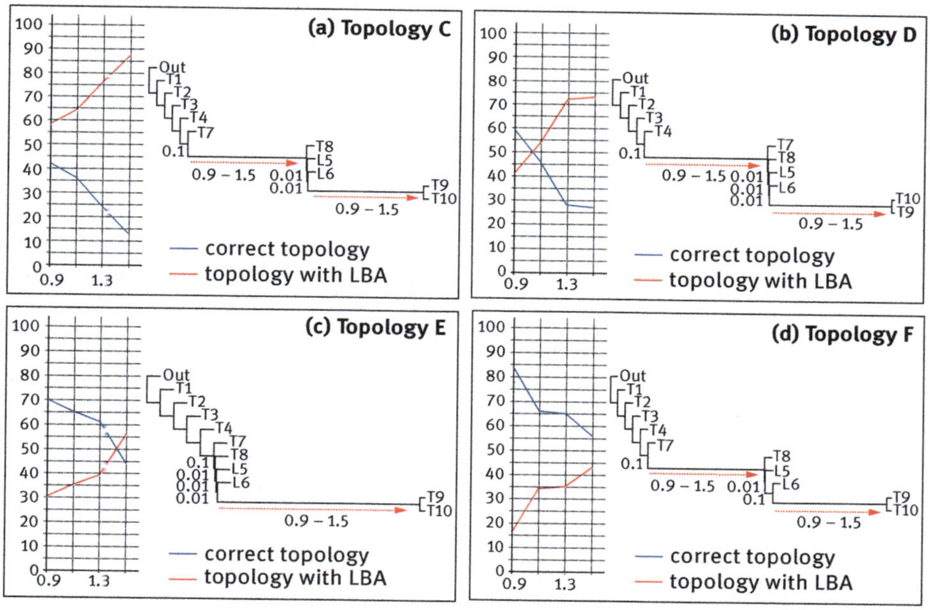

Figure 26.8: Tree reconstruction success of topologies C–F under alignment lengths of 10,000 base positions if model assumptions are identical to the simulated parameters (α = 1.0; θ = 0.3). Y-axis: number of topologies found among 100 repeats. X-axis: length of the long branches (0.9 to 1.5). Short branches were 0.1 if not indicated otherwise. **(a)** Topology C in which two long internal branches are separated by two intermediate branches. **(b)** Topology D in which two long internal branches are separated by three short intermediate branches. **(c)** Topology E in which one of the long branches of Topology D was reduced to a length of 0.1. **(d)** Topology F in which the length of one of the two short intermediate branches of topology C was kept moderately long (0.1). The number of incorrect trees strongly increases for topologies C and D with increasing branch length differences while topologies E and F allow a much higher reconstruction success, especially if one of the two long branches is kept moderately short (Topology F).

26.4 Discussion

For alignment lengths in the range of 2,000 to 10,000 bp, we investigated the success of inference of topologies A and B using correct as well as estimated model parameters (θ and α) under a mixed-distribution model (JC + Γ + I), a Γ-distribution model in which α was estimated without θ (JC + Γ), and without any consideration of ASRV (JC).

Our results show that ML has a much higher reconstruction success if ASRV is taken into account when the underlying data includes rate heterogeneity. This result is not new and agrees with the results of previous studies (Anderson and Swofford, 2004; Huelsenbeck, 1997; Lartillot et al, 2007; Philippe and Germot, 2000; Sullivan, Swofford and Naylor, 1999; Yang, Goldman and Friday, 1994).

The consideration of ASRV through a mixed-distribution model (JC + Γ + I) improves the success of ML tree reconstruction more than a consideration of rate heterogeneity by a Γ-distribution model alone (JC + Γ). The performance of the mixed-distribution model is especially high for the reconstruction of Topology B, in which ML recovers correct topologies under a wide range of branch length conditions (Figures 26.4b and 26.5b). This result is in accordance with the results of Sullivan, Swofford and Naylor (1999) and Anderson & Swofford (2004). Both studies have also shown that ML performs best if the mixed-distribution model is used. This contradicts the widespread opinion that the application of a Γ-distribution model alone is sufficient (e.g. Murienne, Edgecombe and Giribet, 2010; Regier et al., 2010; Rota-Stabelli et al., 2010; Sanderson et al., 2000, Savard et al., 2006). It has to be tested in further analyses whether a mixed-distribution model will perform equally well on empirical data of the type of topology B. However, if branch length differences are very large between long branches (BL2) and short branches (BL1), ML performs poorly even when a mixed-distribution model is included in the tree reconstruction process.

In our analyses, the lower reconstruction success for topologies with very short branch lengths (BL1 = 0.01) cannot be equated with a random choice from a star topology in which ML is expected to “choose” at random from the set of all possible topologies (Bruno and Halpern, 1999; Swofford et al., 2001). This would be the case if information for ancestral states is completely missing. This is not the case in our analyses. Even under large branch length differences ML was able to resolve correct topologies more often than would be expected by chance. Similar, incorrect topologies like the paraphyletic grouping of taxa L5 and L6 as a sister-group (class I effect) occurs more often than can be attributed to chance.

ML does not allow reconstruction of Topology B under correct model assumptions when branch length differences between short branches of BL1 and long branches of BL2 are large (Figure 26.6b), even for long alignments with 100,000 bp. With increased alignment length, the class I effect (symplesiomorphy effect) becomes even stronger after a certain point of branch length differences. The explanation for this phenomenon is the systematic bias. This class of topologies has not been investigated before and cannot be reduced to a 4-taxon example. In the case of a 4-taxon topology, the “Farris Zone” in which two long branches are directly related to each other is a well-known example of reduced ML efficiency. Alignment lengths of 100,000 bp are required to recover trees in the “Farris Zone” correctly with an 85 % probability (Swofford et al., 2001). The ML reconstruction success declines initially for smaller sequence lengths and then improves with increasing sequence length. It can be expected that the topologies of our setups have a similar effect on ML tree inference as found for the “Farris Zone” by Swofford et al. (2001). When model assumptions were correct, ML was able to recover Topology A efficiently even with large branch length differences between BL1 and BL2, if alignment length exceeds 100,000 bp. A decreased overall length of internal branches basal to the small intermediate branch (BL1) as given in the “taxon reduced Topology B” does not reduce the class I effect. The opposite is the case. This implies that a higher number of taxa increases the amount of phylogenetic signal between these taxa and the following long internal branch. The reconstruction success of Topology B (Figure 26.6b) suggests that the class I effect is harder to avoid, which means that alignment lengths may need to be in the millions or even higher. In this case we expect that the class I artifact will occur even for large data such as many bacterial genomes. Since we are currently entering the era of genomics, it would be interesting to investigate the alignment length for which ML will reliably find the correct tree.

One possible explanation for the different reconstruction success between Topologies A and B could be that with Topology B we are dealing with a systematic error. The bidirectional separation of taxa L5 and L6 from all other taxa via long internal branches and the connection of both taxa via very short branches implies little signal evolving between both taxa due to very fast splitting events as well as strong signal loss along the following long internal branch. This increases dramatically the complexity to recover the true branching events in this area of Topology B. Although ML is independent of the root position due the time reversibility of the JC model, there are no signal differences in any direction between the internal nodes of L5 and L6 to another node which would be strong enough to be recovered by ML. Instead, due to the nucleotide patterns produced in the simulation, L5 and L6 will share characters unique for their last common ancestor and absent in T9 and T10 more often than expected by chance (class I effect).

The same effect occurs trying to infer Topologies C and D (Figure 26.8) in which additional nodes connected by short edges were added. Topology C has one additional node in the region between two long internal branches. Though now this region is somewhat longer, the inference of the true tree is still extremely difficult. The explanation is that the number of synapomorphies for (L5 (L6 (T9, T10))) and (L6 (T9, T10)) is still low and probably to a large extent substituted along the following long branch. Topology D has one internal node more than C. The success of ML inference is slightly higher than for Topology C, which can be attributed to the generation of more synapomorphies in this region, but the bias is still present (Figure 26.8b).

A much higher reconstruction success has been found for Topologies E and F (Figure 26.8c and d) in which either one of the internal branches between the two long branches was ten times larger than the very short branches (1.0 contrary to 0.1) or when one of the two long internal branches is missing (reduced to 0.1). For both topologies we have found correct topologies even under very large branch length differences BL1 and BL2. In topology E, the first long branch is missing. This implies that the number of symplesiomorphies for the false clade (L5, L6) is much lower in E than in C or D (Figure 26.8), where they evolve along the basal long branch. In Topology F we obtain more signal for the correct clade (L6, (T9, T10)) than in C or D.

The result for topology E explains why incorrect topologies due to signal erosion along a single long branch (class II effect) are only found under large branch length differences between BL1 and BL2. For example, incorrect topologies have been only found in the majority reconstructions of Topology E when the single long internal branch is 150 times larger than the next internal short branch (BL2 = 1.5, BL1 = 0.01) (Figure 26.8c).

It is also interesting to note that estimates of rate heterogeneity are very accurate when the mixed-distribution model was used ($JC + \Gamma + I$) for tree reconstructions. This high accuracy was found for all simulated branch length differences independent of the reconstruction success of a given topology. Due to the fact that the mixed-distribution model was also used for our data simulations, the use of accurately estimated or correctly given model parameters excludes model misspecification as the source of inaccuracy for our tree reconstructions. In cases of incorrectly resolved topologies (e.g. Figure 26.6b), ML consistency implies that the phylogenetic inaccuracy is more likely caused by a taxon-sampling problem. The explanation that the heuristic ML searches get stuck in local optima is rather unlikely with only 11 taxa and $JC + \Gamma + I$.

It has been shown by Sullivan, Swofford and Naylor (1999) that taxon number is important for a correct estimate of model parameters, because of stochastic errors in small samples. Sullivan et al. have shown that at least a data set of 20 taxa is needed to fulfill this demand. The fact that even a data set of only 11 taxa allows accurate parameter estimates could be explained by longer alignments used in our study. Sullivan, Holsinger and Simon (1996) have shown further that long internal branches can have a strong impact on estimates of the Γ-distribution. However, the correlation between the Γ-distribution and internal branch lengths was based on four-taxon tree simulations and has not been detected in our study.

We have shown in our analyses that the occurrence of long-branch artifacts, especially the class I LBA is not a particular problem of model selection. If only the Γ-distribution model is used, the deficiency of an additional estimation of completely invariant sites will be partially compensated by a lower estimate of the α shape parameter. This results in an increased estimate of the number of positions with low or very low substitution rates and an overestimation of substitution rates for a certain number of invariant sites. However, the reconstruction success with a Γ-distribution model alone is lower in our analyses than with the mixed-distribution model ($\Gamma + I$).

Generally, our results show that the risk of recovering incorrect topologies using ML is dependent on the arrangement of internal nodes and the branch length ratios between them. Although our results base on simulated DNA data, it can be expected that amino acid sequences are also susceptible to LBA if branch length ratios between adjacent branches are extremely large, even though the possibility of obtaining long-branch class III effects decreases with an increasing alphabet of character states. It is also clear from our results that high support for nodes is no guarantee for their correctness. Furthermore, it is important to keep in mind that empirical data can evolve in a much more heterogeneous way than simulated in this study. However, ML is still the most efficient and robust tree reconstruction method currently available. Maximum Parsimony for example would be much more susceptible to LBA (Bruno and Halpern, 1999; Felsenstein, 1978; Sullivan and Swofford, 1997; Sullivan and Swofford, 2001; Swofford et al., 2001; Yang, Goldman and Friday, 1995; Yang, 1997) and thus is entirely inappropriate for molecular data.

Patrick Kück and J. Wolfgang Wägele

27 Topological bias of maximum-likelihood trees inferred from star phylogenies in the event of correct and incorrect model assumptions

Abstract: The theoretical reliability of Maximum Likelihood methods often fails in practice due to the finite size of data sets and because of peculiarities of the software, even when correct substitution models are used. In addition, systematic errors inherent to the method may occur.

To detect biases, we simulated star phylogenies. In the case of star-like divergences, an unbiased tree reconstruction method will choose a best resolved binary tree by chance. Concerning 4-taxon trees, each of the three possible topologies should be chosen with the same probability. Deviations from the expected equal frequency of occurrence of topologies indicate a bias of the method. Based on this assumption, we analyzed the influence of topological bias on maximum-likelihood tree reconstructions under a broad range of different branch length conditions, model parameter assumptions, and alignment lengths. Our analyses show that the Maximum Likelihood method can be influenced by topological bias at certain branch length differences even if model parameter assumptions have been chosen correctly. Furthermore, our analyses show that the topological bias obtained from tree reconstructions with correct model assumptions is more distinct for alignments with 100,000 base positions than for 10,000 base positions.

The quantitative approach based on the resolution of star-like trees seems to be a good starting point to test the influence of long-branch attraction and chosen model assumptions. The artifactual attraction of strongly derived taxa found in our analyses due to oversimplified model assumptions confirms the commonly accepted assumption that an exclusion of among-site rate variation increases the probability of topological bias in maximum-likelihood tree reconstructions when underlying data is heterogeneous. However, the occurrence of topological bias under correct model assumptions and the increased probability of grouping strongly derived taxa with increasing sequence lengths imply that this bias is a systematic error. A better adaption of model assumptions can only delay the occurrence of errors in maximum-likelihood tree reconstructions for a certain range of branch length conditions.

27.1 Introduction

The reliability of phylogenetic tree reconstruction methods depends on data quality and on the accordance between real evolutionary processes and tree model assumptions. The larger the deviation between the chosen model assumptions and the evolutionary processes that shaped the underlying data, the higher the topological bias in inferred trees. Starting from the assumption that empirical data result from heterogeneous evolutionary processes, model assumptions which do not consider among-site rate variation (ASRV) are clearly oversimplified and increase the probability for topological bias regarding the attraction of strongly derived taxa due to the accumulation of undetected random character similarities shared by these taxa (Felsenstein, 1978; Huelsenbeck and Hillis, 1993; Kuhner and Felsenstein 1994; Gaut and Lewis, 1995; Russo, Takezaki, and Nei, 1996; Huelsenbeck, 1997; Bruno and Halpern, 1999; Swofford et al., 2001; Kolaczkowski and Thornton, 2004; Bergsten, 2005; Kück et al. 2012). This applies especially to the reconstruction of deeper taxon relationships, because the underestimation of substitution rates becomes progressively worse with increasing evolutionary distances (Felsenstein, 1978; 1984; Huelsenbeck and Hillis, 1993; Kuhner and Felsenstein 1994; Gaut and Lewis, 1995; Huelsenbeck, 1997; Philippe and Laurent, 1998; Bruno and Halpern, 1999; Kück et al. 2012). Therefore, tree reconstruction methods like Maximum Parsimony, which do not consider ASRV, are unsuitable to reconstruct correct topologies from heterogeneous data. Maximum Likelihood (ML) is certainly more robust against model violations and more efficient than other tree reconstruction methods (Felsenstein, 1978; Huelsenbeck and Hillis, 1993; Kuhner and Felsenstein 1994; Gaut and Lewis, 1995; Yang, 1996; Huelsenbeck, 1997; Bruno and Halpern, 1999; Sullivan, Swofford, and Naylor, 1999; Swofford et al., 2001; Anderson and Swofford, 2004; Kolaczkowski and Thornton, 2004; Kelchner and Thomas, 2006; Fischer and Steel, 2009; Kück et al. 2012). Even oversimplified likelihood models can find the correct tree if branch lengths are well balanced (Yang, Goldman, and Friday, 1994). However, despite the preeminence of ML in comparison to other tree reconstruction methods, it was shown by Kück et al. (2012) that ML can fail to recover the correct tree from strongly heterogeneous data under certain tree shapes and branch length differences if model assumptions are chosen correctly. Kück et al. (2012) have shown further that ML was even unable to find the correct tree under these conditions if underlying data comprises sequence lengths of 100,000 base positions. The authors concluded that their observations are based on a systematic error.

To understand the influence of systematic bias on ML tree reconstruction processes in greater detail, we used a quantitative approach based on the resolution of star-like trees, which was first introduced by Bruno and Halpern (1999). These authors have shown that the influence of topological bias on a tree method in favor of a specific tree can be directly obtained from the distribution deviations of binary trees estimated for star-like phylogenies. If underlying data evolves from a star-like process, an unbiased method has no other choice but to choose randomly a binary tree from the possible alternative topologies (Bruno and Halpern, 1999; Swofford et al., 2001). Concerning the 4-taxon case, each of the three possible topologies should be chosen to an equal extent if the number of repeated tree reconstructions is large enough. Thus, Bruno and Halpern (1999) quantitatively defined the topological bias of a method in the 4-taxon case as the fraction of tree reconstructions in which a specific tree is chosen, minus the expected one-third.

Based on their definition of topological bias, Bruno and Halpern (1999) reanalyzed previously published results of Yang (1997) in which an oversimplified ML model led to a better tree reconstruction success in 4-taxon cases than when using the correct model in cases with two strongly derived taxa as sister-groups (the so called 'Farris-Zone' effect). Their reanalysis suggests that the out-performance of the correct model assumptions is most probably caused by a topological bias. The oversimplified model assumptions favor the sister-group relationship between these two taxa. This observation supports the results of other studies (Felsenstein, 1978; Gaut and Lewis, 1995; Huelsenbeck, 1997; Siddall, 1998; Sullivan, Swofford, and Naylor, 1999; Swofford et al., 2001; Pol and Siddall, 2001). Bruno and Halpern (1999) could also show that this topological bias gets progressively worse for longer sequences while the correct model gives relatively unbiased estimates.

However, the topologies investigated by Bruno and Halpern (1999) on the basis of Yang (1997) consist of less strongly pronounced branch length differences between short and long branches than the study of Kueck et al. (2012). In the study of Bruno and Halpern (1999), branch length differences between short and long branches in which the correct model gives relatively unbiased estimates consist of rather one long branch and three more or less short branches. The occurrence of two equally distinct long branches was not considered. To check the occurrence of topological bias under correct and incorrect model assumptions, we analyzed the influence of topological bias under a much broader range of branch length conditions (Figure 27.1). Furthermore, the sequence lengths used by Bruno and Halpern (1999) ranged only from 2,000 to 8,000 base positions. To obtain a more accurate statement about the occurrence of topological bias, we used alignments of 10,000 to 100,000 base positions.

27.2 Methods

27.2.1 Simulations

We designed three sets of simulations (Figure 27.1) to test the randomness of the Maximum Likelihood selection process regarding the three possible resolved 4-taxon topologies drawn from star-like phylogenies. Simulation setup A was characterized by a stepwise elongation of two branches (Figure 27.1a), simulation setup B by a stepwise elongation of a single branch (Figure 27.1b), and simulation setup C by a stepwise elongation of all four branches (Figure 27.1c). While all four branches were equally elongated in the third setup, remaining branches of the first two simulation sets were kept short (Figures 27.1a, 27.1b). Branch lengths reflect the number of expected substitution rates per site for corresponding lineages and thus the amount of deviations between taxa. We increased branch lengths from 0.1 to 1.5 in steps of 0.2. Therefore, branch length ratios between short and long branches ranged from one to 15 in the first two setups. All alignments were generated with INDELible v.1.01 (Fletcher and Yang, 2009) using the Jukes–Cantor model (JC) of sequence evolution and a mixed-distribution model of Γ + I for ASRV. All data were simulated with ASRV, shape parameter α = 1.0, and a proportion of invariant sites ρinv = 0.3. ASRV was modeled using a continuous Γ-rate distribution while indel events were not simulated. For all three simulation setups, we simulated for each branch length combination 1,000 sequence alignments with a length of 10,000 base positions. Setup A (stepwise elongation of two branches) was further simulated with increased alignment lengths of 100,000 base positions under otherwise equal conditions. The JC model has been chosen for the simulations (i) since it is better understood than any other model of sequence evolution and (ii) to keep the model parameter space as small as possible. Due to the simple assumptions of the JC model (each base in the sequence has an equal probability of changing, which results in equal frequency of the four bases), the reconstruction success of ML is directly linked to the simulated branch length conditions, sequence lengths, and among-site rate variation parameters used in each ML analysis.

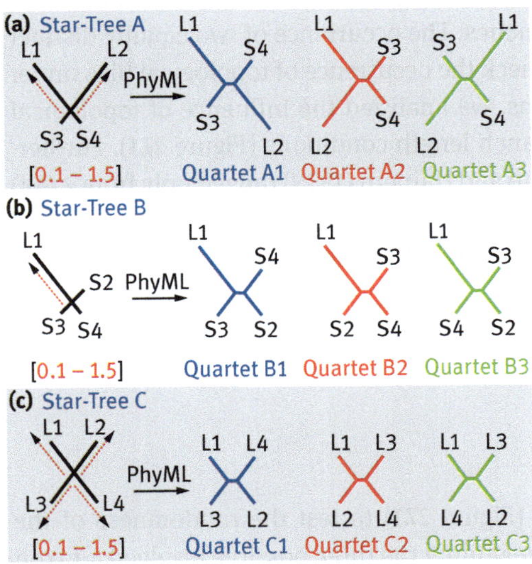

Figure 27.1: Three simulation setups for topological bias tests on 4-taxon star-like trees and corresponding binary trees. **(a)** Simulation setup A: Stepwise elongation of two branches (L1 & L2), **(b)** Simulation setup B: stepwise elongation of one single branch (L1), and **(c)** stepwise elongation of all four branches (L1 to L4). Elongation of branch lengths spans from 0.1 to 1.5 (in steps of 0.2) while remaining branch lengths are kept constant at 0.1. Maximum Likelihood inference is conducted 1,000 times for each elongation setup. PhyML tree reconstructions of simulation setup A are conducted for alignment lengths of 10,000 bp and 100,000 bp, tree reconstructions of simulation setups B and C for sequence lengths of 10,000 bp.

27.2.2 Maximum Likelihood analyses

ML trees were calculated with the Jukes–Cantor (JC) model under three different parameter settings using PhyML-3.0-linux64 (Guindon and Gascuel, 2003; Guindon et al. 2010). We analyzed the data either with a mixed-distribution model (JC + Γ + I) or without consideration of among-site rate variation. For the mixed-distribution model we used an α value of either 1.0 (equal to simulated) or 0.75 (lower than simulated). Both α values were used in combination with a proportion of invariant sites set equal to the simulated value (ρinv = 0.3). As approximation to non-ASRV (JC), model parameters were set to α = 100 and ρinv = 0. For the alignment length of 100,000 base positions, tree reconstructions were only performed under the correct model parameters (α = 1.0 and ρinv = 0.3). With the discrete gamma model, the number of relative substitution rate categories was set to four (c = 4) and tree topologies and branch lengths were optimized (heuristic search). Maximum Likelihood analyses were performed and evaluated using a Perl pipeline.

27.3 Results

27.3.1 Simulation setup A

For alignment lengths of 10,000 base positions, the ML reconstruction success for the three possible quartet topologies of simulation setup A (stepwise elongation of two branches) deviates significantly from a random choice at certain branch length conditions for oversimplified and also for correct model assumptions (Figure 27.2a). Without consideration of ASRV, best ML topologies favor a sister-group relationship of both elongated branches in all 1,000 repeat steps of each branch elongation. This results in a strong topological bias of 70 % favoring the grouping of both elongated branches (quartet A2) even if branch length differences are relatively low (elongated branches three times longer than the remaining branches) (Figure 27.2a). If model parameter assumptions are slightly changed by using a higher substitution rate parameter than simulated (α = 0.75 instead of α = 1.0), our analyses result in nearly unbiased estimates (Figure 27.2a). Each of the three possible quartets (A1, A2, A3) is favored by a topological bias of around 10 % at a certain point of branch length conditions (Figure 27.2a). For the largest simulated branch length differences (elongated branches 15 times longer than remaining branches) all three quartet topologies (A1, A2, A3) end up with zero topological bias in any possible direction of tree shape. Inferring trees with correct model assumptions, topological bias starts to occur when the two elongated branches are nine times longer than the short branches whereas no topological bias has been found for smaller branch length differences. The topological bias which has been found under correct model assumptions favors the sister-group relationship of the two long branches (quartet A2) and increases continuously from 5 % to 40 % with increasing branch length differences between long and short branches (Figure 27.2a).

With increased alignment lengths of 100,000 base positions (Figure 27.3), the topological bias which was found for correct model assumptions occurs even for shorter branch length differences than those found with 10,000 base positions. While for alignment lengths of 10,000 base positions the topological bias starts to favor the sister-group relationship of the two elongated branches when both elongated branches are nine times longer than the short branches, the topological bias occurs

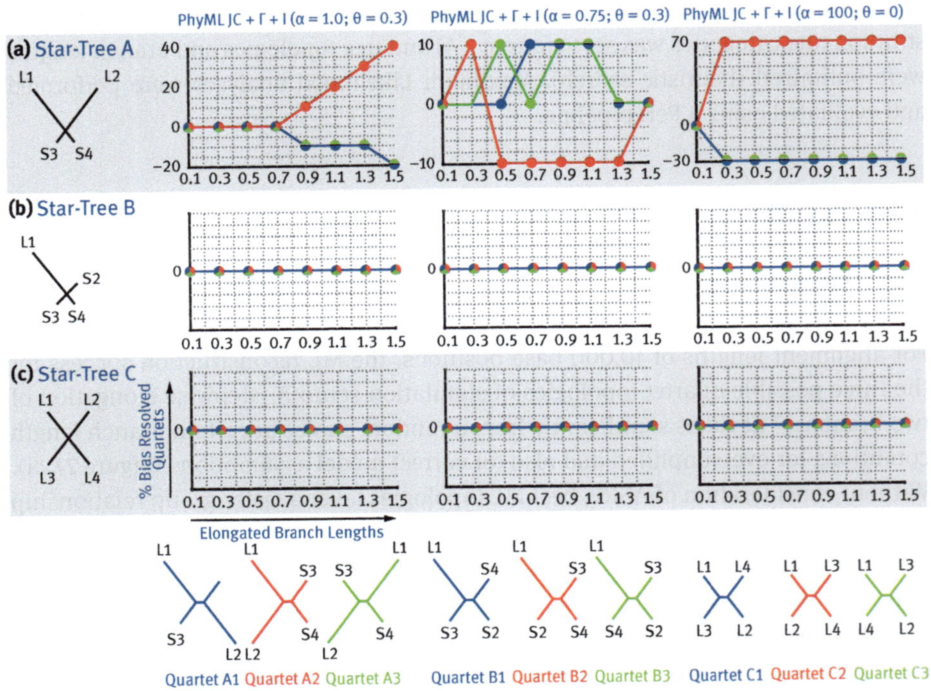

Figure 27.2: Frequencies of topological biases inferred from maximum-likelihood tree reconstructions for sequence lengths of 10,000 bp. Frequency of topological bias for different model parameter settings of **(a)** simulation setup A, **(b)** simulation setup B, and **(c)** simulation setup C. The topological bias is quantified as the fraction of binary tree reconstructions in which a specific binary topology is chosen, minus the expected one-third. While a topological bias value > 0 was found in simulation setup A (stepwise elongations of two branches), the simulation setups B (stepwise elongations of a single branch) and C (equal balanced elongations of all four branches) result in unbiased estimates. Topological bias was found for quartet A2 (red line) under correct model assumptions (α = 1.0; θ = 0.3) if branch lengths of L1 and L2 are larger than 0.7 (bias ≤ 40 %). It is even more pronounced (bias = 70 %) in a much wider range of branch length differences (0.3 to 1.5) when among-site rate variation is completely ignored (α = 100; θ = 0). Nearly unbiased estimates (bias ≤ 10 %) were found when α was set lower than simulated (α = 0.75; θ = 0.3).

already when elongated branches are seven times longer than the short ones if alignment lengths are increased to 100,000 base positions.

The frequency of the topological bias is also higher for 100,000 base positions than for 10,000 base positions (Figure 27.2a and Figure 27.3) and rises to 70 % for the largest branch length differences which is as high as the observed topological bias in analyses without consideration of ASRV (Figure 27.2a and Figure 27.3).

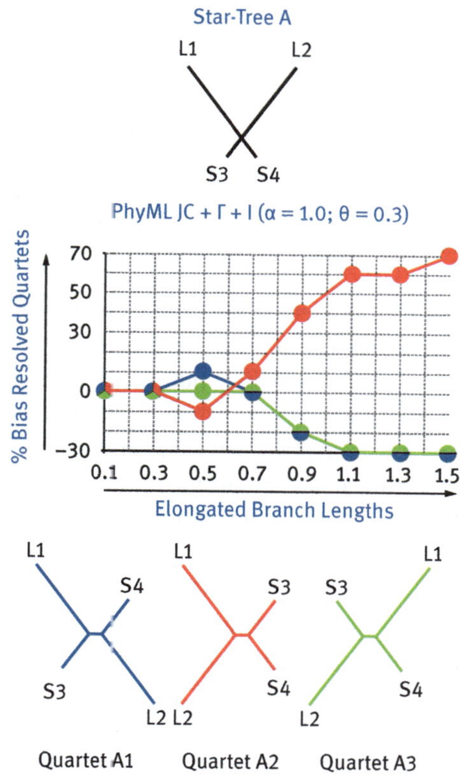

Figure 27.3: Frequencies of topological bias inferred from maximum-likelihood tree reconstructions for sequence lengths of 100,000 bp under correct model assumptions. The topological bias found in tree reconstructions of simulation setup A (Figure 27.1a) is even stronger for alignment lengths of 100,000 bp than for 10,000 bp. While positive long-branch attraction (LBA) bias favoring quartet A2 was found for 10,000 bp if branch length differences are set to ≥ 0.9, quartet A2 was also positively biased for branch length differences of 0.7. Furthermore, while the maximum frequency of positive LBA bias (favoring quartet A2) was 40 % if sequence lengths are set to 10,000 bp, the maximum amount of positive LBA bias (quartet A2) for alignment lengths of 100,000 bp reaches 70 %.

27.3.2 Simulation setups B and C

ML reconstructions of 4-taxon star-like trees which are obtained from the simulation setups B and C (elongation of a single branch and elongation of all four branches) deliver completely unbiased estimates even for large branch length differences and independently of chosen model assumptions (Figures 27.2b and 27.2c).

27.4 Discussion

The quantitative approach on the basis of the resolution of 4-taxon star-like phylogenies first introduced by Bruno and Halpern (1999) seems to be a good starting point to test the reliability of tree building methods in terms of long-branch attraction in conjunction with chosen model assumptions.

Our analyses show that ML can indeed be influenced by topological bias at certain branch length differences, not only when model assumptions are oversimplified, but also when model assumptions are chosen correctly. While no topological bias has been found in any direction of possible tree shapes for simulated data of well-balanced phylogenies (Figure 27.2c) and for phylogenies in which only one branch has been elongated (Figure 27.2b), topological bias occurs in the form of the attraction of two strongly derived taxa when the branch length ratio between short and elongated branches exceed a certain threshold (Figure 27.2a). The point at which this topological bias starts to occur in ML tree reconstructions as well as its frequency depends on the chosen model assumptions (Figure 27.2).

If ASRV is completely ignored by oversimplified model assumptions, topological bias favors the attraction of derived taxa even under relatively short branch length differences (Figure 27.2a). This result is congruent with previously published studies and confirms the commonly accepted assumption that the exclusion of ASRV ignores the accumulation of undetected random character similarities between derived taxa, which seriously enhances the risk of long-branch attraction (Huelsenbeck and Hillis, 1993; Yang, 1996; Gaut and Lewis, 1995; Huelsenbeck, 1995; Bruno and Halpern, 1999; Sullivan and Swofford, 2001; Kelchner and Thomas, 2006; Kück et al. 2012). Furthermore, this result supports the assumption of Bruno and Halpern (1999) that a high reconstruction success of long branched sister-group relationships through oversimplified ML model assumptions in the 'Farris Zone' is most probably due to positive topological bias.

Our results obtained from tree reconstructions under correct model assumptions imply that ML can be influenced by long-branch attraction through topological bias if two branches are clearly longer than the remaining branches and that the occurrence of this bias is not necessarily correlated with deviations of models from the actual evolutionary processes. Furthermore, our analyses show that the frequency of topological biases increases with increasing sequence length. For large branch length differences, in which long branches are 15 times longer than short branches, the frequency of topological biases is equal to inferences without consideration of ASRV (Figure 27.2 and Figure 27.3). Thus, our results support the statement of Kück et al. (2012) that systematic errors can occur in ML analyses.

The extent to which topological bias can influence ML tree reconstructions depends not only on the chosen model assumptions and the sequence lengths of underlying data, but also on phylogenetic signal originating from internal branch lengths. The longer the internal branch between taxa, the larger the signal antagonizing topological bias. When Kück et al. (2012) inferred systematic bias not only caused by random variations but also enforced by shared homologies for topologies in which the internal branch length was set 150 times shorter than the two elongated terminal branches and 10 times shorter than the remaining branches, the antagonizing signal originating from the very short internal branch was very low.

For this reason, we conclude that the topological bias observed in this study is most likely identical with the systematic bias found by Kück et al. (2012). The occurrence of topological bias under correct model assumptions excludes model misspecification as the source of inaccuracy. Therefore, topological bias as inferred in this study seems to be a systematic effect which can lead to incorrect, most probably highly supported biological conclusions. As a result, we have to keep in mind that good 'support values' are no guarantee for the correctness of a tree topology.

The contradiction between our results obtained with correct model assumptions and the conclusion of Bruno and Halpern (1999), who stated that ML delivers nearly unbiased estimates under this condition, can be explained by differences in the chosen branch length ratios. The conditions in which topological bias occurs in our study were not investigated by Bruno and Halpern (1999), whose setup resembles our simulations B and C. These simulations deliver completely unbiased estimates (Figure 27.2b and Figure 27.2c).

A completely different resolution success of ML quartets has been obtained in simulation setup A (Figure 27.2a) when the model parameters are only slightly modified ($\alpha = 0.75$). For this model parameter setting we have found nearly unbiased estimates for all branch length conditions. It seems that to assume a substitution rate that is higher than in the correct model is more suitable for the correction of attraction due to random similarities occurring along long branches. The inflated parameter is most probably better adapted to our simulated data. It has to be tested in further simulations whether this could be a good possibility to overcome topological bias.

Our analyses confirm the statement of previously performed studies (Felsenstein, 1978; Gaut and Lewis, 1995; Yang, 1996; Huelsenbeck, 1997; Bruno and Halpern, 1999; Sullivan, Swofford, and Nalyor, 1999; Siddall, 1998; Pol and Siddall, 2001; Swofford et al., 2001) that an exclusion of ASRV can lead to tree reconstructions which can outperform a better, less biased evolutionary model in cases where underlying data contains strongly derived sister-group lineages (Figure 27.2a). More importantly, our analyses show that even correct ML model assumptions can lead to biased tree reconstructions in the sense of long-branch attraction when underlying data is strongly heterogeneous. The topological bias even gets stronger when sequence lengths are increased from 10,000 base positions to 100,000 base positions (Figure 27.2a and Figure 27.3). This implies that the topological bias is a systematic error and that a better adaptation of model assumptions can only delay the occurrence of errors in ML tree reconstructions for a certain range of branch length conditions. The consideration of a higher substitution rate parameter in ML tree reconstructions may help to overcome this bias and thus reduce long-branch attraction. We have further shown that

ML reconstructions of data sets which include a single derived taxon (Figure 27.1b) or which correspond to well-balanced topologies (Figure 27.1c) are much more robust against topological bias (Figure 27.2).

Unfortunately, empirical phylogenies are rarely “balanced”. We have to keep in mind that empirical data can evolve in a much more heterogeneous way than in our simulations.

Although our results depend on simulated nucleotide data it can be expected that protein sequences are also prone to topological bias if branch lengths differ strongly from each other.

It might well be that this error can be reduced by increasing the number of rate categories in substitution models and adapting them to each edge separately. However, the error we report here is relevant for all hitherto published studies that use the default options in available software.

Ingo Ebersberger and Arndt von Haeseler

28 Exploring phylogenomic data

Abstract: It is almost exactly half a century ago that Zuckerkandl and Pauling (1963) first used sequence differences to infer the evolutionary relationships between a handful of hemoglobins. Since then the relevance of sequence data for phylogeny reconstruction has constantly increased, and meanwhile most if not all evolutionary hypotheses are driven by the analysis of either DNA or proteins. Hand in hand with this development grew also the size of the data being analyzed. Where once individual species had to represent entire phyla, and individual genes an entire genome, now contemporary phylogenomic data sets can easily comprise hundreds of taxa and even more genes. The sheer mass of data now allows a tracing also of very faint evolutionary signals resulting in highly sensitive tree reconstructions. Moreover, the integration of many taxa and different genes into a single reconstruction reduces the influence of individual data points. It facilitates, therefore, a more comprehensive and unbiased reconstruction of the evolutionary past. Meanwhile, the other side of the coin has also appeared. An unreflected compilation and use of phylogenomic data sets can result in statistically very robust albeit wrong trees. Moreover, it is not always clear how to interpret the resulting trees given that subsets of the data can differ in their evolutionary history. A thorough exploration of phylogenomic data sets is therefore necessary to assess their potential and limitation for addressing a particular evolutionary question.

28.1 Introduction

28.1.1 Tree thinking in evolution

Duplication and diversification is a general concept in biology that applies to the evolution of molecules, cells, organisms, and of entire species. In all these cases a parental instance gives rise to typically two co-existing descendants that, at a later point in time, themselves become parents. Following this process through time generates a tree structure of relationships. Changes that occur in the course of this process result in the gradual diversification of the various evolutionary lineages and provide the information for a later reconstruction of their relationships. In reality, however, the original tree structure is often difficult to infer (reviewed in Bapteste, Bouchard and Burian, 2012). This can be mainly attributed to the following reasons: Lateral events, such as crossing-over or gene conversion on the DNA level, horizontal gene transfer, or hybridization between species can interconnect originally separate evolutionary lineages (reviewed in Dagan, 2011). Alternatively, assumptions about the homology of the data used for phylogeny reconstruction may be wrong. And eventually, noise can blur the phylogenetic signal in the data and can cause different sets of genes to support different trees although they, in fact, share the same evolutionary history (reviewed in Philippe et al., 2011a). At first sight, these aspects argue against the use of trees for depicting evolutionary relationships. And consequently, networks have in many cases replaced trees since they facilitate a more comprehensive summary of varying phylogenetic signals in the data (Dagan, 2011; Huson and Scornavacca, 2011). However, at second sight the principle of evolution remains tree-like under all mentioned circumstances, and even in the presence of lateral events. The extreme case can be used to justify this claim: The evolutionary history of an individual nucleotide in a species' genome – we can consider it as the atom of evolution – is always tree-like, because a nucleotide never has more than one ancestor making reticulations on that scale impossible. Only when we analyze the evolutionary relationships of combined characters, such as the sequence of two or more nucleotides we may have to cope with a mixture of different trees. However, for many applications in evolutionary research a network summarizing these trees is not satisfying as it leaves the nature of the underlying relationships in the dark. It is therefore a key task in analyzing and interpreting phylogenomic data to identify subsets in the data that share the same evolutionary history and whose phylogenetic signal to noise ratio facilitates a reliable reconstruction of the past. From these subsets individual trees can be reconstructed and subsequently interpreted. Only then is it possible to differentiate vertical relationships, reflecting the succession of the generations through time, from lateral events, such as horizontal gene transfer and from reconstruction artifacts due to noise in the data.

28.1.2 Reconstructing the evolutionary history of species

The evolutionary relationships of species can be, in principle, reconstructed using characters fulfilling three basal requirements:

- A reliable homology inference across species must be possible.
- A sufficient extent of change during evolution is required.
- The changes must be reconstructable from the contemporary data.

Tree reconstruction and systematic classification of species was originally done by the analysis of morphological features. They are, in many cases, readily accessible and can be directly inspected without the help of sophisticated methods or tools. These obvious advantages are, however, opposed by some disadvantages that come into effect in particular when relationships over larger evolutionary distances are to be reconstructed. Homology for morphological characters becomes harder to establish the more the compared species differ in their body plan. There is no unifying metric to measure the extent of divergence between homologous characters. Individual characters typically do not evolve independently from the rest of the organism and they are in most cases associated with a certain function. Hence, they are subject to selective constraints influencing their change over time. It is especially due to the latter three aspects that there is still no comprehensive (statistical) framework of morphological evolution. As a consequence, phylogenetic interpretations of observed changes are mostly based on the parsimony principle (Farris, 1977).

Molecular sequences both of DNA and of proteins provide a complementary view on the evolutionary history of species (Zuckerkandl and Pauling, 1965). In principle, sequences can be used for phylogeny inference in the same way as morphological characters. However, compared to the latter they provide two key advantages. Firstly, the number of edit steps required to change one sequence into another is a straightforward measure of their evolutionary distance. Secondly, sequences evolve in a well-defined framework where only a very limited number of changes are possible. Within a DNA sequence any nucleotide can at any time be changed only into one of the three other nucleotides. The same is true for proteins, although, due to the five times larger alphabet, the number of possible substitutions for any site is 19. This simple principle makes substitutions amenable for statistical modeling provided only three assumptions about the underlying process are valid:

- It is memory-less, i.e. a character i changes to character j independent of the history of i.
- It is time reversible, i.e. it does not matter whether we model the change $i \rightarrow j$ or the change $j \rightarrow i$.
- It is stationary, i.e. the relative frequencies of the individual characters in the alphabet do not change over time.

Jukes and Cantor (1969) were one of the first to specify a model of nucleotide substitution assigning one rate for all changes and setting the frequencies to ¼ for each of the four nucleotides. Bishop and Friday extended this initial model sixteen years later to the amino acid alphabet (Bishop and Friday, 1985). Meanwhile, numerous derivates of these initial models exist, hosting a varying number of free parameters to describe the evolution of both DNA and protein sequences in a biologically more realistic way (Aris-Brosou and Rodrigue, 2012). Such explicit models of sequence evolution, irrespective of whether they model DNA or protein evolution, provide a crucial advantage over morphological data in evolutionary research. They are ubiquitously applicable, because DNA and proteins belong to the inventory of any living organism. Moreover, they embed the process of character change into a full probabilistic framework. Thus, a standardized interpretation of phylogenetically informative data becomes possible. The question about homology of characters now translates into the question of whether two sequences are more similar to each other than would be expected by chance. With the help of the evolutionary model the observed distance between two homologous sequences can be corrected for undetectable changes due to multiple substitutions at the same site. Finally, a likelihood can be computed that one sequence has evolved into another in a given time t allowing to search for the tree that most likely generated the data. Note that the likelihood can also be used to assess the goodness of fit of the evolutionary model to the sequence data. It is due to these advantages that sequences are now in many cases favored over morphological characters for resolving the evolutionary relationships of species. However, detecting phylogenetic relationships from sequence data is not the ultima ratio. Some problems like non-independent evolution between different sites in the sequences still pose major challenges to a model-based analysis of sequences.

28.1.3 From phylogenetics to phylogenomics

The first evolutionary relationships reconstructed from molecular sequences were based on the analysis of six hemoglobin sequences from human and gorilla (Zuckerkandl and Pauling, 1962), of 7 vertebrate Cytochrome C sequences (Margoliash, 1963), and of 20 cytochrome oxidases II (Fitch and Margoliash, 1967). Sequences used in later studies were chosen from a somewhat larger list, including the large subunit of the RNA polymerase (RPB1, (Denton, McConaughy, and Hall, 1998), EF1 α (Roger et al., 1999), the actins (Bhattacharya and Ehlting, 1995), tubulins (Keeling and Doolittle, 1996), HSP70 (Gupta, Golding, and Singh, 1994), and most prominently rRNAs (Olsen and Woese, 1993). These sequences share some desirable properties for phylogeny reconstruction. They convey basal functionalities to cellular life, and are hence present over a broad phylogenetic range. As a result they facilitate a comprehensive taxon sampling over the entire tree of life. They preserve phylogenetic information even over very large evolutionary distances allowing the reconstruction of even very deep phylogenies. As a prime example, the analysis of rRNA molecules suggested that all cellular life can be placed into one of the three domains of life, eubacteria, archaea and eukaryotes (Embley, Hirt, Williams, 1994). Finally, the above-mentioned molecules by and large support congruent trees when used on the same set of taxa. This suggests that any of these molecules provides a picture of organismal history within the limits of its distribution and the inherent uncertainties of the reconstructed trees (Olsen, Woese, 1993).

However, with increasing numbers of phylogenetic analyses individual trees with at least partially overlapping taxon sets supported mutually exclusive evolutionary scenarios for some parts of the reconstructed tree (Penny, Foulds, and Hendy, 1982; Keeling and Doolittle, 1996; Aguinaldo et al., 1997; Adoutte et al., 2000; Martin and Burg, 2002; Castresana, 2007). One of the underlying causes was acknowledged already in the early days of sequence based phylogeny reconstruction. The amount of information required for resolving certain relationships may exceed the information contained in a single sequence (Zuckerkandl and Pauling, 1965). Hence, the trees reconstructed from individual genes do not necessarily reflect the true evolutionary history of the sequences. Additional causes of distortions are: Individual markers may in some parts of the tree truly differ from the species phylogeny, e.g. due to lineage-specific event such as lateral gene transfer, gene duplications and subsequent reciprocal losses of the duplication products (Martin and Burg, 2002), or due to the different sorting of ancestral polymorphisms (Ruvolo, 1997). Basing phylogenetic inferences on the joint analysis of a greater number of homologous proteins is one way to cope with these issues. This is expected to enhance the phylogenetic signal in the data (e.g. Rokas et al., 2003a; Gadagkar, Rosenberg, and Kumar, 2005) and, hence, should facilitate a more sensitive reconstruction of the evolutionary relationships among species. As a further benefit, gene specific effects should level out as the number of considered genes gets large. Despite being straightforward, the realization of this approach is hindered by the effort it takes to experimentally determine the sequence of even an individual gene from a set of new species. In many cases, multi-gene analyses in comprehensive phylogenetic studies were, therefore, limited to varying combinations of the traditional phylogenetic markers, for which standardized procedures for their retrieval exist (e.g. Sekimoto et al., 2011). It was only the technical advances in DNA sequencing and the increasing availability of complete genome sequences for individual species that alleviated this situation.

With a complete gene set for a species at hand, retrieval of sequences no longer constitutes a bottleneck. And as a result, the numbers of genes integrated into data sets for phylogeny reconstruction substantially increased (e.g. Rokas et al., 2003a; Wolf, Rogozin, and Koonin, 2004; Fitzpatrick et al., 2006; Huerta-Cepas et al., 2007). In fact, the size of the data set is now basically limited only by the representation of homologs to individual genes in the taxa under study and by computational limits in data analysis. Under this condition it is no problem to compile a large collection of randomly selected genes for the tree reconstruction. In the resulting data, gene-specific effects are alleviated and the phylogenetic signal shared by many or even all genes are enhanced (Gatesy and Baker, 2005). Both size and randomness of the data are the hallmarks of the transition from phylogenetics to phylogenomics.

28.1.4 Phylogenomics – The ultima ratio?

Trees reconstructed from phylogenomic data sets are – given the wealth of data – typically statistically very well supported (e.g. Fitzpatrick et al., 2006; Kuramae et al., 2006; Dunn et al., 2008). For the cases where nonetheless different data sets result in mutually exclusive trees, a considerable problem emerges (Jeffroy et al., 2006; Philippe et al., 2011a; Kumar et al., 2012). Each tree represents one evolutionary history, and we face the dilemma of having to choose one to represent the species relationships, unless we are willing to sacrifice the hypothesis of tree-like evolution. Since all trees earn significant support by the data, additional information is needed for explaining the different topologies and making the correct choice. If the collections of genes in the competing data sets are comprehensive, biological explanations such as incomplete lineage sorting or horizontal gene transfer can be ruled out. The data sets should reflect consistently the strongest phylogenetic signal in the genomes of the analyzed taxa. Problems in the compilation of the data sets, such as errors in the ortholog prediction, may however, play a role. And hardest to account for are artifacts during tree reconstructions due to wrong interpretations of the phylogenetic signal. However, they are a common source of incongruence.

28.1.5 Artifacts during phylogeny reconstruction

Felsenstein (1978) was the first to show that unequal amounts of change in adjacent lineages can cause maximum parsimony to result in wrong trees by attracting the long-branch taxa to each other although they are not their next relatives (Felsenstein, 1978; Hendy and Penny, 1989). This phenomenon has been coined long-branch attraction (LBA). At the same time Felsenstein showed that the confidence in the wrong tree increases with longer sequences, i.e. the parsimony is in such instance *statistically inconsistent*. However, maximum parsimony can also be inconsistent for trees that evolve according to a molecular clock (Hendy and Penny, 1989). Subsequently, it turned out that distance-based methods for tree reconstructions, as well as Maximum Likelihood, are similarly misled when an inappropriate model of sequence evolution is used (Lockhart et al., 1996; Philippe and Laurent, 1998; Bruno and Halpern, 1999; Sullivan and Swofford, 2001; Brinkmann et al., 2005; Lockhart and Steel, 2005). However, for the latter approaches the inconsistency disappears once the correct model is specified, while for maximum parsimony (based on its original definition) no such escape route is possible.

In real life, model misspecification is always present, thus the following two problems in tree reconstruction have still not been satisfactorily solved.

First, such trees are hard to reconstruct where long (external) branches are connected by, in relation, only short internal branches. Since a sparse taxon sampling, especially of evolutionarily only distantly related taxa, promotes such trees, taxon sampling has been identified as one possible cause of tree reconstruction artifacts (Hendy and Penny, 1989; Lecointre et al., 1993; Graybeal, 1998; Hillis et al., 2003; Lockhart and Steel, 2005; Philippe et al., 2011a) but see (Rosenberg and Kumar, 2003; Rokas and Carroll, 2005)). Similarly, the presence of individual taxa with a particularly high evolutionary rate promotes such artifacts (e.g. Brinkmann et al., 2005). Second, the choice of the evolutionary model and its fit to the data becomes increasingly important. In particular, tree reconstruction artifacts need to be expected when basic assumptions of the evolutionary model are violated, e.g., homogeneity of the substitution process or compositional homogeneity of the sequences across the tree.

28.1.6 Why are phylogenomic trees sometimes hard to interpret?

Given the various factors that can affect – and compromise – phylogenomic analyses, a careful biological interpretation of the resulting trees is advised. Before jumping to conclusions one should consider potential sources of error:

- Does the data set contain sufficient phylogenetic signal to reproduce the species tree?
- Is there any reason for assuming tree reconstruction artifacts to occur?
- Does the statistical significance reflect biological significance?
- Are there additional data that can challenge or support the findings from the tree?

These four points leave us with a typical chicken-and-egg problem. If we were to know the true evolutionary relationships for a set of species, it would be straightforward investigating and addressing the extent of phylogenetic signal and tree reconstruction artifacts. In turn, if we could be confident that the phylogenetic signal in the data suffices and that no artifact impairs tree reconstruction, we could be sure about the reconstructed relationships. So far mainly general guidelines exist that can be used to build and explore a particular phylogenomic data set, and to judge the relevance of the tree reconstructed from it.

28.2 Compiling phylogenomic data sets

28.2.1 Orthology inference

The general scope of phylogenomic studies is the inference of species relationships from the analysis of many if not all genes encoded in the species' genomes. To make this inference meaningful, the evolutionary histories of the sequences should closely match the corresponding species phylogeny. While sequence homology, i.e. the existence of a last common ancestor sequence, is necessary to achieve this, it is not sufficient. More specifically, it is required that the sequences were separated by the split of the two species from their last common ancestor (orthologs; Fitch, (1970)) rather than being separated by a gene duplication event (paralogs). Correspondingly, internal nodes of gene trees inferred from orthologs reflect in general speciation events (see Ebersberger et al., (2007), and references therein for exceptions) while those inferred from paralogs reflect gene duplication events that are not informative about the species phylogeny.

Orthology inference for a set of homologous sequences is the cornerstone to infer correct species trees from gene trees. Unfortunately orthology is a relative term, which is, in principle, defined only for pairs of species relative to a third taxon. This relativity makes it difficult to identify orthologs in multi-taxon sets.

28.2.1.1 Identification of orthologs in complete gene sets

Various computational approaches exist for identifying orthologous sequences in a collection of species with completely sequenced genomes (see Chen et al., 2007). Three main criteria are typically used: (a) congruence of the gene tree and the species tree (e.g. Zmasek and Eddy, 2002), (b) identical genomic position of the candidates (e.g. Dewey, 2011), and (c) a reciprocal highest sequence similarity (e.g. Chervitz et al., 1998). In phylogenomics, sequence similarity based approaches are most widely applied as they are easy to implement and scalable even to large data sets. The general principle is quickly outlined: For a given protein in species *A* the sequence with the highest similarity score is determined in species *B*, the *hit sequence*. Typically, BLAST (Altschul et al., 1997) is used for this purpose, although in principle any other search tool assessing local similarity between sequences can be used (e.g. Jensen et al., 2008; Ebersberger, Strauss, and von Haeseler, 2009). A reciprocity criterion is then used to conclude from sequence similarity on orthology. For the hit protein the highest scoring sequence in species *A* is determined. If the hit protein from species *A* in the second search is the same as the query sequence from species *A* in the first search, then and only then is orthology between the two sequences from *A* and *B* assumed.

Meanwhile, numerous variations of the so-called reciprocal best hit (RBH) approach have been published. It is now possible to generate orthologous groups containing more than one protein per species (Remm, Storm, and Sonnhammer, 2001). This accounts for the fact that gene duplications occurring after the split of two species result in (in-)paralogs that are, however, co-orthologous to their counterpart in the other species. Two proteins in a species that show a lesser divergence between each other than when either is compared to the ortholog in the other species are typically considered as co-orthologs. Furthermore, sequences from more than two species can be included into an ortholog group (e.g. Tatusov et al., 2000; Li, Stoeckert, and Roos, 2003; Jensen et al., 2008; Roth, Gonnet, and Dessimoz, 2008). Extending the RBH approach is one possibility. One simply computes the RBH for all possible pairs of species and the sequence family of interest. This results in a graph where nodes represent orthologs/in-paralogs in the different species and edges are present when two nodes are identified as orthologs. Thus, if for *n* species (genes) the resulting graph is fully connected then the entire set of sequences represents one orthologous group. However, this strategy is unduly stringent, as it would enforce that every protein has to be present in all analyzed species. Thus, various methods relax the completeness requirement. To automatically resolve the resulting many-to-many orthologous relationships and to remove any non-orthologous proteins, different clustering algorithms are applied (e.g. Li, Stoeckert, and Roos, 2003; Jensen et al., 2008). While these graph-based approaches are highly inclusive – they hardly miss an ortholog if it is present – they suffer from a considerable amount of false positive predictions (Chen et al., 2007). This should be kept in mind when compiling phylogenomic data sets based on these methods.

Recently, an alternative method has been suggested, InParanoid-TC (Ebersberger, Strauss, and von Haeseler, 2009), that extends pairwise orthology predictions using InParanoid to more than two species by applying a transitive closure argument (Figure 28.1). This method makes use of the fact that orthology has the transitivity property, i.e. if A1 and B1, and B1 and C1 are pairwise orthologs it follows that A1 and C1 are orthologs too, if and only if species A and B are more closely related to each other than either is to species C (Ebersberger, Strauss, and von Haeseler, 2009).

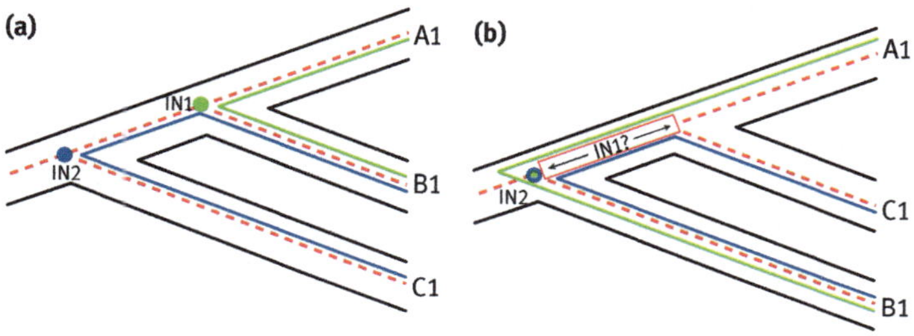

Figure 28.1: Transitivity of orthology inference. The figure depicts the evolutionary relationships of 3 species, A, B, and C (black tree) and the inferred relationships of the corresponding genes A1, B1, and C1 (red dashed line). IN1 and IN2 are the inferred common ancestors in the gene tree connecting A1, B1, and C1. Assume that pairwise orthology prediction has identified two ortholog pairs, A1 and B1 (green line), and B1 and C1 (blue line). **(a)** If species A and B are their closest relative, it follows that the orthologs A1 and B1 share a common ancestor (IN1) to the exclusion of C1. Thus, from the observation that C1 is an ortholog of B1 it follows that it is also an ortholog of A1. **(b)** If, however, species C is more closely related to A this inference is not possible, since no information exists about IN1.

28.2.1.2 Identification of orthologs in incomplete gene sets

We have described in the previous section that it is methodologically straightforward to compile orthologous groups from species whose entire genome has been sequenced. Similarly straightforward is their integration into phylogenomic data sets. Despite the recent increase in taxa with completely sequenced genomes (see www.diark.com for an approximate inventory), still additional sequence data are necessary to get a comprehensive taxon sampling (cf. Section 28.1.5; Artifacts during phylogeny reconstruction). Transcriptome data (**E**xpressed **S**equence **T**ags), i.e. sequences representing the mRNA, are the main source for this additional information (Hughes et al., 2006; Philippe and Telford, 2006; Roeding et al., 2007; Dunn et al., 2008). In the most inclusive studies more than half of the taxa are represented by ESTs (e.g. Ebersberger et al., 2012). However, the use of ESTs for phylogeny reconstruction entails two problems: First, it is not guaranteed that an RBH pair resulting from a search in an incomplete gene set is also the best hit pair in the entire genome, because only a subset was searched. Second, the coding sequence of a gene can be tagged by two or more non-overlapping ESTs. Programs implicitly designed for analyzing full-length sequences allow for more than one sequence per species in an ortholog cluster, only when the intra-specific sequence similarity is higher than the sequence similarity between species (Remm, Storm, and Sonnhammer, 2001; Li, Stoeckert, and Roos, 2003; Roth, Gonnet, and Dessimoz, 2008). However, the intra-specific similarity between non-overlapping ESTs derived from the same gene is approximately that of two randomly chosen and unrelated sequences. Thus, the programs will discard potential phylogenetic information.

Currently, few standardized approaches exist to overcome these issues. Hejnol et al. (2009) follow a graph-based approach as a basis for orthology inference. Nodes in these graphs represent proteins and the edges represent pairwise sequence similarities inferred by BLAST (Altschul et al., 1997). Clusters of homologous sequences are then automatically identified with a Markov clustering algorithm (Dongen, 2000). The sequences from each cluster are subsequently aligned and a phylogenetic tree is computed. This tree is then subjected to a series of pruning steps to obtain subtrees in which each taxon is represented only by a single sequence. The sequences at the leaves of these subtrees are then considered orthologous (Hejnol et al., 2009). Although this approach allows the identification of orthologs in incomplete gene sets, it cannot take into account that more than one sequence can represent the same gene. Ebersberger et al. (2009) presented an alternative strategy and re-introduce the reciprocal hit criterion into the ortholog search in incomplete data (Ebersberger, Strauss, and von Haeseler, 2009). To accomplish this, the ortholog search problem was slightly reformulated. Rather than predicting an ortholog for each and every sequence in an EST data set, they extend pre-defined groups of orthologs, so-called core-orthologs with sequences from the EST set. Figure 28.2 outlines the principle of this approach.

In brief, orthologous sequences from a pre-defined set of primer taxa are used for building and training a profile-hidden Markov model (pHMM; Eddy, (1998)). Moreover, a set of ESTs from a search taxon, assembled e.g. using MIRA (Chevreux et al., 2004), is translated in all six reading frames. HaMStR then compiles a set of ortholog candidates. To this end, hmmsearch from the hmmer3 package (www.janelia.org) identifies sequences in the crude EST translations showing a significant similarity to the pHMM. In the subsequent evaluation phase each candidate is taken as query for a BLAST search against the protein set of one of the primer taxa, the so-called reference taxon. In the case that the highest scoring BLAST hit is the same sequence that is represented in the core-ortholog group, the reciprocity criterion is fulfilled and the candidate is accepted as an ortholog. Otherwise it is discarded. It is now clear that HaMStR can accept more than one sequence per taxon to extend the core-ortholog set. To fit the particular needs of EST data, HaMStR has a further step built into the candidate evaluation. For a set of sequences fulfilling the reciprocity criterion HaMStR globally aligns each sequence against the reference protein and determines the alignment score. A greedy algorithm is then used to identify the highest scoring

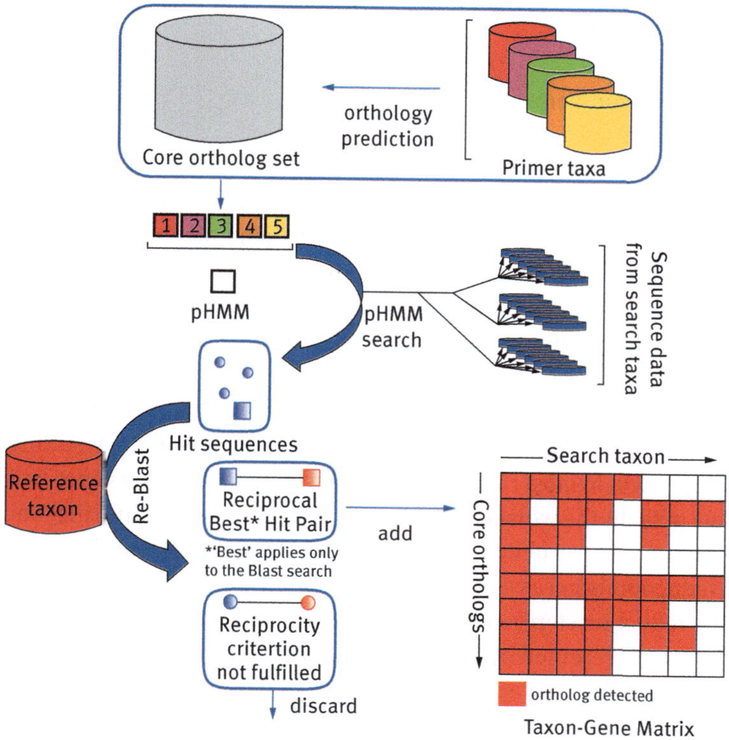

Figure 28.2: Workflow of the HaMStR ortholog search.

set of EST sequences that align to non-overlapping parts of the reference protein. The corresponding sequences can be concatenated and subsequently analyzed together. The remainder of the candidates is discarded. The use of HaMStR is easy and there is only one sensitive step in the analysis. From the available primer taxa the one should be chosen as reference taxon for the re-BLAST that is most closely related to the search taxon. Only then it is likely that the accepted sequence is orthologous to all of the sequences in the core-ortholog group (cf. Figure 28.1). If the evolutionary relationships of the search species to the reference taxa are unknown, or when a more stringent orthology prediction is desired, more than one or even all primer taxa can be specified as reference taxa. A candidate is only then accepted as an ortholog when the reciprocity criterion is fulfilled for all reference taxa.

HaMStR can be used with both ESTs and full-length protein sequences and achieves independent of the type of input data a very low false positive rate of 3%. However, the accuracies of these predictions especially for the EST data do not come for free. We could show that particularly such ESTs escape the ortholog detection that cover less than 20% of a protein's coding sequence (Ebersberger, Strauss, and von Haeseler, 2009).

28.3 The taxon-gene matrix

28.3.1 Generation of the taxon-gene matrix

The various methods for orthology inference provide the input to generate a taxon-gene matrix (TGM) as the basis for a phylogenomic analysis. Basically, one can distinguish between an undirected and a directed approach to generate the TGM. In the *undirected approach* all sequence data from different species subsets are concatenated in one taxon-gene matrix. In other words, once a set of taxa with the corresponding sequences has been defined, every gene finds its way into the matrix for which an ortholog was identified in at least one other taxon. This procedure matches most closely the pretension of phylogenomics to build trees from random data collections (Delsuc, Brinkmann, and Philippe, 2005; Telford, 2007). However, the resulting matrices – though huge – will in most cases only be sparsely filled, as many genes will be represented in few taxa. In the worst case, this TGM can be separated into disjoint submatrices and then the reconstruction of one phylogeny is not possible. Thus, an a posteriori matrix reduction is necessary to obtain a data set that is suitable for tree reconstruction. In the reduction step genes are removed from the TGM that only occur in few taxa and also taxa represented by a few genes only are deleted.

An alternative to the undirected approach is *the directed* approach or core-gene approach (eg. Susko et al., 2006). Here, the taxon-gene matrix, and typically also the preceding ortholog search, is a priori restricted to a subset of pre-selected genes. This allows adapting the set of phylogenetic markers to the particular phylogenetic problem under study (Townsend and Leuenberger, 2011; Ebersberger et al., 2012). A ubiquitous representation of genes in the taxa of interest is the most common selection criterion of core-genes. One example for a core-gene set, used for reconstructing animal phylogenies, are the ribosomal proteins (e.g. Witek et al., 2009; Nesnidal et al., 2010). Their essential role in eukaryotic mRNA translation suggests that the same or at least largely overlapping sets of proteins are present in all animals. As a consequence, a densely filled taxon-gene matrix is expected, at least when analyzing taxa with completely sequenced and annotated genomes. The advantage of the ribosomal proteins for constructing phylogenomic data sets is contrasted by at least two disadvantages. First, the proteins are functionally tightly correlated, which is at odds with the presumed randomness of gene selection in phylogenomics analyses. Second, the number of genes involved is with only 78 (Nesnidal et al., 2010) considerably low, and thus taxa for which only incomplete data – in particular small EST collections – are available have a high chance not to occur in the TGM. More comprehensive core-sets that are compiled based solely on the phylogenetic distribution of genes, and irrespective of their function, should overcome such drawbacks. The core-orthologs introduced in the context of HaMStR (cf. previous section) represent one such alternative. Meanwhile a number of phylogenomic studies with a focus on the integration of EST data include between a couple of hundred to a few thousand genes in the TGM (e.g. Roeding et al., 2007; Simon et al., 2009; Meusemann et al., 2010; Ebersberger et al., 2012). However, the size of these inclusive core-gene sets typically requires an a posteriori reduction of the taxon-gene matrix, similar to the undirected approach.

28.3.2 Matrix reduction: Final selection of taxa and genes

In many cases it is impractical to directly use the TGM that emerged from the ortholog searches for the phylogeny reconstructions. As already pointed out the TGM is typically sparsely filled. The gaps in the matrix result either from taxa for which only incomplete gene sets are available, e.g. those taxa that are represented by ESTs, or from the restricted phylogenetic distribution of individual genes (a biological phenomenon one cannot change). It has been shown that missing data interferes with an accurate phylogeny reconstruction (Sanderson, McMahon, and Steel, 2010). Thus, different strategies have been proposed to reduce the extent of missing data in taxon-gene matrices by removing taxa and genes that are only weakly represented. Mining the data for maximal bicliques, i.e. for completely connected bipartite sub-graphs, is one approach to the problem of matrix reduction. To this end, the taxon-gene matrix is represented as a graph consisting of two sets of nodes. One set represents the genes and the other set represents the taxa. An edge is drawn between a 'gene-node' and a 'taxon-node' when an ortholog to the gene has been identified in the corresponding taxon. The goal is to identify maximal subgraphs (bicliques) such that each gene-node is connected to each taxon-node and vice versa. Unfortunately, the problem of maximal biclique identification is NP complete (Sanderson et al., 2003), but efficient solutions exist (e.g. Sanderson et al., 2003; Liu, Sim, and Li, 2006). In phylogenomics practice the biclique approach is, however, not the method of choice. The criterion of full connectivity is for most applications too stringent and too few taxa and genes would remain for analysis. Alternative methods range from simple iterative steps of taxon and gene removal and re-addition until a certain threshold of connectivity is reached (e.g. Roeding et al., 2007; Simon et al., 2009; Ebersberger et al., 2012) to more elaborate procedures (e.g. Roure, Rodriguez-Ezpeleta, and Philippe, 2007), some of which try to integrate also the phylogenetic information content of individual genes into the selection procedure (Meusemann et al., 2010).

Irrespective of the exact procedure of matrix reduction, the main criterion for an inclusion of a gene into the final taxon-gene matrix remains the fraction of taxa it is represented in. Using a fungal phylogenomics data set where about half of the taxa were represented by EST data, we have investigated to what extent this criterion affects the proposed randomness of gene selection for phylogenomic studies (cf. Ebersberger et al., 2012). As can be seen in Figure 28.3 this introduces a surprisingly strong bias. Gene sets that are selected according to their representation in EST taxa are highly enriched in slowly evolving genes. Moreover, the genes are also functionally highly correlated with a preponderance of genes involved in the primary cellular metabolism

(Ebersberger et al., 2012). The likely origin of these biases lies in the nature of the EST data. Particularly genes with a high and evolutionary conserved expression level will be represented in EST collections from highly diverged taxa. In fact it was shown that the evolutionary rate of a gene is negatively correlated with its expression rate (Drummond et al., 2005). This explains the selection bias towards slowly evolving genes. Furthermore, that genes involved in the primary cellular metabolism are predominantly selected has a similar reason. Such genes are likely to act as housekeeping genes and will be sampled independent of the tissue the ESTs have been derived from (Ebersberger et al., 2012).

These results from the previous analysis show that the assumption of phylogenomic data sets as collections of randomly selected genes cannot be taken as granted. Neither of these biases reduces per se the value of phylogenomic studies. However, the effects need to be taken into account when interpreting the outcomes of their analysis. Especially the confinement to slowly evolving genes may become relevant, e.g. when attempting to resolve phylogenies with short internal branches (Ebersberger et al., 2012).

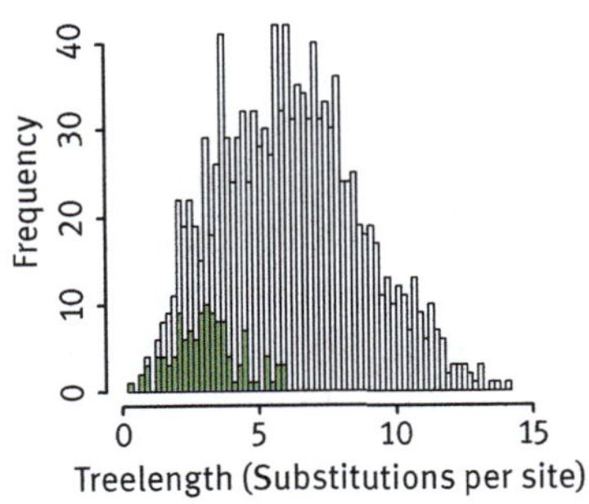

Figure 28.3: EST taxon guided selection of genes favors slowly evolving genes. We have computed for each gene in a core-ortholog set a maximum-likelihood tree and used the total tree length as a proxy of the gene's evolutionary rate (histogram in grey). The green histogram shows the distribution for those genes that were selected during reduction of the TGM using their representation in EST taxa as a criterion.

28.4 Phylogeny reconstruction

28.4.1 Phylogenomics reconstruction from many genes

Once the available data for a taxon set of interest have been collected the reconstruction of the phylogenetic relationship among the taxa follows an established workflow:

First, the sequences are collected into a multiple sequence alignment. After applying the corresponding procedures, one ends up with either one multiple sequence alignment, where all ortholog regions from the different taxa are concatenated, or one starts phylogenetic inference for each multiple sequence alignment from an ortholog region separately. The first approach is also known as the supermatrix method whereas the later infers the trees for each alignment independently and then combines the resulting trees to produce one species trees (supertree). Both methods have their pros and cons (Haeseler, 2012) but we will not enter into this discussion here.

No matter if we start with a supermatrix or with a collection of individual multiple sequence alignments, the next step in the phylogenetic inference pipeline is the reconstruction of the phylogenetic relationships.

28.4.2 Selecting appropriate evolutionary models

All methods to infer the phylogenetic relationships from a multiple sequence alignment make implicit or explicit assumptions about the nature of nucleotide or amino acid changes. Some assumptions are necessary to make the models of change mathematically tractable. For example, it is widely assumed that the sites in a sequence do evolve independently of each other (see Section 28.1.2), an assumption that is clearly violated by most biological sequences. However, tree-reconstruction becomes much easier, because we can treat each column in the alignment as one independent sample from the underlying evolutionary process. As a consequence, this assumption persists in phylogeny inference. Other simplifications, for example reflected in the first evolutionary models like the Jukes–Cantor model, can lead to a systematic bias in the reconstructed trees (cf. Section 28.1.5; Artifacts during phylogeny reconstruction). As a consequence, we observe in recent years a dramatic increase in the number of models of change and also an increase in their complexity. This development aims at a more realistic description of the evolutionary change. For DNA sequences the general time reversible model (GTR) is frequently used (Yang, 1994) together with the assumption of different rates across sites (Yang, 1993; Gu, Fu, and Li, 1995). But many more complexities can be included, like heterogeneity and change in evolutionary models in a tree (Tuffley and Steel, 1998; Foster, 2004). We can also allow invariant sites, i.e. positions in the sequence that are not allowed to change, to maintain the function of the molecule. Thus, a full zoo of models is already available for DNA sequences, and for amino acids the models are even more parameter rich.

The toolbox of realistic models of sequence evolution is well filled, although not perfect. However, the naïve assumption that the model with the most parameters (most realistic) will always produce the best tree and will avoid the inconsistency trap is counterbalanced by the fact that the parameters need to be estimated from the data. The estimation of many parameters typically increases the variance of the estimate and thus it is more likely to estimate a value that is not close to the true value. In other words, the resulting tree may be wrong due to the variability in our estimates.

In phylogenetic inferences we have to deal with the tradeoff between a realistic model (parameter rich) with less precise parameter estimates, and less realistic models (parameter poor) with better parameter estimates. Fortunately, statistics can help here. The simple models are nested within the most complex models. Imagine the general reversible model with six rate parameters for each of the six possible changes between pairs of nucleotides. Then the Jukes–Cantor model, that assumes the same rate for each pair, is a special case of the GTR model; it is nested in the GTR case. For each model we can also compute the likelihood, i.e. the probability to observe the data under the specified model. The likelihood of the parameter rich models is always larger than or equal to the likelihood of the parameter poor models. But sometimes this difference is not significant and in such situations one prefers the parameter poor model. From an epistemological point of view, such a decision is reasonable, it selects the most parsimonious model to explain the data.

Consequently, tools have been developed that allow researchers to select the best model from a collection of models (Posada and Crandall, 1998; Abascal, Zardoya, and Posada, 2005; Posada, 2008). One caveat remains, the selection of a model does not necessarily imply that this model adequately describes the observed multiple sequence alignment. This is the case when the most complex model is selected. Then it is not clear whether another yet more complex model will fit better. Note, we only compare between models. A pessimistic viewer would say that the selected model is the least bad model.

Currently only very few tools are available to determine if the model really describes the data. In the early 1990s Navidi, Churchill, and von Haeseler, (1991) and Goldman (1993a; 1993b) introduced the well-known likelihood ratio statistics into phylogenetic inference. Here the likelihood of a phylogenetic tree is compared to the so-called unconstrained likelihood of the alignment. If the "distance" between the two likelihoods exceeds a critical value, then one concludes that the tree does not describe the multiple sequence alignment sufficiently. While such methods have their merits, researchers want to understand where in the data the deviation occurs. To overcome these limitations only recently a new way of looking at the evolution of DNA sequence was suggested (Klaere, Gesell, and von Haeseler, 2008; Nguyen, Klaere, and von Haeseler, 2011; Nguyen, Klaere, and Haeseler, 2012) that addresses this problem. We will briefly outline the idea.

28.4.2.1 The MISFITS approach

Imagine a multiple DNA-sequence alignment for n taxa then one can theoretically generate 4^n different patterns that represent the columns of the alignment. For simplicity, we ignore positions that contain gaps. How often each of the exponentially many possible patterns actually occurs in a multiple sequence alignment is summarized in the so-called frequency pattern vector. This pattern vector is also used to infer the maximum-likelihood tree. Once the maximum-likelihood tree together with the best model of sequence evolution is found, it is possible to predict the expected number for each pattern. Thus, it is now possible to compare the expected pattern likelihood vector and to compare it to the observed pattern frequency vector. To accommodate the different scales it is easier to take the logarithms. Figure 28.4 displays a typical result from such an analysis. For a perfect fit the white circles should be close to the diagonal. Circles above the diagonal represent patterns that occur too often in the data, whereas circles below the diagonal are indicative for underrepresented patterns in the alignment. To account for stochastic effects one typically constructs a confidence band around the diagonal line. The dotted lines represent the 95% Gold confidence region for multivariate proportions (Nguyen, Klaere, and von Haeseler, 2011). D^+ and D^- denote the sets of site patterns that are significantly over- and underrepresented, respectively. Hence, the patterns in D^+ and D^- are not adequately represented by the model. However, because we have already selected the best model, there is little hope to find a model that explains the observed alignment better.

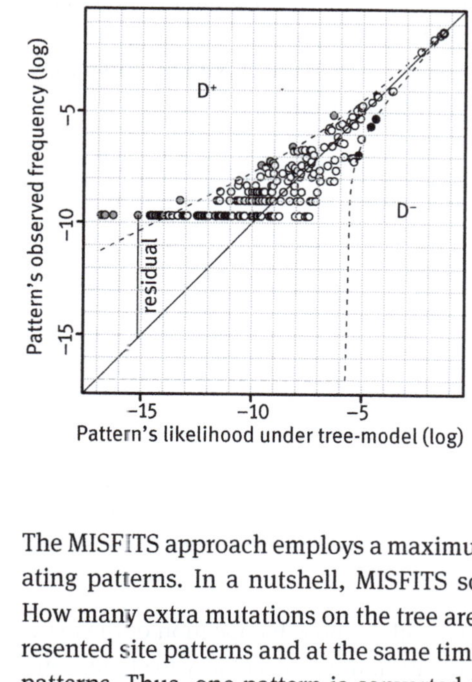

Figure 28.4: Observed versus expected pattern frequencies.

The MISFITS approach employs a maximum parsimony approach to explain the deviating patterns. In a nutshell, MISFITS solves the following optimization problem: How many extra mutations on the tree are required to reduce the number of overrepresented site patterns and at the same time increase the number of underrepresented patterns. Thus, one pattern is converted into another pattern employing a minimal number of changes.

To understand the algorithm it is necessary to recapitulate the Kimura three-parameter model (Kimura, 1981) that distinguishes transitions σ_1 ($A \leftrightarrow G$, $C \leftrightarrow T$) and two types of transversions σ_2 ($A \leftrightarrow C$, $G \leftrightarrow T$), σ_3 ($A \leftrightarrow T$, $G \leftrightarrow C$). With this notation the meaning of extra mutations on a tree becomes clear:

Figure 28.5 explains this for the simple example of a rooted two taxa tree.

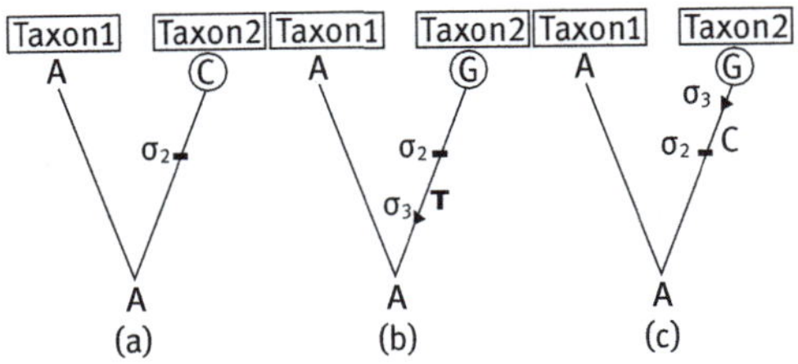

Figure 28.5: Placing an extra substitution on a tree.

Figure 28.5a displays the known evolutionary history of pattern AC. The ancestral character state was an A and on the branch leading to taxon 2 a σ_2 substitution occurred. The extra substitution (σ_3) put on the branch leading to taxon 2 occurs either before s_2 (Figure 28.5b) or after σ_2 (Figure 28.5c). Independently of the precise location of the extra mutation, the resulting pattern is always the same (AG). This can be generalized to any evolutionary history. The bottom line is always the same: The exact series of substitution events along the branch of a tree is not relevant if one wants to put one extra substitution on the branch. More importantly it was shown that this works for all branches in the tree (Klaere, Gesell, and von Haeseler, 2008).

To convert one pattern into another for a given tree then the maximum parsimony method is applicable (Fitch, 1971). Figure 28.6 gives an example.

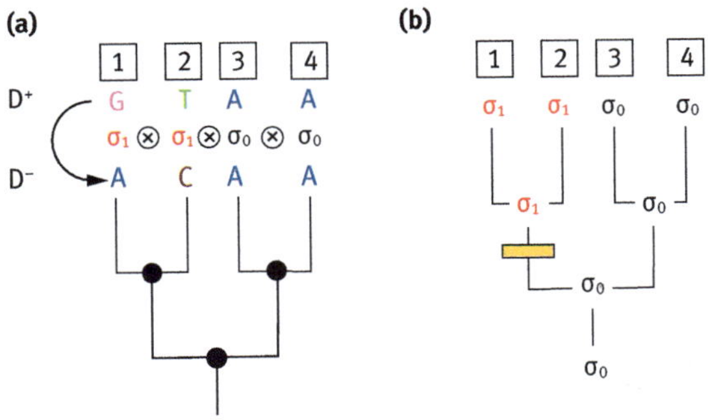

Figure 28.6: Converting the pattern GTAA into ACAA with minimal number of extra substitutions.

To compute the number of extra substitutions and the possible location of the extra substitutions, compute for each taxon the substitution type necessary to change one nucleotide into the other. Thus, for taxon 1 substitution type σ_1 will change the G into an A, σ_1 will also change the T into a C in taxon 2, for the remaining taxa no substitution is necessary, we call this type σ_0. Now the leaf labels of the tree are replaced with the substitution type and σ_0 is also placed at the root (Figure 28.6b). For this leaf labeled 'tree' the Fitch parsimony algorithm works, however instead of using the nucleotides as the underlying alphabet the substitution types σ_0, σ_1, σ_2, and σ_3 constitute the alphabet. Thus, the minimal number m_{min} of extra substitutions is quickly computed. The example in Figure 28.6 needs one extra substitution on the branch leading to taxon 1 and 2 (yellow bar).

To assess the significance of this minimal number a parametric bootstrap test is applied, that provides the empirical distribution of the minimal number of extra mutations under the assumed model of sequence evolution and the phylogenetic tree. Nguyen et al. (2011) showed that the test performs similar to the Goldman test, with the advantage of pointing to the sites in the alignment where the model is not adequate.

In a next, however, yet unexplored step, one needs to detect regions in the alignment where deviations from the model cluster. Such regions would then split from the rest of the alignment and would be subject to another phylogenetic analysis. For example, trying to identify a more appropriate model.

While MISFIT provides a goodness of fit test, the theory of putting extra substitutions on a tree can also be used to systematically introduce deviations from an evolutionary model into an alignment (Nguyen, Klaere, and Haeseler, 2012). By this one can disturb an ideal system in the smallest possible way, an approach that was very successful in physics. As it turns out, for already very mild deviations from the true evolutionary model, tree reconstruction methods can produce very different results.

In summary, this indicates that selecting an appropriate evolutionary model for tree reconstruction is a key step in phylogenomic analyses. However, it also needs to be taken into account that even the best fitting of the available models is not the true model. As a consequence, the reconstructed tree may differ from the true tree due to model misspecifications.

28.4.3 Tree reconstruction and inference of species trees from gene trees

The ultimate goal of phylogenomics studies is the reconstruction of reliable sequence trees from the compiled data set(s), and subsequently the deduction of the species tree from the sequence trees. Ortholog searches, the selection of taxa and genes for tree reconstruction, and the choice of the best-fitting evolutionary model form the basis of the analysis. Once these steps are accomplished, one can choose from a plethora of criteria, algorithms and methods to reconstruct sequence trees from the data. We will not even attempt to review the wealth of existing approaches. Instead, we will focus on one particular aspect in phylogenomics that has been recently referred to as the 'tyranny of total evidence' (Narechania et al., 2012). The problem originates from how the different phylogenetic signals conveyed by the individual genes in a data set are integrated in the tree reconstruction. It is common practice to use supermatrix or total evidence strategies for this purpose. Here, the orthologous sequences for one gene are aligned first, and the resulting alignments are then concatenated to form a superalignment or supermatrix. The supermatrix serves then as input for tree reconstruction and a single tree is inferred from the data. The advantage is that weak but consistent phylogenetic signals across the genes are reinforced by the concatenation. In many cases this results in more accurate trees than, for example, supertree methods where trees are reconstructed from individual alignments and the resulting trees are combined into one (super)tree at a later step (e.g. Kupczok, Schmidt, and von Haeseler, 2010). However, the restriction to a single tree also has a problematic side effect. It implies that all genes or at least their majority share the same phylogeny. In some instances this may indeed be the case. In many others it is an unwanted and undue oversimplification of the true evolutionary relationships within the data. Recent analyses in humans and yeast have revealed that the genealogies of individual genes can substantially differ from that of the species they reside in (Huerta-Cepas et al., 2007; Marcet-Houben and Gabaldon, 2009). And the issue is substantially more severe for the prokaryotes where horizontal gene transfer is believed to be frequent. For a review see Degnan and Rosenberg (2009). Supertrees would be the obvious solution to capture the variability in the phylogenetic signal. However, this poses the problem that individual alignments frequently do not contain sufficient phylogenetic signal for an accurate tree reconstruction.

This makes it hard to distinguish true differences between gene tree topologies from artificial ones. To overcome this limitation it is necessary to combine the advantages of supertree approaches with those of the supermatrices. We will briefly outline three methods. Leigh et al. (2008) developed a stepwise procedure of supermatrix construction, *concaterpillar*, where at each iteration only such alignments are concatenated that do not significantly differ in their phylogenetic signal. The procedure is repeated until no more concatenations are possible. Thus, rather than ending up with a single supermatrix, *concaterpillar* can produce several partial supermatrices each reflecting a different signal in the data. Reconstructing trees from each of the partial supermatrices gives then a good overview of the various evolutionary histories represented by the genes. Obviously, this approach makes, per se, no statement about which of the supported sequence trees – if any – represents the species tree. More recently, Narechania et al. (2012) presented a tool, *radical*, that also concatenates individual alignments iteratively. In contrast to *concaterpillar* the individual alignments are added in a random order to build up only a single supermatrix. *Radical* then tries to assess the variation of the phylogenetic signal among different data subsets by computing a tree each after n concatenation steps, where n is a user-defined integer value. Moreover, the procedure of stepwise addition and tree reconstruction can be repeated multiple times using different starting alignments. This generates a large collection of sequence trees based on differently sized random partitions of the data. Analyzing and integrating the information provided by the individual trees can then explore the varying evolutionary signals in the data. For example, the support values for a given split can be assessed from the number of genes required to recover this split and the fraction of sequence trees where it is present. The authors note that even splits that achieve high bootstrap support with the full data may actually turn out as only weakly supported by a *radical* analysis (Narechania et al., 2012). Unfortunately, it is not always clear how to interpret such a finding. On the one hand it can indicate that different subsets of the data have different evolutionary histories and it just happens that one dominates, maybe even by chance, in the total evidence analysis. On the other hand one may face the situation where a certain phylogenetic signal is so weak in the data that a critical mass of genes is required to lift this signal above the noise level (cf. Gatesy, Baker, 2005). Obviously, the conclusions drawn from such an analysis essentially depend on which of the alternative scenarios apply. It is often that this issue is only solved over time. If independent analyses on different – and in the better case non-overlapping – sequence sets consistently reconstruct the same evolutionary tree it is safe to assume that this tree depicts the dominant phylogenetic signal for the chosen taxon set. An inference of the species tree from the sequence tree is, therefore, admissible. If, however, even partly independent data sets of comparable size significantly disagree in the supported phylogeny, any decisive species tree inference should await clarification of the issue. The past discussion concerning the validity of the Ecdysozoa hypothesis (reviewed in Telford et al., 2008) and the ongoing discussions about the early evolutionary histories of the metazoa (reviewed in Philippe et al., 2011a) illustrate the latter case.

28.4.4 The criterion of tree consistency

Given the wealth of sequence data for many taxa, consistency support of an evolutionary hypothesis – as we have it described above – is nowadays feasible already within a single study. Ebersberger et al. (2012) have chosen the kingdom of fungi as a demonstrative example. The analysis was based on fungal sequences from 99 genomes and 109 EST projects complemented with data from relevant outgroups. Ortholog searches were performed with HaMStR (Ebersberger, Strauss, and von Haeseler, 2009) using three different core-ortholog sets: One, consisting of evolutionarily old genes that were present already in the common ancestor of all fungi, was used for analyzing the backbone of the fungal tree, and two further sets were dedicated to more refined analyses of the Basidiomycota and Pezizomycotina subtrees, respectively. Genes and taxa for the tree reconstructions were selected according to different combinations of three criteria: (i) representation of a gene in EST taxa, (ii) identification of a gene as single copy gene, and (iii) evolutionary rate of a gene. This produced 13 in part completely independent data sets between 45 and 162 genes in size, each shedding light on fungal relationships. Trees were then built with up to four different tree reconstruction methods per data set. Supermatrices served as input for maximum-likelihood and maximum parsimony tree reconstruction, as well as for Bayesian tree search. For the supertree reconstructions they computed a maximum-likelihood tree for each individual alignment, summarized the trees using matrix representation with parsimony and then computed a supertree from this matrix. Overall, the diversity of the phylogenomic data sets and of the tree reconstruction methods appear a good reflection of what typically accumulates over time in independent studies with the same focus. However, having all data at hand, one can compare the resulting phylogenies and choose to trust only such splits that could be reproduced with independent data and independent methods. The resulting phylogeny is shown in Figure 28.7. This shows a single species tree and reticulations are only used to highlight regions where the data (or the methods) so far do not suffice to decisively choose between alternative evolutionary hypotheses. The question remains how this relates to the criticism of Narechania et al. (2012) concerning the 'tyranny of total evidence'. We believe that the answer is hidden in the detail, i.e. in the precise question asked. Our approach is based on two working hypotheses: First, we believe that the evolution of the eukaryotes is by and large tree-like (see the start of this chapter). Second, we assume that the dominating phylogenetic signal in the data is the best approximation of the evolutionary relationships of the analyzed taxa that we can get. Against this background, the problem of species tree inference boils down to reliably identifying the strongest phylogenetic signal in the data, and not just picking up one strong signal by chance or even being misled by artifacts. With the data at hand, it is likely that this task was accomplished.

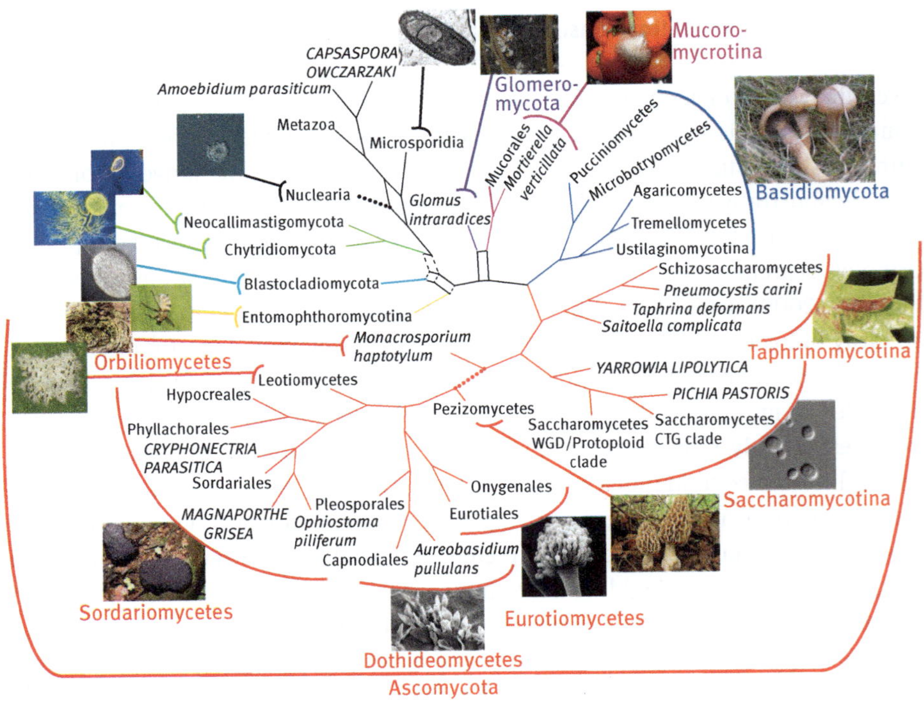

Figure 28.7: A consistency based phylogeny of the fungi. Modified from (Ebersberger et al. 2012).

Once a species phylogeny is available, it forms the necessary basis for further investigating in detail the evolutionary histories of the genes within the species. Any significant deviations of gene trees from the species tree, as they are e.g. caused by incomplete lineage sorting of ancestral alleles or horizontal gene transfer, completes then the picture of evolutionary events resulting in the formation of contemporary species from their common ancestors. If, however, the basal assumption of tree-like evolution is in doubt, networks rather than trees should be used to depict the relationships between the taxa under study (e.g. Kunin et al., 2005).

Acknowledgments

This work was supported by the German Research Foundation (DFG) within the Priority Project “Deep Metazoan Phylogeny” (SPP 1174) grant HA1628/9.

References

Abascal F, Zardoya R, Posada D. ProtTest: Selection of best-fit models of protein evolution. Bioinformatics 2005;21:2104–5.

Abdo DA, Fromont J, McDonald JI. Strategies, patterns and environmental cues for reproduction in two temperate haliclonid sponges. Aquatic Biol 2008;1:291–302.

Adam Z, Turmel M, Lemieux C, Sankoff D. Common intervals and symmetric difference in a model-free phylogenomics, with an application to streptophyte evolution. J Comp Biol 2007;14:436–45.

Adami C. Information theory in molecular biology. Phys Life Rev 2004;1:3–22.

Adoutte A, Balavoine G, Lartillot N, Lespinet O, Prud'homme B, de Rosa R. The new animal phylogeny: Reliability and implications. Proc Natl Acad Sci USA 2000;97:4453–6.

Adrianov AV, Malakhov VV. Cephalorhyncha of the World Ocean. Moscow, Russia: KMK Scientific Press, 1999.

Adrianov AV, Malakhov VV. Cephalorhynchia – A New Phylum of the Animal Kingdom. Moscow, Russia: KMK Scientific Press, 1995.

Adrianov AV, Malakhov VV. Priapulida: Structure, Development, Phylogeny, and Classification. Moscow, Russia: KMK Scientific Press, 1996.

Aguileta G, Marthey S, Chiapello H, et al. Assessing the performance of single-copy genes for recovering robust phylogenies. Syst Biol 2008;57:613–27.

Aguinaldo AMA, Turbeville JM, Linford LS, et al. Evidence for a clade of nematodes, arthropods and other moulting animals. Nature 1997;387:489–93.

Ahlrichs W. Ultrastruktur und Phylogenie von *Seison nebaliae* (Grube 1859) und *Seison annulatus* (Claus 1876). Hypothesen zu phylogenetischen Verwandtschaftsverhältnissen innerhalb der Bilateria. Göttingen, Germany: Cuvillier Verlag, 1995.

Ahlrichs WH. Epidermal ultrastructure of *Seison nebaliae* and *Seison annulatus*, and a comparison of epidermal structures within the Gnathifera. Zoomorphology 1997;117:41–8.

Ahlrichs WH. Spermatogenesis and ultrastructure of the spermatozoa of *Seison nebaliae* (Syndermata). Zoomorphology 1998;118:255–61.

Ahnelt P. Chaetognatha. Chapter 40. X. Minor Coelomatic Phyla. In: Bereiter-Hahn J, Matolsky AG, Richards KS, editors. Biology of the Integument 1. Invertebrates. Berlin, Germany: Springer, 1984:746–55.

Ahnelt P. Das Coelom der Chaetognathen. Dissertation, University of Vienna, Austria, 1980.

Akesson B. The embryology of the polychaete *Eunice kobiensis*. Acta Zool (Stockholm) 1967;48:141–92.

Akesson B. The embryology of *Tomopteris helgolandica*. Acta Zool (Stockholm) 1962;43:135–99.

Alam MS, Kurabayashi A, Hayashi Y, et al. Complete mitochondrial genomes and novel gene rearrangements in two dicroglossid frogs, *Hoplobatrachus tigerinus* and *Euphlyctis hexadactylus*, from Bangladesh. Genes Genet Syst 2010;85:219–32.

Alberti G, Heethoff M, Norton RA, Schmelze S, Seniczak A, Seniczak S. Fine structure of the gnathosoma of *Archaeozetes longisetosus* Aoki (Acari: Oribatida, Trhypochthoniidae). J Morphol 2011;272:1025–79.

Alberti G, Peretti AV. Fine structure of male genital system and sperm in Solifugae does not support a sistergroup relationship with Pseudoscorpiones (Arachnida). J Arachnol 2002;30:268–74.

Aleshin VV, Mikhailov KV, Konstantinova AV, et al. On the phylogenetic position of insects in the pancrustacean clade. Molecular Biology 2009;43:804–18.

Aleshin VV, Petrov NB. Molecular evidence of regression in evolution of metazoa. Zhurnal Obshchei Biologii 2002;63:195–208.

Aleshin VV, Vladychenskaya NS, Kedrova OS, Milyutina IA, Petrov NB. Phylogeny of invertebrates deduced from 18S rRNA comparisons. Mol Biol 1995;29:843–55.

Allman GJ. Monograph of the fresh-water Polyzoa, including all the known species, both British and foreign. London, UK: The Ray Society, 1856.

Altenhoff AM, Schneider A, Gonnet GH, Dessimoz C. OMA 2011: Orthology inference among 1000 complete genomes. Nucleic Acids Res 2010:1–6.

Altschul SF, Gish W, Miller W, Myers EW, Lipman DJ. Basic local alignment search tool. J Mol Biol 1990;215:403–10.

Altschul SF, Madden TL, Schaffer AA, et al. Gapped BLAST and PSI-BLAST: A new generation of protein database search programs. Nucleic Acids Res 1997;25:3389–402.

Amemiya CT, Alföldi J, Lee, AP, et al. The African coelacanth genome provides insights in tetrapod evolution. Nature 2013;496:311–6.

Amin OM. Key to the families and subfamilies of Acanthocephala, with the erection of a new class (Polyacanthocephala) and a new order (Polyacanthorhynchida). J Parasitol 1987;73:1216–9.

Anderson DT. The embryology of the polychaete *Scoloplos armiger*. Q J Microsc Sci 1959;100:89–166.

Anderson DT. The comparative embryology of the Polychaeta. Acta Zool (Stockholm) 1966;47:1–42.

Anderson DT. Embryos, fate maps, and the phylogeny of arthropods. In: Gupta AP., ed. Arthropod Phylogeny. New York, NY, USA: Van Nostrand, 1979:59–106.

Anderson DT. Embryology and Phylogeny in Annelids and Arthropods. Oxford, UK: Pergamon Press, 1973.

Anderson DT. Embryology. In: Abele LG., ed. The Biology of Crustacea, Vol 2, Embryology, Morphology and Genetics. New York, NY, USA: Academic Press, 1982:1–41.

Anderson E. Oocyte differentiation in the sea urchin, *Arbacia punctulata*, with particular reference to the origin of cortical granules and their participation in the cortical reaction. J Cell Biol 1968;37:514–39.

Anderson FE, Córdoba AJ, Thollesson M. Bilaterian phylogeny based on analyses of a region of the sodium-potassium ATPase α-subunit gene. J Mol Evol 2004;58:252–68.

Anderson FE, Swofford DL. Should we be worried about long-branch attraction in real data sets? Investigations using metazoan 18S rDNA. Mol Phylogenet Evol 2004;33:440–51.

Andreasen K., Baldwin B G. Unequal evolutionary rates between annual and perennial lineages of checker mallows (Sidalcea, Malvaceae): Evidence from 18S-26S rDNA internal and external transcribed spacers. Mol Biol Evol 2001;18:936–44.

Andrew DR, Brown SM, Strausfeld NJ. The minute brain of the copepod *Tigriopus californicus* supports a complex ancestral ground pattern of the tetraconate cerebral nervous systems. J Comp Neurol 2012;520:3446–70.

Andrew DR. A new view of insect-crustacean relationships II. Inferences from expressed sequence tags and comparisons with neural cladistics. Arthr Struct Dev 2011;40:289–302.

Angelini DR, Kaufman TC. Insect appendages and comparative ontogenetics. Dev Biol 2005;286:57–77.

Anstey RL. Bryozoans. In: McNamara KJ, ed. Evolutionary Trends. London, UK: Belhaven Press, 1990:232–52.

Aouacheria A, Geourjon C, Aghajari N, et al. Insights into early extracellular matrix evolution: spongin short chain collagen-related proteins are homologous to basement membrane type IV collagens and form a novel family widely distributed in invertebrates. Mol Biol Evol 2006;23:2288–302.

Appeltans W, Ahyong ST, Anderson et al. The magnitude of global marine species diversity. Curr Biol 2012a;22:2189–202.

Appeltans W, Bouchet P, Boxshall GA, , et al. World Register of Marine Species, 2012b. (Accessed Mai 18, 2013, at http://www.marinespecies.org).

Aramant R, Elofsson R. Distribution of monoaminergic neurons in the nervous system of non-malacostracan crustaceans. Cell Tiss Res 1976;166:1–24.

Arendt D, Denes AS, Jekely G, Tessmar-Raible K. The evolution of nervous system centralization. Phil Trans R Soc Lond B Biol Sci 2008;363(1496):1523–8.

Arendt D, Nübler-Jung K. Dorsal or ventral: Similarities in fate maps and gastrulation patterns in annelids, arthropods and chordates. Mech Dev 1997;61:7–21.

Arendt D, Nübler-Jung K. Inversion of dorsoventral axis? Nature 1995;371:26.

Arendt D, Tessmar K, de Campos-Bapista MI, Dorresteijn A, Wittbrodt J. Development of pigment-cup eyes in the polychaete *Platynereis dumerlii* and evolutionary conservation of larval eyes in Bilateria. Development 2002;129:1143–54.

Aris-Brosou S, Rodrigue N. The essentials of computational molecular evolution. Methods Mol Biol 2012;855:111–52.

Arnaud F, Bamber RN. The biology of Pycnogonida. Adv Mar Biol 1987;24:1–96.

Arnaud J, Brunet M, Casanova J-P, Mazza J, Pasqualini V. Morphology and ultrastructure of the gut in *Spadella cephaloptera* (Chaetognatha). J Morph 1996;228:27–44.

Aronowicz J, Lowe CJ. Hox gene expression in the hemichordate *Saccoglossus kowalevskii* and the evolution of deuterostome nervous systems. Integr Comp Biol 2006;46:890–901.

Assis LCS, Rieppel O. Are monophyly and synapomorphy the same or different? Revisiting the role of morphology in phylogenetics. Cladistics 2011;27:94–102.

Atkinson HJ, Isaac RE, Harris PD, Sharpe CM. FMRFamide like immunoreactivity within the nervous system of the nematodes *Panagrellus redivivus*, *Caenorhabditis elegans* and *Heterodera glycines*. J Zool 1988;216:663–71.

Attems C. Myriapoda. In: Kükenthal W, Krumach T, eds. Handbuch der Zoologie IV. Berlin, Germany: de Gruyter Verlag, 1926.

Audouin JV, Milne Edwards H. Recherches pour servir a l'histoire naturelle du littoral de la France, ou Recueil de mémoires sur l'anatomie, la physiologie, la classification et les moeurs des animaux de nos côtes; ouvrage accompagné de planches faites d'après nature. Paris, France: Crochard, 1834.

Averof M, Akam M. Insect-crustacean relationships: Insights from comparative developmental and molecular studies. Phil Trans R Soc Lond B Biol Sci 1995a;347:293–303.

Averof M, Akam M. Hox genes and the diversification of insect and crustacean body plans. Nature 1995b;376:420–3.

Averof M, Patel NH. Crustacean appendage evolution associated with changes in Hox gene expression. Nature 1997;388:682–6.

Ax P. Basic Phylogenetic Systematization of the Metazoa. In: Fernholm B, Bremer K, Jornvall H, ed. The Hierarchy of Life. Amsterdam, the Netherlands: Elsevier, 1989:229–45.

Ax P. Das System der Metazoa I: ein Lehrbuch der phylogenetischen Systematik. Stuttgart, Germany: Gustav Fischer Verlag, 1995.

Ax P. Das System der Metazoa II: ein Lehrbuch der phylogenetischen Systematik. Stuttgart, Germany: Gustav Fischer Verlag, 1999.

Ax P. Das System der Metazoa III: ein Lehrbuch der phylogenetischen Systematik. Heidelberg, Germany: Spektrum Akademischer Verlag, 2001.

Ax P. Die Gnathostomulida, eine rätselhafte Wurmgruppe aus dem Meeressand. Mainz: Verlag der Akademie der Wissenschaften und der Literatur; in Kommission bei F. Steiner, Wiesbaden, Germany, 1956.

Ax P. Multicellular Animals: Order in nature-System made by man. Vol. III. Berlin, Germany: Springer, 2003.

Ax P. Multicellular Animals: a new approach to the phylogenetic order in nature. Vol. I. Berlin: Springer 1996.
Axtell MJ, Snyder JA, Bartel DP. Common functions for diverse small RNAs of land plants. Plant Cell 2007;19:1750–69.
Babbitt CC, Patel NH. Relationships within the Pancrustacea: Examining the influence of additional malacostracan 18s and 28s rDNA. In: Koenemann S, Jenner R, eds. Crustacea and Arthropod Relationships, Boca Ratón, FL, USA: CRC Press, 2005:275–94.
Baccari S, Renaud-Mornant J. Etude du système nerveux de *Derocheilocaris remanei* Delamare et Chappuis 1951 (Crustacea, Mystacocarida). Cah Biol Mar 1974;15:589–604.
Bäcker H, Fanenbruck M, Wägele JW. A forgotten homology supporting the monophyly of Tracheata: The subcoxa of insects and myriapods. Zool Anz 2008;247:185–207.
Bader M Ohlebusch, E. Sorting by weighted reversals, transpositions, and inverted transpositions. J Comp Biol 2007;14:615–36.
Bader M, Abouelhoda M, Ohlebusch E. A fast algorithm for the multiple genome rearrangement problem with weighted reversals and transpositions. BMC Bioinf 2008;9:516.
Baker MB, Shachak M, Brand S. Settling behavior of the desert isopod *Hemilepistus reaumuri* in response to variation in soil moisture and other environmental cues. Israel J Zool 1998;44:345–54.
Baldauf SL, Palmer JD. Animals and fungi are each other's closest relatives: Congruent evidence from multiple proteins. Proc Natl Acad Sci USA 1993;90:11558–62.
Ballard JWO, Olsen GJ, Faith DP, Odgers WA, Rowell DM, Atkinson PW. Evidence from 12S ribosomal RNA sequences that onychophorans are modified arthropods. Science 1992;258:1345–8.
Banerjee-Basu S, Baxevanis AD. Molecular evolution of the homeodomain family of transcription factors. Nucl Acids Res 2001;29:3258–69.
Bard J. Ontologies: Formalising biological knowledge for bioinformatics. BioEssays 2003;25:501–6.
Bard JBL, Rhee SY. Ontologies in biology: Design, applications and future challenges. Nature Reviews Genetics 2004;5:213–22.
Barthélémy RM, Jule Y, Da Prato J-L, Liberge M. Expression of serotonin and enkephalins in calanoid copepods (Crustacea): An immunohistochemical study. J Plankt Res 2006;28:1047–53.
Bartolomaeus T. Die Leibeshöhlenverhältnisse und Nephridialorgane der Bilateria – Ultrastruktur, Entwicklung und Evolution. Habilitationsschrift, Universität Göttingen, Germany, 1993.
Bartolomaeus T. On the ultrastructure of the coelomic lining in the Annelida, Sipuncula and Echiura. Microfauna marina 1994;9:171–220.
Bartolomaeus T. Structure and development of the nephridia of *Tomopteris helgolandica* (Annelida). Zoomorphology 1997;117:1–11.
Bartolomaeus T. Ultrastructure and development of the nephridia in *Anaitides mucosa* (Annelida, Polychaeta). Zoomorphology 1989;109:15–32.
Bartolomaeus T. Chaetogenesis in polychaetous Annelida – significance for annelid systematics and the position of the Pogonophora. Zoology 1998;100:348–64.
Bartolomaeus T. Structure, function and development of segmental organs in Annelida. Hydrobiologia 1999;402:21–37.
Bartolomaeus T. Ultrastructure and formation of the body cavity lining in *Phoronis muelleri* (Phoronida, Lophophorata). Zoomorphology 2001;120:135–48.
Bartolomaeus T, Ax P. Protonephridia and Metanephridia – their relation within the Bilateria. Z Zool Syst Evolutionsforsch 1992;30:21–45.
Bartolomaeus T, Quast B. Structure and development of nephridia in Annelida and related taxa. In: Bartolomaeus T, Purschke G, eds. Morphology, molecules, evolution and phylogeny in Polychaeta and related taxa. Hydrobiologia 2005;535–536:139–65.

Bartolomaeus T, Ruhberg H. Ultrastructure of the body cavity lining in embryos of *Epiperipatus biolleyi* (Onychophora, Peripatidae) – a comparison with annelid larvae. Invertebr Biol 1999;118:165–74.

Bartolomaeus T, Purschke G, Hausen H. Polychaete phylogeny based on morphological data– a comparison of current attempts. In: Bartolomaeus T, Purschke G, eds. Morphology, molecules, evolution and phylogeny in Polychaeta and related taxa. Hydrobiologia 2005;535–536:341–56.

Bartolomaeus T, Quast B, Koch M. Nephridial development and body cavity formation in *Artemia salina* (Crustacea: Branchiopoda): No evidence for any transitory coelom. Zoomorphology 2009;247–62.

Bassler RS. Bryozoa. In: Moore RC, ed. Treatise on Invertebrate Paleontology. Lawrence, KS, USA: University of Kansas Press, 1953.

Bauer M, Schuster SM, Sayood K. The average mutual information profile as a genomic signature. BMC Bioinf 2008;9:48.

Baurain D, Brinkmann H, Philippe H. Lack of resolution in the animal phylogeny: Closely spaced cladogeneses or undetected systematic errors? Mol Biol Evol 2007;24:6–9.

Bauzà-Ribot MM, Jaume D, Juan C, Pons J. The complete mitochondrial genome of the subterranean crustacean *Metacrangonyx longipes* (amphipoda): A unique gene order and extremely short control region. Mitochondr DNA 2009;20:88–99.

Bavestrello G, Puce S, Cerrano C, Zocchi E, Boero F. The problem of seasonality of benthic hydroids in temperate waters. Chemistry and Ecology 2006;22:197–205.

Bayer D, Diaconis P. Trailing the dovetail shuffle to its lair. Ann Appl Prob 1992;2:294–313.

Beard J. On the life-history and development of the genus *Myzostoma*. Mitt zool Stat Neapel 1884;5:544–80.

Bechly G, Brauckmann C, Zessin W, Gröning E. New results concerning the morphology of the most ancient dragonflies (Insecta: Odonatoptera) from the Namurian of Hagen-Vorhalle (Germany). Z Zool Syst Evolutionsforsch 2001;39:209–26.

Beckenbach AT. Mitochondrial genome sequences of nematocera (lower diptera): Evidence of rearrangement following a complete genome duplication in a winter crane fly. Genome Biol Evol 2012;4:89–101.

Becker E. Zum Bau und zur Genese des coxotrochanteralen Teiles des Ateloceratenbeines. Zool Anz 1923;57:137–44.

Beckers P, Faller S, Loesel R. Lophotrochozoan neuroanatomy: An analysis of the brain and nervous system of *Lineus viridis* (Nemertea) using different staining techniques. Front Zool 2011;8:17.

Beckers, P, Loesel, R, Bartolomaeus T (2013) The nervous systems of basally branching Nemertea (Palaeonemertea). PLoS one 8:e66137.

Bejerano G, Pheasant M, Makunin I, et al. Ultraconserved elements in the human genome. Science 2004;304:1321–5.

Belinky F, Cohen O, Huchon D. Large-scale parsimony analysis of metazoan indels in protein-coding genes. Mol Biol Evol 2010;27:441–51.

Bell JJ. The functional roles of marine sponges. Estuar Coast Shelf Sci 2008;79:341–53.

Bellonci G. Sistema nervoso e organi dei sensi dello *Sphaeroma serratum*. Atti dell'Accademia nazionale dei Lincei (Roma) 1881;278:90–103.

Belshaw R, Quicke DL. A molecular phylogeny of the Aphidiinae (Hymenoptera: Braconidae). Mol Phylogenet Evol 1997;7:281–93.

Bembenek JN, Richie CT, Squirrell JM, Campbell JM, Eliceiri KW, Poteryaev D, Spang A, Golden A, White JG. Cortical granule exocytosis in C. elegans is regulated by cell cycle components including separase. Development 2007;134:3837–48.

Bender JA, Pollack AJ, Ritzmann RE. Neural activity in the central complex of the insect brain is linked to locomotor changes. Curr Biol 2010;20:921–6.

Benesch R. Zur Ontogenie und Morphologie von *Artemia salina* L. Zool Jahrb Abt Anat Ontog 1969;86:307–458.
Benham WB. Archiannelida, Polychaeta, and Myzostomaria. In: Harmer SF, Shipley AE, eds. The Cambridge Natural History. London, UK: Macmillan and Co, Ltd, 1896:241–344.
Benzid D, Morris C, Barthélémy RM. First detection of serotonin in the nervous system of the marine calanoid copepod *Centropages typicus*. J Mar Biol Assoc UK 2005;85:71–2.
Bérard S, Bergeron A, Chauve C, Paul C. Perfect sorting by reversals is not always difficult. IEEE-ACM T Comput Bi 2007;4:4–16.
Berezikov E. Evolution of microRNA diversity and regulation in animals. Nat Rev Genet 2011;12:846–60.
Bergeron A, Chauve C, de Montgolfier F, Raffinot M. Computing common intervals of k permutations, with applications to modular decomposition of graphs. SIAM J Discrete Math 2008;22:1022–39.
Bergeron A, Chauve C, Gingras Y. Formal models of gene clusters. In A Zelikovsky and II Mandoiu, eds. Bioinformatics Algorithms: Techniques and Applications. Hoboken, NJ, USA: John Wiley & Sons, Inc., 2007;177–202.
Bergeron A, Stoye J. On the similarity of sets of permutations and its applications to genome comparison. J Comp Biol 2006;13:1340–54.
Bergmann A, Rust J. Morphology, palaeobiology, and phylogeny of *Oryctocaris balssi* gen. nov. (Arthropoda), a phyllocarid from the Lower Devonian Hunsrück Slate (Germany). J Syst Palaeontol 2013, DOI:10.1080/14772019.2012.750630.
Bergsten J, Miller KB. Acilius phylogeny (Coleoptera: Dytiscidae), problems with long-branch attraction and morphological intersexual coevolution. Cladistics 2004;20:76–7.
Bergsten J. A review of long-branch attraction. Cladistics 2005;21:163–93.
Bergter A, Paululat A. Pattern of body-wall muscle differentiation during embryonic development of *Enchytraeus coronatus* (Annelida: Oligochaeta; Enchytraeidae). J Morphol 2007;268:537–49.
Bergter A, Hunnekuhl VS, Schniederjans M, Paululat A. Evolutionary aspects of pattern formation during clitellate muscle evelopment. Evol Dev 2007;9:602–17.
Bernhart SH, Hofacker IL, Will S, Gruber AR, Stadler PF. RNAalifold: Improved consensus structure prediction for RNA alignments. BMC Bioinf 2008;9:474.
Bernt M. Gene order rearrangement methods for the reconstruction of phylogeny. Dissertation, Universität Leipzig, Germany, 2010.
Bernt M, Middendorf M. A method for computing an inventory of metazoan mitochondrial gene order rearrangements. BMC Bioinf 2011;12(Suppl 9):S6.
Bernt M, Merkle D, Middendorf M. A parallel algorithm for solving the reversal median problem. In: Wyrzykowski R, Dongarra J, Meyer N, Waśniewski J, eds. Parallel Processing and Applied Mathematics. Lecture Notes in Computer Science, Berlin, Germany: Springer, 2005;3911:1089–96.
Bernt M, Merkle D, Middendorf M. Genome rearrangement based on reversals that preserve conserved intervals. IEEE-ACM T Comput Bi 2006a;3:275–88.
Bernt M, Merkle D, Middendorf M. The reversal median problem, common intervals and mitochondrial gene orders. In; Berthold MR, Glen RC, Fischer I, eds. Computational Life Sciences. Lecture notes in computer science, Berlin, Germany: Springer, 2006b;4216:52–63.
Bernt M, Merkle D, Middendorf M. Using median sets for inferring phylogenetic trees. Bioinformatics 2007;23:129–35.
Bernt M, Merkle D, Middendorf M. Solving the preserving reversal median problem. IEEE-ACM T Comput Bi 2008a;5:332–47.
Bernt M, Merkle D, Middendorf M. An algorithm for inferring mitogenome rearrangements in a phylogenetic tree. In: Nelson CE, Vialette S, eds. Comparative Genomics. Lecture notes in computer science, Berlin, Germany: Springer, 2008b;5267:143–57.

Bernt M, Bleidorn C, Braband A, et al. A comprehensive analysis of metazoan mitochondrial genomes and animal phylogeny. Mol Phylogenet Evol 2013a;69:352–64.

Bernt M, Braband A, Schierwater B, Stadler PF. Genetic aspects of mitochondrial genome evolution. Mol Phylogenet Evol 2013b;69:328–38.

Bernt M, Braband A, Middendorf M, Misof B, Rota-Stabelli O, Stadler PF. Bioinformatics methods for the comparative analysis of metazoan mitochondrial genome sequences. Mol Phylogenet Evol 2013d;69:320–7.

Bernt M, Chen K-Y, Chen M-C, et al. Finding all sorting tandem duplication random loss operations. J Discrete Math 2011;9:32–48.

Bernt M, Donath A, Jühling F, et al. MITOS: Improved *de novo* metazoan mitochondrial genome annotation. Mol Phylog Evol 2013c;69:313–9.

Bernt M, Merkle D, Ramsch K, et al. CREx: Inferring genomic rearrangements based on common intervals. Bioinformatics 2007;23:2957–8.

Bertrand S, Escriva H. Evolutionary crossroads in developmental biology: Amphioxus. Development 2011;138:4819–30.

Beutel RG, Gorb S. A revised interpretation of the evolution of attachment structures in hexapoda with special emphasis on Mantophasmatodea. Arthropod Syst Phyl 2006;64:3–25.

Beutel RG, Gorb S. Ultrastructure of attachment specializations of hexapods (Arthropoda): evolutionary patterns inferred from a revised ordinal phylogeny. J Zool Syst Evol Res 2001;39:177–207.

Beutel RG, Kristensen NP. Morphology and insect systematics in the era of phylogenomics. Arthropod Struct Dev 2012;41:303–5.

Beutel RG, Pohl HW. Endopterygote systematics–where do we stand and what is the goal (Hexapoda, Arthropoda)? Syst Entomol 2006;31:202–19.

Beutel R, Weide D. Cephalic anatomy of *Zorotypus hubbardi* (Hexapoda: Zoraptera): new evidence for a relationship with Acercaria. Zoomorphology 2005;124:121–36.

Beutel RG, Friedrich F, Hörnschemeyer T, Pohl H, Hünefeld F, Beckmann F, Meier R, Misof B, Whiting MF, Vilhelmsen L. Morphological and molecular evidence converge upon a robust phylogeny of the megadiverse Holometabola. Cladistics 2011;27:341–55.

Beutel RG, Kristensen NP, Pohl H. Resolving insect phylogeny: The significance of cephalic structures of the Nannomecoptera in understanding endopterygote relationships. Arthropod Struct Dev 2009;38:427–60.

Bhattacharya D, Ehlting J. Actin Coding Regions: Gene family evolution and use as a phylogenetic marker. Arch Protistenkunde 1995;145:155–64.

Bieri R. Systematics of the Chaetognatha. In: Bone Q, Kapp H, Pierrot-Bults AC, eds. The Biology of Chaetognaths. New York, NY, USA: Oxford University Press, 1991:122–36.

Bininda-Emonds O. transAlign: Using amino acids to facilitate the multiple alignment of protein-coding DNA sequences. BMC Bioinf 2005;6:156.

Bishop MJ, Friday AE. Evolutionary trees from nucleic acids and protein sequences. Proc R Soc Lond B Biol Sci 1985;226:271–302.

Bitsch C, Bitsch J. Internal anatomy and phylogenetic relationships among apterygote insects (Hexapoda). Ann Soc Entomol Fr 1998;34:339–63.

Bitsch C, Bitsch J. Phylogenetic relationships of basal hexapods among the mandibulate arthropods: A cladistic analysis based on comparative morphological characters. Zool Scripta 2004;33:511–50.

Blackstone NW. A new look at some old animals. PLOS Biol 2009;7:e7.

Blair C, Murphy RW. Recent trends in molecular phylogenetic analysis: Where to next? J Hered 2011;102:130–8.

Blanke A, Greve C, Wipfler B, Beutel RG, Holland BR, Misof B. The identification of concerted convergence in insect heads corroborates palaeoptera. Syst Biol 2013;62:250–63.
Blanke A, Wipfler B, Letsch H, Koch M, Beckmann F, Beutel R, Misof B. Revival of Palaeoptera – head characters support a monophyletic origin of Odonata and Ephemeroptera (Insecta). Cladistics 2012;28:560–81.
Bleidorn C. The role of character loss in phylogenetic reconstruction as exemplified for the Annelida. J Zool Syst Evol Res 2007;45:299–307.
Bleidorn C, Podsiadlowski L, Bartolomaeus T. The complete mitochondrial genome of the orbiniid polychaete *Orbinia latreilli* (Annelida, Orbiniidae) – a novel gene order for Annelida and implications for annelid phylogeny. Gene 2006;370:96–103.
Bleidorn C, Vogt L, Bartolomaeus T. A contribution to sedentary polychaete phylogeny using 18S rRNA sequence data. J Syst Zool Evol Res 2003a;41:186–95.
Bleidorn C, Vogt L, Bartolomaeus T. New insights into polychaete phylogeny (Annelida) inferred from 18S rDNA sequences. Mol Phylogenet Evol 2003b;29:279–88.
Bleidorn C, Hill N, Erséus C, Tiedemann R. On the role of character loss in orbiniid phylogeny (Annelida): Molecules vs. morphology. Mol Phylogenet Evol 2009b;52:57–69.
Bleidorn C, Lanterbecq D, Eeckhaut I, Tiedemann R. A PCR survey of Hox genes in the myzostomid *Myzostoma cirriferum*. Dev Gen Evol 2009c;219:211–6.
Bleidorn C, Eeckhaut I, Podsiadlowski L, et al. Mitochondrial genome and nuclear sequence data support Myzostomida as part of the annelid radiation. Mol Biol Evol 2007;24:1690–701.
Bleidorn C, Podsiadlowski L, Zhong M, et al. On the phylogenetic position of Myzostomida: Can 77 genes get it wrong? BMC Evol Biol 2009a;9:150.
Boero F, Schierwater B, Piraino S. Cnidarian milestones in metazoan evolution. Integr Comp Biol 2007;47:693–700.
Böger H. Versuch über das phylogenetische System der Porifera. Meyniana 1983;40:143–54.
Bone Q, Duvert M. Locomotion and Buoyancy. In: Bone Q, Kapp H, Pierrot-Bults AC, eds. The Biology of Chaetognaths. New York, NY, USA: Oxford University Press, 1991:32–44.
Bone Q, Goto T. The Nervous System. In: Bone Q, Kapp H, Pierrot-Bults AC, eds. The Biology of Chaetognaths. New York, NY, USA: Oxford University Press, 1991:18–31.
Bone Q, Pulsford A. The sense organs and ventral ganglion of *Sagitta* (Chaetognatha). Acta Zool (Stockholm) 1984;65:209–20.
Bone Q, Kapp H, Pierrot-Bults AC. The Biology of Chaetognaths. New York, NY, USA: Oxford University Press, 1991.
Bone Q, Grimmelikhuijzen CLP, Pulsford A, Ryan KP. Possible transmitter functions of acetylcholine and an RFamide-like substance in *Sagitta* (Chaetognatha). Proc R Soc Lond B Biol Sci 1987;230:1–14.
Boore JL. Animal mitochondrial genomes. Nucleic Acids Res 1999;27:1767–80.
Boore JL. Complete mitochondrial genome sequence of Urechis caupo, a representative of the phylum Echiura. BMC Genomics 2004;5:67.
Boore JL. The duplication/random loss model for gene rearrangement exemplified by mitochondrial genomes of deuterostome animals. In: Sankoff D, Nadeau JH, eds. Comparative Genomics: Empirical and Analytical Approaches to Gene Order Dynamics, Map Alignment and the Evolution of Gene Families. Computational Biology Series, Dordrecht, The Netherlands: Kluwer Academic Publishers, 2000;1:133–47.
Boore JL. The use of genome-level characters for phylogenetic reconstruction. Trends Ecol Evol 2006;21:439–46.
Boore JL, Brown WM. Big trees from little genomes: Mitochondrial gene order as a phylogenetic tool. Curr Opin Genet Dev 1998;8:668–74.

Boore JL, Brown WM. Mitochondrial genomes of *Galathealinum*, *Helobdella* and *Platynereis*: Sequence and gene arrangements comparisons indicate that Pogonophora is not a phylum and Annelida and Arthropoda are not sister taxa. Mol Biol Evol 2000;17:87–106.

Boore JL, Fuerstenberg SI. Beyond linear sequence comparisons: The use of genome-level characters for phylogenetic reconstruction. Phil Trans R Soc Lond B Biol Sci 2008;363:1445–51.

Boore JL, Staton JL. The mitochondrial genome of the sipunculid *Phascolopsis gouldii* supports its association with Annelida rather than Mollusca. Mol Biol Evol 2002;19:127–37.

Boore JL, Daehler LL, Brown WM. Complete sequence, gene arrangement, and genetic code of mitochondrial DNA of the cephalochordate *Branchiostoma floridae* (Amphioxus). Mol Biol Evol 1999;16:410–8.

Boore JL, Lavrov DV, Brown WM. Gene translocation links insects and crustaceans. Nature 1998;392:667–8.

Boore JL, Medina M, Rosenberg LA. Complete sequences of two highly rearranged molluscan mitochondrial genomes, those of the scaphopod *Graptacme eborea* and of the bivalve *Mytilus edulis*. Mol Biol Evol 2004;21:1492–503.

Bornberg-Bauer E, Huylmans AK, Sikosek T. How do new proteins arise? Curr Opin Struct Biol 2010;20:390–6.

Börner C. Beiträge zur Morphologie der Arthropoden. I. Ein Beitrag zur Kenntnis der Pedipalpen. Zoologia 1904;42:1–147.

Botting JP, Muir LA, Xiao S, Li X, Lin J-P. Evidence for spicule homology in calcareous and siliceous sponges: Bimineralic spicules in *Lenica* sp. from the Early Cambrian of South China. Lethaia 2012;45:463–75.

Boudreaux HB. Arthropod Phylogeny with Special Reference to Insects. New York, NY, USA: Wiley, 1979.

Bourlat SJ, Hejnol A. Quick guide: Acoels. Curr Biol 2009;19:R279–80.

Bourlat SJ, Rota-Stabelli O, Lanfear R, Telford MJ. The mitochondrial genome structure of *Xenoturbella bocki* (phylum Xenoturbellida) is ancestral within the deuterostomes. BMC Evol Biol 2009;9:107.

Bourlat SJ, Nielsen C, Lockyer AE, Littlewood DTJ, Telford MJ. *Xenoturbella* is a deuterostome that eats molluscs. Nature 2003;424:925–8.

Bourlat SJ, Juliusdottir T, Lowe CJ, et al. Deuterostome phylogeny reveals monophyletic chordates and the new phylum Xenoturbellida. Nature 2006;444:85–8.

Bourque G, Pevzner PA. Genome-scale evolution: Reconstructing gene orders in the ancestral species. Genome Res 2002;12:26–36.

Boury-Esnault N, Rützler K. Thesaurus of sponge morphology. Smithson Contrib Zool 1997;596:1–55.

Boury-Esnault N, Efremova S, Bezac C, Vacelet J. Reproduction of a hexactinellid sponge: first description of gastrulation by cellular determination in the Porifera. Invertebr Reprod Dev 1999;35:187–201.

Boury-Esnault N, Ereskovsky A, Bezac C, Tokina D. Larval development in the Homoscleromorpha (Porifera, Demospongiae). Invertebr Biol 2003;122:187–202.

Boute N, Exposito JY, Boury-Esnault N, et al. Type IV collagen in sponges, the missing link in basement membrane ubiquity. Biol Cell 1996;88:37–44.

Boxshall G. Comparative limb morphology in major crustacean groups: The coxa-basis joint in postmandibular limbs. In: Fortey RA, Thomas RH, eds. Arthropod Relationships, London: Chapman and Hall, 1998:155–67.

Boxshall GA. Copepoda. In: Harrison FW, Humes AG, eds. Microscopic Anatomy of Invertebrates. Vol. 9. New York, NY, USA: Wiley-Liss. 1992:347–84.

Boyan GS, Williams JLD, Herbert Z. Fascicle switching generates a chiasmal neuroarchitecture in the embryonic central body of the grasshopper *Schistocerca gregaria*. Arthr Struct Dev 2008;37:539–44.

Boyan GS, Williams JLD. Embryonic development of the insect central complex: Insights from lineages in the grasshopper and *Drosophila*. Arthr Struct Dev 2011;40:334–48.

Braband A, Cameron SL, Podsiadlowski L, Daniels SR, Mayer G. The mitochondrial genome of the onychophoran *Opisthopatus cinctipes* (Peripatopsidae) reflects the ancestral mitochondrial gene arrangement of Panarthropoda and Ecdysozoa. Mol Phylogenet Evol 2010b;57:285–92.

Braband A, Podsiadlowski L, Cameron SL, Daniels S, Mayer G. Extensive duplication events account for multiple control regions and pseudo-genes in the mitochondrial genome of the velvet worm *Metaperipatus inae* (Onychophora, Peripatopsidae). Mol Phylogenet Evol 2010a;57:293–300.

Bracken HD, De Grave S, Toon A, Felder DL, Crandall, KA. Phylogenetic position, systematic status, and divergence time of the Procarididea (Crustacea: Decapoda). Zool Scripta 2010;39:198–212.

Bracken HD, Toon A, Felder DL et al. The decapod tree of life: Compiling the data and moving toward a consensus of decapod evolution. Arthropod Syst Phyl 2009;67:99–116.

Brady SG, Danforth BN. Recent intron gain in elongation factor–1alpha of colletid bees (Hymenoptera: Colletidae). Mol Biol Evol 2004;21:691–6.

Braga M, Sagot M-F, Scornavacca C, Tannier E. (2008), 'Exploring the solution space of sorting by reversals, with experiments and an application to evolution. IEEE-ACM T Comput Bi 2008;5:348–56.

Brandenburg J. Die Reusenzelle (Cyrtocyte) des *Dinophilus* (Archiannelida). Zoomorphology 1970;68:83–92.

Brazma A. On the importance of standardisation in life sciences. Bioinformatics 2001;17:113–4.

Brazma A, Hingamp P, Quackenbush J, Sherlock G, Spellman P, et al. Minimum information about a microarray experiment (MIAME)–toward standards for microarray data. Nature Genetics 2001;29:365–71.

Brena C, Akam M. The embryonic development of the centipede *Strigamia maritima*. Developmental Biology 2012;363:290–307.

Brenneis G, Richter S. Architecture of the nervous system in Mystacocarida (Arthropoda, Crustacea): An immunohistochemical study and 3D reconstruction. J Morph 2010;271:169–89.

Brenneis G, Arango C, Scholtz G. Morphogenesis of *Pseudopallene* sp. (Pycnogonida, Callipallenidae) I: embryonic development. Dev Genes Evol 2011a;221:309–28.

Brenneis G, Arango C, Scholtz G. Morphogenesis of *Pseudopallene* sp. (Pycnogonida, Callipallenidae) II: Postembryonic development. Dev Genes Evol 2011b;221:329–50.

Brenneis G, Ungerer P, Scholtz G. The chelifores of sea spiders (Arthropoda, Pycnogonida) are the appendages of the deutocerebral segment. Evol Dev 2008;10:717–24.

Bretfeld G. Zur Anatomie und Embryologie der Rumpfmuskulatur und der abdominalen Anhänge der Collembolen. Zool Jahrb (Anatomie) 1963;80:309–84.

Bridge D, Cunningham CW, DeSalle R, Buss LW. Class-level relationships in the phylum Cnidaria: molecular and morphological evidence. Mol Biol Evol 1995;12:679–89.

Bridge D, Cunningham CW, Schierwater B, DeSalle R, Buss LW. Class-level relationships in the phylum Cnidaria: evidence from mitochondrial genome structure. Proc Natl Acad Sci USA 1992;89:8750–3.

Brien P. Embranchements des tuniciers. In: Grasse PP, ed. Traite de Zoologie, vol IX. Paris, France: Masson et Cie, 1948:553–932.

Briggs RT, Moss BL. Ultrastructure of the coxal gland of the horseshoe crab *Limulus polyphemus*: Evidence for ultrafiltration and osmoregulation. J Morphol 1997;234:233–52.

Brinkmann H, van der Giezen M, Zhou Y, de Raucourt GP, Philippe H. An empirical assessment of long-branch attraction artefacts in deep eukaryotic phylogenomics. Syst Biol 2005;54:743–57.

Brinkmann H, Venkatesh B, Brenner S, Meyer A. Nuclear protein-coding genes support lungfish and not the coelacanth as the closest living relatives of land vertebrates. Proc Natl Acad Sci USA 2004;101:4900–5.

Briscoe AD, White RH. Adult stemmata of the butterfly *Vanessa cardui* express UV and green opsin mRNAs. Cell Tissue Res 2005;319:175–9.

Bromham L, Rambaut A, Harvey PH. Determinants of rate variation in mammalian DNA sequence evolution. J Mol Evol 1996;43:610–21.

Brooke NM, Garcia-Fernandez J, Holland PW. The ParaHox gene cluster is an evolutionary sister of the Hox gene cluster. Nature 1998;392:920–2.

Broughton RE, Dowling TE. Evolutionary dynamics of tandem repeats in the mitochondrial DNA control region of the minnow *Cyprinella spiloptera*. Mol Biol Evol 1997;14:1187–96.

Brower AVZ, Schawaroch V. Three steps of homology assessment. Cladistics 1996;12:265–72.

Brown FD, Prendergast A, Swalla BJ. Man is but a worm: Chordate origins. Genesis 2008;46:605–13.

Brown MR, Sieglaff DH, Rees HH. Gonadal ecdysteroidogenesis in Arthropoda: Occurrence and regulation. Annu Rev Entomol 2009;54:105–25.

Brown S, Wolff G. Fine structural organization of the hemiellipsoid body of the land hermit crab *Coenobita clypeatus*. J Comp Neurol 2012;520:2847–63.

Brown S, Rouse GW, Hutchings P, Colgan D. Assessing the usefullness of histone H3, U2 snRNA and 28S rDNA in analyses of polychaete relationships. Aust J Zool 1999;47:499–516.

Brunet M, Arnaud J, Mazza J. Gut structure and digestive cellular processes in marine Crustacea. Oceanogr Mar Biol Annu Rev 1994;32:335–67.

Bruno WJ, Halpern AL. Topological bias and inconsistency in maximum likelihood using wrong models. Mol Biol Evol 1999;16:564–6.

Brusca GJ, Brusca RC, Gilbert SF. Characteristics of metazoan development. In: Gilbert SF, Raunio AM, eds. Embryology: Constructing the Organism. Sunderland, MA, USA: Sinauer Associates, 1997:3–19.

Brusca RC, Brusca GJ. Invertebrates, 1st edition. Sunderland, MA, USA: Sinauer Associates, 1990.

Brusca RC, Brusca GJ. Invertebrates, 2nd edition. Sunderland, MA, USA: Sinauer Associates, 2003.

Bryson-Richardson RJ, Currie PD. The genetics of vertebrate myogenesis. Nature Rev Genet 2008;9:632–46.

Buckley TR, Simon C, Flook PK, Misof B. Secondary structure and conserved motifs of the frequently sequenced domains IV and V of the insect mitochondrial large subunit rRNA gene. Insect Mol Biol 2000;9:565–80.

Budd GE. A palaeontological solution to the arthropod head problem. Nature 2002;417:271–75.

Budd GE, Telford MJ. The origin and evolution of arthropods. Nature 2009;457:812–7.

Buljan M, Bateman A. The evolution of protein domain families. Biochem Soc Trans 2009;37:751–5.

Bullock TH. The functional organization of the nervous system of Enterpneusta. Biol Bull 1940;79:91–113.

Bullock TH. The giant nerve fiber system in balanoglossids. J Comp Neurol 1944;80:355–67.

Bullock TH. Grades in neural complexity: How large is the span? Integr Comp Biol 2002;42:757–61.

Bullock TH, Horridge GA. Structure and Function in the Nervous Systems of Invertebrates. San Francisco, CA, USA: W.H. Freeman and Company, 1965.

Bulteau L, Fertin G, Rusu I. Sorting by transpositions is difficult. SIAM J Discrete Math 2012; 26:1148-80.

Büning J. The telotrophic ovary known from neuropterida exists also in the myxophagan beetle *Hydroscapha natans*. Dev Genes Evol 2005;215:597–607.

Bunke D. Early development of metanephridia in the caudal budding zone of a clitellate annelid, *Dero digitata* (Naididae): An electron-microscopical study. Acta Zool (Stockholm) 2003;84:87–97.

Bunke D. Ultrastructure of the metanephridial system in *Aeolosoma bengalense* (Annelida). Zoomorphology 1994;114:247–58.

Bunke D. Ultrastructure of the nephrostome in *Enchytraeus albidus* (Annelida, Clitellata). Zoomorphology 1998;118:177–82.

Bunke D. Ultrastructure of the preseptal part of metanephridia in *Nais variabilis* and *Dero digitata* (Annelida, Clitellata). Zoomorphology 2000;120:39–46.

Burger G, Forget L, Zhu Y, Gray MW, Lang BF. Unique mitochondrial genome architecture in unicellular relatives of animals. Proc Natl Acad Sci USA 2003;100:892–7.

Burger G, Yan Y, Javadi P, Lang BF. Group I-intron trans-splicing and mRNA editing in the mitochondria of placozoan animals. Trends Genet 2009.

Burighel P, Cloney RA. Urochordata: Ascidiacea. In: Harrison FW, Ruppert EE, eds. Microscopic Anatomy of Invertebrates. Hemichordata, Chaetognatha, and the Invertebrate Chordates, Vol 15. New York, NY, USA: Willey-Liss, Incorporation, 1997:221–347.

Burighel P, Nunzi MG, Schiaffino S. A comparative study of the organization of the sarcotubular system in ascidian muscle. J Morphol 1977;153:205–24.

Burke RD. Deuterostome neuroanatomy and the body plan paradox. Evolution & Development 2011;13:110–5.

Burris ZP. Larval morphologies and potential developmental modes of eight sea spider species (Arthropoda: Pycnogonida) from the southern Oregon coast. J Mar Biol Assoc UK 2011;91:845–55.

Buschbeck EK. The compound eye of Strepsiptera: Morphological development of larvae and pupae. Arthropod Struct Dev 2005;34:315–26.

Byrne M, Nakajima Y, Chee FC, Burke RD. Apical organs in echinoderm larvae: Insights into larval evolution in the Ambulacraria. Evol Dev 2007;9:432–45.

Caetano-Anollés G. Evolved RNA secondary structure and the rooting of the universal tree of life. J Mol Evol 2002a;54:333–45.

Caetano-Anollés G. Tracing the evolution of RNA structure in ribosomes. Nucl Acids Res 2002b;30:2575–87.

Callaway JC Stuart AE. The distribution of histamine and serotonin in the barnacle's nervous system. Microsc Res Tech 1999;44:94–104.

Camacho C, Coulouris G, Avagyan V, et al. BLAST+: architecture and applications. BMC Bioinf 2009;10:421.

Cameron C, Garey J, Swalla B. Evolution of the chordate body plan: New insights from phylogenetic analyses of deuterostome phyla. Proc Natl Acad Sci USA 2000;97:4469–74.

Cameron RA, Rowen L, Nesbitt R, et al. Unusual gene order and organization of the sea urchin hox cluster. J Exp Zool B Mol Dev Evol 2006b;306B:45–58.

Cameron SL, Beckenbach AT, Dowton M, Whiting MF. Evidence from mitochondrial genomics on interordinal relationships in insects. Arthropod Syst Phyl 2006a;64:27–34.

Cameron SL, Miller KB, D'Haese CA, Whiting MF, Barker SC. Mitochondrial genome data alone are not enough to unambiguously resolve the relationships of Entognatha, Insecta and Crustacea sensu lato (Arthropoda). Cladistics 2004;20:534–57.

Cameron SL, Yoshizawa K, Mizukoshi A, Whiting MF, Johnson KP. Mitochondrial genome deletions and minicircles are common in lice (Insecta: Phthiraptera). BMC Genomics 2011;12:394.

Campbell L, Rota-Stabelli O, Edgecombe GD, Marchioro T, Longhorn SJ, Telford MJ. MicroRNAs and phylogenomics resolve the relationships of Tardigrada and suggest that velvet worms are the sister group of Arthropoda. Proc Natl Acad Sci USA 2011;108:15920–4.

Campo-Paysaa F, Sémon M, Cameron RA, Peterson KJ, Schubert M. MicroRNA complements in deuterostomes: origin and evolution of microRNAs. Evol Dev 2011;13:15–27.

Cannon HG. On the Anatomy of a Marine Ostracod, *Cypridina* (*Doloria*) *levis* Skogsberg, Discovery Reports, 2nd edition, London, UK: Cambridge University Press, 1931:435–82.

Cannon HG. On the development of an estherid crustacean. Phil Trans R Soc Lond B Biol Sci 1924;212:395–430.

Cannon JT, Rychel AL, Eccleston H, Halanych KM, Swalla BJ. Molecular phylogeny of hemichordata, with updated status of deep-sea enteropneusts. Mol Phylogenet Evol 2009;52:17–24.

Capella-Gutiérrez S, Silla-Martínez J, Gabaldón T. trimAl: A tool for automated alignment trimming in large-scale phylogenetic analyses. Bioinformatics 2009;15:1972–3.

Caprara A. The reversal median problem. INFORMS J Comput 2003;15:93–113.

Carapelli A, Liò P, Nardi F, van der Wath E, Frati F. Phylogenetic analysis of mitochondrial protein coding genes confirms the reciprocal paraphyly of Hexapoda and Crustacea. BMC Evol Biol 2007; 7 (Suppl 2):S8.

Carapelli A, Nardi F, Dallai R, Boore J, Lìo P, Frati F. Relationships between hexapods and crustaceans based on four mitochondrial genes. In: Koenemann S, Jenner R, eds. Crustacea and Arthropod Relationships, Boca Ratón, FL, USA: CRC Press, 2005:295–306

Cardona A, Saalfeld S, Schindelin J, et al. TrakEM2 software for neural circuit reconstruction. PLoS ONE 2012;7:6, e38011.

Carefoot TH. Nutrition and Growth of *Ligia pallasii*. Symp Zool Soc London 1984;53:455–67.

Carlson SJ. Phylogenetic relationships among extant brachiopods. Cladistics 1995;11:131–97.

Carmel L, Rogozin IB, Wolf YI, Koonin EV. Patterns of intron gain and conservation in eukaryotic genes. BMC Evol Biol 2007;7:192.

Caron J-B, Conway Morris S, Shu D. Tentaculate fossils from the Cambrian of Canada (British Columbia) and China (Yunnan) interpreted as primitive deuterostomes. PLoS ONE 2010;5:e9586.

Carpenter FM, Burnham L. The Geological Record of Insects. Ann Rev Earth Planet Sci 1985;13:297–314.

Carpenter FM. Treatise on Invertebrate paleontology, Part R, Arthropoda 4, vols. 3 & 4: Superclass Hexapoda. Lawrence, KS, USA: Geological Society of America & University of Kansas, 1992.

Carré D, Djediat C, Sardet C. Formation of a large vasa-positive germ granule and its inheritance by germ cells in the enigmatic chaetognaths. Development 2002;129:661–70.

Cartwright P, Schierwater B, Buss LW. Expression of a Gsx parahox gene, Cnox-2, in colony ontogeny in Hydractinia (Cnidaria: Hydrozoa). J Exp Zool Part B 2006;306:460–9.

Carus JV. Vermes. In: Carus JV, Gaerstecker CEA, eds. Handbuch der Zoologie. Leipzig, Germany: Wilhelm Engelmann Verlag, 1863:422–84.

Casanova J-P, Duvert M, Goto T. Ultrastructural study and ontogenesis of the appendages and related musculature of *Paraspadella* (Chaetognatha). Tissue Cell 2003;35:339–51.

Casanova J-P, Duvert M. Comparative studies and evolution of muscles in chaetognaths. Mar Biol 2002;141:925–38.

Castresana J. Selection of conserved blocks from multiple alignments for their use in phylogenetic analysis. Mol Biol Evol 2000;17:540–52.

Castresana J. Topological variation in single-gene phylogenetic trees. Genome Biology 2007;8:216.

Castresana J, Feldmaier-Fuchs G, Pääbo S. Codon reassignment and amino acid composition in hemichordate mitochondria. Proc Natl Acad Sci USA 1998;95:3703–7.

Castresana J, Feldmaier-Fuchs G, Yokobori S, Satoh N, Pääbo S. The mitochondrial genome of the hemichordate *Balanoglossus carnosus* and the evolution of deuterostome mitochondria. Genetics 1998;150:1115–23.

Castro L, Holland P. Chromosomal mapping of ANTP class homeobox genes in amphioxus: Piecing together ancestral genomes. Evol Dev 2003;5:459–65.

Castro LR, Austin AD, Dowton M. Contrasting rates of mitochondrial molecular evolution in parasitic Diptera and Hymenoptera. Mol Biol Evol 2002;19:1100–13.

Caterino MS, Cho S, Sperling FA. The current state of insect molecular systematics: A thriving Tower of Babel. Annu Rev Entomol 2000;45:1–54.

Cavalier-Smith T. A revised six-kingdom system of life. Biol Rev 1998;73:203–66.

Chaigneau J. Etude ultrastructurale de l'organe cephalique de bellonci des crustaces superieurs. Thèse de Doctorat, Université de Poitiers, France, 1976.

Chaigneau J. L'organe de Bellonci du crustacé isopode *Sphaeroma serratum* (Fabricius): ultrastructure et signification. Z Zellforsch 1971;112:166–87.

Chang JT. Full reconstruction of markov models on evolutionary trees: Identifiability and consistency. Math Biosci 1996;137:51–73.

Chase R, Tolloczko B. Interganglionic dendrites constitute an output pathway from the procerebrum of the snail *Achatina fulica*. J Comp Neurol 1989;283:143–52.

Chase R, Tolloczko B. Tracing neural pathways in snail olfaction: From the tip of the tentacles to the brain and beyond. Micros Res Tech 1993;24:214–30.

Chaudhuri K, Chen K, Mihaescu R, Rao S. (2006), On the tandem duplication-random loss model of genome rearrangement. SODA '06 Proc 17th Annu ACM-SIAM Symp Discrete algorithm. ACM, New York, NY, USA, 2006:564–70.

Chauve C, Tannier E. A methodological framework for the reconstruction of contiguous regions of ancestral genomes and its application to mammalian genomes. PLoS Comput Biol 2008;4:e1000234.

Chen CLA, Chen CP, Chen IM. Sexual and asexual reproduction of the tropical corallimorpharian *Rhodactis (=Discosoma) indosinensis* (Cnidaria, Corallimorpharia) in Taiwan. Zool Stud 1995;34:29–40.

Chen F, Mackey A, Stoeckert J, C, Roos D. OrthoMCL-DB: Querying a comprehensive multi-species collection of ortholog groups. Nucleic Acids Res 2006;34:363–8.

Chen F, Mackey AJ, Vermunt JK, Roos DS. Assessing performance of orthology detection strategies applied to eukaryotic genomes. PLoS ONE 2007;2:e383.

Chen HX, Sundberg P, Norenburg JL, Sun SC. The complete mitochondrial genome of *Cephalothrix simula* (Iwata) (Nemertea: Palaeonemertea). Gene 2009;442:8–17.

Chen HX, Sundberg P, Wu HY, Sun SC. The mitochondrial genomes of two nemerteans, *Cephalothrix* sp. (Nemertea: Palaeonemertea) and *Paranemertes* cf. *peregrina* (Nemertea: Hoplonemertea). Mol Biol Rep 2011;38:4509–25.

Chen J-Y, Huang D-Y, Peng Q-Q, Chi H-M, Wang X-Q, Feng M. The first tunicate from the early Cambrian of South China. Proc Natl Acad Sci USA 2003;100:8314–8.

Chen J, Dzik J, Edgecombe GD, Ramsköld L, Zhou G. A possible early cambrian chordate. Nature 1995;377:720–2.

Chen JY. The sudden appearance of diverse animal body plans during the Cambrian explosion. Int J Dev Biol 2009;53:733–51.

Chen JY, Vannier J, Hunag DY. The origin of crustaceans. New evidence from the Early Cambrian of China. Proc R Soc Lond B Biol Sci 2001;268:2181–7.

Chen JY, Waloszek D, Maas A. A new 'great-appendage' arthropod from the Lower Cambrian of China and homology of chelicerate chelicerae and raptorial antero-ventral appendages. Lethaia 2004;37:3–20.

Chen M, Zou M, Yang L, He S. Basal Jawed Vertebrate Phylogenomics Using Transcriptomic Data from Solexa Sequencing. PLoS ONE 2012;7:e36256.

Cheng CH, Yang CH, Chiu HT, Lu CL. Reconstructing genome trees of prokaryotes using overlapping genes. BMC Bioinf 2010;11:102.

Chervitz SA, Aravind L, Sherlock G, et al. Comparison of the complete protein sets of worm and yeast: Orthology and divergence. Science 1998;282:2022–8.

Chevreux E, Pfisterer T, Drescher B, et al. Using the miraEST assembler for reliable and automated mRNA transcript assembly and SNP detection in sequenced ESTs. Genome Research 2004;14:1147–59.

Chiang HR, Schoenfeld LW, Ruby JG, et al. Mammalian microRNAs: Experimental evaluation of novel and previously annotated genes. Genes Dev 2010;24:992–1009.

Chiaverano L, Mianzan H, Ramirez F. Gonad development and somatic growth patterns of *Olindias sambaquiensis* (Limnomedusae, Olindiidae). Hydrobiologia 2004;530:373–81.

Chien PK, Stephen GC, Healy PL. The role of ultrastructure and physiological differentiation of epithelia in amino acid uptake by the bloodworm *Glycera*. Biol Bull 1972;142:219–35.

Chioda M, Spada F, Eskeland R, Thompson EM. Histone mRNAs do not accumulate during S phase of either mitotic or endoreduplicative cycles in the chordate *Oikopleura dioica*. Mol Cell Biol 2004;24:5391–403.

Chiori R, Jager M, Denker E, et al. Are Hox genes ancestrally involved in axial patterning? Evidence from the hydrozoan *Clytia hemisphaerica* (Cnidaria). PLoS ONE 2009;4:e4231.

Chipman AD, Arthur W, Akam M. Early development and segment formation in the centipede, *Strigamia maritima* (Geophilomorpha). Evol Dev 2004;6:78–89.

Chitwood BG, Chitwood MB. Introduction to Nematology. Baltimore, MD, USA: University Park Press, 1950.

Cho G, Doolittle RF. Intron distribution in ancient paralogs supports random insertion and not random loss. J Mol Evol 1997;44:573–84.

Cho S, Jin SW, Cohen A, Ellis RE. A phylogeny of caenorhabditis reveals frequent loss of introns during nematode evolution. Genome Research 2004;14:1207–20.

Cho S, Mitchell A, Regier JC, et al. A highly conserved nuclear gene for low-level phylogenetics: elongation factor-1 alpha recovers morphology-based tree for heliothine moths. Mol Biol Evol 1995;12:650–6.

Chourrout D, Delsuc F, Chourrout P, et al. Minimal ProtoHox cluster inferred from bilaterian and cnidarian Hox complements. Nature 2006;442:684–7.

Clark RB. Dynamics in metazoan evolution. The origin of the coelom and segments. Oxford, UK: Clarendon Press, 1964.

Clark RB. Systematics and phylogeny: Annelida, Echiura, Sipuncula. In: Florkin M, Scheer BT, eds. Chemical Zoology, Vol. 4. New York, NY, USA: Academic Press, 1969:1–68.

Clarkson ENK. The visual system of trilobites. Paleontology 1979;22:1–22.

Claus C. Zur Kenntnis der Organisation und des feineren Baues der Daphniden und verwandter Cladoceren. Z wiss Zool 1876;17.

Clausen C. *Microphthalmus ephippiophorus* sp.n. (Polychaeta: Hesionidae) and two other *Microphthalmus* species from the Bergen area, Western Norway. Sarsia 1986;71:177–91.

Clausen S, Smith AB. Palaeoanatomy and biological affinities of a Cambrian deuterostome (Stylophora). Nature 2005;438:351–4.

Clement P. The phylogeny of rotifers: molecular, ultrastructural and behavioural data. Hydrobiologia 1993;255/256:527–44.

Clements KD, Gray RD, Choat JH. Rapid evolutionary divergences in reef fishes of the family Acanthuridae (Perciformes: Teleostei). Mol Phylogenet Evol 2003;26:190–201.

Clifton SW, Minx P, Fauron CM, et al. Sequence and comparative analysis of the maize NB mitochondrial genome. Plant Physiol 2004;136:3486–503.

Cobbett A, Wilkinson M, Wills MA. Fossils impact as hard as living taxa in parsimony analyses of morphology. Syst Biol 2007;56:753–66.

Coenen-Stass D. Investigations on microclimate and climatic preference of the desert woodlouse *Hemilepistus reaumuri* (Audouin and Savigny, 1826) (Oniscidea Isopoda Crustacea). Monit Zool Ital Monogr 1989;4:425–34.
Cohen BL, Gawthrop A, Cavalier-Smith T. Molecular phylogeny of brachiopods and phoronids based on nuclear-encoded small subunit ribosomal RNA gene sequences. Phil Trans R Soc Lond B Biol Sci 1998;353:2039–61.
Cohen BL, Weydmann A. Molecular evidence that phoronids are a subtaxon of brachiopods (Brachiopoda: Phoronata) and that genetic divergence of metazoan phyla began long before the early Cambrian. Organ Diver Evol 2005;5:253–73.
Cohen BL. Monophyly of brachiopods and phoronids reconciliation of molecular evidence with Linnaean classification (the subphylum Phoroniformea nov.). Proc R Soc Lond B Biol Sci 2000;267:225–31.
Cohen BL. Rerooting the rDNA gene tree reveals phoronids to be 'brachiopods without shells'; dangers of wide taxon samples in metazoan phylogenetics (Phoronida; Brachiopoda). Zool J Linn Soc 2013;167:82–92.
Colgan DJ, Hutchings PA, Braune M. A multigene framework for polychaete phylogenetic studies. Org Divers Evol 2006;6:220–35.
Collins AG. Phylogeny of Medusozoa and the evolution of cnidarian life cycles. J Evol Biol 2002;15:418–32.
Collins AG, Cartwright P, McFadden CS, Schierwater B. Phylogenetic context and basal metazoan model systems. Integr Comp Biol 2005;45:585–94.
Collins AG, Bentlage B, Matsumoto GI, Haddock SH, Osborn KJ, Schierwater B. Solution to the phylogenetic enigma of Tetraplatia, a worm-shaped cnidarian. Biol Lett 2006b;2:120–4.
Collins AG, Schuchert P, Marques AC, Jankowski T, Medina M, Schierwater B. Medusozoan phylogeny and character evolution clarified by new large and small subunit rDNA data and an assessment of the utility of phylogenetic mixture models. Syst Biol 2006a;55:97–115.
Collins D. Misadventures in the Burgess Shale. Nature 2009;460:952–3.
Collins LJ, Moulton V, Penny D. Use of RNA secondary structure for studying the evolution of RNase P and RNase MRP. J Mol Evol 2000;51:194–204.
Collins TM, Fedrigo O, Naylor GJP. Choosing the best genes for the job: The case for stationary genes in genome-scale phylogenetics. Syst Biol 2005;54:493–500.
Comas I, Moya A, Gonzalez-Candelas F. From phylogenetics to phylogenomics: the evolutionary relationships of insect endosymbiotic gamma-Proteobacteria as a test case. Syst Biol 2007a;56:1–16.
Comas I, Moya A, Gonzalez-Candelas F. Phylogenetic signal and functional categories in Proteobacteria genomes. BMC Evol Biol 2007b;7 Suppl 1:S7.
Conaco C, Bassett DS, Zhou H, et al. Functionalization of a protosynaptic gene expression network. Proc Natl Acad Sci USA 2012;109:10612–8.
Conway Morris S. The Burgess Shale animal *Oesia* is not a chaetognath: A reply to Szaniawski (2005). Acta Palaeont Pol 2009;54:175–9.
Conway Morris S, Caron J-B. Pikaia gracilens Walcott, a stem-group chordate from the Middle Cambrian of British Columbia. Biol Rev 2012;87:480–512.
Conway Morris S, Peel JS. Articulated halkieriids from the Lower Cambrian of north Greenland. Nature 1990;345:802–5.
Conway Morris S, Peel JS. Articulated halkieriids from the Lower Cambrian of North Greenland and their role in early protostome evolution. Phil Trans R Soc Lond B Biol Sci 1995;347:305–58.
Conway Morris S, Peel JS. The earliest annelids: Lower Cambrian polychaetes from the Sirius Passet Lagerstätte, Peary Land, North Greenland. Acta Palaeontol Pol 2008;53:137–48.
Conway Morris S, Whittington HB. The animals of the Burgess Shale. Sci Am 1979;110–20.

Cook CE, Jiménez E, Akam M, Saló E. The Hox gene complement of acoel flatworms, a basal bilaterian clade. Evol Dev 2004;6:154–63.
Cook CE, Smith ML, Telford MJ, Bastianello A, Akam M. Hox genes and the phylogeny of the arthropods. Curr Biol 2001;11:759–63.
Cook CE, Yue Q, Akam M. Mitochondrial genomes suggest that hexapods and crustaceans are mutually paraphyletic. Proc R Soc Lond B Biol Sci 2005;272:1295–304.
Copp AJ, Greene NDE, Murdoch JN. The genetic basis of mammalian neurulation. Nat Rev Genet 2003;4:784–93.
Coulier F, Popovici C, Villet R, Birnbaum D. MetaHox gene clusters. J Exp Zool 2000;288:345–51.
Coulombe-Huntington J, Majewski J. Intron loss and gain in Drosophila. Mol Biol Evol 2007;24:2842–50.
Cran DG. Cortical granules during oocyte maturation and fertilization. J Reprod Fertil 1989;38:49–62.
Crawford NG, Faircloth BC, McCormack JE, Brumfield RT, Winker K, Glenn TC. More than 1000 ultraconserved elements provide evidence that turtles are the sister group of archosaurs. Biol Lett 2012;8:783–6.
Cronin TW. Optical design and evolutionary adaptation in crustacean compound eyes. J Crust Biol 1986;6:1–23.
Cross FR, Buchler NE, Skotheim JM. Evolution of networks and sequences in eukaryotic cell cycle control. Phil Trans R Soc Lond B Biol Sci 2011;366:3532–44.
Crow KD, Stadler PF, Lynch VJ, Amemiya CT, Wagner GP. The fish specific Hox cluster duplication is coincident with the origin of teleosts. Mol Biol Evol 2006;23:121–36.
Crowden C, Sithigorngul P, Brackley P, Guastella J, Stretton AOW. Localization and differential expression of FMRFamide-like immunoreactivity in the nematode *Ascaris suum*. J Comp Neurol 1993;333:455–68.
Csűrös M, Holey JA, Rogozin IB. In search of lost introns. Bioinformatics 2007;23:i87–i96.
Cuffey RJ. An improved classification, based upon numerical-taxonomic analyses, for the higher taxa of entoproct and ectoproct bryozoans. In: Larwood GP, ed. Living and Fossil Bryozoa. New York, NY, USA: Academic Press, 1973:549–64.
Cuffey RJ, Blake DB. Cladistic analysis of the phylum Bryozoa. In: Bigey FP, ed. Bryozoaires actuels et fossiles. Bull Soc Sci Nat l'Ouest France Mém HS 1991;1:97–108.
Cummings MP, Otto SP, Wakeley J. Sampling properties of DNA sequence data in phylogenetic analysis. Mol Biol Evol 1995;12:814–22.
Cunningham CW, Zhu H, Hillis DM. Best fit maximum-likelihood models for phylogenetic inference: Empirical tests with known phylogenies. Evolution 1998;52:978–87.
Dabert M, Witalinski W, Kazmierski A, Olszanowski Z, Dabert J. Molecular phylogeny of acariform mites (Acari, Arachnida): Strong conflict between phylogenetic signal and long-branch attraction artifacts. Mol Phylogenet Evol 2010;56:222–41.
Dacks JB, Marinets A, Doolittle WF, Cavalier-Smith T, Logsdon JM. Analyses of RNA polymerase II genes from free-living protists: Phylogeny, long branch attraction, and the eukaryotic big bang. Mol Biol Evol 2002;19:830–40.
Dagan T. Phylogenomic networks. Trends Microbiol 2011;19:483–91.
Dahdul WM, Balhoff JP, Engeman J, et al. Evolutionary characters, phenotypes and ontologies: Curating data from the systematic biology literature. PLoS ONE 2010a;5:e10708.
Dahdul WM, Lundberg JG, Midford PE, et al. The teleost anatomy ontology: Anatomical representation for the genomics age. Syst Biol 2010b;59:369–83.
Dales RP. Annelids. London, UK: Hutchinson University Library, 1963.
Dales RP. The polychaete stomatodeum and phylogeny. In: Reish DJ, Fauchald K, eds. Essays on polychaetous annelids in memory of Dr. Olga Hartman. Los Angeles, CA, USA: Allan Hancock Foundation, University of Southern California, 1977:525–46.

Dalli D, Wilm A, Mainz I, Steger G. STRAL: Progressive alignment of non-coding RNA using base pairing probability vectors in quadratic time. Bioinformatics 2006;22:1593–9.
Daly M, Brugler MR, Cartwright P, et al. The phylum Cnidaria: A review of phylogenetic patterns and diversity 300 years after Linnaeus. Zootaxa 2007:127–82.
Dandekar P, Talbot P. Perivitelline space of mammalian oocytes: Extracellular matrix of unfertilized oocytes and formation of a cortical granule envelope following fertilization. Mol Reprod Dev 1992;31:135–43.
Danovaro R, Dell'anno A, Pusceddu A, Gambi C, Heiner I, Kristensen RM. The first Metazoa living in permanently anoxic conditions. BMC Biol 2010;8:30.
Darling JA, Reitzel AR, Burton PM, et al. Rising starlet: The starlet sea anemone, *Nematostella vectensis*. Bioessays 2005;27:211–21.
Darling KF, Wade CM, Kroon D, Brown AJL. Planktic foraminiferal molecular evolution and their polyphyletic origins from benthic taxa. Mar Micropal 1997;30:251–66.
Dawson PJ, Gutteridge WE, Gull K. Purification and characterization of tubulin from the parasitic nematode, *Ascaridia galli*. Mol Biochem Parasitol 1983;7:267–77.
de Carvalho MR, Ebach MC. Death of the specialist, rise of the machinist. Hist Phil Life Sci 2009;31:467–70.
de Jong DM, Hislop NR, Hayward DC, et al. Components of both major axial patterning systems of the Bilateria are differentially expressed along the primary axis of a 'radiate' animal, the anthozoan cnidarian *Acropora millepora*. Dev Biol 2006;298:632–43.
de Oliveira Almeida W, Christoffersen ML, de Sousa Amorim D, Eloy ECC. Morphological support for the positioning of Pentastomida and related fossils. Biotemas 2008;21:81–90.
de Pinna MCC. Concepts and tests of homology in the cladistic paradigm. Cladistics 1991;7:367–94.
de Putron SJ, Ryland JS. Effect of seawater temperature on reproductive seasonality and fecundity of *Pseudoplexaura porosa* (Cnidaria: Octocorallia): Latitudinal variation in Caribbean gorgonian reproduction. Invertebr Biol 2009;128:213–22.
de Quatrefages AM. Histoire Naturelle des Annelides, Marine et d'Eau Douce. Annelides et gephyriens. Paris, France: Librairie Encyclopedique de Roret, 1866.
De Robertis EM, Kuroda H. Dorsal-ventral patterning and neural induction in *Xenopus* embryos. Annu Rev Cell Dev Biol 2004;20:285–308.
de Rosa R, Grenier JK, Andreeva T, et al. Hox genes in brachiopods and priapulids and protostome evolution. Nature 1999;399:772–6.
de Velasco B, Mandal L, Mkrtchyan M, Hartenstein V. Subdivision and developmental fate of the head mesoderm in *Drosophila melanogaster*. Dev Genes Evol 2006;216:39–51.
Deans AR, Yoder MJ, Balhoff JP. Time to change how we describe biodiversity. Trends Ecol Evol 2012;27:78–84.
Debasieux P. Les yeux de Crustacés. La Cellule 1944;50:9–122.
Degnan JH, Rosenberg NA. Gene tree discordance, phylogenetic inference and the multispecies coalescent. Trends Ecol Evol 2009;24:332–40.
Dehal P, Boore JL. Two rounds of whole genome duplication in the ancestral vertebrate. PLoS Biol 2005;3: e314.
Delarbre C, Gallut C, Barriel V, Janvier P, Gachelin G. Complete mitochondrial DNA of the hagfish, *Eptatretus burgeri*: the comparative analysis of mitochondrial DNA sequences strongly supports the cyclostome monophyly. Mol Phylogenet Evol 2002;22:184–92.
Dellaporta SL, Xu A, Sagasser S, et al. Mitochondrial genome of *Trichoplax adhaerens* supports placozoa as the basal lower metazoan phylum. Proc Natl Acad Sci USA 2006;103:8751–6.
Delsman HC. Beiträge zur Entwicklungsgeschichte von *Oikopleura dioica*. Verhandelingen uit het Rijksinstituut voor het Onderzoek der Zee 1910;3:1–24.

Delsuc F, Brinkmann H, Chourrout D, Philippe H. Tunicates and not cephalochordates are the closest living relatives of vertebrates. Nature 2006;439:965–8.
Delsuc F, Brinkmann H, Philippe H. Phylogenomics and the reconstruction of the tree of life. Nat Rev Genet 2005;6:361–75.
Delsuc F, Stanhope MJ, Douzery EJ. Molecular systematics of armadillos (Xenarthra, Dasypodidae): Contribution of maximum likelihood and Bayesian analyses of mitochondrial and nuclear genes. Mol Phylogenet Evol 2003;28:261–75.
Delsuc F, Tsagkogeorga G, Lartillot N, Philippe H. Additional molecular support for the new chordate phylogeny. Genesis 2008;46:592–604.
Denes AS, Jékely G, Steinmetz PRH, et al. Molecular architecture of annelid nerve cord supports common origin of nervous system centralization in Bilateria. Cell 2007;129:277–88.
Denton AL, McConaughy BL, Hall BD. Usefulness of RNA polymerase II coding sequences for estimation of green plant phylogeny. Mol Biol Evol 1998;15:1082–5.
Dermauw W, Van Leeuwen T, Vanholme B, Tirry L. The complete mitochondrial genome of the house dust mite *Dermatophagoides pteronyssinus* (Trouessart): a novel gene arrangement among arthropods. BMC Genomics 2009;10:107.
DeSalle R, Schierwater B. An even “newer” animal phylogeny. Bioessays 2008;30:1043–7.
DeSalle R. Animal phylogenomics: Multiple interspecific genome comparisons. Methods Enzymol 2005;395:104–33.
Desjardins P, Morais R. Sequence and gene organization of the chicken mitochondrial genome. A novel gene order in higher vertebrates. J Mol Biol 1990;212:599–634.
Deutsch JS. Are Hexapoda members of the Crustacea? Evidence from comparative developmental genetics. Ann Soc entomol France 2001;37:41–9.
Deuve T. Sur la présence d’un “épipleurite” dans le plan de base du segment des Hexapodes. Bull Soc Entomol France 1994;99:199–210.
Devor EJ. Primate micrornas miR-220 and miR-492 lie within processed pseudogenes. J Hered 2006;97:186–90.
Dewel RA. Colonial origin for Eumetazoa: Major morphological transitions and the origin of bilaterian complexity. J Morphol 2000;243:35–74.
Dewey CN. Positional Orthology: Putting genomic evolutionary relationships into context. Briefings in Bioinf 2001;12:401–12.
Dibb N, Newman A. Evidence that introns arose at proto-splice sites. EMBO J 1989;8:2015–21.
Ding F, Sharma S, Chalasani P, Demidov VV, Broude NE, Dokholyan NV. Ab initio RNA folding by discrete molecular dynamics: From structure prediction to folding mechanisms. RNA 2008;14:1164–73.
Dohle W. Are the insects more closely related to the crustaceans than to the myriapods? Ent Scand Suppl 1997;51:7–16.
Dohle W. Are the insects terrestrial crustaceans? A discussion of some new facts and arguments and the proposal of the proper name ‘Tetraconata’ for the monophyletic unit Crustacea + Hexapoda. Ann Soc Entomol Fr 2001;37:85–103.
Dohle W. Die Embryonalentwicklung von *Glomeris marginata* (Villers) im Vergleich zur Entwicklung anderer Diplopoden. Zool Jahrb Abt Anat Ontog 1964;81:241–310.
Dohle W. Myriapod – insect relationships as opposed to an insect-crustacean sister group relationship. In: Fortey RA, Thomas RH, eds. Arthropod Relationships. London, UK: Chapman and Hall, 1998:305–15.
Dohle W. The ancestral cleavage pattern of the clitellates and its phylogenetic deviations. Hydrobiologia 1999;402:267–83.
Dohle W. Vergleichende Entwicklungsgeschichte des Mesoderm bei Articulaten. Fortschr Zool Syst Evolutionsforsch 1979;1:120–40.

Dohrmann M, Wörheide G. Novel scenarios of early animal evolution – Is it time to rewrite textbooks? Integr Comp Biol 2013; 53:503-11.
Dohrmann M, Janussen D, Reitner J, Collins AG, Wörheide G. Phylogeny and evolution of glass sponges (Porifera: Hexactinellida). Syst Biol 2008;57:388–405.
Dominguez Alonso P, Jefferies RPS. A stem group tunicate – microtomography and virtual fossils of a primitive tunicate. International Congress of Vertebrate Morphology 6, Jena, Germany, 2001.
Donath A, Stadler PF. Split-inducing indels in phylogenomic analysis 2011; Preprint, http://www.bioinf.uni-leipzig.de/Publications/PREPRINTS/11-001.pdf.
Donath A, Findeiβ S, Hertel J, et al. Noncoding RNA. In: Caetano-Anollés G, ed. Evolutionary Genomics and Systems Biology. New York, NY, USA: John Wiley & Sons, Inc. 2010:251–93.
Doncaster L. On the development of *Sagitta*, with notes on the anatomy of adult. Q J Microsc Sci 1903;46:351–98.
Dongen SMv. Graph clustering by flow simulation. PhD thesis, University of Utrecht, The Netherlands, 2000.
Donoghue MJ, Doyle JA, Gauthier J, Kluge AG, Rowe T. The importance of fossils in phylogeny reconstructions. Annu Rev Ecol Syst 1989;20:431–60.
Donoghue PCJ, Purnell MA. Distinguishing heat from light in debate over controversial fossils. Bioessays 2009;31:178–89.
Dordel J, Fisse F, Purschke G, Struck TH. Phylogenetic position of Sipuncula derived from multi-gene and phylogenomic data and its implication for the evolution of segmentation. J Zool Syst Evol Res 2010;48:197–207.
Doshi KJ, Cannone JJ, Cobaugh CW, Gutell RR. Evaluation of the suitability of free-energy minimization using nearest-neighbor energy parameters for RNA secondary structure prediction. BMC Bioinf 2004;5:105.
Dove H, Stollewerk A. Comparative analysis of neurogenesis in the myriapod *Glomeris marginata* (Diplopoda) suggests more similarities to chelicerates than to insects. Development 2003;130:2161–71.
Dowton M, Cameron SL, Dowavic JI, Austin AD, Whiting M. Characterization of 67 mitochondrial tRNA gene rearrangements in the hymenoptera suggests that mitochondrial tRNA gene position is selectively neutral. Mol Biol Evol 2009;26:1607–17.
Doyle JJ. Trees within trees: Genes and species, molecules and morphology. Syst Biol 1997;46:537–53.
Dress AW, Flamm C, Fritzsch G, et al. Noisy: Identification of problematic columns in multiple sequence alignments. Algorithms Mol Biol 2008;3:7.
Dress F, Duvert M. Development of the locomotory muscle of the chaetognath *Sagitta*. 2. Stereological study of fibre growth and differentiation processes. Tissue Cell 1994;26:349–73.
Drosophila 12 Genomes Consortium. Evolution of genes and genomes on the *Drosophila* phylogeny. Nature 2007;450:203–18.
Drummond DA, Bloom JD, Adami C, Wilke CO, Arnold FH. Why highly expressed proteins evolve slowly. Proc Natl Acad Sci USA 2005;102:14338–43.
Drummond FH. The eversible vesicles of *Campodea* (Thysanura). Proc Roy Entomol Soc London A, 1953;28:145–8.
Dubuisson M. Recherches sur la ventilation trachéenne chez les chilopodes et sur la circulation sanguine chez les scutigères. Arch Zool exp gen 1928;67:49–63.
Ducret F. Particularites structurales du systeme optique chez deux Chaetognathes (*Sagitta tasmanica* et *Eukrohnia hamata*) et incidences phylogenetiques. Zoomorphologie 1978;91:201–15.
Dunlop JA. Geological history and phylogeny of Chelicerata. Arthropod Struct Dev 2010;39:124–42.
Dunlop JA, Alberti G. The affinities of mites and ticks: A review. J Zool Syst Evol Res 2008;46:1–18.

Dunlop JA, Arango CP. Pycnogonid affinities: A review. J Zool Syst Evol Res 2005;43:8–21.
Dunlop JA, Braddy SJ. Scorpions and their sister group relationships. In: Fet V, Selden PA, eds. Scorpions 2001, In Memoriam Gary A Polis. Burnham Beeches, Bucks, UK: British Arachnological Society;2001:1–24.
Dunlop JA, Selden PA. Calibrating the chelicerate clock: A paleontological reply to Jeyaprakash and Hoy. Exp Appl Acarol 2009;48:183–97.
Dunlop JA, Kamenz C, Talarico G. A fossil trigonotarbid with a ricinuleid-like pedipalpal claw. Zoomorphology 2009;128:305–13.
Dunlop JA, Krüger J, Alberti G. The sejugal furrow in camel spiders and acariform mites. Arachnol Mitt;2012;43:8–15.
Dunn CW, Hejnol A, Matus DQ et al. Broad phylogenomic sampling improves resolution of the animal tree of life. Nature 2008;452:745–9.
Duò A, Bruggmann R, Zoller S, Bernt M, Gruenig CR. Mitochondrial genome evolution in species belonging to the *Phialocephala fortinii* s.l. – *Acephala applanata* species complex. BMC Genomics 2012;13:166.
Dutilh BE, Snel B, Ettema TJ, Huynen MA. Signature genes as a phylogenomic tool. Mol Biol Evol 2008;25:1659–67.
Duvert M. Étude de la structure et de la fonction de la musculature locomotrice d'un invertébré. Apport de la biologie cellulaire à l'histoire naturelle des chaetognathes. Cuad Invest Biol Bilbao 1989;15:1–30.
Duvert M. A very singular muscle: the secondary muscle of chaetognaths. Phil Trans R Soc Lond B Biol Sci 1991;332:245–60.
Duvert M, Barets AL. Ultrastructural studies of neuromuscular junctions in visceral and skeletal muscles of the chaetognath *Sagitta setosa*. Cell Tiss Res 1983;233:657–69.
Duvert M, Dress F. Development of the locomotory muscle of the chaetognath *Sagitta*. 1. Quantitative and qualitative aspects of the body and muscle tissue development within the phylum. Tissue Cell 1994;26:333–48.
Duvert M, Salat C. Fine structure of muscle and other components of the trunk of *Sagitta setosa* (Chaetognatha). Tissue Cell 1979;11:217–30.
Duvert M, Salat C. The primary body wall musculature in the arrow-worm *Sagitta setosa* (Chaetognatha): an ultrastructural study. Tissue Cell 1980;12:723–38.
Duvert M, Salat C. Ultrastructural and cytochemical studies on the connective tissue of chaetognaths. Tissue Cell 1990;22:865–78.
Duvert M, Salat C. Ultrastructural studies of the visceral muscles of chaetognaths. Acta Zool (Stockholm) 1995;76:75–87.
Duvert M, Savineau JP. Ultrastructural and physiological studies of the contraction of the trunk musculature of *Sagitta setosa* (chaetognath). Tissue Cell 1986;18:937–52.
Duvert M, Bouligand Y, Salat C. The liquid crystalline nature of the cytoskeleton in epidermal cells of the chaetognath *Sagitta*. Tissue Cell 1984;16:469–81.
Duvert M, Gros D, Salat C. Ultrastructural studies of the junctional complex in the musculature of the arrow-worm (*Sagitta setosa*) (Chaetognatha). Tissue Cell 1980;12:1–11.
Duvert M, Perez Y, Casanova J-P. Wound healing and survival of beheaded chaetognaths. J Mar Biol Ass UK 2000;80:891–8.
Duvert M, Savineau JP, Campistron G, Onteniente B 1997. Distribution and role of aspartate in the nervous system of the chaetognath *Sagitta*. J Comp Neurol 1997;380:485–94.
Eakin RM, Westfall JA. Fine structures of the eye of a chaetognath. J Cell Biol 1964;21:115–32.
Easterbrook PJ, Matthews DR. Fate of research studies. J R Soc Med 1992;85:71–6.
Eastham LES. The Embryology of *Pieris rapae*. Organogeny. Phil Trans R Soc Lond B Biol Sci 1931;219:1–50.

Eaton TH. The stem-tail problem and the ancestry of chordates. J Palaeontol 1970;44:969–79.
Ebersberger I, De Matos Simoes R, Kupczok A, et al. A consistent phylogenetic backbone for the fungi. Mol Biol Evol 2012;29:1319–34.
Ebersberger I, Galgoczy P, Taudien S, Taenzer S, Platzer M, von Haeseler A. Mapping human genetic ancestry. Mol Biol Evol 2007;24:2266–76.
Ebersberger I, Strauss S, von Haeseler A. HaMStR: Profile hidden markov model based search for orthologs in ESTs. BMC Evol Biol 2009;9:157.
Eckelbarger KJ, Blades-Eckelbarger PI. Oogenesis in calanoid copepods. Invertebr Repod Dev 2005;47:167–81.
Eckelbarger KJ, Grassle JP. Ultrastructure of the ovary and oogenesis in the polychaete *Capitella jonesi* (Hartman, 1959). J Morphol 1982;171:305–20.
Eckelbarger KJ. Oogenesis and oocytes. In: Bartolomaeus T, Purschke G, eds. Morphology, molecules, evolution and phylogeny in Polychaeta and related taxa. Hydrobiologia 2005;535/536:179–89.
Eddy SR. Profile hidden Markov models. Bioinformatics 1998;14:755–63.
Edgar RC. MUSCLE: multiple sequence alignment with high accuracy and high throughput. Nucleic Acids Res 2004;32:1792–7.
Edgecombe GD. Morphological data, extant Myriapoda, and the myriapod stem-group. Contrib Zool 2004;73:1–57.
Edgecombe GD. Anatomical nomenclature: Homology, standardization and datasets. Zootaxa 2008;1950(1950):87–95.
Edgecombe GD. Palaeontological and molecular evidence linking arthropods, onychophorans, and other Ecdysozoa. Evol Edu Outreach 2009;2:178–90.
Edgecombe GD. Arthropod phylogeny: An overview from the perspectives of morphology, molecular data and the fossil record. Arthropod Struct Dev 2010;39:74–87.
Edgecombe G, Giribet G, Dunn C, et al. Higher-level metazoan relationships: Recent progress and remaining questions. Org Divers Evol 2011;11:151–72.
Edgecombe GD, Wilson GD, Colgan DJ, Gray MR, Cassis G. Arthropod cladistics: Combined analysis of histone H3 and U2 snRNA sequences and morphology. Cladistics 2000;16:155–203.
Edney EB, Spencer JO. Cutaneous respiration in woodlice. J Exp Biol 1955;32:256–69.
Eeckhaut I, Fievez L, Müller MCM. Larval development of *Myzostoma cirriferum* (Myzostomida). J Morphol 2003;258:269–83.
Eeckhaut I, Jangoux M. Fine-structure of the spermatophore and intradermic penetration of sperm cells in *Myzostoma cirriferum* (Annelida, Myzostomida). Zoomorphology 1991;111:49–58.
Eeckhaut I, Jangoux M. Life cycle and mode of infestation of *Myzostoma cirriferum* (Annelida), a symbiotic myzostomid of the comatulid crinoid *Antedon bifida* (Echinodermata). Dis Aquat Org 1993;15:207–17.
Eeckhaut I, McHugh D, Mardulyn P, et al. Myzostomida: a link between trochozoans and flatworms? Proc R Soc Lond B Biol Sci 2000;267:1383–92.
Eeckhaut I. *Mycomyzostoma calcidicola* gen. et sp. nov., the first extant parasitic myzostome infesting crinoid stalks, with a nomenclatural appendix by M.J. Grygier. Species Div 1998;3:89–103.
Eernisse DJ, Albert JS, Anderson FE. Annelida and Arthropoda are not sister taxa: A phylogenetic analysis of spiralian metazoan morphology. Syst Biol 1992;41:305–30.
Eernisse DJ. Arthropod and annelid relationships reexamined. In: Fortey RA, Thomas RH, eds. Arthropod Relationships. London, UK: Chapman and Hall, 1997:43–56.
Ehlers U. The basic organization of the Plathelminthes. Hydrobiologia 1995;305:21–6.
Ehlers U, Ahlrichs W, Lemburg C, Schmidt-Rhaesa A. Phylogenetic systematization of the Nemathelminthes (Aschelminthes). Verh Dtsch Zool Ges 1996;891:8.

Eibye-Jacobsen D. A reevaluation of *Wiwaxia* and the polychaetes of the Burgess Shale. Lethaia 2004;37:317–35.

Eibye-Jacobsen D, Vinther J. Reconstructing the ancestral annelid. J Zool Syst Evol Res 2012;50:85–7.

Eidmann H, Kühlhorn F. Lehrbuch der Entomologie. Hamburg and Berlin: Paul Parey Verlag 1970.

Eisen JA, Fraser CM. Phylogenomics: Intersection of evolution and genomics. Science 2003;300:1706–7.

Eisenbeis, G. Physiological abssorption of liquid water by Collembola: absorptions by the ventral tube at different salinities. J Ins Physiol 1982;28:11–20.

Eitel M, Schierwater B. The phylogeography of the Placozoa suggests a taxon rich phylum in tropical and subtropical waters. Mol Ecol 2010;19:2315–27.

Eitel M, Guidi L, Hadrys H, Balsamo M, Schierwater B. New insights into placozoan sexual reproduction and development. PLoS ONE 2011;6:e19639.

Eitel M, Osigus HJ, DeSalle R, Schierwater B. Global diversity of the Placozoa. PLoS ONE 2013;8:e57131.

Elofsson R, Dahl E. The optic neuropils and chiasmata of Crustacea. Z Zellforsch 1970;107:343–60.

Elofsson R, Hessler RR. Central nervous system of *Hutchinsoniella macracantha* (Cephalocarida). J Crust Biol 1990;10:423–39.

Elofsson R, Hessler RR. The tegumental sensory organ and nervous system of *Derocheilocaris typica* (Crustacea: Mystacocarida). Arthropod Struct Dev 2005;34:139–52.

Embley TM, Hirt RP, Williams DM. Biodiversity at the molecular level: The domains, kingdoms and phyla of life. Phil Trans R Soc Lond B Biol Sci 1994;345:21–33.

Ender A, Schierwater B. Placozoa are not derived cnidarians: Evidence from molecular morphology. Mol Biol Evol 2003;20:130–4.

Endo K, Noguchi Y, Ueshima R, Jacobs HT. Novel repetitive structures, deviant protein-encoding sequences and unidentified ORFs in the mitochondrial genome of the brachiopod *Lingula anatina*. J Mol Evol 2005;61:36–53.

Engel MS, Grimaldi DA. A winged *Zorotypus* in Miocene amber from the Dominican Republic (Zoraptera: Zorotypidae), with discussion on relationships of and within the order. Acta Geológica Hispánica 2000;35:149–64.

Erber A, Riemer D, Bovenschulte M, Weber K. Molecular phylogeny of metazoan intermediate filament proteins. J Mol Evol 1998;47:751–62.

Ereskovsky AV, Tokina DB. Asexual reproduction in homoscleromorph sponges (Porifera; Homoscleromorpha). Mar Biol 2007;151:425–34.

Ereskovsky AV, Konyukov PY, Tokina DB. Morphogenesis accompanying larval metamorphosis in *Plakina trilopha* (Porifera, Homoscleromorpha). Zoomorphology 2010;129:21–31.

Ereskovsky AV, Renard E, Borchiellini C. Cellular and molecular processes leading to embryo formation in sponges: Evidences for high conservation of processes throughout animal evolution. Dev Genes Evol 2013;223:5–22.

Eriksson JB, Tait NN. Early development in the velvet worm *Euperipatoides kanangrensis* Reid 1996 (Onychophora: Peripatopsidae). Arthropod Struct Dev 2012;41:483–93.

Erixon P, Svennblad B, Britton T, Oxelman B. Reliability of bayesian posterior probabilities and bootstrap frequencies in phylogenetics. Syst Biol 2003;52:665–73.

Erpenbeck D, Wörheide G. On the molecular phylogeny of sponges (Porifera). Zootaxa 2007:107–26.

Erpenbeck D, Breeuwer J, Parra-Velandia F, van Soest R. Speculation with spiculation? – Three independent gene fragments and biochemical characters versus morphology in demosponge higher classification. Mol Phylogenet Evol 2006;38:293–305.

Erséus C, Kallersjo M. 18S rDNA phylogeny of Clitellata (Annelida). Zool Sripta 2004;33:187–96.

Erséus C. Phylogeny of oligochaetous Clitellata. In: Bartolomaeus T, Purschke G, eds. Morphology, molecules, evolution and phylogeny in Polychaeta and related taxa. Hydrobiologia 2005;535/536:357–72.
Ertas B, von Reumont BM, Wägele J, Misof B, Burmester T. Hemocyanin suggests a close relationship of Remipedia and Hexapoda. Mol Biol Evol 2009;26:2711–8.
Erwin DH, Laflamme M, Tweedt SM, Sperling EA, Pisani D, Peterson KJ. The Cambrian conundrum: Early divergence and later ecological success in the early history of animals. Science 2011;334:1091–7.
Evans HE. Insect Biology: A Textbook of Entomolgy. Reading, MA, USA: Addison Wesley Publishing Company, 1984.
Ewen-Campen B, Donoughe S, Clarke DN, Extavour CG. Germ cell specification requires zygotic mechanisms rather than germ plasm in a basally branching insect. Curr Biol 2013;23:1–8.
Ewing HE. The legs and leg-bearing segments of some primitive arthropod groups, with notes on leg-segmentation in the Arachnida. Smithson Miscell Coll 1928;11:1–41.
Extavour CG, Akam M. Mechanisms of germ cell specification across the metazoans: epigenesis and preformation. Development 2003;130:5869–84.
Fahrein K, Masta SE, Podsiadlowski L. The first complete mitochondrial genome sequences of Amblypygi (Chelicerata: Arachnida) reveal conservation of the ancestral arthropod gene order. Genome 2009;52:456–66.
Fahrein K, Talarico G, Braband A, Podsiadlowski L. The complete mitochondrial genome of *Pseudocellus pearsei* (Chelicerata: Ricinulei) and a comparison of mitochondrial gene rearrangements in Arachnida. BMC Genomics 2007;8:386.
Fahrein, K., Arabi, J., Braband, A., and Podsiadlowski, L. (unpublished work). Rearrangements and mitochondrial control region variation in spiders. University of Bonn, Germany.
Fahrenbach WH. The biology of a harpacticoid copepod. La Cellule 1962;62:303–76.
Fahrenbach WH. The visual system of the horseshoe crab *Limulus polyphemus*. Intern Rev Cytol 1975;41:285–349.
Faircloth BC, McCormack JE, Crawford NG, Harvey MG, Brumfield RT, Glenn TC. Ultraconserved elements anchor thousands of genetic markers spanning multiple evolutionary timescales. Syst Biol 2012;61:717–26.
Faivovich J. On RASA. Cladistics 2002;18:324–33.
Faller S, Todt C, Rothe BH, Schmidt-Rhaesa A, Loesel R. Comparative neuroanatomy of Caudofoveata, Solenogastres, Polyplacophora, and Scaphopoda (Mollusca) and its phylogenetic implications. Zoomorphology 2012;131:149–70.
Fanenbruck M. Die Kopfanatomie der Mandibulata. Eine anatomische Studie des Skelettmuskelsystems zur Klärung der Verwandtschaftsverhältnisse von Krebsen, Tausendfüssern und Insekten. Saarbrücken, Germany: Südwestdeutscher Verlag für Hochschulschriften, 2009.
Fanenbruck M, Harzsch S. A brain atlas of *Godzilliognomus frondosus* Yager, 1989 (Remipedia, Godzilliidae) and comparison with the brain of *Speleonectes tulumensis* Yager, 1987 (Remipedia, Speleonectidae): Implications for arthropod relationships. Arthropod Struct Dev 2005;34:343–78.
Fanenbruck M, Harzsch S, Wägele JW. The brain of the Remipedia (Crustacea) and an alternative hypothesis on their phylogenetic relationships. Proc Natl Acad Sci USA 2004;101:3868–73.
Farias IP, Orti G, Sampaio I, Schneider H, Meyer A. The cytochrome b gene as a phylogenetic marker: the limits of resolution for analyzing relationships among cichlid fishes. J Mol Evol 2001;53:89–103.
Farris JS. Phylogenetic analysis under Dollo's Law. Syst Biol 1977;26:77–88.
Farris SM, Roberts NS. Coevolution of generalist feeding ecologies and gyrencephalic mushroom bodies in insects. Proc Natl Acad Sci USA 2005;102:17394–9.

Fauchald K. Polychaete Phylogeny: A problem in protostome evolution. Syst Zool 1974;23:493–506.
Fauchald K. The polychaete worms. Definitions and keys to the orders, families and genera. Los Angeles: Natural History Museum of Los Angeles County. Science Series, 1977.
Fauchald K, Rouse G. Polychaete systematics: Past and present. Zool Sripta 1997;26:71–138.
Faulkner J. Marine natural products. Nat Prod Rep 2002;19:1–48.
Fauvel P. Polychètes Errantes. Faune de France 1923;5:1–488.
Fauvel P. Polychètes Sédentaires. Addenda aux Errantes, Archiannélides, Myzostomaires. Faune de France 1927;16:1–494.
Fauvel P. The Fauna of India, including Pakistan, Ceylon, Burma and Malaya. Annelida. Polychaeta. Allahabad, India: The Indian Press, 1953.
Feiner N, Ericsson R, Meyer A, Kuraku S. Revisiting the origin of the vertebrate Hox14 by including ilts relict sarcopterygian members. J Exp Zool Part B 2011;316:515–25.
Felgenhauer BE, Abele LG, Felder DL. Remipedia. In: Harrison FW, Humes AG eds. Microscopic Anatomy of Invertebrates Vol. 9. New York, NY, USA: Wiley-Liss. 1992:225–47.
Felsenstein J. Maximum likelihood and minimum-steps methods for estimating evolutionary trees from data on discrete characters. Syst Zool 1973;22:240–9.
Felsenstein J. Cases in which parsimony or compatibility methods will be positively misleading. Syst Zool 1978;27:401–10.
Felsenstein J. Evolutionary trees from DNA sequences: A maximum likelihood approach. J Mol Evol 1981;17:368–76.
Felsenstein J. Distance methods for inferring phylogenies: A justification. Evolution 1984;38:16–24.
Felsenstein J. Confidence limits on phylogenies: An approach using the bootstrap. Evolution 1985;39:783–91.
Felsenstein J. Inferring Phylogenies. Sunderland, MA, USA: Sinauer Associates, 2004.
Felsenstein J, Sober E. Parsimony and likelihood: An exchange. Syst Zool 1986;35:617–26.
Ferrara F, Paoli P, Taiti S. An original respiratory structure in the xeric genus *Periscyphis* Gerstaecker, 1873 (Crustacea: Oniscidea: Eubelidae). Zool Anz 1996;235:147–56.
Ferrier DE, Minguillión C. Evolution of the Hox/ParaHox gene clusters. Int J Dev Biol 2003;47:605–11.
Ferrier DEK, Holland PWH. Ancient origin of the Hox gene cluster. Nat Rev Genet 2001;2:33–8.
Fertin G, Labarre A, Rusu I, Tannier E, Vialette S. Combinatorics of Genome Rearrangements. Cambridge, MA, USA: MIT Press.
Feustel H. Untersuchungen über die Exkretion bei Collembolen. Z wiss Zool 1958;161:209–38.
Fidalgo ML Biology of the freshwater shrimp *Atyephyra desmaresti* Millet (Decapoda, Natantia) in the river Douro, Portugal II. Feeding rate and assimilation efficiency. Publ Instit Zool “Dr. Augusto Nobre” 1990;223:1–19.
Field KG, Olsen GJ, Lane DJ, et al. Molecular phylogeny of the animal kingdom. Science 1988;239:748–53.
Figeac M, Varré J-S. Sorting by reversals with common intervals. In: Jonassen I, KimJ, eds. Algorithms in Bioinformatics. Lecture notes in computer science. Berlin, Germany: Springer, 2006;3240:26–37.
Fischer AHL, Arendt D. Mesoteloblast-like mesodermal stem cells in the polychaete annelid *Platynereis dumerilii* (Nereididae). J Exp Zool B 2013;320:95–104.
Fischer M, Steel M. Sequence length bounds for resolving a deep phylogenetic divergence. J Theor Biol 2009;256:247–52.
Fitch WM, Margoliash E. Construction of phylogenetic trees. Science 1967;155:279–84.
Fitch WM. Distinguishing homologous from analogous proteins. Syst Zool 1970;19:99–113.
Fitch WM. Toward defining course of evolution – minimum change for a specific tree topology. Syst Zool 1971;20:406–16.

Fitzpatrick D, Logue M, Stajich J, Butler G. A fungal phylogeny based on 42 complete genomes derived from supertree and combined gene analysis. BMC Evol Biol 2006;6:99.

Fleck G, Ullrich B, Brenk M, et al. A phylogeny of anisopterous dragonflies (Insecta, Odonata) using mtRNA genes and mixed nucleotide/doublet models. J Zool Syst Evol Res 2008;46:310–22.

Fletcher W, Yang Z. INDELible: A flexible simulator of biological sequence evolution. Mol Biol Evol 2009;26:1879–88.

Flook PK, Rowell CHF. The effectiveness of mitochondrial rRNA gene sequences for the reconstruction of the phylogeny of an insect order (Orthoptera). Mol Phylogenet Evol 1997;8:177–92.

Foata J, Dezfuli BS, Pinelli B, Marchand B. Ultrastructure of spermiogenesis and spermatozoon of *Leptorhynchoides plagicephalus* (Acanthocephala, Palaeacanthocephala), a parasite of the sturgeon *Acipenser naccarii* (Osteichthyes, Acipenseriformes). Parasitol Res 2004;93:582–9.

Foelix RF. Biology of Spiders, 3rd edition. New York, NY, USA: Oxford University Press, 2011.

Fong JJ, Fujita MK. Evaluating phylogenetic informativeness and data-type usage for new protein-coding genes across Vertebrata. Mol Phylogenet Evol 2011;61:300–7.

Fontaneto D, Jondelius U. Broad taxonomic sampling of mitochondrial cytochrome c oxidase subunit I does not solve the relationships between Rotifera and Acanthocephala. Zool Anz 2011;250:80–5.

Forslund K, Sonnhammer ELL. Predicting protein function from domain content. Bioinformatics 2008;24:1681–7.

Forst CV, Schulten K. Phylogenetic analysis of metabolic pathways. J Mol Evol 2001;52:471–89.

Foster PG. Modeling compositional heterogeneity. Syst Biol 2004;53:485–95.

Foster PG, Hickey DA. Compositional bias may affect both DNA-based and protein-based phylogenetic reconstructions. J Mol Evol 1999;48:284–90.

Fransen ME. Ultrastructure of coelomic organization in Polychaeta. PhD thesis, University of North Carolina, Chapel Hill, NC, USA, 1980a.

Fransen ME. Ultrastructure of coelomic organization in annelids. I. Archiannelids and other small polychaetes. Zoomorphology 1980b;95:235–49.

Fransen ME. The role of ECM in the development of invertebrates: A phylogeneticist's view. In: Hawkes S, Wang LJ, eds. Extracellular Matrix. New York, NY, USA: Academic Press, 1982:177–81.

Fransen ME. Coelomic and vascular system. In: Westheide W, Hermans CO, eds. The ultrastructure of the Polychaeta. Microfauna marina 1988;4:199–213.

Freeman R, Ikuta T, Wu M, et al. Identical Genomic Organization of Two Hemichordate Hox Clusters. Curr Biol 2012;22:2053–8.

Freudenstein J. Characters, states and homology. Syst Biol 2005;54:965–73.

Fried C, Hordijk W, Prohaska SJ, Stadler CR, Stadler PF. The footprint sorting problem. J Chem Inf Comp Sci 2004;44:332–8.

Friedemann K, Wipfler B, Bradler S, Beutel RG. On the head morphology of *Phyllium* and the phylogenetic relationships of Phasmatodea (Insecta). Acta Zoologica 2012;93:184–99.

Friedrich F, Beutel RG. Goodbye Halteria? The thoracic morphology of Endopterygota (Insecta) and its phylogenetic implications. Cladistics 2010;26:579–612.

Friedrich M, Tautz D. Ribosomal DNA phylogeny of the major extant arthropod classes and the evolution of myriapods. Nature 1995;376:165–7.

Friedrich M, Tautz D. An episodic change of rDNA nucleotide substitution rate has occurred during the emergence of the insect order Diptera. Mol Biol Evol 1997;14;644–53.

Friedrich M, Tautz D. Arthropod rDNA phylogeny revisited: A consistency analysis using Monte Carlo simulation. Ann Soc Entomol Fr 2001;37:21–40.

Fritsch M, Richter S. The formation of the nervous system during larval development in *Triops cancriformis* (Bosc) (Crustacea, Branchiopoda): An immunohistochemical survey. J Morph 2010;271:1457–81.

Fritsch M, Richter S. Nervous system development in Spinicaudata and Cyclestherida (Crustacea, Branchiopoda) – Comparing two different modes of indirect development by using an event pairing approach. J Morph 2012;273:672–95.

Fritzsch G, Böhme MU, Thorndyke M, et al. PCR Survey of *Xenoturbella bocki* Hox Genes. J Exp Zool Part B 2008;310B:278–84.

Fritzsch G, Schlegel M, Stadler PF. Alignments of mitochondrial genome arrangements: Applications to metazoan phylogeny. J Theor Biol 2006;240:511–20.

Fu X, Adamski M, Thompson EM. Altered miRNA repertoire in the simplified chordate, *Oikopleura dioica*. Mol Biol Evol 2008;25:1067–80.

Fuchs J, Iseto T, Hirose M, Sundberg P, Obst M. The first internal molecular phylogeny of the animal phylum Entoprocta (Kamptozoa). Mol Phylogenet Evol 2010;56:370–9.

Fuchs J, Obst M, Sundberg P. The first comprehensive molecular phylogeny of Bryozoa (Ectoprocta) based on combined analyses of nuclear and mitochondrial genes. Mol Phylogenet Evol 2009;52,225–33.

Fuellen G. Evolution of gene regulation: On the road towards computational inferences. Briefings in Bioinf 2011;12:122–31.

Fuellen G, Wägele JW, Giegerich R. Minimum conflict: a divide-and-conquer approach to a phylogeny estimation. Bioinformatics 2001;17:1168–78.

Fukami-Kobayashi K, Minezaki Y, Tateno Y, Nishikawa K. A tree of life based on protein domain organizations. Mol Biol Evol 2007;24:1181–9.

Füller H. Elektronenmikroskopische Untersuchungen der Malphighischen Gefäße von *Lithobius forficatus* (L.). Z wiss Zool 1966;173:191–216.

Funch P, Sorensen MV, Obst M. On the phylogenetic position of Rotifera – have we come any further? Hydrobiologia 2005;546:11–28.

Furlong RF, Holland PWH. Bayesian Phylogenetic Analysis Supports Monophyly of Ambulacralia and of Cyclostomes. Zool Sci 2002;19:593–9.

Furnestin M-L. Contribution a l'étude histologique des Chaetognathes. Rev Trav Inst Pèches Marit 1967;31:383–92.

Gadagkar SR, Rosenberg MS, Kumar S. Inferring species phylogenies from multiple genes: concatenated sequence tree versus consensus gene tree. J Exp Zool Part B 2005;304:64–74.

Gai Y, Song D, Sun H, Yang Q, Zhou K. The complete mitochondrial genome of *Symphylella* sp. (Myriapoda: Symphyla): Extensive gene order rearrangement and evidence in favor of Progoneata. Mol Phylogenet Evol 2008;49:574–85.

Galant R, Walsh CM, Carroll SB. Hox repression of a target gene: extradenticle-independent, additive action through multiple monomer binding sites. Development 2002;129:3115–26.

Galliot B, de Vargas C, Miller D. Evolution of homeobox genes: Q 50 Paired-like genes founded the Paired class. Dev Genes Evol 1999;209:186–97.

Garcia-Fernandez J. The genesis and evolution of homeobox gene clusters. Nat Rev Genet 2005;6:881–92.

García-Varela M, Cummings MP, Perez-Ponce de Leon G, Gardner SL, Laclette JP. Phylogenetic analysis based on 18S ribosomal RNA gene sequences supports the existence of class Polyacanthocephala (Acanthocephala). Mol Phylogenet Evol 2002;23:288–92.

García-Varela M, Nadler SA. Phylogenetic relationships among Syndermata inferred from nuclear and mitochondrial gene sequences. Mol Phylogenet Evol 2006;40:61–72.

García-Varela M, Pérez-Ponce de León G, de la Torre P, Cummings MP, Sarma SS, Laclette JP. Phylogenetic relationships of Acanthocephala based on analysis of 18S ribosomal RNA gene sequences. J Mol Evol 2000;50:532–40.

Gardiner SL, Rieger RM. Rudimentary cilia in muscle cells of annelids and echinoderms. Cell Tissue Res 1980;213:247–52.

Gardiner SL. Polychaeta: General Organization, Integument, Musculature, Coelom, and Vascular System. In: Harrison FW, Gardiner SL, eds. Microscopic Anatomy of Invertebrates, Vol. 7. New York, NY, USA: Wiley-Liss, 1992:19–52.

Gardner PP, Daub J, Tate J, et al. Rfam: Wikipedia, clans and the "decimal" release. Nucleic Acids Res 2011;39(Database issue):D141–5.

Gardner PP, Giegerich R. A comprehensive comparison of comparative RNA structure prediction approaches. BMC Bioinf 2004;5:140.

Garey JR, Near TJ, Nonnemacher MR, Nadler SA. Molecular evidence for Acanthocephala as a subtaxon of Rotifera. J Mol Evol 1996;43:287–92.

Garey JR, Schmidt-Rhaesa A, Near TJ, Nadler SA. The evolutionary relationships of rotifers and acanthocephalans. In: Wurdak E, Wallace R, Segers H, eds. Rotifera VIII: A comparative approach. Hydrobiologia 1998; 387/388:83–91.

Garstang W. The morphology of the Tunicata, and its bearings on the phylogeny of the Chordata. Q J Microsc Sci 1928;72:51–187.

Gatesy J, Baker RH. Hidden likelihood support in genomic data: Can forty-five wrongs make a right? Syst Biol 2005;54:483–92.

Gauchat D, Mazet F, Berney C, et al. Evolution of Antp-class genes and differential expression of *Hydra* Hox/paraHox genes in anterior patterning. Proc Natl Acad Sci USA 2000;97:4493.

Gaucher EA, Miyamoto MM. A call for likelihood phylogenetics even when the process of sequence evolution is heterogeneous. Mol Phylogenet Evol 2005;37:928–31.

Gaunt MW, Miles MA. An insect molecular clock dates the origin of the insects and accords with palaeontological and biogeographic land marks. Mol Biol Evol 2002;19:748–61.

Gaut S, Lewis PO. Success of maximum likelihood phylogeny inference in the four-taxon case. Mol Biol Evol 1995;12:152–62.

Gazave E, Lapébie P, Ereskovsky A, et al. No longer Demospongiae: Homoscleromorpha formal nomination as a fourth class of Porifera. Hydrobiologia 2012;687:3–10.

Gazi M, Sultana T, Min G-S, et al. The complete mitochondrial genome sequence of *Oncicola luehei* (Acanthocephala: Archiacanthocephala) and its phylogenetic position within Syndermata. Parasitol Int 2012;61:307–16.

Geary TG, Nulf SC, Favreau MA, et al. Three beta-tubulin cDNAs from the parasitic nematode *Haemonchus contortus*. Mol Biochem Parasitol 1992b;50:295–306.

Geary TG, Price DA, Bowman JW, et al. Two FMRFamide-like peptides from the free-living nematode *Panagrellus redivivus*. Peptides 1992a;13:209–14.

Gegenbaur C. Versuch eines Systemes der Medusen mit Beschreibung neuer oder wenig gekannter Formen; zugleich ein Beitrag zur Kenntniss der Fauna des Mittelmeeres. Z wiss Zool 1857;8:202–73.

Gemballa S. Myoseptenarchitektur und Rumpfmuskulatur der Actinopterygii – ein vergleichend-anatomischer Ansatz zum Verstaendnis der undulatorischen Lokomotion. Verh Ges Ichthyol 1998;1:29–58.

Gemballa S, Weitbrecht GW, Sanchéz-Villagra MR. The myosepta in *Branchiostoma lanceolatum* (Cephalochordata):3D reconstruction and microanatomy. Zoomorphology 2003;122:169–79.

Geoffroy Saint-Hilaire E. Considerations generales sur la vertebre. Mem Mus d'Hist Nat 1822;9:89–119.

Gerberding, M. Germ band formation and early neurogenesis of *Leptodora kindti* (Cladocera): First evidence for neuroblasts in the entomostracan crustaceans. Invert Reprod Dev 1997;32:63–73.

Gerhart J, Lowe C, Kirschner M. Hemichordates and the origin of chordates. Curr Opin Genet Dev 2005;15:461–7.

Gerhart J. Inversion of the chordate body axis: Are there alternatives? Proc Natl Acad Sci USA 2000;97:4445–8.

Gerlach D, Wolf M, Dandekar T, Müller T, Pokorny A, Rahmann S. Deep metazoan phylogeny. In Silico 2007;7:151–4.

Ghiradelli E. Chaetognaths: Two Unsolved Problems: The Coelom and their Affinities. In: Lanzavecchia G, Valvassori R, Candia Carnevali MD, editors. Body Cavities: Function and Phylogeny. UZISelected Symposia and Monographs, Vol 8. Modena, Italy: Mucci Editore, 1995:167–85.

Ghirardelli E. Some aspects of the biology of the chaetognaths. Adv Arch Biol 1968;6:271–357.

Gibb GC, Kardailsky O, Kimball RT, Braun EL, Penny D. Mitochondrial genomes and avian phylogeny: complex characters and resolvability without explosive radiations. Mol Biol Evol 2007;24:269–80.

Gibson T, Blok VC, Phillips MS, et al. The mitochondrial subgenomes of the nematode *Globodera pallida* are mosaics: Evidence of recombination in an animal mitochondrial genome. J Mol Evol 2007;64:463–71.

Gilbert SF. Arthropods: The Crustaceans, Spiders, and Myriapods. In: Gilbert SF, Raunio AM., eds. Embryology: Constructing the Organism. Sunderland, MA, USA: Sinauer Associates, 1997:237–58.

Gilula NB, Branton D, Satir P. The septate junction: a structural basis for intercellular coupling. Proc Natl Acad Sci USA 1970;67:213–20.

Giribet G. Current advances in the phylogenetic reconstruction of metazoen evolution. A new paradigm for the Cambrian Explosion? Mol Phylogenet Evol 2002;24:345–57.

Giribet G. Molecules, development and fossils in the study of metazoan evolution; Articulata versus Ecdysozoa revisited. Zoology 2003;106:303–26.

Giribet G. A new dimension in combining data? The use of morphology and phylogenomic data in metazoan systematics. Acta Zoologica 2010;91:11–9.

Giribet G, Edgecombe GD. Reevaluating the arthropod tree of life. Annu Rev Entomol 2012;57:167–86.

Giribet G, Ribera C. A review of arthropod phylogeny: New data based on ribosomal DNA sequences and direct character optimization. Cladistics 2000;16:204–31.

Giribet G, Edgecombe G, Wheeler WC. Arthropod phylogeny based on eight molecular loci and morphology. Nature 2001;413:157–61.

Giribet G, Distel DL, Polz M, Sterrer W, Wheeler WC. Triploblastic relationships with emphasis on the acoelomates and position of Gnathostomulida, Cycliophora, Plathelminthes and Chaetognatha: A combined approach to 18S rDNA sequences and morphology. Syst Biol 2000;49:539–62.

Giribet G, Sørensen MV, Funch P, Kristensen RM, Sterrer W. Investigations into the phylogenetic position of Micrognathozoa using four molecular loci. Cladistics 2004;20:1–13.

Giribet G, Richter S, Edgecombe GD, Wheeler WC. The position of crustaceans within Arthropoda – evidence from nine molecular loci and morphology. In: Koenemann S, Jenner R, eds. Crustacea and arthropod relationships. Boca Ratón, FL, USA: CRC Press, 2005:307–52.

Giribet G, Dunn CW, Edgecombe GD, Hejnol A, Martindale MQ, Rouse GW. Assembling the Spiralian Tree of Life. In: Telford MJ, Littlewood DTJ, eds. Animal Evolution: Genes, Genomes, Fossils and Trees. Oxford, UK: Oxford University Press, 2009:52–64.

Gissi C, Pesole G, Mastrototaro F, Iannelli F, Guida V, Griggio F. Hypervariability of ascidian mitochondrial gene order: Exposing the myth of deuterostome organelle genome stability. Mol Biol Evol 2010;27:211–5.

Glasby CJ, Hutchings PA, Fauchald K, et al. Class Polychaeta. In: Beesly PL, Ross GJ, Glasby CJ, eds. Polychaetes & Allies: The Southern Synthesis. Fauna of Australia. Vol. 4A Polychaeta, Myzostomida, Pogonophora, Echiura, Sipuncula. Melbourne, Australia: CSIRO Publishing, 2000.

Gleizer L, Stent GS. Developmental origin of segmental identity in the leech mesoderm. Development 1993;117:177–89.

Glenner H. Cypris metamorphosis, injection and earliest internal development of the rhizocephalan *Loxothylacus panopaei* (Gissler). Crustacea: Cirripedia: Rhizocephala: Sacculinidae. J Morph 2001;249:43–75.

Glenner H, Hebsgaard M. Phylogeny and evolution of life history strategies of the parasitic barnacles (Crustacea, Cirripedia, Rhizocephala). Mol Phylog Evol 2006;41:528–38.

Glenner H, Thomsen PF, Hebsgaard MB, Sorensen MV, Willerslev E. Evolution: The origin of insects. Science 2006;314:1883–4.

Goetze E. Elongation factor 1-alpha in marine copepods (Calanoida: Eucalanidae): Phylogenetic utility and unique intron structure. Mol Phylogenet Evol 2006;40:880–6.

Goldman N. Simple diagnostic statistical tests of models for DNA substitution. J Mol Evol 1993a;37:650–61.

Goldman, N. Statistical tests of models of DNA substitution. J Mol Evol 1993b;36:182–98.

Goldschmidt R. Das Nervensystem von *Ascaris lumbricoides* und *megalocephala*. Ein Versuch, in den Aufbau eines einfachen Nervensystems einzudringen. I. Teil. Z wiss Zool 1908;40:73–136.

Goldschmidt R. Das Nervensystem von *Ascaris lumbricoides* und *megalocephala*. Ein Versuch, in den Aufbau eines einfachen Nervensystems einzudringen. II. Teil. Z wiss Zool 1909;42:307–58.

Goldschmidt R. Das Nervensystem von *Ascaris lumbricoides* und *megalocephala*. Ein Versuch, in den Aufbau eines einfachen Nervensystems einzudringen. III. Teil. In: Anonymus, ed. Festschrift zum sechzigsten Geburtstage Richard Hertwigs. Jena, Germany: Gustav Fischer Verlag, 1910:255–354.

Goloboff PA, Catalano SA, Mirande JM, et al. Phylogenetic analysis of 73 060 taxa corroborates major eukaryotic groups. Cladistics 2009;25:211–30.

Gols E, Ernsting G, van Straalen NM. Paternity analysis in a hexapod (*Orchesella cincta*; Collembola) with indirect sperm transfer. J Insect Behav 2004;17:317–28.

Gooding RU. *Lightiella incisa* sp. nov. (Cephalocarida) from the West Indies. Crustaceana 1963;5:293–314.

Goodrich ES. On the nephridia of the Polychaeta: Part II. *Glycera* and *Goniada*. Q J Microsc Sci 1898;41:439–57.

Goodrich ES. The study of nephridia and genital ducts since 1895. Q J Microsc Sci 1945;86:113–391.

Goto A, Kitamura K, Shimizu T. Cell lineage analysis of pattern formation in the *Tubifex* embryo. I. Segmentation in the mesoderm. Int J Dev Biol 1999;43:317–27.

Goto A, Kitamura K, Arai A, Shimizu T. Cell fate analysis of teloblasts in the *Tubifex* embryo by intracellular injection of HRP. Dev Growth Differ 1999;41:703–13.

Goto T. Fertilization process in the arrow worm *Spadella cephaloptera* (Chaetognatha). Zool Sci 1999;16:109–14.

Goto T, Yoshida, M. Oriented light reactions of the arrow worm *Sagitta crassa* Tokioka. Biol Bull 1981;160:419–30.

Goto T, Yoshida, M. The role of the eye and CNS components in phototaxis of the arrow worm, *Sagitta crassa* Tokioka. Biol Bull 1983;164:82–92.

Goto T, Yoshida M. Photoreception in Chaetognatha. In: Ali MA, ed. Photoreception and Vision in Invertebrates. New York,NY, USA: Plenum Publishing Corporation, 1984:727–42.

Goto T, Yoshida, M. The mating sequence of the benthic arrowworm *Spadella schizoptera*. Biol Bull 1985;169:328–33.

Goto T, Yoshida M. Nervous System in Chaetognatha. In: Ali MA, ed. Nervous Systems in Invertebrates. New York, NY, USA: Plenum Publishing Corporation, 1987:461–81.

Goto T, Yoshida M. Histochemical demonstration of a rhodopsin-like substance in the eye of the arrow-worm, *Spadella schizoptera* (Chaetognatha). Exp Biol 1988;48:1–4.

Goto T, Yoshida M. Growth and reproduction of the benthic arrowworm *Paraspadella gotoi* (Chaetognatha) in laboratory culture. Invert Repro Dev 1997;32:201–7.
Goto T, Takasu N, Yoshida M. A unique photoreceptive structure in the arrowworms *Sagitta crassa* and *Spadella schizoptera* (Chaetognatha). Cell Tiss Res 1984;235:471–8.
Goto T, Terazaki M, Yoshida M. Comparative morphology of the eyes of *Sagitta* (Chaetognatha) in relation to depth of habitat. Exp Biol 1989;48:95–105.
Goto T, Katayama-Kumoi Y, Tohyama M, Yoshida M. Distribution and development of the serotonin-and RFamide-like immunoreactive neurons in the arrowworm, *Paraspadella gotoi* (Chaetognatha). Cell Tiss Res 1992;267:215–22.
Gowri-Shankar V, Jow H. PHASE: A software package for phylogenetics and sequence evolution. release 2.0. 2006.
Gowri-Shankar V, Rattray M. A reversible jump method for Bayesian phylogenetic inference with a nonhomogeneous substitution model. Mol Biol Evol 2007;24:1286–99.
Graff L. Das Genus *Myzostoma* (F.S. Leuckart). Leipzig, Germany: Wilhelm Engelmann Verlag, 1877.
Grande C, Templado J, Zardoya R. Evolution of gastropod mitochondrial genome arrangements. BMC Evol Biol 2008;8:61.
Grandjean F. Observations sur les Acariens (1re série). Bull Mus Nat Hist Nat 1935;7:119–26.
Grassle JF, Maciolek NJ. Deep-sea species richness: Regional and local diversity estimates from quantitative bottom samples. American Naturalist 1992;139:313–41.
Graur D, Li WH. Fundamentals of molecular evolution. Sunderland, MA, USA: Sinauer Associates, 2000.
Gray J. Jim Grey on eScience: A Transformed Scientific Method. In: Hey T, Transley S, Tolle K, eds. The Fourth Paradigm: Data-Intensive Scientific Discoveries. Redmond, WA, USA: Microsoft Research, 2009:XVII-XXXI.
Graybeal A. Evaluating the phylogenetic utility of genes: A search for genes informative about deep divergences among vertebrates. Syst Biol 1994;43:174–93.
Graybeal A. Is it better to add taxa or characters to a difficult phylogenetic problem? Syst Biol 1998;47:9–17.
Grell KG. Embryonalentwicklung bei *Trichoplax adhaerens* F. E. Schulze. Naturwissenschaften 1971;58:570.
Grell KG. Eibildung und Furchung von *Trichoplax adhaerens* F. E. Schulze (Placozoa). Z Morph Tiere 1972;73:297–314.
Grell KG. Reproduction of Placozoa. In: Engels W, ed. Advances in Invertebrate Reproduction. Amsterdam, The Netherlands: Elsevier, 1984:541–6.
Grell KG, Benwitz G. Die Ultrastruktur von *Trichoplax adhaerens* F. E. Schulze. Cytobiologie 1971;4:216–40.
Grell KG, Benwitz G. Elektronenmikroskopische Beobachtungen über das Wachstum der Eizelle und die Bildung der "Befruchtungsmembran" von *Trichoplax adhaerens* F. E. Schulze (Placozoa). Z Morph Tiere 1974;79:295–310.
Grimaldi DA. 400 million years on six legs: On the origin and early evolution of Hexapoda. Arthropod Struct Dev 2010;39:191–203.
Grimaldi DA, Engel MS. Evolution of the Insects. New York, NY, USA: Cambridge University Press, 2005.
Grimmer JC, Holland ND. The structure of a sessile, stalkless crinoid (*Holopus rangii*). Acta Zool (Stockholm) 1990;71:61–7.
Grimmer JC, Holland ND, Hayami I. Fine structure of the stalk of an isocrinid sea lily (*Metacrinus rotundus*) (Echinodermata, Crinoidea). Zoomorphology 1985;105:39–50.
Grimmer JC, Holland ND, Kubota H. The fine structure of the stalk of the pentacrinoid larva of a feather star, *Comanthus japonica* (Echinodermata: Crinoidea). Acta Zoologica 1984;65:41–58.

Grobe P. Larval development, the origin of the coelom and the phylogenetic relationships of the Phoronida. Dissertation, Freie Universität Berlin, Germany, 2008.
Grobe P, Vogt L. MorphDBase – a morphological description database. J Morphol 2008;269:1478–9.
Groth JG, Barrowclough GF. Basal divergences in birds and the phylogenetic utility of the nuclear RAG-1 gene. Mol Phylogenet Evol 1999;12:115–23.
Grube AE. Die Familien der Anneliden, Arch Naturgesch 1850;16 :249–364.
Gruber AR, Koper-Emde D, Marz M, et al. Invertebrate 7SK snRNAs. J Mol Evol 2008;66:107–15.
Grygier MJ. Class Myzostomida. In: Beesley P L, Ross GJB, Glasby CJ, eds. Melbourne Polychaetes and Allies: The Southern Synthesis. Fauna of Australia. Vol 4A Polychaeta, Myzostomida, Pogonophora, Echiura, Sipuncula. Melbourne, Australia: CSIRO Publishing, 2000:297–330.
Gu X, Fu YX, Li WH. Maximum likelihood estimation of the heterogeneity of substitution rate among nucleotide sites. Mol Biol Evol 1995;12:546–57.
Guindon S, Gascuel O. A simple, fast, and accurate algorithm to estimate large phylogenies by maximum likelihood. Syst Biol 2003;52:696–704.
Guindon S, Dufayard JF, Lefort V, et al. PhyML 3.0: New algorithms and methods to estimate maximum-likelihood phylogenies: Assessing the performance of PhyML 3.0. Syst Biol 2010;59:307–21.
Gullan PJ, Cranston PS. The Insects: An Outline of Entomology. Malden, MA, USA: Blackwell Publishing, 2000.
Gupta A, Larson SD, Condit C, et al. Toward an Ontological Database for Subcellular Neuroanatomy. In: Hainaut JL, Rundensteiner EA, Kirchberg M, et al.,eds. Advances in Conceptual Modeling – Foundations and Applications. Lecture notes in computer science. Berlin, Germany: Springer, 2007; 4802:66–73.
Gupta RS, Golding GB, Singh B. HSP70 phylogeny and the relationship between archaebacteria, eubacteria, and eukaryotes. J Mol Evol 1994;39:537–40.
Gwilliam GF, Cole ES. The morphology of the central nervous system of the barnacle, *Semibalanus cariosus* (Pallas). J Morph 1979;159:297–310.
Gwilliam GF. Neurobiology of Barnacles. In: Southward AJ, ed. Barnacle Biology. Rotterdam, The Netherlands: Balkema, 1987:191–211.
Haas F, Kukalova-Peck J. Dermaptera hindwing structure and folding: New evidence for familial ordinal and superordinal relationships within Neoptera (Insecta). Eur J Entomol 2001;98:445–509.
Haas F, Waloszek D, Hartenberger R. *Devonohexapodus bocksbergensis*, a new marine hexapod from the Lower Devonian Hunsrück Slates, and the origin of Atelocerata and Hexapoda. Org Divers Evol 2002;3:39–54.
Hadrys H, Simon S, Kaune B, Schmitt O, Schoner A, Jakob W, Schierwater B. Isolation of hox cluster genes from insects reveals an accelerated sequence evolution rate. PLoS ONE 2012;7:e34682.
Hadrys T, DeSalle R, Sagasser S, Fischer N, Schierwater B. The Trichoplax PaxB gene: a putative Proto-PaxA/B/C gene predating the origin of nerve and sensory cells. Mol Biol Evol 2005;22:1569–78.
Haeckel E. Generelle Morphologie der Organismen. Allgemeine Grundzüge der organischen Formen-Wissenschaft, mechanisch begründet durch die von Charles Darwin reformirte Descendenz-theorie. Berlin, Germany: Reimer Verlag, 1866.
Haen KM, Lang BF, Pomponi SA, Lavrov DV. Glass sponges and bilaterian animals share derived mitochondrial genomic features: a common ancestry or parallel evolution? Mol Biol Evol 2007;24:1518–27.
Haendel MA, Gkoutos GV, Lewis SE, Mungall CJ. Uberon: Towards a comprehensive multi-species anatomy ontology. Nature Precedings 2009;713:207 (doi:10.1038/npre.2009.3592.1).
Haeseler Av. Do we still need supertrees? BMC Biol 2012;10:13.

Haffer K. Zur Morphologie der Malacostraca: Der Kaumagen der Mysidacea im Vergleich zu dem verschiedener Peracarida und Eucarida. Helgoländer wissensch Meeresunters 1965;12:156–206.

Hafner MS, Sudman PD, Villablanca FX, Spradling TA, Demastes JW, Nadler SA. Disparate rates of molecular evolution in cospeciating hosts and parasites. Science 1994;265:1087–90.

Halanych KM. The phylogenetic position of the pterobranch hemichordates based on 18S rDNA sequence data. Mol Phylogenet Evol 1995;4:72–6.

Halanych KM. Testing hypotheses of chaetognath origins: Long branches revealed by 18S ribosomal DNA. Syst Biol 1996a;45:223–46.

Halanych KM. Convergence in the feeding apparatuses of lophophorates and pterobranch hemichordates revealed by 18S rDNA: An interpretation. Biol Bull 1996b;190:1–5.

Halanych KM. The new view of animal phylogeny. Annu Rev Ecol Evol Syst 2004;35:229–56.

Halanych KM, Bacheller JD, Aguinaldo AMA, Liva SM, Hillis DM, Lake JA. Evidence from 18S ribosomal DNA that the lophophorates are protostome animals. Science 1995;267:1641–3.

Halcrow K. Ultrastructural features of the funnel of *Gammarus oceanicus* (Amphipoda). J Crust Biol 2001;21:631–9.

Hall KA, Hutchings PA, Colgan DJ. Further phylogenetic studies of the Polychaeta using 18S rDNA sequence data. J Mar Biol Ass UK 2004;84:949–60.

Hamel JF, Becker P, Eeckhaut I, Mercier A. Exogonadal oogenesis in a temperate holothurian. Biol Bull 2007;213:101–9.

Hammersen F, Staudte HW. Beiträge zum Feinbau der Blutgefäße von Invertebraten. I. Die Ultrastruktur des Sinus lateralis von *Hirudo medicinalis*. Z Zellforsch Mikrosk Anat 1969;100:215–50.

Han KL, Braun EL, Kimball RT, et al. Are transposable element insertions homoplasy free?: An examination using the avian tree of life. Syst Biol 2011;60:375–86.

Hanelt B, Van-Schyndel D, Adema CM, Lewis LA, Loker ES. The phylogenetic position of Rhopalura ophiocomae (Orthonectida) based on 18S ribosomal DNA sequence analysis. Mol Biol Evol 1996;13:1187–91.

Hanlon RT, Messenger JB. Cephalopod Behaviour. Cambridge, UK: Cambridge University Press, 1996.

Hannenhalli S, Pevzner PA. Transforming cabbage into turnip: Polynomial algorithm for sorting signed permutations by reversals. In: Leighton TF, Borodin A, eds. Proceedings of the 27th annual ACM symposium on Theory of Computing. New York, NY, USA: AMC, 1995:178–89.

Hanson B, Sugden A, Alberts B. Making data maximally available. Science 2011;331:649.

Hanström B. Untersuchungen über das Gehirn, insbesondere die Sehganglien der Crustaceen. Ark Zool Stockholm 1924;16:1–119.

Hanström B. Vergleichende Anatomie des Nervensystems der wirbellosen Tiere unter Berücksichtigung seiner Funktion. Berlin, Germany: Springer Verlag, 1928.

Hanström B. Neue Untersuchungen der Sinnesorgane und Nervenzentren der Crustaceen. I. Zool Jahrb Abt Anat Ontog Tiere 1931;23:81–236.

Hanström B. Neue Untersuchungen der Sinnesorgane und Nervenzentren der Crustaceen. III. Zool Jahrb Abt Anat Ontog Tiere 1934;58:101–44.

Hanström B. "Inkretorische Organe". Sinnesorgane und Nervensystem des Kopfes einiger niederer Insektenordnungen. Kungl Svensk Vetenskapsakad Handl Ser. III 1940; 18:1-266.

Hanström B. The brain, the sense organs, and the incretory organs of the head in the Crustacea Malacostraca. Lunds Universitets Årsskrift 1947;43:1–49

Harbison GR. On the classification and evolution of the Ctenophora. In: Conway Morris S, George JD, Gibson R, and Platt HM, eds. The Origins and Relationships of Lower Invertebrates. Oxford, UK: Oxford University Press, 1985:78–100.

Harley CM, Ritzmann RE. Electrolytic lesions within central complex neuropils of the cockroach brain affect negotiation of barriers. J Exp Biol 2010;231:2851–64.

Harrat A, Ihsan S, Schoeller-Raccaud J. Development of the suboesophageal body during embryogenesis without diapause in *Locusta migratoria* (Linnaeus) (Orthoptera: Acrididae). Int J Insect Morphol Embryol 1999;28:27–39.

Harris TJC, Tepass U. Adherens junctions: From molecules to morphogenesis. Nature Rev Mol Cell Biol 2010;11:502–14.

Harrison PJH, Sandeman DC. Morphology of the nervous system of the barnacle larva (*Balanus amphritre* Darwin) revealed by light and electron microscopy. Biol Bull 1999;197:144–58.

Hartline DK, Christie AE. Immunohistochemical mapping of histamine, dopamine, and serotonin in the central nervous system of the copepod *Calanus finmarchicus* (Crustacea; Maxillopoda; Copepoda). Cell Tiss Res 2010;341:49–71.

Hartmann G. Klassen und Ordnungen des Tierreichs. In: Gruner, HE, ed. Arthropoda, I. Abteilung: Crustacea, 2. Buch, IV. Teil, 2 Ostracoda, , Leipzig, Germany: Akademische Verlagsgesellschaft Geest & Portig KG, 1967:217–408.

Hartmann S, Vision T. Using ESTs for phylogenomics: Can one accurately infer a phylogenetic tree from a gappy alignment? BMC Evol Biol 2008;8:95.

Hartmann S, Helm C, Nickel B, et al. Exploiting gene families for phylogenomic analysis of myzostomid transcriptome data. PLoS ONE 2012;7:e29843.

Harzsch S. The phylogenetic significance of crustacean optic neuropils and chiasmata: A re-examination. J Comp Neurol 2002a;453:10–21.

Harzsch S. Neurobiologie und Evolutionsforschung: “Neurophylogenie” und die Stammesgeschichte der Euarthropoda. Neuroforum 2002b;4:267–73.

Harzsch S. Ontogeny of the ventral nerve cord in malacostracan crustaceans: A common plan for neuronal development in Crustacea, Hexapoda and other Arthropoda? Arthropod Struct Dev 2003a;32:17–37.

Harzsch S. Evolution of identified arthropod neurons: The serotonergic system in relation to *engrailed*-expressing cells in the embryonic ventral nerve cord of the American lobster *Homarus americanus* Milne Edwards, 1873 (Malacostraca, Pleocyemata, Homarida). Dev Biol 2003b;258:44–56.

Harzsch S. Phylogenetic comparison of serotonin-immunoreactive neurons in representatives of the Chilopoda, Diplopoda, and Chelicerata: Implications for arthropod relationships. J Morph 2004;259:198–213.

Harzsch S. Neurophylogeny: architecture of the nervous system and a fresh view on arthropod phylogeny. Comp Integr Biol 2006;46:162–94.

Harzsch S. The architecture of the nervous system provides important characters for phylogenetic reconstructions: Examples from the Arthropoda. Spec Phyl Evol 2007;1:33–57.

Harzsch S, Dawirs RR. A developmental study of serotonin-immunoreactive neurons in the larval central nervous system of the spider crab *Hyas araneus* (Decapoda, Brachyura). Invertebr Neurosci 1995;1:53–65.

Harzsch S, Glötzner J. An immunohistochemical study of structure and development of the nervous system in the brine shrimp *Artemia salina* Linnaeus, 1758 (Branchiopoda, Anostraca) with remarks on the evolution of the arthropod brain. Arthropod Struct Dev 2002;30:251.

Harzsch S, Hafner G. Evolution of eye development in arthropods: phylogenetic aspects. Arthropod Struct Dev 2006;35:319–40.

Harzsch S, Hansson BS. Brain architecture in the terrestrial hermit crab *Coenobita clypeatus* (Anomura, Coenobitidae), a crustacean with a good aerial sense of smell. BMC Neurosci 2008;9:58.

Harzsch S, Müller CHG. A new look at the ventral nerve centre of *Sagitta*: Implications for the phylogenetic position of Chaetognatha (arrow worms) and the evolution of the bilaterian nervous system. Front Zool 2007;4:14.

Harzsch S, Waloszek D. Serotonin-immunoreactive neurons in the ventral nerve cord of Crustacea: A character to study aspects of arthropod phylogeny. Arthropod Struct Dev 2000;29:307–22.

Harzsch S, Waloszek D. Neurogenesis in the developing visual system of the branchiopod crustacean *Triops longicaudatus* (LeConte, 1846): Corresponding patterns of compound-eye formation in Crustacea and Insecta? Dev Genes Evol 2001;211:37–43.

Harzsch S, Wanninger A. Evolution of invertebrate nervous systems: The Chaetognatha as a case study. Acta Zool (Stockholm) 2009;91:35–41.

Harzsch S, Melzer RR, Müller CHG. Eye development in Myriapoda: Implications for arthropod phylogeny. Norw J Entomol 2006;53:177–80.

Harzsch S, Melzer RR, Müller CHG. Mechanisms of eye development and evolution of the arthropod visual system: The lateral eyes of Myriapoda are not modified insect ommatidia. Org Divers Evol 2007;7:20–32.

Harzsch S, Müller CHG, Wolf H. From variable to constant cell numbers: Cellular characteristics of the arthropod nervous system argue against a sister-group relationship of Chelicerata and "Myriapoda" but favour the Mandibulata concept. Dev Gen Evol 2005;215:53–68.

Harzsch S, Miller J, Benton J, Dawirs RR, Beltz B. Neurogenesis in the thoracic neuromeres of two crustaceans with different types of metamorphic development. J Exp Biol 1998;201:2465–79.

Harzsch S, Müller CHG, Rieger V, Perez Y, Sintoni S, Sardet C, Hansson B. Fine structure of the ventral nerve centre and interspecific identification of individual neurons in the enigmatic Chaetognatha. Zoomorphology 2009;128:53–73.

Hassanin A. Phylogeny of arthropoda inferred from mitochondrial sequences: Strategies for limiting the misleading effects of multiple changes in pattern and rates of substitution. Mol Phylogenet Evol 2006;38:100–16.

Hasse C, Rebscher N, Reiher W, et al. Three consecutive generations of nephridia occur during development of *Platynereis dumerilii* (Annelida, Polychaeta). Dev Dyn 2010;239:1967–76.

Haszprunar G, Schander C, Halanych KM. Relationships of Higher Molluscan Taxa. In: Ponder WF, Lindberg DR, eds. Phylogeny and Evolution of the Mollusca. Berkeley, CA, USA: University of California Press, 2008:19–32.

Haszprunar G. The Mollusca: Coelomate turbellarians or mesenchymate annelids? In: Taylor J, ed. Origin and Evolutionary Radiation of the Mollusca. Oxford, UK: Oxford University Press, 1996a:1–28.

Haszprunar G. Plathelminthes and Plathelminthomorpha – paraphyletic taxa. J Zool Syst Evol Res 1996b;34:41–8.

Hatschek B. Lehrbuch der Zoologie, 3. Lieferung. Jena, Germany: G. Fischer, 1891.

Hatschek B. Studien über die Entwicklungsgeschichte der Anneliden. Ein Beitrag zur Morphologie der Bilaterien. Arb Zool Inst Univ Wien Stat Triest 1878;1:1–128.

Hatschek B. Zur Entwicklung des Kopfes von *Polygordius*. Arb Zool Inst Univ Wien Stat Triest 1885;6:109–210.

Haug C. Head shield evolution and development in the Arthropoda. Dissertation, Universität Ulm, Germany, 2011.

Haug JT, Briggs DEG, Haug, C. Morphology and function in the Cambrian Burgess Shale megacheiran arthropod *Leanchoilia superlata* and the application of a descriptive matrix. BMC Evol Biol 2012;12:162.

Haug JT, Maas A, Haug C, Waloszek D. Evolution of crustacean appendages. In: Watling L, Thiel M, eds. The Natural History of the Crustacea Vol. 1: Functional Morphology and Diversity. Oxford UK, Oxford University Press, 2012b:34–73.

Haug JT, Mayer G, Haug C, Briggs DEG. A carboniferous non-onychophoran lobopodian reveals long-term survival of a Cambrian morphotype. Curr Biol 201a2;22:1673–5.
Haupt J. Kutikula-Strukturen der Pseudoculi von Pauropoden und Proturen. Naturwissenschaften 1972;59:80–1.
Haupt J. Die Ultrastruktur des Pseudoculus von *Allopauropus* (Pauropoda) und die Homologie der Schläfenorgane. Z Morph Tiere 1973;76:173–91.
Haupt J. Phylogenetic aspects of recent studies on myriapod sense organs. In: Blower JG, ed. Myriapod Biology. New York, NY, USA: Academic Press. 1979:391–406.
Hausdorf B, Helmkampf M, Nesnidal M, Bruchhaus I. Phylogenetic relationships within the lophophorate lineages (Ectoprocta, Brachiopoda and Phoronida). Mol Phylogenet Evol 2010;55:1121–7.
Hausdorf B, Helmkampf M, Meyer A, et al. Spiralian phylogenomics supports the resurrection of Bryozoa comprising Ectoprocta and Entoprocta. Mol Biol Evol 2007;24:2723–9.
Hausen H. Chaetae and chaetogenesis in polychaetes (Annelida). In: Bartolomaeus T, Purschke G, eds. Morphology, molecules, evolution and phylogeny in Polychaeta and related taxa. Hydrobiologia 2005;535/536:37–52.
Hay-Schmidt A. The evolution of the serotonergic nervous system. Phil Trans R Soc Lond B Biol Sci 2000;267:1071–9.
Hazkani-Covo E. Mitochondrial insertions into primate nuclear genomes suggest the use of numts as a tool for phylogeny. Mol Biol Evol 2009;26:2175–9.
Hazkani-Covo E, Zeller RM, Martin W. Molecular poltergeists: Mitochondrial DNA copies (numts) in sequenced nuclear genomes. PLoS Genet 2010;6:e1000834.
Heber S, Stoye J. Algorithms for finding gene clusters. In: Gascuel O, Moret BME, eds. Algorithms in Bioinformatics. Lecture Notes on Computer Science. Heidelberg, Germany: Springer, 2001; 2149:252-63.
Hebert PDN, Remigio E, Colbourne JK, Taylor DJ, Wilson CC. Accelerated molecular evolution in halophilic crustaceans. Evolution 2002;56:909–26.
Heckmann R, Kutsch W. Common neural „bauplan“ in Tracheata? Innervation of the dorsal longitudinal muscles. In: Elsner N, Roth G, eds. Brain – Perception – Cognition, Stuttgart, Germany: Thieme Verlag, 1990:39.
Hedges S, Blair J, Venturi M, Shoe J. A molecular timescale of eukaryote evolution and the rise of complex multicellular life. BMC Evol Biol 2004;4:2.
Heimberg AM, Cowper-Sal-lari R, Sémon M, Donoghue PCJ, Peterson KJ. MicroRNAs reveal the interrelationships of hagfish, lampreys, and gnathostomes and the nature of the ancestral vertebrate. Proc Natl Acad Sci USA 2010;107:19379–83.
Heimberg AM, Sempere LF, Moy VN, Donoghue PCJ, Peterson K. MicroRNAs and the advent of vertebrate morphological complexity. Proc Natl Acad Sci USA 2007;105:2946–50.
Heiner I, Schmidt-Rhaesa A, and Kristensen RM. Loricifera. In: Schmidt-Rhaesa A, ed. Handbook of Zoology, Gastrotricha, Cycloneuralia and Gnathifera. Berlin, Germany: De Gruyter, 2012: in press.
Heinze S, Homberg U. Maplike representation of celestial *e*-vector orientations in the brain of an insect. Science 2007;315:995–7.
Heisenberg M. Mushroom body memoir: From maps to models. Nat Rev Neurosci 2003;4:266–75.
Hejnol A. A twist in time–the evolution of spiral cleavage in the light of animal phylogeny. Integr Comp Biol 2010;50:695–706.
Hejnol A, Martindale MQ. Acoel development supports a simple planula-like urbilaterian. Phil Trans R Soc Lond B Biol Sci 2008a;363:1493–501.
Hejnol A, Martindale MQ. Acoel development indicates the independent evolution of the bilaterian mouth and anus. Nature 2008b;456:382–7.

Hejnol A, Martindale MQ. The mouth, the anus and the blastopore – open questions about questionable openings. In: Telford MJ, Littlewood DTJ, eds. Animal Evolution: Genes, Genomes, Fossils and Trees. Oxford, UK: Oxford University Press, 2009:33–40.

Hejnol A, Schnabel R. What a couple of dimensions can do for you: Comparative developmental studies using 4D microscopy–examples from tardigrade development. Integr Comp Biol 2006;46:151–61.

Hejnol A, Obst M, Stamatakis A, et al. Assessing the root of bilaterian animals with scalable phylogenomic methods. Proc Roy Soc B 2009;276:4261–70.

Helfenbein KG, Boore JL. The mitochondrial genome of *Phoronis architecta* – comparisons demonstrate that phoronids are lophotrochozoan protostomes. Mol Biol Evol 2004;21:153–7.

Helfenbein KG, Brown WM, Boore JL. The complete mitochondrial genome of the articulate brachiopod *Terebratalia transversa*. Mol Biol Evol 2001;18:1734–44.

Helfenbein KG, Fourcade HM, Vanjani RG, Boore JL. The mitochondrial genome of *Paraspadella gotoi* is highly reduced and reveals that chaetognaths are a sister group to protostomes. Proc Natl Acad Sci USA 2004;101:10639–43.

Helm C, Bernhart SH, Höner zu Siederdissen C, Nickel B, Bleidorn C. Deep sequencing of small RNAs confirms an annelid affinity of Myzostomida. Mol Phylogenet Evol 2012;64:198–203.

Helm C, Weigert A, Meyer G, Bleidorn C. Myoanatomy of *Myzostoma cirriferum* (Annelida, Myzostomida): Implications for the evolution of the myzostomid body plan. J Morphol 2013, 274:456–66.

Helmkampf M, Bruchhaus I, Hausdorf B. Multigene analysis of lophophorate and chaetognath phylogenetic relationships. Mol Phylogenet Evol 2008a;46:206–14.

Helmkampf M, Bruchhaus I, Hausdorf B. Phylogenomic analyses of lophophorates (brachiopods, phoronids and bryozoans) confirm the Lophotrochozoa concept. Proc R Soc Lond B Biol Sci 2008b;275:1927–33.

Hendy MD, Penny D. A framework for the quantitative study of evolutionary trees. Syst Zool 1993;38:297–309.

Hennecke R, Gellissen G, Spindler KD. Haemocyanin synthesis in the crayfish, *Astacus leptodactylus*: in vitro production by the midgut gland after addition of oxygen binding modulators and exogenous copper. Z Naturforsch 1991;46c:163–5.

Hennig W. Grundzüge einer Theorie der phylogenetischen Systematik. Berlin, Germany: Deutscher Zentralverlag, 1950.

Hennig W. Die Stammesgeschichte der Insekten. Frankfurt am Main, Germany: Verlag Dr. Waldemar Kramer, 1969.

Herlyn H, Ehlers U. Ultrastructure and function of the pharynx of *Gnathostomula paradoxa* (Gnathostomulida). Zoomorphology 1997;117:135–45.

Herlyn H, Ehlers U. Organisation of the praesoma in *Acanthocephalus anguillae* (Acanthocephala, Palaeacanthocephala) with special reference to the muscular system. Zoomorphology 2001;121:13–8.

Herlyn H, Piskurek O, Schmitz J, Ehlers U, Zischler H. The syndermatan phylogeny and the evolution of acanthocephalan endoparasitism as inferred from 18S rDNA sequences. Mol Phylogenet Evol 2003;26:155–64.

Hermsen EJ, Hendricks JR. W(h)ither fossils? Studying morphological character evolution in the age of molecular sequences 1. Ann Missouri Bot Garden 2008;95:72–100.

Herranz M, Pardos F, Boyle MJ. Comparative morphology of serotonin-like immunoreactive elements in the central nervous system of kinorhynchs (Kinorhyncha, Cyclorhagida). J Morphol 2013;274:258–74.

Hertel J, Lindemeyer M, Missal K, et al. The expansion of the metazoan microRNA repertoire. BMC Genomics 2006;7:25.

Hertwig O. Die Chaetognathen. Monatl Jenaer Z Med Nat 1880;14:196–311.
Hertwig O, Hertwig R. Die Coelomtheorie. Jena, Germany: Gustav Fischer, 1881.
Hertzel G. Die Segmentation des Keimstreifens von *Lithobius forficatus* (L.) (Myriapoda, Chilopoda). Zool Jahrb Abt Anat Ontog 1984;112:369–86.
Hessler RR. A new species of Mystacocarida from Maine. Vie et Milieu Ser. A 1969;20:105–116.
Hessler RR. The structural morphology of walking mechanisms in eumalacostracan crustaceans. Phil Trans R Soc Lond B Biol Sci 1982;296:245–98.
Hessler RR. Reflections on the position of Cephalocarida. Acta Zool (Stockholm) 1992;73:315–6.
Hessler RR, Elofsson R. Segmental podocytic excretory glands in the thorax of *Hutchinsoniella macracantha* (Cephalocarida). J Crustacean Biol 1995;15:61–9.
Hessler RR, Sanders HL. *Derocheilocaris typicus* Pennak & Zinn (Mystacocarida) revisited. Crustaceana 1965;11:141–55.
Hessler RR, Sanders HL. Faunal diversity in the deep sea. Deep Sea Research 1967;14:65–78.
Hessler RR, Sanders HL. Two new species of *Sandersiella* (Cephalocarida), including one from the deep sea. Crustaceana 1973;24:181–96.
Hessler RR, Yager J. Skeletomusculature of trunk segments and their limbs in *Speleonectes tulumensis* (Remipedia). J Crust Biol 1998;18:111–9.
Hessling R. Metameric organisation of the nervous system in developmental stages of *Urechis caupo* (Echiura) and its phylogenetic implications. Zoomorphology 2002;121:221–34.
Hessling R. Novel aspects of the nervous system of *Bonellia viridis* (Echiura) revealed by the combination of immunohistochemistry, confocal laser-scanning microscopy and three-dimensional reconstruction. Hydrobiologia 2003;496:225–39.
Hessling R, Westheide W. Are Echiura derived from a segmented ancestor? Immunohistochemical analysis of the nervous system in developmental stages of *Bonellia viridis*. J Morphol 2002;252:100–13.
Heuer CM, Loesel R. Immunofluorescence analysis of the internal brain anatomy of *Nereis diversicolor* (Polychaeta, Annelida). Cell Tissue Res 2008;331:713–24.
Heuer CM, Loesel R. Three-dimensional reconstruction of mushroom body neuropils in the polychaete species *Nereis diversicolor* and *Harmothoe areolata* (Phyllodocida, Annelida). Zoomorphology 2009;128:219–26.
Heuer CM, Müller CHG, Todt C, Loesel R. Comparative neuroanatomy suggests repeated reduction of neuroarchitectural complexity in Annelida. Front Zool 2010;7:13.
Heymans M, Singh AK. Deriving phylogenetic trees from the similarity analysis of metabolic pathways. Bioinformatics 2003;19 Suppl 1:i138–46.
Heymons R. Beiträge zur Morphologie und Entwicklungsgeschichte der Rhynchoten. Nova Acta. Abh kaiserl Leopoldinisch-Carolinischen dt Akad Naturf 1899;74:349–456.
Heymons R. Die Entwicklungsgeschichte der Scolopender. Zoologica (Stuttgart) 1901;33:1–244.
Hickman CP, Roberts LS, Larson A, l'Anson H, Eisenhour DJ. Zoologie. München, Germany: Pearson, 2008.
Hickson RE, Simon C, Perrey SW. The performance of several multiple-sequence alignment programs in relation to secondary–structure features for an rRNA sequence. Mol Biol Evol 2000;17:530–9.
Higgs PG. RNA secondary structure: physical and computational aspects. Q Rev Biophys 2000;33:199–253.
Hilken G. Vergleich von Tracheensystemen unter phylogenetischem Aspekt. Verh naturwiss Ver Hamburg (NF) 1998;37:5–94.
Hill N, Leow A, Bleidorn C, et al. Analysis of phylogenetic signal in protostomial intron patterns using Mutual Information. Theory Biosci 2013; 132:93-104.
Hiller M, Findeiß S, Lein S, et al. Conserved introns reveal novel transcripts in *Drosophila melanogaster*. Genome Res 2009;19:1289–300.

Hiller M, Nikolajewa S, Huse K, et al. TassDB: A database of alternative tandem splice sites. Nucleic Acids Res 2007;35:D188–92.
Hillis DM, Dixon MT. Ribosomal DNA: Molecular evolution and phylogenetic inference. Q Rev Biol 1991;66:411–53.
Hillis DM, Wiens JJ. Molecules versus morphology in systematics: Conflicts, artifacts, and misconceptions. In: Wiens JJ, ed. Phylogenetic Analysis of Morphological Data. Washington: Smithsonian Institution Press, 2000:1–19.
Hillis DM, Bull JJ, White ME, Badgett MR, Molineux IJ. Experimental phylogenetics: Generation of a known phylogeny. Science 1992;255:589–92.
Hillis DM, Pollock DD, McGuire JA, Zwickl DJ. Is sparse taxon sampling a problem for phylogenetic inference? Syst Biol 2003;52:124–6.
Hochberg R. Comparative immunohistochemistry of the cerebral ganglion in Gastrotricha: An analysis of FMRFamide-like immunoreactivity in *Neodasys cirritus* (Chaetonotida), *Xenodasys riedli* and *Turbanella* cf *hyalina* (Macrodasyida). Zoomorphology 2007;126:245–64.
Hochberg R, Litvaitis MK. A muscular double helix in Gastrotricha. Zool Anz 2001a;240:61–8.
Hochberg R, Litvaitis MK. The muscular system of *Dactylopodola baltica* and other macrodasyidan gastrotrichs in a functional and phylogenetic perspective. Zool Sripta 2001b;30:325–36.
Hochberg R, Litvaitis MK. Organization of muscles in *Chaetonotida paucitubulatina* (Gastrotricha). Meiofauna Marina 2003;12:47–58.
Hochsmann M, Voss B, Giegerich R. Pure multiple RNA secondary structure alignments: a progressive profile approach. IEEE-ACM T Comput Bi 2004;1:53–62.
Høeg J, Pérez-Losada M, Glenner H, Kolbasov GA, Crandall KA. Evolution of morphology, ontogeny and life cycles within the Crustacea Thecostraca. Arthropod Syst Phylog 2009a;67:199–217.
Høeg JT, Achituv Y, Chan BK, Chan K, Jensen PG, Pérez-Losada M. Cypris morphology in the barnacles *Ibla* and *Paralepas* (Crustacea: Cirripedia Thoracica): implications for cirripede evolution. J Morphol 2009b;270:241–55.
Hoeh WR, Stewart DT, Sutherland BW, Zouros E. Cytochrome c oxidase sequence comparisons suggest an unusually high rate of mitochondrial DNA evolution in *Mytilus* (Mollusca: Bivalvia). Mol Biol Evol 1996;13:418–21.
Hofacker IL, Fekete M, Stadler PF. Secondary structure prediction for aligned RNA sequences. J Mol Biol 2002;319:1059–66.
Hofacker IL, Fontana W, Stadler PF, Bonhoeffer LS, Tacker M, Schuster P. Fast Folding and Comparison of RNA Secondary Structures. Monatsh Chem 1994;125:167–88.
Holland ND. Epidermal cells. Chapter 41. XI. Echinodermata. In: Bereiter-Hahn J, Matoltsy AG, Richards KS, editors. Biology of the Integument 1. Invertebrates. Berlin, Germany: Springer Verlag, 1984:756–74.
Holland ND, Holland LZ. Serotonin-containing cells in the nervous system and other tissues during ontogeny of a lancelet, *Branchiostoma floridae*. Acta Zool (Stockholm) 1993;74:195–204.
Holland PWH. Beyond the Hox: How widespread is homeobox gene clustering? J Anat 2001;199:13–23.
Holland PWH, Takahashi T. The evolution of homeobox genes: Implications for the study of brain development. Brain Res Bull 2005;66:484–90.
Holmgren N. Zur vergleichenden Anatomie des Gehirns von Polychaeten, Onychophoren, Xiphosuren, Arachniden, Crustaceen, Myriapoden und Insekten. K Svenska Vetens Akad Handl 1916;56:1–303.
Holton TA, Pisani D. Deep genomic-scale analyses of the metazoa reject coelomata: Evidence from single- and multigene families analyzed under a supertree and supermatrix paradigm. Genome Biol Evol 2010;2:310–24.

Homberg U. Neurotransmitters and neuropeptides in the brain of the locust. Microsc Res Tech 2002;56:189–209.

Homberg U. Evolution of the central complex in the arthropod brain with respect to visual system. Arthropod Struct Dev 2008;37:347–62.

Homberg U, Hildebrand JG. Histamine-immunoreactive neurones in the midbrain and subesophageal ganglion of the sphinx moth *Manduca sexta*. J Comp Neurol 1991;307:647–57.

Hoofer SR, Reeder SA, Hansen EW, van den Bussche RA. Molecular phylogenetics and taxonomic review of noctilionoid and vespertilionoid bats (Chiroptera: Yangochiroptera). J Mammal 2003:809–21.

Hooper JNA, Van Soest RWM. Systema Porifera: A Guide to the Supraspecific Classification of Sponges and Spongiomorphs (Porifera). New York, NY, USA: Plenum, 2002.

Horie T, Shinki R, Ogura Y, Kusakabe TG, Satoh N, Sasakura Y. Ependymal cells of chordate larvae are stem-like cells that form the adult nervous system. Nature 2011;469:525–8.

Horst R. Die Gephyrea gesammelt während der zwei ersten Fahrten des "Willem Barents". II. Priapulidae. Niederl Arch Zool 1881;1 (Suppl):13–38.

Horwitz HR, Chalfie M, Trent C, Sulston JE, Evans PD. Serotonin and octopamine in the nematode *Caenorhabditis elegans*. Science 1982;216:1012–4.

Hosfeld B. The relationship between the rostrum and the organ of Bellonci in copepods: an ultrastructural stidy of the rostrum of *Canuella perplexa*. Zool Anz 1995;234:175–90.

Hosfeld B, Schminke HK. Discovery of segmental extranephridial podocytes in Harpacticoida (Copepoda) and Bathynellacea (Syncarida). J Crustacean Biol 1997;17:13–20.

Hovmöller R, Pape T, Källersjö M. The Palaeoptera problem: Basal pterygote phylogeny inferred from 18S and 28S rDNA sequences. Cladistics 2002;18:313–23.

Howison M, Sinnott-Armstrong NA, Dunn CW. BioLite, a lightweight bioinformatics framework with automated tracking of diagnostics and provenance. Proceedings of the 4th USENIX Workshop on the Theory and Practice of Provenance (TaPP'12). Berkeley, CA, USA: USENIX Association, 2012.

Huang X, Madan A. CAP3: A DNA sequence assembly program. Genome Res 1999;9:868–77.

Huang YL, Huang CC, Tang CY, Lu CL. SoRT2: A tool for sorting genomes and reconstructing phylogenetic trees by reversals, generalized transpositions and translocations. Nucleic Acids Res 2010;38:W221–7.

Huckstorf K, Kosok G, Seyfarth EA, Wirkner CS. The hemolymph vascular system in *Cupiennius salei* (Araneae: Ctenidae). Zool Anz 2013;252:76–87.

Hudelot C, Gowri-Shankar V, Jow H, Rattray M, Higgs PG. RNA-based phylogenetic methods: Application to mammalian mitochondrial RNA sequences. Mol Phylogenet Evol 2003;28:241–52.

Hudson CM, Conant GC. Expression level, cellular compartment and metabolic network position all influence the average selective constraint on mammalian enzymes. BMC Evol Biol 2011;11:89.

Huelsenbeck JP. Performance of phylogenetic methods in simulation. Syst Biol 1995;44:17–48.

Huelsenbeck JP. Is the Felsenstein zone a fly trap? Syst Biol 1997;46:69–74.

Huelsenbeck JP. Systematic bias in phylogenetic analysis: Is the Strepsiptera problem solved ? Syst Biol 1998;47:519–37.

Huelsenbeck JP, Crandall KA. Phylogeny estimation and hypotheses testing using maximum likelihood. Annu Rev Ecol Syst 1997;28:43766.

Huelsenbeck JP, Hillis DM. Success of phylogenetic methods in the four-taxon case. Syst Zool 1993;42:247–64.

Huelsenbeck JP, Lander KM. Frequent inconsistency of parsimony under a simple model of cladogenesis. Syst Biol 2003;52:641–8.

Huelsenbeck JP, Ronquist F. MRBAYES: Bayesian inference of phylogenetic trees. Bioinformatics 2001;17:754–5.
Huerta-Cepas J, Dopazo H, Dopazo J, Gabaldón T. The human phylome. Genome Biology 2007;8:R109.
Hughes AL, Friedman R. A phylogenetic approach to gene expression data: Evidence for the evolutionary origin of mammalian leukocyte phenotypes. Evol Dev 2009;11:382–90.
Hughes J, Longhorn SJ, Papadopoulou A, Theodorides K, De Riva A, Mejia-Chang M, Foster PG, Vogler AP. Dense taxonomic EST sampling and its applications for molecular systematics of the Coleoptera (Beetles). Mol Biol Evol 2006;23:268–78.
Hughes J, Vogler AP. The phylogeny of acorn weevils (genus Curculio) from mitochondrial and nuclear DNA sequences: The problem of incomplete data. Mol Phylogenet Evol 2004;32:601–15.
Hui JHL, Raible F, Korchagina N, Dray N, Samain S, Magdelenat G, et al. Features of the ancestral bilaterian inferred from *Platynereis dumerilii* ParaHox genes. BMC Biol 2009;7:43.
Humbert W. Ultrastructure des nephrocytes cephaliques et abdominaux chez *Tomocerus minor* (Lubbock) et *Lepidocyrtus curvicollis* Bourlet (Collemboles). Int J Insect Morphol Embryol 1975;4:307–18.
Humbert-David N, Garrone R. A six-armed, tenascin-like protein extracted from the Porifera *Oscarella tuberculata* (Homosclerophorida). Eur J Biochem1993;216:255–60.
Hummel J, Keshvari N, Weckwerth W, Selbig J. Species-specific analysis of protein sequence motifs using mutual information. BMC Bioinformatics 2005;6:164.
Hunnekuhl VS, Wolff C. Reconstruction of cell lineage and spatiotemporal pattern formation of the mesoderm in the amphipod crustacean *Orchestia cavimana*. Dev Dyn 2012;241:697–717.
Hunnekuhl VS, Bergter A, Purschke G, Paululat A. Development and embryonic pattern of body wall musculature in the crassiclitellate *Eisenia andrei* (Annelida, Clitellata). J Morphol 2009;270:1122–36.
Huson DH, Bryant D. User Manual for Splits Tree4 V4. Tübingen 2010. (http://ab.inf.uni-tuebingen.de/data/software/splitstree4/download/V4_11_3/manual.pdf).
Huson DH, Scornavacca C. A survey of combinatorial methods for phylogenetic networks. Genome Biol Evol 2011;3:23–35.
Hutchinson GE. "Homage to Santa Rosalina or why are there so many kinds of animals?" Am Nat 1959;93:145–59.
Huys R, Boxshall GA. Copepod evolution. London: The Ray Society, 1991.
Hwang UW, Friedrich M, Tautz D, Park CJ, Kim W. Mitochondrial protein phylogeny joins myriapods with chelicerates. Nature 2001;413:154–7.
Hyman LH. The Invertebrates, Vol 1. Protozoa through Ctenophora. New York,NY, USA: McGraw-Hill, 1940.
Hyman LH. The Invertebrates, Vol. 2. Platyhelminthes and Rhynchocoela: The Acoelomate Bilateria. New York, NY, USA: McGraw-Hill, 1951.
Hyman LH. The Invertebrates, Vol.5. Chaetognatha. Smaller Coelomate groups. New York, NY, USA: McGraw Hill Book Company, 1959.
Ibrahim MM. Grundzüge der Organbildung im Embryo von *Tachycines* (Insecta: Saltatoria). Zool Jahrb Abt Anat Ontog 1958;76:541–94.
Ikuta T. Evolution of Invertebrate Deuterostomes and Hox/ParaHox Genes. Genomics Proteomics Bioinf 2011;9:77–96.
Imms, AD. A General Textbook of Entomology, including the Anatomy, Physiology, Development and Classification of Insects, 4th edition. London, UK: Methuen & Co, 1938.
Inagaki Y, Susko E, Naomi MF, Roger AJ. Covarion shifts cause a long-branch attraction artifact that units microsporidia and archaebacteria in EF-1α phylogenies. Mol Biol Evol 2004;21:1340–9.

Inoue JG, Miya M, Tsukamoto K, Nishida M. Complete mitochondrial DNA sequence of *Conger myriaster* (Teleostei: Anguilliformes): Novel gene order for vertebrate mitochondrial genomes and the phylogenetic implications for anguilliform families. J Mol Evol 2001;52:311–20.
Inoue JG, Miya M, Tsukamoto K, Nishida M. Evolution of the deep-sea gulper eel mitochondrial genomes: Large-scale gene rearrangements originated within the eels. Mol Biol Evol 2003;20:1917–24.
Inward D, Beccaloni G, Eggleton P. Death of an order: A comprehensive molecular phylogenetic study confirms that termites are eusocial cockroaches. Biol Lett 2007;3:331–5.
Irimia M, Roy SW. Spliceosomal introns as tools for genomic and evolutionary analysis. Nucleic Acids Res 2008;36:1703–12.
Irimia M, Tena JJ, Alexis M, et al. Extensive conservation of ancient microsynteny across metazoans due to cis-regulatory constraints. Genome Res 2012;22:2356–7.
Irvine SQ, Chaga O, Martindale MQ. Larval ontogenetic stages of *Chaetopterus*: Developmental heterochrony in the evolution of Chaetopterid Polychaetes. Biol Bull 1999;197:319–31.
Ishiwata K, Sasaki G, Ogawa J, Miyata T, Su ZH. Phylogenetic relationships among insect orders based on three nuclear protein-coding gene sequences. Mol Phylogenet Evol 2010;58:169–80.
Israelsson O ... and molluscan embryogenesis. Nature 1997;390:32.
Ivanov AV. *Trichoplax adhaerens* Schulze and the question of origin of Metazoa. Dok Akad Nauk Sssr 1973;211:1469–71.
Iwanoff PP. Die Entwicklung der Larvalsegmente bei den Anneliden. Z Morphol Ökol 1928;10:62–161.
Jagdale GB, Gordon R. Distribution of FMRF-amide-like peptide in the nervous system of a mermithid nematode, *Romanomermis culicivorax*. Parasitology 1994;80:467–73.
Jager M, Murienne J, Clabaut C, Deutsch J, Le Guyader H, Manuel M. Homology of arthropod anterior appendages revealed by Hox gene expression in a sea spider. Nature 2006;441:506–8.
Jägersten G. Zur Kenntnis der Morphologie, Entwicklung und Taxonomie der Myzostomida. Nova Acta Reg Soc Sci Ups 1940;11:1–84.
Jägersten G. On the morphology and behaviour of pelagosphaera larvae (Sipunculida). Zool Bidr Upps 1963;36:27–35.
Jägersten G. Evolution of the Metazoan Life Cycle. New York, NY, USA: Academic Press, 1972.
Jakob W, Schierwater B. Changing hydrozoan bauplans by silencing Hox-like genes. PLoS one 2007;2:e694.
Jakob W, Sagasser S, Dellaporta S, Holland P, Kuhn K, Schierwater B. The Trox-2 Hox/ParaHox gene of Trichoplax (Placozoa) marks an epithelial boundary. Dev Genes Evol 2004;214:170–5.
Jang KH, Hwang UW. Complete mitochondrial genome of *Bugula neritina* (Bryozoa, Gymnolaemata, Cheilostomata): phylogenetic position of Bryozoa and phylogeny of lophophorates within the Lophotrochozoa. BMC Genomics 2009;10:167.
Janssen R, Budd GE. Gene expression suggests conserved aspects of Hox gene regulation in arthropods and provides additional support for monophyletic Myriapoda. EvoDevo 2010;1:4.
Janssen R, Damen WGM. Diverged and conserved aspects of heart formation in a spider. Evol Dev 2008;10:155–65.
Janssen R, Damen WGM, Budd GE. Expression of collier in the praemandibular segment of myriapods: Support for the traditional Atelocerata concept or a case of convergence? BMC Ecol Biol 2011;11:50.
Jarvis KJ, Haas F, Whiting MF. Phylogeny of earwigs (Insecta: Dermaptera) based on molecular and morphological evidence: reconsidering the classification of Dermaptera. Syst Entomol 2005;30:442–53.
Jarzembowski EA, Ross AJ. Insect origination and extinction in the Phanerozoic. Geol Soc London Spec Publ 1996;102:65–78.

Jebram D. The importance of different growth directions in the Phylactolaemata and Gymnolaemata for reconstructing the phylogeny of the Bryozoa. In: Larwood GP, ed. Living and Fossil Bryozoa. New York, NY, USA: Academic Press, 1973:565–76.
Jefferies RPS. The ancestry of the vertebrates. Cambridge, MA, USA: Cambridge University Press; 1986.
Jefferies RPS. Two types of bilaterial symmetry in the Metazoa: Chordate and bilaterian. In: Bock GR, Marsch J, eds. Biological Asymmetry and Handedness. Chichester, UK: Wiley, 1991:94–127.
Jeffroy O, Brinkmann H, Delsuc F, Philippe H. Phylogenomics: The beginning of incongruence? Trends Genet 2006;22:225–31.
Jenner RA. Boolean logic and character state identity: Pitfalls of character coding in metazoan cladistics. Contrib Zool 2002;71:67–91.
Jenner RA. Unleashing the force of cladistics? Metazoan phylogenetics and hypothesis testing. Integr Comp Biol 2003;43:207–18.
Jenner RA. Towards a phylogeny of the Metazoa: evaluating alternative phylogenetic positions of Platyhelminthes, Nemertea, and Gnathostomulida, with a critical reappraisal of cladistic characters. Contrib Zool 2004a;73:3–163.
Jenner RA. The scientific status of metazoan cladistics: Why current research practice must change. Zool Scr 2004b;33:293–310.
Jenner RA. Accepting partnership by submission? Morphological phylogenetics in a molecular millennium. Syst Biol 2004c;53:333–42.
Jenner RA. When molecules and morphology clash: Reconciling conflicting phylogenies of the Metazoa by considering secondary character loss. Evol Dev 2004d;6:372–8.
Jenner RA. Higher-level crustacean phylogeny: consensus and conflicting hypotheses. Arthropod Struct Dev 2010;39:143–53.
Jenner RA, Ní Dhubhghaill C, Ferla MP, Wills MA. Eumalacostracan phylogeny and total evidence: limitations of the usual suspects. BMC Evol Biol 2009;9:21.
Jennings RM, Halanych KM. Mitochondrial genomes of *Clymenella torquata* (Maldanidae) and *Riftia pachyptila* (Siboglinidae): evidence for conserved gene order in Annelida. Mol Biol Evol 2005;22:210–22.
Jennison BL. Gametogenesis and reproductive-cycles in the sea-anemone *Anthopleura elegantissima* (Brandt, 1835). Can J Zool 1979;57:403–11.
Jensen LJ, Julien P, Kuhn M, et al. eggNOG: automated construction and annotation of orthologous groups of genes. Nucleic Acids Res 2008;36:D250–4.
Jeram AJ, Selden PA, Edwards D. Land animals in the Silurian: arachnids and myriapods from Shropshire, England. Science 1990;250:658–61.
Jermiin LS, Ho SYW, Ababneh F, Robinson J, Larkum AWD. The biasing effect of compositional heterogeneity on phylogenetic estimates may be underestimated. Syst Biol 2004;53:638–43.
Jeyaprakash A, Hoy MA. First divergence time estimate of spiders, scorpions, mites and ticks (subphylum: Chelicerata) inferred from mitochondrial phylogeny. Exp Appl Acarol 2009;47:1–18.
Jimenez-Guri E, Paps J, García-Fernàndez J, Saló E. Hox and ParaHox genes in Nemertodermatida, a basal bilaterian clade. Int J Dev Biol 2006;50:675–9.
Jintsu Y, Uchifune T, Machida R. Structural features of eggs of the basal phasmatodean *Timema monikensis* Vickery and Sandoval, 1998 (Insecta: Phasmatodea: Timematidae). Arthropod Syst Phylogeny 2010;68:71–8.
Jochmann R, Schmidt-Rhaesa A. New ultrastructural data from the larva of *Paragordius varius* (Nematomorpha). Acta Zool (Stockholm) 2007;88:137–44.
Joffe BI, Kotikova EA. Nervous system of *Priapulus caudatus* and *Halicryptus spinulosus* (Priapulida). Proc Zool Inst USSR Acad Sci 1988;183:52–77.

Johannsen O, Butt FH. Embryology of insects and myriapods, Vol. 1. New York, NY, USA: McGraw-Hill, 1941.

Johansson US, Parsons TJ, Irestedt M, Ericson PGP. Clades within the 'higher land birds', evaluated by nuclear DNA sequences. J Zool Syst Evol Res 2001;39:37–51.

John CC. Habits, structure and development of *Spadella cephaloptera*. Q J Microsc Sci 1933;75:625–96.

Johnson, CD, Reinitz, CA, Sithigorngul, P, Stretton, AOW. Neuronal localization of serotonin in the nematode Ascaris suum. J Comp Neurol 1996;367:352–60.

Joly R, Descamps M. Histology and ultrastructure of the myriapod brain. In: Gupta A, ed. Arthropod Brain: Its Evolution, Development, Structure and Functions. New York, NY, USA: Wiley, 1987:135–57.

Jondelius U, Ruiz-Trillo I, Baguñà J, Riutort M. The Nemertodermatida are basal bilaterians and not members of the Platyhelminthes. Zool Scr 2002;31:201–15.

Jones M, Gantenbein B, Fet V, Blaxter M. The effect of model choice on phylogenetic inference using mitochondrial sequence data: lessons from the scorpions. Mol Phylogenet Evol 2007;43:583–95.

Jöns D. Zur Biologie und Ökologie von *Ligia oceanica* (L.) in der westlichen Ostsee. Kieler Meeresforsch Sonderh 1965;21:203–7.

Joseph JM. On the identification and investigation of homologous gene families, with particular emphasis on the accuracy of multidomain families. PhD thesis, Carnegie Mellon University, Pittsburgh, PA, USA, 2012.

Jow H, Hudelot C, Rattray M, Higgs PG. Bayesian phylogenetics using an RNA substitution model applied to early mammalian evolution. Mol Biol Evol 2002;19:1591–601.

Jühling F, Pütz J, Florentz C, Stadler PF. Armless mitochondrial tRNAs in Enoplea (Nematoda). RNA Biol 2012b;9:1161–6.

Jühling F, Pütz J, Bernt M, et al. Improved systematic tRNA gene annotation allows new insights into the evolution of mitochondrial tRNA structures and into the mechanisms of mitochondrial genome rearrangements. Nucleic Acids Res 2012a;40:2833–45.

Kadner D, Stollewerk A. Neurogenesis in the chilopod *Lithobius forficatus* suggests more similarities to chelicerates than to insects. Dev Genes Evol 2004;214:367–79.

Kamenz C, Staude A, Dunlop JA. Sperm carriers in Silurian sea scorpions. Naturwissenschaften 2011;98:889–96.

Kamm K, Schierwater B. Ancient complexity of the non-Hox ANTP gene complement in the anthozoan *Nematostella vectensis*. Implications for the evolution of the ANTP superclass. J Exp Zool Part B 2006;306B:589–96.

Kamm K, Schierwater B, Jakob W, Dellaporta SL, Miller DJ. Axial patterning and diversification in the cnidaria predate the Hox system. Curr Biol 2006;16:920–6.

Kanehisa M, Goto S, Hattori M, et al. From genomics to chemical genomics: New developments in KEGG. Nucleic Acids Res 2006;34:D354–7.

Kapp H. Morphology and Anatomy. In: Bone Q, Kapp H, Pierrot-Bults AC, editors. The Biology of Chaetognaths. Oxford, UK: Oxford University Press, 1991:5–17.

Kapp H. The unique embryology of Chaetognatha. Zool Anz 2000;239:263–6.

Kapp H. Chaetognatha, Pfeilwürmer. In: Westheide W, Rieger R, eds. Spezielle Zoologie. Teil 1. Einzeller und Wirbellose Tiere. Stuttgart, Germany: Gustav Fischer, 2007:898–904.

Karol KG, Arumuganathan K, Boore JL, et al. Complete plastome sequences of *Equisetum arvense* and *Isoetes flaccida*: implications for phylogeny and plastid genome evolution of early land plant lineages. BMC Evol Biol 2010;10:321.

Karpov SA, Leadbeater BSC. Cytoskeleton structure and composition in choanoflagellates. J Eukaryot Microbiol 1998;45:361–7.

Kato C. Ultrastruktur der Kopfnieren (head kidneys) von sedentären Polychaeten und ihre Bedeutung für die Phylogenie der Annelida. Dissertation, Rheinische-Friedrich-Wilhelms Universität Bonn, Germany, 2013.

Kato C, Lehrke J, Quast B. Ultrastructure and phylogenetic significance of the head kidneys in *Thalassema thalassemum* (Echiura, Thalassematinae). Zoomorphology 2011;130:97–106.

Katoh K, Toh H. Recent developments in the MAFFT multiple sequence alignment program. Brief Bioinform 2008a;9:286–98.

Katoh K, Toh H. Improved accuracy of multiple ncRNA alignment by incorporating structural information into a MAFFT-based framework. BMC Bioinformatics 2008b;9:212–24.

Kaufman ZS. Some problems of regressive evolution. Biol Bull 2008;35:318–26.

Kaul S, Stach T. Ontogeny of the collar cord: neurulation in the hemichordate *Saccoglossus kowalevskii*. J Morphol 2010;271:1240–59.

Kayal E, Lavrov DV. The mitochondrial genome of *Hydra oligactis* (Cnidaria, Hydrozoa) sheds new light on animal mtDNA evolution and cnidarian phylogeny. Gene 2008;410:177–86.

Kayal E, Roure B, Philippe H, Collins AG, Lavrov DV. Cnidarian phylogenetic relationships as revealed by mitogenomics. BMC Evol Biol 2013;13:5.

Kayal E, Bentlage B, Collins AG, Kayal M, Pirro S, Lavrov DV. Evolution of linear mitochondrial genomes in medusozoan cnidarians. Genome Biol Evol 2012;4:1–12.

Keeling P, Doolittle W. Alpha-tubulin from early-diverging eukaryotic lineages and the evolution of the tubulin family. Mol Biol Evol 1996;13:1297–305.

Kelchner SA, Thomas MA. Model use in phylogenetics: nine key questions. Trends Ecol Evol 2006;22:87–94.

Kennedy M, Paterson AM, Morales JC, Parsons S, Winnington AP, Spencer HG. The long and short of it: branch lengths and the problem of placing the New Zealand short-tailed bat, Mystacina. Mol Phylogenet Evol 1999;13:405–16.

Kenning M, Müller C, Wirkner CS, Harzsch S. The Malacostraca from a neurophylogenetic perspective: new insights from brain architecture in *Nebalia herbstii* Leach, 1814 (Leptostraca, Phyllocarida). Zool Anz 2013;252:319–36.

Kenrick P, Crane PR. The origin and early evolution of plants on land. Nature 1997;389:33–9.

Keskinen E, Takaku Y, Meyer-Rochow, B, Hariyama, T. Postembryonic eye growth in the seashore isopods *Ligia exotica* (Crustacea, Isopoda). Biol Bull 2002;202:223–31.

Khodabandeh S, Charmantier G, Blasco C, Grousset E, Charmantier-Daures M. Ontogeny of the antennal glands in the crayfish *Astacus leptodactylus* (Crustacea, Decapoda): anatomical and cell differentiation. Cell Tissue Res 2005;319:153–65.

Kieneke A, Riemann O, Ahlrichs WH. Novel implications for the basal internal relationships of Gastrotricha revealed by an analysis of morphological characters. Zool Scr 2008;37:429–60.

Kilian EF. Stamm Porifera, Schwämme. In: Gruner H-E, ed. Lehrbuch der speziellen Zoologie. Band I (Wirbellose Tiere) 1.Teil: Einführung, Protozoa, Placozoa, Porifera. 5th ed. Jena, Germany: Gustav Fischer, 1993:251–288.

Kilpert F, Podsiadlowski L. The complete mitochondrial genome of the common sea slater, *Ligia oceanica* (Crustacea, Isopoda) bears a novel gene order and unusual control region features. BMC Genomics 2006;7:241.

Kilpert F, Held C, Podsiadlowski L. Multiple rearrangements in mitochondrial genomes of Isopoda and phylogenetic implications. Mol Phylogenet Evol 2012;64:106–17.

Kim JH. General inconsistency conditions for maximum parsimony: effects of branch lengths and increasing numbers of taxa. Syst Biol 1996;45:363–74.

Kimber MJ, Fleming CC, Bjourson AJ, Halton DW, Maule AG. FMRFamide-related peptides in potato cyst nematodes. Mol Biochem Parasit 2001;116:199–208.

Kimble M, Coursey Y, Ahmad N, Hinsch GW. Behavior of the yolk nuclei during embryogenesis, and development of the midgut diverticulum in the horseshoe crab *Limulus polyphemus*. Invertebr Biol 2002;121:365–77.

Kimura M. Estimation of evolutionary distances between homologous nucleotide sequences. Proc Natl Acad Sci USA 1981;78:454–8.

King N, Westbrook M, Young S, et al. The genome of the choanoflagellate *Monosiga brevicollis* and the origin of metazoans. Nature 2008;451:783–8.

Kirsch R, Richter S. The nervous system of *Leptodora kindtii* (Branchiopoda, Cladocera) surveyed with confocal scanning microscopy (cLSM), including general remarks on the branchiopod neuromorphological ground pattern. Arthropod Struct Dev 2007;36:143–56.

Kishinouye K. On the development of Araneina. J Coll Sci Imp Univ Tokyo 1891;4:55–88.

Kitching IJ, Forey PL, Humphries CJ, Williams DM. Cladistics : the theory and practice of parsimony analysis. Oxford, UK: Oxford University Press, 1998.

Kjer KM. Use of rRNA secondary structure in phylogenetic studies to identify homologous positions: an example of alignment and data presentation from the frogs. Mol Phylogenet Evol 1995;4:314–30.

Kjer KM. Aligned 18S and insect phylogeny. Syst Biol 2004;53:506–14.

Kjer KM, Baldridge GD, Fallon AM. Mosquito large subunit ribosomal-RNA – simultaneous alignment of primary and secondary structure. BBA–Gene Struct Expr 1994;1217:147–55.

Kjer KM, Carle FL, Litman J, Ware J. A molecular phylogeny of Hexapoda. Arthropod Syst Phylogeny 2006;65:35–44.

Klaere S, Gesell T, von Haeseler A. The impact of single substitutions on multiple sequence alignments. Phil Trans R Soc Lond B Biol Sci 2008;363:4041–7.

Klag J, Ksiazkiewicz-Kapralska M. Embryonic hemocytes and body cavity formation in Tetrodontophora bielanensis (Insecta, Collembola). In: Dallai R, ed. 3rd International Seminar on Apterygota, Siena, Italy: 1989:215–9.

Klass KD. Possible homologies in the proventriculi of Dicondylia (Hexapoda) and Malacostraca (Crustacea). Zool Anz 1998;237:43–58.

Klass KD. The male abdomen of the relic termite *Mastotermes darwiniensis* (Insecta: Isoptera: Mastotermitidae). Zool Anz 2000;239:231–62.

Klass KD, Kristensen NP. The ground plan and affinities of hexapods: recent progress and open problems. Ann Soc entomol Fr 2001;37:265–98.

Klass KD, Zompro O, Kristensen NP, Adis J. Mantophasmatodea: a new insect order with extant members in the Afrotropics. Science 2002;296:1456–9.

Klingel, H. Vergleichende Verhaltensbiologie der Chilopoden *Scutigera coleoptrata* L. ("Spinnenassel") und *Scolopendra cingulata* Latreille (Skolopender). Z Tierpsychol 1960;17:11–30.

Knoll HJ. Untersuchungen zu Entwicklungsgeschichte von *Scutigera coleoptrata* L. (Chilop.). Zool Jahrb Abt Anat Ontogenie Tiere 1974;92:47–132.

Knudsen V, Caetano-Anollés G. NOBAI: A web server for character coding of geometrical and statistical features in RNA structure. Nucleic Acids Res 2008;36:85–90.

Koch M. Mandibular mechanisms and the evolution of hexapods. Ann Soc entomol Fr 2001;37:129–74.

Kocot KM, Cannon JT, Todt C, et al. Phylogenomics reveals deep molluscan relationships. Nature 2011;477:452–6.

Koenemann S. Post-embryonic development of remipede crustaceans. Evol Dev 2007;9:117–21.

Koenemann S, Iliffe TM, Yager J. *Kaloketos pilosus*, a new species and genus of Remipedia (Crustacea) from the Turks and Caicos islands. Zootaxa 2004;618:1–12.

Koenemann S, Jenner RA, Hoenemann M, Stemme T, von Reumont BM. Arthropod phylogeny revisited, with a focus on crustacean relationships. Arthropod Struct Dev 2010;39:88–110.

Koenemann S, Schram FR, Bloechl A, Iliffe T, Hoenemann M, Held C. Post-embryonic development of remipede crustaceans. Evol Dev 2007;9:117–21.

Koenemann S, Olesen J, Alwes F, et al. The post-embryonic development of Remipedia (Crustacea): additional results and new insights. Dev Genes Evol 2009;219:131–45.

Kohler HJ. Embryologische Untersuchungen an Copepoden: Die Entwicklung von *Lernaeocera branchialis* L. 1767 (Crustacea: Copepoda, Lernaeoidea, Lernaeidae). Zool Jahrb Abt Anat Ontogenie Tiere 1976;95:448–504.

Kojima T. The mechanism of *Drosophila* leg development along the proximodistal axis. Dev Growth Differ 2004;46:115–29.

Kolaczkowski B, Thornton JW. Performance of maximum parsimony and likelihood phylogenetics when evolution is heterogeneous. Nature 2004;431:980–4.

Kolaczkowski B, Thornton JW. A mixed branch length model of heterotachy improves phylogenetic accuracy. Mol Biol Evol 2008;25:1054–66.

Kollmann M, Huetteroth W, Schachtner J. Brain organization in Collembola (springtails). Arthropod Struct Dev 2011;40:304–16.

Kon T, Nohara M, Yamanoue Y, Fujiwara Y, Nishida M, Nishikawa T. Phylogenetic position of a whale-fall lancelet (Cephalochordata) inferred from whole mitochondrial genome sequences. BMC Evol Biol 2007;7:127.

Kopp A, True JR. Phylogeny of the oriental *Drosophila melanogaster* species group: a multilocus reconstruction. Syst Biol 2002;51:786–805.

Kornblihtt AR, de la Mata M, Fededa JP, Muñoz MJ, Nogués G. Multiple links between transcription and splicing. RNA 2004;10:1489–98.

Korschelt EE, Heider K. Vergleichende Entwicklungsgeschichte der Tiere, 2nd ed. Jena, Germany: Gustav Fischer, 1936.

Kowalewsky AO. Entwicklungsgeschichte der einfachen Ascidien. Mem Acad Imp Sci St Petersb Ser 7 1866;10:1–19.

Kozmikova I, Smolikova J, Vlcek C, Kozmik Z. Conservation and diversification of an ancestral chordate gene regulatory network for dorsoventral patterning. PLoS ONE 2011;6:e14650.

Kozomara A, Griffiths-Jones S. miRBase: Integrating microRNA annotation and deep-sequencing data. Nucleic Acids Res 2011;39:D152–7.

Krakauer DC, Collins JP, Erwin D, et al. The challenges and scope of theoretical biology. J Theor Biol 2011;276:269–76.

Kraus O. Zur phylogenetischen Stellung und Evolution der Chelicerata. Entomol Ger 1976;3:1–12.

Kraus O. "Myriapoda" and the ancestry of the hexapoda. Ann Soc entomol Fr 2001;37:105–27.

Kraus O, Brauckmann C. Fossil giants and surviving dwarfs. Arthropleurida and Pselaphognatha (Atelocerata: Diplopoda): Characters, phylogenetic relationships and construction. Verh naturwissensch Ver Hamburg 2003;40:5–50.

Kraus O, Kraus M. Phylogenetic system of the Tracheata (Mandibulata): on "Myriapoda" – Insecta interrelationships, phylogenetic age and primary ecological niches. Verh naturwissensch Ver Hamburg 1994;34:5–31.

Krauss V, Pecyna M, Kurz K, Sass H. Phylogenetic mapping of intron positions: a case study of translation initiation factor eIF2γ. Mol Biol Evol 2005;22:74–84.

Krauss V, Thümmler C, Georgi F, Lehmann J, Stadler P, Eisenhardt C. Near intron positions are reliable phylogenetic markers: an application to holometabolous insects. Mol Biol Evol 2008;25:821–30.

Krebes L, Bastrop R. The mitogenome of *Gammarus duebeni* (Crustacea Amphipoda): a new gene order and non-neutral sequence evolution of tandem repeats in the control region. Comp Biochem Phys D 2012;7:201–11.

Krenn H. Evidence from mouthpart structure on interordinal relationships in endopterygota? Arthropod Syst Phylogeny 2006;65:7–14.

Krieger J, Sandeman RE, Sandeman DC, Hansson BS, Harzsch S. Brain architecture of the largest living land arthropod, the giant robber crab *Birgus latro* (Crustacea, Anomura, Coenobitidae): Evidence for a prominent central olfactory pathway? Front Zool 2010;7:25.

Krieger J, Sombke A, Seefluth F, Kenning M, Hansson BS, Harzsch S. Comparative brain architecture of the European shore crab *Carcinus maenas* (Brachyura) and the common hermit crab *Pagurus bernhardus* (Anomura). Cell Tissue Res 2012;348:47–69.

Kristensen NP. Phylogeny of insect orders. Annu Rev Entomol 1981;26:135–57.

Kristensen NP. Phylogeny of extant hexapods. In: ID Naumann, Lawrence JF, Nielsen ES, Spradberry JP, Taylor RW, Whitten MJ, Littlejohn MJ, eds. The Insects of Australia: A Textbook for Students and Research Workers. Melbourne, Australia: CSIRO, Melbourne Univ. Press, 1991b:125–40.

Kristensen NP. Phylogeny of endopterygote insects, the most successful lineage of living organisms. Eur J Entomol 1999;96:237–53.

Kristensen RM. Loricifera. In: Harrison FW, Ruppert EE, eds. Microscopic Anatomy of Invertebrates, Vol. 4 Aschelminthes. New York, USA: Wiley-Liss, 1991a:351–75.

Kristensen RM. An introduction to Loricifera, Cycliophora, and Micrognathozoa. Integr Comp Biol 2002;42:641–51.

Kristensen RM, Higgins, RP. Kinorhyncha. In: Harrison FW, Ruppert EE, eds. Microscopic Anatomy of Invertebrates, Vol. 4 Aschelminthes. New York, USA: Wiley-Liss, 1991:377–404.

Kristensen RM, Funch P. Micrognathozoa: A new class with complicated jaws like those of Rotifera and Gnathostomulidae. J Morphol 2000;246:1–49.

Kristof A, Wollesen T, Wanninger A. Segmental mode of neural patterning in Sipuncula. Curr Biol 2008;18:1129–32.

Kubatko LS, Degnan JH. Inconsistency of phylogenetic estimates from concatenated data under coalescence. Syst Biol 2007;56:17–24.

Kubo T, Newton KJ. Angiosperm mitochondrial genomes and mutations. Mitochondrion 2008;8:5–14.

Kück P. Homology assessment in molecular phylogenetics. Evaluation, improvement, and influence of data quality on tree reconstruction. Dissertation, Rheinische Friedrich-Wilhelms-Universität Bonn, Germany, 2012.

Kück P, Mayer C, Wägele J-W, Misof B. Long branch effects distort maximum likelihood phylogenies in simulations despite selection of the correct model. PLoS ONE 2012;7:e36593.

Kück P, Meusemann K, Dambach J, et al. Parametric and non-parametric masking of randomness in sequence alignments can be improved and leads to better resolved trees. Front Zool 2010;7:10–21.

Kugler J, Kerner P, Bouquet J-M, Jiang D, Di Gregorio A. Evolutionary changes in the notochord genetic toolkit: A comparative analysis of notochord genes in the ascidian Ciona and the larvacean Oikopleura. BMC Evol Biol 2011;11:21.

Kühl G, Rust J. Devonohexapodus bocksbergensis is a synonym of *Wingertshellicus backesi* (Euarthropoda) – no evidence for marine hexapods living in the Devonian Hunsrück Sea. Org Divers Evol 2009;9:215–31.

Kühl G, Briggs DEG, Rust J. A great-appendage arthropod with a radial mouth from the Lower Hunsrück Slate, Germany. Science 2009;323:771–3.

Kühl G, Bergmann A, Dunlop JA, Garwood RJ, Rust, J. Redescription and palaeobiology of *Palaeoscorpius devonicus* Lehmann, 1944 from the Lower Devonian Hunsrück Slate of Germany. Palaeontology 2012;55:775–87.

Kuhl W. Chaetognatha. In: Bronn HG, ed. Klassen und Ordnungen des Tierreiches. Bd 4, Buch 2, Teil 1. Leipzig, Germany: Akademische Verlagsgesellschaft, 1938.

Kuhl W, Kuhl G. Die Dynamik der Frühentwicklung von *Sagitta setosa*. Helgoland Wiss Meer 1965;12:260–301.

Kuhlmann D. Laboratory studies on the feeding behavior of the chaetognaths *Sagitta setosa* J. Müller and *S. elegans* Verrill with special reference to fish eggs and larvae as food organisms. Kieler Meeresforsch 1977;25:163–71.

Kuhn K, Streit B, Schierwater B. Homeobox genes in the cnidarian *Eleutheria dichotoma*: evolutionary implications for the origin of antennapedia-class (HOM/Hox) genes. Mol Phylogenet Evol 1996;6:30–8.

Kuhner MK, Felsenstein J. A simulation comparison of phylogeny algorithms under equal and unequal evolutionary rates. Mol Biol Evol 1994;11:459–68.

Kukalova-Peck J. Fossil history and the evolution of hexapod structures. In: Naumann ID, Lawrence JF, Nielsen ES, Spradberry JP, Taylor RW, Whitten MJ, Littlejohn MJ, eds. The Insects of Australia: A Textbook for Students and Researcher Workers. Melbourne, Australia: CSIRO Melbourne University Press, 1991:141–79.

Kukalova-Peck J, Lawrence JF. Relationships among coleopteran suborders and major endoneopteran lineages: evidence from hind wing characters. Eur J Entomol 2004;101:95–144.

Kumar S, Filipski AJ, Battistuzzi FU, Kosakovsky Pond SL, Tamura K. Statistics and truth in phylogenomics. Mol Biol Evol 2012;29:457–72.

Kumar S, Skjaeveland A, Orr RJS, et al. AIR: A batch-oriented web program package for construction of supermatrices ready for phylogenomic analyses. BMC Bioinformatics 2009;10:357.

Kumazawa Y. Mitochondrial genomes from major lizard families suggest their phylogenetic relationships and ancient radiations. Gene 2007;388:19–26.

Kunin V, Goldovsky L, Darzentas N, Ouzounis CA. The net of life: reconstructing the microbial phylogenetic network. Genome Res 2005;15:954–9.

Kupczok A, Schmidt HA, von Haeseler A. Accuracy of phylogeny reconstruction methods combining overlapping gene data sets. Algorithm Mol Biol 2010;5:37.

Kuper M. Ultrastrukturuntersuchungen der Segmentalorgane, der Spermien und der Brutpflege-strukturen innerhalb der Syllidae (Annelida: Polychaeta). Dissertation, Universität Osnabrück, Germany, 2001.

Kurabayashi A, Sumida M, Yonekawa H, Glaw F, Vences M, Hasegawa M. Phylogeny, recombination, and mechanisms of stepwise mitochondrial genome reorganization in mantellid frogs from Madagascar. Mol Biol Evol 2008;25:874–91.

Kurabayashi A, Yoshikawa N, Sato N, et al. Complete mitochondrial DNA sequence of the endangered frog *Odorrana ishikawae* (family Ranidae) and unexpected diversity of mt gene arrangements in ranids. Mol Phylogenet Evol 2010;56:543–53.

Kuraku S. Palaeophylogenomics of the vertebrate ancestor – impact of hidden paralogy on hagfish and lamprey gene phylogeny. Integr Comp Biol 2010;50:124–9.

Kuraku S, Meyer A, Kuratani S. Timing of genome duplications relative to the origin of the vertebrates: Did cyclostomes diverge before or after? Mol Biol Evol 2009;26:47–59.

Kuraku S, Hoshiyama D, Katoh K, Suga H, Miyata, T. Monophyly of lampreys and hagfishes supported by nuclear DNA–coded genes. J Mol Evol 1999;49:729–35.

Kuramae EE, Robert V, Snel B, Weiss M, Boekhout T. Phylogenomics reveal a robust fungal tree of life. FEMS Yeast Res 2006;6:1213–20.

Kusche K, Hembach A, Hagner-Holler S, Gebauer W, Burmester T. Complete subunit sequences, structure and evolution of the 6 × 6-mer hemocyanin from the common house centipede, Scutigera coleoptrata. Eur J Biochem 2003;270:2860–8.

Kutsch W, Breidbach O. Homologous structures in the nervous system of Arthropoda. In: Advances in Insect Physiology Vol.24. New York, NY, USA: Academic Press, 1994:1–113.

Kwa MS, Veenstra JG, Van Dijk M, Roos MH. Beta-tubulin genes from the parasitic nematode Haemonchus contortus modulate drug resistance in *Caenorhabditis elegans*. J Mol Biol 1995;246:500–10.

Labandeira CC, Beall BS, Hueber FM. Early insect diversification: Evidence from a Lower Devonian bristletail from Québec. Science 1988;242:913–6.

Lacalli TC. Protochordate body plan and the evolutionary role of larvae: old controversies resolved? Can J Zool 2005;83:216–24.

Lacalli TC. The emergence of the chordate body plan: some puzzles and problems. Acta Zool (Stockholm) 2010;9:4–10.

Lallier FH, Walsh PJ. Activities of uricase, xanthine oxidase, and xanthine dehydrogenase in the hepatopancreas of aquatic and terrestrial crabs. J Crustacean Biol 1991;11:506–12.

Lamarck JBPAdM. Système des animaux sans vertèbres: ou, Tableau général des classes, des ordres et des genres de ces animaux. Présentant leurs caractères essentiels et leur distribution, d'après la considération de leurs rapports naturels et de leur organisation, et suivant l'arrangement établi dans les galeries du Muséum d'hist. naturelle, parmi leurs dépouilles conservées; précédé du discours d'ouverture de cours de zoologie donné dans le Muséum national d'histoire naturelle l'an 8 de la République. Paris, France: Deterville 1801.

Lamarck JPBAdM. Histoire naturelle des animaux sans vertèbres. Vol 5. Paris, France: Deterville, Verdiere, 1818.

Lamsdell JC. Revised systematics of Palaeozoic 'horseshoe crabs' and the myth of monophyletic Xiphosura. Zool J Linn Soc – Lond 2013;167:1–27.

Land MF. Optics and Vision in Invertebrates. In: Autrum H, Bagger-Sjöbäck D, Kornhuber HH, eds. Handbook of Sensory Physiology. Berlin, Germany: Springer, 1981:471–590.

Lang K. Monographie der Harpacticiden, 2 Vol., Stockholm, Sweden: Nordiska Bokhandeln, 1948.

Langworthy K, Helluy S, Benton J, Beltz B. Amines and peptides in the brain of the American lobster: immunocytochemical localization patterns and implications for brain function. Cell Tissue Res 1997;288:191–206.

Lankester ER. The enterocoela and coelomocoela. In: Lankester ER., ed. A Treatise on Zoology, Vol. II, London, UK: A & C Black, 1900:1–37.

Lanterbecq D, Rouse GW, Eeckhaut I. Bodyplan diversification in crinoid-associated myzostomes (Myzostomida, Protostomia). Invertebr Biol 2009;128:283–301.

Lanterbecq D, Rouse GW, Eeckhaut I. Evidence for cospeciation events in the host-symbiont system involving crinoids (Echinodermata) and their obligate associates, the myzostomids (Myzostomida, Annelida). Mol Phylogenet Evol 2010;54:357–71.

Lanterbecq D, Bleidorn C, Michel S, Eeckhaut I. Locomotion and fine structure of parapodia in *Myzostoma cirriferum* (Myzostomida). Zoomorphology 2008;127:59–68.

Lanterbecq D, Rouse GW, Milinkovitch MC, Eeckhaut I. Molecular phylogenetic analyses indicate multiple independent emergences of parasitism in Myzostomida (Protostomia). Syst Biol 2006;55:208–27.

Lapraz F, Besnardeau L, Lepage T. Patterning of the dorsal-ventral axis in echinoderms: Insights into the evolution of the BMP-Chordin signaling network. PLoS Biol 2009;7:e1000248.

Larink O. Zur Entwicklungsgeschichte von *Petrobius brevistylis* (Thysanura, Insecta). Helgoland Wiss Meer 1969;19:111–55.

Larkin MA, Blackshields G, Brown NP, et al. Clustal W and clustal X version 2.0. Bioinformatics 2007;23:2947–8.

Larroux C, Fahey B, Degnan SM, Adamski M, Rokhsar DS, Degnan BM. The NK homeobox gene cluster predates the origin of Hox genes. Curr Biol 2007;17:706–10.

Larroux C. Ancient genes reveal our precambrian ancestor. Australas Sci 2011;2:24–7.

Lartillot N, Philippe H. A Bayesian mixture model for across-site heterogeneities in the amino-acid replacement process. Mol Biol Evol 2004;21:1095–109.

Lartillot N, Brinkmann H, Philippe H. Suppression of long-branch attraction artefacts in the animal phylogeny using a site-heterogeneous model. BMC Evol Biol 2007;7(suppl 1):S4.

Lartillot N, Blanquart S, Lepage T. PhyloBayes 2.3 – a Baysiean software for phylogenic reconstruction using mixture models. University of Montreal, 2008.

Lartillot N, Rodrigue N, Stubbs D, Richer J. PhyloBayes MPI. Phylogenetic reconstruction with infinite mixtures of profiles in a parallel environment. Syst Biol 2013;62:611–5.

Laughlin SB. Energy as a constraint on the coding and processing of sensory information. Curr Opin Neurobiol 2001;11:475–80.

Laurie GW, Leblond CP. Basement membrane nomenclature. Nature 1985;313:372.

Lauterbach KE. Zum Problem der Monophylie der Crustacea. Verh naturwissensch Ver Hamburg 1983;26:293–320.

Lavrov DV. Key transitions in animal evolution: A mitochondrial DNA perspective. Integr Comp Biol 2007;47:734–43.

Lavrov DV, Brown WM. Trichinella spiralis mtDNA: a nematode mitochondrial genome that encodes a putative ATP8 and normally structured tRNAS and has a gene arrangement relatable to those of coelomate metazoans. Genetics 2001;157:621–37.

Lavrov DV, Lang BF. Poriferan mtDNA and animal phylogeny based on mitochondrial gene arrangements. Syst Biol 2005;54:651–9.

Lavrov DV, Boore JL, Brown WM. Complete mtDNA sequences of two millipedes suggest a new model for mitochondrial gene rearrangements: duplication and nonrandom loss. Mol Biol Evol 2002;19:163–9.

Lavrov DV, Brown WM, Boore JL. A novel type of RNA editing occurs in the mitochondrial tRNAs of the centipede Lithobius forficatus. Proc Natl Acad Sci USA 2000;97:13738–42.

Lavrov DV, Brown WM, Boore JL. Phylogenetic position of the Pentastomida and (pan)crustacean relationships. Proc R Soc Lond B: Biol Sci 2004;271:537–44.

Lavrov DV, Pett W, Voigt O, et al. Mitochondrial DNA of *Clathrina clathrus* (Calcarea, Calcinea): six linear chromosomes, fragmented rRNAs, tRNA editing, and a novel genetic code. Mol Biol Evol 2013;30:865–80.

Lawrence JF, Ślipiński A, Seago AE, Thayer MK, Newton AF, Marvaldi AE. Phylogeny of the Coleoptera based on morphological characters of adults and larvae. Ann Zool 2011;61:1–217.

Layton DM, Bundschuh R. A statistical analysis of RNA folding algorithms through thermodynamic parameter perturbation. Nucleic Acids Res 2005;33:519–24.

Le SQ, Gascuel O. An improved general amino acid replacement matrix. Mol Biol Evol 2008;25:1307–20.

Le TH, Blair D, Agatsuma T, et al. Phylogenies inferred from mitochondrial gene orders – a cautionary tale from the parasitic flatworms. Mol Biol Evol 2000;17:1123–5.

Leach L, Trudgill DL, Gahan, PB. Immunocytochemical localization of neurosecretory amines and peptides in the free-living nematode, *Goodeyus ulmi*. Histochem J 1987;19:471–5.

Leasi F, Rouse GW, Sørensen MV. A new species of *Paraseison* (Rotifera: Seisonacea) from the coast of California, USA. J Mar Biol Assoc UK 2011;92:959–65.

Lecointre G, Philippe H, Van Le HL, Le Guyader H. Species sampling has a major impact on phylogenetic inference. Mol Phylogenet Evol 1993;2:205–24.

Leder H. Untersuchungen über den feineren Bau des Nervensystems der Cladoceren. Arbeiten aus den Zoologischen Instituten der Universität Wien. 1915:297–392.

Lee CT, Risom T, Strauss WM. Evolutionary conservation of microRNA regulatory circuits: an examination of microRNA gene complexity and conserved microRNA-target interactions through metazoan phylogeny. DNA Cell Biol 2007;26:209–18.

Lehmann J, Stadler PF, Krauss V. Near intron pairs and the metazoan tree. Mol Phylogenet Evol 2013;66:811–23.

Lehmann J, Eisenhardt C, Stadler P, Krauss V. Some novel intron positions in conserved *Drosophila* genes are caused by intron sliding or tandem duplication. BMC Evol Biol 2010;10:156.

Leigh JW, Susko E, Baumgartner M, Roger AJ. Testing congruence in phylogenomic analysis. Syst Biol 2008;57:104–15.

Lemaire P. Evolutionary crossroads in developmental biology: the tunicates. Development 2011;138:2143–52.

Lemburg C. Ultrastructure of the introvert and associated structures of the larvae of *Halicryptus spinulosus* (Priapulida). Zoomorphology 1995;115:11–29.

Lemburg C. Ultrastrukturelle Untersuchungen an den Larven von *Halicryptus spinulosus* und *Priapulus caudatus* Hypothesen zur Phylogenie der Priapulida und deren Bedeutung für die Evolution der Nemathelminthes. Göttingen, Germany: Cuvillier, 1999.

Lento GM, Hickson RE, Chambers GK, Penny, D. Use of spectral analysis to test hypotheses on the origin of pinnipeds. Mol Biol Evol 1995;12:28–52.

Letsch HO, Kjer KM. Potential pitfalls of modelling ribosomal RNA data in phylogenetic tree reconstruction: evidence from case studies in the Metazoa. BMC Evol Biol 2011;11:146.

Letsch HO, Kück P, Stocsits RR, Misof B. The impact of rRNA secondary structure consideration in alignment and tree reconstruction: simulated data and a case study on the phylogeny of hexapods. Mol Biol Evol 2010;27:2507–21.

Letsch HO, Schmidt C, Kück P, Fleck G, Stocsits RR, Misof B. Simultaneous alignment and folding of 28S rRNA sequences uncovers phylogenetic signal in structure variation. Mol Phylogenet Evol 2009;53:758–71.

Letsch HO, Meusemann K, Wipfler B, Schutte K, Beutel R, Misof B. Insect phylogenomics: results, problems and the impact of matrix composition. Proc R Soc Lond B: Biol Sci 2012;279:3282–90.

Leuckart FS. Versuch einer naturgemässen Eintheilung der Helminthen. Heidelberg, Germany: Neue Akademische Buchhandlung von Karl Gross, 1827.

Leuckart FS. In Beziehung auf den Haarstern (*Comatula*) und *Pentacrinus europaeus*, so wie auf das Schmarotzerthier auf *Comatula*. Notizen aus dem Gebiete der Natur- und Heilkunde gesammelt und mitgetheilt von Dr. L.G.V. Froriep 1836;59:129–31.

Lewontin RC. The genetic basis of evolutionary change. New York, NY, USA: Columbia University Press, 1974.

Leydig F. Ueber *Argulus foliaceus*. Arch mikrosk Anat 1889;23:1–51.

Leys S. Gastrulation in Sponges. In: Stern C, ed. Gastrulation: From Cells to Embryo. Cold Spring Harbor, NY, USA: Cold Spring Harbor Laboratory Press, 2004:23–31.

Leys SP, Riesgo A. Epithelia, an evolutionary novelty of metazoans. J Exp Zool Part B: 2011;318:438–47.

Leys SP, Eerkes-Medrano D. Gastrulation in calcareous sponges: in search of Haeckel's gastraea. Integr Comp Biol 2005;45:342–51.

Li L, Stoeckert CJJ, Roos DS. OrthoMCL: identification of ortholog groups for eukaryotic genomes. Genome Res 2003;13:2178–89.

Lichtneckert R, Reichert H. Insights into the urbilaterian brain: conserved genetic patterning mechanisms in insect and vertebrate brain development. Heredity 2005;94:465–77.

Liley R. Ctenophora. Fiches d'identification du plancton. ICES, 1958:1–5.

Lin CP, Chen MY, Huang JP. The complete mitochondrial genome and phylogenomics of a damselfly, *Euphaea formosa* support a basal Odonata within the Pterygota. Gene 2010;468:20–9.

Lin Y, Rajan V, Moret BM. TIBA: a tool for phylogeny inference from rearrangement data with bootstrap analysis. Bioinformatics 2012;28:3324–5.
Linard B, Thompson J, Poch O, Lecompte O. OrthoInspector: comprehensive orthology analysis and visual exploration. BMC Bioinformatics 2011;12:11.
Lindgreen S, Gardner PP, Krogh A. MASTR: multiple alignment and structure prediction of non-coding RNAs using simulated annealing. Bioinformatics 2007;23:3304–11.
Littlewood DT, Telford MJ, Clough KA, Rohde K. Gnathostomulida – an enigmatic metazoan phylum from both morphological and molecular perspectives. Mol Phylogenet Evol 1998;9:72–9.
Littlewood DTJ, Smith AB, Clough KA, Emson RH. The interrelationships of the echinoderm classes: morphological and molecular evidence. Biol J Linn Soc 1997;61:409–38.
Liu G, Sim K, Li J. Efficient Mining of Large Maximal Bicliques. Data Warehousing and Knowledge Discovery. Berlin, Germany: Springer, 2006:437–48.
Liu G, Seiler H, Wen A, Zars T, Ito K, Wolf R. Distinct memory traces for two visual features in the *Drosophila* brain. Nature 2006;439:551–6.
Liu J, Steiner MM, Dunlop JA, et al. An armoured Cambrian lobopodian from China with arthropod-like appendages. Nature 2011;470:526–30.
Liu K, Raghavan S, Nelesen S, Linder C, Warnow T. Rapid and accurate large-scale coestimation of sequence alignments and phylogenetic trees. Science 2009;324:1561.
Lloyd DG, Calder VL. Multi-residue gaps, a class of molecular characters with exceptional reliability for phylogenetic analyses. J Evolution Biol 1991;4:9–21.
Lockhart P, Steel M. A tale of two processes. Syst Biol 2005;54:948–51.
Lockhart PJ, Cameron SA. Trees for bees. Trends Ecol Evol 2001;16:84–8.
Lockhart PJ, Larkum AW, Steel M, Waddell PJ, Penny D. Evolution of chlorophyll and bacterio-chlorophyll: the problem of invariant sites in sequence analysis. Proc Natl Acad Sci USA 1996;93:1930–4.
Loesel R. Comparative morphology of central neuropils in the brain of arthropods and its evolutionary and functional implications. Acta Biol Hung 2004;55:39–51.
Loesel R. The arthropod brain: retracing six hundred million years of evolution. Arthropod Struct Dev 2005;34:207–9.
Loesel R. Can brain structures help to resolve interordinal relationships in insects? Arthropod Syst Phylogeny 2006;64:101–6.
Loesel R. Neurophylogeny – retracing early metazoan brain evolution. In: Pontarotti P, ed. Evolutionary Biology: Concepts, Biodiversity, Macroevolution, and Genome Evolution. Berlin, Germany: Springer, 2011:169–91.
Loesel R, Heuer CM. The mushroom bodies – prominent brain centers of arthropods and annelids with enigmatic evolutionary origin. Acta Zool (Stockholm) 2010;91:29–34.
Loesel R, Homberg U. Histamine-immunoreactive neurons in the brain of the cockroach *Leucaphaea maderae*. Brain Res 1999;842:408–18.
Loesel R, Nässel DR, Strausfeld NJ. Common design in a unique midline neuropil in the brains of arthropods. Arthropod Struct Dev 2002;31:77–91.
Loesel R, Seyfarth E-A, Bräuning P, Agricola H-J. Neuroarchitecture of the arcuate body in the brain of the spider *Cupiennius salei* (Araneae, Chelicerata) revealed by allatostatin-, proctolin-, and CCAP-immunocytochemistry and its evolutionary implications. Arthropod Struct Dev 2011;40:210–20.
Long JH, Koob-Emunds M, Sinwell B, Koob TJ. The notochord of hagfish *Myxine glutinosa*: visco-elastic properties and mechanical functions during steady swimming. J Exp Biol 2002;205:3819–31.

Longhorn SJ, Pohl HW, Vogler AP. Ribosomal protein genes of holometabolan insects reject the Halteria, instead revealing a close affinity of Strepsiptera with Coleoptera. Mol Phylogenet Evol 2010;55:846–59.
Lopez JV, Yuhki N, Masuda R, Modi W, O'Brien SJ. Numt, a recent transfer and tandem amplification of mitochondrial DNA to the nuclear genome of the domestic cat. J Mol Evol 1994;39:174–90.
Lorenzen S. Phylogenetic aspects of pseudocoelomate evolution. In: Conway Morris S, George JD, Gibson R, Platt HM, eds. The origins and relationships of lower invertebrates. Oxford, UK: Clarendon, 1985:210–23.
Louvet JP. Morphogenèse du pleuropode chez l'embryon de *Carausius morosus* Br. (Phasmida: Lonchodidae): étude au microscope électronique a balayage. Int J Insect Morphol 1976;5:35–49.
Lovett DL, Felder DL. Ontogenetic changes in enzyme distribution and midgut function in developmental stages of *Penaeus setiferus* (Crustacea, Decapoda, Penaeidae). Biol Bull 1990;178:160–74.
Lowe C. Molecular genetic insights into deuterostome evolution from the direct-developing hemichordate *Saccoglossus kowalevskii*. Phil Trans R Soc Lond B Biol Sci 2008;363:1569–78.
Lowe CJ, Terasaki M, Wu M, et al. Dorsoventral patterning in hemichordates: Insights into early chordate evolution. PLoS Biol 2006;4:e291.
Lowe E. On the anatomy of a marine copepod, *Calanus finmarchicus* (Gunnerus). T Roy Soc Edin 1935;58:561–603.
Lowery LA, Sive H. Strategies of vertebrate neurulation and a re-evaluation of teleost neural tube formation. Mech Develop 2004;121:1189–97.
Lucas CH, Lawes S. Sexual reproduction of the scyphomedusa *Aurelia aurita* in relation to temperature and variable food supply. Mar Biol 1998;131:629–38.
Luke GN, Castro LFC, McLay K, Bird C, Coulson A, Holland PWH. Dispersal of NK homeobox gene clusters in amphioxus and humans. P Natl Acad Sci USA 2003;100:5292.
Lüter C. The origin of the coelom in Brachiopoda and its phylogenetic significance. Zoomorphology 2000;120:15–28.
Lüter C, Bartolomaeus T. The phylogenetic position of Brachiopoda – a comparison of morphological and molecular data. Zool Scr 1997;26:245–54.
Lüter C, Grobe P, Bartolomaeus T. Tentaculata (Lophophorata), Tentakulaten. In: Westheide W, Rieger R, eds. Spezielle Zoologie. Teil 1: Einzeller und Wirbellose Tiere. 2nd ed. München, Germany: Elsevier, 2007:764–87.
Lynch C. Jim Gray's fourth paradigm and the construction of the scientific record. In: Hey T, Tansley S, Tolle K, eds. The Fourth Paradigm: Data-Intensive Scientific Discoveries. Redmond, WA, USA: Microsoft Research, 2009:177–83.
Lyons-Weiler J, Hoelzer GA. Escaping from the Felsenstein zone by detecting long branches in phylogenetic data. Mol Phylogenet Evol 1997;8:375–84.
Lyons-Weiler J, Hoelzer GA, Tausch RJ. Relative apparent synapomorphy analysis (RASA) I: The statistical measurement of phylogenetic signal. Mol Biol Evol 1996;13:749–57.
Lyson TR, Sperling EA, Heimberg AM, Gauthier JA, King BL, Peterson KJ. MicroRNAs support a turtle + lizard clade. Biol Letters 2012;8:104–7.
Ma X, Hou X, Edgecombe GD, Strausfeld NJ. Complex brain and optic lobes in an early Cambrian arthropod. Nature 2012;490:258–61.
Maas A, Waloszek D. Cambrian derivatives of the early arthropod stem lineage, pentastomids, tardigrades and lobopodians an 'Orsten'perspective. Zool Anz 2001;240:451–9.
Macey JR, Larson A, Ananjeva NB, Fang Z, Papenfuss TJ. Two novel gene orders and the role of light-strand replication in rearrangement of the vertebrate mitochondrial genome. Mol Biol Evol 1997;14:91–104.

Macey JR, Papenfuss TJ, Kuehl JV, Fourcade HM, Boore JL. Phylogenetic relationships among amphisbaenian reptiles based on complete mitochondrial genomic sequences. Mol Phylogenet Evol 2004;33:22–31.

Mackey LY, Winnepenninckx B, De Wachter R, Backeljau T, Emschermann P, Garey JR. 18S rRNA suggests that Entoprocta are protostomes, unrelated to Ectoprocta. J Mol Evol 1996;42:552–9.

Maddison DR, Maddison WP. Macclade 4. Sunderland, MA, USA: Sinauer Associate Inc., Publishers, 2004.

Maddison DR, Baker MD, Ober KA. Phylogeny of carabid beetles as inferred from 18S ribosomal DNA (Coleoptera: Carabidae). Syst Entomol 1999;24:103–38.

Maddison WP. Gene trees in species trees. Syst Biol 1997;46:523–36.

Maes M. On a cyclic string-to-string correction problem. Inform Process Lett 1990;35:73–8.

Malakhov VV. Structure of nervous system of head end in a free-living marine nematode, *Pontonema vulgare*. Zool Zh 1978;57:645–52

Malakhov VV. Cephalorhyncha, a new type of animal kingdom uniting Priapulida, Kinorhyncha, Gordiacea, and a new system of Aschelminthes worms. Zool Zh 1980;59:485–99

Malakhov VV. The structure of nervous system of the posterior end of the body of the free-living marine nematode *Pontonema vulgare* and the principal plan of the nervous system in nematodes. Zool Zh 1982;61:1481–91

Malakhov VV. Nematodes. Structure, development, classification, and phylogeny. Washington, D.C. USA: Smithsonian Institution Press, 1994.

Maldonado M, Carmona MC, Uriz MJ, Cruzado A. Decline in Mesozoic reef-building sponges explained by silicon limitation. Nature 1999;401:785–8.

Maldonado M, Riesgo A. Intra-epithelial spicules in a homosclerophorid sponge. Cell Tissue Res 2007;328:639–50.

Maldonado M. Choanoflagellates, choanocytes, and animal multicellularity. Invertebr Biol 2004;123:1–22.

Mallatt J, Giribet G. Further use of nearly complete 28S and 18S rRNA genes to classify Ecdysozoa:37 more arthropods and a kinorhynch. Mol Phylogenet Evol 2006;40:772–94.

Mallatt J, Sullivan J. 28S and 18S rDNA sequences support the monophyly of lampreys and hagfishes. Mol Biol Evol 1998;15:1706–18.

Mallat J, Winchell CJ. Testing the new animal phylogeny: first use of combined large-subunit and small sub-unit rRNA gene sequences to classify the protostomes. Mol Biol Evol 2002;19:289–301.

Mallatt J, Winchell CJ. Ribosomal RNA genes and deuterostome phylogeny revisited: more cyclostomes, elasmobranchs, reptiles, and a brittle star. Mol Phylogenet Evol 2007;43:1005–22.

Mallatt J, Craig CW, Yoder MJ. Nearly complete rRNA genes assembled from across the metazoan animals: Effects of more taxa, a structure-based alignment, and paired-sites evolutionary models on phylogeny reconstruction. Mol Phylogenet Evol 2010;55:1–17.

Mallatt J, Craig CW, Yoder MJ. Nearly complete rRNA genes from 371 Animalia: Updated structure-based alignment and detailed phylogenetic analysis. Mol Phylogenet Evol 2012;64:603–17.

Mallatt JM, Garey JR, Shultz JW. Ecdysozoan phylogeny and Bayesian inference: First use of nearly complete 28S and 18S rRNA gene sequences to classify the arthropods and their kin. Mol Phylogenet Evol 2004;31:178–91.

Mann KH. Leeches (Hirundinea): Their structure, physiology, ecology and embryology. Oxford: Pergamon Press, 1962.

Manton SM. On the embryology of a mysid crustacean, *Hemimysis lamornae*. Phil Trans R Soc Lond B Biol Sci 1928;216:363–463.

Manton SM. Mandibular mechanisms and the evolution of arthropods. J Linn Soc London 1964;247:153–87.
Manton SM. Arthropod phylogeny, a modern synthesis. J Zool London 1973;171:111–30.
Manton SM. Uniramian evolution with particular reference to the pleura. In: Camatini M ed. Myriapod Biology. New York, NY, USA: Academic Press, 1979:317–43.
Manuel M, Borchiellini C, Alivon E, Le Parco Y, Vacelet J, Boury-Esnault N. Phylogeny and evolution of calcareous sponges: monophyly of Calcinea and Calcaronea, high level of morphological homoplasy, and the primitive nature of axial symmetry. Syst Biol 2003;52:311–33.
Mao F, Williams D, Zhaxybayeva O, Poptsova M, Lapierre P, Gogarten P, Xu Y. Quartet decomposition server: A platform for analyzing phylogenetic trees. BMC Bioinformatics 2012;13:123.
Marcade I, Cordaux R, Doublet V, Debenest C, Bouchon D, Raimond R. Structure and evolution of the atypical mitochondrial genome of *Armadillidium vulgare* (Isopoda, Crustacea). J Mol Evol 2007;65:651–9.
Marcet-Houben M, Gabaldon T. The tree versus the forest: The fungal tree of life and the topological diversity within the yeast phylome. PLoS ONE 2009;4:e4357.
Marcus E. Bryozoarios marinhos Brasileiros II. Bol Fac Philos Sci Letr Univ São Paulo 1938;IV, Zool. 2:1–196.
Marden JH, Thompson JD. Rowing locomotion by a stonefly that possesses the ancestral condition of co-occuring wings and abdominal gills. Biol J Linn Soc 2003;79:341–9.
Margoliash E. Primary structure and evolution of cytochrome C. Proc Natl Acad Sci USA 1963;50:672–9.
Mark Welch DB. Evidence from a protein-coding gene that acanthocephalans are rotifers. Invertebr Biol 2000;119:17–26.
Markevich GI. SEM observations on *Seison* and phylogenetic relationships of the Seisonidea (Rotifera). Hydrobiologia 1993;255:513–20.
Marlétaz F, Le Parco Y. Careful with understudied phyla: The case of chaetognath. BMC Evol Biol 2008;8:251.
Marlétaz F, Le Parco Y. Phylogeny of animals: Genomes have a lot to say. Adv Mar Genomics 2010;1:119–41.
Marlétaz F, Gyapay G, Le Parco Y. High level of structural polymorphism driven by mobile elements in the *Hox* genomic region of the chaetognath *Spadella cephaloptera*. Gen Biol Evol 2010;2:665–77.
Marlétaz F, Gilles A, Caubit X, et al. Chaetognath transcriptome reveals ancestral and unique features among bilaterians. Gen Biol 2008;9: R94.
Marlétaz F, Martin E, Perez Y, et al. Chaetognath phylogenomics: a protostome with deuterostome-like development. Curr Biol 2006;16: R577–8.
Marques AC, Collins AG. Cladistic analysis of Medusozoa and cnidarian evolution. Invertebr Biol 2004;123:23–42.
Martin AP, Burg TM. Perils of paralogy: using HSP70 genes for inferring organismal phylogenies. Syst Biol 2002;51:570–87.
Martin JL. Branchiopoda. In: Harrison FW, Humes AG, eds. Microscopic Anatomy of Invertebrates Vol. 9. New York: Wiley-Liss. 1992:25–224.
Martin MF. On the morphology and classification of *Argulus* (Crustacea). Proc Zool Soc London 1932;103:771–806.
Martín-Durán JM, Janssen R, Wennberg S, Budd GE, Hejnol A. Deuterostomic development in the protostome *Priapulus caudatus*. Curr Biol 2012;22:1–6.
Martindale MQ, Heijnol A. A developmental perspective: Changes in the position of the blastopore during bilaterian evolution. Dev Cell 2009;17:162–74.

Martindale MQ, Shankland M. Developmental origin of segmental differences in the leech ectoderm: Survival and differentiation of the distal tubule cell is determined by the host segment. Dev Biol 1988;125:290–300.
Martindale MQ, Finnerty JR, Henry JQ. The Radiata and the evolutionary origins of the bilaterian body plan. Mol Phylogenet Evol 2002;24:358–65.
Masci AM, Arighi CN, Diehl AD, et al. An improved ontological representation of dendritic cells as a paradigm for all cell types. BMC Bioinf 2009;10:70.
Mason-Gamer RJK, E. A. Testing for phylogenetic conflict among molecular data sets in the tribe Triticeae (Gramineae). Syst Biol 1996:522–43.
Masta SE. Mitochondrial rRNA secondary structures and genome arrangements distinguish chelicerates: comparisons with a harvestman (Arachnida: Opiliones: *Phalangium opilio*). Gene 2010;449:9–21.
Masta SE, Boore JL. Parallel evolution of truncated transfer RNA genes in arachnid mitochondrial genomes. Mol Biol Evol 2008;25:949–59.
Masta SE, Klann AE, Podsiadlowski L. A comparison of the mitochondrial genomes from two families of Solifugae (Arthropoda: Chelicerata): Eremobatidae and Ammotrechidae. Gene 2008;417:35–42.
Masta SE, McCall A, Longhorn SJ. Rare genomic changes and mitochondrial sequences provide independent support for congruent relationships among the sea spiders (Arthropoda, Pycnogonida). Mol Phylogenet Evol 2010;57:59–70.
Mathew AP. Embryology of *Heterometrus scaber*. Bull Res Inst Travancore 1956;1:1–96.
Mathews DH. Predicting a set of minimal free energy RNA secondary structures common to two sequences. Bioinformatics 2005;21:2246–53.
Matsuda R. Morphology and evolution of the insect thorax. Mem Can Entomol Soc 1970;76:1–483.
Mattei X, Marchand B. The spermatozoa of Acanthocephala and Myzostomida – resemblance and phyletic consequences. C R Acad Sci III-Vie 1987;305:525–9.
Matthysse AG, Deschet K, Williams M, Marry M, White AR, Smith WC. A functional cellulose synthase from ascidian epidermis. Proc Natl Acad Sci USA 2004;101:986–91.
Matus DQ, Copley RR, Dunn CW, et al. Broad taxon and gene sampling indicate that chaetognaths are protostomes. Curr Biol 2006;16: R575–6.
Matus DQ, Halanych KM, Martindale MQ. The *Hox* gene complement of a pelagic chaetognath, *Flaccisagitta enflata*. Int Comp Biol 2007;47:854–64.
Maxmen A, Browne W, Martindale M, Giribet G. Neuroanatomy of sea spiders implies an appendicular origin of the protocerebral segment. Nature 2005;437:1144–8.
Maxmen AB. Marine worm rewrites theory of brain evolution. Nature News 2012; doi:10.1038/nature.2012.10226.
Mayer G. Ultrastructural investigations on mesoderm differentiation in Onychophora. Dissertation, Freie Universität Berlin, Germany, 2005.
Mayer G. Origin and differentiation of nephridia in the Onychophora provide no support for the Articulata. Zoomorphology 2006;125:1–12.
Mayer G, Koch M. Ultrastructure and fate of the nephridial anlagen in the antennal segment of *Epiperipatus biolleyi* (Onychophora, Peripatidae)–evidence for the onychophoran antennae being modified legs. Arthropod Struct Dev 2005;34:471–80.
Mayer G, Tait NN. Position and development of oocytes in velvet worms shed light on the evolution of the ovary in Onychophora and Arthropoda. Zool J Linn Soc 2009;157:17–33.
Mayer G, Whitington PM. Neural development in Onychophora (velvet worms) suggests a step-wise evolution of segmentation in the nervous system of Panarthropoda. Dev Biol 2009a;335:263–75.

Mayer G, Whitington PM. Velvet worm development links myriapods with chelicerates. Proc R Soc Lond B Biol Sci 2009b;276:3571–9.
Mayer G, Ruhberg H, Bartolomaeus T. When an epithelium ceases to exist: an ultrastructural study on the fate of the embryonic coelom in *Epiperipatus biolleyi* (Onychophora, Peripatidae). Acta Zool (Stockholm) 2004;85:163–70.
Mayer G, Bartolomaeus T, Ruhberg H. Ultrastructure of mesoderm in embryos of *Opisthopatus roseus* (Onychophora, Peripatopsidae): Revision of the "long germ band" hypothesis for *Opisthopatus*. J Morphol 2005;263:60–70.
Mayhew PJ. Shifts in hexapod diversification and what Haldane could have said. Proc Roy Soc B 2002;269:969–74.
Mayhew PJ. A tale of two analyses: Estimating the consequences of shifts in hexapod diversification. Biol J Linn Soc 2003;80:23–36.
Mayhew PJ. Why are there so many insect species? Perspectives from fossils and phylogenies. Biol Rev Camb Philos Soc 2007;82:425–54.
Mayrat A, McMahon BR, Tanaka K. (2006). The circulatory system. In: Forest J, v. Vaupel Klein JC, eds. Treatise on Zoology, Vol. 2, The Crustacea. Leiden, TheNetherlands: Brill, 2006:3–84.
Mayrose I, Friedman N, Pupko T. A gamma mixture model better accounts for among site heterogeneity. Bioinformatics 2005;21:151–8.
Mazurie A, Bonchev D, Schwikowski B, Buck GA. Phylogenetic distances are encoded in networks of interacting pathways. Bioinformatics 2008;24:2579–85.
McHugh D. Molecular evidence that echiurans and pogonophorans are derived annelids. Proc Natl Acad Sci USA 1997;94:8006–9.
McHugh D. Molecular phylogeny of the Annelida. Can J Zool 2000;78:1873–84.
McHugh D. Molecular systematics of polychaetes (Annelida). In: Bartolomaeus T, Purschke G, eds. Morphology, molecules, evolution and phylogeny in Polychaeta and related taxa. Hydrobiologia 2005;535/536:309–18.
McKenna DD, Farrell BD. 9-genes reinforce the phylogeny of Holometabola and yield alternate views on the phylogenetic placement of Strepsiptera. PLoS ONE 2010;5:e11887.
McMahon DP, Hayward A, Kathirithamby J. The mitochondrial genome of the 'twisted-wing parasite' Mengenilla australiensis (Insecta, Strepsiptera): A comparative study. BMC Genomics 2009;10:603.
Mecznikow E. Zur Entwicklungsgeschichte von *Myzostomum*. Z wiss Zool 1866;16:236–43.
Medina M, Collins AG, Silberman JD, Sogin ML. Evaluating hypotheses of basal animal phylogeny using complete sequences of large and small subunit rRNA. Proc Natl Acad Sci USA 2001;98:9707–12.
Medina M, Collins AG, Takaoka TL, Kuehl JV, Boore JL. Naked corals: skeleton loss in Scleractinia. Proc Natl Acad Sci USA 2006;103:9096–100.
Meglitsch PA, Schram FR. Invertebrate Zoology. Oxford, UK: Oxford University Press, 1991.
Mei J, Chen B, Yue H, Gui JF. Identification of a C1q family member associated with cortical granules and follicular cell apoptosis in *Carassius auratus gibelio*. Mol Cell Endocrinol 2008;289:67–76.
Melone G, Ferraguti W. The spermatozoon of *Brachionus plicatilis* (Rotifera, Monogononta) with some notes on spermultrastructure in Rotifera. Acta Zool (Stockholm) 1994;75:81–8.
Melone G, Ricci C, Segers H, Wallace RL. Phylogenetic relationships of phylum Rotifera with emphasis on the families of Bdelloidea. Hydrobiologia 1998;387:101–7.
Melzer RR. Persisting stemma neuropils in *Chaoborus crystallinus* (Diptera: Chaoboridae): Development and evolution of a bipartite visual system. J Morphol 2009;270:1524–30.
Melzer RR, Paulus HF. Post-larval development of compund eyes and stemmata of *Chaoborus crystallinus* (De Greer, 1776) (Diptera: Chaoboridae): Stage-specific reconstructions within individual organs of vision. Int J Insect Morph Embryol 1994;23:261–74.

Melzer RR, Petyko Z, Smola U. Photoreceptor axons and optic neuropils in *Lithobius forficatus* (Linnaeus, 1758) (Schilopoda, Lithobiidae). Zool Anz 1997;235:177–82.

Mercurio M, Corriero G, Gaino E. A 3-year investigation of sexual reproduction in *Geodia cydonium* (Jameson 1811) (Porifera, Demospongiae) from a semi-enclosed Mediterranean bay. Mar Biol 2007;151:1491–500.

Metschnikoff VE. Über die systematische Stellung von *Balanoglossus*. Zool Anz 1881;4:139–57.

Mettam C. Functional constraints in the evolution of the Annelida. In: Morris SC, George JD, Gibson R, Platt HM, eds. The origins and relationships of lower invertebrates. The Systematics Association Special Volume, Oxford, UK: Clarendon Press, 1985:297–309.

Metzker ML. Sequencing technologies – the next generation. Nat Rev Genet 2010;11:31–46.

Meusemann K, Von Reumont B, Simon S, et al. A phylogenomic approach to resolve the arthropod tree of life. Mol Biol Evol 2010;27:2451–64.

Meyer A, Todt C, Mikkelsen NT, Lieb B. Fast evolving 18S rRNA sequences from Solenogastres (Mollusca) resist standard PCR amplification and give new insights into mollusk substitution rate heterogeneity. BMC Evol Biol 2010;10:70.

Meyer B, Meusemann K, Misof B. April 2011. MARE: MAtrix Reduction – A tool to select optimized data subsets from supermatrices for phylogenetic inference, http://mare.zfmk.de.

Meyer E. Die Abstammung der Anneliden. Der Ursprung der Metamerie und die Bedeutung des Mesoderms. Biol Zentralbl 1890;10:296–308.

Michaelsen W. Clitellata = Gürtelwürmer. In: Kükenthal W, ed. Handbuch der Zoologie. Berlin, Germany: DeGruyter, 1928–1932:1–118.

Michener WK. Meta-information concepts for ecological data management. Ecol Informatics 2006;1:3–7.

Miller DJ, Ball EE. Animal evolution: *Trichoplax*, trees, and taxonomic turmoil. Curr Biol 2008;18:R1003–R5.

Miller RL, Cnidaria. In: Adiyodi KG, Adiyodi, RG, eds. Reproductive Biology of Invertebrates. Vol. II (Spermatogenesis and Sperm Function). Chichester, UK: John Wiley and Sons,1983:23–73.

Min GS, Kim SH, Kim W. Molecular phylogeny of arthropods and their relatives: Polyphyletic origin of arthropodization. Mol Cells 1998;8:75–83.

Min GS, Park JK. Eurotatorian paraphyly: Revisiting phylogenetic relationships based on the complete mitochondrial genome sequence of *Rotaria rotatoria* (Bdelloidea: Rotifera: Syndermata). BMC Genomics 2009;10:533–45.

Mindell DP, Sorenson MD, Dimcheff DE. Multiple independent origins of mitochondrial gene order in birds. Proc Natl Acad Sci USA 1998;95:10693–7.

Minelli, S. Chilopoda. In: Harrison FW, Rice ME, eds. Microscopic Anatomy of Invertebrates Vol. 12 Onychopohora, Chilopoda, and lesser Protostomata. New York, NY, USA: Wiley-Liss, 1993:11–56.

Minot CS. Cephalic homologies. A contribution to the determination of the ancestry of vertebrates. Am Nat 1897;31:927–43.

Minxiao W, Song S, Chaolun L, Xin S. Distinctive mitochondrial genome of calanoid copepod *Calanus sinicus* with multiple large non-coding regions and reshuffled gene order: Useful molecular markers for phylogenetic and population studies. BMC Genomics 2011;12 73.

Mirkin B, Muchnik I, Smith TF. A biologically consistent model for comparing molecular phylogenies. J Comput Biol 1995;2:493–507.

Mishler BD, Kelch DG. Phylogenomics and early land plant evolution. In: Shaw AJ, Goffinet B, editors, Bryophyte Biology. 2nd ed. Cambridge, UK: Cambridge University Press, 2009:173–97.

Misof B, Anderson CL, Buckley TR, Erpenbeck D, Rickert A, Misof K. An empirical analysis of mt 16S rRNA covarion-like evolution in insects: site-specific rate variation is clustered and frequently detected. J Mol Evol 2002;55:460–9.

Misof B, Fleck G. Comparative analysis of mt LSU rRNA secondary structures of Odonates: structural variability and phylogenetic signal. Insect Mol Biol 2003;12:535–47.

Misof B, Misof K. A Monte Carlo approach successfully identifies randomness in multiple sequence alignments: A more objective means of data exclusion. Syst Biol 2009;58:21–34.

Misof B, Niehuis O, Bischoff I, Rickert A, Erpenbeck D, Staniczek A. Towards an 18S phylogeny of hexapods: accounting for group-specific character covariance in optimized mixed nucleotide/doublet models. Zoology 2007;110:409–29.

Misof B, Niehuis O, Bischoff I, Rickert A, Erpenbeck D, Staniczek A. A hexapod nuclear SSU rRNA secondary-structure model and catalog of taxon-specific structural variation. J Exp Zool Part B 2006;306B:70–88.

Mitchell A, Cho S, Regier JC, Mitter C, Poole RW, Matthews M. Phylogenetic utility of elongation factor-1 alpha in Noctuoidea (Insecta: Lepidoptera): the limits of synonymous substitution. Mol Biol Evol 1997;14:381–90.

Mittmann B. Die Keimstreifbildung und das Expressionsmuster des Homöobox-Gens *Distal-less* bei primär flügellosen Insekten (Collembola und Zygentoma). Sber Ges Naturf Freunde Berlin (NF) 2000;38:93–103.

Mittmann B, Wolff C. Embryonic development and staging of the cobweb spider *Parasteatoda tepidariorum* C. L. Koch, 1841 (syn.: *Achaearanea tepidariorum*; Araneomorphae; Theridiidae). Dev Genes Evol 2012;222:189–216.

Miya M, Satoh TR, Nishida M. The phylogenetic position of toadfishes (order Batrachoidiformes) in the higher ray-finned fish as inferred from partitioned Bayesian analysis of 102 whole mitochondrial genome sequences. Biol J Linn Soc 2005;85:289–306.

Miyamoto H, Machida RJ, Nishida S. Complete mitochondrial genome sequences of the three pelagic chaetognaths *Sagitta nagae, Sagitta decipiens* and *Sagitta enflata*. Comp Biochem Physiol D Gen Prot 2010;5:65–72.

Miyazawa H, Yoshida MA, Tsuneki K, Furuya H. Mitochondrial genome of a Japanese placozoan. Zool Sci 2012;29:223–8.

Møller OS, Olesen J, Avenant-Oldewage A, Thomsen PF, Glenner H. First maxillae suction discs in Branchiura (Crustacea): Development and evolution in light of the first molecular phylogeny of Branchiura, Pentastomidae, and other "Maxillopoda". Arthropod Struct Dev 2008;37:333–46.

Monks S. Phylogeny of the Acanthocephala based on morphological characters. Syst Parasitol 2001;48:81–116.

Monteiro AS, Schierwater B, Dellaporta SL, Holland PWH. A low diversity of ANTP class homeobox genes in Placozoa. Evol Dev 2006;8:174–82.

Montgomery TH. The adult organisation of *Paragordius varius* (Leidy). Zool Jahrb Abt Ontog Anat 1903;18:387–474.

Mooi R, Gill A. Phylogenies without synapomorphies – a crisis in fish systematics: Time to show some character. Zootaxa 2010;2450:26–40.

Moreno E, Permanyer J, Martinez P. The origin of patterning systems in Bilateria – insights from the Hox and ParaHox genes in Acoelomorpha. Gen Prot Bioinf 2011;9:65–76.

Moret B, Tang J, Warnow T. Reconstructing phylogenies from gene-content and gene-order data. In Mathematics of Evolution and Phylogeny. Oxford, UK: Oxford University Press, 2005:321–52.

Morgenstern B. A simple and space-efficient fragment-chaining algorithm for alignment of DNA and protein sequences. Appl Math Lett 2002;15:11–6.

Morgenstern B, Prohaska SJ, Pöhler D, Stadler PF. Multiple sequence alignment with user-defined anchor points. Algorithms Mol Biol 2006;1:6.

Moritz M. Zur Embryonalentwicklung der Phalangiiden (Opiliones; Palpatores) II. Die Anlage und Entwicklung der Coxaldrüse bei *Phalangium opilio* L. Zool Jahrb Abt Anat Ontog 1959;77:229–40.

Morris J, Cardona A, Del Mar De Miguel-Bonet M, Hartenstein V. Neurobiology of the basal platyhelminth *Macrostomum lignano*: map and digital 3D model of the juvenile brain neuropile. Dev Genes Evol 2007;217:569–84.

Morrison D A framework for phylogenetic sequence alignment. Plant Syst Evol 2009;282:127–49.

Morrison DA. Using data-display networks for exploratory data analysis in phylogenetic studies. Mol Biol Evol 2010;27:1044–57.

Mosig A, Hofacker IL, Stadler PF. Comparative analysis of cyclic sequences: Viroids and other small circular RNAs. In Proceedings GCB 2006, Lecture Notes in Informatics P-83, Bonn: Gesellschaft für Informatik, 2006:93–102.

Moura G, Christoffersen ML. The system of the mandibulate arthropods: Tracheata and Remipedia as sister groups, "Crustacea" non-monophyletic. J Comp Biol 1996;1:95–112.

Mueller RL, Boore JL. Molecular mechanisms of extensive mitochondrial gene rearrangement in plethodontid salamanders. Mol Biol Evol 2005;22:2104–12.

Mueller RL, Macey JR, Jaekel M, Wake DB, Boore JL. Morphological homoplasy, life history evolution, and historical biogeography of plethodontid salamanders inferred from complete mitochondrial genomes. Proc Natl Acad Sci USA 2004;101:13820–5.

Müller CHG, Sombke A, Rosenberg J. The fine structure of some bristly millipedes (Penicillata, Diplopoda): Additional support for the homology of mandibulate ommatidia. Arthrop Struct Dev 2007;36:463–76.

Müller CHG, Rieger V, Perez Y, Harzsch S. Immunohistochemical and ultrastructural studies on ciliary sense organs of arrow worms (Chaetognatha). Zoomorphology 2013: DOI 10.1007/s00435-013-0211-6

Müller CHG, Rosenberg J, Richter S, Meyer-Rochow VB. The compound eye of *Scutigera coleoptrata* (Linnaeus, 1758) (Chilopoda: Notostigmophora): An ultrastructural reinvestigation that adds support to the Mandibulata concept. Zoomorphology 2003;122:191–209.

Müller J. Über die Gattungen und Arten der Comatulen. Arch Naturgesch 1841;7:139–48.

Müller K. Incorporating information from length-mutational events into phylogenetic analysis. Mol Phylogenet Evol 2006;38:667–76.

Müller KJ. Crustacea with preserved soft parts from the Upper Cambrian of Sweden. Lethaia 1983;16:93–109.

Müller KJ, Waloszek D. Skaracarida, a new order of Crustacea from the Upper Cambrian of Västergötland, Sweden. Fossils Strata 1985;17:1–65.

Müller KJ, Waloszek D. *Martinssonia elongata* gen. et sp.n., a crustacean-like euarthropod from the Upper Cambrian 'Orsten' of Sweden. Zool Scripta 1986;15:73–92.

Müller M, Homberg U, Kühn A. Neuroarchitecture of the lower division of the central body in the brain of the locust (*Schistocerca gregaria*). Cell Tiss Res 1997;288:159–76.

Müller MCM, Westheide W. Structure of the nervous system of *Myzostoma cirriferum* (Annelida) as revealed by immunohistochemistry and cLSM analyses. J Morphol 2000;245:87–98.

Müller W, Wang X, Kropf K, et al. Silicatein expression in the hexactinellid *Crateromorpha meyeri*: the lead marker gene restricted to siliceous sponges. Cell Tissue Res 2008;333:339–51.

Mundy SP, Taylor PD, Thorpe JP. A reinterpretation of Phylactolaemate phylogeny. In: Larwood GP, Nielsen C, ed. Recent and Fossil Bryozoa. Fredensborg, Denmark: Olsen and Olsen, 1981:185–90.

Murienne J, Edgecombe G, Giribet G. Including secondary structure, fossils and molecular dating in the centipede tree of life. Mol Phylogenet Evol 2010;57:301–13.

Murphy WJ, Pringle TH, Crider TA, Springer MS, Miller W. Using genomic data to unravel the root of the placental mammal phylogeny. Genome Res 2007;17:413–21.

Mutlu E. Diel vertical migration of *Sagitta setosa* as inferred acoustically in the Black Sea. Mar Biol 2006;149:573–84.

Mwinyi A, Bailly X, Bourlat S, Jondelius U, Littlewood DT, Podsiadlowski L. The phylogenetic position of acoela as revealed by the complete mitochondrial genome of *Symsagittifera roscoffensis*. BMC Evol Biol 2010;10:309.

Mwinyi A, Meyer A, Bleidorn C, Lieb B, Bartolomaeus T, Podsiadlowski L. Mitochondrial genome sequence and gene order of *Sipunculus nudus* give additional support for an inclusion of Sipuncula into Annelida. BMC Genomics 2009;10:27.

Nair SG. On the embryology of the isopod *Irona*. J Embryol Exp Morphol 1956;4:1–33.

Nakao T. Electron microscope study of the circulatory system in *Nereis japonica*. J Morphol 1974;144:217–36.

Nakashima K, Nishino A, Hirose E. Forming a tough shell via an intracellular matrix and cellular junctions in the tail epidermis of *Oikopleura dioica* (Chordata: Tunicata: Appendicularia). Naturwissenschaften 2011;98:661–9.

Nakashima K, Yamada L, Satou Y, Azuma J-i, Satoh N. The evolutionary origin of animal cellulose synthase. Dev Genes Evol 2004;214:81–8.

Nansen F. Bidrag til Myzostomernes anatomi og histologi. Bergen, Norway: John Griegs Bogtrykkerl, 1885.

Nardi F, Spinsanti G, Boore JL, Carapelli A, Dallai R, Frati F. Hexapod origins: monophyletic or paraphyletic? Science 2003;299:1887–9.

Narechania A, Baker RH, Sit R, Kolokotronis S-O, DeSalle R, Planet PJ. Random addition concatenation analysis: A novel approach to the exploration of phylogenomic signal reveals strong agreement between core and shell genomic partitions in the cyanobacteria. Gen Biol Evol 2012;4:30–43.

Navidi WC, Churchill GA, von Haeseler A. Methods for inferring phylogenies from nucleic acid sequence data by using maximum likelihood and linear invariants. Mol Biol Evol 1991;8:128–43.

Nawrocki EP, Kolbe DL, Eddy SR. Infernal 1.0: inference of RNA alignments. Bioinformatics 2009;25:1335–7.

Naylor GJ, Brown WM. Structural biology and phylogenetic estimation. Nature 1997;388:527–8.

Near TJ. Acanthocephalan phylogeny and the evolution of parasitism. Integr Comp Biol 2002;42:668–77.

Near TJ, Garey JR, Nadler SA. Phylogenetic relationships of the Acanthocephala inferred from 18S ribosomal DNA sequences. Mol Phylogenet Evol 1998;10:287–98.

Nebelsick, M. Introvert, mouth cone, and nervous system of *Echinoderes capitatus* (Kinorhyncha, Cyclorhagida) and implications for the phylogenetic relationships of Kinorhyncha. Zoomorphology 1993;113:211–32.

Negrisolo E, Minelli A, Valle G. The mitochondrial genome of the house centipede *Scutigera* and the monophyly versus paraphyly of Myriapoda. Mol Biol Evol 2004a;21:770–80.

Negrisolo E, Minelli A, Valle G. Extensive gene order rearrangement in the mitochondrial genome of the centipede *Scutigera coleoptrata*. J Mol Evol 2004b;58:413–23.

Nesnidal MP, Helmkampf M, Bruchhaus I, Hausdorf B. Compositional heterogeneity and phylogenomic inference of metazoan relationships. Mol Biol Evol 2010;27:2095–104.

Nesnidal M, Helmkampf M, Bruchhaus I, Hausdorf B. The complete mitochondrial genome of *Flustra foliacea* (ectoprocta, cheilostomata) – compositional bias affects phylogenetic analyses of lophotrochozoan relationships. BMC Genomics 2011;12:572.

Neuhaus B. Ultrastructure of alimentary canal and body cavity, ground pattern, and phylogenetic relationships of Kinorhyncha. Microfauna marina 1994;9:61–156.

Neuhaus B. Kinorhyncha. In: Schmidt-Rhaesa A, ed. Handbook of Zoology, Gastrotricha, Cycloneuralia and Gnathifera. Berlin, Gemany: De Gruyter, 2012: in press.

Nguyen HD, Yoshihama M, Kenmochi N. New maximum likelihood estimators for eukaryotic intron evolution. PLoS Comput Biol 2005;1:e79.
Nguyen MAT, Klaere S, von Haeseler A. MISFITS: Evaluating the goodness of fit between a phylogenetic model and an alignment. Mol Biol Evol 2011;28:143–52.
Nguyen MAT, Klaere S, von Haeseler A. ImOSM: Intermittent evolution and robustness of phylogenetic methods. Mol Biol Evol 2012;29:663–73.
Nichols R. Gene trees and species trees are not the same. TREE 2001;16:358–64.
Nickel M. Evolutionary emergence of synaptic nervous systems: What can we learn from the non-synaptic, nerveless Porifera? Invertebr Biol 2010;129:1–16.
Nickel M. The pre-nervous system and beyond – poriferan milestones in the early evolution of the metazoan nervous system. In: DeSalle R, Schierwater B, eds. Key Transitions in Animal Evolution. Bocca Raton, FL, USA: CRC Press, 2011:85–126.
Niehuis O, Hartig G, Grath S, et al. Genomic and morphological evidence converge to resolve the enigma of Strepsiptera. Curr Biol 2012;22:1309–13.
Niehuis O, Naumann CM, Misof B. Phylogenetic analysis of Zygaenoidea small-subunit rRNA structural variation implies initial oligophagy on cyanogenic host plants in larvae of the moth genus Zygaena (Insecta : Lepidoptera). Zool J Linn Soc Lond 2006;147:367–81.
Nielsen C. Entoproct life cycles and the entoproct/ectoproct relationship. Ophelia 1971;9:209–341.
Nielsen C. Animal phylogeny in the light of the trochaea theory. Biol J Linn Soc 1985;25:243–99.
Nielsen C. Larval and adult brains. Evol Dev 2005a;7:483–9.
Nielsen C. Trochophora larvae: Cell-lineages, ciliary bands and body regions. 2. Other groups and general discussion. J Exp Zool 2005b;304B:401–47.
Nielsen C. Six major steps in animal evolution: Are we derived sponge larvae? Evol Dev 2008;10:241–57.
Nielsen C. Some aspects of spiralian development. Acta Zool (Stockholm) 2010;91:20–8.
Nielsen, C Animal evolution: interrelationships of the living phyla. 3rd edition. Oxford, UK: Oxford University Press, 2012.
Nielsen C, Hay-Schmidt A. Development of the enteropneust *Ptychodera flava*: Ciliary bands and nervous system. J Morphol 2007;268:551–70.
Nielsen C, Scharff N, Eibye-Jacobsen D. Cladistic analyses of the animal kingdom. Biol J Linn Soc 1996;57:385–410.
Nilsson DE, Osorio D. Homology and parallelism in arthropod sensory processing. In: Fortey RA, Thomas RH eds. Arthropod Relationships. London, UK: Chapman and Hall, 1998:333–47.
Nishida H. Cell lineage analysis in ascidian embryos by intracellular injection of a tracer enzyme. Dev Biol 1987;121:526–41.
Nitsche A, Rose D, Fasold M, Stadler PF. Comparison of splice sites reveals that long non-coding RNAs are evolutionarily well conserved 2012; Submitted.
Noguchi Y, Endo K, Tajima F, Ueshima R. The mitochondrial genome of the brachiopod *Laqueus rubellus*. Genetics 2000;155:245–59.
Nohara M, Nishida M, Miya M, Nishikawa T. Evolution of the mitochondrial genome in Cephalochordata as inferred from complete nucleotide sequences from two *Epigonichthys* species. J Mol Evol 2005;60:526–37.
Nolte C, Ahn Y, Krumlauf R. Evolutionary developmental biology: Hox gene evolution. eLS 2001.
Nomaksteinsky M, Roettinger E, Dufour HD, Chettouh Z, Lowe CJ, Martindale MQ, Brunet J-F. Centralization of the deuterostome nervous system predates chordates. Curr Biol 2009;19:1264–9.
Nordhausen W. Impact of the nemertean *Lineus viridis* on its polychaete prey on an intertidal sandflat. Hydrobiol 1988;156:39–46.
Norén M, Jondelius U. *Xenoturbella*'s molluscan relatives… . Nature 1997;390:31–2.
Nosek J. Biogenesis of the cortical granules in fish oocyte. Histochem J 1984;16:435–7.

Nosenko T, Schreiber F, Adamska M, et al. Deep metazoan phylogeny: When different genes tell different stories. Mol Phylogenet Evol 2013;67:223–33.
Novikova E, Bakalenko N, Nesterenko A, Kulakova M. Expression of Hox genes during regeneration of nereid polychaete *Alitta (Nereis) virens* (Annelida, Lophotrochozoa). EvoDevo 2013;4:14.
Nübler-Jung K, Arendt D. Dorsoventral axis inversion: Enteropneust anatomy links invertebrates to chordates turned upside down. J Zool Syst Evol Res 1999;37:93–100.
Nyhart LK. Biology takes form – animal morphology and the German universities, 1800–1900. Chicago, IL, USA: The University of Chicago Press, 1995.
O'Leary MA, Kaufman S. MorphoBank: Phylophenomics in the "cloud". Cladistics 2011;27:1–9.
Oakley TH, Wolfe JM, Lindgren AR, Zaharoff AK. Phylotranscriptomics to bring the understudied into the fold: monophyletic Ostracoda, fossil placement, and pancrustacean phylogeny. Mol Biol Evol 2013;30:215–33.
Odorico DM, Miller DJ. Internal and external relationships of the Cnidaria: Implications of primary and predicted secondary structure of the 5'-end of the 23S-like rDNA. Proc Roy Soc B Biol Sci 1997;264:77–82.
Ogden TH, Rosenberg M. Multiple sequence alignment accuracy and phylogenetic inference. Syst Biol 2006;55:314–28.
Ogden TH, Whiting MF. The problem with "the Paleoptera Problem:" sense and sensitivity. Cladistics 2003;19:432–42.
Ohta T. Role of random genetic drift in the evolution of interactive systems. J Mol Evol 1997;44:9–14.
Okamura K. Diversity of animal small RNA pathways and their biological utility. Wiley Interdiscip Rev RNA 2012;3:351–68.
Olesen J. Monophyly and phylogeny of Branchiopoda, with focus on morphology and homologies of branchiopod phyllopodous limbs. J Crust Biol 2007;27:165–83.
Olsen GJ, Woese CR. Ribosomal RNA: A key to phylogeny. FASEB Journal: Official Publ Federation Am Soc Exp Biol 1993;7:113–23.
Olsson R. Reissner's Fibre Mechanisms: Some Common Denominators. In: Oksche A, Rodríguez EM, Fernández-Llebrez P, eds. The subcommissural organ. Berlin, Germany: Springer-Verlag, 1993:33–9.
Omilian AR, Taylor DJ. Rate acceleration and long-branch attraction in a conserved gene of cryptic daphniid (Crustacea) species. Mol Biol Evol 2001;18:2201–12.
Omori S, Sato Y, Isobe T, Yukawa M, Murata K. Complete nucleotide sequences of the mitochondrial genomes of two avian malaria protozoa, *Plasmodium gallinaceum* and *Plasmodium juxtanucleare*. Parasitol Res 2007;100:661–4.
Orengo CA, Thornton JM. Protein families and their evolution – a structural perspective. Annu Rev Biochem 2005;74:867–900.
Orrhage L. Anatomische und morphologische Studien über die Polychaetenfamilien Spionidae, Disomidae und Poecilochaetidae. Zool Bidr Uppsala 1964;36:335–405.
Ortega Hernández J, Legg DA, Braddy SJ. The phylogeny of aglaspidid arthropods and the internal relationships within Artiopoda. Cladistics 2013;29:15–45.
Osigus HJ, Eitel M, Schierwater B. Chasing the urmetazoon: Striking a blow for quality data? Mol Phylogenet Evol 2013;66:551–7.
Osorio D, Averof M, Bacon JP. Arthropod evolution: Great brains, beautiful bodies. TREE 1995, 10:449–54.
Ostlund G, Schmitt T, Forslund K, et al. InParanoid 7: New algorithms and tools for eukaryotic orthology analysis. Nucleic Acids Res 2010;38:D196–203.
Ota KG, Kuratani S. Cyclostome embryology and early evolutionary history of vertebrates. Integr Comp Biol 2007;47:329–37.

Ott M, Zola J, Aluru S, Stamatakis A. Large-scale maximum likelihood-based phylogenetic analysis on the IBM BlueGene/L. Proceedings of the ACM/IEEE conference on Supercomputing 2007. Reno, NV, USA: ACM, 2007.

Ovchinnikov S, Masta SE. Pseudoscorpion mitochondria show rearranged genes and genome-wide reductions of RNA gene sizes and inferred structures, yet typical nucleotide composition bias. BMC Evol Biol 2012;12:31.

Overstreet RM, Dyková I, Hawkins WE. Branchiura. In: Harrison FW, Humes AG, eds. Microscopic Anatomy of Invertebrates, Vol. 9. New York, NY, USA: Wiley-Liss, 1992:385–413.

Pacheco MA, Battistuzzi FU, Lentino M, Aguilar RF, Kumar S, Escalante AA. Evolution of modern birds revealed by mitogenomics: Timing the radiation and origin of major orders. Mol Biol Evol 2011;28:1927–42.

Paffenhöfer G-A. The cultivation of an appendicularian through numerous generations. Mar Biol 1973;22:183–5.

Paffenhöfer G-A, Gibson DM. Determination of generation time and asexual fecundity of doliolids (Tunicata, Thaliacea). J Plankton Res 1999;21:1183–9.

Page RD, Charleston MA. From gene to organismal phylogeny: Reconciled trees and the gene tree/species tree problem. Mol Phylogenet Evol 1997;7:231–40.

Palakodeti D, Smielewska M, Graveley BR. MicroRNAs from the planarian Schmidtea mediterranea: a model system for stem cell biology. RNA 2006;12:1640–9.

Palmer JD, Soltis D, Soltis P. Large size and complex structure of mitochondrial DNA in two nonflowering land plants. Curr Genet 1992;21:125–9.

Pamilo P, Nei M. Relationships between gene trees and species trees. Mol Biol Evol 1998;5:568–83.

Pani AM, Mullarkey EE, Aronowicz J, Assimacopoulos S, Grove EA, Lowe CJ. Ancient deuterostome origins of vertebrate brain signalling centres. Nature 2012;483:289–94.

Papillon D, Perez Y, Caubit X, Le Parco Y. Identification of chaetognaths as protostomes is supported by the analysis of their mitochondrial genome. Mol Biol Evol 2004;21:2122–9.

Papillon D, Perez Y, Caubit X, Le Parco Y. Systematics of Chaetognatha under the light of molecular data, using duplicated ribosomal 18S DNA sequences. Mol Phylogenet Evol Mar 2006;38:621–34.

Papillon D, Perez Y, Fasano L, Le Parco Y, Caubit X. *Hox* gene survey in the chaetognath *Spadella cephaloptera*: Evolutionary implications. Dev Genes Evol 2003;213:142–8.

Papillon D, Perez Y, Fasano L, Parco YL, Caubit X. Restricted expression of a median *Hox* gene in the central nervous system of chaetognaths. Dev Genes Evol 2005;215:369–73.

Paps J, Baguñà J, Riutort M. Lophotrochozoa internal phylogeny: new insights from an up-to-date analysis of nuclear ribosomal genes. Proc R Soc Lond B Biol Sci 2009a;276:1245–54.

Paps J, Baguñà J, Riutort M. Bilaterian phylogeny: A broad sampling of 13 nuclear genes provides a new Lophotrochozoa phylogeny and supports a paraphyletic basal Acoelomorpha. Mol Biol Evol 2009b;26:2397–406.

Parida L. Using PQ structures for genomic rearrangement phylogeny. J Comp Biol 2006;13(10):1685–700.

Parikesit AA, Stadler PF, Prohaska SJ. Evolution and quantitative comparison of genome-wide protein domain distributions. Genes 2011;2:912–24.

Park JK, Rho HS, Kristensen RM, Kim W, Giribet G. First molecular data on the phylum Loricifera – an investigation into the phylogeny of Ecdysozoa with emphasis on the positions of Loricifera and Priapulida. Zool Sci 2006;23:943–54.

Park JK, Kim KH, Kang S, Kim W, Eom KS, Littlewood DT. A common origin of complex life cycles in parasitic flatworms: evidence from the complete mitochondrial genome of *Microcotyle sebastis* (Monogenea: Platyhelminthes). BMC Evol Biol 2007;7:11.

Park SJ, Lee YS, Hwang UW. The complete mitochondrial genome of the sea spider *Achelia bituberculata* (Pycnogonida, Ammotheidae): arthropod ground pattern of gene arrangement. BMC Genomics 2007;8:343.

Parsch J, Braverman JM, Stephan W. Comparative sequence analysis and patterns of covariation in RNA secondary structures. Genetics 2000;154:909–21.

Paśko L, Ericson PGP, Elzanowski A. Phylogenetic utility and evolution of indels: A study in neognathous birds. Mol Phylogenet Evol 2011;61:760–71.

Pass G. Antennal circulatory organs in Onychophora, Myriapoda and Hexapoda: Functional morphology and evolution. Zoomorphology 1991;110:145–64.

Passamaneck YJ, Halanych KM. Evidence from *Hox* genes that bryozoans are lophotrochozoans. Evol Dev 2004;6:275–81.

Passamaneck Y, Halanych KM. Lophotrochozoan phylogeny assessed with LSU and SSU data: evidence of lophophorate polyphyly. Mol Phylogenet Evol 2006;40:20–8.

Patel NH. The evolution of arthropod segmentation; insights from comparisons of gene expression patterns. Dev Suppl 1994:201–7.

Patten W, Hazen AP. The development of the coxal gland, branchial cartilages, and genital ducts of *Limulus polyphemus*. J Morphol 1900;16:459–502.

Pattengale ND, Alipour M, Bininda-Emonds ORP, Moret BME, Stamatakis A. How many bootstrap replicates are necessary? J Comput Biol 2010;17:337–54.

Patterson C. Homology in classical and molecular biology. Mol Biol Evol 1988;5:603–25.

Paul DH. A neurophylogenist's view of decapod Crustacea. Bull Mar Sci 1989;45:487–504.

Paulus HF. Eye structure and the monophyly of the Arthropoda. In: Gupta AP, ed. Arthropod Phylogeny. New York, NY, USA: Van Nostrand Reinhold Co., 1979:299–383.

Paulus HF. Phylogeny of the Myriapoda – Crustacea – Insecta: A new attempt using photoreceptor structure. J Zool Syst Evol Res 2000;38:189–208.

Pawlowski J, Bolivar I, Fahrni JF, de Vargas C, Gouy M, Zaninetti L. Extreme differences in rates of molecular evolution of Foraminifera revealed by comparison of ribosomal DNA sequences and the fossil record. Mol Biol Evol 1997;14:498–505.

Pawson DL. Phylum Echinodermata. Zootaxa 2007:1668;749–64.

Pearre S. Growth and Reproduction. In: Bone Q, Kapp H, Pierrot-Bults AC, eds. The biology of chaetognaths. New York, NY, USA: Oxford University Press, 1991:61–75.

Pedersen KJ. Structure and composition of basement membranes and other basal matrix systems in selected invertebrates. Acta Zool (Stockholm) 1991;72:181–201.

Penner O, Grassberger P, Paczuski M. Sequence alignment, mutual information, and dissimilarity measures for constructing phylogenies. PLoS ONE 2011;6:e14373.

Pennerstorfer M, Scholtz G. Early cleavage in *Phoronis muelleri* (Phoronida) displays spiral features. Evol Devol 2012;13:484–500.

Penney D. Does the fossil record of spiders track that of their principal prey, the insects? Trans Roy Soc Edinburgh: Earth Sciences;2004:94:275–81.

Pennings SC, Carefoot TH, Zimmer M, Danko JP, Ziegler A. Feeding preferences of supralittoral isopods and amphipods. Canadian J Zool 2000;78:1918–29.

Penny D, Foulds LR, Hendy MD. Testing the theory of evolution by comparing phylogenetic trees constructed from five different protein sequences. Nature 1982;297:197–200.

Penny D, White WT, Hendy M, Phillips MJ. A bias in ML estimates of branch lengths in the presence of multiple signals. Mol Biol Evol 2008;25:239–42.

Pepato AR, Rocha CEF, Dunlop JA. Phylogenetic position of the actinotrichid mites: Sensitivity to homology assessment under total evidence. BMC Evol Biol 2010;10:235.

Perez Y, Casanova J-P, Mazza J. Changes in the structure and ultrastructure of the intestine of *Spadella cephaloptera* (Chaetognatha) during feeding and starvation experiments. J Exp Mar Biol Ecol 2000;253:1–15.

Perez Y, Casanova J-P, Mazza J. Degrees of vacuolation of the absorptive intestinal cells of five *Sagitta* (Chaetognatha) species: Possible ecophysiological implications. Mar Biol 2001;138:125–33.

Perez Y, Arnaud J, Brunet M, Casanova J-P, Mazza J. Morphological study of the gut in *Sagitta setosa*, *S. serratodentata* and *S. pacifica* (Chaetognatha). Functional implications in digestive processes. J Mar Biol Ass UK 1999;79:1097–109.

Perez Y, Rieger V, Martin E, Müller C, Harzsch S. Neurogenesis in an early protostome relative: progenitor cells in the ventral nerve centre of chaetognath hatchlings are arranged in a highly organized geometrical pattern. J. Exp. Zool. 2013;320:179–93.

Pérez-Losada M, Høeg JT, Crandall K. Remarkable convergent evolution in specialized parasitic Thecostraca (Crustacea). BMC Biology 2009;7:15.

Perseke M, Golombek A, Schlegel M, Struck TH. The impact of mitochondrial genome analyses on the understanding of deuterostome phylogeny. Mol Phylogenet Evol 2013;66:898–905.

Perseke M, Bernhard D, Fritzsch G, Brämmer F, Stadler PF, Schlegel M. Mitochondrial genome evolution in ophiuroidea, echinoidea, and holothuroidea: Insights in phylogenetic relationships of echinodermata. Mol Phylogenet Evol 2010;56:201–11.

Perseke M, Hetmank J, Bernt M, Stadler P, Schlegel M, Bernhard D. The enigmatic mitochondrial genome of *Rhabdopleura compacta* (Pterobranchia) reveals insights into selection of an efficient tRNA system and supports monophyly of Ambulacraria. BMC Evol Biol 2011;11:134.

Perseke M, Fritzsch G, Ramsch K, et al. Evolution of mitochondrial gene orders in echinoderms. Mol Phylogenet Evol 2008;47:855–64.

Perseke M, Hankeln T, Weich B, et al. The mitochondrial DNA of *Xenoturbella bocki*: Genomic architecture and phylogenetic analysis. Theor Biosci 2007;126:35–42.

Peterson KJ. A phylogenetic test of the calcichordate scenario. Lethaia 1995;28:25–38.

Peterson KJ, Eernisse DJ. Animal phylogeny and the ancestry of bilaterians: Inferences from morphology and 18S rDNA gene sequences. Evol Dev 2001;3:170–205.

Peterson KJ, Dietrich MR, McPeek MA. MicroRNAs and metazoan macroevolution: insights into canalization, complexity, and the Cambrian explosion. Bioessays 2009;31:736–47.

Peterson KJ, Cotton JA, Gehling JG, Pisani D. The Ediacaran emergence of bilaterians: Congruence between the genetic and the geological fossil records. Phil Trans R Soc Lond B Biol Sci 2008;363:1435–43.

Peterson KJ, Lyons JB, Nowak KS, Takacs CM, Wargo MJ, McPeek MA. Estimating metazoan divergence times with a molecular clock. Proc Natl Acad Sci USA 2004;101:6536–41.

Petrov NB, Pegova A, Manylov O, Vladychenskaya N, Mugue N, Aleshin V. Molecular phylogeny of Gastrotricha on the basis of a comparison of the 18S rRNA genes: Rejection of the hypothesis of a relationship between Gastrotricha and Nematoda. Mol Biol 2007;41:445–52.

Petrov NB, Vladychenskaya NS. Phylogeny of molting protostomes (Ecdysozoa) as inferred from 18S and 28S rRNA gene sequences. Mol Biol 2005;39:503–13.

Petrunkevitch A, I. A study of Palaeozoic Arachnida. Trans Conn Acad Arts Sci 1949;37:69–315.

Pett W, Ryan JF, Pang K, Mullikin JC, Martindale MQ, Baxevanis AD, Lavrov DV. Extreme mitochondrial evolution in the ctenophore *Mnemiopsis leidyi*: Insight from mtDNA and the nuclear genome. Mitochondr DNA 2011;22:130–42.

Petyko Z, Zimmermann T, Smola U, Melzer RR. Central projections of Tömösvary's organ in *Lithobius forficatus* (Chilopoda, Lithobiidae). Cell Tissue Res 1996;283:331–4.

Pflugfelder O. Entwicklung von *Paraperipatus amboinensis* n. sp. Zool Jahrb Abt Anat Ontog 1948;69:443–92.

Philippe H, Forterre P. The rooting of the universal tree of life is not reliable. J Mol Evol 1999;49:509–23.
Philippe H, Germot A. Phylogeny of eukaryotes based on ribosomal RNA: Long-branch attraction and models of sequence evolution. Mol Biol Evol 2000;17:830–4.
Philippe H, Laurent J. How good are deep phylogenetic trees? Current Opinion in Genetics & Development 1998;8:616–23.
Philippe H, Telford MJ. Large-scale sequencing and the new animal phylogeny. Trends Ecol Evol 2006;21:614–20.
Philippe H, Chenuil A Adoutte A. Can the cambrian explosion be inferred through molecular phylogeny? Development (Suppl) 1994:15–25.
Philippe H, Lartillot N, Brinkmann H. Multigene analyses of bilaterian animals corroborate the monophyly of Ecdysozoa, Lophotrochozoa, and Protostomia. Mol Biol Evol 2005;22:1246–53.
Philippe H, Brinkmann H, Martinez P, Riutort M, Baguñà J, Volff J. Acoel flatworms are not Plathelminthes: Evidence from phylogenomics. PLoS ONE 2007;2:e717.
Philippe H, Snell EA, Bapteste E, Lopez P, Holland PWH, Casane D. Phylogenomics of eukaryotes: impact of missing data on large alignments. Mol Biol Evol 2004;21:1740–52.
Philippe H, Brinkmann H, Copley RR, et al. Acoelomorph flatworms are deuterostomes related to *Xenoturbella*. Nature 2011b;470:255–8.
Philippe H, Brinkmann H, Lavrov DV, et al. Resolving difficult phylogenetic questions: Why more sequences are not enough. PLoS Biol 2011a;9:e1000602.
Philippe H, Derelle R, Lopez P et al. Phylogenomics revives traditional views on deep animal relationships. Curr Biol 2009;19:706–12.
Phillips MJ, Penny D. The root of the mammalian tree inferred from whole mitochondrial genomes. Mol Phylogenet Evol 2003;28:171–85.
Picaud JL, Souty-Grosset C, Martin G. Vitellogenesis in terrestrial isopods: Female specific proteins and their control. Monit Zool Ital Monogr 1989;4:305–31.
Pick KS, Philippe H, Schreiber F, et al. Improved phylogenomic taxon sampling noticeably affects non-bilaterian relationships. Mol Biol Evol 2010;27:1983–7.
Pick L, Heffer A. Hox gene evolution: Multiple mechanisms contributing to evolutionary novelties. Ann N Y Acad Sci 2012;1256:15–32.
Piekarski N, Olsson L. Muscular derivatives of the cranialmost somites revealed by long-term fate mapping in the Mexican axolotl (*Ambystoma mexicanum*). Evol Dev 2007;9:566–78.
Pietsch A, Westheide W. Protonephridial organs in *Myzostoma cirriferum* (Myzostomida). Acta Zool (Stockholm) 1987;68:195–203.
Pioro HL, Stollewerk A. The expression pattern of genes involved in early neurogenesis suggests distinct and conserved functions in the diplopod *Glomeris marginata*. Dev Genes Evol 2006;216:417–30.
Pisani D, Benton MJ, Wilkinson M. Congruence of morphological and molecular phylogenies. Acta Biotheor 2007;55:269–81.
Pisani D, Feuda R, Peterson KJ, Smith AB. Resolving phylogenetic signal from noise when divergence is rapid: A new look at the old problem of echinoderm class relationships. Mol Phylogenet Evol 2012;62:27–34.
Pisani D, Poling LL, Lyons-Weiler M, Hedges SB. The colonization of land by animals: Molecular phylogeny and divergence times among arthropods. BMC Biology 2004;2:1–10.
Piskurek O, Jackson DJ. Tracking the ancestry of a deeply conserved eumetazoan SINE domain. Mol Biol Evol 2011;28:2727–30.
Pleijel F. On character coding for phylogeny reconstruction. Cladistics 1995;11:309–15.
Pocock RI. On the classification of the tracheate Arthropoda. Zool Anz 1893a;16:271–5.

Pocock RI. On some points in the morphology of the Arachnida (s.s.), with notes on the classification of the group. Ann Mag Nat Hist, Ser 6 1893b;11:1–19.

Podar M, Haddock SH, Sogin ML, Harbison GR. A molecular phylogenetic framework for the phylum Ctenophora using 18S rRNA genes. Mol Phylogenet Evol 2001;21:218–30.

Podsiadlowski L, Braband A. The complete mitochondrial genome of the sea spider *Nymphon gracile* (Arthropoda: Pycnogonida). BMC Genomics 2006;7:284.

Podsiadlowski L, Fahrein K. The mitochondrial genome of *Opilio parietinus* (Arachnida: Opiliones). Mitochondr DNA 2010;21:149–50.

Podsiadlowski L, Braband A, Mayer G. The complete mitochondrial genome of the onychophoran *Epiperipatus biolleyi* reveals a unique transfer RNA set and provides further support for the Ecdysozoa hypothesis. Mol Biol Evol 2008;25:42–51.

Podsiadlowski L, Kohlhagen H, Koch M. The complete mitochondrial genome of *Scutigerella causeyae* (Myriapoda: Symphyla) and the phylogenetic position of Symphyla. Mol Phylogenet Evol 2007;45:251–60.

Podsiadlowski L, Braband A, Struck TH, Döhren Jv, Bartolomaeus T. Phylogeny and mitochondrial gene order variation in Lophotrochozoa in the light of new mitogenomic data from Nemertea. BMC Genomics 2009;10:364–77.

Poe S. The effect of taxonomic sampling on accuracy of phylogeny estimation: test case of a known phylogeny. Mol Biol Evol 1998;15:1086–90.

Poe S, Wiens JJ. Character selection and the methodology of morphological phylogenetics. In: Wiens John J ed. Phylogenetic analysis of morphological data. Washington, D.C., USA: Smithsonian Institution Press, 2000:20–36.

Poe S, Swofford DL. Taxon sampling revisited. Nature 1999;398:299–300.

Pol D, Siddall ME. Biases in maximum likelihood and parsimony: A simulation approach to a 10-taxon case. Cladistics 2001;17:266–81.

Pollard SL, Holland PWH. Evidence for 14 homeobox gene clusters in human genome ancestry. Curr Biol 2000;10:1059–62.

Pollock DD, Zwickl DJ, McGuire JA, Hillis DM. Increased taxon sampling is advantageous for phylogenetic inference. Syst Biol 2002;51:664–71.

Pont-Kingdon GA, Okada NA, Macfarlane JL, et al. A coral mitochondrial mutS gene. Nature 1995;375:109–11.

Popadic A, Rusch D, Peterson M, Rogers BT, Kaufman TC. Origin of the arthropod mandible. Nature 1996;380:395.

Popper KR. Logik der Forschung. Tübingen, Germany: Mohr, J. C. B. (Paul Siebeck), 1984.

Por FD, Bromley HJ. Morphology and anatomy of *Maccabeus tentaculatus* (Priapulida: Seticoronaria). J Zool London 1974;173:173–97.

Posada D. jModelTest: Phylogenetic model averaging. Mol Biol Evol 2008;25:1253–6.

Posada D, Crandall KA. MODELTEST: Testing the model of DNA substitution. Bioinformatics 1998;14:817–8.

Potswald HE. Abdominal segment formation in *Spirorbis moerchi* (Polychaeta). Zoomorphology 1981;97:225–45.

Powell S, Szklarczyk D, Trachana K, et al. eggNOG v3.0: Orthologous groups covering 1133 organisms at 41 different taxonomic ranges. Nucleic Acids Res 2011;40:D284–D9.

Pozdnyakov IR, Karpov SA. Flagellar apparatus structure of choanocyte in *Sycon* sp. and its significance for phylogeny of Porifera. Zoomorphology 2013:in press.

Prendini L. Species or supraspecic taxa as terminals in cladistic analysis – groundplans versus exemplars revisited. Syst Biol 2001;50:290–300.

Prochnik SE, Rokhsar DS, Aboobaker AA. Evidence for a microRNA expansion in the bilaterian ancestor. Dev Genes Evol 2007;217:73–7.

Proctor H.C. Indirect sperm transfer in arthropods: Behavioral and evolutionary trends. Ann Rev Entomol 1998;43:153–74.
Prohaska SJ, Fried C, Amemiya CT, Ruddle FH, Wagner GP, Stadler PF. The shark HoxN cluster is homologous to the human HoxD cluster. J Mol Evol 2004;58:212–7.
Proost S, Van Bel M, Sterck L, Billiau K, Van Parys T, Van de Peer Y, Vandepoele K. PLAZA: A comparative genomics resource to study gene and genome evolution in plants. The Plant Cell Online 2009;21:3718–31.
Pross A. Diskussionsbeitrag zur Segmentierung des Cheliceraten-Kopfes. Zoomorphology 1977;86:183–96.
Pupko T, Huchon D, Cao Y, Okada N, Hasegawa M. Combining multiple data sets in a likelihood analysis: Which models are the best? Mol Biol Evol 2002;19:2294–307.
Purschke G. Echiura (Echiurida), Igelwürmer. In: Westheide W, Rieger RM, eds. Spezielle Zoologie, Teil 1: Einzeller und Wirbellose Tiere. Stuttgart: Gustav Fischer, 1996:345–9.
Purschke G. On the ground pattern of Annelida. Org Divers Evol 2002;2:181–96.
Purschke G, Hessling R, Westheide W. The phylogenetic position of the Clitellata and the Echiura – on the problematic assessment of absent characters. J Zool Syst Evol Res 2000;38:165–73.
Purschke G, Westheide W, Rohde D, Brinkhurst RO. Morphological reinvestigation and phylogenetic relationship of *Acanthobdella peledina* (Annelida, Clitellata). Zoomorphology 1993;113:91–101.
Putnam NH, Butts T, Ferrier DEK, et al. The amphioxus genome and the evolution of the chordate karyotype. Nature 2008;453:1064–72.
Qiu Y, Song D, Zhou K, Sun H. The mitochondrial sequences of *Heptathela hangzhouensis* and *Ornithoctonus huwena* reveal unique gene arrangements and atypical tRNAs. J Mol Evol 2005;60:57–71.
Qiu YL, Lee J, Bernasconi-Quadroni F, Soltis DE, et al. The earliest angiosperms: Evidence from mitochondrial, plastid and nuclear genomes. Nature 1999;402:404–7.
Rafael JA, Engel MS. A new species of *Zorotypus* from Central Amazonia, Brazil (Zoraptera: Zorotypidae). Am Mus Novitates 2006:1–11.
Ragsdale EJ, Baldwin JG. Resolving phylogenetic incongruence to articulate homology and phenotypic evolution: A case study from Nematoda. Proc R Soc Lond B Biol Sci 2010;277:1299–307.
Raikova OI, Reuter M, Kotikova EA, Gustafsson MKS. A commissural brain! The pattern of 5-HT immunoreactivity in Acoela (Plathelminthes). Zoomorphology 1998;118:69–77.
Ramírez MJ, Coddington J a, Maddison WP, et al. Linking of digital images to phylogenetic data matrices using a morphological ontology. Syst Biol 2007;56:283–94.
Ramsköld D, Luo S, Wang Y-C, et al. Full-length mRNA-Seq from single-cell levels of RNA and individual circulating tumor cells. Nat Biotech 2012;30:777–82.
Rannala B, Huelsenbeck JP, Yang Z, Nielsen R. Taxon sampling and the accuracy of large phylogenies. Syst Biol 1998;47:702–10.
Rasmussen AS, Janke A, Arnason U. The mitochondrial DNA molecule of the hagfish (*Myxine glutinosa*) and vertebrate phylogeny. J Mol Evol 1998;46:382–8.
Rasmussen M, Kellis M. Accurate gene-tree reconstruction by learning gene- and species-specific substitution rates across multiple complete genomes. Genome Res 2007;17:1932–42.
Rattenbury JC. The embryology of *Phoronopsis viridis*. J Morph 1954;95:289–340.
Rauther M. Das Cerebralganglion und die Leibeshöhle der Gordiiden. Zool Anz 1904;27:606–14.
Rauther M. Beiträge zur Kenntnis der Morphologie und der phylogenetischen Beziehungen der Gordiiden. Z Naturwiss 1905;40:1–94.
Rauther M. Nematomorpha (Saitenwürmer). In: Krumbach T, ed. Handbuch der Zoologie, Vol 2: Vermes Amera. Berlin, Germany: De Gruyter, 1930:403–48.

Rawlings TA, Collins TM, Bieler R. Changing identities: tRNA duplication and remolding within animal mitochondrial genomes. Proc Natl Acad Sci USA 2003;100:15700–5.
Ray DA, Xing J, Salem AH, Batzer MA. SINEs of a nearly perfect character. Syst Biol 2006;55:928–35.
Reed RD, Sperling FA. Interaction of process partitions in phylogenetic analysis: an example from the swallowtail butterfly genus Papilio. Mol Biol Evol 1999;16:286–97.
Reeder J, Steffen P, Giegerich R. Effective ambiguity checking in biosequence analysis. BMC Bioinformatics 2005;6:153.
Rees G. Locomotion of the cercaria of *Parorchis acanthus*, Nicoll and the ultrastructure of the tail. Parasitol 1971;62:489–503.
Reeve MR Complete cycle of development of a pelagic chaetognath in culture. Nature 1970;227:381.
Regier JC, Shultz WJ. Molecular phylogeny of the major arthropod groups indicates polyphyly of crustaceans and a new hypothesis for the origin of hexapods. Mol Biol Evol 1997;14:902–13.
Regier JC, Shultz JW. Molecular phylogeny of arthropods and the significance of the Cambrian explosion for molecular systematics. Am. Zool 1998;38:918–28.
Regier JC, Shultz JW. Elongation factor-2: A useful gene for arthropod phylogenetics. Mol Phylogenet Evol 2001;20, 136–48.
Regier JC, Shultz JW, Ganley AR et al. Resolving arthropod phylogeny: Exploring phylogenetic signal within 41 kb of protein-coding nuclear gene sequence. Syst Biol 2008;57:920–38.
Regier JC, Shultz JW, Kambic RE. Pancrustacean phylogeny: Hexapods are terrestrial crustaceans and maxillopods are not monophyletic. Proc R Soc Lond B Biol Sci 2005;272:395–401.
Regier JC, Shultz JW, Zwick A et al. Arthropod relationships revealed by phylogenomic analysis of nuclear protein-coding sequences. Nature 2010;463:1079–83.
Rehkämper G, Welsch U. On the fine structure of the cerebral ganglion of *Sagitta* (Chaetognatha). Zoomorphology 1985;105:83–9.
Rehkämper G, Storch V, Alberti G, Welsch U. On the fine structure of the nervous system of *Tubiluchus philippinensis* (Tubiluchidae, Priapulida). Acta Zool (Stockholm) 1989;70:111–20.
Rehm P, Borner J, Meusemann K et al. Dating the arthropod tree based on large-scale transcriptome data. Mol Phylogenet Evol 2011;61:880–7.
Rehm P, Pick C, Borner J, Markl J, Burmester T. The diversity and evolution of chelicerate hemocyanins. BMC Evol Biol 2012;12:19.
Reid AL. Review of the Peripatopsidae (Onychophora) in Australia, with comments on peripatopsid relationships. Invertebr Taxonomy 1996;10:663–936.
Reisinger E. Was ist *Xenoturbella*? Z Wiss Zool Abt A 1960;164:188–98.
Reitner J, Mehl D. Monophyly of the Porifera. Verh Natwiss Ver Hambg 1996;36:5–32.
Remane A. Gastrotricha. In: Bronn HG, ed. Bronns Klassen und Ordnungen des Tierreichs, Vol. 4 (II12): Gastrotricha und Kinorhyncha. Leipzig, Germany: Akademische Verlagsges, 1936:1–385.
Remm M, Storm CE, Sonnhammer EL. Automatic clustering of orthologs and in-paralogs from pairwise species comparisons. J Mol Biol 2001;314:1041–52.
Ren F, Tanaka H, Yang Z. An empirical examination of the utility of codon-substitution models in phylogeny reconstruction. Syst Biol 2005;54:808–18.
Renaud-Mornant J, Pochon-Masson J, Chaigneau J. Mise en évidence et ultrastructure d'un organe de Bellonci chez un crustacé mystacocaride. Ann Sci nat Zool Paris Ser 12 1977;19:459–78.
Reuter M, Raikova OI, Gustafsson MKS. Patterns in the nervous and muscle systems in lower flatworms. Belg J Zool 2001;131 (Suppl 1), 47–53.
Reuter M, Raikova OI, Jondelius U, Gustafsson MKS, Maule AG, Halton DW. Organisation of the nervous system in the Acoela: An immunocytochemical study. Tiss Cell 2001;33:119–28.
Reyes A, Gissi C, Pesole G, Saccone C. Asymmetrical directional mutation pressure in the mitochondrial genome of mammals. Mol Biol Evol 1998;15:957–66.

Reynolds PD, Steiner G. Scaphopoda. In: Ponder WF, Lindberg DR, eds. Phylogeny and evolution of the Mollusca. Berkeley, CA, USA: University of California Press, 2008:143–61.
Ricci C. Are lemnisci and proboscis present in the Bdelloidea? Hydrobiologia 1998;387:93–6.
Rice ME. Morphology, behavior, and histogenesis of the Pelagosphera larva of *Phascolosoma agassizii* (Sipuncula). Smiths Contrib Zool 1973;132:1–51.
Rice ME. Sipuncula: Developmental evidence for phylogenetic inference. In: Morris SC, George JD, Gibson R, Platt HM, eds. The origins and relationships of lower invertebrates. Oxford, UK: Oxford University Press, 1985:274–96.
Richter S. The Tetraconata concept: Hexapod-crustacean relationships and the phylogeny of Crustacea. Org Div Evol 2002;2:217–37.
Richter S. Homologies in phylogenetic analyses – concept and tests. Theory Biosci 2005;124:105–20.
Richter S, Edgecombe GD, Wilson GDF. The lacinia mobilis and similar structures – a valuable character in arthropod phylogenetics? Zool Anz 2002;241:339–61.
Richter S, Møller OS, Wirkner CS. Advances in crustacean phylogenetics. Arthr Syst Phylogenet 2009;67:275–86.
Richter S, Loesel R, Purschke G, et al. Invertebrate neurophylogeny: Suggested terms and definitions for a neuroanatomical glossary. Front Zool 2010;7:29.
Riedl RJ. Gnathostomulida from America. Science 1969;163:445–52.
Rieger RM. Monociliated epidermal cells in Gastrotricha: Significance for concepts of early metazoan evolution. J Zool Syst Evol Res 1976;14:198–226.
Rieger RM. Über den Ursprung der Bilateria: die Bedeutung der Ultrastrukturforschung für ein neues Verstehen der Metazoenevolution. Verh Dtsch Zool Ges 1986;79:31–50.
Rieger RM, Lombardi J. Ultrastructure of coelomic lining in echinoderm podia: Significance for concepts in the evolution of muscle and peritoneal cells. Zoomorphology 1987;107:191–208.
Rieger RM, Purschke G. (2005). The coelome and the origin of the annelid body plan. In: Bartolomaeus T, Purschke G, eds. Morphology, molecules, evolution and phylogeny in Polychaeta and related taxa. Hydrobiologia 2005;535/536:127–37.
Rieger RM, Tyler S. Sister-group relationship of Gnathostomulida and Rotifera-Acanthocephala. Invertebr Biol 1995;114:186–8.
Rieger V, Perez Y, Müller CHG, et al. Immunohistochemical analysis and 3D reconstruction of the cephalic nervous system in Chaetognatha: Insights into an early bilaterian brain? Invert Biol 2010;129:77–104.
Rieger V, Perez Y, Müller CHG, Lacalli T, Hansson BS, Harzsch S. Development of the nervous system in hatchlings of *Spadella cephaloptera* (Chaetognatha), and implications for nervous system evolution in Bilateria. Dev Growth Diff 2011;53:740–59.
Riemann F. *Kinonchulus sattleri* n.g.n.sp. (Enoplida, Tripyloidea), an aberrant freeliving nematode from the lower Amazonas. Veröff Inst Meeresforsch Bremerhaven 1972;13:317–26.
Rigaud T, Bouchon D, Souty-Grosset C, Raimond R. Mitochondrial DNA polymorphism, sex ratio distorters and population genetics in the isopod *Armadillidium vulgare*. Genetics 1999;152:1669–77.
Riutort M, Álvarez-Presas M, Lázaro E, Solà E, Paps J. Evolutionary history of the Tricladida and the Platyhelminthes: An up-to-date phylogenetic and systematic account. Int J Dev Biol 2012;56:5–17.
Rivera AS, Hammel JU, Haen KM, et al. RNA interference in marine and freshwater sponges: actin knockdown in Tethya wilhelma and *Ephydatia muelleri* by ingested dsRNA expressing bacteria. BMC Biotech 2011;11:67.
Rodriguez-Ezpeleta N, Brinkmann H, Roure B, Lartillot N, Lang B, Philippe H. Detecting and overcoming systematic errors in genome-scale phylogenies. Syst Biol 2007;56:389–99.

Roeding F, Borner J, Kube M, Klages S, Reinhardt R, Burmester T. A 454 sequencing approach for large scale phylogenomic analysis of the common emperor scorpion (*Pandinus imperator*). Mol Phylogenet Evol 2009;53:826–34.

Roeding F, Hagner-Holler S, Ruhberg H et al. EST sequencing of Onychophora and phylogenomic analysis of Metazoa. Mol Phylogenet Evol 2007;45:942–51.

Roger AJ, Sandblom O, Doolittle WF, Philippe H. An evaluation of elongation factor 1 alpha as a phylogenetic marker for eukaryotes. Mol Biol Evol 1999;16:218–33.

Rogers JS. On the consistency of maximum likelihood estimation of phylogenetic trees from nucleotide sequences. Syst Biol 1997;46:354–7.

Roggen DR, Raski DJ, Jones NO. Further electron microscope observations of *Xiphinema index*. Nematologica 1967;13:1–16.

Rogozin IB, Lyons-Weiler J, Koonin EV. Intron sliding in conserved gene families. Trends Genet 2000;16:430–2.

Rogozin IB, Sverdlov AV, Babenko VN, Koonin EV. Analysis of evolution of exon-intron structure of eukaryotic genes. Brief Bioinform 2005;6:118–34.

Rogozin IB, Wolf YI, Carmel L, Koonin EV. Ecdysozoan clade rejected by genome-wide analysis of rare amino acid replacements. Mol Biol Evol 2007;24:1080–90.

Rogozin IB, Wolf YI, Sorokin AV, Mirkin BG, Koonin EV. Remarkable interkingdom conservation of intron positions and massive, lineage-specific intron loss and gain in eukaryotic evolution. Curr Biol 2003;13:1512–7.

Rohdendorf B, Rasnitsyn A. Historical development of the class Insecta. Moscow: Nauka Press, 1980.

Rokas A, Carroll SB. More genes or more taxa? The relative contribution of gene number and taxon number to phylogenetic accuracy. Mol Biol Evol 2005;22:1337–44.

Rokas A, Carroll SB. Bushes in the tree of life. PLoS Biol 2006;4:e352.

Rokas A, Holland PW. Rare genomic changes as a tool for phylogenetics. Trends Ecol Evol 2000;15:454–9.

Rokas A, Kruger D, Carroll SB. Animal evolution and the molecular signature of radiations compressed in time. Science 2005;310:1933–8.

Rokas A, King N, Finnerty J, Carroll SB. Conflicting phylogenetic signals at the base of the metazoan tree. Evol Dev 2003b;5:346–59.

Rokas A, Williams BL, King N, Carrol SB. Genome-scale approaches to resolving incongruence in molecular phylogenies. Nature 2003a;425:798–804.

Ronquist F, Huelsenbeck JP. MrBayes 3: Bayesian phylogenetic inference under mixed models. Bioinformatics 2003;19:1572–4.

Ronshaugen M, McGinnis N, McGinnis W. Hox protein mutation and macroevolution of the insect body plan. Nature 2002;415:914–7.

Roonwal M. Studies on the embryology of the African migratory locust, *Locusta migratoria* migratorioides. Phil Trans R Soc Lond B Biol Sci 1936;226:391–421.

Rosenberg J. Coxal organs of *Lithobius forficatus* (Myriapoda, Chilopoda). Cell Tissue Res 1983;230:421–30.

Rosenberg MS, Kumar S. Taxon sampling, bioinformatics, and phylogenomics. Syst Biol 2003;52:119–24.

Rosse C, Mejino JLV. A reference ontology for biomedical informatics: The foundational model of anatomy. J Biomed Inf 2003;36:478–500.

Rössler W, Oland LA, Higgins MR, Hildebrand JG, Tolbert LP. Development of a glia-rich axon-sorting zone in the olfactory pathway of the moth *Manduca sexta*. J Neurosci 1999;22:9865–77.

Rota E, Martin P, Erséus C. Soil-dwelling polychaetes: Enigmatic as ever? Some hints on their phylogenetic relationships as suggested by a maximum parsimony analysis of 18S rRNA gene sequences. Cont Zool 2001;70:127–38.

Rota-Stabelli O, Telford MJ. A multi criterion approach for the selection of optimal outgroups in phylogeny: Recovering some support for Mandibulata over Myriochelata using mitogenomics. Mol Phylogenet Evol 2008;48:103–11.

Rota-Stabelli O, Daley AC, Pisani D. Molecular timetrees reveal a Cambrian colonization of land and a new scenario for ecdysozoan evolution. Curr Biol 2013;23:1–7.

Rota-Stabelli O, Campbell L, Brinkmann H et al. A congruent solution to arthropod phylogeny: Phylogenomics, microRNAs and morphology support monophyletic Mandibulata. Proc R Soc Lond B Biol Sci 2011;278:298–306.

Rota-Stabelli O, Kayal E, Gleeson D, et al. Ecdysozoan mitogenomics: evidence for a common origin of the legged invertebrates, the Panarthropoda. Genome Biol Evol 2010;2:425–40.

Roth A, Gonnet G, Dessimoz C. Algorithm of OMA for large-scale orthology inference. BMC Bioinformatics 2008;9:518.

Rothe BH, Schmidt-Rhaesa A. Variation in the nervous system in three species of the genus *Turbanella* (Gastrotricha, Macrodasyida). Meiofauna marina 2008;16:175–84.

Rothe BH, Schmidt-Rhaesa A. Architecture of the nervous system in two *Dactylopodola* species (Gastrotricha, Macrodasyida). Zoomorphology 2009;128:227–46.

Rothe BH, Schmidt-Rhaesa A. Structure of the nervous system in *Tubiluchus troglodytes* (Priapulida). Invertebr Biol 2010a;129:39–58.

Rothe BH, Schmidt-Rhaesa A. *Oregodasys cirratus*, a new species of Gastrotricha (Macrodasyida) from Tenerife (Canary Islands), with a description of the muscular and nervous system. Meiofauna Marina 2010b;18:49–66.

Rothe BH, Kieneke A, Schmidt-Rhaesa A. The nervous system of *Xenotrichula intermedia* and *X velox* (Gastrotricha: Paucitubulatina) by means of immunohistochemistry (IHC) and TEM. Meiofauna Marina 2011;19:71–88.

Rothe BH, Schmidt-Rhaesa A, Kieneke A. The nervous system of *Neodasys chaetonotoideus* (Gastrotricha: *Neodasys*) revealed by combining confocal laserscanning and transmission electron microscopy – evolutionary comparison of neuroanatomy within the Gastrotricha and basal Protostomia. Zoomorphology 2011;130:51–84.

Röttinger E, Lowe CJ. Evolutionary crossroads in developmental biology: hemichordates. Development 2012;139:2463–75.

Roure B, Baurain D, Philippe H. Impact of missing data on phylogenies inferred from empirical phylogenomic datasets. Mol Biol Evol 2012;30:197–214.

Roure B, Rodriguez-Ezpeleta N, Philippe H. SCaFoS: A tool for selection, concatenation and fusion of sequences for phylogenomics. BMC Evol Biol 2007;7:S2.

Rouse GW. Trochophore concepts: Ciliary bands and the evolution of larvae in spiralian Metazoa. Biol J Linn Soc 1999;66:411–64.

Rouse GW. Bias? What bias? The evolution of downstream larval-feeding in animals. Zool Scr 2000;29:213–36.

Rouse GW, Fauchald K. The articulation of annelids. Zool Scr 1995;24:269–301.

Rouse GW, Fauchald K. Cladistics and polychaetes. Zool Scr 1997;26:139–204.

Rouse GW, Pleijel F. Polychaetes Oxford: University Press, 2001.

Rouse GW, Pleijel F. Problems in polychaete systematics. Hydrobiologia 2003; 496:175–89.

Rouse GW, Pleijel F. Reproductive biology and phylogeny of Annelida. Enfield, NH, Jersey, Plymouth: Science Publishers, 2006.

Rouse GW, Pleijel F. Annelida. Zootaxa 2007;1668:245–64.

Rousset V, Pleijel F, Rouse GW, Erséus C, Siddall ME. A molecular phylogeny of annelids. Cladistics 2007;23:41–63.

Roy SW. Intron-rich ancestors. Trends in Genetics 2006;22:468–71.

Roy SW, Gilbert W. Resolution of a deep animal divergence by the pattern of intron conservation. Proc Natl Acad Sci USA 2005;102:4403–8.
Roy SW, Gilbert W. The evolution of spliceosomal introns: patterns, puzzles and progress. Nat Rev Genet 2006;7:211–21.
Roy SW, Irimia M. Rare genomic characters do not support Coelomata: Intron loss/gain. Mol Biol Evol 2008a;25:620–3.
Roy SW, Irimia M. When good transcripts go bad: artifactual RT-PCR 'splicing' and genome analysis. Bioessays 2008b;30:601–5.
Royuela M, Astier C, Grandier-Vazeille X, et al. Immunohistochemistry of chaetognath body wall muscles. Invert Biol 2003;122:74–82.
Ruiz-Trillo I, Riutort M, Littlewood DTJ, Herniou EA, Bagunà J. Acoel Flatworms: Earliest extant bilaterian metazoans, not members of Platyhelminthes. Science 1999;283:1919–23.
Ruiz-Trillo I, Paps J, Loukota M, et al. A phylogenetic analysis of myosin heavy chain type II sequences corroborates that Acoela and Nemertodermatida are basal bilaterians. Proc Natl Acad Sci USA 2002;99:11246–51.
Ruppert EE. Gastrotricha. In: Harrison FW, Ruppert EE, eds. Microscopic Anatomy of Invertebrates, Vol. 4 Aschelminthes. New York, NY, USA: Wiley-Liss, 1991a:41–109.
Ruppert EE. Introduction of the aschelminth phyla: A consideration of mesoderm, body cavities and cuticle. In: Harrison FW, Ruppert EE, eds. Microscopic Anatomy of Invertebrates Vol. 4. New York, NY, USA: Wiley-Liss, 1991b:1–17.
Ruppert EE. Introduction: Microscopic anatomy of the notochord, heterochrony, and chordate evolution. In: Harrison FW, Ruppert EE, eds. Microscopic Anatomy of Invertebrates. Hemichordata, Chaetognatha, and the Invertebrate Chordates, Vol 15. New York, NY, USA: Willey-Liss, Incorporation, 1997:1–13.
Ruppert EE, Carle KJ. Morphology of metazoan circulatory systems. Zoomorphology 1983;103:193–208.
Ruppert EE, Rice ME. Structure, ultrastructure, and function of the terminal organ of a Pelagosphera larva (Sipuncula). Zoomorphology 1983;102:143–63.
Ruppert EE, Smith PR. The functional organisation of filtration nephridia. Biol Rev 1988;63:231–58.
Russo CAM, Takezaki N, Nei M. Efficiencies of different genes and different tree-building methods in recovering a known vertebrate phylogeny. Mol Biol Evol 1996;13:525–36.
Ruvolo M. Molecular phylogeny of the hominoids: Inferences from multiple independent DNA sequence data sets. Mol Biol Evol 1997;14:248–65.
Ryan JF, Mazza ME, Pang K, Matus DQ, Baxevanis AD, Martindale MQ, Finnerty JR. Pre-bilaterian origins of the Hox cluster and the Hox code: Evidence from the sea anemone, Nematostella vectensis. PLoS one 2007;2:e153.
Rzhetsky A. Estimating substitution rates in ribosomal RNA genes. Genetics 1995;141:771–83.
Saeys Y, Rouzé P, Van de Peer Y. In search of the small ones: improved prediction of short exons in vertebrates, plants, fungi and protists. Bioinformatics 2007;23:414–20.
Sagane Y, Zech K, Bouquet J-M, Schmid M, Bal U, Thompson EM. Functional specialization of cellulose synthase genes of prokaryotic origin in chordate larvaceans. Development 2010;137:1483–92.
Salvini-Plawen Lv. Zur Morphologie und Phylogenie der Mollusken: Die Beziehungen der Caudofoveata und der Solenogastres als Aculifera, als Mollusca und als Spiralia. Z wiss Zool 1972;184:205–394.
Salvini-Plawen Lv. Systematic notes on *Spadella* and on the Chaetognatha in general. Z Zool Syst Evol 1986;24:122–8.
Salvini-Plawen Lv. The epineural (vs. gastroneural) cerebral complex of the Chaetognatha. Z Zool Syst Evol 1988;26:425–9.

Salvini-Plawen Lv. On the phylogenetic significance of the aplacophoran Mollusca. Iberus 2003;21:67–97.
Sammler S, Bleidorn C, Tiedemann R. Full mitochondrial genome sequences of two endemic Philippine hornbill species (Aves: Bucerotidae) provide evidence for pervasive mitochondrial DNA recombination. BMC Genomics 2011;12:35.
San Mauro D, Gower DJ, Zardoya R, Wilkinson M. A hotspot of gene order rearrangement by tandem duplication and random loss in the vertebrate mitochondrial genome. Mol Biol Evol 2006;23:227–34.
Sandeman DC, Scholtz G. Ground plans, evolutionary changes and homologies in decapod crustacean brains. In: The nervous system in invertebrates: An evolutionary aspect, O Breidbach, W Kutsch, eds., Basel: Birkenhäuser Verlag,1995:329–47.
Sandeman DC, Sandeman RE, Aitken AR. Atlas of serotonin-containing neurons in the optic lobes and brain of the crayfish *Cherax destructor*. J Comp Neurol 1988;269:465–78.
Sandeman DC, Scholtz G, Sandeman RE. Brain evolution in decapod Crustacea. J Exp Zool 1993;265:112–33.
Sandeman DC, Sandeman RE, Derby C, Schmidt M. Morphology of the brain of crayfish, crabs, and spiny lobsters: A common nomenclature for homologous structures. Biol Bull 1992;183:304–26.
Sanders HL. The Cephalocarida and crustacean phylogeny. Syst Zool 1957;6:112–48.
Sanders KL, Lee MSY. Arthropod molecular divergence times and the Cambrian origin of pentastomids. Syst Biodivers 2010;8:63–74.
Sanderson MJ, McMahon MM, Steel M. Phylogenomics with incomplete taxon coverage: The limits to inference. BMC Evol Biol 2010;10:155.
Sanderson MJ, Driskell AC, Ree RH, Eulenstein O, Langley S. Obtaining maximal concatenated phylogenetic data sets from large sequence databases. Mol Biol Evol 2003;20:1036–42.
Sanderson MJ, Wojciechowski MF, Hu JM, Khan TS, Brady SG. Error, bias, and long-branch attraction in data for two chloroplast photosystem genes in seed plants. Mol Biol Evol 2000;17:782–97.
Sankoff D. Edit distance for genome comparison based on non-local operations. In Combinatorial pattern matching. Vol. 644 of LNCS, Heidelberg: Springer, 1992:121–35.
Sankoff D. Conserved segment statistics and rearrangement inferences in comparative genomics. In Mathematics of evolution and phylogeny. Oxford: Oxford University Press, 2005:236–61.
Sansone S-A, Rocca-Serra P, Tong W, Fostel J, Morrison N, Jones AR. A strategy capitalizing on synergies: The reporting structure for biological investigation (RSBI) Working Group. Omics 2006;10:164–71.
Sasaki T, Takahashi K, Nikaido M, Miura S, Yasukawa Y, Okada N. First application of the SINE (short interspersed repetitive element) method to infer phylogenetic relationships in reptiles: An example from the turtle superfamily Testudinoidea. Mol Biol Evol 2004;21:705–15.
Sato A, Bishop JDD, Holland PWH. Developmental biology of pterobranch hemichordates: History and perspectives. Genesis 2008;46:587–91.
Satoh N. An aboral-dorsalization hypothesis for chordate origin. Genesis 2008;46:614–22.
Satoh N, Tagawa K, Takahashi H. How was the notochord born? Evol Dev 2012;14:56–75.
Satta Y, Klein J, Takahata N. DNA archives and our nearest relative: The trichotomy problem revisited. Mol Phylogenet Evol 2000;14:259–75.
Savard J, Tautz D, Richards S, et al. Phylogenomic analysis reveals bees and wasps (hymenoptera) at the base of the radiation of holometabolous insects. Genome Res 2006;16:1334–8.
Saveliev SV. Evolution of brain development in amphibians. Biol Bull 2009;36:128–38.
Savill NJ, Hoyle DC, Higgs PG. RNA sequence evolution with secondary structure constraints. Comparison of substitution rate models using maximum-likelihood methods. Genetics 2001;157:399–411.

Savineau JP, Duvert M. Physiological and cytochemical studies of Ca in the primary muscle of the trunk of *Sagitta setosa* (Chaetognath). Tissue Cell 1986;18:953–66.
Sawyer JK, Harris NJ, Slep KC, Gaul U, Peifer M. The *Drosophila* afadin homologue *Canoe* regulates linkage of the actin cytoskeleton to adherens junctions during apical constriction. J Cell Biol 2009;186:57–73.
Sbita SJ, Morgan RC, Buschbeck EK. Eye and optic lobe metamorphosis in the sunburst diving beetle, *Thermonectus marmoratus* (Coleoptera: Dytiscidae). Arthropod Struct Dev 2007;36:449–62.
Scharff R. On the skin and nervous system of *Priapulus* and *Halicryptus* Quart J Microsc Sci 1885;25:193–213.
Scharrer E. The fine structure of the retrocerebral organ of *Sagitta* (Chaetognatha). Life Sci 1965;4:923–6.
Scheloske HW. Ist die Unterteilung des Magens der Malacostraca in "Cardia" und "Pylorus" berechtigt? Z Zool Syst Evolutionsforsch 1976;14:253–80.
Scheltema AH. Aplacophora as progenetic aculiferans and the coelomate origin of mollusks as the sister taxon of Sipuncula. Biol Bull 1993;184:57–78.
Schierwater B, Kamm K. The early evolution of Hox genes: A battle of belief? In: Deutsch JS, editor. Hox genes: studies from the 20th to the 21st Century. New York: Springer-Verlag 2010:81–90.
Schierwater B, de Jong D, DeSalle R. Placozoa and the evolution of Metazoa and intrasomatic cell differentiation. Int J Biochem Cell B 2009;41:370–9.
Schierwater B, Kolokotronis SO, Eitel M, DeSalle R. The Diploblast-Bilateria sister hypothesis: Parallel evolution of a nervous system may have been a simple step. Comp Integr Biol 2009b:1–3.
Schierwater B, Kamm K, Srivastava M, Rokhsar D, Rosengarten RD, Dellaporta SL. The early ANTP gene repertoire: Insights from the Placozoan genome. PLoS ONE 2008;3:e2457.
Schierwater B, Eitel M, Jakob W, et al. Concatenated analysis sheds light on early metazoan evolution and fuels a modern "Urmetazoon" hypothesis. PLoS Biol 2009a;7:e1000020.
Schindelin J, Arganda-Carreras I, Frise E, et al. Fiji: an open-source platform for biological-image analysis. Nat Methods 2012;9:676–82.
Schinkmann K, Li C. Localization of FMRFamide-like peptides in *Caenorhabditis elegans*. J Comp Neurol 1992;316:251–60.
Schmidt C, Wägele JW. Morphology and evolution of respiratory structures in the pleopod exopodites of terrestrial Isopoda (Crustacea, Isopoda, Oniscidea). Acta Zool (Stockholm) 2001;82:315–30.
Schmidt D, Wilson MD, Ballester B, et al. Five-vertebrate ChIP-seq reveals the evolutionary dynamics of transcription factor binding. Science 2010;328:1036–40.
Schmidt-Rhaesa A. The nervous system of *Nectonema munidae* and *Gordius aquaticus*, with implications on the ground pattern of the Nematomorpha. Zoomorphology 1996;116:133–42.
Schmidt-Rhaesa A. Phylogenetic relationships of the Nematomorpha – a discussion of current hypothesis. Zool Anz 1998;236:203–16.
Schmidt-Rhaesa A. Two dimensions of biodiversity research exemplified by Nematomorpha and Gastrotricha. Integr Comp Biol 2002;42:633–40.
Schmidt-Rhaesa A. Perplexities concerning the Ecdysozoa: A reply to Pilato et al. Zool Anz 2006;244:205–8.
Schmidt-Rhaesa A. The Evolution of Organ Systems. Oxford, UK: Oxford University Press, 2007.
Schmidt-Rhaesa A. Gastrotricha, Cycloneuralia and Gnathifera, General history and phylogeny. In: Schmidt-Rhaesa A, ed. Handbook of Zoology, Gastrotricha, Cycloneuralia and Gnathifera. Berlin, Germany: De Gruyter, 2012a: in press.
Schmidt-Rhaesa A. Nematomorpha. In: Schmidt-Rhaesa A, ed. Handbook of Zoology, Gastrotricha, Cycloneuralia and Gnathifera. Berlin, Germany: De Gruyter, 2012b: in press.

Schmidt-Rhaesa A, Bartolomaeus T, Lemburg C, Ehlers U, Garey JR. The position of the Arthropoda in the phylogenetic system. J Morphol 1998;238:263–85.

Schmitz EH. Amphipoda. In: Harrison FW, Humes AG, eds. Microscopic Anatomy of Invertebrates vol. 9. New York, NY, USA: Wiley-Liss, 1992:443–528.

Schoeniger M, von Haeseler A. A stochastic model for the evolution of autocorrelated DNA sequences. Mol Phylogenet Evol 1994;3:240–7.

Schoenwolf GC, Smith JL. Mechanisms of neurulation: traditional viewpoint and recent advances. Development 1990;109:243–70.

Scholl G. Die Embryonalentwicklung des Kopfes und Prothorax von *Carausius morosus* Br. (Insecta, Phasmida). Zoomorphology 1969;65:1–142.

Scholl G. Beiträge zur Embryonalentwicklung von *Limulus polyphemus* L. (Chelicerata, Xiphosura). Zoomorphology 1977;86:99–154.

Scholtz G. The Articulata hypothesis – or what is a segment. Org Divers Evol 2002;2:197–215.

Scholtz G. Is the taxon Articulata obsolete? Arguments in favour of a close relationship between annelids and arthropods. In: Legakis A, Sfenthourakis S, Polymeni R, Thessalou-Legaki M, eds. The New Panorama of Animal Evolution, Proceedings of the 18th International Congress of Zoology. Sofia: Pensoft, 2003:489–501.

Scholtz G. Homology and ontogeny: Pattern and process in comparative developmental biology. Theory Biosci 2005;124:121–43.

Scholtz G. Deconstructing morphology. Acta Zool (Stockholm) 2010;91:44–63.

Scholtz G, Edgecombe GD. Heads, hox and the phylogenetic position of trilobites. In: Koenemann S, ed. Crustacean Issues 16, Crustacean and arthropod relationships. Boca Raton, FL, USA: CRC Press;2005:139–65.

Scholtz G, Kamenz C. The book lungs of Scorpiones and Tetrapulmonata (Chelicerata, Arachnida): Evidence for homology and a single terrestrialisation event of a common arachnid ancestor. Zoology 2006;109:2–13.

Scholtz G, Wolff C. Arthropod embryology: Cleavage and germ band development. In: Minelli A, Boxshall G, Fusco G, eds. Arthropod biology and evolution. Berlin, Germany: Springer, 2013:63–89.

Schram FR. Remipedia and Crustacean Phylogeny. In: Schram FR, ed. Crustacean Issues 1: Crustacean Phylogeny. Rotterdam, The Netherlands: Balkema, 1983:23–8.

Schram FR. Crustacea. Oxford, UK: Oxford University Press, 1986.

Schram FR, Hof CHJ. Fossils and interrelationships of major crustacean groups. In: Edgecombe GD, ed. Arthropod fossils and phylogeny. New York, NY, USA: Columbia University Press, 1998:233–302.

Schram FR, Koenemann S. Developmental genetics and arthropod evolution: Part I, on legs. Evol Dev 2001;3:343–54.

Schram FR, Koenemann S. Are the crustaceans monophyletic? In: Cracraft J, Donoghue MJ, eds. Assembling the Tree of Life. Oxford, UK: Oxford University Press, 2004:319–29.

Schram FR, Yager J, Emerson MJ. Remipedia. Part I. Systematics. Mem San Diego Soc Nat Hist 1986;15:1–60.

Schreiber F, Pick K, Erpenbeck D, Wörheide G, Morgenstern B. OrthoSelect: A protocol for selecting orthologous groups in phylogenomics. BMC Bioinformatics 2009;10:219.

Schröder C, Bleidorn C, Hartmann S, Tiedemann R. Occurrence of Can-SINEs and intron sequence evolution supports robust phylogeny of pinniped carnivores and their terrestrial relatives. Gene 2009;448:221–6.

Schroeder PC, Hermans CO. Annelida: Polychaeta. In: Giese AC, Pearse JS, eds. Annelids and Echiurans, Reproductions of Marine Invertebrates, Vol. 3. New York, NY, USA: Academic Press, 1975:1–213.

Schubart CD, Diesel R, Hedges SB. Rapid evolution to terrestrial life in Jamaican crabs. Nature 1998;393,363–5.
Schulze A, Cutler E, Giribet G. Reconstructing the phylogeny of the Sipuncula. In: Bartolomaeus T, Purschke G, eds. Morphology, molecules, evolution and phylogeny in Polychaeta and related taxa. Hydrobiologia 2005;535–536:277–96.
Schulze A, Cutler EB, Giribet G. Phylogeny of sipunculan worms: A combined analysis of four gene regions and morphology. Mol Phylogenet Evol 2007;42:171–92.
Schulze FE. *Trichoplax adhaerens*, nov. gen., nov. spec. Zool Anz 1883;6:92–7.
Scotland RW, Olmstead RG, Bennett JR. Phylogeny reconstruction: The role of morphology. Syst Biol 2003;52:539–48.
Scouras A, Smith JM. The complete mitochondrial genomes of the sea lily *Gymnocrinus richeri* and the feather star *Phanogenia gracilis*: signature nucleotide bias and unique *nad4L* gene rearrangement within crinoids. Mol Phylogenet Evol 2006;39:323–34.
Scouras A, Smith MJ. A novel mitochondrial gene order in the crinoid echinoderm *Florometra serratissima*. Mol Biol Evol 2001;18:61–73.
Scouras A, Beckenbach K, Arndt A, Smith MJ. Complete mitochondrial genome DNA sequence for two ophiuroids and a holothuroid: the utility of protein gene sequence and gene maps in the analyses of deep deuterostome phylogeny. Mol Phylogenet Evol 2004;31:50–65.
Seaver EC, Thamm K, Hill SD. Growth patterns during segmentation in the two polychaete annelids, *Capitella* sp. I and *Hydroides elegans*: Comparisons at distinct life history stages. Evol Dev 2005;7:312–26.
Sedgwick A. The development of *Peripatus capensis*. Part IV. The changes from stage G to birth. Q J Microsc Sci 1888;28:373–96.
Sedgwick A. *Peripatus*. In: Harmer SF, Shipley AE, eds. The Cambridge Natural History, Vol. 5. London, UK: Macmillan, 1895:1–26.
Segovia R, Pett W, Trewick S, Lavrov DV. Extensive and evolutionarily persistent mitochondrial tRNA editing in Velvet Worms (phylum Onychophora). Mol Biol Evol 2011;28:2873–81.
Seifert G. Considerations about the evolution of excretory organs in terrestrial arthropods. In: Camatini M., ed. Myriapod Biology. New York, NY, USA: Academic Press, 1979:354–72.
Seifert G, Rosenberg J. Die Ultrastruktur der Nephrozyten von *Orthomorpha gracilis* (C. L. Koch 1847) (Diplopoda, Strongylosomidae). Zoomorphology 1976;85:23–37.
Sekiguchi K, Yamamichi Y, Costlow JD. Horseshoe crab developmental studies I. Normal embryonic development of *Limulus polyphemus* compared with *Tachypleus tridentatus*. Prog Clin Biol Res 1982;31:53–73.
Sekimoto S, Rochon DA, Long JE, Dee JM, Berbee ML. A multigene phylogeny of Olpidium and its implications for early fungal evolution. BMC Evol Biol 2011;11:331.
Selden PA, Shear WA, Sutton M. Fossil evidence for the origin of spider spinnerets, and a proposed arachnid order. Proc Nat Acad Sci USA 2008;105:20781–5.
Selden PA, Shih C-K, Ren D. A golden orb-weaver spider (Araneae: Nephilidae: *Nephila*) from the Middle Jurassic of China. Biol Lett 2011;7:775–8.
Selden PA, Anderson JM, Anderson HM, Fraser NC. Fossil araneomorph spiders from the Triassic of South Africa and Virginia. J Arachnol 1999;27:401–14.
Semmler H, Wanninger A, Høeg JT, Scholtz G. Immunocytochemical studies on the naupliar nervous system of *Balanus improvisus* (Crustacea, Cirripedia, Thecostraca). Arthropod Struct Dev 2008;37:383–95.
Semper C. Zur Anatomie und Entwicklungsgeschichte der Gattung *Myzostoma* Leuckart. Z wiss Zool 1858;9:48–65.

Sempere LF, Cole CN, McPeek MA, Peterson KJ. The phylogenetic distribution of metazoan microRNAs: insights into evolutionary complexity and constraint. J Exp Zoolog B 2006;306:575–88.
Seo H, Edvardsen R, Maeland A, et al. Hox cluster disintegration with persistent anteroposterior order of expression in *Oikopleura dioica*. Nature 2004;431:67–71.
Sereno PC. Logical basis for morphological characters in phylogenetics. Cladistics 2007;23:565–87.
Serna F, Bolton B, Mackay W. On the morphology of Procryptocerus (Hymenoptera: Formicidae). Some comments and corrigenda. Zootaxa 2011;2923:67–8.
Shain DH. Annelids in modern biology. Hoboken, NJ, USA: John Wiley & Sons, Inc., 2009.
Shan Y, Gras R. 43 genes support the lungfish-coelacanth grouping related to the closest living relative of tetrapods with the Bayesian method under the coalescence model. BMC Res Notes 2011;4.
Shankland M, Seaver EC. Evolution of the bilaterian body plan: What have we learned from annelids? Proc Natl Acad Sci USA 2000;97:4434–7.
Shannon CE. A mathematical theory of communication. Bell Labs Tech J 1948;27:379–423.
Shao R, Kirkness EF, Barker SC. The single mitochondrial chromosome typical of animals has evolved into 18 minichromosomes in the human body louse, *Pediculus humanus*. Genome Res 2009;19:904–12.
Shao Z, Graf S, Chaga OY, Lavrov DV. Mitochondrial genome of the moon jelly *Aurelia aurita* (Cnidaria, Scyphozoa): A linear DNA molecule encoding a putative DNA-dependent DNA polymerase. Gene 2006;381:92–101.
Sharov AG. Basic arthropodan stock with special reference to insects. Oxford, UK: Pergamon Press, 1966.
Shear WA, Edgecombe GD. The geological record and phylogeny of the Myriapoda. Arthr Struct Dev 2010;39:174–90.
Shear WA, Selden PA, Rolfe WDI, Bonamo PM, Grierson JD. New terrestrial arachnids from the Devonian of Gilboa, New York (Arachnida: Trigonotarbida). Am Mus Novit 1987;2901:1–74.
Shedlock AM, Okada N. SINE insertions: Powerful tools for molecular systematics. Bioessays 2000;22:148–60.
Shen X, Ma X, Ren J, Zhao F. A close phylogenetic relationship between Sipuncula and Annelida evidenced from the complete mitochondrial genome sequence of *Phascolosoma esculenta*. BMC Genomics 2009;10:136.
Shen X, Tian M, Meng X, et al. Complete mitochondrial genome of *Membranipora grandicella* (Bryozoa: Cheilostomatida) determined with next-generation sequencing: the first representative of the suborder Malacostegina. Comp Biochem Physiol Part D Genomics Proteomics 2012;7:248–53.
Shigeno S, Sasaki T, Haszprunar G. Central nervous system of *Chaetoderma japonicum* (Caudofoveata, Aplacophora): Implications for diversified ganglionic plans in early molluscan evolution. Biol Bull 2007;213:122–34.
Shimamura M, Yasue H, Ohshima K, et al. Molecular evidence from retroposons that whales form a clade within even-toed ungulates. Nature 1997;388:666–70.
Shimeld SM, Donoghue PC. Evolutionary crossroads in developmental biology: cyclostomes (lamprey and hagfish). Development 2012;139:2091–9.
Shimizu K, Cha J, Stucky GD, Morse DE. Silicatein alpha: cathepsin L-like protein in sponge biosilica. Proc Natl Acad Sci USA 1998;95:6234–8.
Shimizu T. Development in the freshwater oligochaete *Tubifex*. In: Harrison FW, Cowden RR, eds. Developmental biology of freshwater invertebrates. New York, NY, USA: Alan R. Liss, 1982:283–316.

Shimizu T, Nakamoto A. Segmentation in annelids: Cellular and molecular basis for metameric body plan. Zool Sci 2001;18:285–98.
Shimizu T, Kitamura K, Arai A, Nakamoto A. Pattern formation in embryos of the oligochaete annelid *Tubifex*: Cellular basis for segmentation and specification of segmental identity. Hydrobiologia 2001;463:123–31.
Shimodaira H. An approximately unbiased test of phylogenetic tree selection. Syst Biol 2002;51:492–508.
Shimodaira H, Hasegawa M. CONSEL: For assessing the confidence of phylogenetic tree selection. Bioinformatics 2001;17:1246–7.
Shimotori T, Goto T. Establishment of axial properties in the arrow worm embryo, *Paraspadella gotoi* (Chaetognatha): Developmental fate of the first two blastomeres. Zool Sci 1999;16:459–69.
Shimotori T, Goto T. Developmental fates of the first four blastomeres of the chaetognath *Paraspadella gotoi*: Relationship to protostomes. Dev Growth Diff 2001;43:371–82.
Shinn GL. Ultrastructure of somatic tissues in the ovaries of a chaetognath (*Ferosagitta hispida*). J Morph 1992;211:221–41.
Shinn GL, Roberts ME. Ultrastructure of hatchling chaetognaths (*Ferosagitta hispida*): epithelial arrangement of mesoderm and its phylogenetic implications. J Morph 1994;219:143–63.
Shinn GL. Chaetognatha. In: Harrison FW, Ruppert EE, editors. Microscopic Anatomy of Invertebrates, Vol. 15: Hemichordata, Chaetognatha, and the Invertebrate Chordates. New York, NY, USA: Wiley-Liss, 1997:103–220.
Shu D, Chen L, Han J, Zhang X. An Early Cambrian tunicate from China. Nature 2001;411:472–3.
Shu D, Zhang X, Chen L. Reinterpretation of *Yunnanozoon* as the earlies known hemichordate. Nature 1996;380:428–30.
Shultz JW. Evolutionary morphology and phylogeny of Arachnida. Cladistics 1990;6:1–38.
Shultz JW. Muscular anatomy of a whipspider, *Phrynus longipes* (Pocock (Arachnida: Amblypygi), and its evolutionary significance. Zool J Linn Soc 1999;126:81–116.
Shultz JW. Gross muscular anatomy of *Limulus polyphemus* (Xiphosura, Chelicerate) and its bearing on evolution in the Arachnida. J Arachnol 2001;29:283–303.
Shultz JW. A phylogenetic analysis of the arachnid orders based on morphological characters. Zool J Linn Soc 2007;150:221–65.
Shultz JW, Regier JC. Phylogenetic analysis of arthropods using two nuclear protein-encoding genes supports a Crustacea + Hexapoda clade. Proc Biol Sci B 2000;267:1011–9.
Shuster CN, Sekiguchi K. Growing up takes about ten years and eighteen stages. In: Shuster CN, Barlow RB, Brockmann HJ, eds. The American horseshoe crab. Cambridge, UK: Harvard Univ. Press, 2003:103–32.
Siddal ME. Success of parsimony in the four-taxon case: Long branch repulsion by likelihood in the Farris zone. Cladistics 1998;14:209–20.
Siddall ME. Unringing a bell: Metazoan phylogenomics and the partition bootstrap. Cladistics 2010;26:444–52.
Siepel AC, Moret BME. Finding an optimal inversion median: Experimental results. In: Gascuel O, Moret BME, eds. Algorithms in Bioinformatics. Lecture Notes on Computer Science. Heidelberg, Germany: Springer, 2001; 2149:189–203.
Siewing R. Lehrbuch der vergleichenden Entwicklungsgeschichte der Tiere. Hamburg: Parey, 1969.
Siewing R. Das Archicoelomatenkonzept. Zool Jahrb Anat 1980;103,439–82.
Signorovitch AY, Buss LW, Dellaporta SL. Comparative genomics of large mitochondria in placozoans. PLOS Genet 2007;3:e13.
Signorovitch AY, Dellaporta SL, Buss LW. Molecular signatures for sex in the Placozoa. Proc Natl Acad Sci USA 2005;102:15518–22.

Simmons MP, Ochoterena H. Gaps as characters in sequence-based phylogenetic analyses. Syst Biol 2000;49:369–81.
Simmons MP, Randle CP, Freudenstein JV, Wenzel JW. Limitations of relative apparent synapomorphy analysis (RASA) for measuring phylogenetic signal. Mol Biol Evol 2002;19:14–23.
Simon S, Hadrys H. A comparative analysis of complete mitochondrial genomes among Hexapoda. Mol Phylogenet Evol 2013;69:393–403.
Simon S, Schierwater B, Hadrys H. On the value of elongation factor-1alpha for reconstructing pterygote insect phylogeny. Mol Phylogenet Evol 2010;54:651–6.
Simon S, Narechania A, Desalle R, Hadrys H. Insect phylogenomics: exploring the source of incongruence using new transcriptomic data. Genome Biol Evol 2012;4:1295–309.
Simon S, Strauss S, von Haeseler A, Hadrys H. A phylogenomic approach to resolve the basal pterygote divergence. Mol Biol Evol 2009;26:2719–30.
Simpson JT, Wong K, Jackman SD, Schein JE, Jones SJ, Birol I. ABySS: A parallel assembler for short read sequence data. Genome Res 2009;19:1117–23.
Simpson TL. Gamete, embryo, larval development. In: Simpson TL, ed. The cell biology of sponges. New York: Springer-Verlag, 1984:341–413.
Sims GE, Jun S-R, Wu GA, Kim S-H. Whole-genome phylogeny of mammals: Evolutionary information in genic and nongenic regions. Proc Natl Acad Sci USA 2009;106:17077–82.
Sinakevitch I, Douglass JK, Scholtz G, Loesel R, Strausfeld NJ. Conserved and convergent organization in the optic lobes of insects and isopods, with reference to other crustacean taxa. J Comp Neurol 2003;467:150–72.
Singh TR, Tsagkogeorga G, Delsuc F, et al. Tunicate mitogenomics and phylogenetics: peculiarities of the *Herdmania momus* mitochondrial genome and support for the new chordate phylogeny. BMC Genomics 2009;10.
Sinha R, Lenser T, Jahn N, et al. TassDB2 – a comprehensive database of subtle alternative splicing events. BMC Bioinf 2010;11:216.
Siveter DJ, Waloszek D, Williams M. An early Cambrian phosphatocopid crustacean with three-dimensionally preserved soft parts from Shropshire, England. Spec Papers Palaeont 2003;70:9–30.
Siveter DJ, Williams M, Waloszek D. A phosphatocopid crustacean with appendages from the Lower Cambrian. Science 2001;293:479–81.
Siveter JD, Sutton MD, Briggs DEG, Siveter DJ. A new probable stem-lineage crustacean with three-dimensionally preserved soft parts from the Herefordshire (Silurian) Lagerstätte, UK. Proc R Soc Lond B Biol Sci 2007;274:2099–108.
Slater GS, Birney E. Automated generation of heuristics for biological sequence comparison. BMC Bioinformatics 2005;6:31.
Sloan DB, Alverson AJ, Chuckalovcak JP, et al. Rapid evolution of enormous, multichromosomal genomes in flowering plant mitochondria with exceptionally high mutation rates. PLoS Biol 2012;10:e1001241.
Smalheiser NR, Torvik VI. Mammalian microRNAs derived from genomic repeats. Trends Genet 2005;21:322–6.
Smith B, Ph D, Kusnierczyk W, Schober D, Ceusters W. Towards a reference terminology for ontology research and development in the biomedical domain. J Global Optim 2006;48:57–65.
Smith B, Ashburner M, Rosse C, et al. The OBO Foundry: Coordinated evolution of ontologies to support biomedical data integration. Nature Biotechnology 2007;25(11):1251–5.
Smith JJ, Saha NR, Amemiya CT. Genome biology of the cyclostomes and insights into the evolutionary biology of vertebrate genomes. Integr Comp Biol 2010;50:130–7.
Smith JJ, Antonacci F, Eichler EE, Amemiya CT. Programmed loss of millions of base pairs from a vertebrate genome. Proc Natl Acad Sci USA 2009;106:11212–7.

Smith JJ, Baker C, Eichler EE, Amemiya CT. Genetic consequences of programmed genome rearrangement. Curr Biol 2012;22:1524–9.

Smith JJ, Kuraku S, Holt C, et al. Sequencing of the sea lamprey (*Petromyzon marinus*) genome provides insights into vertebrate evolution. Nature Genetics 2013;45:415–21.

Smith MJ, Arndt A, Gorski S, Fajber E. The phylogeny of echinoderm classes based on mitochondrial gene arrangements. J Mol Evol 1993;36:545–54.

Smith MJ, Banfield DK, Doteval K, Gorski S, Kowbel DJ. Nucleotide sequence of nine protein-coding genes and 22 tRNAs in the mitochondrial DNA of the sea star *Pisaster ochraceus*. J Mol Evol 1990;31:195–204.

Smith PR. Development of the blood vascular system in *Sabellaria cementarium* (Annelida, Polychaeta) An ultrastructural investigation. Zoomorphology 1986;106:67–74.

Smith PR, Ruppert EE. Nephridia. In: Westheide W, Hermans CO, eds. The ultrastructure of the Polychaeta. Microfauna marina 1988;4:231–62.

Smith PR, Lombardi J, Rieger RM. Ultrastructure of the body cavity lining in a secondary acoelomate, *Microphthalmus* cf. *listensis* Westheide (Polychaeta: Hesionidae). J Morphol 1986;188:257–71.

Snel B, Huynen MA, Dutilh BE. Genome trees and the nature of genome evolution. Annu Rev Microbiol 2005;59:191–209.

Snir S, Pachter L. Tracing the most parsimonious indel history. J Comput Biol 2011;18:967–986.

Snodgrass RE. The thorax of insects and the articulation of the wings. Proc US Natl Mus Smiths Inst 1909;36:511–95.

Snodgrass RE. Morphology and mechanism of the insect thorax. Smiths Misc Coll 1927;80:1–108.

Snodgrass RE. Principles of Insect Morphology. New York, NY, USA: McGraw-Hill, 1935.

Snodgrass RE. Evolution of the Annelida, Onychophora and Arthropoda. Smiths Misc Coll 1938;97:1–159.

Snodgrass RE. Comparative studies on the jaws of mandibulate arthropods. Smiths Misc Coll 1950;116:1–85.

Snodgrass RE. Comparative studies on the head of Mandibulate Arthropods. Ithaca, New York, NY, USA: Comstock Publ. Comp. Inc., 1951.

Snodgrass RE. Evolution of arthropod mechanisms. Smiths Misc Coll 1958;138:1–77.

Sombke A, Harzsch S, Hansson BS. Organization of deutocerebral neuropils and olfactory behavior in the centipede *Scutigera coleoptrata* (Linnaeus, 1758) (Myriapoda: Chilopoda). Chem Senses 2011;36:43–61.

Sombke A, Lipke E, Kenning M, Müller CHG, Hansson BS, Harzsch S. Comparative analysis of deutocerebral neuropils in Chilopoda (Myriapoda): Implications for the evolution of the arthropod olfactory system and support for the Mandibulata concept. BMC Neurosci 2012;13:1.

Sombke A, Rosenberg J, Hilken G. Chilopoda – nervous system. In: Minelli A, ed. Treatise on Zoology. Anatomy, Taxonomy, Biology. The Myriapoda, Vol 1. Leiden, The Netherlands: Brill, 2011:217–34.

Sørensen MV. On the evolution and morphology of the rotiferan trophi, with a cladistic analysis of Rotifera. J Zool Syst Evol Res 2002;40:129–54.

Sørensen MV. Further structures in the jaw apparatus of *Limnognathia maerski* (Micrognathozoa), with notes on the phylogeny of the Gnathifera. J Morphol 2003;255:131–45.

Sørensen MV, Giribet G. A modern approach to rotiferan phylogeny: Combining morphological and molecular data. Mol Phylogenet Evol 2006;40:585–608.

Sørensen MV, Pardos F. Kinorhynch systematics and biology – an introduction to the study of kinorhynchs, inclusive identification keys to the genera. Meiofauna marina 2008;16:21–73.

Sørensen MV, Sterrer W, Giribet G. Gnathostomulid phylogeny inferred from a combined approach of four molecular loci and morphology. Cladistics 2006;22:32–58.

Sørensen MV, Funch P, Willerslev E, Hansen AJ, Olesen J. On the phylogeny of the Metazoa in the light of Cycliophora and Micrognathozoa. Zool Anz 2000;239:297–318.
Sørensen MV, Hebsgaard MB, Heiner I, Glenner H, Willerslev E, Kristensen RM. New data from an enigmatic phylum: evidence from molecular sequence data supports a sister-group relationship between Loricifera and Nematomorpha. J Zool Syst Evol Res 2008;46:231–9.
Sousa GL, Lenz PH, Hartline DK, Christie AE. Distribution of pigment dispersing hormone- and tachykinin-related peptides in the central nervous system of the copepod crustacean *Calanus finmarchicus*. Gener Comp Endocrinol. 2008;156:454–9.
Spagnuolo A, Ristoratore F, Di Gregorio A, Aniello F, Branno M, Di Lauro R. Unusual number and genomic organization of *Hox* genes in the tunicate *Ciona intestinalis*. Gene 2003;309:71–9.
Spears T, Abele LG. Crustacean phylogeny inferred from 18S rDNA. In: Arthropods Relationships, RA Fortey, RH Thomas, eds. London, UK: Chapman & Hall, 1998:169–87.
Sperling EA, Peterson KJ. microRNAs and metazoan phylogeny: Big trees from little genes. In: Telford MJ, Littlewood DTJ, eds. Animal Evolution – Genomes, Trees and Fossils. Oxford, UK: Oxford University Press, 2009:157–70.
Sperling EA, Peterson KJ, Pisani D. Phylogenetic signal dissection of nuclear housekeeping genes supports the paraphyly of sponges and the monophyly of Eumetazoa. Mol Biol Evol 2009;26:2261–74.
Sperling EA, Pisani D, Peterson KJ. Poriferan paraphyly and its implications for Precambrian paleobiology. In: Vickers-Rich P, Komarower P, eds. The rise and fall of Ediacaran biota. London, UK: Geological Society of London, Special Publications, Vol. 286, 2007:355–68.
Sperling EA, Pisani D, Peterson KJ. Molecular paleobiological insights into the origin of the Brachiopoda. Evol Dev 2011;13:290–303.
Sperling EA, Robinson JM, Pisani D, Peterson KJ. Where's the glass? Biomarkers, molecular clocks, and microRNAs suggest a 200-Myr missing Precambrian fossil record of siliceous sponge spicules. Geobiol 2010;8:24–36.
Sperling EA, Vinther J, Moy VN, et al. MicroRNAs resolve an apparent conflict between annelid systematics and their fossil record. Proc R Soc Lond B Biol Sci 2009;276:4315–22.
Spies T. Structure and phylogenetic interpretation of diplopod eyes (Diplopoda). Zoomorph 1981;98:241–60.
Srivastava M, Begovic E, Chapman J, et al. The *Trichoplax* genome and the nature of placozoans. Nature 2008;454:955–60.
Srivastava M, Simakov O, Chapman J, et al. The *Amphimedon queenslandica* genome and the evolution of animal complexity. Nature 2010;466:720–6.
Stach T. The ontogeny of the notochord of *Branchiostoma lanceolatum*. Acta Zoologica 1999;80:25–33.
Stach T. Microscopic anatomy of developmental stages of *Branchiostoma lanceolatum* (Cephalochordata, Chordata). Bonn Zool Monogr 2000;47:1–111.
Stach T. Comparison of the serotonergic nervous system among Tunicata: Implications for its evolution within Chordata. Org Divers Evol 2005;5:15–24.
Stach T. Ontogeny of the appendicularian *Oikopleura dioica* (Tunicata, Chordata) reveals characters similar to ascidian larvae with sessile adults. Zoomorphology 2007;203–14.
Stach T, Kaul S. The post-anal tail of the enteropneust *Saccoglossus kowalevskii* is a ciliary creeping organ without distinct similarities to the chordate tail. Acta Zool (Stockholm) 2011;92: 150–60.
Stach T, Kirbach A. Larval convergence in a colonial tunicate: The organization of the sarcotubular complex in *Ecteinascidia turbinata* (Perophoridae, Phlebobranchiata, Tunicata, Chordata). Zoomorphology 2009;128:1–11.
Stach T, Turbeville JM. Phylogeny of Tunicata inferred from molecular and morphological characters. Mol Phylogenet Evol 2002;25:408–28.

Stach T, Braband A, Podsiadlowski L. Erosion of phylogenetic signal in tunicate mitochondrial genomes on different levels of analysis. Mol Phylogenet Evol 2010;55:860–70.
Stach T, Gruhl A, Kaul-Strehlow S. The central and peripheral nervous system of *Cephalodiscus gracilis* (Pterobranchia, Deuterostomia). Zoomorphology 2012;131:11–24.
Stach T, Winter J, Bouquet J-M, Chourrout D, Schnabel R. Embryology of a planktonic tunicate reveals traces of sessility. Proc Natl Acad Sci USA 2008;105:7229–34.
Stamatakis A. RAxML-VI-HPC: Maximum likelihood-based phylogenetic analyses with thousands of taxa and mixed models. Bioinformatics 2006;22:2688–90.
Stamatakis A, Hoover P, Rougemont J. A rapid bootstrap algorithm for the RAxML Web servers. Syst Biol 2008;57:758–71.
Staniczek AH. The mandible of silverfish (Insecta: Zygentoma) and mayflies (Ephemeroptera): its morphology and phylogenetic significance. Zool Anz 2000;239:147–78.
Starck D. Embryologie. Stuttgart, Germany: Georg Thieme Verlag, 1975.
Staton JL, Daehler LL, Brown WM. Mitochondrial gene arrangement of the horseshoe crab *Limulus polyphemus* L.: Conservation of major features among arthropod classes. Mol Biol Evol 1997;14:867–74.
Stecher HJ. Zur Organisation und Fortpflanzung von *Pisione remota* (Southern) (Polychaeta, Pisionidae). Z Morphol Ökol Tiere 1968;61:347–410.
Stechmann A, Schlegel M. Analysis of the complete mitochondrial DNA sequence of the brachiopod *Terebratulina retusa* places Brachiopoda within the protostomes. Proc R Soc Lond B Biol Sci 1999;266:2043–52.
Steel M, Hudson D, Lockhart PJ. Invariable sites models and their use in phylogeny reconstruction. Syst Biol 2000;49:225–32.
Stefanovic S, Rice DW, Palmer JD. Long branch attraction, taxon sampling, and the earliest angiosperms: Amborella or monocots? BMC Evolut Biol 2004;4:1–19.
Stegner MEJ, Brenneis G, Richter S. The ventral nerve cord of Cephalocarida (Crustacea) – New insights into the ground pattern of Tetraconata. J Morph 2013; DOI: 10.1002/jmor.20213.
Stegner MEJ, Richter S. Morphology of the brain in *Hutchinsoniella macracantha* (Cephalocarida, Crustacea). Arthr Struct Dev 2011;40:221–43.
Steinauer ML, Nickol BB, Broughton R, Orti G. First sequenced mitochondrial genome from the phylum Acanthocephala (*Leptorhynchoides thecatus*) and its phylogenetic position within Metazoa. J Mol Evol 2005;60:706–15.
Steinmetz PRH, Kraus JEM, Larroux C, et al. Independent evolution of striated muscles in cnidarians and bilaterians. Nature 2012;487:231–4.
Stemme T, Iliffe TM, Bicker G, Harzsch S, Koenemann S. Serotonin immunoreactive interneurons in the brain of the Remipedia: New insights into the phylogenetic affinities of an enigmatic crustacean taxon. BMC Evol. Biol. 2012;12:168.
Stemme T, Iliffe TM, von Reumont BM, Koenemann S, Harzsch S, Bicker G. Serotonin-immunoreactive neurons in the ventral nerve cord of Remipedia (Crustacea) – Support for a sister group relationship of Remipedia and Hexapoda? BMC Evol. Biol. 2013;13:119.
Stenderup J, Olesen J, Glenner H. Molecular phylogeny of the Branchiopoda (Crustacea)–multiple approaches suggest a 'diplostracan' ancestry of the Notostraca. Mol Phylogenet Evol 2006;41:182–94.
Stephan W. The rate of compensatory evolution. Genetics 1996;144:419–26.
Stephen S, Pheasant M, Makunin IV, Mattick JS. Large-scale appearance of ultraconserved elements in tetrapod genomes and slowdown of the molecular clock. Mol Biol Evol 2008;25:402–8.
Sterrer W, Mainitz M, Rieger RM. Gnathostomulida: Enigmatic as ever. In: Conway Morris S, George JD, Gibson RH, Platt M, eds. The origin and relationships of lower invertebrates. Oxford: Clarendon Press, 1985:181–91.

Stevens R, Goble CA, Bechhofer S. Ontology-based knowledge representation for bioinformatics. Briefings in Bioinformatics 2000;1:398–414.

Stewart C, Kural D, Strömberg MP, et al. A comprehensive map of mobile element insertion polymorphisms in humans. PLoS Genet 2011;7:e1002236.

Stickney HL, Barresi MJF, Devoto SH. Somite development in zebrafish. Dev Dyn 2000;219:287–303.

Stiller JW, Hall BD. Long-branch attraction and the rDNA model of early eukaryotic evolution. Mol Biol Evol 1999;16:1270–9.

Stiller JW, Riley J, Hall BD. Are red algae plants? A critical evaluation of three key molecular data sets. J Mol Evol 2001;52:527–39.

Stocsits RR, Letsch H, Hertel J, Misof B, Stadler PF. Accurate and efficient reconstruction of deep phylogenies from structured RNAs. Nucleic Acids Res 2009;37:6184–93.

Stokes MD. Larval locomotion of the lancelet *Branchiostoma floridae*. J Exp Biol 1997;200:1661–80.

Stokes MD, Holland ND. Ciliary hovering in larval lancelets (=Amphioxus). Biol Bull 1995a;188:231–3.

Stokes MD, Holland ND. Embryos and larvae of a lancelet, *Branchiostoma floridae*, from hatching through metamorphosis: Growth in the laboratory and external morphology. Acta Zool (Stockholm) 1995b;76:105–20.

Stollewerk A, Simpson P. Evolution of early development of the nervous system: A comparison between arthropods. BioEssays 2005;27:874–83.

Stollewerk A, Tautz D, Weller M. Neurogenesis in the spider: New insights from comparative analysis of morphological processes and gene expression patterns. Arthr Struct Dev 2003;32:5–16.

Storch V. Zur vergleichenden Anatomie der segmentalen Muskelsysteme und zur Verwandtschaft der Polychaeten-Familien. Z Morphol Ökol Tiere1968;63:251–342.

Storch V. Microscopic anatomy and ultrastructure of the stomach of *Porcellio scaber* (Crustacea, Isopoda). Zoomorphology 1987;106:301–11.

Storch V, Higgins RP, Morse MP. Internal anatomy of *Meiopriapulus fijiensis* (Priapulida). Trans Am Microsc Soc 1989;108:245–61.

Stoye J, Wittler R. A unified approach for reconstructing ancient gene clusters. IEEE-ACM T Comput Bi 2009;6 387–400.

Strathmann RR, Staver JM, Hoffman JR. Risk and the evolution of cell-cycle duration of embryos. Evolution 2002;56:708–20.

Strausfeld NJ. Atlas of an insect brain. Berlin: Springer, 1976.

Strausfeld NJ. Crustacean-insect relationships: The use of brain characters to derive phylogeny amongst segmented invertebrates. Brain Behav Evol 1998;52:186–202.

Strausfeld NJ. A brain region in insects that supervises walking. Prog Brain Res 1999;123:273–84.

Strausfeld NJ. The evolution of crustacean and insect optic lobes and the origins of chiasmata. Arthr Struct Dev 2005;34:235–56.

Strausfeld NJ. Brain organization and the origin of insects: An assessment. Proc R Soc Lond B Biol Sci 2009;276:1929–37.

Strausfeld NJ. Arthropod brains: Evolution, functional elegance, and historical significance. Cambridge, MA, USA: The Belknap Press of Harvard University Press, 2012.

Strausfeld NJ, Andrew DR. A new view of insect-crustacean relationships I. Inferences from neural cladistics and comparative neuroanatomy. Arthr Struct Dev 2011;40:276–88.

Strausfeld NJ, Hirth F. Deep homology of arthropod central complex and vertebrate basal ganglia. Science 2013;340:157.

Strausfeld NJ, Strausfeld CM, Loesel R, Rowell D, Stowe S. Arthropod phylogeny: onychophoran brain organization suggests an archaic relationship with a chelicerate stem lineage. Proc R Soc Lond B Biol Sci 2006a;273:1857–66.

Strausfeld NJ, Strausfeld CM, Stowe S, Rowell D, Loesel R. The organization and evolutionary implications of neuropils and their neurons in the brain of the onychophoran *Euperipatoides rowelli*. Arthropod Struct Dev 2006b;35:169–96.

Strausfeld NJ, Weltzen P, Barth FG. Two visual systems in one brain: Neuropils serving the principal eyes of the spider *Cupiennius salei*. J. Comp. Neurol. 1993;328:63–75.

Strauß J, Zhang Q, Verleyen P, et al. Pigment-dispersing hormone in *Daphnia* interneurons, one type homologous to insect clock neurons displaying circadian rhythmicity. Cell Mol Life Sci 2011;68:3403–23.

Strauss R. The central complex and the genetic dissection of locomotor behavior. Curr Opin Neurobiol 2002;6:633–8.

Strauss R. Control of *Drosophila* walking and orientation behavior by functional subunits localized in different neuropils in the central brain. In: Elsner N, Zimmermann H., eds. Proceedings of the 29th Göttingen Neurobiol Conference. Stuttgart, Germany: Thieme, 2003:206.

Strömberg JO. On the embryology of the isopod *Idotea*. Ark Zool Ser 2 1965;7:421–73.

Strote G, Bonow I, Attah S. The ultrastructure of the anterior end of male *Onchocerca volvulus:* papillae, amphids, nerve ring and first indication of an excretory system in the adult filarial worm. Parasitology 1996;113:71–85.

Struck TH. Progenetic species in polychaetes (Annelida) and problems assessing their phylogenetic affiliation. Integr Comp Biol 2006;46:558–68.

Struck TH. Direction of evolution within Annelida and the definition of Pleistoannelida. J Zool Syst Evol Res 2011;49:340–5.

Struck TH. Phylogeny of Annelida. Zoology Online. Walter de Gruyter, Berlin, 2012. http://www.degruyter.com/view/Zoology/bp_029147-6_1 (published online ahead of print)

Struck TH, Fisse F. Phylogenetic position of Nemertea derived from phylogenomic data. Mol Biol Evol 2008;25:728–36.

Struck TH, Purschke G. The sister group relationship of Aeolosomatidae and Potamodrilidae (Annelida: "Polychaeta") – a molecular phylogenetic approach based on 18S rDNA and Cytochrome Oxidase I. Zool Anz 2005;243:281–93.

Struck TH, Halanych KM, Purschke G. Dinophilidae (Annelida) is most likely not a progenetic Eunicida: Evidence from 18S and 28S rDNA. Mol Phylogenet Evol 2005;37:619–23.

Struck TH, Nesnidal MP, Purschke G, Halanych KM. Detecting possibly saturated positions in 18S and 28S sequences and their influence on phylogenetic reconstruction of Annelida (Lophotrochozoa). Mol Phylogenet Evol 2008;48:628–45.

Struck TH, Paul C, Hill N, et al. Phylogenomic analyses unravel annelid evolution. Nature 2011;471:95–8.

Struck TH, Schult N, Kusen T, et al. Annelida phylogeny and the status of Sipuncula and Echiura. BMC Evol Biol 2007;7:57–67.

Struckmann S, Arauzo-Bravo M, Scholer H, Reinbold R, Fuellen G. ReXSpecies – a tool for the analysis of the evolution of gene regulation across species. BMC Evol Biol 2008;8:111.

Subramoniam T. Spermatophores and sperm transfer in marine crustaceans. Adv Mar Biol 1993;29:129–202.

Sudhaus W. Die Notwendigkeit morphologischer Analysen zur Rekonstruktion der Stammesgeschichte. Species Phyl Evol 2006;1:17–32.

Suga K, Welch DB, Tanaka Y, Sakakura Y, Hagiwara A. Two circular chromosomes of unequal copy number make up the mitochondrial genome of the rotifer *Brachionus plicatilis*. Mol Biol Evol 2008;25:1129–37.

Sullivan J, Swofford DL. Are guinea pigs rodents? The importance of adequate models in molecular phylogenetics. J Mammal Evol 1997;4:77–86.

Sullivan J, Swofford DL. Should we use model-based methods for phylogenetic inference when we know that assumptions about among-site rate variation and nucleotide substitution pattern are violated? Syst Biol 2001;50:723–9.
Sullivan J, Holsinger KE, Simon C. The effect of topology on estimates of among-site rate variation. J Mol Evol 1996;42:308–12.
Sullivan J, Swofford DL, Naylor GJP. The effect of taxon sampling on estimating rate heterogeneity parameters of maximum-likelihood models. Mol Biol Evol 1999;16:1347–56.
Sun M, Shen X, Liu H, Liu X, Wu Z, Liu B. Complete mitochondrial genome of *Tubulipora flabellaris* (Bryozoa: Stenolaemata): the first representative from the class Stenolaemata with unique gene order. Mar Genomics 2011;4:159–65.
Sun M, Wu Z, Shen X, et al. The complete mitochondrial genome of *Watersipora subtorquata* (Bryozoa, Gymnolaemata, Ctenostomata) with phylogenetic consideration of Bryozoa. Gene 2009;439:17–24.
Sunkar R, Jagadeeswaran G. In silico identification of conserved microRNAs in large number of diverse plant species. BMC Plant Biol 2008;8:37.
Susko E, Leigh J, Doolittle WF, Bapteste E. Visualizing and assessing phylogenetic congruence of core gene sets: A case study of the Γ-proteobacteria. Mol Biol Evol 2006;23:1019–30.
Suzuki Y, Glazko GV, Nei M. Overcredibility of molecular phylogenies obtained by Bayesian phylogenetics. Proc Natl Acad Sci USA 2002;99:16138–43.
Švábeník J. Contributions to the anatomy and histology of Nematomorpha. Sitzungsber Ges Wiss Prag Math-Naturwiss Kl 1909;7:1–64.
Švábeník J. Parasitism and metamorphosis of the species *Gordius tolosanus* Duj (Parasitismus a metamorfosa druhu *Gordius tolosanus* Duj). Publ Fac Sci Univ Masaryk 1925;58:1–48.
Svane I. Observations on the long-term population dynamics of the perennial ascidian, *Ascidia mentula* O. F. Müller, on the Swedish west coast. Biol Bull 1984;167:630–46.
Svane I, Young CM. The ecology and behaviour of ascidian larvae. Oceanogr Mar Biol Annu Rev 1989;27:45–90.
Sverdlov A, Rogozin I, Babenko V, Koonin E. Reconstruction of ancestral protosplice sites. Curr Biol 2004;14:1505–8.
Swalla BJ, Smith AB. Deciphering deuterostome phylogeny: molecular, morphological and palaeontological perspectives. Phil Trans R Soc Lond B Biol Sci 2008;363:1557–68.
Swalla BJ, Cameron CB, Corley LS, Garey JR. Urochordates are monophyletic within the deuterostomes. Syst Biol 2000;49:52–64.
Swiatek P. Oogenesis in the leech *Glossiphonia heteroclita* (Annelida, Hirudinea, Glossiphoniidae) II. Vitellogenesis, follicle cell structure and egg shell formation. Tissue Cell 2006;38:263–70.
Swofford DL. PAUP*: Phylogenetic analysis using parsimony (*and other methods). Version 4. Sinauer Associates, Sunderland MA. 2003.
Swofford DL, Waddell PJ, Huelsenbeck JP, et al. Bias in phylogenetic estimation and its relevance to the choice between parsimony and likelihood methods. Syst Biol 2001;50:525–39.
Syed T, Schierwater B. The evolution of the Placozoa: A new morphological model. Senckenbergiana lethaea 2002a;82:315–24.
Syed T, Schierwater B. *Trichoplax adhaerens*: Discovered as a missing link, forgotten as a hydrozoan, re-discovered as a key to metazoan evolution. Vie Milieu 2002b;52:177–87.
Szaniawski H. Chaetognath grasping spines recognized among Cambrian protoconodonts. J Paleontol 1982;56:806–10.
Szaniawski H. New evidence for the protoconodont origin of chaetognaths. Acta Pal Polonica 2002;47:405–19.
Szaniawski H. Cambrian chaetognaths recognized in Burgess Shale fossils. Acta Pal Polonica 2005;50:1–8.

Szaniawski H. Fossil chaetognaths from the Burgess Shale: A reply to Conway Morris. Acta Pal Polonica 2009;54:361–4.

Tabei Y, Kiryu H, Kin T, Asai K. A fast structural multiple alignment method for long RNA sequences. BMC Bioinf 2008;9:33–49.

Takada N, Goto T, Satoh N. Expression pattern of the brachyury gene in the arrow worm *Paraspadella gotoi* (Chaetognatha). Genesis 2002;32:240–5.

Takezaki N, Figueroa F, Zaleska-Rutczynska Z, Takahata N, Klein J. The phylogenetic relationship of tetrapod, coelacanth, and lungfish revealed by the sequences of forty-four nuclear genes. Mol Biol Evol 2004;21:1512–24.

Takezaki N, Nei M. Inconsistency of the maximum parsimony method when the rate of nucleotide substitution is constant. J Mol Evol 1994;39:210–8.

Talavera G, Castresana J. Improvement of phylogenies after removing divergent and ambiguously aligned blocks from protein sequence alignments. Syst Biol 2007;56:564–77.

Talavera G, Vila R. What is the phylogenetic signal limit from mitogenomes? The reconciliation between mitochondrial and nuclear data in the Insecta class phylogeny. BMC Evol Biol 2011;11:315.

Tang KL, Berendzen PB, Wiley EO, Morrisey JF, Winterbottom R, Johnson GD. The phylogenetic relationships of the suborder Acanthuroidei (Teleostei: Perciformes) based on molecular and morphological evidence. Mol Phylogenet Evol 1999;11:415–25.

Tanzer A, Stadler PF. Molecular evolution of a microRNA cluster. J Mol Biol 2004;339:327–35.

Tatián M, Lagger C, Demarchi M, Mattoni C. Molecular phylogeny endorses the relationship between carnivorous and filter-feeding tunicates (Tunicata, Ascidiacea). Zool Sripta 2011;40:603–12.

Tattersall WM, Sheppard EM. Observations on the bipinnaria of the asteroid genus *Luidia*. In: Daniel RJ, ed. James Johnstone Memorial Volume. Liverpool, UK: University of Liverpool Press, 1934:35–61.

Tatusov RL, Galperin MY, Natale DA, Koonin EV. The COG database: A tool for genome-scale analysis of protein functions and evolution. Nucleic Acids Res 2000;28:33–6.

Tautz D. Segmentation. Dev Cell 2004;7:301–12.

Tautz D. Morphologie versus DNA-Sequenzen in der Phylogenie-Rekonstruktion. Spec Phyl Evol 2006;1:9–16.

Taylor MS, Ponting CP, Copley RR. Occurrence and consequences of coding sequence insertions and deletions in Mammalian genomes. Genome Res 2004;14:555–66.

Taylor MW, Thacker RW, Hentschel U. Genetics. Evolutionary insights from sponges. Science 2007;316:1854–5.

Technau U, Steele RE. Evolutionary crossroads in developmental biology: Cnidaria. Development 2011;138:1447–58.

Teeling EC, Madsen O, Murphy WJ, Springer MS, O'Brien SJ. Nuclear gene sequences confirm an ancient link between New Zealand's short-tailed bat and South American noctilionoid bats. Mol Phylogenet Evol 2003;28:308–19.

Teeling EC, Springer MS, Madsen O, Bates P, O'Brien SJ, Murphy WJ. A molecular phylogeny for bats illuminates biogeography and the fossil record. Science 2005;307:580–4.

Tekle YI, Grant JR, Kovner AM, Townsend JP, Katz LA. Identification of new molecular markers for assembling the eukaryotic tree of life. Mol Phylogenet Evol 2010;55:1177–82.

Telford MJ. Phylogenomics. Curr Biol: CB 2007;17:R945–6.

Telford MJ, Copley RR. Improving animal phylogenies with genomic data. Trends Genet 2011;27:186–95.

Telford MJ, Holland P. The phylogenetic affinities of the chaetognaths: a molecular analysis. Mol Biol Evol 1993;10:660–76.

Telford MJ, Wise MJ, Gowri-Shankar V. Consideration of RNA secondary structure significantly improves likelihood-based estimates of phylogeny: Examples from the bilateria. Mol Biol Evol 2005;22:1129–36.

Telford MJ, Lockyer AE, Cartwright-Finch C, Littlewood DTJ. Combined large and small subunit ribosomal RNA phylogenies support a basal position of the acoelomorph flatworms. Proc R Soc Lond B: Biol Sci 2003;270:1077–83.

Telford MJ, Bourlat SJ, Economou A, Papillon D, Rota-Stabelli O. The evolution of the Ecdysozoa. Phil Trans R Soc Lond B Biol Sci 2008;363:1529–37.

Temereva EN, Malakhov VV. The evidence of metamery in adult brachiopods and phoronids. Invert Zool 2011;8:87–101.

Terry MD, Whiting MF. Mantophasmatodea and phylogeny of the lower neopterous insects. Cladistics 2005;21:240–57.

Tetlie OE. Distribution and dispersal history of Eurypterida (Chelicerata). Palaeogeog, Palaeoclim, Palaeoecol 2007;252:557–74.

Teuchert G. The ultrastructure of the marine gastrotrich *Turbanella cornuta* Remane (Macrodasyoidea) and its functional and phylogenetical importance. Zoomorphologie 1977;88:189–246.

Thiel M, Kruse I. Status of the Nemertea as predators in marine ecosystems. Hydrobiol 2001;45:21–32.

Thomas-Chollier M, Ledent V, Leyns L, Vervoort M. A non-tree-based comprehensive study of metazoan Hox and ParaHox genes prompts new insights into their origin and evolution. BMC Evol Biol 2010;10.

Thompson JD, Linard B, Lecompte O, Poch O. A comprehensive benchmark study of multiple sequence alignment methods: Current challenges and future perspectives. PLoS ONE 2011;6:e18093.

Thompson JV. Memoir on the Star-Fish of the genus *Comatula*, demonstrative of the *Pentacrinus europaeus* being the Young of our hidigenous species. New Edinburgh Philosoph J 1836;20:295–300.

Tichy H. Untersuchungen über die Feinstruktur des Tömösváryschen Sinnesorgans von *Lithobius forficatus* L. (Chilopoda) und zur Frage seiner Funktion. Zool Jahrb Abt Anat Ontog 1973;91:93–139.

Tiegs OW. The embryology and affinities of the Symphyla, based on a study of *Hanseniella agilis*. Q J Microsc Sci 1940;82:1–225.

Tiegs OW. The post-embryonic development of *Hanseniella agilis* (Symphyla). Q J Microsc Sci 1945;85:191–328.

Tiegs OW. The development and affinities of the Pauropoda, based on a study of *Pauropus silvaticus*. Q J Microsc Sci 1947;88:165–267(Part I),275–336(Part II).

Tillier ERM, Collins RA. Neighbor joining and maximum-likelihood with RNA sequences – addressing the interdependence of sites. Mol Biol Evol 1995;12:7–15.

Tillier ERM, Collins RA. High apparent rate of simultaneous compensatory base-pair substitutions in ribosomal RNA. Genetics 1998;148:1993–2002.

Timm RW. Observations on the morphology and histological anatomy of a marine nematode, *Leptosomatum acephalatum* Chitwood, 1936: new combination (Enoplidae: Leptosomatinae). Am Midland Nat 1953;49:229–48.

Timmermans M, Roelofs D, Marien J, van Straalen NM. Revealing pancrustacean relationships: Phylogenetic analysis of ribosomal protein genes places Collembola (springtails) in a monophyletic Hexapoda and reinforces the discrepancy between mitochondrial and nuclear markers. BMC Evol Biol 2008;8:1–10.

Todaro MA, Telford MJ, Lockyer AE, Littlewood DTJ. Interrelationships of the Gastrotricha and their place among the Metazoa inferred from 18S rRNA genes. Zool Sripta 2006;35:251–9.
Todd JA. The central role of ctensotomes in bryozoan phylogeny. In: Herrera Cubilla A, Jackson JBC, eds. Proceedings of the 11th International Bryozoology Association Conference. Balboa, Panama: Smithsonian Tropical Research Institute, 2000:104–35.
Todt C, Okusu A, Schander C, Schwabe E. Solenogastres, Caudofoveata, and Polyplacophora. In: Ponder WF, Lindberg DR, eds. Phylogeny and evolution of the Mollusca. Berkeley, CA, USA: University of California Press, 2008:71–96.
Toh Y, Mizutani A. Structure of the visual system of the larva of the tiger beetle (*Cicindela chinensis*). Cell Tissue Res 1994;278:125–34.
Tomer R, Denes AS, Tessmar-Raible K, Arendt D. Profiling by image registration reveals common origin of annelid mushroom bodies and vertebrate pallium. Cell 2010;142:800–9.
Torarinsson E, Havgaard JH, Gorodkin J. Multiple structural alignment and clustering of RNA sequences. Bioinformatics 2007;23:926–32.
Torruella G, Derelle R, Paps J, et al. Phylogenetic relationships within the opisthokonta based on phylogenomic analyses of conserved single–copy protein domains. Mol Biol Evol 2012;29:531–44.
Tourasse NJ, Gouy M. Accounting for evolutionary rate variation among sequence sites consistently changes universal phylogenies deduced from rRNA and protein-coding genes. Mol Phylogenet Evol 1999;13:159–68.
Townsend JP. Profiling phylogenetic informativeness. Syst Biol 2007;56:222–31.
Townsend JP, Leuenberger C. Taxon sampling and the optimal rates of evolution for phylogenetic inference. Syst Biol 2011;60:358–65.
Townsend JP, Lopez-Giraldez F. Optimal selection of gene and ingroup taxon sampling for resolving phylogenetic relationships. Syst Biol 2010;59:446–57.
Townsend JP, Lopez-Giraldez F, Friedman R. The phylogenetic informativeness of nucleotide and amino acid sequences for reconstructing the vertebrate tree. J Mol Evol 2008;67:437–47.
Trachana K, Larsson TA, Powell S, Chen W-H, Doerks T, Muller J, Bork P. Orthology prediction methods: A quality assessment using curated protein families. Bioessays 2011;33:769–80.
Trautwein MD, Wiegmann BM, Beutel R, Kjer KM, Yeates DK. Advances in insect phylogeny at the dawn of the postgenomic era. Annu Rev Entomol 2012;57:449–68.
Triphan T, Poeck B, Neuser K, Strauss R. Visual targeting of motor actions in climbing *Drosophila*. Curr Biol 2010;20:663–8.
Tsagkogeorga G, Turon X, Galtier N, Douzery E, Delsuc F. Accelerated evolutionary rate of housekeeping genes in tunicates. J Mol Evol 2010;71:153–67.
Tsagkogeorga G, Turon X, Hopcroft RR, et al. An updated 18S rRNA phylogeny of tunicates based on mixture and secondary structure models. BMC Evol Biol 2009;9:187.
Tsuzuki S, Sekiguchi S, Kamimura M, Kiuchi M, Hayakawa Y. A cytokine secreted from the suboesophageal body is essential for morphogenesis of the insect head. Mech Dev 2005;122:189–97.
Tsyganov-Bodounov A, Hayward PJ, Porter JS, Skibinski DOF. Bayesian phylogenetics of Bryozoa. Mol Phylogenet Evol 2009;52,904–10.
Tuffley C, Steel M. Modeling the covarion hypothesis of nucleotide substitution. Math Biosci 1998;147:63–91.
Turbeville JM. An ultrastructural analysis of coelomogenesis in the hoplonemertine *Prosorhochmus americanus* and the Polychaete *Magelona* sp. J Morphol 1986;187:51–60.
Turbeville JM, Schulz JR, Raff RA. Deuterostome phylogeny and the sister group of the chordates: evidence from molecules and morphology. Mol Biol Evol 1994;11:648–55.

Turbeville JM, Pfeifer D, Field KG, Raff RA. The phylogenetic status of arthropods, as inferred from 18S rRNA sequences. Mol Biol Evol 1991;8:669–86.

Türk F. Über einige im Golfe von Neapel frei lebende Nematoden. Mitt Zool Stat Neapel 1903;16:1–67.

Tweedie S, Ashburner M, Falls K, et al. FlyBase: enhancing *Drosophila* gene ontology annotations. Nucleic Acids Res 2009;37:D555–9.

Tyler S, Rieger GE. Adhensive organs of the Gastrotricha. I. Duo-gland organs. Zoomorphologie 1980;95:1–15.

Tzetlin A, Purschke G. Fine structure of the pharyngeal apparatus of the pelagosphera larva in *Phascolosoma agassizii* (Sipuncula) and its phylogenetic significance. Zoomorphology 2006;125:109–17.

Tzetlin AB, Filippova AV. Muscular system in polychaetes (Annelida). In: Bartolomaeus T, Purschke G, eds. Morphology, molecules, evolution and phylogeny in Polychaeta and related taxa. Hydrobiologia 2005;535/536:113–26.

Ullmann SL. The origin and structure of the mesoderm and the formation of the coelomic sacs in *Tenebrio molitor* L. (Insecta, Coleoptera). Phil Trans R Soc Lond B Biol Sci 1964;248:245–77.

Ungerer P, Scholtz G. Filling the gap between identified neuroblasts and the neurons in crustaceans adds new support for Tetraconata. Proc R Soc Lond B Biol Sci 2008;275:369–76.

Uriz MJ, Turon X, Becerro MA, Agell G. Siliceous spicules and skeleton frameworks in sponges: Origin, diversity, ultrastructural patterns, and biological functions. Microsc Res Tech 2003;62:279–99.

Utting M, Agricola H-J, Sandeman RE, Sandeman DC. Central complex in the brain of crayfish and its possible homology with that of insects. J Comp Neurol 2000;416:245–61.

Vallès Y, Boore JL. Lophotrochozoan mitochondrial genomes. Integr Comp Biol 2006;46:544–57.

Valles Y, Halanych KM, Boore JL. Group II introns break new boundaries: presence in a bilaterian's genome. PLoS One 2008;3:e1488.

van der Hammen L. An Introduction to Comparative Arachnology. The Hague, The Netherlands: SPB Academic Publishing, 1989.

Van Roy P, Orr PJ, Botting JP, et al. Ordovician faunas of Burgess Shale type. Nature 2010;465:215–18.

Van Soest RW, Boury-Esnault N, Vacelet J, et al. Global diversity of sponges (Porifera). PLoS ONE 2012;7:e35105.

Vannier J, Steiner M, Renvoise E, Hu S-X, Casanova J-P. Early Cambrian origin of modern food webs: Evidence from predator arrow worms. Proc Roy Soc London B 2007;274:627–33.

Vejdovsky F. Zur Morphologie der Gordiiden. Z wiss Zool 1886;43:369–433.

Vejdovsky F. Organogenie der Gordiiden. Z wiss Zool 1894;57:642–703.

Velilla E, Izquierdo D, Rodriguez-Gonzalez E, Lopez-Bejar M, Vidal F, Paramio MT. Distribution of prepubertal and adult goat oocyte cortical granules during meiotic maturation and fertilisation: Ultrastructural and cytochemical study. Mol Reprod Dev 2004;68:507–14.

Venkatesh B, Erdmann MV, Brenner S. Molecular synapomorphies resolve evolutionary relationships of extant jawed vertebrates. Proc Natl Acad Sci USA 2011;98:11382–7.

Venkatesh B, Ning Y, Brenner S. Late changes in spliceosomal introns define clades in vertebrate evolution. Proc Natl Acad Sci USA 1999;96:10267–71.

Veremeichik GN, Shkryl YN, Bulgakov VP, et al. Occurrence of a silicatein gene in glass sponges (Hexactinellida: Porifera). Mar Biotechnol 2011;13:810–9.

Verhoeff KW. Beiträge zur vergleichenden Morphologie des Thorax der Insekten mit Berücksichtigung der Chilopoden. Nova Acta Abhandl kaiserl Leopoldinisch-Carolinischen dt Akad Naturf 1902;81:64–109.

Verhoeff KW. Vergleichend-morphologische Studie über die coxopleuralen Körperteile der Chilopoden, mit besonderer Berücksichtigung der Scolopendromorpha, ein Beitrag zur Anatomie und Systematik derselben, nebst physiologischen und phylogenetischen

Mitteilungen und Ausblicken auf die Insekten. Nova Acta Abhandl kaiserl Leopoldinisch-Carolinischen Dt Akad Naturf 1906;86:351–501.
Verweyen L, Klimpel S, Palm HW. Molecular phylogeny of the Acanthocephala (class Palaeacanthocephala) with a paraphyletic assemblage of the orders Polymorphida and Echinorhynchida. PLoS ONE 2011;6:e28285.
Vinther J, Eibye-Jacobsen D, Harper DAT. An Early Cambrian stem polychaete with pygidial cirri. Biol Lett 2011;7:929–32.
Vogel F, Gemballa S. Locomotory design of 'cyclostome' fishes: spatial arrangement and architecture of myosepta and lamellae. Acta Zool (Stockholm) 2000;81:267–83.
Vogel S. Life in Moving Fluids. The Physical Biology of Flow. Princeton, NJ, USA: Princeton University Press, 1994.
Vogt L. Weighting indels as phylogenetic markers of 18S rDNA sequences in Diptera and Strepsiptera. Org Divers Evol 2002;2:335–49.
Vogt L. Learning from Linnaeus: towards developing the foundations for a general structure concept for morphology. Zootaxa 2008;1950:123–52.
Vogt L. The future role of bio-ontologies for developing a general data standard in biology: Chance and challenge for zoo-morphology. Zoomorphology 2009;128:201–17.
Vogt L. Spatio-structural granularity of biological material entities. BMC Bioinf 2010;11:289.
Vogt L. Signs and terminology: Science caught between language and perception. Bionomina 2011;4:1–41.
Vogt L, Grobe P. Morph-D-Base – Eine online Datenbank für morphologische Daten und Metadaten. GfbS Newsletter 2010;24:29–34.
Vogt L, Bartolomaeus T, Giribet G. The linguistic problem of morphology: structure versus homology and the standardization of morphological data. Cladistics 2010;26:301–25.
Vogt L, Grobe P, Quast B, Bartolomaeus T. Top-level categories of constitutively organized material entities – suggestions for a formal top-level ontology. PLoS ONE 2011;6:e18794.
Vogt L, Grobe P, Quast B, Bartolomaeus T. Accommodating ontologies to biological reality – top-level categories of cumulative-constitutively organized material entities. PLoS ONE 2012;7:e30004.
Voigt O, Collins AG, Pearse VB, Pearse JS, Ender A, Hadrys H, Schierwater B. Placozoa – no longer a phylum of one. Curr Biol 2004;14:R944–5.
von Nordheim H, Schrader A. Ultrastructure and functional morphology of the protonephridia and segmental fenestrated lacunae in *Protodrilus rubropharyngeus* (Polychaeta, Protodrilidae). Helgoländer Meeresunters 1994;48:467–85.
von Reumont BM Molecular insights to crustaecan phylogeny. A status quo of past, present and perspective prospects also covering phylogenomics. Saarbrücken, Germany: Südwestdeutscher Verlag für Hochschulschriften, 2010.
von Reumont, BM, Meid S, Misof B. Aspects of quality and project management in analyses of large scale sequencing data. In: Akyar I. ed. Wide Spectra of Quality Control. Rijeka, Croatia: InTech Europe 2011:71-90 DOI:10.5772/24070.
von Reumont BM, Jenner RA, Wills MA et al. Pancrustacean phylogeny in the light of new phylogenomic data: support for Remipedia as the possible sister group of Hexapoda. Mol Biol Evol 2012;29:1031–45.
von Reumont BM, Meusemann K, Szucsich NU, et al. Can comprehensive background knowledge be incorporated into substitution models to improve phylogenetic analyses? A case study on major arthropod relationships. BMC Evol Biol 2009;9:119.
Wada H, Satoh N. Details of the evolutionary history from invertebrates to vertebrates as deducted from the sequences of 18S rDNA. Proc Natl Acad Sci USA 1994;91:1801–4.
Waddell PJ, Cao Y, Hauf J, Hasegawa M. Using novel phylogenetic methods to evaluate mammalian mtDNA, including amino acid invariant sites LogDet plus site stripping, to detect internal

conflicts in the data, with special reference to the positions of hedgehog, armadillo, and elephant. Syst Biol 1999;48:31–53.
Waeschenbach A, Telford MJ, Porter JS, Littlewood DTJ. The complete mitochondrial genome of *Flustrellidra hispida* and the phylogenetic position of Bryozoa among the Metazoa. Mol Phylogenet Evol 2006;40:195–207.
Waeschenbach A, Taylor PD, Littlewood DTJ. A molecular phylogeny of bryozoans. Mol Phylogenet Evol 2012;62:718–35.
Waeschenbach A, Webster BL, Littlewood DTJ. Adding resolution to ordinal level relationships of tapeworms (Platyhelminthes: Cestoda) with large fragments of mtDNA. Mol Phylogenet Evol 2012;63:834–47.
Wägele JW. On the reproductive biology of *Ceratoserolis trilobitoides* (Crustacea: Isopoda): Latitudinal variation of fecundity and embryonic development. Polar Biology 1987;7:11–24.
Wägele JW. Isopoda. In: Harrison FW, Humes AG, eds. Microscopic Anatomy of Invertebrates Vol. 9. New York, NY, USA: Wiley-Liss, 1992:529–617.
Wägele JW. Rejection of the "Uniramia" hypothesis and implications of the Mandibulata concept. Zool Jahrb Abt Syst 1993;120:253–88.
Wägele J-W. Grundlagen der Phylogenetischen Systematik. München, Germany: Verlag Dr. Friedrich Pfeil, 2001.
Wägele JW. Foundations of Phylogenetic Systematics. München, Germany: Verlag Dr. Friedrich Pfeil, 2005.
Wagner GP. The character concept in evolutionary biology. New York, NY, USA: Academic Press, 2001.
Wägele JW, Mayer C. Visualizing differences in phylogenetic information content of alignments and distinction of three classes of long-branch effects. BMC Evol Biol 2007;7:147.
Wägele JW, Misof B. On quality of evidence in phylogeny reconstruction: A reply to Zrzavý's defence of the 'Ecdysozoa' hypothesis. J Zool Syst Evol Res 2001;39:165–76.
Wagner GP, Amemiya C, Ruddle F. Hox cluster duplications and the opportunity for evolutionary novelties. Proc Natl Acad Sci USA 2003;100:14603–6.
Wagner GP, Stadler PF. Quasi-independence, homology and the unity of type: A topological theory of characters. J Theor Biol 2003;220:505–27.
Wainright PO, Hinkle G, Sogin ML, Stickel SK. Monophyletic origins of the metazoa: An evolutionary link with fungi. Science 1993;260:340–2.
Walker G, Clare AS, Rittschof D, Mensching D. Aspects of the life cycle of *Loxothylacus panopaei* (Gissler), a sacculinid parasite of the mud crab, *Rhithropanopeus harrisii* (Gould): A laboratory study. J Exp Mar Biol Ecol 1992;157:181–93.
Walker MH, Tait NN. Studies on embryonic development and the reproductive cycle in ovoviviparous Australian Onychophora (Peripatopsidae). J Zool 2004;264:333–54.
Wallace RL, Ricci C, Melone G. Through Alice's looking glass: a cladistic analysis of pseudocoelomate anatomy. In: Lanzavecchia G, Valvassori R, Candia Carnevali MD, editors. Body Cavities: Function and Phylogeny. UZI Selected Symposia and Monographs, Vol. 8. Modena, Italy: Mucci Editore, 1995: 1–67.
Wallace RL, Ricci C, Melone G. A cladistic analysis of pseudocoelomate (aschelminth) morphology. Invertebr Biol 1996;115:104–12.
Wallberg A, Curini-Galletti M, Ahmadzadeh A, Jondelius U. Dismissal of Acoelomorpha: Acoela and Nemertodermatida are separate early bilaterian clades. Zool Scripta 2007;36:509–23.
Wallstabe P. Beiträge zur Kenntnis der Entwicklungsgeschichte der Araneinen. Zool Jahrb Abt Anat Ontog 1908;26:683–712.
Walossek D. The Upper Cambrian *Rehbachiella* and the phylogeny of Branchiopoda and Crustacea. Fossils Strata 1993;32:1–202.

Waloszek D. The Upper Cambrian *Rehbachiella*, its larval development, morphology and significance for the phylogeny of Branchiopoda and Crustacea. Hydrobiology 1995;298:1–13.

Waloszek D. On the Cambrian Diversity of Crustacea. In: Schram FR, Vaupel Klein JC v, eds. Crustaceans and the biodiversity crisis. Leiden, The Netherlands: Brill. 1999:3–27.

Waloszek D. Cambrian 'Orsten'-type preserved arthropods and the phylogeny of Crustacea. In: Legakis A, Sfenthourakis S, Polymeni R, Thessalou-Legaki M, eds. The new panorama of animal evolution. Sofia, Bulgaria: Pensoft Publishers, 2003a:69–87.

Waloszek D. The 'Orsten' window – a three-dimensionally preserved Upper Cambrian meiofauna and its contribution to our understanding of the evolution of Arthropoda. Paleontol Res 2003b;7:71–88.

Waloszek D, Dunlop JA. A larval sea spider (Arthropoda: Pycnogonida) from the Upper Cambrian 'Orsten' of Sweden, and the phylogenetic position of pycnogonids. Paleontology 2002;45:421–46.

Waloszek D, Müller KJ. Cambrian 'Orsten'-type arthropods and the phylogeny of Crustacea. In: Fortey RA, Thomas RH, eds. Arthropod relationships. London, UK: Chapman and Hall. 1998a:139–53.

Waloszek D, Müller KJ. Early arthropod phylogeny in light of the Cambrian "Orsten" fossils. In: Edgecombe GD ed. Arthropod fossils and phylogeny. New York, NY, USA: Columbia University Press. 1998b:185–231.

Waloszek D, Repetski JE, Maas A. A new Late Cambrian pentastomid and a review of the relationships of this parasitic group. Trans R Soc Edinburgh Earth Sci 2006;96:163–76.

Waloszek D, Chen J, Maas A, Wang X. Early Cambrian arthropods – new insights into arthropod head and structural evolution. Arthropod Struct Dev 2005;34:189–205.

Waloszek D, Maas A, Chen J, Stein M. Evolution of cephalic feeding structures and the phylogeny of Arthropoda. Palaeogeogr Palaeoclim Palaeoecol 2007;254:273–87.

Walter DE, Proctor HC. 1998. Feeding behaviour and phylogeny: Observations on early derivative Acari. Exp Appl Acarol 1998;22:39–50.

Wang X, Gorlitsky R, Almeida JS. From XML to RDF: How semantic web technologies will change the design of "omic" standards. Nature Biotechnol 2005;23:1099–103.

Wang X, Lavrov DV. Seventeen new complete mtDNA sequences reveal extensive mitochondrial genome evolution within the Demospongiae. PLoS one 2008;3:e2723.

Wang X, Lavrov DV. Gene recruitment – common mechanism in the evolution of transfer RNA gene families. Gene 2011;475:22–9.

Warburg MR. Evolutionary biology of land isopods. Berlin, Germany: Springer, 1993.

Warn JM. Presumed myzostomid infestation of an ordovician crinoid. J Paleontol 1974;48:506–13.

Watterson GA, Ewens WJ, Hall TE, Morgan A. The chromosome inversion problem. J Theor Biol 1982;99:1–7.

Weatherbee SD, Halder G, Kim J, Hudson A, Carroll S. Ultrabithorax regulates genes at several levels of the wing-patterning hierarchy to shape the development of the *Drosophila* haltere. Genes Dev 1998;12:1474–82.

Weatherbee SD, Nijhout HF, Grunert LW, eds. Ultrabithorax function in butterfly wings and the evolution of insect wing patterns. Curr Biol 1999;9:109–15.

Weber H. Die Gliederung der Sternopleuralregion des Lepidopterenthorax. Eine vergleichend morphologische Studie zur Subcoxaltheorie. Z wissensch Zool 1928;131:181–254.

Weber H. Lehrbuch der Entomologie. Stuttgart, Germany: Gustav Fischer Verlag, 1933.

Weber M, Wey-Fabrizius AR, Podsiadlowski L, et al. Phylogenetic analysis of endoparasitic Acanthocephala based on mitochondrial genomes suggests secondary loss of sense organs. Mol Phylogenet Evol 2013;66:182–9.

Weber M, Wey AR, Podsiadlowski L, et al. Phylogenetic analysis of endoparasitic Acanthocephala based on mitochondrial genomes suggests secondary loss of sense organs. Mol Phylogenet Evol 2013;66:182–9.
Weber RE. Functions of invertebrate hemoglobins with special reference to adaptations to environmental hypoxia. Am Zool 1980;20:79–101.
Webster BL, Copley RR, Jenner RA, et al. Mitogenomics and phylogenomics reveal priapulid worms as extant models of the ancestral Ecdysozoan. Evol Dev 2006;8:502–10.
Webster BL, Littlewood DTJ. Mitochondrial gene order change in *Schistosoma* (Platyhelminthes: Digenea: Schistosomatidae). Int J Parasitol 2012;42:313–21.
Webster SG. Peptidergic neurons in barnacles: An immunohistochemical study using antisera raised against crustacean neuropeptides. Biol Bull 1998;195:282–9.
Weckwerth W, Selbig J. Scoring and identifying organism-specific functional patterns and putative phosphorylation sites in protein sequences using mutual information. Biochem Biophys Res Commun 2003;307:516–21.
Weigert A, Helm C, Hausen H, Zakrzewski A-C, Bleidorn C. Expression pattern of Piwi-like genes in adult *Myzostoma cirriferum* (Annelida). Dev Gen Evol 2013;223:329–34.
Weihe U, Milan, M, Cohen SM. *Drosophila* Limb Development. In: Gilbert LI, ed. Development, morphogenesis, molting and metamorphosis. New York, NY, USA: Academic Press, 2009:59–96.
Weir M, Eaton M, Rice M. Challenging the spliceosome machine. Genome Biol 2006;7:R3.
Weisblat DA, Kim SY, Stent G. Embryonic origins of cells in the leech *Helobdella triserialis*. Dev Biol 1984;104:65–85.
Weiss LC, Tollrian R, Herbert Z, Laforsch C. Morphology of the *Daphnia* nervous system: A comparative study on *Daphnia pulex*, *Daphnia lumholtzi*, and *Daphnia longicephala*. J Morph 2012;273(12):1392–405.
Weisz PB. The histological pattern of metameric development in *Artemia salina*. J Morphol 1947;81:45–95.
Welsch U, Storch V. Fine structure of the coelomic epithelium of *Sagitta elegans*. Zoomorphology 1982;100:217–22.
Welsch U, Storch V. Fine structural and histochemical observations on the epidermis and the sensory cells of *Sagitta elegans* (Chaetognatha). Zool Anz 1983;210:34–43.
Wenzel JW, Siddall ME. Noise. Cladistics 1999;15:51–64.
Werner B. Stamm Cnidaria, Nesseltiere. In: Gruner H-E, ed. Lehrbuch der speziellen Zoologie. Band I (Wirbellose Tiere) 2. Teil: Cnidaria, Ctenophora, Mesozoa, Plathelminthes, Nemertini, Entoprocta, Nemathelminthes, Priapulida. 5. Auflage. Stuttgart, Germany: Gustav Fischer Verlag, 1993:11–305.
Westblad E. *Xenoturbella bocki* n.g., n.sp., a peculiar, primitive turbellarian type. Ark Zool 1949;1:3–29.
Westheide W. Ultrastructure of the protonephridia in the dorvilleid polychaete *Apodotrocha progenerans* (Annelida). Zool Sripta 1985;14:273–8.
Westheide W. The nephridia of the interstitial polychaete *Hesionides arenaria* and their phylogenetic significance (Polychaeta, Hesionidae). Zoomorphology 1986;106:35–43.
Westheide W. Progenesis as a principle in meiofauna evolution. J Nat Hist 1987;21:843–54.
Westheide W. The direction of evolution within the Polychaeta. J Nat Hist 1997;31:1–15.
Westheide W, Müller MC. Cinematographic documentation of enchytraeid morphology and reproductive biology. Hydrobiologia 1996;334:263–7.
Westheide W, Rieger R. Spezielle Zoologie. Teil 1. Einzeller und Wirbellose Tiere. Stuttgart, Germany: Gustav Fischer Verlag, 1996.
Westheide W, Watson Russell C. Ultrastructure of chrysopetalid paleal chaetae (Annelida, Polychaeta). Acta Zool (Stockholm) 1992;73:197–202.

Westheide W, McHugh D, Purschke G, Rouse G. Systematization of the Annelida: Different approaches. In: Dorresteijn AWC, Westheide W, eds. Reproductive strategies and developmental patterns in annelids. Hydrobiologia 1999;402:291–307.
Wey-Fabrizius AR, Herlyn H, Rieger B, et al. (submitted): Transcriptome data reveal syndermatan relationships and suggest the evolution of endoparasitism in Acanthocephala via an epizoic stage.
Weyda F. Coxal vesciles of Machilidae. Pedobiologia 1974;14:138–41.
Weygoldt P. Die Embryonalentwicklung des Amphipoden *Gammarus pulex pulex* (L). Zool Jahrb Abt Anat Ontog 1958;77:51–110.
Weygoldt P. Beitrag zur Kenntnis der Ontogenie der Dekapoden: Embryologische Untersuchungen an *Palaemonetes varians* (Leach). Zool Jahrb Abt Anat Ontog 1961;79:223–70.
Weygoldt P. Vergleichend-embryologische Untersuchungen an Pseudoscorpionen II. Das zweite Embryonalstadium von *Lasiochernes pilosus* Ellingsen und *Cheiridium museorum* Leach. Zool Beitr, NF 1964;10:353–68.
Weygoldt P. Untersuchungen zur Embryologie und Morphologie der Geißelspinne *Tarantula marginemaculata* C. L. Koch (Arachnida, Amblypygi, Tarantulidae). Zoomorphology 1975;82:137–99.
Weygoldt P. Significance of later embryogenic stages and head development in arthropod phylogeny. In: Gupta AP, ed. Arthropod Phylogeny. New York, NY, USA: Van Nostrand, 1979:107–36.
Weygoldt P. Whip Spiders (Chelicerata, Amblipygi) – Their biology, morphology and systematics. Stenstrup, Denmark: Apollo Books, 2000.
Weygoldt P, Paulus HF. Untersuchungen zur Morphologie, Taxonomie und Phylogenie der Chelicerata. I. Morphologische Untersuchungen. Z Zool Syst Evolutionsforsch 1979a;17:85–116.
Weygoldt P, Paulus HF. Untersuchungen zur Morphologie, Taxonomie und Phylogenie der Chelicerata. II. Cladogramme und die Entfaltung der Chelicerata. Z Zool Syst Evolutionsforsch 1979b;17:85–116, 177–200.
Whalan S, Battershill C, de Nys R. Sexual reproduction of the brooding sponge *Rhopaloeides odorabile*. Coral Reefs 2007;26:655–63.
Wheeler BM, Heimberg, AM, Moy VN, et al. The deep evolution of metazoan microRNAs. Evol Dev 2009;11:50–68.
Wheeler WC. Sampling, groundplans, total evidence and the systematics of arthropods. In:Fortey RA, Thomas RH, eds. Arthropod relationships. London, UK: Chapman and Hall. 1998:87–97.
Wheeler WC, Cartwright P, Hayashi CY. Arthropod phylogeny: A combined approach. Cladistics 1993;9:1–39.
Wheeler WC, Giribet G, Edgecombe GD. Arthropod systematics. The comparative study of genomic, anatomical, and paleontological information. In: Cracraft J, Donoghue MJ, eds. Assembling the Tree of Life. New York, NY, USA: Oxford University Press 2004:281–95.
Wheeler WC, Hayashi CY. The phylogeny of the extant chelicerate orders. Cladistics 1998;14:173–92.
Wheeler WC, Whiting MF, Wheeler QD, Carpenter JM. The phylogeny of the extant hexapod orders. Cladistics 2001;17:113–69.
Wheeler WM. A contribution to insect embryology. J Morphol 1893;8:1–148.
Wheeler WM. The sexual phases of *Myzostoma*. Mitt zool Stat Neapel 1896;12:227–302.
Whelan S. Spatial and temporal heterogeneity in nucleotide sequence evolution. Mol Biol Evol 2008;25:1683–94.
Whelan S, Goldman N. A general empirical model of protein evolution derived from multiple protein families using a maximum-likelihood approach. Mol Biol Evol 2001;18:691–9.
White JG, Southgate E, Thomson JN, Brenner S. The structure of the nervous system of the nematode *Caenorhabditis elegans*. Phil Trans R Soc Lond B Biol Sci 1986;314:1–340.
Whitfield JB, Kjer KM. Ancient rapid radiations of insects: Challenges for phylogenetic analysis. Ann Rev Entomol 2008;53:449–72.

Whiting MF. Phylogenetic position of the strepsiptera: Review of molecular and morphological evidence. Int J Insect Morphol Embryol 1998;27:53–60.
Whiting MF. Mecoptera is paraphyletic: Mulitple genes and phylogeny of Mecoptera and Siphonaptera. Zool Scripta 2001;31:93–104.
Whiting MF. Phylogeny of the holometabolous insect orders based on 18S ribosomal DNA: When bad things happen to good data. In; DeSalle R, Giribet G, Wheeler WC, eds. Molecular systematics and evolution: Theory and practice. Basel, Switzerland: Birkhäuser Verlag, 2002a:69–83.
Whiting MF. Phylogeny of the holometabolous insect orders: molecular evidence. Zool Scripta 2002b;31:3–15.
Whiting MF, Bradler S, Maxwell T. Loss and recovery of wings in stick insects. Nature 2003;421:264–7.
Whiting MF, Carpenter JC, Wheeler QD, Wheeler WC. The Strepsiptera problem: Phylogeny of the holometabolous insect orders inferred from 18S and 28S ribosomal DNA sequences and morphology. Syst Biol 1997;46:1–68.
Whitington PM, Bacon JP. The organization and development of the arthropod ventral nerve cord: Insights into arthropod relationships. In: Fortey RA, Thomas RH eds. Arthropod Relationships. London,UK: Chapman and Hall, 1998:349–67.
Whitington PM. Conservation versus change in early axongenesis in arthropod embryos: A comparison between myriapods, crustaceans and insects. In: Breidbach O, Kutsch W, eds. The nervous system of invertebrates: an evolutionary and comparative approach. Basel, Switzerland: Birkhäuser, 1995:181–219.
Whitington PM, Harris KL, Leach D. Early axonogenesis in the embryo of a primitive insect, the silverfish *Ctenolepisma longicaudata*. Roux's Arch Dev Biol 1996;205:272–81.
Whitington PM, Leach D, Sandeman R. Evolutionary change in neural development within the arthropods: axongenesis in the embryos of two crustaceans. Development 1993;118:449–61.
Whitington PM, Meier T, King P. Segmentation, neurogenesis and formation of early axonal pathways in the centipede, *Ethmostigmus rubripes* (Brandt). Roux's Arch Dev Biol 1991;199:349–63.
Wiedermann A. Zur Ultrastruktur des Nervensystems bei *Cephalodasys maximus* (Macrodasyida, Gastrotricha). Microfauna marina 1995;10:173–233.
Wiegmann BM, Trautwein MD, Kim JW, et al. Single-copy nuclear genes resolve the phylogeny of the holometabolous insects. BMC Biol 2009;7:34.
Wiens JJ. Character analysis in morphological phylogenetics: problems and solutions. Syst Biol 2001;50:689–99.
Wiens JJ. The role of morphological data in phylogeny reconstruction. Syst Biol 2004;53:653–61.
Wiens JJ, Hollingsworth BD. War of the iguanas: Conflicting molecular and morphological phylogenies and long-branch attraction in iguanid lizards. Syst Biol 2000;49:143–59.
Wiens JJ, Moen DS. Missing data and the accuracy of Bayesian phylogenetics. J Syst Evol 2008;46:307–14.
Wiens JJ, Morrill MC. Missing data in phylogenetic analysis: Reconciling results from simulations and empirical data. Syst Biol 2011;60:719–31.
Wiesmann R. Zur Kenntnis der Anatomie und Entwicklungsgeschichte der Stabheuschrecke *Carausius morosus* Br. III. Entwicklung und Organogenese der Cölomblasen. PhD thesis, ETH Zürich, Switzerland, 1926.
Wilcox TP, Garcia de Leon FJ, Hendrickson DA, Hillis DM. Convergence among cave catfishes: Long-branch attraction and a Bayesian relative rates test. Mol Phylogenet Evol 2004;31:1101–13.
Wiley EO, Lieberman BS. Phylogenetics. Theory and practice of phylogenetic systematics. Hoboken, NJ, USA: John Wiley & Sons, Inc., 2011.

Williams A, Carlson SJ, Brunton CHC, Holmer LE, Popov L. A supra-ordinal classification of the Brachiopoda. Phil Trans R Soc Lond B Biol Sci 1996;351:1171–93.

Williams JLD. Anatomical studies of insect central nervous system: Ground plan of midbrain and an introduction to central complex in locust, *Schistocerca gregaria* (Orthoptera). J Zool 1975;176:67–6.

Willkommen J, Hornschemeyer T. The homology of wing base sclerites and flight muscles in Ephemeroptera and Neoptera and the morphology of the pterothorax of *Habroleptoides confusa* (Insecta: Ephemeroptera: Leptophlebiidae). Arthropod Struct Dev 2007;36:253–69.

Wills MA. A phylogeny of recent and fossil Crustacea derived from morphological characters. In: Fortey RA, Thomas RH eds. Arthropod relationships. London, UK: Chapman and Hall, 1998:189–209.

Wilmer P. Invertebrate relationships: Patterns in animal evolution. Cambridge,UK: Cambridge University Press, 1990.

Wilson CH, Christie AE. Distribution of C-type allatostatin (C-AST)-like immunoreactivity in the central nervous system of the copepod *Calanus finmarchicus*. Gener Comp Endocrinol 2010;2:252–60.

Winchell C, Sullivan J, Cameron CB, Swalla BJ, Mallatt J. Evaluating hypotheses of deuterostome phylogeny and chordate evolution with new LSU and SSU ribosomal DNA data. Mol Biol Evol 2002;19:762–76.

Winnepenninckx B, Van de Peer Y, Backeljau T. Metazoa relationships on the basis of 18S rRNA sequences: A few years later .... Am Zool 1998;38:888–906.

Winther R. Character analysis in cladistics: Abstraction, reification, and the search for objectivity. Acta Biotheor 2009;57:129–62.

Wipfler B, Machida R, Müller B, Beutel RG. On the head morphology of Grylloblattodea (Insecta) and the systematic position of the order, with a new nomenclature for the head muscles of Dicondylia. Syst Entomol 2011;36:241–66.

Wirkner CS, Richter S. Evolutionary morphology of the circulatory system in Peracarida (Malacostraca; Crustacea). Cladistics 2010;26:143–67.

Witek A, Herlyn H, Ebersberger I, Welch DBM, Hankeln T. Support for the monophyletic origin of Gnathifera from phylogenomics. Mol Phylogenet Evol 2009;53:1037–41.

Witek A, Herlyn H, Meyer A, Boell L, Bucher G, Hankeln T. EST based phylogenomics of Syndermata questions monophyly of Eurotatoria. BMC Evol Biol 2008;8:345–55.

Witte U, Barthel D, Tendal O. The reproductive-cycle of the sponge *Halichondria panicea* Pallas (1766) and its relationship to temperature and salinity. J Exp Mar Biol Ecol 1994;183:41–52.

Woese CR. Bacterial Evolution. Microbiol Rev 1987;51:221–71.

Wolf YI, Rogozin IB, Koonin EV. Coelomata and not Ecdysozoa: Evidence from genome-wide phylogenetic analysis. Genome Res 2004;14:29–36.

Wolff C, Hilbrant M. The embryonic development of the Central American wandering spider *Cupiennius salei*. Front Zool 2011;8:15.

Wolff G, Harzsch S, Hansson BS, Brown S, Strausfeld NJ. Neuronal organization of the hemiellipsoid body of the land hermit crab *Coenobita clypeatus*: Correspondence with the mushroom body ground pattern. J Comp Neurol 2012;520:2824–46.

Wollesen T, Loesel R, Wanninger A. Distribution of FMRFamidergic neurons in the central nervous system of the cephalopod mollusc *Idiosepius notoides*. Acta Biol Hun 2008;59:39–51.

Wollesen T, Loesel R, Wanninger A. Pygmy squids and giant brains: Mapping the complex cephalopod CNS by phalloidin staining of vibratome sections and whole-mount preparations. J Neurosci Methods 2009;179:63–7.

Wolstenholme DR. Animal mitochondrial DNA: Structure and evolution. Int Rev Cytol 1992;141:173–216.

Wong JL, Wessel GM. FRAP analysis of secretory granule lipids and proteins in the sea urchin egg. Methods Mol Biol 2008;440:61–76.
Wong KM, Suchard MA, Huelsenbeck JP. Alignment uncertainty and genomic analysis. Science 2008;319:473–6.
Woo HJ, Lee YS, Park SJ et al. Complete mitochondrial genome of a troglobite millipede *Antrokoreana gracilipes* (Diplopoda, Juliformia, Julida), and juliformian phylogeny. Mol Cells 2007;23:182–91.
Woodland JT. A contribution to our knowledge of lepismatid development. J Morphol 1957;101:523–77.
Woollacott R, Pinto R. Flagellar basal apparatus and its utility in phylogenetic analyses of the porifera. J Morphol 1995;226:247–65.
Woolridge TH, Greenwood JG, Greenwood J. A new species of *Haplostylus* (Mysidacea) from sandy beaches on the East Coast of Australia. Crustaceana 1992;63:160–8.
Wörheide G, Hooper JNA. Calcarea from the Great Barrier Reef. 1: Cryptic Calcinea from Heron Island and Wistari Reef (Capricorn-Bunker Group). Mem Queensl Mus 1999;43:859–91.
Wörheide G, Hooper JNA. New species of Calcaronea (Porifera: Calcarea) from cryptic habitats of the southern Great Barrier Reef (Heron Island and Wistari Reef, Capricorn-Bunker Group, Australia). J Nat Hist 2003;37:1–47.
Wörheide G, Dohrmann M, Erpenbeck D, et al. Deep phylogeny and evolution of sponges (Phylum Porifera). Adv Mar Biol 2012;61:1–78.
Worsaae K, Kristensen RM. Evolution of interstitial Polychaeta (Annelida). In: Bartolomaeus T, Purschke G, eds. Morphology, molecules, evolution and phylogeny in Polychaeta and related taxa. Hydrobiologia 2005;535–536:319–40.
Worsaae K, Rouse GW. Is *Diurodrilus* an annelid? J Morphol 2008;269:1426–55.
Wright KA. Nematoda. In: Harrison FW, Ruppert EE, eds. Microscopic Anatomy of Invertebrates, Vol. 4 Aschelminthes. New York, NY, USA: Wiley-Liss, 1991:111–95.
Wu Z, Shen X, Sun Ma, et al. Phylogenetic analyses of complete mitochondrial genome of *Urechis unicinctus* (Echiura) support that echiurans are derived annelids. Mol Phylogenet Evol 2009;52:558–62.
Xian-Guang H, Bergström J, Xiao-Ya M, Jie Z. The Lower Cambrian Phlogites Luo & Hu re-considered. GFF 2006;128:47–51.
Xiang QY, Moody ML, Soltis DE, Fan CZ, Soltis PS. Relationships within Cornales and circumscription of Cornaceae - matK and rbcL sequence data and effects of outgroups and long branches. Mol Phylogenet Evol 2002;24:35–57.
Xiao Y, Zhang Y, Gao T, Yabe M, Sakurai Y. Phylogenetic relationships of the lancelets of the genus *Branchiostoma* in China inferred from mitochondrial genome analysis. Afr J Biotechnol 2008;7:3845–52.
Xie Q, Tian X, Qin Y, Bu W. Phylogenetic comparison of local length plasticity of the small subunit of nuclear rDNAs among all Hexapoda orders and the impact of hyper-length-variation on alignment. Mol Phylogenet Evol 2009;50:310–6.
Yancopoulos S, Attie O, Friedberg R. Efficient sorting of genomic permutations by translocation, inversion and block interchange. Bioinformatics 2005;21:3340–6.
Yandell M, Mungall CJ, Smith C, et al. Large-scale trends in the evolution of gene structures within 11 animal genomes. PLoS Comput Biol 2006;2:e15.
Yang AS. Modularity, evolvability, and adaptive radiations: a comparison of the hemi- and holometabolous insects. Evol Dev 2001;3:59–72.
Yang J., Ortega-Hernández J, Butterfield NJ, Zhang X. Specialized appendages in fuxianhuiids and the head organization of early artropods. Nature 2013;494:468–71.
Yang JS, Lai EC. Alternative miRNA biogenesis pathways and the interpretation of core miRNA pathway mutants. Mol Cell 2011;43:892–903.

Yang S, Doolittle RF, Bourne PE. Phylogeny determined by protein domain content. Proc Natl Acad Sci USA 2005;102:373–8.

Yang Z. Maximum-likelihood estimation of phylogeny from DNA sequences when substitution rates differ over sites. Mol Biol Evol 1993;10:1396–401.

Yang Z. Estimating the pattern of nucleotide substitution. J Mol Evol 1994;39:105–11.

Yang Z. Among-site rate variation and its impact on phylogenetic analyses. Trends Ecol Evol 1996;11:367–72.

Yang Z. How often do wrong models produce better phylogenies? Mol Biol Evol 1997;14:105–8.

Yang Z, Goldman N, Friday AE. Comparison of models for nucleotide substitution used in maximum likelihood phylogenetic estimation. Mol Biol Evol 1994;11:316–24.

Yang Z, Goldman N, Friday AE. Maximum likelihood trees from DNA sequences: A peculiar statistical estimation problem. Syst Biol 1995;44:384–99.

Yasuda E, Goto T, Makabe KW, Satoh N. Expression of actin genes in the arrow worm *Paraspadella gotoi* (Chaetognatha). Zool Sci 1997;14:953–60.

Yoder MJ, Mikó I, Seltmann KC, Bertone MA, Deans AR. A gross anatomy ontology for Hymenoptera. PLoS ONE 2010;5(12):8.

Yokobori S, Lindsay DJ, Tsuchiya K, Yaniagishi A, Maruyama T, Shirna T. Mitochondrial genome structure and evolution in the living fossil vampire squid, *Vampyroteuthis infernalis*, and extant cephalopods. Mol Phylogenet Evol 2007;44:898–910.

Yokobori S, Iseto T, Asakawa S, et al. Complete nucleotide sequences of mitochondrial genomes of two solitary entoprocts, *Loxocorone allax* and *Loxosomella aloxiata*: Implications for lophotrochozoan phylogeny. Mol Phylogenet Evol 2008;47:612–28.

Yoshikura M. Embryological studies on the liphistiid spider, *Heptathela kimurai*. Part II. Kumamoto J Sci B 1955;2:1–86.

Yoshikura M. Comparative embryology and phylogeny of Arachnida. Kumamoto J Sci B 1975;12:71–142.

Yoshizawa K. The Zoraptera problem: evidence for Zoraptera + Embiodea from the wing base. Syst Entomol 2007;32:197–204.

Yoshizawa K. Direct optimization overly optimizes data. Syst Entomol 2010;35:199–206.

Yoshizawa K. Monophyletic Polyneoptera recovered by wing base structure. Syst Entomol 2011;36:377–94.

Yoshizawa K, Johnson KP. Aligned 18S for Zoraptera (Insecta): phylogenetic position and molecular evolution. Mol Phylogenet Evol 2005;37:572–80.

Young JZ. The Anatomy of the nervous system of *Octopus vulgaris*. Oxford, UK: Clarendon Press, 1971.

Young JZ. Computation in the learning system of cephalopods. Biol Bull 1991;180:200–8.

Yuan Z, Sun X, Liu H, Xie J. MicroRNA genes derived from repetitive elements and expanded by segmental duplication events in mammalian genomes. PLoS ONE 2011;6:e17666.

Yusupov MM, Yusupova GZ, Baucom A, et al. Crystal structure of the ribosome at 5.5 angstrom resolution. Science 2001;292:883–96.

Zachvatkin AA [The division of the Acarina into orders and their position in the system of the Chelicerata.] Parazit Sborn 1952;14:5–46 [in Russian].

Zaćwilichowski J. Über die Innervation der Haftapparate der Karpfenlaus *Argulus foliaceus* L. (Branchiura). Bull Int Acad Polo Sci Lett, Sci Math Nat, Série B: Sci Nat (II) 1935;2:145–62.

Zaćwilichowska J. The nervous system of the carplouse *Argulus foliaceus* L. Bull Int Acad Polo Sci Lett, Sci Math Nat, Série B: Sci Nat (II) 1948;1:117–28.

Zakharov EV, Caterino MS, Sperling FA. Molecular phylogeny, historical biogeography, and divergence time estimates for swallowtail butterflies of the genus *Papilio* (Lepidoptera: Papilionidae). Syst Biol 2004;53:193–215.

Zanol J, Halanych KM, Struck TH, Fauchald K. Phylogeny of the bristle worm family Eunicidae (Eunicida, Annelida) and the phylogenetic utility of noncongruent 16S, COI and 18S in combined analyses. Mol Phylogenet Evol 2010;55:660–76.

Zapotosky JE. Fine structure of the larval stage of *Paragordius varius* (Leidy, 1851) (Gordioidea: Paragordidae) I. The preseptum. Proc Helminthol Soc Washington 1974;41:209–21.

Zdobnov E, Bork P. Quantification of insect genome divergence. Trends Genet 2007;23:16–20.

Zelinka C. Die Gastrotrichen. Eine monographische Darstellung ihrer Anatomie, Biologie und Systematik. Z wiss Zool 1889;49:209–384.

Zeng L, Swalla BJ. Molecular phylogeny of the protochordates: Chordate evolution. Can J Zool 2005;83:24–33.

Zerbst-Boroffka I, Haupt J. Morphology and function of the metanephridia in annelids. Fortschr Zool 1975;23:33–47.

Zhang B, Pan X, Cannon CH, Cobb GP, Anderson TA. Conservation and divergence of plant microRNA genes. Plant J 2006;46:243–59.

Zhang J, Wu X, Xie M, Xu X, Li A. The mitochondrial genome of *Polylabris halichoeres* (Monogenea: Microcotylidae). Mitochondr DNA 2011;22:3–5.

Zhang J, Zhou C, Gai Y, Song D, Zhou K. The complete mitochondrial genome of *Parafronurus youi* (Insecta: Ephemeroptera) and phylogenetic position of the Ephemeroptera. Gene 2008;424:18–24.

Zhang TT, Jiang YQ, Zhou H, Yang WX. Ultrastructural observation on genesis and morphology of cortical granules in *Macrobrachium nipponense* (Crustacea, Caridea). Micron 2010a;41:59–64.

Zhang YY, Xuan WJ, Zhao JL, Zhu CD, Jiang GF. The complete mitochondrial genome of the cockroach *Eupolyphaga sinensis* (Blattaria: Polyphagidae) and the phylogenetic relationships within the Dictyoptera. Mol Biol Rep 2010b;37:3509–16.

Zhang Z-Q. Animal biodiversity: An outline of higher-level classification and survey of taxonomic richness. Zootaxa 2011;3148:7–12.

Zheng J, Rogozin IB, Koonin EV, Przytycka TM. Support for the Coelomata clade of animals from a rigorous analysis of the pattern of intron conservation. Mol Biol Evol 2007;24:2583–92.

Zhong Y-F, Holland PWH. HomeoDB2: Functional expansion of a comparative homeobox gene database for evolutionary developmental biology. Evol Dev 2011;13:567–8.

Zhong M, Struck TH, Halanych KM. Phylogenetic information from three mitochondrial genomes of Terebelliformia (Annelida) worms and duplication of the methionine tRNA. Gene 2008;416:11–21.

Zhong M, Hansen B, Nesnidal MP, Golombek A, Halanych KM, Struck TH. Detecting the symplesiomorphy trap: A multigene phylogenetic analysis for terebelliform annelids. BMC Evol Biol 2011;11:369.

Zhong J, Zhang J, Mukwaya E, Wang Y. Reevaluation of deuterostome phylogeny and evolutionary relationships among chordate subphyla using mitogenome data. J Genet Genomics 2009;36:151–60.

Zhou Y, Rodrigue N, Lartillot N, Philippe H. Evaluation of the models handling heterotachy in phylogenetic inference. BMC Evol Biol 2007;7:206–18.

Zhuo D, Madden R, Elela S, Chabot B. Modern origin of numerous alternatively spliced human introns from tandem arrays. Proc Natl Acad Sci USA 2007;104:882–6.

Zmasek CM, Eddy SR. A simple algorithm to infer gene duplication and speciation events on a gene tree. Bioinformatics 2001;17:821–8.

Zmasek CM, Eddy SR. RIO: Analyzing proteomes by automated phylogenomics using resampled inference of orthologs. BMC Bioinf 2002;3:14.

Zrzavý J. The interrelationships of metazoan parasites: A review of phylum-and higher-level hypotheses from recent morphological and molecular phylogenetic analyses. Folia Parasit (Praha) 2001a;48:81–103.

Zrzavý J. Ecdysozoa versus Articulata: Clades, artifacts, prejudices. J Zool Syst Evol Res 2001b;39:159–63.

Zrzavý J. Gastrotricha and metazoan phylogeny. Zool Sripta 2003;32:61–81.

Zrzavý J, Stys P. The basic body plan of arthropods: Insights from evolutionary morphology and developmental biology. J Evol. Biol 1997;10:353–67.

Zrzavý J, Hypsa V, Tietz DF. Myzostomida are not annelids: Molecular and morphological support for a clade of animals with anterior sperm flagella. Cladistics 2001;17:170–98.

Zrzavý J, Hypsa V, Vlaskova M. Arthropod phylogeny: Taxonomic congruence, total evidence and conditional combination approaches to morphological and molecular data sets. In: Fortey R, Thomas R, eds. Arthropod Relationships, London, UK: Chapman & Hall, 1998:97–107.

Zrzavý J, Mihulka S, Kepka P, Bezděk A, Tietz D. Phylogeny of the Metazoa based on morphological and 18S ribosomal DNA evidence. Cladistics 1998;14:249–85.

Zrzavý J, Riha P, Pialek L, Janouskovec J. Phylogeny of Annelida (Lophotrochozoa): Total-evidence analysis of morphology and six genes. BMC Evol Biol 2009;9:189.

Zuckerkandl E, Pauling L. Molecular Disease, Evolution, and Genic Heterogeneity. In: Kasha M, Pullman B, eds. Horizons in Biochemistry. New York, NY, USA: Academic Press, 1962:189–225.

Zuckerkandl E, Pauling L. Molecules as documents of evolutionary history. J Theor Biol 1965;8:357–66.

Zuker M. Mfold web server for nucleic acid folding and hybridization prediction. Nucleic Acids Res 2003;31:3406–15.

Zuker M, Sankoff D. RNA secondary structures and their prediction. B Math Biol 1984;46:591–621.

Zur Strassen O. *Anthraconema*, eine neue Gattung freilebender Nematoden. Zool Jahrb Syst Suppl 1904;7:301–46.

Zwickl DJ, Hillis DM. Increased taxon sampling greatly reduces phylogenetic error. Syst Biol 2002;51:588–98.

Index

Printed and bound by CPI Group (UK) Ltd, Croydon, CR0 4YY
13/09/2026
14960855-0001